Drug Discovery in Cancer Epigenetics

Translational Epigenetics Series
Trygve O. Tollefsbol, Series Editor

Transgenerational Epigenetics
Edited by Trygve O. Tollefsbol, 2014

Personalized Epigenetics
Edited by Trygve O. Tollefsbol, 2015

Epigenetic Technological Applications
Edited by Y. George Zheng, 2015

Epigenetic Cancer Therapy
Edited by Steven G. Gray, 2015

DNA Methylation and Complex Human Disease
By Michel Neidhart, 2015

Epigenomics in Health and Disease
Edited by Mario F. Fraga and Agustin F. Fernández, 2016

DNA Biomarkers and Diagnostics
Edited by José Luis García-Giménez, 2016

Drug Discovery in Cancer Epigenetics
Edited by Gerda Egger and Paola Arimondo, 2016

Drug Discovery in Cancer Epigenetics

Edited by

Gerda Egger

and

Paola Arimondo

ELSEVIER

AMSTERDAM • BOSTON • HEIDELBERG • LONDON
NEW YORK • OXFORD • PARIS • SAN DIEGO
SAN FRANCISCO • SINGAPORE • SYDNEY • TOKYO
Academic Press is an imprint of Elsevier

Academic Press is an imprint of Elsevier
The Boulevard, Langford Lane, Kidlington, Oxford OX5 1GB
225 Wyman Street, Waltham MA 02451

Notices
Knowledge and best practice in this field are constantly changing. As new research and experience broaden our understanding, changes in research methods, professional practices, or medical treatment may become necessary.

Practitioners and researchers may always rely on their own experience and knowledge in evaluating and using any information, methods, compounds, or experiments described herein. In using such information or methods they should be mindful of their own safety and the safety of others, including parties for whom they have a professional responsibility.

To the fullest extent of the law, neither the Publisher nor the authors, contributors, or editors, assume any liability for any injury and/or damage to persons or property as a matter of products liability, negligence or otherwise, or from any use or operation of any methods, products, instructions, or ideas contained in the material herein.

Library of Congress Cataloging-in-Publication Data
A catalog record for this book is available from the Library of Congress

British Library Cataloguing-in-Publication Data
A catalogue record for this book is available from the British Library

ISBN: 978-0-12-802208-5

Publisher: Mica Haley
Acquisition Editor: Catherine Van Der Laan
Editorial Project Manager: Lisa Eppich
Production Project Manager: Melissa Read
Designer: Mark Rogers

Contents

PART III CLASSES OF EPIGENETIC DRUGS

List of Contributors

Paola Arimondo
Unité de Service et de Recherche, CRDPF, Toulouse, France

Dina Arvanitis
INSERM, UMR1048, Institute of Cardiovascular and Metabolic Diseases, University-Paul Sabatier, Toulouse, France

Fréderic Ausseil
Unité de Service et de Recherche, CRDPF, Toulouse, France

Mina Bekheet
Laboratory of Cancer Biology, Department of Oncology, University of Oxford, Oxford, UK

Christopher G. Bell
Department of Twin Research & Genetic Epidemiology, St Thomas' Hospital, King's College London, London, UK; MRC Lifecourse Epidemiology Unit, University of Southampton, Southampton, UK; Academic Unit of Human Development and Health, University of Southampton, Southampton, UK; Epigenomic Medicine, Centre for Biological Sciences, Faculty of Environmental and Natural Sciences, University of Southampton, Southampton, UK

Magdalena Benetkiewicz
Groupe Coopérateur Multidisciplinaire en Oncologie, Paris, France

Kate H. Brettingham-Moore
School of Medicine, University of Tasmania, Hobart, TAS, Australia

Peter J. Brown
Structural Genomics Consortium, University of Toronto, Toronto, ON, Canada

Corey Carter
John P. Murtha Cancer Center, Walter Reed National Military Medical Center/National Cancer Institute, Bethesda, MD, USA

Christophe Cisarovsky
Research Laboratory in Oncology, University of Lausanne, Centre Hospitalier Universitaire Vaudois, Lausanne, Switzerland

Pierre Cordelier
INSERM, UMR1037, Cancer Research Center of Toulouse, University of Toulouse-Paul Sabatier, Toulouse, France

Armand de Gramont
New Drug Evaluation Laboratory, Centre of Experimental Therapeutics and Medical Oncology, Department of Oncology, University of Lausanne, Centre Hospitalier Universitaire Vaudois, Lausanne, Switzerland

Frank J. Dekker
Pharmaceutical Gene Modulation, Groningen Research Institute of Pharmacy, University of Groningen, Groningen, The Netherlands

Yannick Delpu
Skirball Institute of Biomolecular Medicine, New York University Langone Medical Center, New York, NY, USA

Helmut Dolznig
Institute of Medical Genetics, Medical University of Vienna, Vienna, Austria

Marlène Dufresne
INSERM, UMR1037, Cancer Research Center of Toulouse, University of Toulouse-Paul Sabatier, Toulouse, France

Gerda Egger
Clinical Institute of Pathology, Medical University of Vienna, Vienna, Austria

Chantal Etievant
Unité de Service et de Recherche, CRDPF, Toulouse, France

Sandrine Faivre
Medical Oncology, Department of Oncology, University of Lausanne, Centre Hospitalier Universitaire Vaudois, Lausanne, Switzerland

Gary R. Fanger
EpicentRx, Mountain View, CA, USA

Panagis Filippakopoulos
Structural Genomics Consortium, Nuffield Department of Clinical Medicine, University of Oxford, Oxford, UK; Ludwig Institute for Cancer Research, Nuffield Department of Clinical Medicine, University of Oxford, Oxford, UK

Alexandre Gagnon
Département de Chimie, Université du Québec à Montréal, Montréal, QC, Canada

Marion Gayral
INSERM, UMR1037, Cancer Research Center of Toulouse, University of Toulouse-Paul Sabatier, Toulouse, France

Melanie R. Hassler
Clinical Institute of Pathology, Medical University of Vienna, Vienna, Austria

Markus Hengstschläger
Institute of Medical Genetics, Medical University of Vienna, Vienna, Austria

Quanah J. Hudson
CeMM Research Center for Molecular Medicine of the Austrian Academy of Sciences, Vienna, Austria

Jean-Pierre J. Issa
Fels Institute for Cancer Research and Molecular Biology, Temple University School of Medicine, Philadelphia, PA, USA

Manfred Jung
Institute of Pharmaceutical Sciences, Albert-Ludwigs-Universität Freiburg, Freiburg, Germany

Stefan Knapp
Structural Genomics Consortium, Nuffield Department of Clinical Medicine, University of Oxford, Oxford, UK; Institute for Pharmaceutical Chemistry, Johann Wolfgang Goethe-University, Frankfurt am Main, Germany

Nina Kramer
Institute of Medical Genetics, Medical University of Vienna, Vienna, Austria

Stefan Kubicek
CeMM Research Center for Molecular Medicine of the Austrian Academy of Sciences, Vienna, Austria; Christian Doppler Laboratory for Chemical Epigenetics and Antiinfectives, CeMM Research Center for Molecular Medicine of the Austrian Academy of Sciences, Vienna, Austria

Nicholas B. La Thangue
Laboratory of Cancer Biology, Department of Oncology, University of Oxford, Oxford, UK

Dorian Larrieu
INSERM, UMR1037, Cancer Research Center of Toulouse, University of Toulouse-Paul Sabatier, Toulouse, France

Maxime Leroy
Département de Chimie, Université du Québec à Montréal, Montréal, QC, Canada

Niek G.J. Leus
Pharmaceutical Gene Modulation, Groningen Research Institute of Pharmacy, University of Groningen, Groningen, The Netherlands

Marco P. Licciardello
CeMM Research Center for Molecular Medicine of the Austrian Academy of Sciences, Vienna, Austria

Marie Lopez
Unité de Service et de Recherche, CRDPF, Toulouse, France

Michelle Lybeck
EpicentRx, Mountain View, CA, USA

Wolfgang J. Miller
Laboratories of Genome Dynamics, Department of Cell and Developmental Biology, Center of Anatomy and Cell Biology, Medical University of Vienna, Vienna, Austria

Heidi Olzscha
Laboratory of Cancer Biology, Department of Oncology, University of Oxford, Oxford, UK

Arnold L. Oronsky
InterWest Partners, Menlo Park, CA, USA

Bryan T. Oronsky
EpicentRx, Mountain View, CA, USA

Neil C. Oronsky
CFLS, San Jose, CA, USA

Thomas Prebet
Departement d'Hematologie et Unite d'Evaluation Therapeutique en Oncologie/Hematologie, Institut Paoli Calmettes, Marseille, France; Hematology Department, Smilow Cancer Center at Yale University, New Haven, CT, USA

Eric Raymond
Medical Oncology, Department of Oncology, University of Lausanne, Centre Hospitalier Universitaire Vaudois, Lausanne, Switzerland

Noël J.-M. Raynal
Département de Pharmacologie, Université de Montréal and Sainte-Justine University Hospital Research Center, Montréal, QC, Canada

Elisa Redl
Clinical Institute of Pathology, Medical University of Vienna, Vienna, Austria

Tony R. Reid
Moores Cancer Institute, University of California, San Diego, CA, USA

Dina Robaa
Institut für Pharmazie, Martin-Luther-Universität Halle-Wittenberg, Halle, Germany

Martin Scherzer
Institute of Medical Genetics, Medical University of Vienna, Vienna, Austria

Jan J. Scicinski
EpicentRx, Mountain View, CA, USA

Semira Sheikh
Laboratory of Cancer Biology, Department of Oncology, University of Oxford, Oxford, UK

Wolfgang Sippl
Institut für Pharmazie, Martin-Luther-Universität Halle-Wittenberg, Halle, Germany

Mira Stadler
Institute of Medical Genetics, Medical University of Vienna, Vienna, Austria

Phillippa C. Taberlay
Chromatin Dynamic Group, Genomics and Epigenetics Division, Garvan Institute of Medical Research, Darlinghurst, NSW, Australia; Epigenetics Program, Genomics and Epigenetics Division, Garvan Institute of Medical Research, Darlinghurst, NSW, Australia; St. Vincent's Clinical School, Faculty of Medicine, University of New South Wales, Darlinghurst, NSW, Australia

Tirza Timmerman,
Pharmaceutical Gene Modulation, Groningen Research Institute of Pharmacy, University of Groningen, Groningen, The Netherlands

Jérôme Torrisani
INSERM, UMR1037, Cancer Research Center of Toulouse, University of Toulouse-Paul Sabatier, Toulouse, France

Christine Unger
Institute of Medical Genetics, Medical University of Vienna, Vienna, Austria

Thea van den Bosch
Pharmaceutical Gene Modulation, Groningen Research Institute of Pharmacy, University of Groningen, Groningen, The Netherlands

Norbert Vey
Departement d'Hematologie et Unite d'Evaluation Therapeutique en Oncologie/Hematologie, Institut Paoli Calmettes, Marseille, France; Aix-Marseille Université, Marseille, France

Tobias Wagner
Institute of Pharmaceutical Sciences, Albert-Ludwigs-Universität Freiburg, Freiburg, Germany

Stefanie Walter
Institute of Medical Genetics, Medical University of Vienna, Vienna, Austria

Angelika Walzl
Institute of Medical Genetics, Medical University of Vienna, Vienna, Austria

Daniel J. Weisenberger
USC/Norris Comprehensive Cancer Center, University of Southern California, Los Angeles, CA, USA

Preface

The causal involvement of epigenetic pathways in tumor development and progression has been widely accepted and recent findings have nourished our knowledge of tumor biology and opened potential clinical applications dependent on epigenetic aberrations. Deregulation and mutation of epigenetic enzymes as well as global and local changes of epigenetic chromatin modifications are implicated in a variety of malignancies and provide novel therapeutic and diagnostic targets for oncology. There is great interest in developing novel epigenetic drugs targeting chromatin modifiers as well as chromatin reader proteins and some successful recent studies have confirmed the validity of these drugs for cancer therapy.

We were fortunate to gather many experts and renowned authors to contribute to this book and are very thankful for their efforts and valuable contributions. This book intends to provide (i) an introduction into cancer epigenetics and to give a comprehensive overview on, (ii) methods and tools for epigenetic drug development, (iii) classes of epigenetic drugs, (iv) development of diagnostic tools, and (v) clinical implications of epigenetic therapy. Emerging concepts, such as "epi-sensitization," are also presented alongside with indications beyond cancer.

The intended audience includes both basic scientists as well as clinicians. The book is directed toward scientists of the academic and industrial sector, who are aiming to test and develop epigenetic cancer drugs. Importantly, the book will increase the awareness level of epigenetic drugs for oncologists. We envision that this might foster increased and more rapid translation of epigenetic drugs into the clinics.

Gerda Egger and Paola Arimondo

PART

I

INTRODUCTION

CHAPTER

BASIC EPIGENETIC MECHANISMS AND PHENOMENA

1

Melanie R. Hassler[1], Elisa Redl[1], Quanah J. Hudson[2], Wolfgang J. Miller[3] and Gerda Egger[1]

[1]Clinical Institute of Pathology, Medical University of Vienna, Vienna, Austria [2]CeMM Research Center for Molecular Medicine of the Austrian Academy of Sciences, Vienna, Austria [3]Laboratories of Genome Dynamics, Department of Cell and Developmental Biology, Center of Anatomy and Cell Biology, Medical University of Vienna, Vienna, Austria

CHAPTER OUTLINE

G. Egger & P. Arimondo (Eds): Drug Discovery in Cancer Epigenetics. DOI: http://dx.doi.org/10.1016/B978-0-12-802208-5.00001-1

1.1 INTRODUCTION

The history of epigenetics dates back to the early Greek philosopher Aristotle, who considered development as a process that provides form from unformed material within male and female germ cells and involves a dynamic course based on intrinsic factors [1]. During the last few centuries vivid debates were fueled by rather extreme scientific views as to the development of an organism. Preformation as a predictable predetermined and stable process was opposed to epigenesis, which involves morphogenesis and differentiation based on regulatory response to the environment, cellular communication, and dynamic processes [2]. The term epigenetics was coined by Conrad Hal Waddington to describe causal "mechanisms by which the genes of a genotype bring about phenotypic effects" during development [3]. He introduced the concept of "epigenotype" to indicate the complex processes and networks of genetic control linking the genotype and the phenotype [4]. Thus, disruption of genes at early stages of development could affect this epigenotype and have far-reaching consequences on different organs and tissues. He modified this name from Valentin Haecker, who created the term "phenogenetics" to describe visible stages of ontogeny [5]. Being both an embryologist and geneticist he felt there was a need to combine genetics and experimental embryology and the expression "epigenetics" for him was in accordance with the classical concept of epigenesis. In the years and decades to follow, the term "epigenetics" was used to describe a variety of phenomena and mechanisms [6] and up to now no clear definition as to the use of the term has been adopted. The persistence of epigenetic systems to implement different phenotypes and a potential heredity or cellular inheritance was tied to epigenetics by Nanney and Ephrussi, respectively [7,8]. With his new concept of carcinogenesis, Holliday [9] proposed that malignant transformation not only results from genetic mutations but also from epigenetic changes in DNA methylation resulting in heritable alterations in gene expression. This put DNA methylation at center stage as an epigenetic mechanism. The discovery of the pivotal role of DNA methylation for imprinting and X chromosome inactivation (XCI), plus the discovery of histone-modifying enzymes provided a revival of the term "epigenetic" in the 1990s [6,10–12]. Histone modifications were suggested to provide an epigenetic code allowing for both transient regulation of gene expression and long-term epigenetic memory [13,14]. Among epigenetic mechanisms, RNAi was shown to induce not only posttranscriptional gene silencing but was also found to confer chromatin alterations including DNA methylation and histone modifications in plants and in yeast [15]. Another noncoding RNA component, long noncoding RNA (lncRNA), was shown to bridge long-range chromatin silencing and chromatin modification by recruitment of histone-modifying complexes both at the inactive X (Xi)-chromosome or gene-specific loci to promote stable chromatin states [16,17]. Thus, epigenetics, which is in fact a transdisciplinary field, has moved from observing phenomena to elucidating its underlying mechanisms, which include DNA methylation, chromatin modification/remodeling, and noncoding RNA. All three are interrelated and cooperate to maintain the stable epigenome of a given cell lineage in response to internal or external stimuli (Figure 1.1).

A generally well-accepted definition of "modern" epigenetics was coined by Arthur D. Riggs describing epigenetics as "The study of mitotically and/or meiotically heritable changes in gene function that cannot be explained by changes in DNA sequence" [18]. Currently, the term "epigenetics" is en vogue and used in very different scientific communities, including ethics and

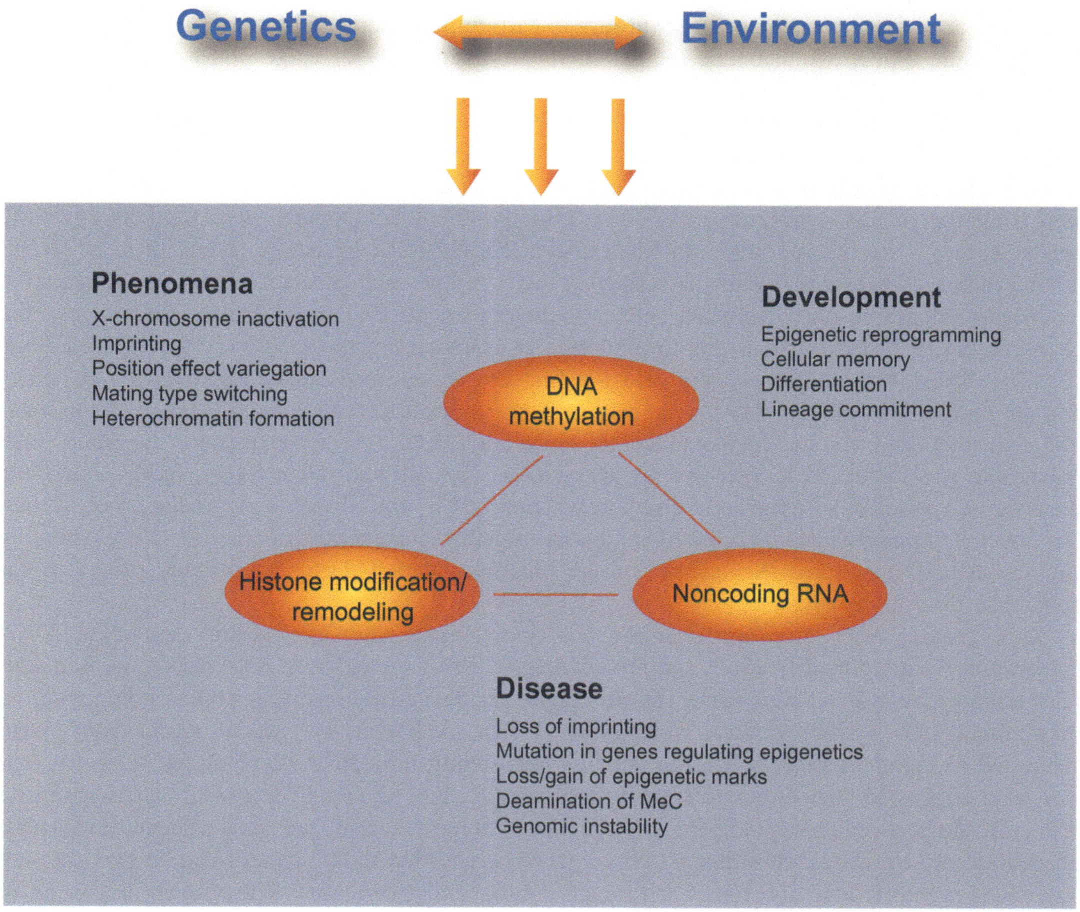

FIGURE 1.1

Epigenetic mechanisms and their role in different processes. Epigenetic mechanisms include DNA methylation, histone modification, and remodeling, as well as noncoding RNA. All three are interrelated and are involved in different phenomena and developmental processes, and also in a variety of diseases. Changes in epigenomic signatures can be due to genetic alterations in epigenetic pathway proteins and to exposure to different environmental stimuli.

social sciences or religious studies and the debate on "nature versus nurture" has been newly sparked by discussions of whether environmental factors can alter the epigenome and be transmitted to subsequent generations. Such discussions and studies are usually based on unresolved questions and often lack biological evidence. Clearly, epigenetic mechanisms play a central role for development, cellular differentiation, and homeostasis and their deregulation is associated with diverse disease states including cancer. As discussed in more detail in Chapter 2, epigenetic alterations in cancer can be the result of mutations in genes involved in epigenetic regulation and very often occur as a consequence of exposure to harmful environmental factors.

One prime aspect of epigenetic regulation is the reversibility of epigenetic marks that can be modulated by chemical compounds, which makes it an important target for developing new therapeutic drugs, which will be addressed in great detail in the following chapters of this book. Starting in the 1960s, antineoplastic activity of 5-azacytidine (5-azaCR) and 5-aza-2′deoxycytidine (5-azaCdR) was shown against leukemia in mice and the first clinical studies were initiated in 1967 in Europe [19–21]. Initially, these compounds were used as cytotoxic agents as they were incorporated into RNA (5-azaCR) and DNA (5-azaCR, 5-azaCdR) and inhibit DNA synthesis. Jones and Taylor then showed that cytidine analogs induced (trans)differentiation of mouse embryonic and fibroblast cells into myotubes, striated muscle cells, adipocytes, and chondrocytes, and that this was due to inhibition of DNA methylation of newly synthesized DNA [22–24]. These two inhibitors were also the first epigenetic drugs to be approved by the FDA and are currently in clinical use for patients with myelodysplastic syndromes (MDSs), acute myeloid leukemia, and chronic myelomonocytic leukemia [25,26]. At approximately the same time, *N*-butyrate was identified to alter histone acetylation [27–29], and interestingly, HDAC inhibitors were the second class of epigenetic drugs to be approved by the FDA for the treatment of cutaneous T-cell lymphoma [30]. The list of inhibitors targeting epigenetic machineries is ever-growing and has recently attracted the interest of the pharmaceutical industry to develop programs for epigenetic drug discovery.

This chapter will give an overview on basic epigenetic mechanisms including DNA methylation, histone modification, and remodeling as well as noncoding RNA-based mechanisms. Furthermore, we will address some classic epigenetic phenomena such as dosage compensation, imprinting, and position effect variegation (PEV). Although epigenetic mechanisms were suggested to maintain gene expression patterns persistently through subsequent cell generations it was also suggested that environmental factors could cause dynamic epigenetic alterations and result in diseases. Epidemiological studies have suggested that the environment can cause phenotypic effects and directly impact on epigenetics, and that these changes can be heritable not only during mitotic cell divisions but through subsequent generations. We will discuss some of the current data available analyzing potential inter- and transgenerational epigenetic effects and disease.

1.2 BASIC EPIGENETIC MECHANISMS

In eukaryotic cells DNA is not “naked” but exists as an intimate complex with specialized proteins called histones, which together with DNA comprise chromatin. Nuclear DNA is spooled around nucleosomal units consisting of small histone proteins in order to fit into the nucleus [31]. Chromatin exists in two forms: the less condensed euchromatin that is associated with transcriptionally active regions and the highly condensed, usually transcriptionally inactive heterochromatin. Chromatin can be modified by posttranslational histone modifications, histone variants, energy-dependent chromatin-remodeling steps that mobilize or alter nucleosome structures, and noncoding RNA that can influence chromatin structure. Furthermore, DNA itself can be modified covalently by methylation of cytosines, usually of CpG dinucleotides.

1.2.1 DNA METHYLATION

DNA methylation is an epigenetic modification that is correlated with gene repression and is known to play an important role in gene regulation, development, and tumorigenesis. It consists of the addition of a methyl group to the carbon at position 5 of cytosine residues in the DNA template. In mammals it usually occurs at CpG dinucleotides but symmetric, asymmetric and non-CpG methylation is also known in embryonic stem cells (ESC), *Neurospora crassa*, and plants [32–34]. In mammalian somatic tissues approximately 70% of all CpG sites within the DNA are methylated and DNA methylation distribution shows enrichment in noncoding regions and interspersed repetitive elements [35] but not in CpG islands of active genes [36]. CpG islands were first defined in 1987 by Gardiner-Garden and Frommer as 200 bp stretches of DNA with a G + C content of at least 50% and an observed CpG/expected CpG excess of 0.6 [37]. CpG islands are often associated with promoters and usually lack DNA methylation, thus allowing gene expression. Approximately 60% of human genes have CpG island promoters. However, methylation of CpG islands can reinforce silencing of genes, for example, for genes on the Xi chromosome or for some imprinted genes. Additionally, in cancer cells, genes are often aberrantly silenced by CpG island methylation.

DNA methylation is catalyzed by DNA methyltransferases (DNMTs), which can either catalyze *de novo* methylation (DNMT3A/B) at novel sites or maintenance methylation (DNMT1) following DNA replication [38]. DNMT1 is specific for CpG and its preferred substrate is hemimethylated DNA, DNA methylated at CpG on one strand. Inactivation of *Dnmt1* in mouse ESC results in genome-wide loss of CpG methylation, indicating that this DNMT is necessary for stable maintenance of DNA methylation [39]. However, Lei et al. [40] showed that *de novo* methylation of proviral DNA introduced into ES cells was not catalyzed by DNMT1 but another unknown DNMT. Besides DNMT1, three candidate proteins that could potentially encode additional DNMTs were found. DNMT2 had minimal DNMT activity *in vitro* and deletion of *Dnmt2* did not alter the level of DNA methylation [41]. However, DNMT2 acts as an RNA methyltransferase that methylates tRNAs [42]. In contrast to DNMT1, DNMT3A and DNMT3B did not show preference for hemimethylated DNA *in vitro* [43] and both genes were necessary for *de novo* methylation of proviral genomes and repetitive elements in embryos and ESC [44]. In mice, inactivation of *Dnmt3a* and *Dnmt3b* results in early embryonic lethality and loss of one gene causes postnatal or embryonic lethality. Furthermore, in humans, mutations of *DNMT3B* are associated with the ICF syndrome, a rare condition characterized by immunodeficiency, centromeric instability, and facial abnormalities [45].

But how does DNA methylation interfere with gene expression? First, the presence of methyl groups interferes with the binding ability of transcription factors that are crucial for transcriptional activation. Transcription factors often recognize CG-rich motifs and several of these are unable to bind to methylated CpG sequences [46]. Second, proteins that bind to methyl-CpGs can also repress gene expression at methylated sites. These proteins were found by performing band-shift assays using random methylated DNA sequences as probes [47]. MeCP1, a DNA–protein complex that is specific for methylated DNA, was discovered in a variety of mammalian cell types. However, the first individual methyl-CpG-binding protein to be purified and cloned was MeCP2. Upon the presence of a methyl-CpG-binding domain four other members of this family, MBD1, MBD2, MBD3, and MBD4, could be identified [48]. Furthermore, two other structural domains are currently known to bind methylated DNA: the SET and RING finger-associated domain, found in UHRF1 and UHRF2 and zinc fingers, found in Kaiso and Kaiso-like proteins [49]. Methyl-CpG-binding proteins recruit a variety of proteins including histone deacetylases (HDACs) and chromatin-remodeling factors that are responsible for transcriptional repression [50] and can also interact with transcription factors, such as MeCP2 with TFIIB [51].

1.2.2 DNA DEMETHYLATION

Studies of cellular reprogramming have demonstrated that differentiated cellular states can be radically altered, suggesting that DNA methylation may be reversible in mammalian cells. Today it is known that not only passive DNA demethylation due to reduction in activity or absence of DNMTs, but also active demethylation catalyzed by special enzymes takes place within a cell. Active DNA demethylation plays an important role in early mammalian development as well as in tissue-specific differentiation but is also observed in adult cells. One of the most prominent processes that involve rapid active DNA demethylation in adult cells is the activity-dependent DNA demethylation of brain-derived neurotrophic factor and fibroblast growth factor 1 promoters in postmitotic neurons [52]. 5-Methylcytosine (5mC) can be hydroxylated to 5-hydroxymethylcytosine (5hmC) by members of the ten–eleven translocation (TET) enzyme family. 5hmC can further be oxidized to 5-formylcytosine or 5-carboxylcytosine. Furthermore, 5mC or 5hmC can be deaminated to 5-methyluracil or 5-hydroxymethyluracil by members of the AID/APOBEC enzyme family. Finally, the intermediates produced by TET and AID/APOBEC are replaced by the uracil DNA glycosylase [53] family of base excision repair glycosylases, that mediate DNA repair [54].

1.2.3 HISTONE MODIFICATIONS

The basic chromatin unit, or nucleosome, consists of a protein octamer containing two molecules of each canonical (or core) histone (H2A, H2B, H3, H4), around which 147 bp of DNA are wrapped. The core histones consist of a globular domain and flexible aminoterminal "histone tails" [55]. These tails, particularly, those of H3 and H4, are accessible for modifications, such as phosphorylation, acetylation, methylation, and ubiquitylation, which can be correlated to both transcriptional activation or repression. In general, active marks include acetylation, arginine methylation, and lysine methylation, such as H3K4 and H3K36, while repressive marks include methylation of H3K9, H3K27, and H4K20. However, the globular domains of core histones can also be modified [56] (Figure 1.2).

But how do histone modifications work in general? First, histone modifications can act in *cis* by directly altering the chromatin structure by disrupting the contact between adjacent nucleosomes or between histones and DNA, for example, by charge changes [59]. Second, histone modifications can also act in *trans* and can be recognized by specific binding proteins, so-called "reader" proteins, which can further recruit chromatin-modifying complexes and modulate chromatin structure [60].

1.2.3.1 Histone Acetylation and Deacetylation

Histone acetylation, catalyzed by histone acetyltransferases (HATs), can directly alter chromatin structure through the loss of positive charges within histones, resulting in reduced interactions of histones with negatively charged DNA and increased accessibility of DNA-binding sites [61]. Reader

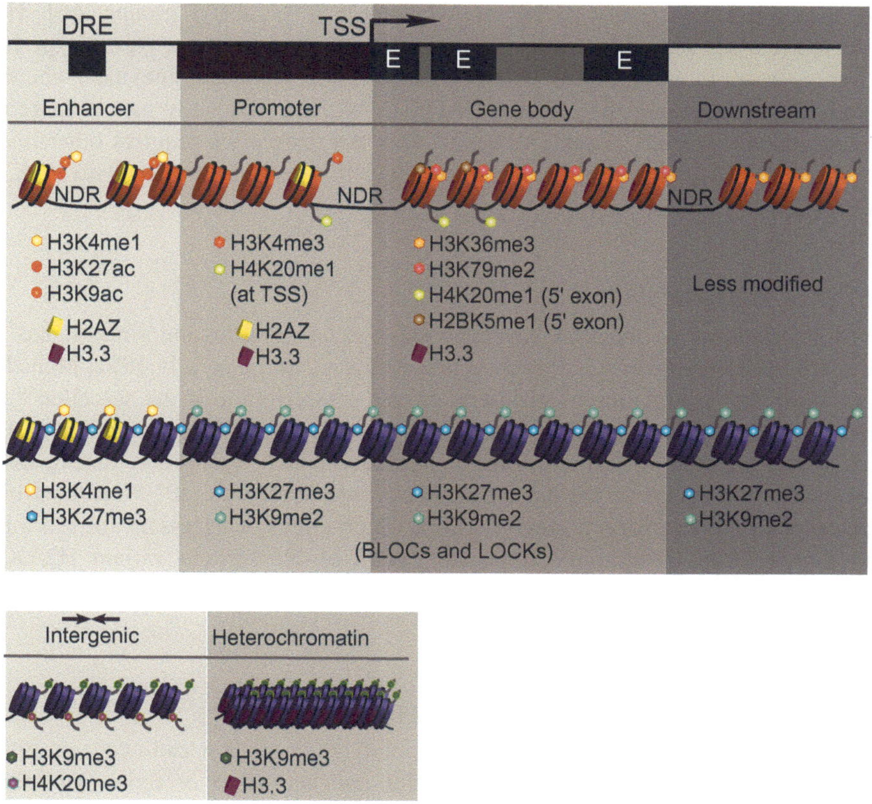

FIGURE 1.2

Histone modification and nucleosome occupancy determine different genomic regions. Dependent on their activity state, genes are marked by characteristic chromatin modifications, nucleosome density, and histone variants, which together reflect the on/off state of a gene (top). Note that only selected histone modification marks are illustrated. NDR indicates nucleosome-depleted regions in enhancers, promoters and 3′ regions of active genes. BLOCs and LOCKs designate large silent gene-rich regions marked by repressive histone marks H3K27me3 and H3K9me2, respectively [57,58]. (Bottom) Chromatin composition including histone modifications, grade of compaction, and histone variants is indicated for intergenic regions and heterochromatin. Inverted arrows on top indicate inverted repeats. These regions show generally high levels of DNA methylation and are associated with repressive histone marks. Red, nucleosomes of active genes; blue, nucleosomes found in repressed genes, intergenic regions or heterochromatin; DRE, distal response element; TSS, transcription start site; E, exon.

proteins with bromo-like domains recognize and bind acetylated histones. These proteins can either be basal transcription factors, HATs, such as p300 or GCN5 or part of large chromatin-associating/altering complexes such as the ATP-dependent remodeling complex SWI/SNF, which further alter the chromatin structure [62]. The same principle applies to histone methylation, which is recognized by chromo-like domains, and phosphorylation, which is recognized by 14-3-3 proteins [63,64].

Histone deacetylation is catalyzed by HDACs that remove acetyl groups [65]. HDACs are grouped into four classes according to their homology with yeast proteins: type I, type II, and type IV HDACs are referred to as "classical" HDACs and are zinc-dependent enzymes, whereas type III or SIR2-related HDACs require the cofactor NAD. HDACs are frequently part of large multisubunit complexes, which target the enzyme to genes. This targeting often requires other histone interacting proteins, for example, the Sin3 corepressor, which is targeted by Rpd3 to H3K36me sites and suppresses transcription in yeast [66,67].

1.2.3.2 Histone Phosphorylation

Histone phosphorylation was the first characterized histone modification since it has long been understood that kinases regulate signal transduction pathways in the cell. In 1991, Mahadevan et al. were able to show that stimulation of proliferation and transcription of the so-called "immediate-early" genes in cells correlate with histone H3 phosphorylation. Phosphorylation of H3S10 has been shown to be required for chromosome condensation and segregation during mitosis [68]. H3S10 phosphorylation occurs at the onset of mitosis, interferes with HP1-H3K9me3 binding and leads to release of heterochromatin protein 1 (HP1), which might be necessary for full mitotic chromatin condensation [69]. Furthermore, phosphorylation of the histone variant H2AX has been shown to be associated with double-strand DNA breaks, where it helps to recruit repair proteins to the site of the break [70].

1.2.3.3 Histone Methylation and Demethylation

Histone methylation is the most complex histone modification since it can occur on either lysines or arginines and is associated with either transcriptional activation or repression. Furthermore, lysines can be mono- (me1), di- (me2), or tri- (me3) methylated and arginines can be monomethylated or symmetrically or asymmetrically dimethylated.

Enzymes of three distinct families can catalyze lysine methylation of histones: the PRMT1 family, whose substrate is arginine, the SET domain containing protein family, and the non-SET domain proteins DOT1/DOT1L, which methylate lysine residues [71,72].

In general, methylation of H3K4, H3K36, and H3K79 is linked to transcriptional activation, however, depending on the methylation state and the genomic location, the same modification may lead to different effects [59,73]. In contrast, methylation of H3K9, H3K27, H3K64, and H4K20, as well as methylation of the linker histone H1, has been implicated in transcriptional repression [73]. Specific reader proteins bind to the methylation sites and alter gene expression. These proteins contain one of three distinct methyl lysine recognition domains: the so-called chromo, tudor or PHD repeat domains [74]. Furthermore, H3K79me3 and H4K20me are associated with DNA repair [75].

For a long time it was not clear whether histone demethylation takes place in the cell or not. Shi et al. [76] discovered the protein LSD1, an enzyme that removes methyl groups specifically from H3K4. It can associate with different complexes and depending on the complex it can also change its demethylation specificity [77]. Today several families of histone demethylases are known, which act on various substrates and therefore play different roles in gene activation and repression [78].

1.2.3.4 Chromatin-Remodeling Complexes and Histone Variants

Another major mechanism to alter chromatin and nucleosome composition in a noncovalent manner is the recruitment of ATP-dependent chromatin-remodeling complexes. These protein complexes can be generally categorized into two families: the SNF2H or ISWI family, which mobilizes nucleosomes and the Brahma or SWI/SNF family, which alters the structure of nucleosomes [79].

Furthermore, core histones can be exchanged in an ATP-dependent manner by specialized histone variants. Histone variants are associated with specific expression states, genomic localization, and species-distribution patterns. They change the structural and functional properties of nucleosomes by affecting chromatin remodeling and histone modifications. There are two important histone variants of H3: H3.3, which marks transcriptionally active genes [80] and CenpA, which is found in centromeric chromatin and is essential for centromeric function and chromosome segregation [81].

Histone H2A shows three histone variants. The presence of H2A.Z correlates with transcriptional activity, while H2A.X senses DNA damage and is therefore crucial for DNA repair [82]. MacroH2A specifically associates with the Xi chromosomes in mammals and is an epigenetic regulator of key developmental genes.

1.2.4 NONCODING RNAs

Within the last few years different classes of noncoding RNAs have been identified, which can either cause transcriptional or posttranscriptional gene silencing or directly impact on chromatin structure by recruiting chromatin-modifying complexes.

RNA interference was originally discovered in *Caenorhabditis elegans*, where exogenously introduced double-stranded RNA molecules are able to silence the expression of homologous sequences within the genome [83]. RNAi and RNAi-mediated transcriptional and post-transcriptional silencing is best understood in *Schizosaccharomyces pombe*, where deletion of one of the genes encoding for Argonaute, Dicer or RNA-dependent RNA polymerase proteins leads to loss of heterochromatic gene silencing, reduced H3K9 methylation at centromeric repeats and accumulation of noncoding RNAs [84–86]. Furthermore, RNAi has also been shown to be required for heterochromatin assembly in *Schizosaccharomyces pombe* and it was shown that pericentromeric siRNA accumulation requires the H3K9 methyltransferase Clr4 [87]. In higher organisms, for example, *Drosophila melanogaster* mutations of Argonaute proteins lead to a decrease of H3K9me as well as HP1 binding [88] and mouse cell lines with mutations in Dicer show loss of silencing at centromeres [89]. Furthermore, gene silencing triggered by RNAi can be long term and heritable, as demonstrated in *Caenorhabditis elegans*, where four chromatin-remodeling factors are necessary for maintaining the silent state of genes [90].

In higher organisms microRNAs (miRNA), a class of small noncoding RNAs, are essential for the regulation of early differentiation programs. For example, *HOX* genes encode a family of transcriptional regulators important for patterning along body axes. Expression of *HOX* genes is regulated by different molecular mechanisms including nuclear dynamics, RNA processing, translational regulation, and miRNAs [91]. Most of the miRNAs that regulate *HOX* gene expression are encoded within the *HOX* cluster [92]. In mammals there are four *HOX* gene clusters, which contain five genes encoding miRNA-10 and miRNA-96 [93].

An additional class of noncoding RNAs constitute lncRNAs, which are able to recruit histone-modifying enzymes using RNAi-independent strategies to particular genomic sites [94,95]. The most prominent lncRNA in mammals so far is *XIST* that mediates XCI and was shown to target the Polycomb repressive complex 2 (PRC2) including the H3K27 methyltransferase EZH2 to chromatin [96,97]. Other lncRNAs that recruit PRC2 and confer silencing include *HOTAIR, RepA,* and *Kcnq1ot1* amongst others [94]. Generally, lncRNAs appear to work as scaffolds to enable complex formation and recruitment to distinct genomic loci. A group of lncRNAs transcribed from enhancers can act as transcriptional activators in *cis* via DNA looping and recruitment of coactivators. An example for this is the lncRNA *HOTTIP,* which recruits a H3K4 methyltransferase complex to activate transcription of *HOXA* homeobox genes [98].

1.3 EPIGENETIC (RE)PROGRAMMING

Generating a broad range of different cell types in a highly ordered and reproducible manner is one of the most remarkable hallmarks of complex genomes [99]. In mammals ~25,000 genes contain the information for the development of ~200 different cell types. During development of multicellular organisms different cells and tissues acquire specific gene expression programs, which are mainly regulated by epigenetic modifications such as DNA methylation and histone modification [36,100,101]. These epigenetic mechanisms propagate appropriate patterns of gene expression and can be heritable but potentially reversible [102].

Cells of multicellular organisms can be functionally divided into two major groups: totipotent reproductive germ cells, necessary for the transmission of genetic information to the next generation, and differentiated somatic cells. For somatic cells epigenetic marks become fixed once the cell differentiates [100] and developmental processes in these cells are usually associated with a progressive loss of developmental potential. However, in cells undergoing dedifferentiation (e.g., cancer cells), or transdifferentiation, reprogramming can also take place in differentiated somatic cells [103].

Major reprogramming and global erasure of epigenetic marks take place at two time points in the mammalian life cycle: during early embryonic development after fertilization and in primordial germ cells (PGCs), where totipotency or pluripotency is restored and parental imprints are erased (Figure 1.3).

1.3.1 EPIGENETIC ASYMMETRY IN THE ZYGOTE

At fertilization, the parental genomes are in different stages of the cell cycle and contain divergent epigenetic marks and chromatin composition [100]. The paternal genome is single copy (1C), packed densely with protamines instead of histones, whereas the maternal genome is arrested in metaphase II (2C) and contains histones. Upon fertilization and zygote formation protamines in the paternal genome are rapidly replaced with histones that lack H3K9me2 and H3K27me3, followed by an active, genome-wide loss of DNA methylation. In contrast, the maternal pronucleus contains histones with H3K9me2 and H3K27me3 marks that were acquired during oocyte growth and does not lose 5mC at this early stage, but rather by a passive mechanism during subsequent

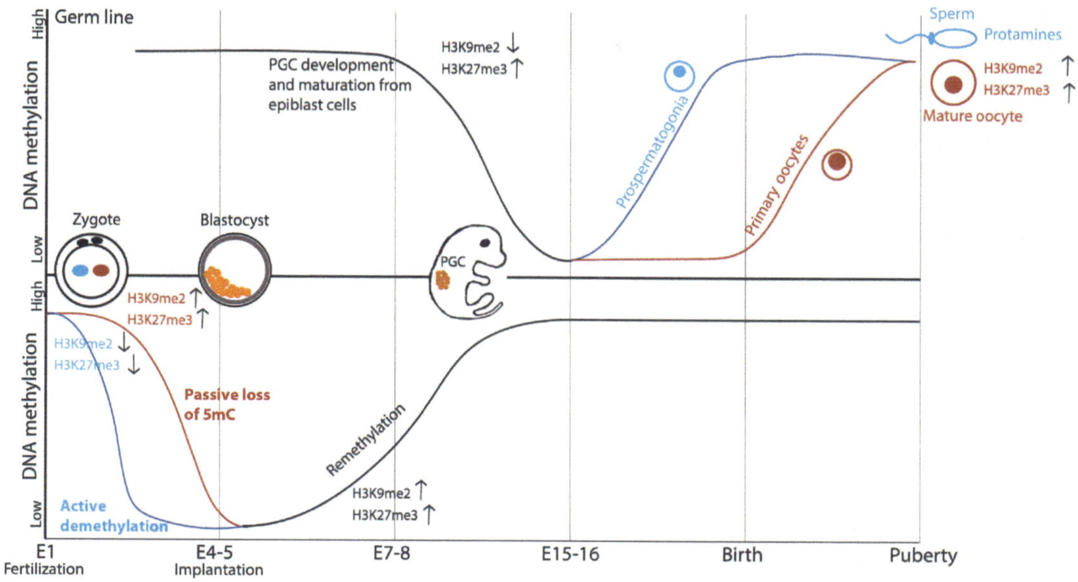

FIGURE 1.3

Epigenetic reprogramming. After fertilization, the paternal genome (bottom blue line) becomes actively demethylated rapidly in the zygote, while the maternal genome (bottom red line) is demethylated by a passive mechanism. Both genomes are remethylated at the time of implantation. During embryonic development, another reprogramming event is occurring in germ cell development in PGCs, were genome-wide DNA methylation is followed by remethylation in male germ cells (top blue line) in prospermatogonia and in female germ cells (top red line) after birth. Both reprogramming events are associated also with changes in repressive histone modification patterns as indicated by methylation changes in histone H3K27 and H3K9.

cell divisions. This asymmetry in paternal and maternal epigenomes can be detected up to the four-cell state. However, the functional importance of this asymmetry in the zygote is not clear [99]. Gill et al. [104] hypothesized that chromatin modifications established in the gametes may be required for proper embryonic development, and it could be shown that loss of early zygotic paternal demethylation perturbs the activation of several pluripotency-associated genes and impairs development [105].

During preimplantation development a passive loss of 5mC can be detected in the maternal genome until the blastocyst state when the inner cell mass acquires high levels of 5mC, H3K9me2, and H3K27me3. In contrast to the paternal active DNA demethylation catalyzed by TET enzymes, it has been hypothesized that DNA demethylation in the maternal genome occurs passively through exclusion of the maintenance methyltransferase DNMT1 from the nucleus [106]. During active paternal demethylation the maternal genome is protected by DPPA3 (also called STELLA) from TET3 oxidation [107]. DPPA3 is essential for embryonic development during preimplantation and loss of *Dppa3* leads to embryonic lethality in mice [108]. Furthermore, although DNA demethylation is a global process during preimplantation development, some genomic sequences, like imprinted loci, centromeric heterochromatin, and intracisternal A particle (IAP) retrotransposons maintain DNA methylation during preimplantational development, which is also mediated by DPPA3.

All in all, an elaborate temporal and spatial epigenetic modification program seems to be necessary for zygote development and to exploit the entire potential of the genome. However totipotency is lost as cleavage progresses [99]. Histone modifications as well as DNA methylation are necessary for regulation of lineage induction and defects in DNA and histone methyltransferases can lead to impaired differentiation in ESC [109]. Polycomb repressive complexes are required for the bivalency of key regulator developmental genes in a yet transcriptionally poised state, where activating and repressing histone marks are present at the same time, ensuring lineage flexibility.

1.3.2 REPROGRAMMING IN THE GERMLINE

A second major epigenetic reprogramming event takes place during development of germ cells in the early embryo. PGCs are formed from epiblast cells by signaling molecules that are produced by extraembryonic ectoderm and primary endoderm [110]. This process is associated with a loss of 5mC and H3K9me2 and an increase of H3K27me3, a more flexible histone modification that allows rapid activation of poised developmental regulator genes [111]. Global demethylation of PCGs occurs between E11.5 and E12.5 of mouse development and leads to erasure of methylation in imprinted genes and single copy genes [112] and to reactivation of the X chromosome in females [113]. However, there is evidence that not all epigenetic marks are erased in imprinted genes in PGCs. The paternal *H19* and the maternal *Snrpn* alleles are demethylated during PGC development but later during spermatogenesis and oogenesis they need to be methylated again. This *de novo* methylation of formerly methylated genes seems to happen at an earlier stage than *de novo* methylation of originally unmethylated genes [114,115], indicating that other epigenetic marks are not completely erased during PGC development and provide signals for *de novo* methylation [100]. To distinguish between somatic cells and PGCs several markers can be used, which reflect the different genetic programs of PGC and somatic cell specification. PGCs were found to express several pluripotency genes, including *Sox2* and *Oct4*, which were not expressed in the neighboring somatic cells, indicating that PGCs exhibit pluripotency, which is lost in somatic cells. On the other hand some genes, including *Hoxb1* and *Hoxa1*, which are important for cell differentiation and morphogenesis are significantly upregulated in somatic cells, but not expressed in PGCs, indicating that PGCs repress somatic cell fate [116].

1.3.3 INDUCED PLURIPOTENCY

Briggs and King [117] performed the first nuclear transfer experiments with cells isolated from late blastula stage frog embryos from *Rana pipiens* that developed into complete embryos when transplanted into enucleated oocytes. Subsequently, Gurdon and colleagues succeeded in producing sexually mature frogs from nuclear transplantation of adult nuclei in *Xenopus laevis* [118]. Meanwhile, it could be demonstrated that adult somatic cell nuclei can be reprogrammed by nuclear transfer from different species including mammals [119]. These studies indicated on one hand that developmental restrictions are due to reversible epigenetic modifications [119] and on the other hand that the oocyte must contain factors that mediate the reprogramming of adult cells into an embryonic state [103].

However, the big breakthrough was the discovery of transcription factors that can induce pluripotency in somatic cells to generate induced pluripotent stem (iPS) cells without the need for oocytes. Takahashi and Yamanaka [120] identified four transcription factors, OCT4 (O), SOX2 (S), KLF4 (K), and cMYC (M) that were able to reprogram adult mouse fibroblasts into ES-like iPS cells. The advantages of transcription-factor-induced reprogramming lies in its simplicity and robustness. By ectopic expression of OSKM a broad range of different cell types can easily be reprogrammed to pluripotency. In their ideal state iPS cells are functionally indistinguishable from ESC, can form chimeras and teratomas, and are able to differentiate into cells of all three germ layers [121]. However, the induction of pluripotency upon OSKM expression requires a latency period of 1–2 weeks and occurs only in 1% of the starting cells. Furthermore, it has been shown that the differentiation state of a somatic cell can influence the efficiency of iPS cell generation and that compounds that inhibit DNA methylation and histone modification can increase the efficiency [122].

1.4 GENOMIC IMPRINTING AS A MODEL OF EPIGENETIC SILENCING

Monoallelic expression can occur randomly for specific clusters of genes, for example, for B- or T-cell receptor genes [123], and for up to 10–24% of individual genes per cell [123,124]. Genetic differences between alleles that affect the activity of promoters or *cis*-regulatory elements can also lead to monoallelic or strongly biased allelic expression that may affect many genes in human [125]. In contrast, parental-specific monoallelic or imprinted expression is relatively rare, with only approximately 0.5% of genes (125 in mouse) showing imprinted expression in mammals (http://igc.otago.ac.nz, [126]). In spite of their small number, many imprinted genes play important roles in development and growth, as was already demonstrated in the 1980s by experiments that showed early embryonic lethality of androgenic and gynogenic embryos, indicating that both the maternally and paternally inherited genomes together are required for normal development [127,128]. Following this, a number of developmental disorders have been shown to be caused by defects at specific imprinted gene loci, such as the chromosome 15q11-13 imprinted region, where paternal mutations can cause Prader–Willi syndrome and maternal mutations can cause Angelman syndrome [129]. In such diseases a nonmutated allele is present, but epigenetically silenced by the imprinting mechanism, opening the possibility for treatment by reactivation of the silent allele with epigenetic drugs.

Genomic imprinting is a classical epigenetic process in mammals controlled by DNA methylation that is deposited in genomic regions called imprint control elements (ICEs) in either the male or female germline by the *de novo* DNA methyltransferases DNMT3A and DNMT3L [130–132] (Figure 1.4). Differential DNA methylation is then maintained on the same parental allele in all somatic cells in the organism by DNMT1 [133]. The majority of reported imprinted genes in mice lie near to one of the 24 reported gametic differentially methylated regions (gDMRs) [134,135], seven of which have been demonstrated to be ICEs by genetic deletion that resulted in loss of

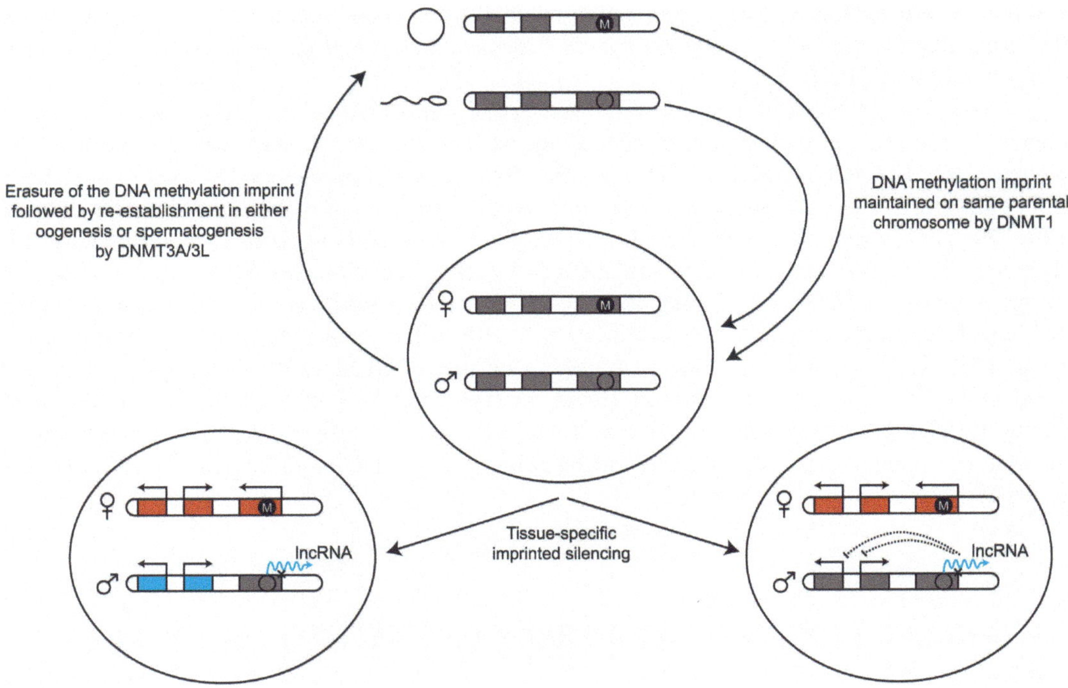

FIGURE 1.4

Regulation of imprinted expression. Imprinted silencing is controlled by differential DNA methylation on the ICE that is established during either oogenesis or spermatogenesis by the DNMT3A/3L *de novo* DNMTs complex and maintained on the same parental chromosome by DNMT1. The unmethylated ICE then acts to silence a cluster of imprinted genes in *cis*, in the most common mechanism by acting as a promoter for a lncRNA that then causes imprinted silencing. Although, differential DNA methylation of the ICE is present in all somatic cells, and the lncRNA is also expressed in most cell types, many imprinted genes are only susceptible to imprinted silencing in some cell types or developmental stages, and hence show tissue-specific imprinted expression.

imprinted expression of nearby genes [10]. This indicates that most imprinted genes are in co-regulated clusters controlled by an ICE, that ranges in size from 3 to 12 genes spread over 100–3700 kilobases (kb) [10]. DNA methylation is associated with gene silencing, but paradoxically many imprinted genes are expressed from the same allele as where the ICE is methylated. This is explained by the mechanism by which imprinted silencing is achieved. Different mechanisms of imprinted silencing via the ICE have been shown, but the most common mechanism described so far for four of the seven defined ICEs (the *Igf2r*, *Kcnq1*, *Gnas*, and Prader–Willi clusters) is by lncRNA-mediated silencing [136–139]. In these cases the ICE is associated with the promoter of the lncRNA and silenced on the maternal allele by DNA methylation. The lncRNA is then expressed and silences imprinted genes in *cis* on the paternal allele. Similarly, in the *Igf2* cluster where imprinted silencing is not controlled by an lncRNA, the imprinted protein-coding genes *Igf2* and *Ins2* are expressed from the paternal allele where the ICE is methylated. In this cluster the insulator protein CTCF binds the unmethylated maternal ICE blocking *Igf2* from accessing enhancers, while CTCF binding is prevented by methylation on the paternal allele allowing the enhancers to activate *Igf2* expression [140,141]. Hence, in imprinted silencing DNA, methylation acts to repress the repressor allowing the expression of imprinted genes. For a limited number of imprinted genes DNA methylation can then play a secondary role to reinforce silencing directly on the silent promoter, forming a somatic DMR [10].

Genomic imprinting is a phenomenon with a clear phenotype, making it a useful model to investigate aspects of epigenetic gene regulation. With the advent of high-throughput sequencing technologies, thousands of lncRNAs have been recently identified [142], but imprinted lncRNAs remain among the few lncRNAs that have so far been demonstrated to be functional by genetic experiments. lncRNAs could act by their transcription alone or via their RNA product to regulate gene expression. Posttranscriptional knockdown by RNA interference-based approaches will not affect the function of lncRNAs that act via their transcription, while genetic deletions of lncRNAs could also remove *cis*-regulatory elements such as enhancers, making neither strategy ideal for investigating lncRNA function. Therefore, in the imprinting field lncRNAs have been truncated by introducing a polyadenylation cassette to investigate their function and mechanism of action [136–139]. Using this approach, the lncRNA *Airn* that silences *Igf2r* in mouse was truncated to various lengths, demonstrating that *Airn* need only be transcribed across the *Igf2r* promoter to cause silencing due to transcriptional interference [143]. In contrast, the *Airn* lncRNA product was shown to be necessary to silence the nonoverlapped gene *Slc22a3* by recruiting the repressive EHMT H3K9 dimethyltransferase to its promoter [144]. Similarly, the *Kcnq1ot1* lncRNA responsible for silencing genes in the *Kcnq1* cluster is associated with EHMT2 and the PRC2 H3K27trimethyltransferase complex, and imprinted expression of some genes in the *Kcnq1* cluster is lost when these complexes are disrupted, indicating that *Kcnq1ot1* may also cause silencing by recruiting repressive histone-modifying complexes [145]. The experimental approaches developed by studying imprinted lncRNAs, and the silencing mechanisms uncovered, may be applied to investigate the function and mechanism of action of lncRNAs outside of imprinted regions.

The majority of imprinted genes show tissue-specific imprinted expression [146], indicating that they may have a dose-dependent function in specific tissues. This indicates that the imprinted silencing mechanism may act tissue-specifically, which has been hypothesized to occur, for example, by imprinted lncRNAs blocking tissue-specific enhancer activity [147]. Open questions in well-studied imprinted clusters, and the likely possibility that other imprinted clusters may achieve imprinted silencing in different ways, makes genomic imprinting a fertile field of study to further elucidate mechanisms of epigenetic gene silencing.

1.5 DOSAGE COMPENSATION IN MAMMALS

The difference in the distribution of the sex chromosomes in mammals requires a mechanism of gene dose balancing to adjust for the double amount of X-linked genes in females. H.J. Muller was the first to develop the concept of dosage compensation in *Drosophila* following his studies on X-linked eye pigment genes [148]. While in *Drosophila* the dosage difference is equaled by transcriptional upregulation of the genes on the single male X chromosome [149], mammals have adopted a silencing mechanism that affects one of the two female X chromosomes [150]. This process involves random selection of one of the two Xs for inactivation, which is remarkably illustrated by the fur color mosaicism of calico and tortoiseshell cats. In these cats the different colored patches are directly reflecting the expression pattern resulting from random XCI of two different alleles of fur color encoded on the X chromosome (http://www.bio.miami.edu/dana/dox/calico.html). XCI utilizes all major epigenetic silencing mechanisms including DNA methylation, histone modification, and incorporation of histone variants, changes in nuclear architecture and noncoding RNA. These mechanisms act together to generate the Xi chromosome, which exists as facultative heterochromatin, is late replicating and condensed during interphase, and is represented by the Barr body in female somatic cells. Together, these structural characteristics are associated with transcriptional silencing of about 1000 X-linked genes.

XCI involves several important steps including "sensing" (more than one X chromosome is present), "counting" (only one X chromosome per cell should be active) and "choice" (random selection of one X chromosome for inactivation) [11]. The onset of XCI is associated with the expression of a noncoding RNA called *Xist*, which covers and marks the X chromosome in *cis* and is essential for the initiation of silencing [151,152]. It is transcribed from the X inactivation center, a key locus for the initiation and regulation of XCI, next to other noncoding RNAs such as the Xist repressor *Tsix*. *Xist* expression is imprinted and only expressed from the paternal X in marsupials and in the early zygote and extraembryonic tissues in rodents, whereas it appears to be randomly expressed from either X chromosomes in other eutherian mammals including human. *Xist* expression and coating of the X chromosome in *cis* represents the first step in XCI. The exact mechanism of interaction between *Xist* and the X chromosome is not clear, but it is suggested to involve several interaction domains. XCI initiation events also include loss of active histone marks (H3K9Ac, H3K4me2, H3K4me3), global H4 hypoacetylation, loss of RNA polymerase II, and transcriptional silencing of X-linked genes. The *Xist* transcript might also play a role for the propagation of silencing by recruitment of histone modification complexes such as PRC2, that deposit repressive marks (H3K27me3, H3K9me3, H4K20me1, H2AK119ub1). Interestingly, H3K9me and H3K27me are regionally enriched at nonoverlapping regions [153] and the association of repressive histone modification complexes is found early during XCI but diminished later on, suggesting a role for these in establishment rather than maintenance of the silenced state. The chromodomain-containing reader protein Cdyl was recently identified as a factor binding to the Xi and was suggested to propagate the silencing via recruitment of the histone methyltransferase G9a [154]. Following these early chromatin changes, the Xi chromosome is decorated with the histone variant macroH2A and DNA methylation of X-linked CpG islands, which is essential for the maintenance of a stable silenced X chromosome [11].

The X chromosome includes genes that are important for sexual dimorphism and reproduction, but also for neuronal development, general intelligence, and social cognition [155]. Intriguingly, almost 30% of known genes that are linked to mental retardation are encoded on the X chromosome [155]. Generally, XCI has important implications for different diseases, because mosaic inactivation of X-linked genes can dampen the phenotype of mutations in female patients, while skewed X inactivation can be associated with worse disease outcomes of X-linked disorders [156]. On the other hand, escape from silencing or mutations in escape genes are involved in disease including cancer [157]. Depending on the cell types and tissues some genes regularly escape

XCI in humans. Besides genes located within the pseudoautosomal (PAR) regions, which are homologous between the X and Y chromosomes, genes outside the PAR regions with Y homologs (e.g., *KDM5c, UTX*) but also genes without Y-linked homologs can escape inactivation resulting in higher expression levels in females, although their exact biological function has been unresolved [11]. Interestingly, the H3K27me3 demethylase UTX was recently identified as a gender-specific tumor suppressor in T-cell acute lymphoblastic leukemia in males [158].

Somatic loss of the Barr body, the Xi chromosome, in cancer was already reported in early studies and was also found in aging tissues in mice [159,160]. A role for loss of the Xi in cancer was indicated by silencing of *Xist* in the blood compartment of mice, being correlated with a highly aggressive myeloproliferative neoplasm and MDS [161]. However, only recently a report by Chaligne et al. [162] demonstrated for the first time that loss of the Xi is due to epigenetic "erosion" rather than genetic loss in the context of breast cancer.

Taken together, these findings highlight the essential role for proper XCI for development and disease, which are often associated with gender-specific prevalence.

1.6 PEV IN *DROSOPHILA*

In addition to the concept of dosage compensation of X-chromosomal linked genes, Muller [163] made in the same publication a second seminal discovery named PEV. Using X-rays as a mutagen, Muller observed a group of unusual phenotypes, in which the eye of *Drosophila* flies was variegating, with some patches of red and some patches of white facets. In contrast to classic loss-of-function mutations that strictly follow Mendelian rules, the variegating nature of the mosaic, so-called "*white-mottled*" eye phenotype of PEV mutants suggested that the X-chromosomal *white* gene, encoding for the red pigment of wild-type flies, itself was not damaged at all, since some facets in the eye remained red, and flies with entirely red eyes could be recovered, again using X-rays as a mutagen. Muller suggested that in this kind of mutation the *white* gene stays intact (not mutated), but has been clonally silenced in some of the facet cells (white), whereas in others it was normally expressed (red). Subsequent studies on somatic giant chromosomes of larval salivary glands showed that the PEV eye phenotype is caused by an inversion or rearrangement with one breakpoint within the pericentric heterochromatin, and one breakpoint adjacent to the *white* gene. Hence this phenomenon is referred to as PEV since the variegating phenotype is caused by a change in the position of the gene within the chromosome and not by a mutation of the gene itself [163].

Today, virtually every *Drosophila* gene that has been examined in an appropriate chromosomal rearrangement has been shown to variegate, and rearrangements involving the pericentric heterochromatin of any chromosome can lead to PEV [164]. This phenomenon is not limited to *Drosophila* but PEV has subsequently also been observed in a variety of organisms, including yeasts (recently reviewed in Ref. [165]), flies, and mammals (recently reviewed in Refs [166,167]). Generally, the phenotype of PEV results whenever a euchromatic gene is juxtaposed with heterochromatin by any kind of rearrangement or transposition. When heterochromatin packaging spreads across the heterochromatin/euchromatin border, it shuts off transcription of the affected reporter gene in a stochastic, that is, variegating, pattern. The strength of PEV, however, can be modified by a variety of extrinsic as well as intrinsic factors [168] (and recently reviewed in Ref. [169]). The first group shown to affect the extent of variegation extrinsically is environmental factors such as temperature acting during development, and the latter represent factors like the amount of heterochromatin available within the genome. High temperatures from 25 to 29 °C during larval development result in suppression of variegation (loss of silencing, red eyes), whereas lower temperatures (e.g., 18 °C) cause enhancement of variegation (increase in silencing, white eyes). The presence of additional Y chromosomes (XXY females and XYY males) that mainly consist of heterochromatin can titrate-off heterochromatic silencing factors from the PEV reporter site and hence act as strong suppressors of PEV, whereas flies without a Y chromosome (X0) act as strong enhancers [170].

In addition to these chromosomal modifiers of PEV mentioned above, hypersensitive reporter genes such as *white-mottled* were intensely studied in *Drosophila* in order to screen for dominant second-site mutations of modifier genes that act either as suppressors or enhancers of PEV. Since then many genetic dominant suppressors and enhancers of PEV have been isolated [171,172] (and recently reviewed in Ref. [169]). These second-site mutations were induced by chemical mutagens that cause point mutations or small insertions/deletions by transposons or irradiation, but that do not impact the chromosome rearrangement responsible for the PEV phenotype. By this genetic approach two groups of PEV modifier genes were extracted from the fly genome. The first class are genes that, when mutated, result in a loss of silencing (reversion to red eyes) and hence act as suppressors of the variegation, thereby named Su(var). The second class are genes that when mutated result in an increase in silencing (white), hence acting as enhancer of variegation, consequently named E(var). In *Drosophila* about 150 loci have been identified from such screens, and of these approximately 30 modifiers of PEV have been studied in detail (recently summarized in Ref. [169]).

In recent years further genetic screens for dominant PEV mutations as well as bioinformatic analysis for extracting homologs of E(VAR) and SU(VAR) proteins not only of *Drosophila* but also mammals, have identified many conserved epigenetic factors including the histone H3 lysine 9 methyltransferase SU(VAR)3-9. As one of the many examples, the heterochromatin protein HP1a binds H3K9me2/3 and interacts with SU(VAR)3-9, creating a core memory system. Thereby, genetic, molecular, and biochemical analyses of PEV that were initiated in the *Drosophila* model system have contributed many key findings concerning establishment and maintenance of heterochromatin with concomitant gene silencing.

1.7 TRANSGENERATIONAL AND INTERGENERATIONAL EPIGENETIC INHERITANCE

Transgenerational epigenetic effects are defined as changes in the offspring's phenotype that are propagated over several generations and that are not due to changes in the genomic DNA sequence [173]. This concept represents a significant alteration regarding the inheritance of phenotypic traits across generations when opposed to Mendelian inheritance, which strictly correlates phenotypes to genotypes. However, it revives Lamarckian ideas by stating that characteristics (i.e., phenotypes) that were acquired or "learned" due to external influences can be passed on to the next generation without having an impact on the genome sequence.

Before giving an overview of the most well-established examples, we want to point out some important discussion points in the field of transgenerational inheritance. Although it is now known that transgenerational epigenetic effects are common in plants and worms, the occurrence of these phenomena in mammals is still to some extent subject to debate. This is due to the fact that mammalian PGCs and the zygote undergo efficient reprogramming at distinct developmental steps, which leads to nearly complete erasure of all inherited epigenetic marks [174]. Through this mechanism, totipotency of the early embryo is achieved, but also epigenetic changes acquired by parents are erased. However, during pregnancy, epigenetic alterations not only affect the mother and the fetus, but also the fetus's PGCs, which will eventually give rise to the grandchildren. Considering this concomitant exposure of three generations during pregnancy, it is important to note that many of the currently termed "transgenerational" effects in mammals are observed when pregnant mothers, their unborn offspring, and the latters' PGCs are exposed to certain external stimuli, but that these effects often cease in the subsequent generation [175]. If the effect on the phenotype thus does not persist longer than in the grandchildren of the affected generation, the effect on the phenotype should rather be considered as an intergenerational and not a transgenerational effect.

Indeed, one of the first studies in mice demonstrating the influence of epigenetic alterations on offsprings' phenotype is due to an intergenerational effect. In these studies, it was observed that mice with *agouti* (brown) coat color give rise to offspring with a range of colors from brown to yellow, but also mottled (both yellow and brown patches) colored mice are seen. This characteristic coat color shift in *agouti* offspring is dependent on the methylation status of a murine retrotransposon (the so-called IAP) upstream of the *agouti* gene locus: if the IAP retrotransposon is unmethylated, mice develop a yellow coat, whereas a methylated IAP leads to silencing of the *agouti* gene and a brown (agouti) coat [176]. The reason why parents do not always pass on their own coat colors but produce offspring with varying degrees of yellow, brown, or mottled color is that the erasure and the following setting of the methylation mark at the retrotransposon are variable during germline reprogramming, owing to the stochastic methylation of the retrotransposon in the early embryo. Interestingly, the methylation status can be influenced by maternal nutrition or environmental exposure during early development. For example, when pregnant dams are fed with methyl supplements, the coat color ratios of their offspring can be shifted toward the brown (or methylated IAP) phenotype [177]. However, there is no cumulative increase in brown color across several generations of methyl supplemented yellow dams, suggesting that the methylated IAP locus is not inherited transgenerationally through the female [178].

A real transgenerational effect in mammals was observed in rats upon treatment with the anti-androgenic compound vinclozolin. Female rats produced male offspring with abnormal spermatogenesis (reduced cell number and viability) and altered DNA methylation patterns when they were exposed to vinclozolin during pregnancy. The effects on male spermiogenesis were stably transferred to at least four subsequent generations through the male germline, although it is possible that vinclozolin caused both epigenetic and genetic changes [179].

Besides chemical substances, nutrition and parental behavior can also influence the offspring's phenotypes potentially via epigenetic mechanisms. In humans, a remarkable transgenerational or intergenerational effect has been observed in children of women who were pregnant during the Dutch famine in the Second World War. Offspring of women who suffered from undernourishment during pregnancy were found to be at higher risk of developing poorer glucose tolerance, and the grandchildren of these women were found to have increased neonatal adiposity [180]. Additionally, another study performed on people who were prenatally exposed to the Dutch famine showed that children of prenatally undernourished fathers, but not mothers, were more obese than children of parents who had not been undernourished prenatally [181]. Similar effects regarding glucose tolerance and obesity are obtained in mouse models of maternal undernutrition [182,183]. In rats, a chronic high-fat diet in fathers leads to female offspring with ß-cell dysfunction, and methylation changes in genes that showed a significant alteration in expression were detected, indicating that epigenetic modifications could be passed on through paternal gametes in this setting [184]. In a *Drosophila* model of paternal-diet-induced intergenerational metabolic reprogramming, paternal sugar intake could control adiposity and reprogram chromatin subtypes in offspring via Polycomb and H3K9me3-centric regulators in the germline and zygote [185]. Epigenetic behavioral reprogramming can also be influenced after birth, as it was shown in rats that maternal licking and grooming of pubs had profound effects on their behavior in adult life and was associated with the promoter methylation status of the glucocorticoid receptor in the hippocampus, which is an important area for memory and learning [186].

In plants, transgenerational epigenetic effects are relatively common, as the plant germline arises from somatic cells which are exposed to developmental and environmental cues, and plants, which are sessile organisms primed to respond quickly to these cues, are under selective pressure to develop ways to pass on their experiences to the next generation [173]. For example, in the plant *Linaria vulgaris*, promoter DNA methylation at the *Lcyc* gene determines flower symmetry. Methylation is passed on from one generation to the next, but can occasionally get lost, leading to a change from radial to bilateral symmetry [187]. In *Arabidopsis*, increased temperature inhibits siRNA biogenesis, and this siRNA inhibition can be observed in subsequent generations grown under normal conditions [188]. Also, DMRs in *Arabidopsis* have been shown to constitute epigenetic variants that are stably transmitted over multiple generations and that are associated with complex traits, such as flowering time and root length [189]. Epigenetic inheritance of transposable elements and their silencing by DNA methylation, piRNA, or chromatin remodeling, which can be passed on over several generations and were first described by Barbara McClintock in maize, also contribute to typical transgenerational effects in plants [173,190].

Another species in which transgenerational phenomena have been described is the nematode *Caenorhabditis elegans*. Examples include worms that were exposed to certain olfactory stimuli during their development showing a higher ability to migrate toward these stimuli in adult life and also laying significantly more eggs [191]. This process, known as "olfactory imprinting," can be transmitted over more than 40 generations and is a way for the worms to pass on the memory of a favorable environment to their offspring. Likewise, deficiencies in H3K4me3 chromatin-remodeling proteins in parental worms were reported to extend the lifespan of their offspring up to the third generation [192]. In another study, starvation-induced arrest of *Caenorhabditis elegans* led to the expression of small RNA molecules targeting genes with tasks in nutrition that were inherited through at least three generations, and the third generation of starved animals also showed an increased lifespan [193].

Transgenerational epigenetic mechanisms are thought to be controlled by chromatin-based mechanisms, such as DNA methylation or chromatin-remodeling proteins mediated by noncoding or coding RNA are in general considered as their main mediators, but more studies are needed to develop a profound understanding of the exact nature of these processes [173]. Additionally, the fact that transgenerational mechanisms are often involved in the regulation of transposable elements, repeat elements or transgenes makes it also likely that they might have originally evolved as germline defense strategies. Compared to the impact of the genotype, their influence on the phenotype is in most cases of transient nature, but reflects temporal and flexible possibilities for organisms and their offspring to respond to their surroundings via epigenetic mechanisms.

1.8 EPIGENETICS AND DISEASE

Although each cell in a living organism contains the same DNA sequence, gene expression patterns are characteristic for individual cell types. This is due to epigenetic modifications determining which parts of the genome are turned on and which parts are turned off. If epigenetic modifications are—due to genetic mutations, stress responses or, other external stimuli—altered in such a way that the cell shifts its specific balance between active and repressed gene transcription toward dysfunctional expression patterns, the consequences for the cell are defective responses and cellular disorders.

Indeed, deregulated epigenetic mechanisms can contribute to a large extent in the developmental course of specific diseases, as is especially demonstrated in the example of monozygotic twins. Although monozygotic twins share the same DNA sequence, they do not always develop the same diseases or present with the same clinical symptoms of a specific disease. In these cases, epigenetic changes can serve as a possible explanation for the discordant phenotype in people who share similar DNA sequence alterations. Epigenetic mechanisms in disorders where genetic and environmental factors have been shown to play a role will be discussed in the context of a set of neurodegenerative diseases such as Alzheimer or Parkinson and autoimmune diseases like systemic lupus erythematosus (SLE), diabetes mellitus type 1, or multiple sclerosis (MS). Other diseases leading to a disruption of epigenetic mechanisms in the cell are clearly monogenetic and can be attributed to the mutation or epigenetic modification of one single gene, some of which are listed in Table 1.1.

1.8.1 SELECTED MONOGENETIC DISEASES

A genetic disorder that is predominantly found in males and interestingly displays more severe symptoms in each passing generation is *Fragile X syndrome*. Fragile X syndrome is a genetic disorder that results from a mutation of the *FMR1* gene located on the X chromosome, thus it mostly affects male individuals [194]. Patients physically present with an elongated face, protruding ears, flat feet, large testes, and low muscle tone and show behavioral characteristics such as stereotypic movements or limited social interaction [195]. The disease is in most cases due to a genetic mutation followed by an epigenetic silencing mechanism in the promoter region of the *FMR1* gene: an increase in the number of CGG trinucleotide repeats in the promoter region of the *FMR1* gene is observed, which leads to methylation of the CGG expansion at the *FMR1* promoter and subsequent epigenetic silencing of the gene [196]. The Fragile X syndrome is also an example for "anticipation" or the Sherman paradox, meaning that the effects of the disease are more pronounced in each passing generation, which is a clear deviation from Mendel's rules. An explanation for this phenomenon is that male carriers in early generations pass on their premutation (between 50 and 200 CGG repeats) on the X chromosome to all of their daughters, however, males with a full mutation

Table 1.1 Selected Monogenetic Diseases

Disorder	Gene	Mutation	Effect of Mutation	Phenotype
Fragile X syndrome (X-linked: males)	*FMR1*	Increase of CGG trinucleotide repeats in *FMR1* promoter	Silencing of *FMR1* gene	Elongated face, protruding ears, larger testes, stereotypic movements
Rubinstein–Taybi syndrome	*CREBBP*, *EP300*	Point mutations	Loss of *CREBBP/EP300* histone acetylase activity	Broad thumbs and toes, small head, small testes, mental disability
Rett syndrome (X-linked: females)	*MECP2*	Point mutations	Loss of methyl-CpG-binding protein 2	Stereotypic hand movements, autistic-like behavior, breathing irregularities, developmental regression
ATRX (X-linked: males)	*ATRX*	Point mutations	SWI/SNF remodeling and DNA methylation changes at repeat sequences	Facial and urogenital abnormalities, alpha-thalassemia, mental retardation
Prader–Willi syndrome (paternally inherited)	*SNRPN*, *NECDIN*, snoRNAs	Point mutations on paternal allele, maternal uniparental disomy, imprinting defect, translocation	Loss of *SNRPB/NECDIN*/snoRNA function	Poor muscle tone, neurocognitive defects, hyperphagia, obesity
Angelman syndrome (maternally inherited)	*UBE3A*	Point mutations on maternal allele, paternal uniparental disomy, imprinting defect, translocation	Loss of ubiquitin E3 ligase activity	Developmental delay, balance disorders, happy-puppet demeanor
Beckwith–Wiedemann syndrome	*IGF-2*, *CDKN1C*	Point mutations, uniparental disomy, imprinting defects	Overexpression of paternally expressed *IFG-2*, no expression of maternally expressed *CDKN1C*	Overgrowth, congenital abnormalities, abdominal wall defects, increased embryonic malignancies

only pass on premutations because male germ cells with full mutations are not viable. Carrier females can pass on full mutations and premutations. In females with premutations, the length of the CGG repeat frequently increases during meiosis, which makes it possible that they pass on a full mutation to their children [197].

Another disorder that leads to incorrect histone acetylation patterns and subsequent dysfunctional gene expression during development is *Rubinstein–Taybi syndrome* (RTS). RTS is morphologically characterized by broad thumbs, broad first toes, a small head, cryptorchidism (small genitals) in males, distinct facial features, and mental disability [198]. It was shown to be due to mutations in the *CREBBP* (*CREB Binding Protein*, *CBP*) gene or in the *EP300* gene, both of which have HAT function and thus contribute to chromatin regulation and to proper gene expression [199,200]. In RTS, one copy of these genes is lost, which reduces the amount of either CREBBP/CBP or EP300 protein leading to dysfunctional development.

A disorder mainly occurring in female patients is *Rett syndrome* (RS). In RS, patients show developmental regression regarding motoric and intellectual abilities starting from the age of 6 to 18 months. Typical hand stereotypes, autistic-like behavior, and breathing irregularities are also characteristic for the disorder [201,202]. Its onset is due to mutations in the *MECP2* gene on the X chromosome, which codes for the methyl-CpG-binding protein 2 [203]. The protein binds to methylated CpG DNA and is essential for mature nerve cells, probably by regulating proper gene transcription in these cells [204,205]. Total lack of protein function is embryonically lethal, which is the reason why only females, who have another functional protein on their second X chromosome, but not males, display a disease phenotype.

Another mental retardation syndrome that is X-linked and thus predominantly affects males is *alpha-thalassemia mental retardation syndrome* (*ATRX*). Physical characteristics of these patients include small head size, widely spaced eyes, a depressed nasal bridge, large lips, coarse facial features, urogenital abnormalities, and alpha-thalassemia. Moderate to severe mental retardation is also encountered in affected individuals [206]. The *ATRX* gene belongs to the family of SWI/SNF chromatin-remodeling complexes and its absence results in changes in DNA methylation patterns at repeat sequences, which may indicate a link between chromatin remodeling, DNA methylation, and gene expression during developmental processes [207].

Prader–Willi and *Angelman syndromes* are two sister-syndromes that show parent-of-origin-dependent transmission, meaning that Prader–Willi syndrome is exclusively inherited from the father and Angelman syndrome from the mother. While patients with Prader–Willi syndrome are characterized by poor muscle tone, neurocognitive defects, hyperphagia and obesity, individuals with Angelman syndrome show developmental delays, balance disorders, and a uniquely happy demeanor. Both syndromes are due to mutations in a region of chromosome 15 (15q11-q13) [208–210]. The reason why these syndromes show parent-of-origin-dependent inheritance is that this region is imprinted on both alleles, a term which is used for the occurrence of parent-of-origin specific gene silencing by DNA methylation, meaning that expression of the genes in this region occurs from either the paternal or the maternal allele, as the other one is silenced by DNA methylation. Thus, if the paternal allele is mutated or if two maternal copies are inherited, Prader–Willi syndrome results, and if the maternal allele is mutated or if two paternal copies are inherited, affected individuals will develop Angelman syndrome. Also, if the imprinting process per se is disturbed and the imprinting patterns are altered, that is, if the paternal allele shows the imprinting configuration of the maternal one and vice versa, the respective phenotypes develop. Regarding the genes mutated in the diseases, it has been shown for Angelman syndrome that about 10–15% of cases are caused by loss-of-function mutations in the ubiquitin E3 ligase gene (*UBE3A*), whereas for Prader–Willi syndrome deletions in the *SNRPN* and *NECDIN* genes and in clusters of snoRNAs have been identified to contribute to the disease phenotype [211–214].

Another syndrome linked to imprinting defects is *Beckwith–Wiedemann syndrome.* Patients presenting with this condition show overgrowth, congenital abnormalities, abdominal wall defects, and an increased incidence of childhood embryonal malignancies [215]. Mutations, uniparental disomy, and imprinting defects have been shown to occur at a region on chromosome 11 (11p15), leading to overexpression of the *IGF-2* gene, which is paternally expressed, or no active copy of *CDKN1C* encoding p57^{kip2}, which is maternally expressed, and subsequent development of Beckwith–Wiedemann symptoms, but also deregulation of other genes in this region are thought to contribute to the phenotype [216].

1.8.2 SELECTED NEURODEGENERATIVE DISEASES

Alzheimer's disease is among the most frequent neurodegenerative diseases to date, mainly affecting people over 65 years [217]. Its signs and symptoms comprise dementia, neuronal loss, and altered synaptic connectivity caused by the extracellular accumulation of amyloid-β peptide, which is aberrantly produced from the amyloid precursor protein (APP) [218]. Mutations in certain genes, as well as environmental factors, predispose to disease development, the latter likely inducing epigenetic alterations that contribute to disease progression [219,220]. Epigenetic modifications have also been implicated in Alzheimer disease by the detection of dysregulated DNA methylation patterns in the entorhinal and dorsolateral prefrontal cortices of Alzheimer patients, and by results from a primate model of Alzheimer, which showed decreased DNMT activity and increased APP mRNA expression in the cortex [221–224]. Also, histone acetylation is overall increased in brains of Alzheimer patients and histone modification patterns are dysregulated, which might be due to a product of APP cleavage that interacts with the HAT TIP60 thus increasing acetylation [225]. In other models, APP overexpression resulted in reduced H3 and H4 acetylation and decreased CBP levels, and treatment with HDAC inhibitors restored histone acetylation and rescued memory deficits in a mouse model of Alzheimer's disease, indicating that HDAC inhibitors may be useful drugs in certain clinical settings of the disease [226,227].

Parkinson's disease is a neurodegenerative disorder of the elderly that is clinically characterized by slow movement, rest tremor, rigidity, and neuropsychiatric symptoms [228]. Its pathophysiology is due to cell death of dopaminergic cells in the substantia nigra of the basal ganglia of the brain, which control voluntary motor movements [229]. No specific cause for the death of these cells has been found to date, but genetic variants, environmental effects, and mitochondrial dysfunction are under debate, and epigenetic mechanisms also seem to play a role in disease modulation [230]. One of the genes implicated in Parkinson's disease, α-synuclein (*SNCA*), is regulated by DNA methylation, and DNA hypomethylation of *SNCA* intron 4 has been detected in brains of Parkinson patients [231]. Furthermore, accumulated SNCA prevents histone acetylation and gene expression at certain genes important for mitochondrial function [232,233]. Thus, a new focus in disease modulation is the possibility of delaying or alleviating symptoms by targeting epigenetic modifications in dopaminergic neurons with HDAC inhibitors. For example, the HDAC inhibitor trichostatin A was able to rescue mitochondrial dysfunction and neuronal death in neuroblastoma cells, which were treated with the neurotoxin MPP +, indicating that alteration of DNA methylation or histone acetylation might also have a positive influence on cell survival [234].

Huntington disease is a type of late-onset neurodegenerative disorder that leads to progressive cognitive impairment, motor dysfunctions such as chorea and dystonia, and psychotic symptoms. The onset of the disease is usually between 35 and 40 years of age. It is autosomal-dominantly inherited and its development is due to a trinucleotide repeat expansion in the Huntingtin encoding gene *HTT*, which leads to the production of a protein with multiple glutamine residues that aggregates and forms intracellular inclusions. Repeat numbers increase from each generation to the next, which accelerates disease onset [235]. Although the disease is due to a genetic mutation, epigenetic alterations have been detected as well, for example, it has been reported that Huntingtin protein can interfere with the acetyltransferase domains of CBP and p300/CBP-associated factor, which inhibits HAT activity and CREB-mediated transcription [236]. Interestingly, an increase in histone acetylation levels by HDAC inhibitors was accompanied by a reduction in motor dysregulation and in neuronal deficits in mouse models of the disease, making HDAC inhibition a promising possible strategy to alleviate disease symptoms in affected individuals [237]. Huntingtin can also associate with PRC2, which is responsible for trimethylation of H3K27 and gene repression, and embryos lacking Huntingtin have lower H3K27 methylation levels [238]. Another histone modification that was shown to be increased in the disease is monoubiquitylation of histone H2A (uH2A), which might be due to a disrupted interaction between Huntingtin and the hPRC1L E3 ubiquitin ligase complex [239]. Indeed, increased uH2A levels were linked to increased H3K9 methylation and gene repression. These data suggest that a disturbed balance between histone acetylation, methylation, and ubiquitylation is an important contributing factor in disease progression, and inhibition of epigenetic modifiers might help to delay disease onset and symptoms.

1.8.3 SELECTED AUTOIMMUNE DISEASES

SLE is an autoimmune disease characterized by the production of antinuclear and other self-reacting antibodies, which lead to inflammatory processes in a variety of organs such as skin, kidney, liver, blood vessels, joints, and the nervous system [240]. It is caused by both genetic and environmental factors, and several studies have shown that epigenetic mechanisms, particularly impaired T-cell DNA methylation, contribute to the pathogenesis of the disease [241]. The two drugs procainamide and hydralazine, an antiarrhythmic agent and a smooth-muscle relaxant, which can both inhibit DNA methylation, were shown to cause a lupus-like disease in a genetically predisposed subset of patients, and treatment of murine T cells with the DNMT inhibitor 5-azaCR also led to lupus-like symptoms in affected mice [242–247]. T cells from patients with active lupus have hypomethylated DNA and reduced DNMT1 expression due to a defect in the ERK pathway, and demethylated CD4 T cells were found to be sufficient to cause autoimmunity and activating lupus in predisposed people [241,248,249]. Furthermore, demethylation and subsequent overexpression of genes found on the epigenetically silenced X chromosome in females might also contribute to the fact that SLE more often affects females than males [250]. For example, demethylation and overexpression have been specifically detected for the *CD40L* gene on the X chromosome in women but not in men suffering from SLE, indicating that demethylation of the Xi chromosome might predispose women to SLE [251].

Rheumatoid arthritis (*RA*) is a systemic autoimmune disease with a prevalence of 1% of the population worldwide [252]. RA patients suffer from chronic inflammation of synovial joints, joint damage, and pain-induced reduced mobility. Monozygotic twins show a concordance of 12–30%, and certain HLA alleles are significantly associated with disease susceptibility, demonstrating the influence of genetic factors in RA [253]. An immune response to triggering agents at the target tissues is thought to present the initiating event in RA, but less is known about epigenetic mechanisms in the disease [254]. It has been reported that synovial fibroblasts without genetic mutations show altered gene expression and demonstrate DNA demethylation and increased expression of normally methylated repetitive DNA elements in RA patients [255]. Additionally, hypermethylation of regulatory and death-associated proteins and aberrant histone and protein deacetylation induced by HDACs could be detected, and the use of HDAC inhibitors as alternative therapeutic options in RA is currently under debate [254,256].

In diabetes mellitus type 1 (T1DM), which is characterized by autoimmune destruction of insulin-producing β-cells in the pancreas, affected individuals show polyuria (excessive urination), polydipsia (excessive drinking), ketoacidosis, fatigue, and weight loss due to the inability of the cells to take up glucose in the absence of insulin [257]. The trigger for the disease is unknown, but comprises genetic as well as environmental factors. Monogenetic twins show a concordance rate of 30–50%, and several screens have identified genetic susceptibility loci such as certain HLA alleles and the insulin gene locus [258]. Regarding epigenetic alterations, analysis of DNA methylation in T1DM patients revealed several differentially methylated CpG sites that were linked to the disease, among which were CpG sites close to the insulin promoter gene and sites that correlated with the time of onset of nephropathy, which is a severe complication in T1DM patients [259,260]. Also, H3K9me2 levels were found to be increased at high-risk genes for T1DM such as the *CLTA4* gene in lymphocytes from T1DM patients compared to healthy controls, and hyperglycemia was shown to affect NFkappaB promoter histone methylation [261,262].

MS is a neurodegenerative autoimmune disease that is caused by genetic and environmental factors. In MS, oligodendrocytes, which produce myelin, and myelin sheaths of nerve cells are destroyed by inflammatory processes, which leads to autonomic, visual, motor and sensory neuronal dysfunction [263]. As observed in other autoimmune diseases, genetic factors such as certain HLA alleles are associated with MS and environmental factors such as sunlight exposure, geographical location, or dietary intake seem to influence disease manifestation, indicating that epigenetic mechanisms might contribute to disease susceptibility [264–266]. However, DNA methylation studies in CD4 lymphocytes isolated from three monozygotic twin pairs that had no concordant MS phenotype did not reveal significant methylation changes in susceptibility genes, suggesting that DNA methylation does not play a major role in susceptibility [53]. Regarding histone modifications, histone citrullination has been reported to be increased in patients with MS compared to individuals without MS, and aberrant citrullination of myelin proteins was associated with myelin sheath instability and increased proteolysis leading to oligodendrocyte apoptosis [267–270]. However, additional studies with larger patient cohorts are clearly needed to elucidate the effects of epigenetic mechanisms in the etiology of the disease.

REFERENCES

[1] Aristotle. Generation of animals. Cambridge, MA: Harvard University Press; 1979.

[2] Maienschein J. Epigenesis and preformationism. In: Edward N. Zalta, editor. The Stanford Encyclopedia of Philosophy. Spring 2012 Edition, Available at: <http://plato.stanford.edu/archives/spr2012/entries/epigenesis/>.

[3] Waddington CH. The epigenotype. 1942. Int J Epidemiol 2012;41(1):10–13.

[4] Waddington CH. The pupal contraction as an epigenetic crisis in *Drosophila*. Proc Zool Soc Lond 1942; A111:181–8.

[5] Haecker V. Entwicklungsgeschichtliche Eigenschaftsanalyse (Phänogenetik). Jena: G. Fischer; 1918.

[6] Haig D. Commentary: the epidemiology of epigenetics. Int J Epidemiol 2012;41(1):13–16.

[7] Nanney DL. Epigenetic control systems. Proc Natl Acad Sci USA 1958;44(7):712–17.
[8] Ephrussi B. The cytoplasm and somatic cell variation. J Cell Physiol Suppl 1958;52(Suppl. 1):35–49.
[9] Holliday R. A new theory of carcinogenesis. Br J Cancer 1979;40(4):513–22.
[10] Barlow DP, Bartolomei MS. Genomic imprinting in mammals. Cold Spring Harb Perspect Biol 2014;6 (2):a018382.
[11] Chaligne R, Heard E. X-chromosome inactivation in development and cancer. FEBS Lett 2014;588 (15):2514–22.
[12] Rothbart SB, Strahl BD. Interpreting the language of histone and DNA modifications. Biochim Biophys Acta 2014;1839(8):627–43.
[13] Jenuwein T, Allis CD. Translating the histone code. Science 2001;293(5532):1074–80.
[14] Turner BM. Histone acetylation and an epigenetic code. BioEssays: News Rev Mol Cell Dev Biol 2000;22(9):836–45.
[15] Denli AM, Hannon GJ. RNAi: an ever-growing puzzle. Trends Biochem Sci 2003;28(4):196–201.
[16] Lee JT. Lessons from X-chromosome inactivation: long ncRNA as guides and tethers to the epigenome. Genes Dev 2009;23(16):1831–42.
[17] Peschansky VJ, Wahlestedt C. Non-coding RNAs as direct and indirect modulators of epigenetic regulation. Epigenetics: Off J DNA Methylation Soc 2014;9(1):3–12.
[18] Berger SL, Kouzarides T, Shiekhattar R, Shilatifard A. An operational definition of epigenetics. Genes Dev 2009;23(7):781–3.
[19] Sorm F, Vesely J. The activity of a new antimetabolite, 5-Azacytidine, against lymphoid leukaemia in Ak mice. Neoplasma 1964;11:123–30.
[20] Sorm F, Vesely J. Effect of 5-aza-2′-deoxycytidine against leukemic and hemopoietic tissues in AKR mice. Neoplasma 1968;15(4):339–43.
[21] Von Hoff DD, Slavik M, Muggia FM. 5-Azacytidine. A new anticancer drug with effectiveness in acute myelogenous leukemia. Ann Int Med 1976;85(2):237–45.
[22] Constantinides PG, Taylor SM, Jones PA. Phenotypic conversion of cultured mouse embryo cells by aza pyrimidine nucleosides. Dev Biol 1978;66(1):57–71.
[23] Taylor SM, Jones PA. Multiple new phenotypes induced in 10T1/2 and 3T3 cells treated with 5-azacytidine. Cell 1979;17(4):771–9.
[24] Jones PA, Taylor SM. Cellular differentiation, cytidine analogs and DNA methylation. Cell 1980;20 (1):85–93.
[25] Gore SD, Jones C, Kirkpatrick P. Decitabine. Nat Rev Drug Discov 2006;5(11):891–2.
[26] Kaminskas E, Farrell AT, Wang YC, Sridhara R, Pazdur R. FDA drug approval summary: azacitidine (5-azacytidine, Vidaza) for injectable suspension. Oncologist 2005;10(3):176–82.
[27] Riggs MG, Whittaker RG, Neumann JR, Ingram VM. n-Butyrate causes histone modification in HeLa and friend erythroleukaemia cells. Nature 1977;268(5619):462–4.
[28] Sealy L, Chalkley R. The effect of sodium butyrate on histone modification. Cell 1978;14(1):115–21.
[29] Vidali G, Boffa LC, Bradbury EM, Allfrey VG. Butyrate suppression of histone deacetylation leads to accumulation of multiacetylated forms of histones H3 and H4 and increased DNase I sensitivity of the associated DNA sequences. Proc Natl Acad Sci USA 1978;75(5):2239–43.
[30] Grant S, Easley C, Kirkpatrick P. Vorinostat. Nat Rev Drug Discov 2007;6(1):21–2.
[31] Kornberg RD. Chromatin structure: a repeating unit of histones and DNA. Science 1974;184(4139): 868–71.
[32] Pinney SE. Mammalian non-CpG methylation: stem cells and beyond. Biology 2014;3(4):739–51.
[33] Selker EU, Cambareri EB, Garrett PW, Haack KR, Jensen BC, Schabtach E. DNA methylation and control of genome organization in *Neurospora crassa*. Gene 1988;74(1):109–11.
[34] Vanyushin BF, Kirnos MD. DNA methylation in plants. Gene 1988;74(1):117–21.

[35] Ehrlich M, Gama-Sosa MA, Huang LH, Midgett RM, Kuo KC, McCune RA, et al. Amount and distribution of 5-methylcytosine in human DNA from different types of tissues of cells. Nucl Acids Res 1982;10(8):2709–21.
[36] Bird A. DNA methylation patterns and epigenetic memory. Genes Dev 2002;16(1):6–21.
[37] Gardiner-Garden M, Frommer M. CpG islands in vertebrate genomes. J Mol Biol 1987;196(2):261–82.
[38] Bestor TH. The DNA methyltransferases of mammals. Hum Mol Genet 2000;9(16):2395–402.
[39] Li E, Bestor TH, Jaenisch R. Targeted mutation of the DNA methyltransferase gene results in embryonic lethality. Cell 1992;69(6):915–26.
[40] Lei H, Oh SP, Okano M, Juttermann R, Goss KA, Jaenisch R, et al. *De novo* DNA cytosine methyltransferase activities in mouse embryonic stem cells. Development 1996;122(10):3195–205.
[41] Okano M, Xie S, Li E. Dnmt2 is not required for *de novo* and maintenance methylation of viral DNA in embryonic stem cells. Nucl Acids Res 1998;26(11):2536–40.
[42] Goll MG, Kirpekar F, Maggert KA, Yoder JA, Hsieh CL, Zhang X, et al. Methylation of tRNAAsp by the DNA methyltransferase homolog Dnmt2. Science 2006;311(5759):395–8.
[43] Okano M, Xie S, Li E. Cloning and characterization of a family of novel mammalian DNA (cytosine-5) methyltransferases. Nat Genet 1998;19(3):219–20.
[44] Okano M, Bell DW, Haber DA, Li E. DNA methyltransferases Dnmt3a and Dnmt3b are essential for *de novo* methylation and mammalian development. Cell 1999;99(3):247–57.
[45] Ehrlich M. The ICF syndrome, a DNA methyltransferase 3B deficiency and immunodeficiency disease. Clin Immunol 2003;109(1):17–28.
[46] Watt F, Molloy PL. Cytosine methylation prevents binding to DNA of a HeLa cell transcription factor required for optimal expression of the adenovirus major late promoter. Genes Dev 1988;2(9):1136–43.
[47] Hendrich B, Guy J, Ramsahoye B, Wilson VA, Bird A. Closely related proteins MBD2 and MBD3 play distinctive but interacting roles in mouse development. Genes Dev 2001;15(6):710–23.
[48] Bird AP, Wolffe AP. Methylation-induced repression—belts, braces, and chromatin. Cell 1999;99 (5):451–4.
[49] Joulie M, Miotto B, Defossez PA. Mammalian methyl-binding proteins: what might they do? BioEssays: News Rev Mol Cell Dev Biol 2010;32(12):1025–32.
[50] Wade PA, Gegonne A, Jones PL, Ballestar E, Aubry F, Wolffe AP. Mi-2 complex couples DNA methylation to chromatin remodelling and histone deacetylation. Nat Genet 1999;23(1):62–6.
[51] Singh J, Saxena A, Christodoulou J, Ravine D. MECP2 genomic structure and function: insights from ENCODE. Nucl Acids Res 2008;36(19):6035–47.
[52] Martinowich K, Hattori D, Wu H, Fouse S, He F, Hu Y, et al. DNA methylation-related chromatin remodeling in activity-dependent BDNF gene regulation. Science 2003;302(5646):890–3.
[53] Baranzini SE, Mudge J, van Velkinburgh JC, Khankhanian P, Khrebtukova I, Miller NA, et al. Genome, epigenome and RNA sequences of monozygotic twins discordant for multiple sclerosis. Nature 2010;464 (7293):1351–6.
[54] Bhutani N, Burns DM, Blau HM. DNA demethylation dynamics. Cell 2011;146(6):866–72.
[55] Luger K, Mader AW, Richmond RK, Sargent DF, Richmond TJ. Crystal structure of the nucleosome core particle at 2.8 A resolution. Nature 1997;389(6648):251–60.
[56] Allis CD, Jenuwein T, Reinberg D. Epigenetics. Cold Spring Harbor, NY: Cold Spring Harbor Laboratory Press; 2007.
[57] Pauler FM, Sloane MA, Huang R, Regha K, Koerner MV, Tamir I, et al. H3K27me3 forms BLOCs over silent genes and intergenic regions and specifies a histone banding pattern on a mouse autosomal chromosome. Genome Res 2009;19(2):221–33.
[58] Wen B, Wu H, Shinkai Y, Irizarry RA, Feinberg AP. Large histone H3 lysine 9 dimethylated chromatin blocks distinguish differentiated from embryonic stem cells. Nat Genet 2009;41(2):246–50.

[59] Izzo A, Schneider R. Chatting histone modifications in mammals. Brief Funct Genom 2010;9(5–6):429–43.
[60] Strahl BD, Allis CD. The language of covalent histone modifications. Nature 2000;403(6765):41–5.
[61] Vettese-Dadey M, Grant PA, Hebbes TR, Crane- Robinson C, Allis CD, Workman JL. Acetylation of histone H4 plays a primary role in enhancing transcription factor binding to nucleosomal DNA *in vitro*. EMBO J 1996;15(10):2508–18.
[62] Neely KE, Hassan AH, Brown CE, Howe L, Workman JL. Transcription activator interactions with multiple SWI/SNF subunits. Mol Cell Biol 2002;22(6):1615–25.
[63] Adams-Cioaba MA, Min J. Structure and function of histone methylation binding proteins. Biochem Cell Biol 2009;87(1):93–105.
[64] Winter S, Fischle W, Seiser C. Modulation of 14-3-3 interaction with phosphorylated histone H3 by combinatorial modification patterns. Cell Cycle 2008;7(10):1336–42.
[65] Kurdistani SK, Grunstein M. Histone acetylation and deacetylation in yeast. Nat Rev Mol Cell Biol 2003;4(4):276–84.
[66] Carrozza MJ, Li B, Florens L, Suganuma T, Swanson SK, Lee KK, et al. Histone H3 methylation by Set2 directs deacetylation of coding regions by Rpd3S to suppress spurious intragenic transcription. Cell 2005;123(4):581–92.
[67] Yang XJ, Seto E. Collaborative spirit of histone deacetylases in regulating chromatin structure and gene expression. Curr Opin Genet Dev 2003;13(2):143–53.
[68] Prigent C, Dimitrov S. Phosphorylation of serine 10 in histone H3, what for? J Cell Sci 2003;116(Pt 18):3677–85.
[69] Dormann HL, Tseng BS, Allis CD, Funabiki H, Fischle W. Dynamic regulation of effector protein binding to histone modifications: the biology of HP1 switching. Cell Cycle 2006;5(24):2842–51.
[70] Morrison AJ, Shen X. DNA repair in the context of chromatin. Cell Cycle 2005;4(4):568–71.
[71] Stallcup MR. Role of protein methylation in chromatin remodeling and transcriptional regulation. Oncogene 2001;20(24):3014–20.
[72] Zhang Y, Reinberg D. Transcription regulation by histone methylation: interplay between different covalent modifications of the core histone tails. Genes Dev 2001;15(18):2343–60.
[73] Hublitz P, Albert M, Peters AH. Mechanisms of transcriptional repression by histone lysine methylation. Int J Dev Biol 2009;53(2–3):335–54.
[74] Yap KL, Zhou MM. Keeping it in the family: diverse histone recognition by conserved structural folds. Crit Rev Biochem Mol Biol 2010;45(6):488–505.
[75] Hunt CR, Ramnarain D, Horikoshi N, Iyengar P, Pandita RK, Shay JW, et al. Histone modifications and DNA double-strand break repair after exposure to ionizing radiations. Radiat Res 2013;179(4):383–92.
[76] Shi Y, Lan F, Matson C, Mulligan P, Whetstine JR, Cole PA, et al. Histone demethylation mediated by the nuclear amine oxidase homolog LSD1. Cell 2004;119(7):941–53.
[77] Metzger E, Wissmann M, Yin N, Muller JM, Schneider R, Peters AH, et al. LSD1 demethylates repressive histone marks to promote androgen-receptor-dependent transcription. Nature 2005;437(7057):436–9.
[78] Mosammaparast N, Shi Y. Reversal of histone methylation: biochemical and molecular mechanisms of histone demethylases. Ann Rev Biochem 2010;79:155–79.
[79] Eisen JA, Sweder KS, Hanawalt PC. Evolution of the SNF2 family of proteins: subfamilies with distinct sequences and functions. Nucl Acids Res 1995;23(14):2715–23.
[80] Ahmad K, Henikoff S. The histone variant H3.3 marks active chromatin by replication-independent nucleosome assembly. Mol Cell 2002;9(6):1191–200.

[81] Howman EV, Fowler KJ, Newson AJ, Redward S, MacDonald AC, Kalitsis P, et al. Early disruption of centromeric chromatin organization in centromere protein A (Cenpa) null mice. Proc Natl Acad Sci USA 2000;97(3):1148–53.
[82] Redon C, Pilch D, Rogakou E, Sedelnikova O, Newrock K, Bonner W. Histone H2A variants H2AX and H2AZ. Curr Opin Genet Dev 2002;12(2):162–9.
[83] Fire A, Xu S, Montgomery MK, Kostas SA, Driver SE, Mello CC. Potent and specific genetic interference by double-stranded RNA in *Caenorhabditis elegans*. Nature 1998;391(6669):806–11.
[84] Volpe T, Martienssen RA. RNA interference and heterochromatin assembly. Cold Spring Harb Perspect Biol 2011;3(9):a003731.
[85] Volpe TA, Kidner C, Hall IM, Teng G, Grewal SI, Martienssen RA. Regulation of heterochromatic silencing and histone H3 lysine-9 methylation by RNAi. Science 2002;297(5588):1833–7.
[86] Reinhart BJ, Bartel DP. Small RNAs correspond to centromere heterochromatic repeats. Science 2002;297(5588):1831.
[87] Noma K, Sugiyama T, Cam H, Verdel A, Zofall M, Jia S, et al. RITS acts in cis to promote RNA interference-mediated transcriptional and post-transcriptional silencing. Nat Genet 2004;36(11): 1174–80.
[88] Pal-Bhadra M, Leibovitch BA, Gandhi SG, Chikka MR, Bhadra U, Birchler JA, et al. Heterochromatic silencing and HP1 localization in *Drosophila* are dependent on the RNAi machinery. Science 2004;303 (5658):669–72.
[89] Kanellopoulou C, Muljo SA, Kung AL, Ganesan S, Drapkin R, Jenuwein T, et al. Dicer-deficient mouse embryonic stem cells are defective in differentiation and centromeric silencing. Genes Dev 2005;19(4):489–501.
[90] Vastenhouw NL, Brunschwig K, Okihara KL, Muller F, Tijsterman M, Plasterk RH. Gene expression: long-term gene silencing by RNAi. Nature 2006;442(7105):882.
[91] Mallo M, Alonso CR. The regulation of Hox gene expression during animal development. Development 2013;140(19):3951–63.
[92] Yekta S, Tabin CJ, Bartel DP. MicroRNAs in the Hox network: an apparent link to posterior prevalence. Nat Rev Genet 2008;9(10):789–96.
[93] Lund AH. miR-10 in development and cancer. Cell Death Differ 2010;17(2):209–14.
[94] Holoch D, Moazed D. RNA-mediated epigenetic regulation of gene expression. Nat Rev Genet 2015;16(2):71–84.
[95] Yang L, Froberg JE, Lee JT. Long noncoding RNAs: fresh perspectives into the RNA world. Trends Biochem Sci 2014;39(1):35–43.
[96] Kohlmaier A, Savarese F, Lachner M, Martens J, Jenuwein T, Wutz A. A chromosomal memory triggered by Xist regulates histone methylation in X inactivation. PLoS Biol 2004;2(7):E171.
[97] Plath K, Fang J, Mlynarczyk-Evans SK, Cao R, Worringer KA, Wang H, et al. Role of histone H3 lysine 27 methylation in X inactivation. Science 2003;300(5616):131–5.
[98] Wang KC, Yang YW, Liu B, Sanyal A, Corces-Zimmerman R, Chen Y, et al. A long noncoding RNA maintains active chromatin to coordinate homeotic gene expression. Nature 2011;472(7341):120–4.
[99] Cantone I, Fisher AG. Epigenetic programming and reprogramming during development. Nat Struct Mol Biol 2013;20(3):282–9.
[100] Morgan HD, Santos F, Green K, Dean W, Reik W. Epigenetic reprogramming in mammals. Hum Mol Genet 2005;14(Spec No 1):R47–58.
[101] Li E. Chromatin modification and epigenetic reprogramming in mammalian development. Nat Rev Genet 2002;3(9):662–73.
[102] Surani MA. Reprogramming of genome function through epigenetic inheritance. Nature 2001;414 (6859):122–8.

[103] Hochedlinger K, Plath K. Epigenetic reprogramming and induced pluripotency. Development 2009;136 (4):509–23.
[104] Gill ME, Erkek S, Peters AH. Parental epigenetic control of embryogenesis: a balance between inheritance and reprogramming? Curr Opin Cell Biol 2012;24(3):387–96.
[105] Gu TP, Guo F, Yang H, Wu HP, Xu GF, Liu W, et al. The role of Tet3 DNA dioxygenase in epigenetic reprogramming by oocytes. Nature 2011;477(7366):606–10.
[106] Santos F, Hendrich B, Reik W, Dean W. Dynamic reprogramming of DNA methylation in the early mouse embryo. Dev Biol 2002;241(1):172–82.
[107] Nakamura T, Liu YJ, Nakashima H, Umehara H, Inoue K, Matoba S, et al. PGC7 binds histone H3K9me2 to protect against conversion of 5mC to 5hmC in early embryos. Nature 2012;486 (7403):415–19.
[108] Nakamura T, Arai Y, Umehara H, Masuhara M, Kimura T, Taniguchi H, et al. PGC7/Stella protects against DNA demethylation in early embryogenesis. Nat Cell Biol 2007;9(1):64–71.
[109] Fisher CL, Fisher AG. Chromatin states in pluripotent, differentiated, and reprogrammed cells. Curr Opin Genet Dev 2011;21(2):140–6.
[110] Yoshimizu T, Obinata M, Matsui Y. Stage-specific tissue and cell interactions play key roles in mouse germ cell specification. Development 2001;128(4):481–90.
[111] Margueron R, Reinberg D. The Polycomb complex PRC2 and its mark in life. Nature 2011;469 (7330):343–9.
[112] Lee J, Inoue K, Ono R, Ogonuki N, Kohda T, Kaneko-Ishino T, et al. Erasing genomic imprinting memory in mouse clone embryos produced from day 11.5 primordial germ cells. Development 2002;129(8):1807–17.
[113] Chuva de Sousa Lopes SM, Hayashi K, Shovlin TC, Mifsud W, Surani MA, McLaren A. X chromosome activity in mouse XX primordial germ cells. PLoS Genet 2008;4(2):e30.
[114] Davis TL, Trasler JM, Moss SB, Yang GJ, Bartolomei MS. Acquisition of the H19 methylation imprint occurs differentially on the parental alleles during spermatogenesis. Genomics 1999;58(1):18–28.
[115] Lucifero D, Mann MR, Bartolomei MS, Trasler JM. Gene-specific timing and epigenetic memory in oocyte imprinting. Hum Mol Genet 2004;13(8):839–49.
[116] Saitou M, Barton SC, Surani MA. A molecular programme for the specification of germ cell fate in mice. Nature 2002;418(6895):293–300.
[117] Briggs R, King TJ. Transplantation of living nuclei from blastula cells into enucleated frogs' eggs. Proc Natl Acad Sci USA 1952;38(5):455–63.
[118] Gurdon JB, Elsdale TR, Fischberg M. Sexually mature individuals of *Xenopus laevis* from the transplantation of single somatic nuclei. Nature 1958;182(4627):64–5.
[119] Hochedlinger K, Jaenisch R. Nuclear transplantation: lessons from frogs and mice. Curr Opin Cell Biol 2002;14(6):741–8.
[120] Takahashi K, Yamanaka S. Induction of pluripotent stem cells from mouse embryonic and adult fibroblast cultures by defined factors. Cell 2006;126(4):663–76.
[121] Bock C, Kiskinis E, Verstappen G, Gu H, Boulting G, Smith ZD, et al. Reference maps of human ES and iPS cell variation enable high-throughput characterization of pluripotent cell lines. Cell 2011;144 (3):439–52.
[122] Saha K, Jaenisch R. Technical challenges in using human induced pluripotent stem cells to model disease. Cell Stem Cell 2009;5(6):584–95.
[123] Mostoslavsky R, Alt FW, Rajewsky K. The lingering enigma of the allelic exclusion mechanism. Cell 2004;118(5):539–44.
[124] Deng Q, Ramskold D, Reinius B, Sandberg R. Single-cell RNA-seq reveals dynamic, random monoallelic gene expression in mammalian cells. Science 2014;343(6167):193–6.

[125] Leung D, Jung I, Rajagopal N, Schmitt A, Selvaraj S, Lee AY, et al. Integrative analysis of haplotype-resolved epigenomes across human tissues. Nature 2015;518(7539):350–4.
[126] Morison IM, Ramsay JP, Spencer HG. A census of mammalian imprinting. Trends Genet 2005;21(8):457–65.
[127] McGrath J, Solter D. Completion of mouse embryogenesis requires both the maternal and paternal genomes. Cell 1984;37(1):179–83.
[128] Surani MA, Barton SC, Norris ML. Development of reconstituted mouse eggs suggests imprinting of the genome during gametogenesis. Nature 1984;308(5959):548–50.
[129] Glenn CC, Driscoll DJ, Yang TP, Nicholls RD. Genomic imprinting: potential function and mechanisms revealed by the Prader-Willi and Angelman syndromes. Mol Hum Reprod 1997;3(4):321–32.
[130] Bourc'his D, Xu GL, Lin CS, Bollman B, Bestor TH. Dnmt3L and the establishment of maternal genomic imprints. Science 2001;294(5551):2536–9.
[131] Kaneda M, Okano M, Hata K, Sado T, Tsujimoto N, Li E, et al. Essential role for *de novo* DNA methyltransferase Dnmt3a in paternal and maternal imprinting. Nature 2004;429(6994):900–3.
[132] Jia D, Jurkowska RZ, Zhang X, Jeltsch A, Cheng X. Structure of Dnmt3a bound to Dnmt3L suggests a model for *de novo* DNA methylation. Nature 2007;449(7159):248–51.
[133] Li E, Beard C, Jaenisch R. Role for DNA methylation in genomic imprinting. Nature 1993;366(6453):362–5.
[134] Proudhon C, Duffie R, Ajjan S, Cowley M, Iranzo J, Carbajosa G, et al. Protection against *De Novo* methylation is instrumental in maintaining Parent-of-Origin methylation inherited from the gametes. Mol Cell 2012;47(6):909–20.
[135] Xie W, Barr CL, Kim A, Yue F, Lee AY, Eubanks J, et al. Base-resolution analyses of sequence and parent-of-origin dependent DNA methylation in the mouse genome. Cell 2012;148(4):816–31.
[136] Meng L, Person RE, Beaudet AL. Ube3a-ATS is an atypical RNA polymerase II transcript that represses the paternal expression of Ube3a. Hum Mol Genet 2012;21(13):3001–12.
[137] Sleutels F, Zwart R, Barlow DP. The non-coding air RNA is required for silencing autosomal imprinted genes. Nature 2002;415(6873):810–13.
[138] Williamson CM, Ball ST, Dawson C, Mehta S, Beechey CV, Fray M, et al. Uncoupling antisense-mediated silencing and DNA methylation in the imprinted gnas cluster. PLoS Genet 2011;7(3):e1001347.
[139] Mancini-Dinardo D, Steele SJ, Levorse JM, Ingram RS, Tilghman SM. Elongation of the Kcnq1ot1 transcript is required for genomic imprinting of neighboring genes. Genes Dev 2006;20(10):1268–82.
[140] Bell AC, Felsenfeld G. Methylation of a CTCF-dependent boundary controls imprinted expression of the Igf2 gene. Nature 2000;405(6785):482–5.
[141] Hark AT, Schoenherr CJ, Katz DJ, Ingram RS, Levorse JM, Tilghman SM. CTCF mediates methylation-sensitive enhancer-blocking activity at the H19/Igf2 locus. Nature 2000;405(6785):486–9.
[142] Derrien T, Johnson R, Bussotti G, Tanzer A, Djebali S, Tilgner H, et al. The GENCODE v7 catalog of human long noncoding RNAs: analysis of their gene structure, evolution, and expression. Genome Res 2012;22(9):1775–89.
[143] Latos PA, Pauler FM, Koerner MV, Şenergin HB, Hudson QJ, Stocsits RR, et al. Airn transcriptional overlap, but not its lncRNA products, induces imprinted Igf2r silencing. Science 2012;338(6113):1469–72.
[144] Nagano T, Mitchell JA, Sanz LA, Pauler FM, Ferguson-Smith AC, Feil R, et al. The air noncoding RNA epigenetically silences transcription by targeting G9a to chromatin. Science 2008;322(5908):1717–20.
[145] Terranova R, Yokobayashi S, Stadler MB, Otte AP, van Lohuizen M, Orkin SH, et al. Polycomb group proteins Ezh2 and Rnf2 direct genomic contraction and imprinted repression in early mouse embryos. Dev Cell 2008;15(5):668–79.

[146] Prickett AR, Oakey RJ. A survey of tissue-specific genomic imprinting in mammals. Mol Genet Genom 2012;287(8):621–30.
[147] Pauler FM, Barlow DP, Hudson QJ. Mechanisms of long range silencing by imprinted macro non-coding RNAs. Curr Opin Genet Dev 2012;22(3):283–9.
[148] Muller HJ, League BB, Offermann CA. Effects of dosage changes of sex-linked genes, and the compensatory effect of other gene differences between male and female. Anatom Record (Suppl) 1931;51:110.
[149] Mukherjee AS, Beermann W. Synthesis of ribonucleic acid by the X-chromosomes of *Drosophila melanogaster* and the problem of dosage compensation. Nature 1965;207(998):785–6.
[150] Lyon MF. Gene action in the X-chromosome of the mouse (*Mus musculus* L.). Nature 1961; 190:372–3.
[151] Brockdorff N, Ashworth A, Kay GF, McCabe VM, Norris DP, Cooper PJ, et al. The product of the mouse Xist gene is a 15 kb inactive X-specific transcript containing no conserved ORF and located in the nucleus. Cell 1992;71(3):515–26.
[152] Brown CJ, Ballabio A, Rupert JL, Lafreniere RG, Grompe M, Tonlorenzi R, et al. A gene from the region of the human X inactivation centre is expressed exclusively from the inactive X chromosome. Nature 1991;349(6304):38–44.
[153] Chadwick BP, Willard HF. Multiple spatially distinct types of facultative heterochromatin on the human inactive X chromosome. Proc Natl Acad Sci USA 2004;101(50):17450–5.
[154] Escamilla-Del-Arenal M, da Rocha ST, Spruijt CG, Masui O, Renaud O, Smits AH, et al. Cdyl, a new partner of the inactive X chromosome and potential reader of H3K27me3 and H3K9me2. Mol Cell Biol 2013;33(24):5005–20.
[155] Skuse DH. X-linked genes and mental functioning. Hum Mol Genet 2005;14(Spec No 1):R27–32.
[156] Van den Veyver IB. Skewed X inactivation in X-linked disorders. Semin Reprod Med 2001;19 (2):183–91.
[157] Lee JT, Bartolomei MS. X-inactivation, imprinting, and long noncoding RNAs in health and disease. Cell 2013;152(6):1308–23.
[158] Van der Meulen J, Sanghvi V, Mavrakis K, Durinck K, Fang F, Matthijssens F, et al. The H3K27me3 demethylase UTX is a gender-specific tumor suppressor in T-cell acute lymphoblastic leukemia. Blood 2015;125(1):13–21.
[159] Barr ML, Moore KL. Chromosomes, sex chromatin, and cancer. Proc Canadian Cancer Conf, 2. 1957. p. 3–16.
[160] Wareham KA, Lyon MF, Glenister PH, Williams ED. Age related reactivation of an X-linked gene. Nature 1987;327(6124):725–7.
[161] Yildirim E, Kirby JE, Brown DE, Mercier FE, Sadreyev RI, Scadden DT, et al. Xist RNA is a potent suppressor of hematologic cancer in mice. Cell 2013;152(4):727–42.
[162] Chaligne R, Popova T, Mendoza-Parra MA, Saleem MA, Gentien D, Ban K, et al. The inactive X chromosome is epigenetically unstable and transcriptionally labile in breast cancer. Genome Res 2015.
[163] Muller HJ. Types of visible variations induced by X-rays in *Drosophila*. J Genet 1930;22(3):299–334.
[164] Girton JR, Johansen KM. Chromatin structure and the regulation of gene expression: the lessons of PEV in *Drosophila*. Adv Genet 2008;61:1–43.
[165] Allshire RC, Ekwall K. Epigenetic regulation of chromatin states in schizosaccharomyces pombe. Cold Spring Harbor Perspect Biol 2015;7(7):a018770.
[166] Blewitt M, Whitelaw E. The use of mouse models to study epigenetics. Cold Spring Harb Perspect Biol 2013;5:a017939.
[167] Brockdorff N, Turner BM. Dosage compensation in mammals. Cold Spring Harbor Perspect Biol 2015;7(3):a019406.

[168] Spofford JB. Position-effect variegation in *Drosophila.* In: Ashburner MN, editor. The Genetics and Biology of *Drosophila.* New York, NY: Academic Press; 1976. p. 955–1018.
[169] Elgin SC, Reuter G. Position-effect variegation, heterochromatin formation, and gene silencing in Drosophila. Cold Spring Harbor Perspect Biol 2013;5(8):a017780.
[170] Dimitri P, Pisano C. Position effect variegation in *Drosophila melanogaster*: relationship between suppression effect and the amount of Y chromosome. Genetics 1989;122(4):793–800.
[171] Grigliatti T. Position-effect variegation—an assay for nonhistone chromosomal proteins and chromatin assembly and modifying factors. Methods Cell Biol 1991;35:587–627.
[172] Reuter G, Spierer P. Position effect variegation and chromatin proteins. BioEssays: News Rev Mol Cell Dev Biol 1992;14(9):605–12.
[173] Heard E, Martienssen RA. Transgenerational epigenetic inheritance: myths and mechanisms. Cell 2014;157(1):95–109.
[174] Monk M, Boubelik M, Lehnert S. Temporal and regional changes in DNA methylation in the embryonic, extraembryonic and germ cell lineages during mouse embryo development. Development 1987;99 (3):371–82.
[175] Grossniklaus U, Kelly WG, Kelly B, Ferguson-Smith AC, Pembrey M, Lindquist S. Transgenerational epigenetic inheritance: how important is it? Nat Rev Genet 2013;14(3):228–35.
[176] Morgan HD, Sutherland HG, Martin DI, Whitelaw E. Epigenetic inheritance at the agouti locus in the mouse. Nat Genet 1999;23(3):314–18.
[177] Cropley JE, Suter CM, Beckman KB, Martin DI. Germ-line epigenetic modification of the murine A vy allele by nutritional supplementation. Proc Natl Acad Sci USA 2006;103(46):17308–12.
[178] Waterland RA, Travisano M, Tahiliani KG. Diet-induced hypermethylation at agouti viable yellow is not inherited transgenerationally through the female. FASEB J 2007;21(12):3380–5.
[179] Anway MD, Cupp AS, Uzumcu M, Skinner MK. Epigenetic transgenerational actions of endocrine disruptors and male fertility. Science 2005;308(5727):1466–9.
[180] Painter RC, Osmond C, Gluckman P, Hanson M, Phillips DI, Roseboom TJ. Transgenerational effects of prenatal exposure to the Dutch famine on neonatal adiposity and health in later life. BJOG 2008;115 (10):1243–9.
[181] Veenendaal MV, Painter RC, de Rooij SR, Bossuyt PM, van der Post JA, Gluckman PD, et al. Transgenerational effects of prenatal exposure to the 1944–45 Dutch famine. BJOG 2013;120(5):548–53.
[182] Jimenez-Chillaron JC, Hernandez-Valencia M, Reamer C, Fisher S, Joszi A, Hirshman M, et al. Beta-cell secretory dysfunction in the pathogenesis of low birth weight-associated diabetes: a murine model. Diabetes 2005;54(3):702–11.
[183] Jimenez-Chillaron JC, Isganaitis E, Charalambous M, Gesta S, Pentinat-Pelegrin T, Faucette RR, et al. Intergenerational transmission of glucose intolerance and obesity by in utero undernutrition in mice. Diabetes 2009;58(2):460–8.
[184] Ng SF, Lin RC, Laybutt DR, Barres R, Owens JA, Morris MJ. Chronic high-fat diet in fathers programs β-cell dysfunction in female rat offspring. Nature 2010;467(7318):963–6.
[185] Ost A, Lempradl A, Casas E, Weigert M, Tiko T, Deniz M, et al. Paternal diet defines offspring chromatin state and intergenerational obesity. Cell 2014;159(6):1352–64.
[186] Weaver IC, Cervoni N, Champagne FA, D'Alessio AC, Sharma S, Seckl JR, et al. Epigenetic programming by maternal behavior. Nat Neurosci 2004;7(8):847–54.
[187] Cubas P, Vincent C, Coen E. An epigenetic mutation responsible for natural variation in floral symmetry. Nature 1999;401(6749):157–61.
[188] Zhong SH, Liu JZ, Jin H, Lin L, Li Q, Chen Y, et al. Warm temperatures induce transgenerational epigenetic release of RNA silencing by inhibiting siRNA biogenesis in Arabidopsis. Proc Natl Acad Sci USA 2013;110(22):9171–6.

[189] Cortijo S, Wardenaar R, Colome-Tatche M, Gilly A, Etcheverry M, Labadie K, et al. Mapping the epigenetic basis of complex traits. Science 2014;343(6175):1145–8.
[190] Schmitz RJ, Ecker JR. Epigenetic and epigenomic variation in Arabidopsis thaliana. Trends Plant Sci 2012;17(3):149–54.
[191] Remy JJ. Stable inheritance of an acquired behavior in *Caenorhabditis elegans*. Curr Biol 2010;20(20): R877–8.
[192] Greer EL, Maures TJ, Ucar D, Hauswirth AG, Mancini E, Lim JP, et al. Transgenerational epigenetic inheritance of longevity in *Caenorhabditis elegans*. Nature 2011;479(7373):365–71.
[193] Rechavi O, Houri-Ze'evi L, Anava S, Goh WS, Kerk SY, Hannon GJ, et al. Starvation-induced transgenerational inheritance of small RNAs in *C. elegans*. Cell 2014;158(2):277–87.
[194] Harrison CJ, Jack EM, Allen TD, Harris R. The fragile X: a scanning electron microscope study. J Med Genet 1983;20(4):280–5.
[195] Hagerman RJ, Van Housen K, Smith AC, McGavran L. Consideration of connective tissue dysfunction in the fragile X syndrome. Am J Med Genet 1984;17(1):111–21.
[196] Verkerk AJ, Pieretti M, Sutcliffe JS, Fu YH, Kuhl DP, Pizzuti A, et al. Identification of a gene (FMR-1) containing a CGG repeat coincident with a breakpoint cluster region exhibiting length variation in fragile X syndrome. Cell 1991;65(5):905–14.
[197] Fu YH, Kuhl DP, Pizzuti A, Pieretti M, Sutcliffe JS, Richards S, et al. Variation of the CGG repeat at the fragile X site results in genetic instability: resolution of the Sherman paradox. Cell 1991;67 (6):1047–58.
[198] Rubinstein JH, Taybi H. Broad thumbs and toes and facial abnormalities. A possible mental retardation syndrome. Am J Dis Child 1963;105:588–608.
[199] Petrij F, Giles RH, Dauwerse HG, Saris JJ, Hennekam RC, Masuno M, et al. Rubinstein-Taybi syndrome caused by mutations in the transcriptional co-activator CBP. Nature 1995;376(6538):348–51.
[200] Roelfsema JH, White SJ, Ariyürek Y, Bartholdi D, Niedrist D, Papadia F, et al. Genetic heterogeneity in Rubinstein-Taybi syndrome: mutations in both the CBP and EP300 genes cause disease. Am J Hum Genet 2005;76(4):572–80.
[201] Rett A. On a unusual brain atrophy syndrome in hyperammonemia in childhood. Wien Med Wochenschr 1966;116(37):723–6.
[202] Hagberg B, Aicardi J, Dias K, Ramos O. A progressive syndrome of autism, dementia, ataxia, and loss of purposeful hand use in girls: Rett's syndrome: report of 35 cases. Ann Neurol 1983;14(4): 471–9.
[203] Amir RE, Van den Veyver IB, Wan M, Tran CQ, Francke U, Zoghbi HY. Rett syndrome is caused by mutations in X-linked MECP2, encoding methyl-CpG-binding protein 2. Nat Genet 1999;23 (2):185–8.
[204] Lewis JD, Meehan RR, Henzel WJ, Maurer-Fogy I, Jeppesen P, Klein F, et al. Purification, sequence, and cellular localization of a novel chromosomal protein that binds to methylated DNA. Cell 1992;69 (6):905–14.
[205] Mullaney BC, Johnston MV, Blue ME. Developmental expression of methyl-CpG binding protein 2 is dynamically regulated in the rodent brain. Neuroscience 2004;123(4):939–49.
[206] Villard L, Fontes M. Alpha-thalassemia/mental retardation syndrome, X-Linked (ATR-X, MIM #301040, ATR-X/XNP/XH2 gene MIM #300032). Eur J Hum Genet 2002;10(4):223–5.
[207] McDowell TL, Gibbons RJ, Sutherland H, O'Rourke DM, Bickmore WA, Pombo A, et al. Localization of a putative transcriptional regulator (ATRX) at pericentromeric heterochromatin and the short arms of acrocentric chromosomes. Proc Natl Acad Sci USA 1999;96(24):13983–8.
[208] Ledbetter DH, Riccardi VM, Airhart SD, Strobel RJ, Keenan BS, Crawford JD. Deletions of chromosome 15 as a cause of the Prader-Willi syndrome. N Engl J Med 1981;304(6):325–9.

[209] Magenis RE, Brown MG, Lacy DA, Budden S, LaFranchi S. Is Angelman syndrome an alternate result of del(15)(q11q13)? Am J Med Genet 1987;28(4):829–38.
[210] Nicholls RD, Knoll JH, Butler MG, Karam S, Lalande M. Genetic imprinting suggested by maternal heterodisomy in nondeletion Prader-Willi syndrome. Nature 1989;342(6247):281–5.
[211] Kishino T, Lalande M, Wagstaff J. UBE3A/E6-AP mutations cause Angelman syndrome. Nat Genet 1997;15(1):70–3.
[212] Matsuura T, Sutcliffe JS, Fang P, Galjaard RJ, Jiang YH, Benton CS, et al. *De novo* truncating mutations in E6-AP ubiquitin-protein ligase gene (UBE3A) in Angelman syndrome. Nat Genet 1997;15 (1):74–7.
[213] Tsai TF, Jiang YH, Bressler J, Armstrong D, Beaudet AL. Paternal deletion from Snrpn to Ube3a in the mouse causes hypotonia, growth retardation and partial lethality and provides evidence for a gene contributing to Prader-Willi syndrome. Hum Mol Genet 1999;8(8):1357–64.
[214] Meguro M, Mitsuya K, Nomura N, Kohda M, Kashiwagi A, Nishigaki R, et al. Large-scale evaluation of imprinting status in the Prader-Willi syndrome region: an imprinted direct repeat cluster resembling small nucleolar RNA genes. Hum Mol Genet 2001;10(4):383–94.
[215] Weksberg R, Smith AC, Squire J, Sadowski P. Beckwith-Wiedemann syndrome demonstrates a role for epigenetic control of normal development. Hum Mol Genet 2003;12(Spec No 1):R61–8.
[216] Joyce JA, Lam WK, Catchpoole DJ, Jenks P, Reik W, Maher ER, et al. Imprinting of IGF2 and H19: lack of reciprocity in sporadic Beckwith-Wiedemann syndrome. Hum Mol Genet 1997;6(9):1543–8.
[217] Burns A, Iliffe S. Alzheimer's disease. BMJ 2009;338:b158.
[218] Crews L, Masliah E. Molecular mechanisms of neurodegeneration in Alzheimer's disease. Hum Mol Genet 2010;19(R1):R12–20.
[219] Sierksma AS, van den Hove DL, Steinbusch HW, Prickaerts J. Major depression, cognitive dysfunction and Alzheimer's disease: is there a link? Eur J Pharmacol 2010;626(1):72–82.
[220] Lahiri DK, Maloney B, Zawia NH. The learn model: an epigenetic explanation for idiopathic neurobiological diseases. Mol Psychiatry 2009;14(11):992–1003.
[221] Wu J, Basha MR, Brock B, Cox DP, Cardozo-Pelaez F, McPherson CA, et al. Alzheimer's disease (AD)-like pathology in aged monkeys after infantile exposure to environmental metal lead (Pb): evidence for a developmental origin and environmental link for AD. J Neurosci 2008;28(1):3–9.
[222] Bollati V, Galimberti D, Pergoli L, Dalla Valle E, Barretta F, Cortini F, et al. DNA methylation in repetitive elements and Alzheimer disease. Brain Behav Immun 2011;25(6):1078–83.
[223] Siegmund KD, Connor CM, Campan M, Long TI, Weisenberger DJ, Biniszkiewicz D, et al. DNA methylation in the human cerebral cortex is dynamically regulated throughout the life span and involves differentiated neurons. PLoS One 2007;2(9):e895.
[224] Mastroeni D, Grover A, Delvaux E, Whiteside C, Coleman PD, Rogers J. Epigenetic changes in Alzheimer's disease: decrements in DNA methylation. Neurobiol Aging 2010;31(12):2025–37.
[225] Cao X, Südhof TC. A transcriptionally [correction of transcriptively] active complex of APP with Fe65 and histone acetyltransferase Tip60. Science 2001;293(5527):115–20.
[226] Rouaux C, Jokic N, Mbebi C, Boutillier S, Loeffler JP, Boutillier AL. Critical loss of CBP/p300 histone acetylase activity by caspase-6 during neurodegeneration. EMBO J 2003;22(24):6537–49.
[227] Francis YI, Fà M, Ashraf H, Zhang H, Staniszewski A, Latchman DS, et al. Dysregulation of histone acetylation in the APP/PS1 mouse model of Alzheimer's disease. J Alzheimers Dis 2009;18 (1):131–9.
[228] Lang AE, Obeso JA. Challenges in Parkinson's disease: restoration of the nigrostriatal dopamine system is not enough. Lancet Neurol 2004;3(5):309–16.
[229] Braak H, Del Tredici K, Rüb U, de Vos RA, Jansen Steur EN, Braak E. Staging of brain pathology related to sporadic Parkinson's disease. Neurobiol Aging 2003;24(2):197–211.
[230] Coppedè F. Genetics and epigenetics of Parkinson's disease. Sci World J 2012;2012: 489830.

[231] Jowaed A, Schmitt I, Kaut O, Wüllner U. Methylation regulates alpha-synuclein expression and is decreased in Parkinson's disease patients' brains. J Neurosci 2010;30(18):6355–9.
[232] Kontopoulos E, Parvin JD, Feany MB. Alpha-synuclein acts in the nucleus to inhibit histone acetylation and promote neurotoxicity. Hum Mol Genet 2006;15(20):3012–23.
[233] Siddiqui A, Chinta SJ, Mallajosyula JK, Rajagopolan S, Hanson I, Rane A, et al. Selective binding of nuclear alpha-synuclein to the PGC1alpha promoter under conditions of oxidative stress may contribute to losses in mitochondrial function: implications for Parkinson's disease. Free Radic Biol Med 2012;53(4):993–1003.
[234] Zhu M, Li WW, Lu CZ. Histone decacetylase inhibitors prevent mitochondrial fragmentation and elicit early neuroprotection against MPP + . CNS Neurosci Ther 2014;20(4):308–16.
[235] Bates GP. History of genetic disease: the molecular genetics of Huntington disease—a history. Nat Rev Genet 2005;6(10):766–73.
[236] Sugars KL, Brown R, Cook LJ, Swartz J, Rubinsztein DC. Decreased cAMP response element-mediated transcription: an early event in exon 1 and full-length cell models of Huntington's disease that contributes to polyglutamine pathogenesis. J Biol Chem 2004;279(6):4988–99.
[237] Thomas EA, Coppola G, Desplats PA, Tang B, Soragni E, Burnett R, et al. The HDAC inhibitor 4b ameliorates the disease phenotype and transcriptional abnormalities in Huntington's disease transgenic mice. Proc Natl Acad Sci USA 2008;105(40):15564–9.
[238] Seong IS, Woda JM, Song JJ, Lloret A, Abeyrathne PD, Woo CJ, et al. Huntingtin facilitates polycomb repressive complex 2. Hum Mol Genet 2010;19(4):573–83.
[239] Kim MO, Chawla P, Overland RP, Xia E, Sadri-Vakili G, Cha JH. Altered histone monoubiquitylation mediated by mutant huntingtin induces transcriptional dysregulation. J Neurosci 2008;28(15):3947–57.
[240] Rahman A, Isenberg DA. Systemic lupus erythematosus. N Engl J Med 2008;358(9):929–39.
[241] Altorok N, Sawalha AH. Epigenetics in the pathogenesis of systemic lupus erythematosus. Curr Opin Rheumatol 2013;25(5):569–76.
[242] Dubroff LM, Reid RJ. Hydralazine-pyrimidine interactions may explain hydralazine-induced lupus erythematosus. Science 1980;208(4442):404–6.
[243] Lee BH, Yegnasubramanian S, Lin X, Nelson WG. Procainamide is a specific inhibitor of DNA methyltransferase 1. J Biol Chem 2005;280(49):40749–56.
[244] Yung R, Chang S, Hemati N, Johnson K, Richardson B. Mechanisms of drug-induced lupus. IV. Comparison of procainamide and hydralazine with analogs *in vitro* and *in vivo*. Arthritis Rheum 1997;40(8):1436–43.
[245] ten Veen JH, Feltkamp TE. Studies on drug induced lupus erythematosus in mice. I. Drug induced anti-nuclear antibodies (ANA). Clin Exp Immunol 1972;11(2):265–76.
[246] Cannat A, Seligmann M. Induction by isoniazid and hydrallazine of antinuclear factors in mice. Clin Exp Immunol 1968;3(1):99–105.
[247] Yung RL, Quddus J, Chrisp CE, Johnson KJ, Richardson BC. Mechanism of drug-induced lupus. I. Cloned Th2 cells modified with DNA methylation inhibitors *in vitro* cause autoimmunity *in vivo*. J Immunol 1995;154(6):3025–35.
[248] Fali T, Le Dantec C, Thabet Y, Jousse S, Hanrotel C, Youinou P, et al. DNA methylation modulates HRES1/p28 expression in B cells from patients with Lupus. Autoimmunity 2014;47(4):265–71.
[249] Jeffries MA, Dozmorov M, Tang Y, Merrill JT, Wren JD, Sawalha AH. Genome-wide DNA methylation patterns in CD4 + T cells from patients with systemic lupus erythematosus. Epigenetics: Off J DNA Methylation Soc 2011;6(5):593–601.
[250] Hewagama A, Gorelik G, Patel D, Liyanarachchi P, McCune WJ, Somers E, et al. Overexpression of X-linked genes in T cells from women with lupus. J Autoimmun 2013;41:60–71.
[251] Lu Q, Wu A, Tesmer L, Ray D, Yousif N, Richardson B. Demethylation of CD40LG on the inactive X in T cells from women with lupus. J Immunol 2007;179(9):6352–8.

[252] Gibofsky A. Epidemiology, pathophysiology, and diagnosis of rheumatoid arthritis: a synopsis. Am J Manag Care 2014;20(7 Suppl):S128–35.
[253] McInnes IB, Schett G. The pathogenesis of rheumatoid arthritis. N Engl J Med 2011;365(23):2205–19.
[254] Sánchez-Pernaute O, Ospelt C, Neidhart M, Gay S. Epigenetic clues to rheumatoid arthritis. J Autoimmun 2008;30(1–2):12–20.
[255] Kuchen S, Seemayer CA, Rethage J, von Knoch R, Kuenzler P, Beat AM, et al. The L1 retroelement-related p40 protein induces p38delta MAP kinase. Autoimmunity 2004;37(1):57–65.
[256] Takami N, Osawa K, Miura Y, Komai K, Taniguchi M, Shiraishi M, et al. Hypermethylated promoter region of DR3, the death receptor 3 gene, in rheumatoid arthritis synovial cells. Arthritis Rheum 2006;54(3):779–87.
[257] Atkinson MA, Maclaren NK. The pathogenesis of insulin-dependent diabetes mellitus. N Engl J Med 1994;331(21):1428–36.
[258] Ounissi-Benkalha H, Polychronakos C. The molecular genetics of type 1 diabetes: new genes and emerging mechanisms. Trends Mol Med 2008;14(6):268–75.
[259] Fradin D, Le Fur S, Mille C, Naoui N, Groves C, Zelenika D, et al. Association of the CpG methylation pattern of the proximal insulin gene promoter with type 1 diabetes. PLoS One 2012;7(5):e36278.
[260] Bell CG, Teschendorff AE, Rakyan VK, Maxwell AP, Beck S, Savage DA. Genome-wide DNA methylation analysis for diabetic nephropathy in type 1 diabetes mellitus. BMC Med Genom 2010;3:33.
[261] Miao F, Smith DD, Zhang L, Min A, Feng W, Natarajan R. Lymphocytes from patients with type 1 diabetes display a distinct profile of chromatin histone H3 lysine 9 dimethylation: an epigenetic study in diabetes. Diabetes 2008;57(12):3189–98.
[262] Brasacchio D, Okabe J, Tikellis C, Balcerczyk A, George P, Baker EK, et al. Hyperglycemia induces a dynamic cooperativity of histone methylase and demethylase enzymes associated with gene-activating epigenetic marks that coexist on the lysine tail. Diabetes 2009;58(5):1229–36.
[263] Frohman EM, Racke MK, Raine CS. Multiple sclerosis—the plaque and its pathogenesis. N Engl J Med 2006;354(9):942–55.
[264] Kragt J, van Amerongen B, Killestein J, Dijkstra C, Uitdehaag B, Polman C, et al. Higher levels of 25-hydroxyvitamin D are associated with a lower incidence of multiple sclerosis only in women. Mult Scler 2009;15(1):9–15.
[265] Oksenberg JR, Baranzini SE, Sawcer S, Hauser SL. The genetics of multiple sclerosis: SNPs to pathways to pathogenesis. Nat Rev Genet 2008;9(7):516–26.
[266] Pitzalis M, Zavattari P, Murru R, Deidda E, Zoledziewska M, Murru D, et al. Genetic loci linked to type 1 diabetes and multiple sclerosis families in Sardinia. BMC Med Genet 2008;9:3.
[267] Mastronardi FG, Wood DD, Mei J, Raijmakers R, Tseveleki V, Dosch HM, et al. Increased citrullination of histone H3 in multiple sclerosis brain and animal models of demyelination: a role for tumor necrosis factor-induced peptidylarginine deiminase 4 translocation. J Neurosci 2006;26(44):11387–96.
[268] Pritzker LB, Joshi S, Harauz G, Moscarello MA. Deimination of myelin basic protein. 2. Effect of methylation of MBP on its deimination by peptidylarginine deiminase. Biochemistry 2000;39(18):5382–8.
[269] Shanshiashvili LV, Kalandadze IV, Ramsden JJ, Mikeladze DG. Adhesive properties and inflammatory potential of citrullinated myelin basic protein peptide 45-89. Neurochem Res 2012;37(9):1959–66.
[270] Harauz G, Ishiyama N, Hill CM, Bates IR, Libich DS, Farès C. Myelin basic protein-diverse conformational states of an intrinsically unstructured protein and its roles in myelin assembly and multiple sclerosis. Micron 2004;35(7):503–42.

CHAPTER

CANCER EPIGENETICS

2

Kate H. Brettingham-Moore[1] **and Phillippa C. Taberlay**[2,3,4]

[1]*School of Medicine, University of Tasmania, Hobart, TAS, Australia* [2]*Chromatin Dynamic Group, Genomics and Epigenetics Division, Garvan Institute of Medical Research, Darlinghurst, NSW, Australia* [3]*Epigenetics Program, Genomics and Epigenetics Division, Garvan Institute of Medical Research, Darlinghurst, NSW, Australia* [4]*St. Vincent's Clinical School, Faculty of Medicine, University of New South Wales, Darlinghurst, NSW, Australia*

CHAPTER OUTLINE

2.1 BACKGROUND

A repertoire of epigenetic mechanisms is damaged as a collective unit in cancers; inclusive of both known and presumably as-yet undefined epigenetic marks. To date, epigenetic aberrations in cancers can be typically described as changes in DNA methylation, histone modifications, histone variant usage, small and noncoding RNA molecules and nucleosome positions while, more recently, it has become clear that an additional layer of epigenetic damage probably occurs as a consequence of genetic mutations in chromatin-modifying proteins (Figure 2.1). A major role for epigenetic mechanisms in cancer development is one that is widely accepted, but the underlying molecular

G. Egger & P. Arimondo (Eds): Drug Discovery in Cancer Epigenetics. DOI: http://dx.doi.org/10.1016/B978-0-12-802208-5.00002-3

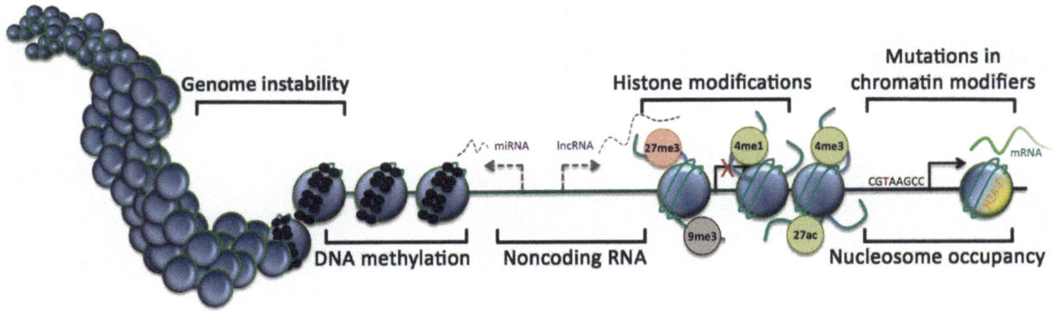

FIGURE 2.1

Epigenetic hallmarks of the cancer genome. The entire epigenome is damaged in cancer cells including DNA methylation, histone modifications (e.g., H3K27me3 or H3K4me3), histone variants (e.g., H2A.Z), nucleosome positions, and noncoding RNA (ncRNA), whose expression can be disrupted and/or whose epigenetic activity can be dysregulated. A global loss of DNA methylation and changes in nucleosome positions contribute to overall genomic instability, while genetic mutations in chromatin modifiers are likely to impact severely on the epigenome. Nucleosome, blue circle; DNA methylation, small black circle; transcriptional start site, arrow; inactive transcriptional start site, red cross; histone 3 lysine 27 trimethylation, 27me3 (red, repressed); histone 3 lysine 9 trimethylation, 9me3 (gray, silenced); histone 3 lysine 4 monomethylation, 4me1 (green, active); histone 3 lysine 4 trimethylation (green, active); histone 3 lysine 27 acetylation, 27ac (green, active).

etiology remains to be fully elucidated. In particular, the high level of concordant changes emphasizes the intricacy of epigenetic control over the genome but highlights the necessity to identify the primary epigenetic events leading to cancer onset and progression. This task remains a challenge as we must now collate and evaluate ever-increasing amounts of sequencing data, which continue to identify novel epigenetic defects as well as emphasizing roles for epigenetic changes that appear in almost all cancers. Here, we describe some well-characterized aberrations, as well as more recent advances in the field of cancer epigenetics.

2.2 DNA METHYLATION

Tumor progression is driven by abnormal transcriptional profiles resulting from both genetic and epigenetic changes. Aging and environmental contributors to cancer progression may cause changes to the epigenome, which in turn influence expression of the genome. Mapping methylation profiles of normal versus tumor cells has provided a great deal of insight into the potential epigenetic mechanisms driving cancer progression. Epigenetic reprogramming via DNA methylation in cancer cells is typically characterized by global hypomethylation along with promoter-specific hypermethylation [1–3]. Global hypomethylation results in oncogene activation and genomic instability, increasing the rate of mutation, resulting in cells accumulating even more genetic damage. Meanwhile locus-specific promoter hypermethylation is involved in silencing tumor-suppressor genes (TSGs). These genes normally protect the cell from cancer progression and unrestricted proliferation. Both TSG alleles must be functionally inactivated in order for cancer to progress and in the case of familial cancers, aberrant TSG promoter methylation can contribute to the functional loss of the second TSG, acting as the "second hit." Loss of heterozygosity in TSG alleles removes the protective function of the TSG product. In addition, while global hypomethylation is a characteristic feature of the cancer epigenome there is also recent evidence for long-range epigenetic silencing in cancer [4]. Certain regions of the genome are more prone to hypermethylation, for example Polycomb target genes [5]. Differential methylation of certain regions may also be attributed to replication timing and attachment to the nuclear lamina [6]. Genetic lesions can contribute to DNA methylation changes and vice versa. For example, genetic changes can trigger epigenetic mutations by impacting levels or activity of DNA methyltransferases (DNMT). Meanwhile, an epigenetic lesion, such as hypermethylation and silencing of a TSG, can lead to chromosomal instability and genetic mutations.

The epimutations acquired during tumorigenesis alter the epigenetic landscape enabling growth advantage. It is important to note that cancer cells acquire thousands of different mutations and epimutations during cancer progression and there is a strong distinction between driver and passenger epigenetic changes. Some methylation marks for example have a direct impact on the expression of an oncogene or TSG. Other methylation changes are simply seeded from these initial methylation or mutational events and result in altered expression of genes not essential to tumor cell growth and evolution. Nonetheless, the alteration of DNA methylation and the resultant transcriptome helps to drive transformation.

Abnormal hyper- and hypomethylation are evident in cancer. Dysregulation of DNA methylation appears to be an early event in tumorigenesis triggering chromosomal instability and tumor progression. Early and late epigenetic events in colorectal cancer have been well characterized and have provided insight into the role of DNA methylation during cancer progression. Epigenetic studies in colorectal cancer have demonstrated both hyper- and hypomethylation in the early stages of normal colon epithelium transitioning to adenocarcinoma [7,8]. Localized promoter CpG island methylation has been detected in normal tissue, aberrant crypt foci, and adenomas, with evidence demonstrating epigenetic silencing of TSGs. Genome-wide methylation changes appear to occur early in disease progression and are often highly heterogeneous [8]. Most methylation changes occur before progression to cancer, perhaps consolidating the transcriptome necessary to drive transformation.

Inactivation of DNA mismatch repair (MMR) pathways via DNA methylation can promote genetic instability. Defects in the MMR pathway are reflected by variation in DNA microsatellite length between cells. This genetic instability is referred to as microsatellite instability (MSI) and often has implications in terms of prognosis and response to therapy. MSI malignancies tend to respond well to chemotherapy and radiotherapy with increased DNA damage following genotoxic insult pushing cells toward apoptosis. MSI in colorectal cancer is observed in both familial (hereditary nonpolyposis colorectal cancer or Lynch syndrome) and sporadic disease. MSI tumors account for about 15% of sporadic colorectal cancer cases and are characterized by genomic instability, accumulating mutations at a significant rate. This can be attributed to silencing of genes involved in MMR pathway such as methylation of MutL homolog 1 (*MLH1*). *MLH1* loss is associated with MSI developing secondarily. The relatively high frequency of gene hypermethylation in colorectal cancer has been used to identify a subgroup termed CpG Island Methylator Phenotype (CIMP). CIMP accounts for around 20% of colorectal cancer cases. It is characterized by numerous features such as BRAF mutation; the concurrent hypermethylation of marker genes *RUNX3, SOCS1, NEUROG1, IGF2,* and *CACNA1G* [9]; female patients and proximal tumor location.

DNA methylation patterns have been used to clinically stratify and identify molecular subtypes in various cancers. High-throughput sequencing technologies and methylation arrays have expanded understanding of the cancer epigenome in recent years. Such technologies have been used successfully to identify methylation patterns associated with different subtypes of cancer. DNA methylation has been extensively studied in myelodysplastic syndrome (MDS) and acute myeloid leukemia (AML). DNA methylation plays a significant role in leukemic transformation with silencing of TSG alleles via this mechanism evident in immature MDS cells. A key study performed analysis of methylation signatures across AML patients and demonstrated that methylation patterns could be used to divide patients into 16 different subgroups [10]. Different subtypes emerging from different molecular and evolutionary pathways are reflected in these methylation subgroupings. These included distinct methylation signatures for common chromosomal rearrangements associated with AML such as t8:21 and inv16. The fact that DNA methylation patterns can accurately reflect genetic aberrations and subgroupings demonstrates how closely the genome and epigenome are intertwined. This also raises the question as to whether the genetic change seeds the corresponding methylation pattern, or whether perhaps an epigenetic lesion contributed to the related genetic instability and mutation. In addition to the ability to define subgroups using methylation profiling in AML, a common 45-gene methylation signature was shared across subgroups indicating a general role in AML progression and provides insight into an epigenetic pathway linked to hematopoietic transformation. This methylation pattern impacted TSGs along with genes involved in transcription and nuclear import [10].

Distinct DNA methylation signatures have been associated with various hallmark features of cancer such as inflammation, cell cycle progression, angiogenesis, and metastasis. Alterations in DNA methylation are associated with chronic inflammation [5]. For example, increased DNA methylation is linked to *Helicobacter pylori* (*H. pylori*)-associated gastric cancer [11]. Gastric cancer associated with *H. pylori* infection can be prevented through treatment with 5-aza-2′-deoxycytidine in a rodent model [12]. This result indicates that preventing methylation changes induced by inflammation can inhibit the progression from the environmental signal, chronic inflammation, to cancer. Methylation signatures have also been associated with human papilloma virus (HPV) positive and negative tumors in oropharyngeal squamous cell carcinoma (reviewed by van Kempen et al. in Ref. [13]). This demonstrates the role of DNA methylation changes during oncogenic transformation.

DNA methylation plays a pathogenic role in tumor cell evolution. A subpopulation of tumor cells has stem-like properties driven by epigenetic reprogramming. These cancer stem cells have increased potential for self-renewal and tumor initiation. Alterations in DNA methylation can induce pluripotency with cancer stem/progenitor cells driven by abnormal methylation programs. Epigenetic plasticity is crucial for cancer cells, which undergo an epithelial to mesenchymal transition (EMT) to metastasize, followed by a mesenchymal to epithelial transition to enable colonization at distal sites. It is crucial to remember that tumor evolution and the resultant heterogeneity means that within a population of cancer cells many subclones and therefore cancer epigenomes exist. This variation within a single tumor complicates treatment considerably with resistant subclones contributing to disease recurrence and progression.

2.3 HISTONE MODIFICATIONS

Posttranslational modification of histone proteins, or indeed, of proteins with the capacity to modify chromatin, can be associated with both active and repressive regions of the genome.

The exact residue and type of modification that decorates histones is often correlated with a panel of complementary marks that are easily identified as "active" or "repressive" combinations. Methylation can be both active and repressive, and histones can also be acetylated, ubiquitinated, phosphorylated, and sumoylated, amongst others. The "histone code" [14] explains the coexistence of some marks but disparities do occur. For example, the co-occurrence of both histone 3 lysine 4 trimethyl (H3K4me3) and histone 3 lysine 27 trimethyl (H3K27me3) at certain promoters has been observed predominantly in embryonic stem cells [15,16] and interestingly, these promoters exhibit a low level of expression [16]. Moreover, there has been a rapid rise in the sequence-based histone modification data, meaning that we have new knowledge of previously uncharacterized marks and greater insight regarding the normal, cell-type-specific distribution of other marks and then, how these are disrupted in cancer.

Not surprisingly, histone modifications associated with both active and repressive chromatin are implicated in cancers and these can occur simultaneously and amongst the other epigenetic changes. It is equally significant that a mark is lost from a particular site along the genome as much as it gained (e.g., loss of H3K27ac from an active enhancer can lead to inactivity and acquisition of a poised state [17]) or the combination with which it exists (e.g., H3K4me1 can coexist with DNA methylation at enhancers repressed by a nucleosome in cancer [18]). The altered histone code of cancer cells is associated with direct and indirect subsequent changes such as additional reprogramming of the histone landscape and concomitant changes of other epigenetic marks (e.g., DNA methylation). All of these changes can be ultimately attributed to the atypical activity of chromatin "writers" or abnormal interactions with histone modification "readers" and "erasers." It should also be considered that in normal cells, healthy competition between "writers" must be decided since the same histone tail residue can be modified in multiple ways; for example, methylation or acetylation can mark histone 3 lysine 27 to yield H3K27me (mono-, di-, or tri-) or H3K27ac, respectively. The result of this particular competition has been partly explained in embryonic stem cells, with the repressive H3K27me3 mark found to be dominant when all components of the methyltransferase machinery (i.e., Polycomb repressive complex) are intact. Therefore, the levels of expression of the catalytic enzyme complexes and their availability probably dictate the amount and distribution of histone marks. This is applicable to normal and cancer cells but can be further complicated by mutations that directly impair the "writers," "readers," and "erasers," as we discuss below.

The H3K4me3 and H3K27me3 modifications are the most widely studied and demonstrate the damage, though ironically, and the potential utility of histone modification disturbances in cancers. For example, both H3K4me3 and H3K27me3 characterize sets of genes that allow the subtyping of malignant, "tumor-sustaining" and chemoresistant ovarian tumors [19]. The H3K27me3 modification is typically but not always [20] exclusive of DNA methylation, though both are markers of gene inactivity and interestingly, genes that carry H3K27me3 in normal cells are more prone to aberrant DNA methylation events in prostate [21] and colon cancer [22] cells. The catalytic subunits of complexes catalyzing H3K27me3 and H3K4me3 (enhancer of zeste 2 and mixed-lineage leukemia, respectively) are amongst the most frequently mutated in many cancers including prostate, breast, colon, skin, lung, and blood cancers [23], suggesting that these could be putative driver mutations leading to widespread epigenetic aberrations in cancers.

2.4 NUCLEOSOME POSITIONS AND HIGHER-ORDER STRUCTURES

Tools to map open chromatin and nucleosome positions are constantly evolving and most have now been adapted so that the entire genome can be interrogated, much like DNA methylation with whole-genome bisulfite sequencing [24] and histone modifications through chromatin immunoprecipitation sequencing [25]. These techniques include DNaseI hypersensitivity sequencing (DNaseI-Seq [26]), micrococcal nuclease sequencing (MNase-Seq [27]), formaldehyde-assisted interrogation of regulatory elements sequencing (FAIRE-Seq [28]), assay for transposase-accessible chromatin sequencing (ATAC-Seq [29]), and nucleosome occupancy and methylation sequencing (NOMe-Seq [18,30]), which directly measures nucleosome positions and simultaneously assays DNA methylation. However, most of these assays are currently costly and/or require a substantial amount of genomic input for sequencing, meaning that few studies have interrogated nucleosome positions or accessibility on a global scale in cancers. In addition, the importance of the nucleosome has been overlooked until relatively recently.

Many general features of cancer cells could be explained by atypical nucleosome dynamics. For example, an overall loss of DNA methylation in cancer cells is associated with genome instability, which could be due to concomitant loss of nucleosomes (i.e., an increased number of nucleosome-depleted regions). This would also be consistent with fewer histones as we age [31,32] and the knowledge that DNMT enzymes require nucleosomes for selective anchoring and activity [33,34]. A potential order of events could be hypothesized from these and observations that nucleosome occupancy is the mode of silencing key pluripotent genes prior to DNA methylation of these loci during stem cell differentiation [35]. The same order has been suggested at gene promoters dysregulated in colon adenoma (*CDKN2B* and *CDH1* [36]), whose inactivity can be associated with nucleosome occupancy in the absence of DNA methylation. Nucleosome dynamics are also observed in cancer cell lines treated with 5-aza-2′-deoxycytidine [37,38], reflecting the fact that a reverse cooperation occurs between these two epigenetic mechanisms during gene reactivation. Outside of promoter regions hypermethylation is accompanied by nucleosome occupancy of insulators and enhancers in breast and prostate cancer cell lines [18] demonstrating that in cancers chromatin is deliberately organized, or reorganized, at a genome-wide scale. These changes could have important therapeutic implications, as indicated by a comparison of endocrine-responsive versus endocrine-resistant breast cancer cell lines that revealed distinct differences in chromatin accessibility linked to pathways involving NOTCH in resistant breast cancer cells [39].

Assessment of open and closed chromatin or nucleosome positions in primary tumors is extremely limited, particularly genome-wide. FAIRE-seq has been successfully used to map accessibility in matched normal compared to renal cell carcinoma samples (clear cell [40]) with tumors tending to be more accessible. The specific loss of nucleosomes (i.e., a gain in accessible chromatin) is predominantly associated with DNA hypomethylation. Interestingly, a subset of these nucleosome repositioning events was found to be common in all samples, indicative of a key event in tumor establishment or progression with the remaining appearing in ~50% of tumors. Again, these data are consistent with overall loss of hypomethylation and increased genome instability that is accompanied by focal gains in methylation and increased nucleosome occupancy at promoters and distal regulatory elements in cancer cells.

Nucleosome depletion is an important feature of the epigenome and one that must be precisely regulated so that transcription factors and other chromatin-modulating proteins bind appropriately.

Indeed, nucleosome-depleted regions are largely cell-type-specific [18] and very few DNase I hypersensitive sites are ubiquitous in cell lines [41], as are MNase accessible sites in primary cells [42]. Interestingly, the chromatin state, in combination with cell-type-specific histone marks and replication timing patterns, can provide an explanation for almost 90% of the mutational variance across the genomes of melanoma, multiple myeloma, lung adenocarcinoma, liver cancer, colorectal cancer, glioblastoma, esophageal adenocarcinoma, and lung squamous cell carcinoma [43]. The rate of mutations, for example C-to-T mutations, is positively correlated with local chromatin with repressive marks and inaccessibility of local chromatin regions harboring these mutations. This work builds on the previous discovery that open chromatin and protein binding can be linked to structural breakpoints in prostate, breast, ovarian, colorectal, and head and neck cancers [44]. The utility of techniques such as ATAC-seq and FAIRE-seq, and the importance of chromatin remodeling changes in cancer are also evident in a temporal study of Ras-induced tumor development. Here, chromatin conformation was found to be more open in early developing tumors, and substantially more accessible in late-stage tumor development [45]. Together, these data further highlight that nucleosomes, thus chromatin accessibility, could determine the aggressiveness of cancer. The degree of nucleosome movement, depletion or compaction, may facilitate new mutations that could serve as a potential driver of disease progression.

Nucleosomes form part of the higher-order chromatin structure of cells, which also includes the folding and looping of DNA, as occurs when enhancers are brought into close proximity to promoters. Relatively little is known about the impact of global nucleosome repositioning on the interactions between regulatory elements, or vice versa in cancers. It is clear from epigenetic profiling that active enhancers in normal cells are DNase I hypersensitive [26,46] and, interestingly, that permissive but not active enhancers are also nucleosome-depleted [47]. Indeed, subtypes of enhancers carry their own unique signatures reflecting their cellular activity [17,48]. The epigenetic remodeling of enhancers and insulators in cancer includes the insertion of a nucleosome into previously vacant genomic regions [18,47] and the inability to be activated following the forced expression of master regulatory factors [47]. These findings are consistent with results from cloning experiments demonstrating that a hypersensitive region is insufficient to confer promoter luciferase activity [49]. Presumably, open chromatin at the promoter and the presence of other epigenetic marks prevent promiscuous enhancer activity in cell types where these enhancers are normally silenced. Indeed, enhancers can become disengaged from their promoters in cancers, as observed at the 8q24 region in prostate cancer [50] or genome-wide in colon cancer cells [51]. In breast cancer, the long-range loop linking *IGFBP3* promoter to its enhancer is disrupted but new interactions are formed, including those that may activate *EGFR* and *BCAS*, which is particularly prone to chromosomal translocation [52]. These are significant studies that emphasize that understanding nucleosome positions and higher-order chromatin structures will be an important area of continued research.

2.5 NONCODING RNAs

Noncoding DNA comprises a significant proportion of the human genome and until relatively recently these sequences were considered junk. It is now evident that, not only are these sequences extensively transcribed [53], they have critical roles to play in gene regulation and in turn cancer progression. Among these nonprotein-coding sequences are microRNAs (miRNAs) and long noncoding RNAs (lncRNAs). Noncoding RNAs (ncRNAs) play a role in regulating gene expression via translational repression and protein binding. An ever-growing list of ncRNAs has been implicated in tumor initiation and progression (reviewed in Ref. [54]). Two mechanisms can trigger tumorigenic ncRNA alterations, either genetic mutation resulting in sequence changes or epigenetic mechanisms which can impact the level of ncRNA transcribed. Altered expression or sequence mutation in ncRNAs impacts target binding and therefore function. Once again these epigenetic players exert their tumorigenic effects via dysregulation of oncogene and TSG pathways. The distinction between driver and passenger mutations in these noncoding regions must be made. However, it is clear that some of these miRNAs and lncRNAs exert significant influence on cellular transformation.

2.5.1 microRNAs

miRNAs are 18–25 bases in length and are involved in posttranscriptional regulation via mRNA degradation and translational repression. miRNAs bind the 3′ UTR of the target mRNA and require near perfect base complementarity. miRNAs are involved in all aspects of cancer initiation and progression including apoptosis, cell cycle progression, invasion and metastasis, and angiogenesis [55,56]. Changes in miRNA sequence or expression can impact molecular pathways responsible for each of the hallmark features cancer cells acquire during transformation [57]. Interestingly, approximately half the miRNAs in the human genome are located in either fragile sites of the genome or regions previously linked to cancer [58]. miRNAs associated with cancer can exert their influence on oncogenic or tumor-suppressive pathways. It is important to note that some miRNAs can be both oncogenic or tumor-suppressive in their function, depending on cellular context and the targets expressed in a particular cell type.

Studies comparing normal versus tumor tissue have revealed miRNA signatures associated with cancer. miRNAs are dysregulated in tumor tissues compared to normal tissue. Conflicting evidence has demonstrated miRNAs to be downregulated or upregulated in cancer tissues [59,60]. It is now evident that the miRNA signatures detected in cancer tissues are dependent on the miRNA panel and cancer type studied. An miRNA expression signature associated with a number of solid tumors (including breast, pancreatic, colon) demonstrated a large number of overexpressed miRNAs including *miR-21, miR-92,* and *miR-155*. The predicted targets for the miRNAs were significantly enriched for oncogenes and TSGs [60].

Oncogenic miRNAs are upregulated in cancer and promote cell proliferation, survival, invasion, and angiogenesis. A cluster of six oncogenic miRNAs located in a region frequently amplified in B-cell lymphoma, *miR-17-92*, display increased expression in a variety of cancer types. The *miR-17-92* cluster is regulated by c-Myc and acts on a number of targets including *p21*, *PTEN*, and *HIF1α*.This region has consequently been implicated in cell cycle progression and angiogenesis [56]. Invasion and metastasis are key properties of malignant disease and a number of oncogenic miRNAs have been demonstrated to promote these features in cancer cells. *miR-21* is well studied and frequently upregulated in numerous cancer types including breast, colon, lung, pancreatic, stomach, and prostate cancers [60]. This miRNA is antiapoptotic and associated with increased metastatic potential by targeting phosphatase and tensin homolog (*PTEN*) [61].

Tumor-suppressive miRNAs are downregulated in cancer or have a sequence mutation which hinders normal activity. Loss of tumor-suppressive miRNAs increases features associated with transformation including proliferation, survival, and invasiveness. Indeed, widespread downregulation of miRNAs is observed in a number of cancer types including prostate cancer [62]. Loss of the *let-7* family of tumor-suppressive miRNAs is associated with pluripotency and increased self-renewal capacity. The *let-7* family of miRNAs regulates the expression of Ras, a GTPase and oncogene frequently mutated in cancer. The 3′ UTR of the Ras gene contains multiple *let-7* sites [63]. Loss of *let-7a, let-7b*, and *let-7g* promotes cell motility and is associated with increased risk of metastasis [64]. In addition to this, *let-7* levels can alter radiosensitivity with increased expression improving radiation response in an *in vitro* model [65].

Two tumor-suppressive miRNAs, *miR-15a* and *miR-16-1*, map to the 13q14 locus, a region deleted in up to 65% of B-cell chronic lymphocytic leukemia cases. These miRNAs control cell proliferation, survival, and invasion via targeted suppression of multiple oncogenes including the antiapoptotic BCL2 [66]. Absence of these two miRNAs results in increased tumor cell survival and proliferation. This region is also deleted in prostate cancer [67] with levels of *miR-15a* and *miR-16-1* significantly decreased in advanced disease. The direct role of these miRNAs in transformation in prostate cancer has been demonstrated by experimental knockdown of *miR-15a* and *miR-16-1* in normal prostate epithelial cells. Restoring *miR-16-1* expression in prostate cancer cell lines inhibits cell proliferation. Synthetic *miR-16* could also reduce bone metastasis in a murine xenograft model [68].

EMT is key to cancer cells moving from the primary tumor site and becoming invasive and metastatic. miRNAs have been linked to epithelial plasticity required for EMT in cancer. A reduction in *miR-200* levels is associated with EMT in cancer. Inhibition of *miR-200* is associated with reduced E cadherin expression, vimentin upregulation, and increased motility [69]. The *miR-200* family of miRNAs has been implicated in a number of cancer types as having a tumor-suppressive function. *miR-200b* and *miR-200c* have been demonstrated to be downregulated in gastric cancer when compared to normal tissue [70]. These miRNAs act as tumor suppressors by targeting the DNA methyltransferases DNMT3A and DNMT3B and the transcription factor, Sp1. Ultimately *miR-200b* and *miR-200c* promote genetic instability via global hypomethylation attributed to dampening DNMT expression [70].

Similar to DNA methylation signatures, miRNA signatures can be used to divide cancers into subgroupings and often reflect the genotype of the tumor. miRNA profiles can not only indicate the origin of the tumor tissue but also subtypes within a single tumor type [59]. In the case of acute lymphocytic leukemia (ALL), shared miRNA patterns between tumors reflect the translocation present [59]. This may be attributed to the miRNA transcriptional profile being dictated by the methylation profile. Alternatively, miRNAs may alter methylation or the genetic lesion may impact miRNA profiles. miRNA expression profiles have also been reported to change over time during cancer progression. In the case of cervical cancer, HPV infection can alter the miRNA expression program as cells transition from normal epithelium to invasive cancer. This is attributed to HPV oncoproteins E6 and E7 altering the activation of transcription factors E2F and c-Myc and therefore their miRNA targets [71]. The ability of miRNA to reflect genotype, DNA methylation, and change expression over time demonstrates the ever-increasing layers of epigenetic complexity in cancer cell evolution.

2.5.2 LONG NONCODING RNAs

Many cancer-associated SNPs lie beyond the protein-coding sequences indicating that the noncoding sequences play a significant role in tumorigenesis. lncRNAs are 200 nucleotides or more in length. These ncRNAs have been shown to interact with proteins to alter binding and function via conformational change. lncRNAs have diverse roles in the cell, with the ability to influence chromatin remodeling, transcriptional activation, protein inhibition, and activation and posttranscriptional modifications. With the ability to bind transcriptional activators or repressors and take control of epigenetic machinery, lncRNAs have a significant impact on gene expression and chromatin structure. lncRNAs can also act as miRNA sponges and behave as decoys for miRNA binding. They operate at specific loci and play a role in pre-mRNA splicing, protein activities, degradation, and translation. They may act in *cis* or *trans* by regulating transcript abundance at local or distal sites. The process of transcribing these noncoding regions may also alter the epigenetic landscape and genetic stability. As with other epigenetic mechanisms promoting tumor formation, lncRNAs exert their influence on oncogenic and TSG pathways. They have been implicated in cell proliferation, apoptosis, angiogenesis, and metastasis. In contrast to miRNAs, most lncRNAs appear to be poorly conserved and while selective pressures have enabled lncRNAs to evolve it is clear they have an important role in regulating numerous cellular pathways.

Aberrant expression of the large intergenic ncRNA HOX transcript antisense intergenic RNA (*HOTAIR*) has been detected in a variety of cancers. Overexpression of *HOTAIR* is linked to more aggressive disease and prediction of survival outcomes in breast, colon, liver, and pancreatic cancer [72]. *HOTAIR* increases proliferation, invasion, and metastasis via interaction with Polycomb Repressive Complex 2 (PRC2). The resulting alterations to H3K27me3 mediate the repression of numerous genes including *HOXD, PRG1*, and *EPHA1*. These genes have wide-ranging functions from involvement in angiogenesis to encoding hormone receptors, clearly demonstrating the broad tumorigenic potential of alterations to this lncRNA. Tumor-suppressive lncRNAs include PTEN pseudogene 1 (*PTENpg1*). This lncRNA acts as a sponge, competing for miRNAs targeting *PTEN* [73]. *PTENpg1* is targeted by the same miRNAs directed against *PTEN* mRNA, acting as a decoy. This lncRNA is lost in a number of cancers including prostate. *GAS5* is another lncRNA with tumor-suppressive function. Downregulation in cancer cells is associated with increased cell proliferation and inhibition of apoptosis. *GAS5* has been reported to target a range of oncogenic transcripts including cyclin-dependent kinases, *miR-21* and *E2F1* [74]. It is consequently associated with poor prognosis in a number of cancers.

While most lncRNAs are poorly conserved, transcribed ultraconserved regions (*T-UCR*) are a class of lncRNA that display 100% sequence conservation in orthologous regions for humans, rats, and mice. These transcripts have been reported as dysregulated in numerous cancer types and are often found in fragile sites and cancer-associated regions of the genome. Malignant cells display distinct *T-UCR* signatures in comparison to normal tissue [75]. They also appear to have tumor-type-specific roles displaying opposing deregulation in different tumor types [76].

2.5.3 OTHER NONCODING RNAs

Other ncRNAs emerging as regulators of tumorigenesis include the human ortholog of Piwi-interacting RNAs (piRNAs). piRNAs are involved in germline stem cell maintenance and maintain genome integrity by silencing transposons. These ncRNAs are dysregulated in cancer [77]. piRNAs may contribute to cancer progression via oncogenic or tumor-suppressor roles or by promoting epigenetic profiles associated with cancer stem cells. Small nucleolar RNAs (snoRNAs) have also been implicated in cancer. snoRNAs are responsible for processing rRNA, however their role in the cell has been expanded. Differential regulation of snoRNAs has been detected in breast, prostate, lung, and brain cancer and reports demonstrate that they can behave as oncogenes or TSGs. For example the snoRNA, *snoRNA42* acts in an oncogenic manner in nonsmall-cell lung cancer and suppression reduces tumorigenicity *in vitro* [78].

2.6 MUTATION OF EPIGENETIC ENZYMES

Epigenetic enzymes play a wide range of roles in modifying the epigenome and therefore gene expression patterns in cells. In cancer these epigenetic modifiers are often altered and the change in activity has a significant impact on the resulting epigenetic landscape. Epigenetic enzymes can be divided into several broad categories including chromatin-remodeling enzymes, histone modifiers, DNMT, and enzymes involved in processing ncRNAs. Sequence mutation or dysregulated expression of these enzymes can dramatically alter the transcriptome of the cell. Epigenetic enzymes are frequently mutated in pediatric cancers [79] highlighting their ability to fast-track tumorigenesis.

2.6.1 CHROMATIN-REMODELING ENZYMES

Chromatin-remodeling enzymes help to shape the epigenetic landscape necessary for day-to-day cellular activities such as transcription and DNA replication. Dysregulation of these enzymes leads to the accumulation of epigenetic abnormalities necessary for cancer initiation and progression. Five major families of ATP-dependent chromatin-remodeling enzymes exist: SWI/SNF, ISWI, Nurd/Mi/CHD, SWR1, and INO80. The most frequently mutated chromatin-remodeling enzyme in cancer is the SWI/SNF complex. Comprising of 10–15 subunits, a broad spectrum of mutations is evident [80]. Mutation and altered expression of SWI/SNF complex subunits is linked to a variety of cancers but the scope of epigenetic consequences occurring as a result of genetic mutations in SWI/SNF is yet to be fully elucidated [81]. *SWI/SNF* mutations are found in up to 20% of cancers and the majority of these are inactivating mutations, indicating a tumor-suppressive function for this ATPase [80,82]. While the major role of SWI/SNF is gene activation it is also involved in double-strand DNA repair and lineage determination, therefore functional loss of this ATPase can increase sensitivity to DNA damage and disrupt differentiation programs. SWI/SNF activity has been linked to metastatic potential. Inhibition of SWI/SNF function in normal cells impacts invasiveness and increased expression of the BRG1 ATPase has been linked with more advanced-stage disease.

The chromodomain helicase DNA binding (CHD) family of chromatin-remodeling enzymes has been implicated in a number of cancers. CHD members are involved in transcription, double-strand break repair, replication, and recombination. This role is key in preventing cancer initiation and progression by maintaining genetic stability. Frameshift mutations are evident in CHD genes in colorectal and gastric cancer with MSI [83]. Another closely related family member is the nucleosome remodeling and deacetylase (NuRD) complex. Similar to CHD, this complex includes the Mi-2 ATPase and is an ATP-dependent chromatin remodeler coupled with histone-modifying properties. Evidence demonstrates that this chromatin-remodeling enzyme interacts with DNMTs and localizes to TSGs to promote silencing *in vitro* [84]. NuRD plays a key role in modulating pluripotency and dysregulation of this enzyme can reprogram cells and help drive tumorigenesis.

2.6.2 HISTONE MODIFIERS

Enzymes responsible for catalyzing the addition or removal of groups from histone tails help to establish chromatin structure and accessibility. Altered expression or function of histone-modifying enzymes results in aberrant chromatin architecture and gene expression programs. Posttranslational histone modifications identified to date include methylation, acetylation, sumoylation, phosphorylation, ADP-ribosylation, ubiquitylation, deimination, and proline isomerization. These modifications impact histone contact between neighboring nucleosomes and DNA. Histone deacetylases (HDACs) are often overexpressed in cancer, resulting in increased acetylation levels and transcriptional activity. HDAC1, HDAC2, and HDAC3 are frequently upregulated in colorectal cancer and increased expression is linked to poor prognosis. Altered histone acetyltransferase (HAT) activity has also been reported in cancer. Dysregulation of HATs triggers aberrant histone acetylation signatures and transcriptional profiles. A number of fusion proteins reported in ALL and AML are the product of chromosomal translocations of HATs, for example the MLL/CBP fusion protein.

The histone methyltransferase PRC2 is involved in promoting H3K27 trimethylation and repression of genes involved in differentiation, thereby maintaining epigenetic silencing during development and cancer. PRC2 dysregulation in cancer promotes pluripotency and suppresses cell differentiation. Enhancer of zeste homolog 2 (EZH2) is the catalytic subunit of this complex. Overexpression or increased activity of EZH2 is linked to silencing of TSGs, EMT, increased migration and tumor progression. Histone demethylases have also been implicated in cancer. Overexpression of the histone demethylases KDM4A, KDM4B, and KDM4C is observed in numerous cancer types including lung, colorectal, and breast. This group of enzymes has broad roles in cancer. KDM4A can form complexes with nuclear receptors such as the estrogen receptor (ER) to target activity to ER-responsive genes [85] and downregulation in ER-positive breast cancer cells can slow cell proliferation. KDM4B expression is increased under hypoxic conditions, typical of the tumor microenvironment. KDM4 enzymes can also act on nonhistone proteins adding another layer of regulation of tumorigenic targets.

2.6.3 DNA METHYLTRANSFERASES

As discussed previously, cancer cells exhibit abnormal methylomes displaying global hypomethylation along with locus-specific hypermethylation. These abnormal methylation profiles are driven by defects in the DNA methyltransferases DNMT1, DNMT3A, and DNMT3B. These enzymes catalyze the addition of methyl groups to hemimethylated and unmethylated CpG dinucleotides. With a role in methylation maintenance and *de novo* methylation events these enzymes are key contributors to the aberrant methylation profiles observed in cancer. DNA methylation patterns are maintained during DNA replication and cell division by DNMT1. This methyltransferase also appears to play a role in DNA repair pathways with DNMT1 recruited to sites of DNA damage [86]. Consequently, aberrations in DNMT1 levels or activity can have a significant impact on the methylation profile of the cell and accrual of mutations.

Subgroups of AML share distinct methylation signatures and this may be attributed to mutation in *DNMT3A* [87]. A high frequency of *DNMT3A* mutations is observed in AML, with approximately 20% of AML patients having some form of mutation. Variants in *DNMT3A* are linked to poor prognosis [88]. Mutations in this enzyme are also associated with MDS. *DNMT3B* is frequently mutated in cancer. A *DNMT3B* splice variant, *DNMT3B4*, lacking DNMT motifs, has been demonstrated to be overexpressed in hepatocellular carcinoma [89]. The increased expression of this nonfunctional methyltransferase was associated with DNA hypomethylation in pericentromeric satellite regions, indicating the ability of DNMT3B4 to destabilize the genome. DNMT have also been implicated in regulation of telomere length [90]. Loss of these enzymes increases telomeric recombination and telomere length changes associated with cancer.

2.6.4 ncRNA MACHINERY

Enzymes involved in processing ncRNA exert a significant influence on the gene expression programs of the cell. Disrupted miRNA biogenesis is evident in numerous malignancies and this can often be attributed to dysregulation of the ncRNA machinery, Dicer and Drosha. Genetic mutations or epimutations to enzymes responsible for processing miRNAs alter the corresponding levels of miRNA within the cell and in turn levels of miRNA targets. Loss or gain of Drosha and Dicer activity accounts for the dysregulation of mature miRNAs observed in cancer. Cancerous cells exhibit both increased and decreased Dicer expression. Dicer expression, and subsequently miRNA expression, is reduced in ovarian cancer and is linked to poor prognosis [91]. Meanwhile Dicer upregulation in prostate adenocarcinoma [92] correlates with the corresponding increase in miRNA. Drosha is an RNase III enzyme located on chromosome 5p, which is amplified in approximately half of advanced cases of cervical squamous cell carcinomas. Drosha copy number gain is frequently observed in cervical squamous cell carcinoma. The resulting increased expression of Drosha contributes to the dysregulated miRNA profiles in cervical cancer [93]. Altered Drosha expression is correlated with tumor progression and can be used as a predictor of patient prognosis and recurrence of disease.

2.7 CONCLUSION

An explosion of genome-wide data has revealed a vast and increasingly complex epigenetic landscape in cancer cells (Figure 2.2). In particular, the changes outside of promoter regions and the contribution of higher-order chromatin structures point to potential new epigenetic mechanisms underlying the initiation and development of cancer. This new knowledge enables identification of new targets that may be useful for therapeutic intervention but the challenge remains how best to deliver the drugs so that they are most specific and effective. Our continued study must also carefully consider how components of the epigenetic machinery (e.g., DNMTs and histone-modifying complexes) interact with each other and ensure that epigenetic marks are not considered as individual entities. This will be important as we seek to understand the key drivers of such heterogeneous and evolving diseases.

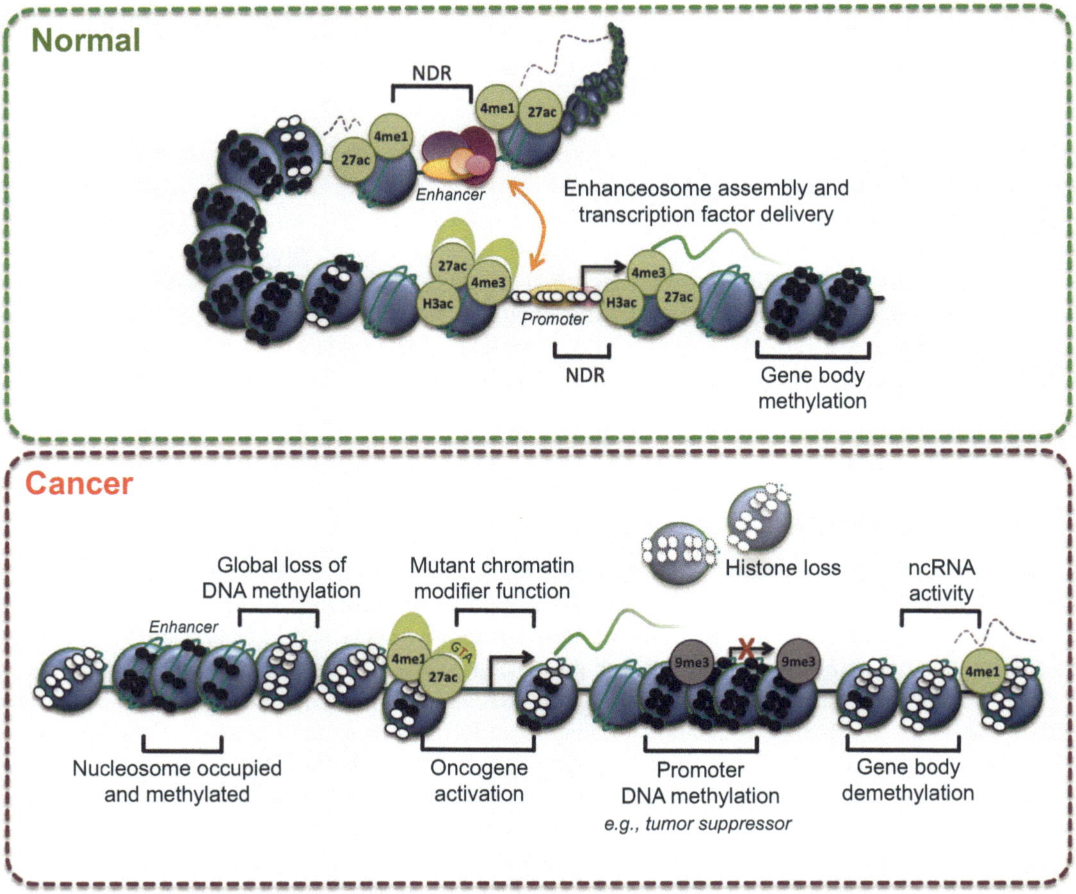

FIGURE 2.2

In normal cells, active genes are marked with a complement of correlating histone modifications (e.g., H3K4me3, H3K27ac, and H3ac) by chromatin-modifying complexes and a nucleosome is absent from the −1 position upstream of the transcriptional start site to create a NDR, which overlaps the unmethylated promoter. However, a patch of heavy methylation in the gene body is positively correlated with gene expression. At the enhancer, which also contains a distinct set of active histone modifications (e.g., H3K4me1 and H3K27ac) and a clear NDR, the enhanceosome is assembled and transcription factors can be delivered to the cognate promoter through DNA looping and enhancer–promoter interactions. Often, the enhancers and other intergenic regions are able to express noncoding RNAs (e.g., lncRNA or miRNA), which have epigenetic activity. In cancer cells, both the promoter and enhancer regions of atypically silenced genes (e.g., TSGs) are occupied by nucleosomes, become marked by repressive histone modifications (e.g., H3K9me3) and are methylated, especially those containing CpG island promoters. Gene body methylation is often lost concomitant with gene silencing, contributing to the overall decrease in DNA methylation observed in cancer cells. This is accompanied by loss of histone proteins, and therefore probably total nucleosome number and composition, contributing to genomic instability seen in tumors. This likely leads to open chromatin near oncogenes that are ordinarily silenced, but gain a NDR and active histone marks. This may be actioned by chromatin-modifying proteins carrying mutations and therefore, that recognize distinct binding sites or partners from their wild-type counterparts. The mutant proteins may also interfere with the epigenetic machinery to contribute to widespread epigenetic remodeling in cancers. Nucleosome, blue circle; nucleosome-depleted region, NDR; DNA methylation, small black circle; unmethylated DNA, small white circle; transcriptional start site, arrow; inactive transcriptional start site, red cross; histone 3 lysine 9 trimethylation, 9me3 (gray, silenced); histone 3 lysine 4 monomethylation, 4me1 (green, active); histone 3 lysine 4 trimethylation (green, active); histone 3 lysine 27 acetylation, 27ac (green, active); histone 3 acetylated (H3ac).

REFERENCES

[1] Esteller M. Epigenetic gene silencing in cancer: the DNA hypermethylome. Hum Mol Genet 2007;16 Spec No 1:R50–9.

[2] Ehrlich M. DNA hypomethylation in cancer cells. Epigenomics 2009;1(2):239–59.

[3] Gama-Sosa MA, Slagel VA, Trewyn RW, Oxenhandler R, Kuo KC, Gehrke CW, et al. The 5-methylcytosine content of DNA from human tumors. Nucleic Acids Res 1983;11(19):6883–94.

[4] Frigola J, Song J, Stirzaker C, Hinshelwood RA, Peinado MA, Clark SJ. Epigenetic remodeling in colorectal cancer results in coordinate gene suppression across an entire chromosome band. Nat Genet 2006;38(5):540–9.

[5] Hahn MA, Hahn T, Lee DH, Esworthy RS, Kim BW, Riggs AD, et al. Methylation of polycomb target genes in intestinal cancer is mediated by inflammation. Cancer Res 2008;68(24):10280–9.

[6] Berman BP, Weisenberger DJ, Aman JF, Hinoue T, Ramjan Z, Liu Y, et al. Regions of focal DNA hypermethylation and long-range hypomethylation in colorectal cancer coincide with nuclear lamina-associated domains. Nat Genet 2012;44(1):40–6.

[7] Goelz SE, Vogelstein B, Hamilton SR, Feinberg AP. Hypomethylation of DNA from benign and malignant human colon neoplasms. Science 1985;228(4696):187–90.

[8] Luo Y, Wong CJ, Kaz AM, Dzieciatkowski S, Carter KT, Morris SM, et al. Differences in DNA methylation signatures reveal multiple pathways of progression from adenoma to colorectal cancer. Gastroenterology 2014;147(2):418–429 e8.

[9] Weisenberger DJ, Siegmund KD, Campan M, Young J, Long TI, Faasse MA, et al. CpG island methylator phenotype underlies sporadic microsatellite instability and is tightly associated with BRAF mutation in colorectal cancer. Nat Genet 2006;38(7):787–93.

[10] Figueroa ME, Lugthart S, Li Y, Erpelinck-Verschueren C, Deng X, Christos PJ, et al. DNA methylation signatures identify biologically distinct subtypes in acute myeloid leukemia. Cancer Cell 2010;17(1):13–27.

[11] Maekita T, Nakazawa K, Mihara M, Nakajima T, Yanaoka K, Iguchi M, et al. High levels of aberrant DNA methylation in *Helicobacter pylori*-infected gastric mucosae and its possible association with gastric cancer risk. Clin Cancer Res 2006;12(3 Pt 1):989–95.

[12] Niwa T, Toyoda T, Tsukamoto T, Mori A, Tatematsu M, Ushijima T. Prevention of *Helicobacter pylori*-induced gastric cancers in gerbils by a DNA demethylating agent. Cancer Prev Res (Phila) 2013;6(4):263–70.

[13] van Kempen PM, van Bockel L, Braunius WW, Moelans CB, van Olst M, de Jong R, et al. HPV-positive oropharyngeal squamous cell carcinoma is associated with TIMP3 and CADM1 promoter hypermethylation. Cancer Med 2014;3(5):1185–96.

[14] Strahl BD, Allis CD. The language of covalent histone modifications. Nature 2000;403(6765):41–5.

[15] Azuara V, Perry P, Sauer S, Spivakov M, Jorgensen HF, John RM, et al. Chromatin signatures of pluripotent cell lines. Nat Cell Biol 2006;8(5):532–8.

[16] Bernstein BE, Mikkelsen TS, Xie X, Kamal M, Huebert DJ, Cuff J, et al. A bivalent chromatin structure marks key developmental genes in embryonic stem cells. Cell 2006;125(2):315–26.

[17] Rada-Iglesias A, Bajpai R, Swigut T, Brugmann SA, Flynn RA, Wysocka J. A unique chromatin signature uncovers early developmental enhancers in humans. Nature. 2011;470(7333):279–83.

[18] Taberlay PC, Statham AL, Kelly TK, Clark SJ, Jones PA. Reconfiguration of nucleosome-depleted regions at distal regulatory elements accompanies DNA methylation of enhancers and insulators in cancer. Genome Res 2014;24(9):1421–32.

[19] Chapman-Rothe N, Curry E, Zeller C, Liber D, Stronach E, Gabra H, et al. Chromatin H3K27me3/H3K4me3 histone marks define gene sets in high-grade serous ovarian cancer that distinguish malignant, tumour-sustaining and chemo-resistant ovarian tumour cells. Oncogene 2013;32(38):4586–92.

[20] Statham AL, Robinson MD, Song JZ, Coolen MW, Stirzaker C, Clark SJ. Bisulfite sequencing of chromatin immunoprecipitated DNA (BisChIP-seq) directly informs methylation status of histone-modified DNA. Genome Res 2012;22(6):1120–7.

[21] Gal-Yam EN, Egger G, Iniguez L, Holster H, Einarsson S, Zhang X, et al. Frequent switching of Polycomb repressive marks and DNA hypermethylation in the PC3 prostate cancer cell line. Proc Natl Acad Sci USA 2008;105(35):12979–84.
[22] Widschwendter M, Fiegl H, Egle D, Mueller-Holzner E, Spizzo G, Marth C, et al. Epigenetic stem cell signature in cancer. Nat Genet 2007;39(2):157–8.
[23] Chi P, Allis CD, Wang GG. Covalent histone modifications—miswritten, misinterpreted and mis-erased in human cancers. Nat Rev Cancer 2010;10(7):457–69.
[24] Lister R, Pelizzola M, Dowen RH, Hawkins RD, Hon G, Tonti-Filippini J, et al. Human DNA methylomes at base resolution show widespread epigenomic differences. Nature 2009;462(7271):315–22.
[25] Barski A, Cuddapah S, Cui K, Roh TY, Schones DE, Wang Z, et al. High-resolution profiling of histone methylations in the human genome. Cell 2007;129(4):823–37.
[26] Song L, Crawford GE. DNase-seq: a high-resolution technique for mapping active gene regulatory elements across the genome from mammalian cells. Cold Spring Harb Protoc 2010;2010(2): pdb prot5384.
[27] Zhang Z, Pugh BF. High-resolution genome-wide mapping of the primary structure of chromatin. Cell 2011;144(2):175–86.
[28] Giresi PG, Kim J, McDaniell RM, Iyer VR, Lieb JD. FAIRE (Formaldehyde-Assisted Isolation of Regulatory Elements) isolates active regulatory elements from human chromatin. Genome Res 2007;17(6):877–85.
[29] Buenrostro JD, Giresi PG, Zaba LC, Chang HY, Greenleaf WJ. Transposition of native chromatin for fast and sensitive epigenomic profiling of open chromatin, DNA-binding proteins and nucleosome position. Nat Methods 2013;10(12):1213–18.
[30] Kelly TK, Liu Y, Lay FD, Liang G, Berman BP, Jones PA. Genome-wide mapping of nucleosome positioning and DNA methylation within individual DNA molecules. Genome Res 2012;22(12):2497–506.
[31] Feser J, Truong D, Das C, Carson JJ, Kieft J, Harkness T, et al. Elevated histone expression promotes life span extension. Mol Cell 2010;39(5):724–35.
[32] Hu Z, Chen K, Xia Z, Chavez M, Pal S, Seol JH, et al. Nucleosome loss leads to global transcriptional up-regulation and genomic instability during yeast aging. Genes Dev 2014;28(4):396–408.
[33] Jeong S, Liang G, Sharma S, Lin JC, Choi SH, Han H, et al. Selective anchoring of DNA methyltransferases 3A and 3B to nucleosomes containing methylated DNA. Mol Cell Biol 2009;29(19):5366–76.
[34] Sharma S, De Carvalho DD, Jeong S, Jones PA, Liang G. Nucleosomes containing methylated DNA stabilize DNA methyltransferases 3A/3B and ensure faithful epigenetic inheritance. PLoS Genet 2011;7(2):e1001286.
[35] You JS, Kelly TK, De Carvalho DD, Taberlay PC, Liang G, Jones PA. OCT4 establishes and maintains nucleosome-depleted regions that provide additional layers of epigenetic regulation of its target genes. Proc Natl Acad Sci USA 2011;108(35):14497–502.
[36] Hesson LB, Sloane MA, Wong JW, Nunez AC, Srivastava S, Ng B, et al. Altered promoter nucleosome positioning is an early event in gene silencing. Epigenetics 2014;9(10):1422–30.
[37] Lin JC, Jeong S, Liang G, Takai D, Fatemi M, Tsai YC, et al. Role of nucleosomal occupancy in the epigenetic silencing of the MLH1 CpG island. Cancer Cell 2007;12(5):432–44.
[38] Yang X, Noushmehr H, Han H, Andreu-Vieyra C, Liang G, Jones PA. Gene reactivation by 5-aza-2′-deoxycytidine-induced demethylation requires SRCAP-mediated H2A.Z insertion to establish nucleosome depleted regions. PLoS Genet 2012;8(3):e1002604.
[39] Magnani L, Brunelle M, Gevry N, Lupien M. Chromatin landscape and endocrine response in breast cancer. Epigenomics 2012;4(6):675–83.
[40] Simon JM, Hacker KE, Singh D, Brannon AR, Parker JS, Weiser M, et al. Variation in chromatin accessibility in human kidney cancer links H3K36 methyltransferase loss with widespread RNA processing defects. Genome Res 2014;24(2):241–50.

[41] Thurman RE, Rynes E, Humbert R, Vierstra J, Maurano MT, Haugen E, et al. The accessible chromatin landscape of the human genome. Nature 2012;489(7414):75–82.
[42] Valouev A, Johnson SM, Boyd SD, Smith CL, Fire AZ, Sidow A. Determinants of nucleosome organization in primary human cells. Nature 2011;474(7352):516–20.
[43] Polak P, Karlic R, Koren A, Thurman R, Sandstrom R, Lawrence MS, et al. Cell-of-origin chromatin organization shapes the mutational landscape of cancer. Nature 2015;518(7539):360–4.
[44] Grzeda KR, Royer-Bertrand B, Inaki K, Kim H, Hillmer AM, Liu ET, et al. Functional chromatin features are associated with structural mutations in cancer. BMC Genomics 2014;15:1013.
[45] Davie K, Jacobs J, Atkins M, Potier D, Christiaens V, Halder G, et al. Discovery of transcription factors and regulatory regions driving *in vivo* tumor development by ATAC-seq and FAIRE-seq open chromatin profiling. PLoS Genet 2015;11(2):e1004994.
[46] Xi H, Shulha HP, Lin JM, Vales TR, Fu Y, Bodine DM, et al. Identification and characterization of cell type-specific and ubiquitous chromatin regulatory structures in the human genome. PLoS Genet 2007;3 (8):e136.
[47] Taberlay PC, Kelly TK, Liu CC, You JS, De Carvalho DD, Miranda TB, et al. Polycomb-repressed genes have permissive enhancers that initiate reprogramming. Cell 2011;147(6):1283–94.
[48] Zentner GE, Tesar PJ, Scacheri PC. Epigenetic signatures distinguish multiple classes of enhancers with distinct cellular functions. Genome Res 2011;21(8):1273–83.
[49] Crawford GE, Holt IE, Whittle J, Webb BD, Tai D, Davis S, et al. Genome-wide mapping of DNase hypersensitive sites using massively parallel signature sequencing (MPSS). Genome Res 2006;16 (1):123–31.
[50] Du M, Yuan T, Schilter KF, Dittmar RL, Mackinnon A, Huang X, et al. Prostate cancer risk locus at 8q24 as a regulatory hub by physical interactions with multiple genomic loci across the genome. Hum Mol Genet 2015;24(1):154–66.
[51] Blattler A, Yao L, Witt H, Guo Y, Nicolet CM, Berman BP, et al. Global loss of DNA methylation uncovers intronic enhancers in genes showing expression changes. Genome Biol 2014;15(9):469.
[52] Zeitz MJ, Ay F, Heidmann JD, Lerner PL, Noble WS, Steelman BN, et al. Genomic interaction profiles in breast cancer reveal altered chromatin architecture. PLoS One 2013;8(9):e73974.
[53] Consortium EP, Birney E, Stamatoyannopoulos JA, Dutta A, Guigo R, Gingeras TR, et al. Identification and analysis of functional elements in 1% of the human genome by the ENCODE pilot project. Nature 2007;447(7146):799–816.
[54] Hauptman N, Glavac D. MicroRNAs and long non-coding RNAs: prospects in diagnostics and therapy of cancer. Radiol Oncol 2013;47(4):311–18.
[55] Liu Z, Zhu J, Cao H, Ren H, Fang X. miR-10b promotes cell invasion through RhoC-AKT signaling pathway by targeting HOXD10 in gastric cancer. Int J Oncol 2012;40(5):1553–60.
[56] Dews M, Homayouni A, Yu D, Murphy D, Sevignani C, Wentzel E, et al. Augmentation of tumor angiogenesis by a Myc-activated microRNA cluster. Nat Genet 2006;38(9):1060–5.
[57] Olson P, Lu J, Zhang H, Shai A, Chun MG, Wang Y, et al. MicroRNA dynamics in the stages of tumorigenesis correlate with hallmark capabilities of cancer. Genes Dev 2009;23(18):2152–65.
[58] Calin GA, Sevignani C, Dumitru CD, Hyslop T, Noch E, Yendamuri S, et al. Human microRNA genes are frequently located at fragile sites and genomic regions involved in cancers. Proc Natl Acad Sci USA 2004;101(9):2999–3004.
[59] Lu J, Getz G, Miska EA, Alvarez-Saavedra E, Lamb J, Peck D, et al. MicroRNA expression profiles classify human cancers. Nature 2005;435(7043):834–8.
[60] Volinia S, Calin GA, Liu CG, Ambs S, Cimmino A, Petrocca F, et al. A microRNA expression signature of human solid tumors defines cancer gene targets. Proc Natl Acad Sci USA 2006;103(7):2257–61.

[61] Zhang JG, Wang JJ, Zhao F, Liu Q, Jiang K, Yang GH. MicroRNA-21 (miR-21) represses tumor suppressor PTEN and promotes growth and invasion in non-small cell lung cancer (NSCLC). Clin Chim Acta 2010;411(11–12):846–52.

[62] Ozen M, Creighton CJ, Ozdemir M, Ittmann M. Widespread deregulation of microRNA expression in human prostate cancer. Oncogene 2008;27(12):1788–93.

[63] Johnson SM, Grosshans H, Shingara J, Byrom M, Jarvis R, Cheng A, et al. RAS is regulated by the let-7 microRNA family. Cell 2005;120(5):635–47.

[64] Hu X, Guo J, Zheng L, Li C, Zheng TM, Tanyi JL, et al. The heterochronic microRNA let-7 inhibits cell motility by regulating the genes in the actin cytoskeleton pathway in breast cancer. Mol Cancer Res 2013;11(3):240–50.

[65] Oh JS, Kim JJ, Byun JY, Kim IA. Lin28-let7 modulates radiosensitivity of human cancer cells with activation of K-Ras. Int J Radiat Oncol Biol Phys 2010;76(1):5–8.

[66] Cimmino A, Calin GA, Fabbri M, Iorio MV, Ferracin M, Shimizu M, et al. miR-15 and miR-16 induce apoptosis by targeting BCL2. Proc Natl Acad Sci USA 2005;102(39):13944–9.

[67] Wolter H, Gottfried HW, Mattfeldt T. Genetic changes in stage pT2N0 prostate cancer studied by comparative genomic hybridization. BJU Int 2002;89(3):310–16.

[68] Takeshita F, Patrawala L, Osaki M, Takahashi RU, Yamamoto Y, Kosaka N, et al. Systemic delivery of synthetic microRNA-16 inhibits the growth of metastatic prostate tumors via downregulation of multiple cell-cycle genes. Mol Ther 2010;18(1):181–7.

[69] Park SM, Gaur AB, Lengyel E, Peter ME. The miR-200 family determines the epithelial phenotype of cancer cells by targeting the E-cadherin repressors ZEB1 and ZEB2. Genes Dev 2008;22(7):894–907.

[70] Tang H, Deng M, Tang Y, Xie X, Guo J, Kong Y, et al. miR-200b and miR-200c as prognostic factors and mediators of gastric cancer cell progression. Clin Cancer Res 2013;19(20):5602–12.

[71] Zheng ZM, Wang X. Regulation of cellular miRNA expression by human papillomaviruses. Biochim Biophys Acta 2011;1809(11–12):668–77.

[72] Gupta RA, Shah N, Wang KC, Kim J, Horlings HM, Wong DJ, et al. Long non-coding RNA HOTAIR reprograms chromatin state to promote cancer metastasis. Nature 2010;464(7291):1071–6.

[73] Poliseno L, Salmena L, Zhang J, Carver B, Haveman WJ, Pandolfi PP. A coding-independent function of gene and pseudogene mRNAs regulates tumour biology. Nature 2010;465(7301):1033–468.

[74] Sun M, Jin FY, Xia R, Kong R, Li JH, Xu TP, et al. Decreased expression of long noncoding RNA GAS5 indicates a poor prognosis and promotes cell proliferation in gastric cancer. BMC Cancer 2014;14:319.

[75] Calin GA, Liu CG, Ferracin M, Hyslop T, Spizzo R, Sevignani C, et al. Ultraconserved regions encoding ncRNAs are altered in human leukemias and carcinomas. Cancer Cell 2007;12(3):215–29.

[76] Peng JC, Shen J, Ran ZH. Transcribed ultraconserved region in human cancers. RNA Biol 2013;10 (12):1771–7.

[77] Cheng J, Guo JM, Xiao BX, Miao Y, Jiang Z, Zhou H, et al. piRNA, the new non-coding RNA, is aberrantly expressed in human cancer cells. Clin Chim Acta 2011;412(17–18):1621–5.

[78] Mei YP, Liao JP, Shen J, Yu L, Liu BL, Liu L, et al. Small nucleolar RNA 42 acts as an oncogene in lung tumorigenesis. Oncogene 2012;31(22):2794–804.

[79] Huether R, Dong L, Chen X, Wu G, Parker M, Wei L, et al. The landscape of somatic mutations in epigenetic regulators across 1,000 paediatric cancer genomes. Nat Commun 2014;5:3630.

[80] Kadoch C, Hargreaves DC, Hodges C, Elias L, Ho L, Ranish J, et al. Proteomic and bioinformatic analysis of mammalian SWI/SNF complexes identifies extensive roles in human malignancy. Nat Genet 2013;45(6):592–601.

[81] Skulte KA, Phan L, Clark SJ, Taberlay PC. Chromatin remodeler mutations in human cancers: epigenetic implications. Epigenomics 2014;6(4):397–414.

[82] Shain AH, Pollack JR. The spectrum of SWI/SNF mutations, ubiquitous in human cancers. PLoS One 2013;8(1):e55119.
[83] Kim MS, Chung NG, Kang MR, Yoo NJ, Lee SH. Genetic and expressional alterations of CHD genes in gastric and colorectal cancers. Histopathology 2011;58(5):660−8.
[84] Cai Y, Geutjes EJ, de Lint K, Roepman P, Bruurs L, Yu LR, et al. The NuRD complex cooperates with DNMTs to maintain silencing of key colorectal tumor suppressor genes. Oncogene 2014;33 (17):2157−68.
[85] Berry WL, Janknecht R. KDM4/JMJD2 histone demethylases: epigenetic regulators in cancer cells. Cancer Res 2013;73(10):2936−42.
[86] Ha K, Lee GE, Palii SS, Brown KD, Takeda Y, Liu K, et al. Rapid and transient recruitment of DNMT1 to DNA double-strand breaks is mediated by its interaction with multiple components of the DNA damage response machinery. Hum Mol Genet 2011;20(1):126−40.
[87] Singh RR, Bains A, Patel KP, Rahimi H, Barkoh BA, Paladugu A, et al. Detection of high-frequency and novel DNMT3A mutations in acute myeloid leukemia by high-resolution melting curve analysis. J Mol Diagn 2012;14(4):336−45.
[88] Ley TJ, Ding L, Walter MJ, McLellan MD, Lamprecht T, Larson DE, et al. DNMT3A mutations in acute myeloid leukemia. N Engl J Med 2010;363(25):2424−33.
[89] Saito Y, Kanai Y, Sakamoto M, Saito H, Ishii H, Hirohashi S. Overexpression of a splice variant of DNA methyltransferase 3b, DNMT3b4, associated with DNA hypomethylation on pericentromeric satellite regions during human hepatocarcinogenesis. Proc Natl Acad Sci USA 2002;99(15):10060−5.
[90] Gonzalo S, Jaco I, Fraga MF, Chen T, Li E, Esteller M, et al. DNA methyltransferases control telomere length and telomere recombination in mammalian cells. Nat Cell Biol 2006;8(4):416−24.
[91] Merritt WM, Lin YG, Han LY, Kamat AA, Spannuth WA, Schmandt R, et al. Dicer, Drosha, and outcomes in patients with ovarian cancer. N Engl J Med 2008;359(25):2641−50.
[92] Chiosea S, Jelezcova E, Chandran U, Acquafondata M, McHale T, Sobol RW, et al. Up-regulation of dicer, a component of the MicroRNA machinery, in prostate adenocarcinoma. Am J Pathol 2006;169 (5):1812−20.
[93] Muralidhar B, Goldstein LD, Ng G, Winder DM, Palmer RD, Gooding EL, et al. Global microRNA profiles in cervical squamous cell carcinoma depend on Drosha expression levels. J Pathol 2007;212(4):368−77.

PART

METHODS AND TOOLS FOR EPIGENETIC DRUG DEVELOPMENT

II

CHAPTER

DRUG DISCOVERY METHODS

3

Marie Lopez[1], Maxime Leroy[2], Chantal Etievant[1], Fréderic Ausseil[1], Alexandre Gagnon[2] and Paola Arimondo[1]

[1]Unité de Service et de Recherche, CRDPF, Toulouse, France [2]Département de Chimie, Université du Québec à Montréal, Montréal, QC, Canada

CHAPTER OUTLINE

G. Egger & P. Arimondo (Eds): Drug Discovery in Cancer Epigenetics. DOI: http://dx.doi.org/10.1016/B978-0-12-802208-5.00003-5

3.1 INTRODUCTION

Epigenetics controls gene expression without changing the DNA base sequence. Epigenetic modifications concern nucleosome positioning, histone posttranslational modifications, DNA methylation, and noncoding RNAs. Abnormalities of the epigenetic landscape contribute to cancer initiation and progression [1,2]. Epigenetics represents a great promise for the discovery of new therapies to fight cancer [3] (cf. Chapter 2), since the chemical modifications of histones and DNA (epigenetic marks) are reversible, in contrast to genetic mutations that are irreversible. Thus, even if the epigenetic regulators are mutated, their downstream consequences on the marks and their regulation can be reversed and the gene expression pattern of the cells reprogrammed [4]. In particular, DNA methylation and the chemical modifications of histones H3 and H4 are major explored therapeutic targets. Epigenetic regulation is orchestrated by (i) *writers*, enzymes that "catalyze" the introduction of epigenetic marks such as the histone methyltransferases (HMT) and acetyltransferases (HAT), and the C5 DNA methyltransferases (DNMT); (ii) *erasers*, enzymes that "remove" epigenetic marks, such as the histone demethylases (HDM) and deacetylases (HDAC), and the Ten-Eleven Translocation methylcytosine dioxygenase 1 (TET) and Isocitrate DeHydrogenase (IDH) enzymes that participate to the removal of methyl groups on DNA; and (iii) *readers*, proteins that "read" the marks and activate downstream signaling, such as methyl-binding proteins (MDB), bromodomains (BRDs), chromodomains, Tudor, PHD domains (cf. Chapter 1). Chemical agents can potentially target each of these actors. In the last 10 years, five epigenetic drugs were approved by the food and drug administration (FDA) to treat hematological cancers (azacitidine and decitabine targeting the DNMTs; vorinostat, romidepsin, and beleodaq targeting the HDACs) and more drugs targeting HMT, HDM, IDH, and the BRD are in clinical trials. Here we describe and discuss the chemical and biological strategies that led to the discovery of inhibitors of the epigenetic regulators. First, we describe rational and random approaches to identify the starting point for chemical optimization. Second, examples are used to illustrate the chemical approaches that resulted in potent lead compounds. Third, we discuss the biological assays that have been set up for preclinical evaluation and drug–target engagement. Importantly, new pharmacological concepts were developed since the effect of these inhibitors is not a direct cytotoxicity, as it is the case for the classical anticancer agents or targeted therapies, but rather a reprogramming of the cancer cells that can eventually result in cell death [5–7]. Several parameters need to be redefined: the time of response—the effects are observed long after treatment or after repeated treatments; the efficient doses—doses that are able to reprogram cells but are not immediately cytotoxic; and the markers of response. The discovery of the first epigenetic drugs opened the road to new pharmacological models.

3.2 HIT DISCOVERY

Several chemical strategies were used to find the chemical starting point (hit) for the development of epidrugs (Figure 3.1): random screening (*in silico* or *in vitro*), fragment- and structure-based approaches, design of hybrid inhibitors, drug repurposing, and exploration of natural products.

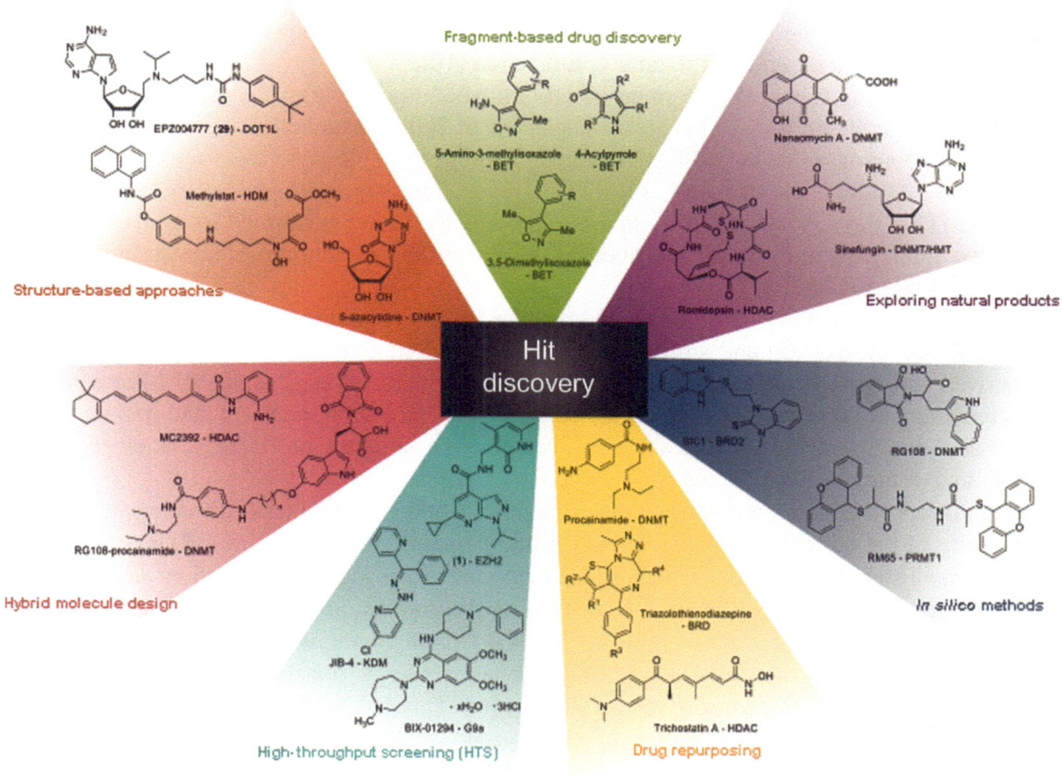

FIGURE 3.1

Main hit discovery strategies applied to epigenetic targets. The chemical structures of the main described examples are shown.

3.2.1 *IN SILICO* SCREENING AND MOLECULAR DOCKING

In silico approaches are efficient methods for hit identification as they allow very large compound libraries to be screened at a relative low cost. They also provide suggestions about the mode of interaction between the molecules and their potential target. These methods require a 3D structure of the target or a homology model and cannot be a substitute for the biological evaluation of such identified compounds. Therefore, computational methods provide valuable starting points for hit identification by exploring novel scaffolds that are structurally divergent from existing active molecules.

RG108 was the first non-nucleoside analog DNMT inhibitor reported. The DNMTs catalyze the transfer of a methyl group from the *S*-adenosyl-L-methionine (SAM) cofactor to the C5-position of cytosine in a CpG context. CpG methylation of promoters induces the silencing of the corresponding genes: in cancer, several genes, including tumor suppressor genes, are silenced by promoter methylation (cf. Chapter 7). RG108 was identified *in silico* from a homology model of DNMT1 out of 1990 molecules of the NCI database [8]. The same groups also reported other starting points based on the same methodology, which were recently reviewed [9]. However these RG108-related molecules lack specificity and still require optimization [10].

In silico screening was also performed on HMTs, which transfer a methyl group from the SAM to a lysine or arginine on histone tails (cf. Chapter 9). In 2007, Spannhoff et al. conducted a virtual screening on homology models of PRMT1, human arginine methyltransferase 1 methylating histone 4 at arginine 3 (H4R3), and of RmtA, PRMT1 homolog from *Aspergillus nidulans*, on the basis of the rat PRMT3 X-ray crystal structure. *In vitro* testing was further performed to validate the molecules. Two main structures were described as potent PRMT inhibitors: Allatodapsone [11] and RM65 [12] (Figure 3.1).

In 2013, Di Micco et al. identified three selective HDAC compounds. The deacetylation of histones, catalyzed by HDACs and sirtuins, is involved in chromatin compaction and gene silencing and its overexpression is oncogenic (HDAC inhibitors are detailed in Chapter 8). After a first virtual screening and docking calculation associated with structural analysis and rationalization, *in vitro* assays allowed the identification of one selective inhibitor of HDAC2, a second one of HDAC8, and a third of HDAC3 and HDAC8 [13].

Ito et al. reported the identification of BRD ligands using computational screening in the acetyl lysine cavity of the corresponding BRD2-BD1 bromodomain crystal structure. BRDs are readers of the acetyl modification of histones and play an important role in signal transductions and the modulation of gene expression (cf. Chapter 10). Following the computational screening and a Lipinski's rule-based selection, 192 hits out of more than 600,000 compounds were further analyzed by surface plasmon resonance then by fluorescence resonance energy transfer in cells to evaluate the interaction between BRD2 and acetylated H4K12 [14]. This led to the selection of the molecule BIC1 (Figure 3.1) and to the report of its crystal structure in complex with BRD2-BD1. More recently, virtual screening on BRD4 was used to identify four new BRD ligand scaffolds for which the crystal structure was solved [15].

3.2.2 FRAGMENT-BASED DRUG DISCOVERY

Fragment-based drug discovery (FBDD) consists of the evaluation of interactions of small chemical entities at high concentrations with the target protein. Whereas the molecular weight of tested compounds in high-throughput screening (HTS) is commonly limited to 500 g/mol and the expected affinity is in the nanomolar range, the fragments analyzed in FBDD have much smaller size (typically 150–250 g/mol) and a micromolar affinity with the target is sufficient to be considered as potent fragments.

FBDD was applied with success in the field of epigenetics, especially to target BRD that lack catalytic activity. Hewings et al. explored methyl-bearing heterocycles as small fragments from commercially available sources and identified 3,5-dimethylisoxazole (Figure 3.1) as an acetyl lysine-mimicking ligand [16]. Micromolar affinity was reported for the aminomethylisoxazole fragment (Figure 3.1) and participated to the optimization of the drug JQ1 (see Section 3.3.2.1). *In silico* fragment identification was also reported for BRD4 targeting. Out of 7 million compounds, Lucas et al. identified 22 potential candidates and further evaluation by isothermal titration calorimetry allowed the identification of 4-acyl pyrrole XD14 (Figure 3.1), 5-isoxazole XD15, the poisonous alkaloid colchiceine (XD1), and its analog colchicine (XD25) as micromolar BRD4 ligands [17].

Selected compounds from fragment libraries were screened directly by X-ray crystallography, leading to new BRD4 interacting scaffolds such as 2-thiazolidinones [18], providing an interesting starting point for structure-based lead optimization (LO).

3.2.3 STRUCTURE-BASED APPROACHES

Hit discovery by structure-based approaches consists of analyzing the natural protein substrate, the cofactor or mechanism intermediate and mimicking this structure with chemical compounds in order to optimize enzyme recognition. This approach is widely used in the discovery of epigenetic enzymes and led to highly potent molecules. Indeed, among the five FDA-approved epigenetic drugs, four of them (azacitidine, decitabine, vorinostat, and belinostat) were designed via a structure-based approach.

Two DNMT inhibitors are used in the clinic: 5-azacytidine (5azaC; azacitidine, Vidaza®) (Figure 3.1) and 5-azadeoxycytidine (5azadC; decitabine, Dacogen™) approved by the FDA in 2004 and 2006, respectively. Both inhibitors are analogs of cytidine, the substrate of the methylation reaction and were first designed as antimetabolic drugs prior to their description as DNMT inhibitors [19,20]. In addition, SAM donates the methyl group into the DNMT catalytic pocket, leaving SAH as coproduct of the reaction. Isakovic et al. developed constrained homocysteine derivatives as SAH analogs. (2*S*,4*S*)-4-Mercaptopyrrolidine-2-carboxylic acid was reported to be a good replacement for the SAH and modulation of its substitution pattern showed an increase in activity against either DNMT1 or DNMT3B [21].

Suberoylanilidehydroxamic acid (SAHA, vorinostat) is a FDA-approved HDAC inhibitor (Zolinza®). Reuben et al. first observed the cellular activity of hybrid polar compounds such as hexamethylenebisacetamide (HMBA) [22]. This observation was the starting point for the design of HDAC inhibitors, leading to SAHA [23]. Interestingly, the target for SAHA was only determined later by structural similarity to trichostatin A (TSA), another HDAC inhibitor [24]. More recently, modulations of the SAHA aromatic part and its linker group led to the identification of novel active molecules. Among them belinostat [25] received FDA approval in 2014.

Same approaches were applied for the identification of lysine demethylase (KDM) inhibitors (cf. Chapter 9). The jumonji C domain-containing KDMs were targeted via molecules mimicking the substrate and cofactor, leading to methylstat (Figure 3.1), a conjugate of fragments mimicking lysine and α-ketoglutarate [26]. The *in vitro* activity of this molecule is in the micromolar range and it is selective for trimethylated lysine. The design of a peptide mimicking the H3K4 histone tail was accomplished to prepare inhibitors of lysine-specific demethylase 1 (LSD1), a flavin-dependent monoamine oxidase, which demethylates mono- and demethylated lysines, specifically on H3K4 and H3K9 [27]. Because of its overexpression in a number of human cancers, LSD1 has emerged as an important target for development of antitumor agents. Lohse et al. studied various truncated H3K9 histone tails to determine the minimal peptide length for protein recognition. Upon addition of a uracil moiety as a promiscuous iron chelator, the affinity of the molecules dramatically increased and allowed the identification of three peptide inhibitors of PHF8 (ARTKQTARK$(Me)_2$-STGGK(Ac)), JmjD2A (ARK$(Me)_3$-STGGK), and JmjD2C (ARK$(Me)_3$-S) [28].

In the case of HMTs, mechanism-based hit identification approaches also gave successful results. In the search for specific inhibitors of DOT1L, a H3K79 methyltransferase, Daigle et al. focused their efforts on SAM and SAH structures guided by the enzyme crystal structure, and discovered EPZ004777 (Figure 3.1), a very selective inhibitor and highly potent *in vitro* ($IC_{50} = 0.4$ nM) and in mouse xenograft model (Section 3.3.4.1) [29].

3.2.4 HYBRID INHIBITOR DESIGN

The design of hybrid inhibitors consists of the association of at least two active molecules in order to increase the potency and/or the selectivity of the inhibitor. For example, the conjugation of two described inhibitors of DNMTs, RG108 [8] and procainamide [30], resulted in a more active compound (Figure 3.1). Procainamide, a weak DNA binder, was used to target the CpG-rich regions and increase the local concentration of the RG108 inhibitor at these sites, increasing its inhibition potency [31]. Another example is represented by MC2392 (Figure 3.1), a hybrid compound between entinostat (an HDAC inhibitor of the benzamide family) and all-*trans* retinoic acid (ATRA) [32].

3.2.5 DRUG REPURPOSING

Drug repurposing, which consists of using existing drugs in the identification of hit compounds against other targets, is a growing hit discovery approach in pharmaceutical companies, as illustrated by sildenafil citrate (Viagra®) [33] and thalidomide [34]. The main advantage resides in the fact that the molecules have already passed several critical steps in the drug development pipeline including toxicity evaluation, thus limiting the failure rate and reducing the costs of drug development.

TSA (Figure 3.1), a microbial metabolite first reported as an antifungal, anti-infectious molecule, was identified more than 20 years ago as an HDAC inhibitor [35]. TSA inhibits several HDAC isoforms in the nanomolar range and exhibits antiproliferative properties [36]. Valproic acid (VPA) is an antiepileptic drug found to have strong anti-HDAC activity, inducing cell differentiation in various cancer cell lines [36]. Drug repurposing was also applied for the identification of DNMT inhibitors. Procaine (Figure 3.1), used as a local anesthetic, and procainamide, its amide analog, both showed anti-DNMT activity [30,37], similarly to hydralazine, an antihypertensive drug [38]. By structure homology, olsalazine, an anti-inflammatory drug, was identified as a hypomethylating agent in a cellular context [39]. Currently, hydralazine and VPA, both coming from drug repurposing and both orally available, entered clinical trials for myelodysplastic syndrome (MDS) and cutaneous T-cell lymphoma [40].

Based on target structural similarities, three molecules developed as flavin adenine dinucleotide (FAD)-dependent monoamine oxidase inhibitors were tested as inhibitors of KDMs. Pargyline [41], first used as an antihypertensive agent, phenelzine [42], a compound with antidepressant and anxiolytic effects, and tranylcypromine, an antidepressant, exhibited anti-LSD1 activity [43]. Triazolothienodiazepines (Figure 3.1) were used for anti-inflammatory purposes before being established as the first BRD ligands with antitumor properties [44]. Based on this scaffold, the first potent inhibitor of BRD JQ1 was discovered [45] (cf. Chapter 10).

3.2.6 EXPLORING NATURAL PRODUCTS

Natural products are an extremely rich source for the identification of new chemical scaffolds. Epigenetic drugs are no exception to the rule. Nanaomycin A (Figure 3.1), curcumin, epigallocatechin gallate (EGCG), genistein, and romidepsin (Istodax®) (Figure 3.1) are natural products described to target epigenetic factors [46], together with two more recently described molecules, kazinol Q [47] and antroquinonol D [48]. However for many, such as EGCG [49], chaetocin [50], psammaplin [51], or curcumin [52], their mode of action and selectivity is controversial. Sinefungin (Figure 3.1), a metabolite produced by bacteria *Streptomyces* sp. mimics SAH and inhibits several methyltransferases; its derivatives exhibit potent and selective inhibition [53,54].

3.2.7 HIGH-THROUGHPUT SCREENING

HTS has been the source of an increasing number of approved drugs in the past 10 years [55]. This drug discovery approach relies on the testing of large compound collections with no or poor information regarding their ability to interact with the target. The expected hit rate generally varies from 0% to 0.001–0.5% of the total number of screened compounds. This approach is particularly interesting when little information is known about the target structure or when multiprotein pathways, cell functions, or phenotypes are considered [56]. Importantly, HTS experiments must be followed by extensive validation analysis in order to eliminate false positives, promiscuous and PAINS (pan assay interference compounds) substructures [57]. The quality, diversity or specificity of the molecules contained in the libraries are of primary importance and strongly influence the output of the screening process.

The search for inhibitors of epigenetic targets has been extensively investigated by HTS, resulting in the use of many different technologies (Table 3.1). These technologies are based on different assay principles and read-outs [58].

Approaches like LANCE® and HTRF™ are based on time-resolved fluorescence resonance energy transfer (TR-FRET) in homogeneous format [76,77]. A TR-FRET assay, detecting the H3K4me0 mark with the lanthanide chelate excite (LANCE) technology, was developed for the HTS and pharmacological profiling of a 14,000 compound library targeting LSD1 (KDM1) [78].

The AlphaLISA/AlphaScreen (developed by Perkin Elmer) is a bead-based homogeneous chemiluminescent detection technology (Alpha = amplified luminescent proximity homogeneous assay) [79]. It was used in the discovery of chemical scaffolds targeting HATs. Out of 243,000 compounds, compound Ctx-1 was shown to bind in the Acetyl-CoA site of the monocytic leukemia zinc finger (MOZ) enzyme and described as the first reversible selective small-molecule inhibitor of MOZ HAT activity. This technology, applied to measure the H3K9me2 mark, was used to identify potent inhibitors of G9A, also known as EHMT2, which methylates H3K9 and participates to maintain heterochromatin. This enzyme is overexpressed in human cancers and knockdown of G9a inhibits cancer cell growth. Using this approach, the potent nanomolar inhibitor A-366 was

Table 3.1 HTS Technologies Used to Identify Hits Against Epigenetic Targets

Assay Technology	Hit Molecules	Target Class	Chemical Library	References
Cell-based assays				
Fluor de Lys assay	KD5170	HDACs	600,000 compounds	[59]
Yeast survival: URA3 gene reporter	Sirtinol, A3, M15	Sirtuins (Sir2p)	1600 compounds	[60]
P53-reporter assay in T22-DFos-RGC lacZ murine cells	Tenovin-6	Sirtuins (Sirt1, Sirt2)	30,000 compounds	[61]
Fluorescence quenching (AMC)	Indole series	Sirtuins (Sirt1)	280,000 compounds	[62]
Fluorescence microscopy	JIB-04	HDM–KDM5	2080 compounds from NCI's diversity set	[63]
HCS single-cell analysis: H3K27me3	5 active compounds	HDM–JMJD3	87,500 compounds	[64]
Enzymatic-based assay				
Alphascreen/AlphaLISA				
H4 biotinylated histone peptide	Ctx-1	HAT–MOZ	243,000 compounds	[65]
H3K9me2 mark	A-366	HMT–G9a	Chemical diversity subset of AbbVie Inc.	[66]
H3K4me mark	PBIT	HDM–JARID1B	15,134 molecules and natural products	[67]
H3K56ac mark	3 actives	HAT–Rtt109	225,000 synthetic molecules, natural products, repurposing, and peptide mimetics	[57]
SPA technology				
Flashplate				
Histone acetylation, tritiated acetyl-CoA	CCT077791	HAT–p300/PCAF	69,000 compounds	[68]
PCR2, tritiated SAM	CPI-169	HMT–EZH2	150,000 compounds (Costellation library)	[69]
PRC2, tritiated SAM	EPZ005687	HMT–EZH2	175,000 compounds (Epizyme library)	[70]
DNA methylation, tritiated SAM	–	DNMT1–DNMT3A	–	[71]
Beads				
H3K27me3 mark	GSK926; GSK343	HMT–EZH2	2,000,000 compounds (GSK library)	[72]
DELPHIA H3K9me2 immunofluoroassay	BIX-01294	HMT–G9a	125,000 compounds	[73]
Restriction enzyme-coupled assay	Dichlone	DNMT3A/L	1120 compounds (MNHN Paris library)	[74]
	Quinone-like	DNMT1	2320 molecules	[75]

obtained [66]. The AlphaScreen technology was also applied to HDMs and a collection of 15,134 molecules of various origins, including natural products, was screened against H3K4 histone demethylase JARID1B with an *in vitro* assay based on an anti-H3K4me1 antibody [67]. This screen led to the identification of 90 *in vitro* validated hits, among which PBIT (2-(4(4-methylphenyl)-1,2-benzisothiazol-3(2H)-one), a benzisothiazolinone derivative, was shown to prevent the JARID1B overexpression-induced decrease of H3K4me3 mark in HeLa cells.

Immunoassays were widely applied and cell-based assays, including high-content phenotypic analysis, were used to screen large compound collections [73,80,81]. Using a Dissociation Enhanced Lanthanide FluoroImmuno Assay (DELFIA) based on a recombinant GST-G9a, scientists from Boehringer Ingelheim in collaboration with the Research Institute of Molecular Pathology (IMP) in Vienna, conducted a HTS campaign on a 125,000-compound library and identified BIX-01294 (Figure 3.1) as a G9a inhibitor with modest potency [73] (Figure 3.1) that served as a starting point for further optimization (see Section 3.3.2.1).

Other cell-based assays were specifically developed. For example, sirtinol was identified as an inhibitor of sirtuins (a class III HDAC subgroup) among a 1600-compound collection with a lethality assay in yeast [60]. A strain containing a URA3 reporter gene integrated into a transcriptionally silent telomeric region was used. Sir2p, the founding member of the sirtuin family, represses URA3 expression at this locus. A small-molecule inhibitor of Sir2p allows expression of URA3, resulting in cell death in the presence of 5-FOA. Sirtinol was shown to inhibit human SIRT2 deacetylase activity in the micromolar range (IC_{50} of 38 ± 2 μM) and did not show any measurable inhibition of HDAC I activity. Using a *p53*-reporter cell-based assay, a screen of 30,000 drug-like small molecules from the Chembridge DIVERS set led to the identification of the Tenovin molecules (Tenovin 1–6) as inhibitors of SIRT1 and SIRT2 targets [61].

Other HTS assays are based on microscopy technologies. A high-content imaging screening assay based on single-cell multiparameter measurements against JMJD3 was developed and validated using a library of 87,500 molecules [64]. Another cell-based automated fluorescence microscopy assay was used to screen a 2080-compound NCI's diversity set [63]. The assay consisted of mammary adenocarcinoma cells containing a stably integrated GFP-estrogen receptor transgene under the control of the CMV promoter, which is subject to epigenetic regulation. A pyridine hydrazone molecule named JIB-04 (Figure 3.1) was shown to induce a dose-dependent GFP induction associated to KDM5 activity inhibition *in vitro*, in cells and in tumors.

Many enzymatic assays are based on the measurement of fluorescence. The Fluor de Lys protocol was applied to find HDAC inhibitors by using an acetylated substrate, leading to the production of a fluorescent signal when deacetylated. This led to the discovery of KD5170, a mercaptoketone inhibitor with a nonhydroxamate-based structure. The KD5170 was shown to present nanomolar scale *in vitro* HDACs and cell proliferation inhibitions associated to tumor growth inhibition when used as monotherapy or combined with docetaxel [59]. A large series of 280,000 molecules was screened using a fluorimetric assay in which a modified acetyl lysine conjugated to aminomethylcoumarin (AMC) is deacetylated by sirtuins SIRT1 in the presence of the cofactor nicotinamide adenine dinucleotide (NAD) [62]. The addition of a proteolytic developer releases the fluorescent AMC. Using this assay, an indole molecule series was identified with a nanomolar range potency (IC_{50} between 50 and 100 nM) and poor activity against SIRT2, SIRT3, and HDACs.

In the case of DNA methylation, fluorescence assays were coupled to DNA cleavage by restriction enzymes that are sensitive to C5 DNA methylation. A multistep screening process designed to identify DNMT1 inhibitors has been described [75]. It associated a DNMT1 restriction enzyme-coupled fluorescence assay to a counter screen against the restriction enzyme alone. DNA intercalators were eliminated from the hit list and differential scanning fluorimetry (DSF) assay validated direct DNMT1 binding. The Microsource Spectrum collection of 2320 compounds was screened using this process and nine molecules, mainly quinone-like substructures and chlorobenzene compounds, with potencies ranging from 300 nM to 11 μM, were identified. Another HTS based on the same principle but targeting the DNMT3A/L complex was proposed and led to the identification of nitroflavanone compounds among a set of 114 flavones derivatives [82]. This active series affected zebrafish embryo development by inducing a short-axis phenotype similar to the reference inhibitor 5azaC to be homogeneous, and a global demethylation of the genome. A new screening of a 1120 chemical library from the Museum national d'Histoire naturelle in Paris (France) with the same assay led to the identification of 12 new molecules with potencies ranging from 800 nM to 23 μM [74].

Finally, radioactive assays were also developed using tritiated SAM and measuring the amount of radioactive histone or DNA substrate after the enzymatic reaction. Homogeneous tests were developed and the radioactivity was measured with the Scintillation Proximity Assay (SPA) developed by Perkin-Elmer by using either plates (FlashPlates) or beads (Table 3.1). Very large HTS campaigns were performed on the polycomb repressive complex 2 (PRC2). Enhancer of Zeste Homologue 2 (EZH2), the enzymatic subunit of PRC2, is responsible of the trimethylation of lysine 27 on histone H3 (H3K27), a transcriptionally repressive epigenetic mark associated with a number of cancers. Approximately 2,000,000 molecules were screened by GlaxoSmithKline Pharmaceuticals using SPA-beads and mixing together substrates, ^{3}H-SAM, PRC2, and compounds [72]. SAM-competitive and cell-active GSK926 and GSK343 molecules were derived from these screening hits as described in Section 3.3.1 [83] (Figure 3.1, compound **1**). This type of assay was also used to screen 150,000 compounds from the unbiased Constellation compound collection [69]. In this case, detection was ensured by the flashplate support instead of SPA beads. (see Section 3.3.3.1) to obtain a highly potent compound (IC_{50} of 0.032 μM) inhibiting the H3K27me3 mark in cancer cells [84]. A 175,000 compound subset of a chemical diversity library was tested against this target by Epizyme [70]. After chemical structure optimization (see Section 3.3.3.1) this work led to the discovery of the EPZ005687.

In the case of DNA methylation, a flashplate-based assay developed with the objective of providing both high-throughput performances and insights in compound mechanism of action was used to carry out SAM and DNA competition studies of sinefugine, zebularine, and a chloronitroflavanone compound [71] and gave new insights into the mechanism of inhibition of the SGI-1027 family [85].

Another flashplate-based screening of a 69,000 diverse compound collection, measuring the transfer of a tritiated acetyl group from acetyl-CoA to histone enabled the identification of the isothiazolone-derived structure CCT077791 as an inhibitor of both p300/cyclic AMP-responsive element binding protein-associated factor (PCAF) and P300 [68]. Complementary analysis showed that CCT077791 inhibited acetylation level and proliferation of HCT116 and HT29 cell lines.

Other types of assays were used to find the hit for HAT inhibition. A screening of 225,000 chemical sources including synthetic molecules, natural products, repurposing and peptidomimetic compounds against the Rtt109 lysine acetyltransferase was performed by measuring the CoA production from acetyl-CoA by the acetylation reaction. The extent of CoA was detected with the thiol-sensitive probe 7-diethylamino-3-(49-maleimidyl-phenyl)-4-methylcoumarin (CPM). This screen led to the discovery of three validated inhibitors [57].

3.3 LEAD OPTIMIZATION AND IDENTIFICATION

Following the identification of a molecule capable of modulating the activity of an epigenetic target, a process aimed at improving its intrinsic and cell-potency, selectivity, biophysical properties and pharmacokinetic profiles is usually initiated. Depending on the potency of the inhibitor, this process is either called "hit-to-lead" or "lead optimization (LO)." Examples of programs taken from the literature and related to the field of epigenetics will be used to illustrate these processes. A general review on LO has been published elsewhere [86].

3.3.1 LEAD OPTIMIZATION IN THE ABSENCE OF STRUCTURAL INFORMATION

In cases where little structural information regarding the target is available, the optimization of the compound's activity often relies on classical structure–activity relationship (SAR) studies. These studies consist of systematically exploring every sector of the molecule in order to identify groups that are essential for activity, portions that can be deleted, and sites that can be modified without affecting the potency. Optimal modifications are then combined with the objective of obtaining an additive or synergistic effect that translates into a major improvement in potency. In order to accelerate LO, this process can also be supported by combinatorial chemistry or guided by a homology model of the biological target built from a similar enzyme for which a crystallographic structure has been resolved (see Section 3.2.1).

3.3.1.1 Lead optimization guided by SAR data

Following an HTS campaign against the PRC2 complex as described above (Section 3.2.7), chemists at GlaxoSmithKline Pharmaceuticals identified compound **1** (Figure 3.1) as a potent inhibitor of EZH2 with a measured K_i^{app} of 149 nM [83]. In order to elucidate the mode of action (MoA) of this inhibitor, the IC_{50} was determined at different ratios of [SAM]/K_m. The observed pattern showed a loss of potency with increasing SAM concentrations, suggesting that **1** is a SAM-competitive EZH2 inhibitor. Further characterization demonstrated that **1** is not competing against the H3 peptide substrate or the nucleosome. Classical SAR studies then led to the replacement of the cyclopropyl group at the 6-position of the azaindolyl central core with a 2-*N*-methylpiperazinyl-pyridine and the removal of the nitrogen atom at the 7-position, providing GSK926 (**2**), which showed an improved K_i^{app} of 7.9 nM (Figure 3.2A). The introduction of a propyl chain at the 4-position of the pyridone, combined with an isomeric left-hand-side piperazinylpyridine, afforded GSK343 (**3**, $K_i^{app} = 1.2$ nM). Profiling of this compound showed a 1000-fold selectivity over 17 other SAM-dependent enzymes and a dose-dependent reduction of the H3K27me3 marks in HCC1806 breast cells. However, due to a poor rat pharmacokinetic profile (PK), this compound was not tested *in vivo*.

In another example, using a recombinant wild-type assay, scientists at Epizyme Inc. identified compound **4** as a potent PRC2 inhibitor by screening a 175,000-compound collection in a HTS

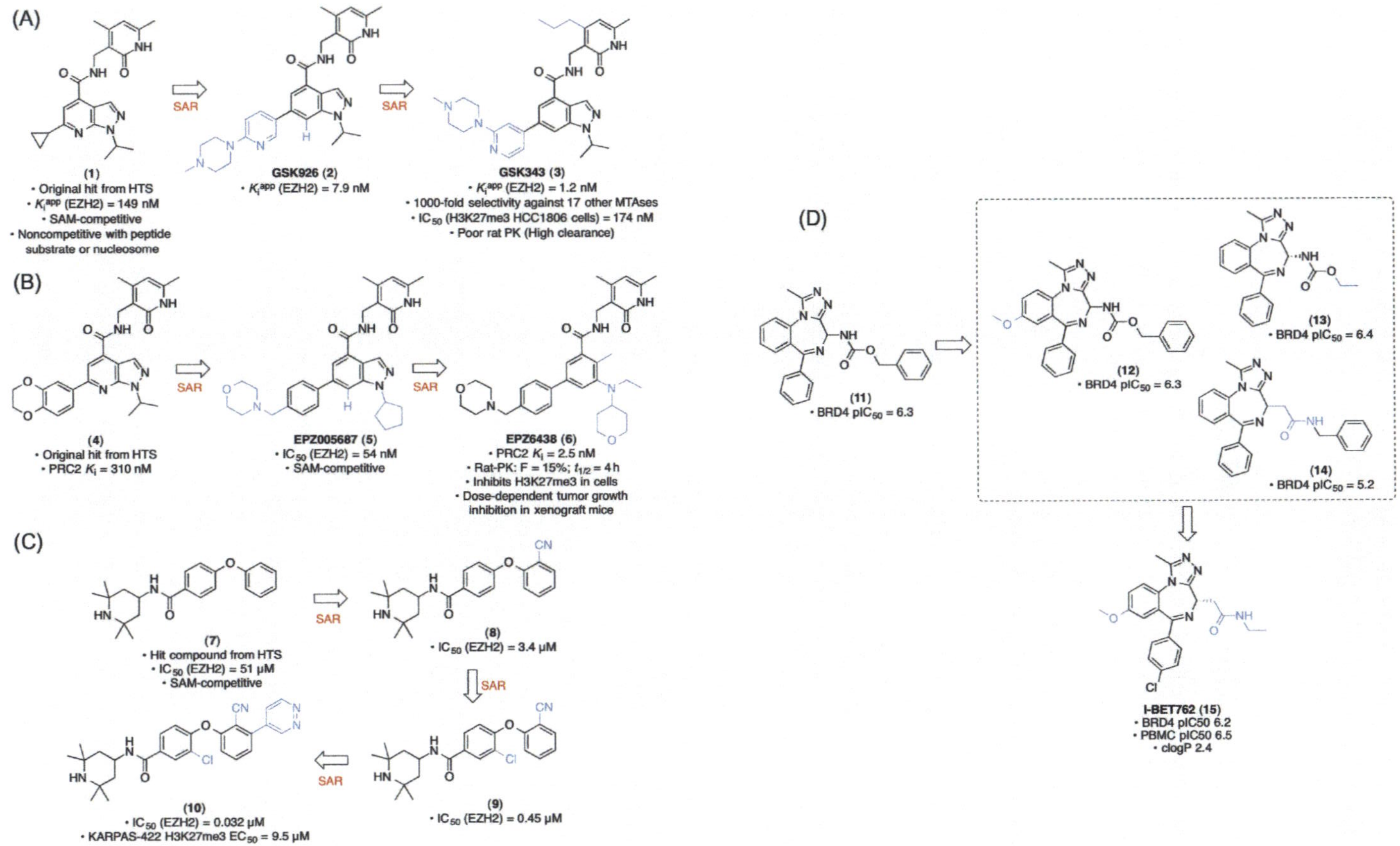

FIGURE 3.2

Lead optimization via SAR studies. (A) Discovery of GSK343, a potent EZH2 inhibitor, by iterative SAR studies. (B) Identification of EPZ6438, a potent inhibitor of the Polycomb Repressive Complex 2 (PRC2), via SAR studies. (C) Identification of compound **10**, a cell-active low-nanomolar inhibitor of EZH2, by sequential SAR modifications on the hit compound **7**. (D) Identification of I-BET762, a low nanomolar inhibitor of BRD4, by sequential structural modifications of **11**. The main parts of the molecules or substituents that are modified during the optimization process are depicted in color.

format (Section 3.2.7 and Figure 3.2B). Substantial improvement in potency was obtained by introducing a morpholinomethylene unit on the left-hand side of the molecule, by removing the nitrogen from the azaindazolyl central core and by replacing the *N*-isopropyl group with a cyclopentyl fragment, as shown in EPZ005687 (**5**, $K_i = 80$ nM). A plot of IC_{50} values as a function of SAM concentrations pointed toward a SAM-competitive mechanism [70]. Opening of the indazolyl central core led to EPZ6438 (**6**, $K_i = 2.5$ nM) where a phenyl ring was used as a scaffold onto which key pharmacophores found in EPZ005687 were attached. EPZ6438 was found to have superior potency and pharmacokinetic profile over EPZ005687 and was shown to reduce the H3K27me3 levels in SMARCB1-deficient MRT cells. In addition, oral dosing at 250 and 500 mg/kg in SCID mice bearing G401 xenografts showed that EPZ6438 is well tolerated and that it practically eliminates the G401 tumors [87,88]. Epizyme has entered into a collaboration with Eisai to develop EPZ6438 (named E7438 by Eisai). Epizyme released data from the phase 1 dose escalation study (March 3, 2015) demonstrating that EPZ6438 shows activity in B-cell non-Hodgkin lymphomas (NHL) and malignant rhabdoid tumors. At a dose up to 800 mg BID, four patients out of 10 achieved a partial or complete response. On March 12, 2015, Epizyme announced that it reacquired the rights from Eisai for EPZ6438 and that it planned to conduct a phase 2 study in approximately 150 patients with NHL and in adults with INI1-deficient rhabdoid tumors as well as a phase 1 study in children with INI1-deficient tumors.

Starting from compound **7** (identified from a HTS campaign as described above, Figure 3.2C), Constellation Pharmaceuticals reported the discovery of **10**, a low-nanomolar SAM-competitive inhibitor of EZH2 [84]. Since no structural information was available, the optimization of this compound was performed via iterative SAR modifications. The first improvement in potency was observed upon introducing a cyano group at the *ortho*-position of the right-hand-side (RHS) phenyl ring (**8**, $IC_{50} = 3.4$ μM). The submicromolar barrier was breached upon adding a chloride at the *ortho*-position of the central phenyl ring, as shown in **9** ($IC_{50} = 0.45$ μM). Re-optimization of the RHS phenyl ring led to compound **10** where a pyridazine ring is present at the 3-position ($IC_{50} = 32$ nM). When KARPAS-422 cells were treated with **10**, a dose- and time-dependent reduction in global H3K27me3 levels was observed [69]. However, due to poor permeability and high efflux ratio, the development of this compound was not further pursued. A model of the cofactor binding site of EZH2 was recently reported in order to elucidate the binding of such small-molecule inhibitors of this target [89].

In 2013, GlaxoSmithKline (GSK) published the structure of I-BET762 (**15**, Figure 3.2D), a novel and potent inhibitor of BRD4 [90–92]. BRDs contain two consecutive BRDs, which recognize the acetyl lysine (Ac-K) residues in histone tails and mediate the protein–protein interactions between BRDs and histones. Small molecules mimicking the Ac-K residues in histone tails may inhibit these protein–protein interactions and lead to modulation of the functions of BRDs. The GSK research program started with the goal of identifying compounds that can upregulate apolipoprotein A-1 (ApoA1) without any knowledge of BRD proteins as the molecular targets. This was performed using a cell-based luciferase reporter assay leading to the identification of **11**, which showed potent induction of the ApoA1 reporter gene with an EC_{170} of 0.22 μM. This compound was considered to be a suitable starting point for medicinal chemistry optimization since it contained a 1,4-benzodiazepine, a motif found in many marketed pharmaceutical agents (see Section 3.2.5 and Figure 3.1). SAR studies enabled the identification of modifications that provided better affinity or efficiency against BRD4, such as a methoxy group on the left-hand-side phenyl ring (compound **12**) and an ethyl carbamate on the lateral chain (compound **13**). However, because the triazolobenzo- and triazolothienodiazepines were susceptible to hydrolysis, removal of the nitrogen atom at position 3 of the benzodiazepine ring was performed, leading to compound **14**. Even though this modification prevented the hydrolysis of the central core, lower potency was observed, leading the authors to combine all three modifications, affording I-BET762 (**15**). This compound was found to have excellent potency against BRD4 as well as high solubility in physiologically relevant media, low plasma protein binding, good passive permeability, excellent metabolic stability, lack of mutagenicity in Ames tests, and good oral bioavailability in mice, rats, dogs, and monkeys.

These four examples demonstrate how the potency of a hit compound possessing moderate activity against a desired target can be improved in the absence of any structural information by simply performing iterative modifications on the molecule. Other examples of LO programs using a SAR approach have been reported (such as [93–95]) and are described in the respective chapters.

3.3.1.2 Lead optimization based on docking or modeling studies

To facilitate the optimization of a lead compound, a homology model of the biological target can be built from another enzyme belonging to the same class. For example, scientists from the University of North Carolina (UNC) in collaboration with the Structural Genomics Consortium (SGC) in Toronto reported in 2013 the discovery of UNC1999 (**17**, Figure 3.3), an EZH2 inhibitor with low nanomolar potency. This compound was designed from docking of the known inhibitor EPZ005687 (**5**) into an EZH2 homology model which was generated from the X-ray crystal structure of GLP (G9a-like protein), a H3K9 mono- and dimethylase. Compounds that combined structural features of EPZ005687 (**5**) and GSK126 (**16**) [96,97] were synthesized and tested against EZH2 using a radioactive biochemical assay, leading to UNC1999. This compound was found to be SAM-competitive with a K_i of 4.6 nM, but noncompetitive with the H3 peptide substrate. Moreover, a reduction in the H3K27 marks was observed in MCF10A cells, with an IC_{50} of 124 nM [98].

3.3.2 LEAD OPTIMIZATION USING STRUCTURAL INFORMATION

A cocrystal of the inhibitor with the biological target provides highly valuable information about the binding environment of the molecule. This information can then be utilized to design novel molecules that can interact with residues in the vicinity of the molecule, fill empty pockets, displace water molecules, or induce conformational changes of the binding pocket. These second-generation inhibitors can then be cocrystallized with the target to generate structural information that can serve as a new starting point for LO.

3.3.2.1 Lead optimization using a structure-based design approach

Following a HTS campaign (Figure 3.1) as described in Section 3.2.7, BIX-01294 (**18**) was identified as a G9a inhibitor with modest nanomolar potency [73]. This compound was used in 2009 by Jin and coworkers as a starting point for lead discovery [99]. An X-ray crystal structure of BIX-01294 in complex with G9a revealed that the *N*-benzyl portion of the aminopiperidinyl group was solvent exposed and that the 7-methoxy group of the quinazoline central core was pointing toward a lysine-binding channel. Using this information, Jin and coworkers designed compound **19** (UNC0224), which showed a dramatic improvement in potency (Figure 3.4A). Further optimization

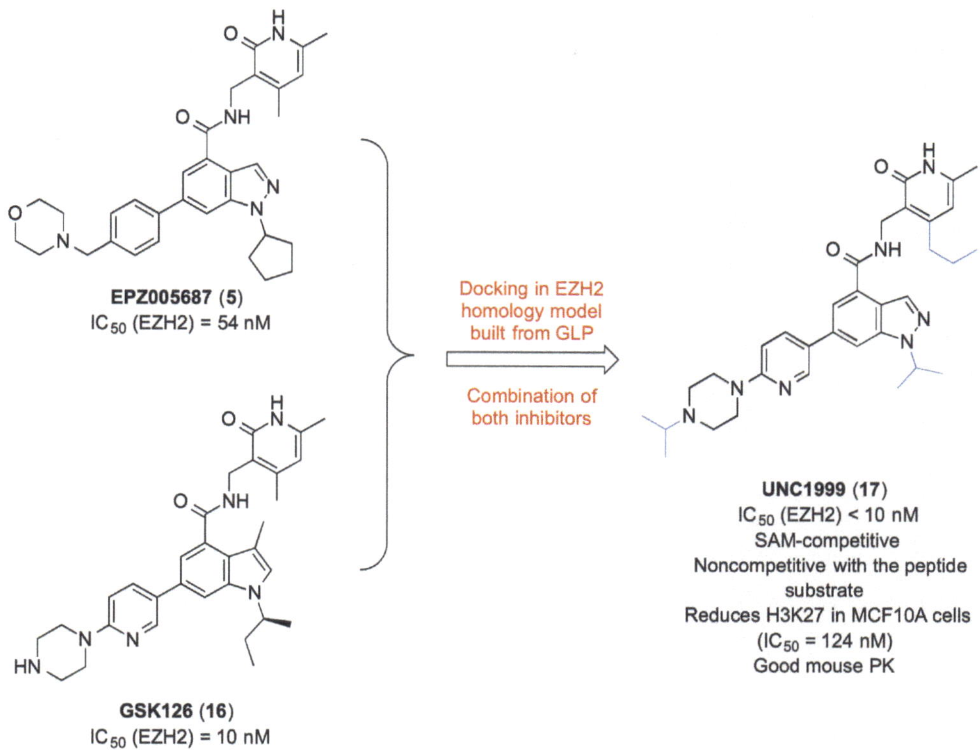

FIGURE 3.3

Lead optimization based on docking or modeling studies. Discovery of UNC1999, a potent EZH2 inhibitor, by docking studies of EPZ00587 and GSK126. The main parts of the molecules or substituents that are modified during the optimization process are depicted in color.

of the 7-dimethylaminopropoxy chain led to UNC0321 (**20**) which possesses an inhibitory constant of 63 pM and an IC_{50} of 9 nM against G9a [100]. However, due to poor cell potency, every area of the molecule was reoptimized in order to balance the lipophilicity of the molecule and obtain adequate passive membrane permeability that would translate into acceptable activity in cells. This exercise resulted in UNC0646 (**21**) that showed an IC_{50} of 6 nM in S-Adenosyl-homocystein hydrolase (SAHH)-coupled assay and of 26 nM in a cell assay [101]. Another solution to the issue of poor cell activity of UNC0321 consisted of replacing the homopiperazine with a cyclohexyl unit and the *N*-methyl group on the 4-aminopiperazine with an isopropyl fragment. These modifications, performed concomitantly with the reinstallment of a 7-aminopropyl side chain, provided UNC0638 (**22**), which showed similar intrinsic potency as UNC0224 [102]. The investigation of the Michaelis–Menten kinetic parameters suggested that this compound competes with the peptide substrate. In addition, UNC0638 was found to inhibit cellular H3K9 dimethylation in MDA-MB-231 cells. Since UNC0638 had a poor pharmacokinetic profile, the difluoro derivative **23** (UNC0642) was prepared, resulting in an improved PK profile over UNC0638, supporting the hypothesis that the cyclohexyl group was metabolized by the CYP450s [103].

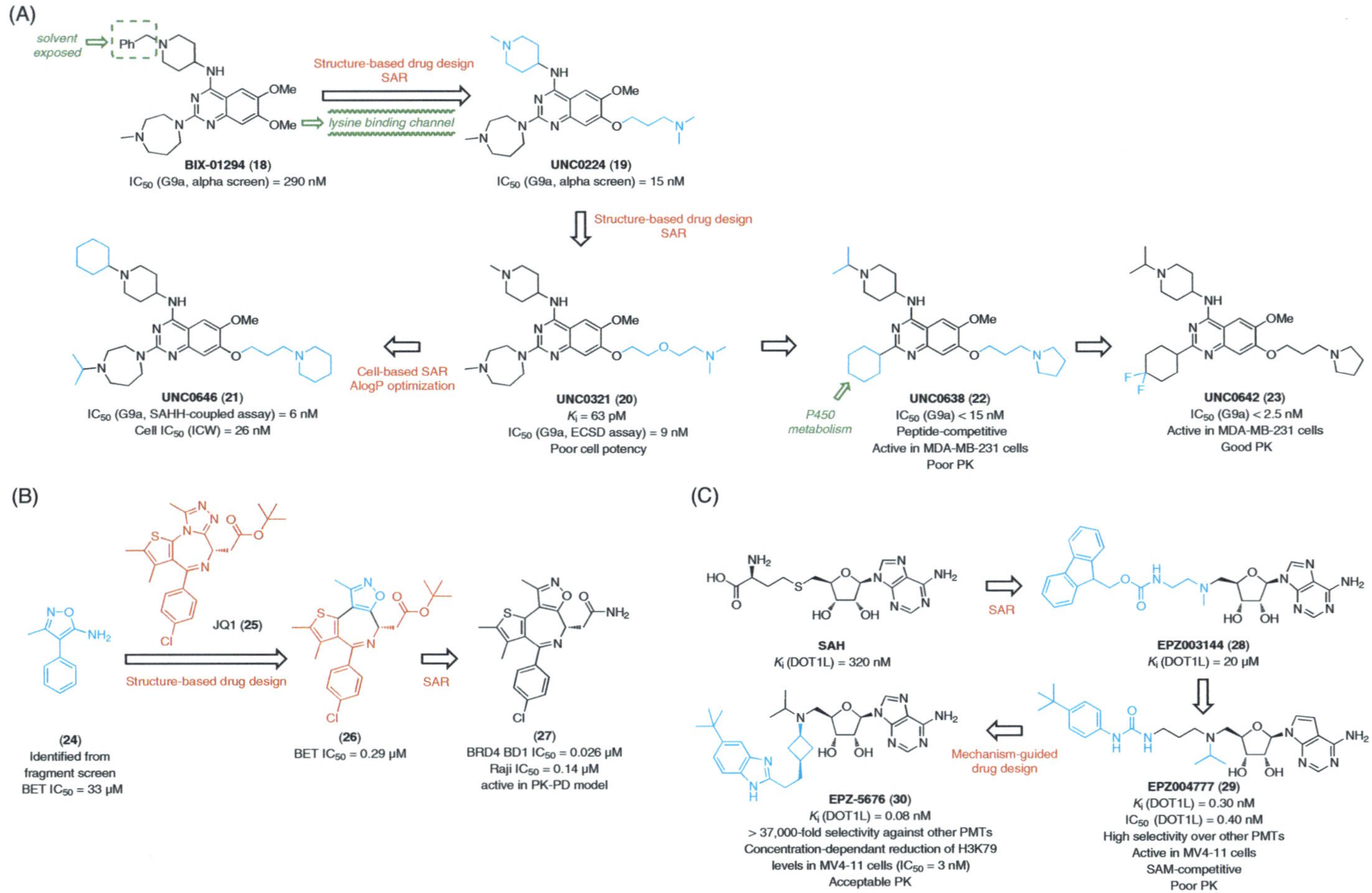

FIGURE 3.4

Lead optimization using structural information. (A) Discovery of UNC0642 from BIX-01294 via a combination of structure-based design, *in vitro* and cell SAR, optimization of lipophilicity, and PK studies. (B) Discovery of **27**, a potent BRD4 inhibitor, by a combination of fragment-based drug discovery and structure-based design. (C) Identification of EPZ-5676, a potent inhibitor of DOT1L. The main parts of the molecules or substituents that are modified during the optimization process are depicted in color.

Constellation Pharmaceuticals reported in 2013 the identification of **24**, a micromolar inhibitor of BRD4, using a fragment-based approach [104] (Figure 3.1). Cocrystallization of this molecule with BRD4 BD1 showed that the amino-isoxazole was binding to the endogenous acetylated lysine site. Moreover, it appeared that the same important interactions involved between JQ1 (**25**) and BRD4 were present in **24** (Figure 3.4B). Consequently, the synthesis and biochemical evaluation of **26** was accomplished, providing an IC_{50} of 290 nM against BET. Since **26** was found to be 10-fold less active than JQ1, the SAR exploration shifted toward the *tert*-butyl ester chain, leading to the identification of **27** as a low-nanomolar inhibitor of BRD4 ($IC_{50} = 26$ nM). Further profiling showed that **27** possessed rat and dog PK profiles suitable for *in vivo* experiments and that **27** inhibited MYC mRNA expression in Raji cells.

Other examples of optimization of inhibitors of BRDs, LSD1, DOT1L, and WDR5 using a structure-based design approach are reported in the literature (see corresponding chapters of this book).

3.3.2.2 Lead optimization using a mechanism- or ligand-based approach

As described in Section 3.2.3, lead compounds can be generated using the natural enzyme substrate or cofactor as a starting point. The objective of this approach consists of generating molecules that have better affinities with the enzyme than the natural substrate and that can interfere with the enzymatic reaction. In one example, reported by Epizyme Inc., novel aminonucleoside inhibitors of DOT1L were designed by replacing the homocysteine moiety of SAH with simple dimethylamines, leading to the identification of EPZ003144 (**28**), which showed a K_i of 20 μM against DOT1L (Figure 3.4C) [105]. A library of compounds focusing on the appended 5′-amine chain was then prepared, enabling the replacement of the Fmoc moiety by a *tert*-butylphenylurea, the installation of a *N*-isopropyl group, and the replacement of the adenine by a deazapurine, as shown in EPZ004777 (**29**). These modifications led to a dramatic improvement in potency, as illustrated by the K_i value of 300 pM. Crystallographic analysis of EPZ004777 with DOT1L showed that the inhibition was likely driven by an induced-fit mechanism where important conformational changes were induced by the sterically hindered *tert*-butylphenylurea [106]. Due to poor pharmacokinetic properties, EPZ004777 was unsuitable for clinical development, thus warranting further optimization of the inhibitor, leading to EPZ-5676 (**30**), a potent DOT1L inhibitor with improved potency and drug-like properties suitable for clinical evaluation [107]. In May 2014, EPZ-5676 was in phase 1b studies to evaluate its safety, pharmacokinetics, and pharmacodynamics with escalating doses and to conduct a preliminary assessment of its antileukemia activity in patients (adults and children) with acute leukemias bearing rearrangements of the MLL gene.

Using a ligand-based approach and starting from the 21 amino acid H3K4 binding region, Kumarasinghe and Woster reported the synthesis and characterization of potent cyclic peptide inhibitors of LSD1 [108]. Additional examples of LO by structure-based design have also been reported for inhibitors of Jumonji C domain-containing HDMs (JHDMs), PRMTs, HMTs, and PRMTs and are cited in the corresponding chapters of this book.

3.3.3 LEAD GENERATION USING OTHER APPROACHES

In this section, other approaches such as parallel or combinatorial chemistry, transposition of fragments from other series, introduction of group isosteres or surrogates, and identification of lead compounds from literature molecules ("lit-to-lead") and by "target hopping" will be described.

3.3.3.1 Lead optimization using parallel or combinatorial chemistry tools

The use of parallel combinatorial chemistry tools can expedite the discovery of potent inhibitors of epigenetic targets. This approach consists of synthesizing very rapidly a high number of analogs of the hit compound where single-point modifications are introduced. Alternatively, many points of diversity can be simultaneously explored using a matrix-like format, allowing the fast identification of combinations that lead to additive or synergistic effects.

CREB-binding proteins (CBP) are lysine acetyl transferases that possess a BRD and a lysine acetyltransferase (KAT) domain. These enzymes play a key role as transcriptional coactivators in human cells and are implicated in cell cycle arrest, senescence, and apoptosis following their recruitment by K382 mutated *p53* genes. Therefore, inhibition of CBP/BRD has potential clinical applications. In one example, Brennan et al. reported the use of parallel chemistry to optimize the activity and selectivity of **31a** [109], a modest inhibitor of CBP (Figure 3.5A) [110]. An X-ray crystal structure of the dimethylisoxazole **31b**, a reported inhibitor of CBP/BRD4, in complex with the CBP/BRD suggested two potential areas where additional groups could interact with the target, potentially leading to improvements in potency and selectivity. On the basis of this analysis, it was anticipated that analogs of compound **31b**, which possessed substitution at N-1 and C-2 may interact with those identified regions. Preparation of 83 analogs of **31b** in a library-format using a Suzuki cross-coupling reaction between heterocycle **31c** and boronic acid pinacol ester **31d** allowed the fast identification of inhibitors with better affinity for CBP, such as **32** which showed a ΔT_m of 4.5 °C against CBP and of 3.2 °C against BRD4 by DSF. Finally, traditional SAR studies led to the identification of **33** as a potent and highly selective CBP inhibitor. Results from a luciferase reporter assay for P53 induction suggested that **33** inhibits the CBP coactivation of *p53* target genes in cells.

In 2014, Bradner et al. identified UMB-32 (**35**) as a new submicromolar probe for BRD4 [111] (Figure 3.5A). After identifying the 3,5-dimethyl-1,2,4-triazole as a recurrent scaffold in BRD4 inhibitors such as in JQ1 (**25**), the synthesis of a focused library using this biasing element was accomplished. The first library enabled the identification of **34** that permitted the selection of the *para*-substitution pattern on the left-hand-side aryl group. In order to expand the chemical diversity, a library was prepared using a multicomponent reaction involving isonitrile **36**, aminoheterocycle **37**, and aldehyde **38**. A perfluoroalkyl phase tag was utilized on aldehyde **38** to facilitate the purification of the products using fluorous solid-phase extraction (F-SPE) techniques. A subsequent Suzuki reaction between sulfonate **39** and 3,5-dimethylisoxazole boronic acid pinacol ester **40** afforded a set of derivatives of **41**, leading to the identification of **35**, which possessed an IC_{50} of 637 nM against BRD4 and a cellular potency in BRD4-dependent cell lines of 724 nM.

A general review of the application of combinatorial chemistry in drug discovery has also been published [112].

3.3.3.2 Lead optimization by transposition of key pharmacophores from known inhibitors

In 2014, AbbVie, in collaboration with the SGC group in Toronto, reported the discovery of hit compound **42**, an inhibitor of G9a with nanomolar potency (Figure 3.5B). This compound, which was identified from a screening of a subset of the company's compound collection using a peptide-based AlphaLISA assay (Section 3.2.7), served as a starting point for LO. Based on the similarity of the dimethoxy unit of **42** with UNC0638 (**22**), the pyrrolidinylpropyl chain of UNC0638 was

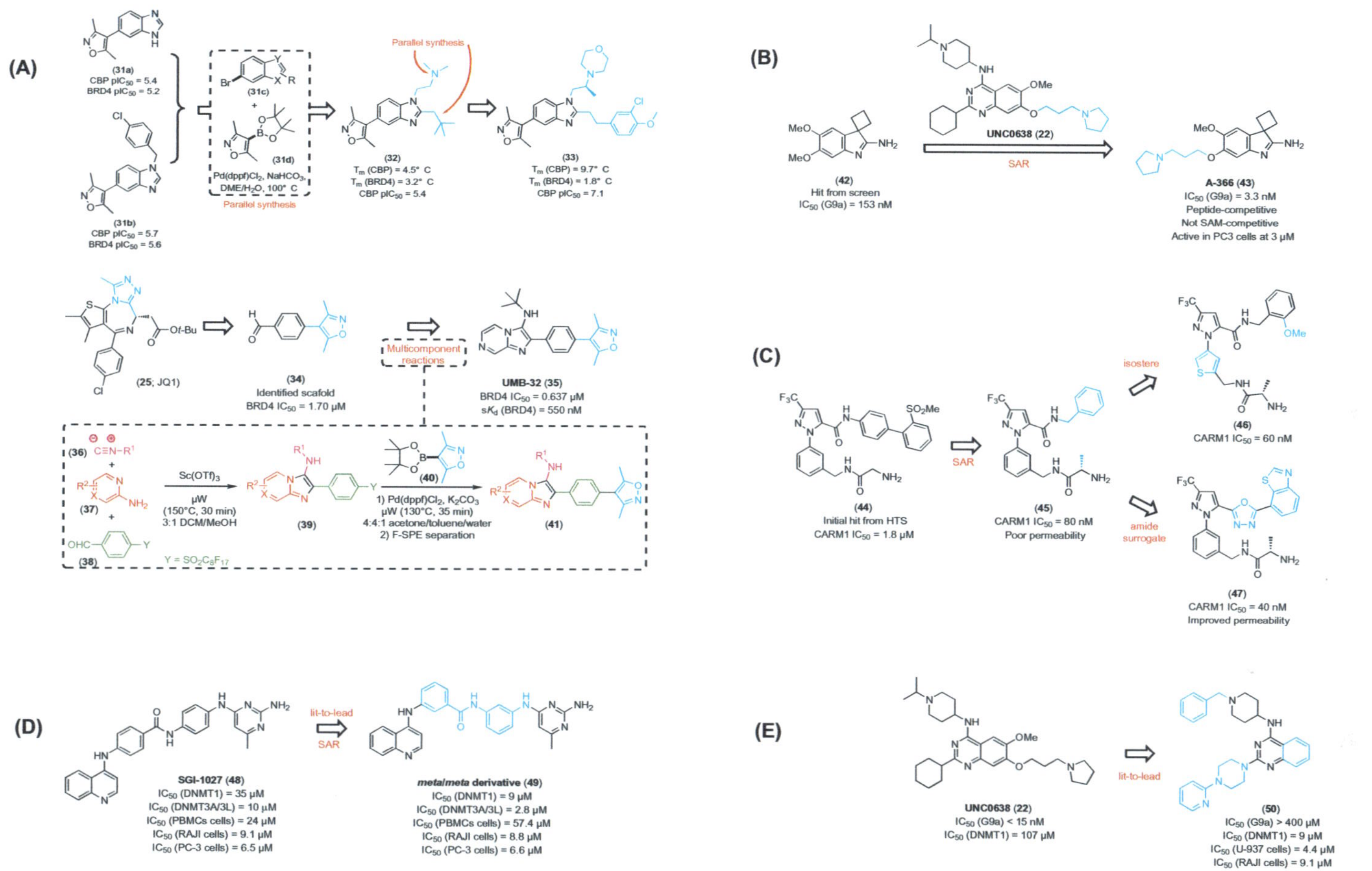

FIGURE 3.5

Lead optimization using other approaches. (A) Use of parallel or combinatorial chemistry tools. Identification of compound **34**, a nanomolar inhibitor of CBP, using a parallel synthesis approach and identification of compound UMB-32, a micromolar probe for BRD4, by SAR using multicomponent reactions. (B) Upon transposition of key pharmacophores from known inhibitors. Development of A-366, a potent G9 inhibitor, by transposition of an aminopropyl chain from UNC0638 to the hit compound **42**. (C) Upon introduction of group isosteres or surrogates. Discovery of **46** and **47**, two low-nanomolar inhibitors of CARM1, by transformation of **45** using group isosteres and surrogates. (D) Discovery of the *meta/meta* derivative **49** of SGI-1027 via a combination of lit-to-lead and SAR approaches. (E) Discovery of **50**, a micromolar inhibitor of DNMT1, starting from the G9a inhibitor UNC0638 and using a lit-to-lead approach.

transposed onto **42**, affording A-366 (**43**), which showed an IC_{50} of 3.3 nM against G9a. Further characterization suggested that, similarly to UNC0638, A-366 is a peptide-competitive inhibitor and does not compete with the cofactor SAM. This mode of action was further supported by a cocrystal of **43** with G9a. Additionally, A-366 was found to reduce the cellular levels of H3K9me2 in PC3 cells at a concentration of 3 μM [66].

3.3.3.3 Lead optimization by introduction of group isosteres or surrogates

PRMT4, also known as coactivator-associated arginine methyltransferase 1 (CARM1), catalyzes the methylation of arginine residues in proteins such as histone H3, p300/CBP, U1C, SAP49, CA150, HuR, and PABP. CARM1 has been shown to be upregulated during the progression of prostate cancer and therefore represents a potentially valuable epigenetic target for drug development. In 2008, scientists from Bristol-Myers Squibb reported the identification of **44** as a low-micromolar CARM1 inhibitor from a HTS screening campaign [113]. SAR studies indicated that truncation of the biphenyl moiety was tolerated and that the introduction of a methyl group alpha to the amino acetamide group provided an improvement in potency, as indicated by compound **45** (Figure 3.5C). In order to identify alternative series of inhibitors of CARM1, scientists from MethylGene Inc. explored different replacements for the phenyl ring in **45** and found that a thiazolyl scaffold was providing similar potency (compound **46**) [114]. Because compound **45** showed poor permeability, evaluation of amide surrogates was also performed, leading to **47** where an oxadiazole ring replaces the amide moiety. This compound provided a twofold improvement in potency over **45** and, more importantly, was found to be cell-permeable [115]. Compounds **46** and **47** demonstrate the use of group isosteres to generate new lead compounds.

A general review of bioisosteres and scaffold hopping in medicinal chemistry has been published [116].

3.3.3.4 Lead identification starting from literature compounds ("lit-to-lead")

The identification of hit or lead compounds is usually accomplished by screening a large collection of compounds as described in Section 3.2. However, one strategy that is less exploited consists of using a compound from the literature as a starting point for LO. This approach, which is sometimes called "lit-to-lead," presents the advantage of utilizing a compound for which a proof-of-concept has already been demonstrated and where biophysical properties have been optimized. In one example, Mai and coworkers performed the optimization of SGI-1027 (**48**), a micromolar DNA DNMT inhibitor, through classical SAR studies and found that a substantial improvement in potency could be obtained with the so-called *meta/meta* isomer **49** (Figure 3.5D). That compound was found to be potent at single-digit micromolar concentrations against a panel of cancer cells. Compound **49** is currently one of the most potent nonnucleoside DNMT inhibitors known [117].

This example demonstrates how lead compounds can be obtained from literature molecules.

3.3.3.5 Lead generation via target hopping approaches

During the LO stage of an inhibitor, the selectivity of the molecule against a selected epigenetic target is usually evaluated by measuring its inhibitory activity against a panel of related enzymes. This process can sometimes lead to the unexpected identification of weak inhibitors of other epigenetic targets. This approach is sometimes referred as "target hopping." Mai and coworkers took note of the fact that UNC0638 (**22**), an inhibitor of G9a, showed weak activity against DNMT1. Using this compound as a starting point for LO, they identified, after extensive SAR studies, derivative **50** as a low-micromolar inhibitor of DNMT1 (Figure 3.5E) [118].

3.4 PHARMACOLOGICAL PROFILING AND DRUG–TARGET ENGAGEMENT

Epigenetic targets represent a novel class of anticancer targets that are different in various aspects from more classical ones addressed by chemotherapy or targeted therapies [119]. First, many epigenetic targets are multidomain proteins that belong to protein complexes. While this offers several opportunities for drug design and single-gene or multifunctional complex modulation, it raises the question of what is the best assay to identify and validate the pharmacological activity of a compound. Indeed, a biochemical screen may not be relevant for the cellular context in multiprotein complexes. In addition, demonstrating the cellular reprogramming effects of an epigenetic compound may take several days and these effects might result from the interaction of the molecule with multiple targets [120]. All these considerations influence the strategy implemented in preclinical studies to prove that a lead compound efficiently induces an antitumor effect by interacting with its target.

Here we illustrate these issues by using as example two types of clinically relevant epigenetic drugs as model examples, for example, DNMT inhibitors, the first generation of epigenetic drugs, and HMT inhibitors, representing a novel generation of epigenetic drugs.

3.4.1 PHARMACOLOGICAL PROFILING OF DNMT INHIBITORS

DNA methylation, which is the most stable epigenetic mark, is catalyzed by DNMTs [121]. Two families of DNMTs have been identified to methylate DNA: (i) DNMT1, responsible for DNA methylation maintenance by methylating the newly synthesized DNA strand, and (ii) DNMT3A and DNMT3B responsible for *de novo* DNA methylation [122]. In cancer cells, a global loss of DNA methylation (gene bodies and intergenic regions) associated with increased genomic instability is accompanied by hypermethylation of many gene promoters including tumor suppressor genes, thus resulting in specific gene silencing in tumors. Mutations in the catalytic domain of DNMT3A and DNMT1 have been reported in hematological tumors but their impact on methylation changes is not clear. In contrast, the expression of DNMT1, DNMT3B, or DNMT3A is increased in various solid tumors thus rendering DNMTs attractive therapeutic targets [123]. The two FDA-approved DNMT inhibitors, azacitidine and decitabine, are nucleoside analogs that are incorporated into DNA and form an irreversible complex with the enzyme that triggers its proteasomal degradation. However, their chemical instability limits their use, and recently a prodrug of 5azadC, SGI-110, presenting a better pharmacokinetic profile than its parent molecule, entered clinical trials (cf. Chapter 7). Here we will focus on these nucleoside inhibitors used in the clinic to illustrate how preclinical studies first demonstrated the demethylating effect of DNMTIs and optimized their use and second enlarged their therapeutic applications.

While these molecules were originally synthesized as cytotoxic agents in the 1960s their DNA methylation inhibitory and differentiating properties were first described in the 1980s [20]. It took then more than 30 years to obtain their clinical approval. The clinical success of these molecules is associated with various groups of clinicians who pushed for the design of lower-dose regimens in which the drugs are given over several days and multiple cycles. These regimens were associated with hypomethylation activity, encouraging response rates, and a more manageable toxicity profile [124].

As mentioned above, a consequence of the incorporation of nucleoside analogs into genomic DNA is the depletion of the pool of DNMTs that induces a progressive loss of DNA methylation and re-expression of silenced genes. Such a reprogramming effect is observed at low doses of DNMT inhibitors, while at higher doses the drugs induce DNA damage responses and cytotoxicity [125]. Thus, dual effect of nucleoside-like DNMTIs is in part responsible for their pharmacological activities. The importance of using low doses of DNMTIs has been clearly illustrated in preclinical studies in the work published in 2011 by Tsai et al. [126]. Using transient exposure to clinically relevant nanomolar doses of 5azaC or 5azadC they demonstrated that both drugs induced long-term antitumor effect against leukemia cells (Kasumi-1, KG1) *in vitro* (clonogenicity) and *in vivo* (engraftment assay), with no or little cell cycle changes or DNA damages. Similar treatments inhibited the tumorigenic properties of breast cancer cells, either cell lines or patients' tumor samples *in vitro* (anchorage-independent growth) and *in vivo* (xenografted tumors). The DNA demethylation together with the re-expression of genes such as *CDKN2A*, *p14ARF*, and *CDKN1A* in leukemia cells were associated with antitumor effects. Thus, exposure to low doses and delayed measurement of pharmacological effects are needed to reveal the reprogramming effect of DNMT inhibitors.

However, there is little correlation between the demethylating effect and the therapeutic effect of these drugs in MDS, acute myeloid leukemia (AML), or chronic myelomonocytic leukemia [127]. Therefore, it remains difficult to prove that the antitumor activity of these drugs is strictly due to their demethylating effect. Several reasons have been proposed for this [128]: (i) the concomitant hypermethylation and hypomethylation events observed in tumors may impair the predictive power of global methylation measurements; (ii) mutations in epigenetic regulators of chromatin configuration (i.e., EZH2) or of the production of 5-hydroxymethylcytosines (TET2, IDH1/2) may influence the response to DNA demethylating agents; (iii) alterations of the proapoptotic and/or the immunoregulatory effects induced by these agents as well as changes in their metabolism may also impact their antitumor activity; and (iv) these compounds also induce DNA damage, not just DNA demethylation. Consequently, while DNA demethylating effects of DNMT nucleoside inhibitors can be demonstrated in tumor cells, consolidated data linking their antitumor response to their demethylating properties are still awaited.

3.4.1.1 Special considerations

As emphasized above, the relationship between the drug sensitivity to DNMT nucleoside inhibitors and their demethylating activity remains to be clarified and indeed predictive markers for response to DNMTIs are lacking. The identification of such markers will depend on better understanding of the oncogenic pathways deregulated in tumors that may contribute to their drug sensitivity. Recently, molecular mutations ($TET2^{Mut}$ and/or $DNMT3A^{Mut}$) have been proposed as predictive markers for overall *response rate* to DNMT nucleoside inhibitors in myelodysplasia and related neoplasms [129]. Another approach proposed by Lund et al. [130] combines whole-genome bisulfite sequencing and gene expression data to identify a set of genes whose methylation and silencing are reversed by the demethylating agents, and that could allow their therapeutic effects in AML cell lines to be followed.

One very important concept that has emerged from the pharmacological characterization of DNMTIs is the concept of using low doses. An additional concept has been the combination of the reprogramming at low doses of DNMT nucleoside inhibitors with other epigenetic, chemotherapeutic or immunologic treatments to improve antitumor efficacy without increasing toxicity (cf. Chapter 15). This was first demonstrated in nonsmall-cell lung cancer by Juergens et al. [131] and is now being investigated in both hematological [132] and solid tumors [133].

All these pharmacological features will have to be reassessed with the novel nonnucleosidic DNMTIs, such as catalytic inhibitors of the DNMTs [7].

3.4.2 PHARMACOLOGICAL PROFILING OF HMT INHIBITORS

HMTs form a large family of more than 60 enzymes that catalyze the methylation of histones on lysine (HKMTs) and arginine (PRMTs) residues using SAM as cofactor (cf. Chapter 9). In the past decade, genetic alterations or overexpression of different HMTs have been associated with various cancers. Among them, EZH2 plays an oncogenic driver role in specific lymphoma subtypes. Here we will describe the preclinical evaluation of EZH2 inhibitors to illustrate the specificities and challenges of second-generation epigenetic drug discovery.

EZH2, the only catalytic component of the PRC2 complex, trimethylates H3K27 but cannot function without interacting at least with EED and SUZ12, two other members of PRC2 [134]. In the 2010s, sequencing studies revealed the existence of an oncogenic mutation of tyrosine 641 (Y641) within the Suppressor of variegation, Enhancer of Zeste, and Trithorax (SET) domain of EZH2 in about 20–30% of diffuse large B-cell lymphoma (DLBCL), a follicular lymphoma (FL) [135]. This mutation is an heterozygous activating mutation that alters substrate specificity and recognition of monomethylated H3K27me1 by the wild-type proteins and accelerates its conversion to H3K27me3. Around the same period, inactivating mutations have been identified in myeloid neoplasms [136], proving that either activating or inactivating EZH2 can lead to cancer, depending on the pathological context. Considerable efforts have been invested in identifying and characterizing potent EZH2 inhibitors as described in the sections above.

Once the lead candidates were identified, cellular experiments showing the specific decrease of the H3K27me3 histone mark versus a panel of other marks were conducted to prove the on-target effects of these inhibitors. These effects were obtained at nanomolar range of the inhibitors after 2–4 days of treatment. EZH2 inhibitors preferentially killed DLBCL lymphoma cell lines bearing EZH2 mutations. Moreover, strong transcriptional activation (including PRC2-regulated genes) was observed in the sensitive mutated cells, while minimal changes occurred in wild-type EZH2 cells. Interestingly, using the second generation of pyridone derivative (EPZ6438, Figure 3.2B), Knutson et al. [88] showed that the time of antiproliferative effects in sensitive mutated cells coincided with the time of optimal H3K27me3 inhibition (4 days). In wild-type cells, even if the compound inhibited the mark, no antiproliferative effect was noted. Thus modifying the histone mark is not sufficient to predict an efficient antiproliferative effect of EZH2 inhibitors. Moreover, *in vivo* preclinical studies [88] clearly demonstrated that repeated EPZ6438 oral treatment of mice bearing xenografted EZH2-mutant DLBCL induced sustained or complete regression. Interestingly, target inhibition, that is, H3K27me3 decrease, was measurable by ELISA in lymphoma xenografts as well as in nontumor tissues (skin, PBMCs, BM cells) thus offering a potential noninvasive pharmacodynamic biomarker for the clinical trials.

3.4.2.1 Special considerations

These important data confirm the interest of targeting dominant gain-of-function mutations of epigenetic targets to selectively eliminate cancer cells. Still, compensatory mechanisms or co-occurring genetic alterations affecting other epigenetic regulators, such as the H3K27 demethylases KDM6A/UTX or apoptotic pathways effectors, may impact positively or negatively the antitumor effect of these EZH2 inhibitors. Moreover, a different latency period can be noticed in sensitive cell lines prior to fully obtaining an antitumor effect [137] and heterogeneous cytotoxic or cytostatic responses can be observed depending on the cell line used [70]. As a consequence, the role of EZH2 in cancer appears dependent on the cellular context and not only on the mutational status of the enzyme, but also on its interplay with other regulators of gene expression [87]. This has been recently illustrated in pediatric malignant rhabdoid tumors by the demonstration that specific inactivating mutation of the switch/sucrose nonfermentable (SWI/SNF) complex subunit SMARCB1 induces a dependency on EZH2 enzymatic activity and a pronounced sensitivity to EZH2 inhibitors [87]. Such addiction, defined as "induced essentiality," can take part of antitumor activity of EZH2 inhibitors in different cellular contexts.

Therefore, as emphasized for DNMTIs, understanding the determinants of the cellular sensitivity to EZH2 inhibitors remains critical to optimize and better predict the pharmacological response to these novel molecules.

3.4.3 DRUG–TARGET ENGAGEMENT

While the above-mentioned pharmacological approaches strongly support the interaction of epigenetic drugs with their targets, the mechanism underling their activity in cells remains to be elucitated. As very recently reviewed by Durham and Blanco [138], being able to establish the correlation between *in vitro* activity, *in vivo* efficacy, and target engagement is crucial for compound selection and the identification of drug–effect biomarkers during the lead discovery process. The following examples illustrate different approaches that have been used for this purpose with a particular emphasis on DNMTIs and EZH2 inhibitors.

The ICM (*In vivo* Complex of Methylase) assay is based on detecting DNMT-DNA adducts in living cells and was used to demonstrate that azacitidine and decitabine induce the formation of DNMT-DNA irreversible adducts with all three catalytic DNMTs in cells [139]. Another interesting demonstration that the two azanucleosides block DNMT1 was given by Schermelleh et al. [140] by using fluorescent bleaching experiments in cells on DNMT1 fused to green fluorescent protein, proving the drug-induced engagement of DNMTs.

Recently [141], EZH2 target engagement in cells has been shown by cellular thermal shift assay. This assay is based on the principle of thermodynamic stabilization of a protein as a result of ligand binding. The shift in thermal stability is estimated by measurement of the amount of remaining soluble target protein by western blotting at different temperatures for ligand-treated and control samples. Using this approach, the authors demonstrated that CPI-360 derivative induced time-dependent EZH2 thermostabilization, demonstrating its engagement in cells. This was reversible and was achieved irrespective of the EZH2 status of the leukemia cells. Moreover, the global

H3K27me3 and H3K27me2 levels were reduced in a dose-dependent manner. The recovery of the mark after compound removal was progressive, consistent with reversible suppression of EZH2 catalytic activity.

Various technologies can be used to prove target engagement in living cells such as the chemoproteomic technique referred as activity-based protein profiling (ABPP) [142]. ABPP is based on the use of chemical probes that are specific ligands of a protein target. The probes contain a photoreactive group to trap probe–target interactions in cells and a hooking-group to attach a tag for target enrichment prior to mass-spectrometry (MS)-based identification. Such technology has been pioneered by the group of Cravatt [143] for HDAC inhibitors, which demonstrated that SAHA, originally considered as a pan HDAC inhibitor, only interacts with four individual HDACs in a cellular context.

These studies not only provide sensitive methods to map on-target interactions and propose drug–effect biomarkers but can be used for predicting off-target effects in living cells.

3.5 GENERAL CONCLUSIONS

The concept of epigenetic therapy of cancer was pushed forward by the approval of the first generation of epigenetic drugs (DNMT and HDAC inhibitors) and by the demonstration that epigenetic deregulations are implicated in the development of many different types of tumors. The epigenetic nature of their antitumor effects has emerged on the course of their pharmacological characterization proving that these effects were not the consequence of direct cytotoxicity, as shown with classical chemotherapy and/or the targeted therapies, but resulted from cell reprogramming through the transcriptional modulation of epigenetically silenced genes.

The field of epigenetic drug discovery is now focused on few selected targets, mostly histone-modifying enzymes and readers, for whom activating mutations, translocations, or induced essentiality have been described [123]. In the light of the two examples described above, specificities and challenges linked to epigenetic drug discovery can be emphasized.

First, many of the aberrant epigenetic regulators are proteins, mostly enzymes that can be targeted with pharmacological inhibitors. However, the existing compound libraries may not provide good chemical starting points for these new targets. Thus computational approaches based on structural determination of these proteins represent helpful strategies to address these targets.

Second, the inhibition of the aberrant epigenetic regulators can result in specific effects on cancer cells. This requires repeated treatments with low doses and longtime exposure to observe the expected biological responses, that is, reprogramming effects. This requires adapting the traditional pharmacological models both *in vitro* and *in vivo*.

Third, given the complexity of cancer cell epigenome regulation, a big challenge remains to determine the underlying genetic or epigenetic changes that predispose tumors to inhibitor sensitivity. This is crucial to better select the tumor types, and thus the patients, who will respond to the treatment. Moreover, this knowledge will help to better predict the combinations with other treatments such as other epigenetic drugs, chemotherapy, or immunotherapy.

Fourth, because of the crosstalk between the different epigenetic regulators in determining epigenetic landscape, the demonstration of the engagement of a novel molecule with its intended target is of particular importance to fully interpret drug effects and predict its efficacy.

ACKNOWLEDGMENTS

Support for this work was provided to P.B.A. by Centre National de la Recherche Scientifique (CNRS) [ATIP], Région Midi-Pyrénées [Equipe d'Excellence and FEDER CNRS/Région Midi-Pyrénées], and Fondation InNaBioSanté.

REFERENCES

[1] Feinberg AP. Phenotypic plasticity and the epigenetics of human disease. Nature 2007;447(7143): 433–40.
[2] Sharma S, Kelly TK, Jones PA. Epigenetics in cancer. Carcinogenesis 2010;31(1):27–36.
[3] Wee S, Dhanak D, Li H, Armstrong SA, Copeland RA, Sims R, et al. Targeting epigenetic regulators for cancer therapy. Ann NY Acad Sci 2014;1309:30–6.
[4] You JS, Jones PA. Cancer genetics and epigenetics: two sides of the same coin? Cancer Cell 2012;22 (1):9–20.
[5] Azad N, Zahnow CA, Rudin CM, Baylin SB. The future of epigenetic therapy in solid tumours—lessons from the past. Nat Rev Clin Oncol 2013;10(5):256–66.
[6] Ahuja N, Easwaran H, Baylin SB. Harnessing the potential of epigenetic therapy to target solid tumors. J Clin Invest 2014;124(1):56–63.
[7] Erdmann A, Halby L, Fahy J, Arimondo PB, Targeting DNA. methylation with small molecules: what's next? J Med Chem 2015. Available from: http://dx.doi.org/10.1021/jm500843d.
[8] Brueckner B, Garcia Boy R, Siedlecki P, Musch T, Kliem HC, Zielenkiewicz P, et al. Epigenetic reactivation of tumor suppressor genes by a novel small-molecule inhibitor of Human DNA methyltransferases. Cancer Res 2005;65(14):6305–11.
[9] Medina-Franco JL, Méndez-Lucio O, Dueñas-González A, Yoo J. Discovery and development of DNA methyltransferase inhibitors using in silico approaches. Drug Discov Today 2015;20(5):569–77 http://dx.doi.org/10.1016/j.drudis.2014.12.007
[10] Asgatay S, Champion C, Marloie G, Drujon T, Senamaud-Beaufort C, Ceccaldi A, et al. Synthesis and evaluation of analogues of *N*-phthaloyl-L-tryptophan (RG108) as inhibitors of DNA methyltransferase 1. J Med Chem 2014;57(2):421–34.
[11] Spannhoff A, Heinke R, Bauer I, Trojer P, Metzger E, Gust R, et al. Target-based approach to inhibitors of histone arginine methyltransferases. J Med Chem 2007;50(10):2319–25.
[12] Spannhoff A, Machmur R, Heinke R, Trojer P, Bauer I, Brosch G, et al. A novel arginine methyltransferase inhibitor with cellular activity. Bioorg Med Chem Lett 2007;17(15):4150–3.
[13] Di Micco S, Chini MG, Terracciano S, Bruno I, Riccio R, Bifulco G. Structural basis for the design and synthesis of selective HDAC inhibitors. Bioorg Med Chem 2013;21(13):3795–807.
[14] Ito T, Umehara T, Sasaki K, Nakamura Y, Nishino N, Terada T, et al. Real-time imaging of histone H4K12–specific acetylation determines the modes of action of histone deacetylase and bromodomain inhibitors. Chem Biol 2011;18(4):495–507.
[15] Vidler LR, Filippakopoulos P, Fedorov O, Picaud S, Martin S, Tomsett M, et al. Discovery of novel small-molecule inhibitors of BRD4 using structure-based virtual screening. J Med Chem 2013;56 (20):8073–88.
[16] Hewings DS, Wang M, Philpott M, Fedorov O, Uttarkar S, Filippakopoulos P, et al. 3,5-Dimethylisoxazoles act as acetyl-lysine-mimetic bromodomain ligands. J Med Chem 2011;54(19): 6761–70.

[17] Lucas X, Wohlwend D, Hügle M, Schmidtkunz K, Gerhardt S, Schüle R, et al. 4-acyl pyrroles: mimicking acetylated lysines in histone code reading. Angew Chem Int Ed Engl 2013;52(52):14055–9.
[18] Zhao L, Cao D, Chen T, Wang Y, Miao Z, Xu Y, et al. Fragment-based drug discovery of 2-thiazolidinones as inhibitors of the histone reader BRD4 bromodomain. J Med Chem 2013; 56(10):3833–51.
[19] Constantinides PG, Jones PA, Gevers W. Functional striated muscle cells from non-myoblast precursors following 5-azacytidine treatment. Nature 1977;267(5609):364–6.
[20] Jones PA, Taylor SM. Cellular differentiation, cytidine analogs and DNA methylation. Cell 1980;20 (1):85–93.
[21] Isakovic L, Saavedra OM, Llewellyn DB, Claridge S, Zhan L, Bernstein N, et al. Constrained (L-)-*S*-adenosyl-L-homocysteine (SAH) analogues as DNA methyltransferase inhibitors. Bioorg Med Chem Lett 2009;19(10):2742–6.
[22] Reuben RC, Wife RL, Breslow R, Rifkind RA, Marks PA. A new group of potent inducers of differentiation in murine erythroleukemia cells. Proc Natl Acad Sci USA 1976;73(3):862–6.
[23] Marks PA, Breslow R. Dimethyl sulfoxide to vorinostat: development of this histone deacetylase inhibitor as an anticancer drug. Nat Biotechnol 2007;25(1):84–90.
[24] Richon VM, Emiliani S, Verdin E, Webb Y, Breslow R, Rifkind RA, et al. A class of hybrid polar inducers of transformed cell differentiation inhibits histone deacetylases. Proc Natl Acad Sci USA 1998;95 (6):3003–7.
[25] Plumb JA, Finn PW, Williams RJ, Bandara MJ, Romero MR, Watkins CJ, et al. Pharmacodynamic response and inhibition of growth of human tumor xenografts by the novel histone deacetylase inhibitor PXD101. Mol Cancer Ther 2003;2(8):721–8.
[26] Luo X, Liu Y, Kubicek S, Myllyharju J, Tumber A, Ng S, et al. A selective inhibitor and probe of the cellular functions of Jumonji C domain-containing histone demethylases. J Am Chem Soc 2011;133 (24):9451–6.
[27] Culhane JC, Szewczuk LM, Liu X, Da G, Marmorstein R, Cole PA. A mechanism-based inactivator for histone demethylase LSD1. J Am Chem Soc 2006;128(14):4536–7.
[28] Lohse B, Nielsen AL, Kristensen JBL, Helgstrand C, Cloos PAC, Olsen L, et al. Targeting histone lysine demethylases by truncating the histone 3 tail to obtain selective substrate-based inhibitors. Angew Chem Int Ed Engl 2011;50(39):9100–3.
[29] Daigle SR, Olhava EJ, Therkelsen CA, Majer CR, Sneeringer CJ, Song J, et al. Selective killing of mixed lineage leukemia cells by a potent small-molecule DOT1L inhibitor. Cancer Cell 2011;20 (1):53–65.
[30] Villar-Garea A, Fraga MF, Espada J, Esteller M. Procaine is a DNA-demethylating agent with growth-inhibitory effects in human cancer cells. Cancer Res 2003;63(16):4984–9.
[31] Halby L, Champion C, Sénamaud-Beaufort C, Ajjan S, Drujon T, Rajavelu A, et al. Rapid synthesis of new DNMT inhibitors derivatives of procainamide. Chem Bio Chem 2012;13(1):157–65.
[32] De Bellis F, Carafa V, Conte M, Rotili D, Petraglia F, Matarese F, et al. Context-selective death of acute myeloid leukemia cells triggered by the novel hybrid retinoid-HDAC inhibitor MC2392. Cancer Res 2014;74(8):2328–39.
[33] Ghofrani HA, Osterloh IH, Grimminger F. Sildenafil: from angina to erectile dysfunction to pulmonary hypertension and beyond. Nat Rev Drug Discov 2006;5(8):689–702.
[34] Silverman WA. The schizophrenic career of a "monster drug". Pediatrics 2002;110(2):404–6.
[35] Yoshida M, Horinouchi S, Beppu T. Trichostatin A and trapoxin: novel chemical probes for the role of histone acetylation in chromatin structure and function. BioEssays 1995;17(5):423–30.
[36] Vigushin DM, Ali S, Pace PE, Mirsaidi N, Ito K, Adcock I, et al. Trichostatin A is a histone deacetylase inhibitor with potent antitumor activity against breast cancer *in vivo*. Clin Cancer Res 2001;7(4):971–6.

[37] Lee BH, Yegnasubramanian S, Lin X, Nelson WG. Procainamide is a specific inhibitor of DNA methyltransferase 1. J Biol Chem 2005;280(49):40749–56.
[38] Singh N, Dueñas-González A, Lyko F, Medina-Franco JL. Molecular modeling and molecular dynamics studies of hydralazine with human DNA methyltransferase 1. Chem Med Chem 2009;4(5):792–9.
[39] Méndez-Lucio O, Tran J, Medina-Franco JL, Meurice N, Muller M. Toward drug repurposing in epigenetics: olsalazine as a hypomethylating compound active in a cellular context. Chem Med Chem 2014;9 (3):560–5.
[40] Dueñas-Gonzalez A, Coronel J, Cetina L, González-Fierro A, Chavez-Blanco A, Taja-Chayeb L. Hydralazine–valproate: a repositioned drug combination for the epigenetic therapy of cancer. Expert Opin Drug Metab Toxicol 2014;10(10):1433–44.
[41] Lee H, Jung K, Kim S, Choi M, Chai Y. Effects of pargyline on cellular proliferation in human breast cancer cells. Mol Cell Toxicol 2012;8(4):393–9.
[42] Schmidt DMZ, McCafferty DG. *trans*-2-Phenylcyclopropylamine is a mechanism-based inactivator of the histone demethylase LSD1. Biochemistry 2007;46(14):4408–16.
[43] Schenk T, Chen WC, Gollner S, Howell L, Jin L, Hebestreit K, et al. Inhibition of the LSD1 (KDM1A) demethylase reactivates the all-*trans*-retinoic acid differentiation pathway in acute myeloid leukemia. Nat Med 2012;18(4):605–11.
[44] Miyoshi S, Ooike S, Iwata K, Hikawa H, Sugahara K. Antitumour Agent Patent WO/2009/084693; 2009.
[45] Filippakopoulos P, Qi J, Picaud S, Shen Y, Smith WB, Fedorov O, et al. Selective inhibition of BET bromodomains. Nature 2010;468(7327):1067–73.
[46] Cherblanc FL, Davidson RWM, Di Fruscia P, Srimongkolpithak N, Fuchter MJ. Perspectives on natural product epigenetic modulators in chemical biology and medicine. Nat Prod Rep 2013;30(5):605–24.
[47] Weng J-R, Lai IL, Yang H-C, Lin C-N, Bai L-Y. Identification of Kazinol Q, a natural product from Formosan plants, as an inhibitor of DNA methyltransferase. Phytother Res 2014;28(1):49–54.
[48] Wang S-C, Lee T-H, Hsu C-H, Chang Y-J, Chang M-S, Wang Y-C, et al. Antroquinonol D, isolated from *Antrodia camphorata*, with DNA demethylation and anticancer potential. J Agric Food Chem 2014;62(24):5625–35.
[49] Lee WJ, Shim J-Y, Zhu BT. Mechanisms for the inhibition of DNA methyltransferases by tea catechins and bioflavonoids. Mol Pharmacol 2005;68(4):1018–30.
[50] Cherblanc FL, Chapman KL, Brown R, Fuchter MJ. Chaetocin is a nonspecific inhibitor of histone lysine methyltransferases. Nat Chem Biol 2013;9(3):136–7.
[51] Godert AM, Angelino N, Woloszynska-Read A, Morey SR, James SR, Karpf AR, et al. An improved synthesis of psammaplin A. Bioorg Med Chem Lett 2006;16(12):3330–3.
[52] Schneider-Stock R, Ghantous A, Bajbouj K, Saikali M, Darwiche N. Epigenetic mechanisms of plant-derived anticancer drugs. Front Biosci 2012;17:129–73.
[53] Zheng W, Ibáñez G, Wu H, Blum G, Zeng H, Dong A, et al. Sinefungin derivatives as inhibitors and structure probes of protein lysine methyltransferase SETD2. J Am Chem Soc 2012;134 (43):18004–14.
[54] Devkota K, Lohse B, Liu Q, Wang M-W, Stærk D, Berthelsen J, et al. Analogues of the natural product sinefungin as inhibitors of EHMT1 and EHMT2. ACS Med Chem Lett 2014;5(4):293–7.
[55] Macarron R, Banks MN, Bojanic D, Burns DJ, Cirovic DA, Garyantes T, et al. Impact of high-throughput screening in biomedical research. Nat Rev Drug Discov 2011;10(3):188–95.
[56] Mayr LM, Bojanic D. Novel trends in high-throughput screening. Curr Opin Pharmacol 2009;9 (5):580–8.
[57] Dahlin JL, Sinville R, Solberg J, Zhou H, Han J, Francis S, et al. A cell-free fluorometric high-throughput screen for inhibitors of Rtt109-catalyzed histone acetylation. PLoS One 2013;8(11):e78877.

[58] Eglen RM, Reisine T. Screening for compounds that modulate epigenetic regulation of the transcriptome: an overview. J Biomol Screen 2011;16(10):1137–52.
[59] Hassig CA, Symons KT, Guo X, Nguyen P-M, Annable T, Wash PL, et al. KD5170, a novel mercaptoketone-based histone deacetylase inhibitor that exhibits broad spectrum antitumor activity *in vitro* and *in vivo*. Mol Cancer Ther 2008;7(5):1054–65.
[60] Grozinger CM, Chao ED, Blackwell HE, Moazed D, Schreiber SL. Identification of a class of small molecule inhibitors of the sirtuin family of NAD-dependent deacetylases by phenotypic screening. J Biol Chem 2001;276(42):38837–43.
[61] Lain S, Hollick JJ, Campbell J, Staples OD, Higgins M, Aoubala M, et al. Discovery, *in vivo* activity, and mechanism of action of a small-molecule p53 activator. Cancer Cell 2008;13(5):454–63.
[62] Napper AD, Hixon J, McDonagh T, Keavey K, Pons J-F, Barker J, et al. Discovery of indoles as potent and selective inhibitors of the deacetylase SIRT1. J Med Chem 2005;48(25):8045–54.
[63] Wang L, Chang J, Varghese D, Dellinger M, Kumar S, Best AM, et al. A small molecule modulates Jumonji histone demethylase activity and selectively inhibits cancer growth. Nat Commun 2013;4:2035.
[64] Mulji A, Haslam C, Brown F, Randle R, Karamshi B, Smith J, et al. Configuration of a high-content imaging platform for hit identification and pharmacological assessment of JMJD3 demethylase enzyme inhibitors. J Biomol Screen 2012;17(1):108–20.
[65] Falk H, Connor T, Yang H, Loft KJ, Alcindor JL, Nikolakopoulos G, et al. An efficient high-throughput screening method for MYST family acetyltransferases, a new class of epigenetic drug targets. J Biomol Screen 2011;16(10):1196–205.
[66] Sweis RF, Pliushchev M, Brown PJ, Guo J, Li F, Maag D, et al. Discovery and development of potent and selective inhibitors of histone methyltransferase G9a. ACS Med Chem Lett 2014;5(2):205–9.
[67] Sayegh J, Cao J, Zou MR, Morales A, Blair LP, Norcia M, et al. Identification of small molecule inhibitors of Jumonji AT-rich interactive domain 1B (JARID1B) histone demethylase by a sensitive high throughput screen. J Biol Chem 2013;288(13):9408–17.
[68] Stimson L, Rowlands MG, Newbatt YM, Smith NF, Raynaud FI, Rogers P, et al. Isothiazolones as inhibitors of PCAF and p300 histone acetyltransferase activity. Mol Cancer Ther 2005;4(10):1521–32.
[69] Garapaty-Rao S, Nasveschuk C, Gagnon A, Chan EY, Sandy P, Busby J, et al. Identification of EZH2 and EZH1 small molecule inhibitors with selective impact on diffuse large B cell lymphoma cell growth. Chem Biol 2013;20(11):1329–39.
[70] Knutson SK, Wigle TJ, Warholic NM, Sneeringer CJ, Allain CJ, Klaus CR, et al. A selective inhibitor of EZH2 blocks H3K27 methylation and kills mutant lymphoma cells. Nat Chem Biol 2012;8(11):890–6.
[71] Gros C, Chauvigné L, Poulet A, Menon Y, Ausseil F, Dufau I, et al. Development of a universal radioactive DNA methyltransferase inhibition test for high-throughput screening and mechanistic studies. Nucleic Acids Res 2013;41(19):e185.
[72] Diaz E, Machutta CA, Chen S, Jiang Y, Nixon C, Hofmann G, et al. Development and validation of reagents and assays for EZH2 peptide and nucleosome high-throughput screens. J Biomol Screen 2012;17(10):1279–92.
[73] Kubicek S, O'Sullivan RJ, August EM, Hickey ER, Zhang Q, Teodoro ML, et al. Reversal of H3K9me2 by a small-molecule inhibitor for the G9a histone methyltransferase. Mol Cell 2007;25(3):473–81.
[74] Ceccaldi A, Rajavelu A, Ragozin S, Sénamaud-Beaufort C, Bashtrykov P, Testa N, et al. Identification of novel inhibitors of DNA methylation by screening of a chemical library. ACS Chem Biol 2013;8(3):543–8.
[75] Fagan RL, Wu M, Chédin F, Brenner C. An ultrasensitive high throughput screen for DNA methyltransferase 1-targeted molecular probes. PLoS One 2013;8(11):e78752.

[76] Monovich L, Vega RB, Meredith E, Miranda K, Rao C, Capparelli M, et al. A novel kinase inhibitor establishes a predominant role for protein kinase D as a cardiac class IIa histone deacetylase kinase. FEBS Lett 2010;584(3):631–7.
[77] Degorce F, Card A, Soh S, Trinquet E, Knapik GP, Xie B. HTRF: a technology tailored for drug discovery—a review of theoretical aspects and recent applications. Curr Chem Genomics 2009;3:22–32.
[78] Wang C, Caron M, Burdick D, Kang Z, Auld D, Hill WA, et al. A sensitive, homogeneous, and high-throughput assay for lysine-specific histone demethylases at the H3K4 site. Assay Drug Dev Technol 2012;10(2):179–86.
[79] Bielefeld-Sevigny M. AlphaLISA immunoassay platform—the "no-wash" high-throughput alternative to ELISA. Assay Drug Dev Technol 2009;7(1):90–2.
[80] Cheng D, Yadav N, King RW, Swanson MS, Weinstein EJ, Bedford MT. Small molecule regulators of protein arginine methyltransferases. J Biol Chem 2004;279(23):23892–9.
[81] Luense S, Denner P, Fernández-Montalván A, Hartung I, Husemann M, Stresemann C, et al. Quantification of histone H3 Lys27 trimethylation (H3K27me3) by high-throughput microscopy enables cellular large-scale screening for small-molecule EZH2 inhibitors. J Biomol Screen 2015;20 (2):190–201.
[82] Ceccaldi A, Rajavelu A, Champion C, Rampon C, Jurkowska R, Jankevicius G, et al. C5-DNA methyltransferase inhibitors: from screening to effects on zebrafish embryo development. Chem Bio Chem 2011;12(9):1337–45.
[83] Verma SK, Tian X, LaFrance LV, Duquenne C, Suarez DP, Newlander KA, et al. Identification of potent, selective, cell-active inhibitors of the histone lysine methyltransferase EZH2. ACS Med Chem Lett 2012;3(12):1091–6.
[84] Nasveschuk CG, Gagnon A, Garapaty-Rao S, Balasubramanian S, Campbell R, Lee C, et al. Discovery and optimization of tetramethylpiperidinyl benzamides as inhibitors of EZH2. ACS Med Chem Lett 2014;5(4):378–83.
[85] Gros C, Fleury L, Nahoum V, Faux C, Valente S, Labella D, et al. New insights on the mechanism of quinoline-based DNA methyltransferase inhibitors. J Biol Chem 2015;290(10):6293–302.
[86] Jorgensen WL. Efficient drug lead discovery and optimization. Acc Chem Res 2009;42(6):724–33.
[87] Knutson SK, Warholic NM, Wigle TJ, Klaus CR, Allain CJ, Raimondi A, et al. Durable tumor regression in genetically altered malignant rhabdoid tumors by inhibition of methyltransferase EZH2. Proc Natl Acad Sci USA 2013;110(19):7922–7.
[88] Knutson SK, Kawano S, Minoshima Y, Warholic NM, Huang K-C, Xiao Y, et al. Selective inhibition of EZH2 by EPZ-6438 leads to potent antitumor activity in EZH2-mutant non-Hodgkin lymphoma. Mol Cancer Ther 2014;13(4):842–54.
[89] Kalinić M, Zloh M, Erić S. Structural insights into binding of small molecule inhibitors to Enhancer of Zeste Homolog 2. J Comput Aided Mol Des 2014;28(11):1109–28.
[90] Chung C-W, Coste H, White JH, Mirguet O, Wilde J, Gosmini RL, et al. Discovery and characterization of small molecule inhibitors of the BET family bromodomains. J Med Chem 2011;54(11):3827–38.
[91] Mirguet O, Gosmini R, Toum J, Clément CA, Barnathan M, Brusq J-M, et al. Discovery of epigenetic regulator I-BET762: lead optimization to afford a clinical candidate Inhibitor of the BET bromodomains. J Med Chem 2013;56(19):7501–15.
[92] Zhao Y, Yang C-Y, Wang S. The making of I-BET762, a BET bromodomain inhibitor now in clinical development. J Med Chem 2013;56(19):7498–500.
[93] Kabro A, Lachance H, Marcoux-Archambault I, Perrier V, Dore V, Gros C, et al. Preparation of phenylethylbenzamide derivatives as modulators of DNMT3 activity. Med Chem Comm 2013;4(12):1562–70.
[94] Srimongkolpithak N, Sundriyal S, Li F, Vedadi M, Fuchter MJ. Identification of 2,4-diamino-6,7-dimethoxyquinoline derivatives as G9a inhibitors. Med Chem Comm 2014;5(12):1821–8.

[95] Ma A, Yu W, Li F, Bleich RM, Herold JM, Butler KV, et al. Discovery of a selective, substrate-competitive inhibitor of the lysine methyltransferase SETD8. J Med Chem 2014;57(15):6822–33.
[96] McCabe MT, Ott HM, Ganji G, Korenchuk S, Thompson C, Van Aller GS, et al. EZH2 inhibition as a therapeutic strategy for lymphoma with EZH2-activating mutations. Nature 2012;492(7427):108–12.
[97] Van Aller GS, Pappalardi MB, Ott HM, Diaz E, Brandt M, Schwartz BJ, et al. Long residence time inhibition of EZH2 in activated polycomb repressive complex 2. ACS Chem Biol 2014;9(3):622–9.
[98] Konze KD, Ma A, Li F, Barsyte-Lovejoy D, Parton T, MacNevin CJ, et al. An orally bioavailable chemical probe of the lysine methyltransferases EZH2 and EZH1. ACS Chem Biol 2013;8(6):1324–34.
[99] Liu F, Chen X, Allali-Hassani A, Quinn AM, Wasney GA, Dong A, et al. Discovery of a 2,4-diamino-7-aminoalkoxyquinazoline as a potent and selective inhibitor of histone lysine methyltransferase G9a. J Med Chem 2009;52(24):7950–3.
[100] Liu F, Chen X, Allali-Hassani A, Quinn AM, Wigle TJ, Wasney GA, et al. Protein lysine methyltransferase G9a inhibitors: design, synthesis, and structure activity relationships of 2,4-diamino-7-aminoalkoxy-quinazolines. J Med Chem 2010;53(15):5844–57.
[101] Liu F, Barsyte-Lovejoy D, Allali-Hassani A, He Y, Herold JM, Chen X, et al. Optimization of cellular activity of G9a inhibitors 7-aminoalkoxy-quinazolines. J Med Chem 2011;54(17):6139–50.
[102] Vedadi M, Barsyte-Lovejoy D, Liu F, Rival-Gervier S, Allali-Hassani A, Labrie V, et al. A chemical probe selectively inhibits G9a and GLP methyltransferase activity in cells. Nat Chem Biol 2011;7(8):566–74.
[103] Liu F, Barsyte-Lovejoy D, Li F, Xiong Y, Korboukh V, Huang X-P, et al. Discovery of an *in vivo* chemical probe of the lysine methyltransferases G9a and GLP. J Med Chem 2013;56(21):8931–42.
[104] Gehling VS, Hewitt MC, Vaswani RG, Leblanc Y, Côté A, Nasveschuk CG, et al. Discovery, design, and optimization of isoxazole azepine BET inhibitors. ACS Med Chem Lett 2013;4(9):835–40.
[105] Basavapathruni A, Jin L, Daigle SR, Majer CRA, Therkelsen CA, Wigle TJ, et al. Conformational adaptation drives potent, selective and durable inhibition of the human protein methyltransferase DOT1L. Chem Biol Drug Des 2012;80(6):971–80.
[106] Yu W, Chory EJ, Wernimont AK, Tempel W, Scopton A, Federation A, et al. Catalytic site remodelling of the DOT1L methyltransferase by selective inhibitors. Nat Commun 2012;3:1288.
[107] Daigle SR, Olhava EJ, Therkelsen CA, Basavapathruni A, Jin L, Boriack-Sjodin PA, et al. Potent inhibition of DOT1L as treatment of MLL-fusion leukemia. Blood 2013;122(6):1017–25.
[108] Kumarasinghe IR, Woster PM. Synthesis and evaluation of novel cyclic peptide inhibitors of lysine-specific demethylase 1. ACS Med Chem Lett 2014;5(1):29–33.
[109] Hay D, Fedorov O, Filippakopoulos P, Martin S, Philpott M, Picaud S, et al. The design and synthesis of 5- and 6-isoxazolylbenzimidazoles as selective inhibitors of the BET bromodomains. Med Chem Comm 2013;4(1):140–4.
[110] Hay DA, Fedorov O, Martin S, Singleton DC, Tallant C, Wells C, et al. Discovery and optimization of small-molecule ligands for the CBP/p300 bromodomains. J Am Chem Soc 2014;136(26):9308–19.
[111] McKeown MR, Shaw DL, Fu H, Liu S, Xu X, Marineau JJ, et al. Biased multicomponent reactions to develop novel bromodomain inhibitors. J Med Chem 2014;57(21):9019–27.
[112] Kennedy JP, Williams L, Bridges TM, Daniels RN, Weaver D, Lindsley CW. Application of combinatorial chemistry science on modern drug discovery. J Comb Chem 2008;10(3):345–54.
[113] Purandare AV, Chen Z, Huynh T, Pang S, Geng J, Vaccaro W, et al. Pyrazole inhibitors of coactivator associated arginine methyltransferase 1 (CARM1). Bioorg Med Chem Lett 2008;18(15):4438–41.
[114] Allan M, Manku S, Therrien E, Nguyen N, Styhler S, Robert M-F, et al. *N*-Benzyl-1-heteroaryl-3-(trifluoromethyl)-1H-pyrazole-5-carboxamides as inhibitors of co-activator associated arginine methyltransferase 1 (CARM1). Bioorg Med Chem Lett 2009;19(4):1218–23.

[115] Huynh T, Chen Z, Pang S, Geng J, Bandiera T, Bindi S, et al. Optimization of pyrazole inhibitors of coactivator associated arginine methyltransferase 1 (CARM1). Bioorg Med Chem Lett 2009;19(11):2924–7.

[116] Brown N. Bioisosteres and scaffold hopping in medicinal chemistry. Mol Inf 2014;33(6–7):458–62.

[117] Valente S, Liu Y, Schnekenburger M, Zwergel C, Cosconati S, Gros C, et al. Selective non-nucleoside inhibitors of human DNA methyltransferases active in cancer including in cancer stem cells. J Med Chem 2014;57(3):701–13.

[118] Rotili D, Tarantino D, Marrocco B, Gros C, Masson V, Poughon V, et al. Properly substituted analogues of BIX-01294 lose inhibition of G9a histone methyltransferase and gain selective anti-DNA methyltransferase 3A activity. PLoS One 2014;9(5):e96941.

[119] Campbell RM, Tummino PJ. Cancer epigenetics drug discovery and development: the challenge of hitting the mark. J Clin Invest 2014;124(1):64–9.

[120] Chik F, Szyf M. Effects of specific DNMT gene depletion on cancer cell transformation and breast cancer cell invasion; toward selective DNMT inhibitors. Carcinogenesis 2011;32(2):224–32.

[121] Jurkowska RZ, Jurkowski TP, Jeltsch A. Structure and function of mammalian DNA methyltransferases. Chem Bio Chem 2011;12(2):206–22.

[122] Gros C, Fahy J, Halby L, Dufau I, Erdmann A, Gregoire J-M, et al. DNA methylation inhibitors in cancer: recent and future approaches. Biochimie 2012;94(11):2280–96.

[123] Mair B, Kubicek S, Nijman SMB. Exploiting epigenetic vulnerabilities for cancer therapeutics. Trends Pharmacol Sci 2014;35(3):136–45.

[124] Issa J-PJ. Optimizing therapy with methylation inhibitors in myelodysplastic syndromes: dose, duration, and patient selection. Nat Clin Pract Oncol 2005;2(Suppl. 1):S24–9.

[125] Palii SS, Van Emburgh BO, Sankpal UT, Brown KD, Robertson KD. DNA methylation inhibitor 5-aza-2′-deoxycytidine induces reversible genome-wide DNA damage that is distinctly influenced by DNA methyltransferases 1 and 3B. Mol Cell Biol 2008;28(2):752–71.

[126] Tsai H-C, Li H, Van Neste L, Cai Y, Robert C, Rassool FV, et al. Transient low doses of DNA-demethylating agents exert durable antitumor effects on hematological and epithelial tumor cells. Cancer Cell 2012;21(3):430–46.

[127] Ren J, Singh BN, Huang Q, Li Z, Gao Y, Mishra P, et al. DNA hypermethylation as a chemotherapy target. Cell Signal 2011;23(7):1082–93.

[128] Voso MT, Santini V, Fabiani E, Fianchi L, Criscuolo M, Falconi G, et al. Why methylation is not a marker predictive of response to hypomethylating agents. Haematologica 2014;99(4):613–19.

[129] Traina F, Visconte V, Elson P, Tabarroki A, Jankowska AM, Hasrouni E, et al. Impact of molecular mutations on treatment response to DNMT inhibitors in myelodysplasia and related neoplasms. Leukemia 2014;28(1):78–87.

[130] Lund K, Cole JJ, VanderKraats ND, McBryan T, Pchelintsev NA, Clark W, et al. DNMT inhibitors reverse a specific signature of aberrant promoter DNA methylation and associated gene silencing in AML. Genome Biol 2014;15(8):406.

[131] Juergens RA, Wrangle J, Vendetti FP, Murphy SC, Zhao M, Coleman B, et al. Combination epigenetic therapy has efficacy in patients with refractory advanced non-small cell lung cancer. Cancer Discov 2011;1(7):598–607.

[132] Clozel T, Yang S, Elstrom RL, Tam W, Martin P, Kormaksson M, et al. Mechanism-based epigenetic chemosensitization therapy of diffuse large B-cell lymphoma. Cancer Discov 2013;3(9):1002–19.

[133] Flis S, Gnyszka A, Flis K. DNA methyltransferase inhibitors improve the effect of chemotherapeutic agents in SW48 and HT-29 colorectal cancer cells. PLoS One 2014;9(3):e92305.

[134] Helin K, Dhanak D. Chromatin proteins and modifications as drug targets. Nature 2013;502(7472):480–8.

[135] Morin RD, Johnson NA, Severson TM, Mungall AJ, An J, Goya R, et al. Somatic mutations altering EZH2 (Tyr641) in follicular and diffuse large B-cell lymphomas of germinal-center origin. Nat Genet 2010;42(2):181–5.

[136] Ernst T, Chase AJ, Score J, Hidalgo-Curtis CE, Bryant C, Jones AV, et al. Inactivating mutations of the histone methyltransferase gene EZH2 in myeloid disorders. Nat Genet 2010;42(8):722–6.

[137] McCabe MT, Graves AP, Ganji G, Diaz E, Halsey WS, Jiang Y, et al. Mutation of A677 in histone methyltransferase EZH2 in human B-cell lymphoma promotes hypertrimethylation of histone H3 on lysine 27 (H3K27). Proc Natl Acad Sci USA 2012;109(8):2989–94.

[138] Durham TB, Blanco M-J. Target engagement in lead generation. Bioorg Med Chem Lett 2015;25 (5):998–1008.

[139] Liu K, Wang YF, Cantemir C, Muller MT. Endogenous assays of DNA methyltransferases: evidence for differential activities of DNMT1, DNMT2, and DNMT3 in mammalian cells *in vivo*. Mol Cell Biol 2003;23(8):2709–19.

[140] Schermelleh L, Spada F, Easwaran HP, Zolghadr K, Margot JB, Cardoso MC, et al. Trapped in action: direct visualization of DNA methyltransferase activity in living cells. Nat Methods 2005;2(10):751–6.

[141] Bradley WD, Arora S, Busby J, Balasubramanian S, Gehling VS, Nasveschuk CG, et al. EZH2 inhibitor efficacy in non-Hodgkin's lymphoma does not require suppression of H3K27 monomethylation. Chem Biol 2014;21(11):1463–75.

[142] Cravatt BF, Wright AT, Kozarich JW. Activity-based protein profiling: from enzyme chemistry to proteomic chemistry. Annu Rev Biochem 2008;77:383–414.

[143] Simon GM, Niphakis MJ, Cravatt BF. Determining target engagement in living systems. Nat Chem Biol 2013;9(4):200–5.

CHAPTER

PRECLINICAL CANCER MODELS WITH THE POTENTIAL TO PREDICT CLINICAL RESPONSE

4

Martin Scherzer, Nina Kramer, Christine Unger, Angelika Walzl, Stefanie Walter, Mira Stadler, Markus Hengstschläger and Helmut Dolznig

Institute of Medical Genetics, Medical University of Vienna, Vienna, Austria

CHAPTER OUTLINE

4.1 INTRODUCTION

Compared to cancer incidence rates, which were rising about 10% during the last 40 years in the United States, death rates of cancer patients dropped only by roughly 15% [1] despite the massive gain of knowledge in tumor biology in this time period. Even to date, the drug development process is far from being efficient and only about 5% of all anticancer new molecular entities (NMEs) succeed in gaining approval after clinical testing [2]. Reasons for these enormous fail rates are

G. Egger & P. Arimondo (Eds): Drug Discovery in Cancer Epigenetics. DOI: http://dx.doi.org/10.1016/B978-0-12-802208-5.00004-7

predominantly the lack of efficacy in humans (60%) and safety problems (30%) [3]. Therefore, despite some success with targeted therapies in some tumor entities, such as lymphomas or breast cancer, highly effective treatment strategies for advanced cancers are generally still missing.

It is discussed that one of the reasons for the poor efficacy of NMEs in humans is due to the compound selection process during drug screening. Novel anticancer drugs are either identified in biochemical assays or selected and engineered for high efficacy in conventional cell cultures in 2D using a limited number of established tumor cell lines (Figure 4.1A). Cell culture in 2D was essential to reveal basic principles and greatly helped to develop modern concepts in cell biology and tumorigenesis [4–6]. However, cell culture on flat surfaces has a major drawback; that is, spatial parameters such as tissue architecture, close cell/cell contact, cell/matrix attachment, diffusion gradients, and barriers (Figure 4.1B) and physiological properties, such as specific signaling and biochemical networks are absent under these artificial conditions. Cells grown in 3D scaffolds or as sphere-like structures (multicellular spheroids) faithfully recapitulate the characteristics of tissues or organs [7–9] and cancers [10–13] as demonstrated by similar gene expression profiles

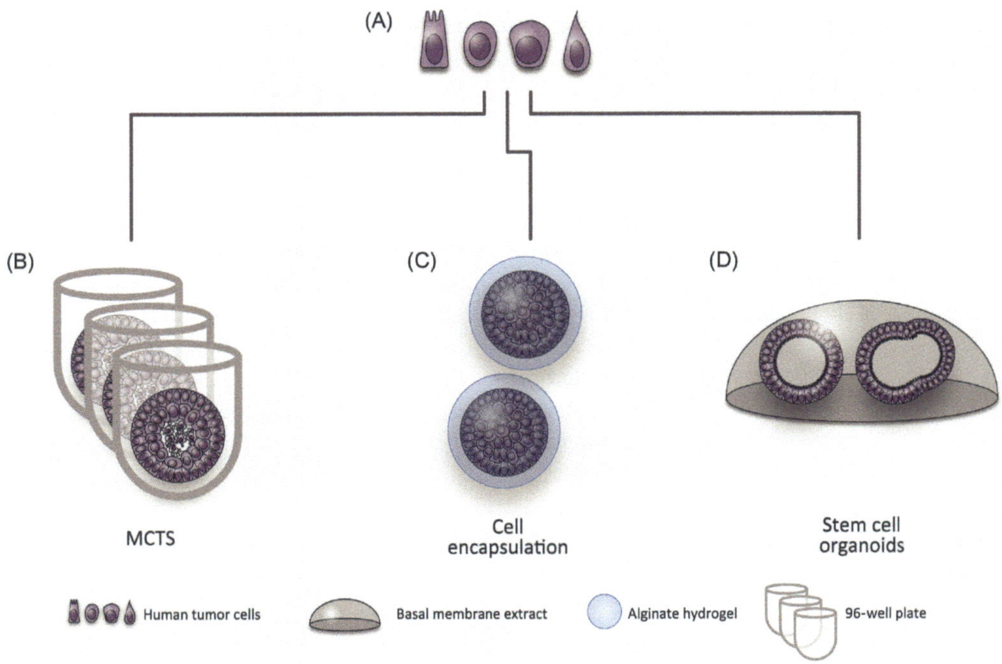

FIGURE 4.1

Homotypic 3D *in vitro* cancer models. (A) Tumor cell heterogeneity can be established in (B) multicellular tumor spheroids (MCTS) in which gradients of oxygen, nutrients, metabolites, and anticancer drugs generate different zones containing proliferative, quiescent, hypoxic, apoptotic, and necrotic cells. (C) Three-dimensional cell aggregates can also be established by encapsulating single-cell suspensions in alginate hydrogels before gelling. (D) Tumor cell heterogeneity in the patient is maintained in organoid cultures of cancer cells in basal membrane extracts in the presence of factors supporting stem cell expansion.

compared to tumors *in vivo* [14]. Hence, these 3D models appear to be more appropriate models in cellular drug screening approaches.

In addition, results from subcutaneous mouse xenograft cancer models, the main *in vivo* preclinical test systems for drug response, were often used for "go or no go" decisions toward proceeding into clinical trials. However, these models have been described to be unreliable predictors of drug response in humans [15–17]. The subcutaneous models have been and are still commonly used, because tumor cells can be easily injected beneath the skin, tumor growth is fast and can be monitored with caliper measurements over time, and therefore, reduced tumor growth or shrinkage can be easily measured to determine drug response in the living animal. The fast growth rate of cancer cell lines in the subcutaneous setting limits the establishment of a prominent mouse stromal response. However, it seems that an organ-specific tumor microenvironment is of essential importance for tumor progression, metastatic potential, drug response, and resistance mechanisms [18]. Hence, growing human tumor cells derived from other organs than skin also fail to experience the proper environmental cues in subcutaneous mouse xenograft models, do not progress properly, and most likely display inadequate response to treatment. In carcinomas, the intimate environment of epithelial cancer cells is built up by fibroblasts, endothelial cells, pericytes, and inflammatory cells [19] and a specific extracellular matrix (ECM). These cells form the tumor stroma, which was shown to be able to promote cancer progression [20,21] and induce invasiveness, eventually leading to metastasis [22]. Moreover, the tumor stroma alters the vulnerability of cancer cells to anticancer drug treatment [23–25]. The nature of the microenvironment-derived resistance mechanisms and the molecular pathways behind them are currently being investigated [26] but remain in many cases rather elusive.

However, there is strong evidence that epigenetic mechanisms are involved in the tumor stroma crosstalk as the microenvironment can influence DNA methylation in cancer cells as demonstrated for glioma [27], squamous cell carcinoma [28], and ovarian cancer [29]. Vice versa, the cancer cells were shown to change epigenetic patterns in the tumor microenvironment. For example, myofibroblast transdifferentiation of stellate cells during liver metastasis involved epigenetic reprogramming [30]. Moreover, breast cancer cells induced the expression of the metallopeptidase ADAMTS1 in cancer-associated fibroblasts by decreasing H2K27 methylation at the ADAMTS1 promoter, thereby enhancing tumor cell invasiveness [31]. Therefore, inhibitors specifically targeting epigenetic programs being changed by tumor–stroma interactions might be relevant strategies to interfere with tumor progression and/or development of resistance to therapy in addition to therapies targeting aberrant signal transduction pathways in cancer.

There is no doubt that there is an urgent need for screening models recapitulating tumor–stroma interaction and for more elaborate preclinical human cancer models [32–35] in order to recapitulate physiological mechanisms acting *in vivo*, such as, for example, chromatin remodeling, but not being taken into account for novel therapy approaches.

4.2 REMODELING HUMAN CANCERS AND THEIR USE AS PRECLINICAL MODELS

One major aspect for the development of efficient drugs in humans is to better understand human cancer biology. This holds true for signal transduction pathways, transcriptional networks, homotypic and heterotypic cell communication, epigenetic mechanisms, as well as drug response and resistance development. As all solid tumors display heterogeneity it is crucial to investigate and understand the topologic and cell-type-specific makeup of individual cancer subregions. Separation in distinct parts can be achieved by laser-guided dissection (laser capture microdissection) [36–38], digestion of the tissue and subsequent flow cytometric [39] or magnetic selection strategies [40]. Of note, in prostate carcinoma, stroma expression profiling was able to predict relapse [41] and poor prognosis in colorectal cancer [42]. Such experiments are indispensable to better understand human cancer biology and the reciprocal impact of cancer and stromal cells on each other. As a side effect, expression profiling of patient material can serve as a "status quo" *ex vivo* standard to corroborate the quality of human tumor models.

Once separated, cells of different cancer regions (e.g., established cell lines, primary cultures of epithelial or stromal source) can be used in *in vitro* 3D cancer models (see Figure 4.2) or animal experiments (see Figure 4.3) for partial reconstruction of the human tumor. Many of these *in vitro* or *in vivo* assays are already used as preclinical models and will be described in this chapter referring to primary literature and outstanding reviews describing each model.

4.2.1 PRECLINICAL *IN VITRO* MODELS

4.2.1.1 Spheroid models

Cells grown as aggregates, so-called multicellular tumor spheroids (MCTS), have been used for many decades to mimic 3D tissue structures *in vitro* (Figure 4.1B). Compared to cells grown as 2D monolayers, MCTS establish tight cell–cell interactions and are subjected to nutrient and oxygen gradients, which generate necrotic, hypoxic, quiescent, and proliferative zones from the inner spheroid core to the surface [43,44] (reviewed in Refs [45–47]). Importantly, gradients of anticancer drugs from the outside of the MCTS to the inner core are also established. Increased resistance (termed multicellular resistance) compared to 2D cultures to chemo- and radiotherapy was frequently reported in these MCTS models [48–50]. High-throughput suitability of MCTS models was demonstrated with a cellular high-content-based screening approach using colon, breast, and prostate MCTS to identify drugs that specifically induce cell death in the spheroid core [51].

Homotypic MCTS already encompass fundamental physiological aspects of tumor biology. However, more sophisticated models are comprised of cocultures of cancer cells with stromal cells such as fibroblasts, endothelial, and immune cells (Figure 4.2A). Here, two or more different cell types are mixed and during spheroid formation heterotypic aggregates are formed (Figure 4.2B). Heterotypic spheroids of human colorectal cancer cells and mouse fibroblasts were used to examine the influence of knockdown of a mitotic kinesin-like motor protein [52]. Cancer cells in 2D were more resistant to the motor protein depletion, whereas the knockdowns failed to grow in the heterotypic 3D model [53].

Alternatively, preformed spheroids can be incubated with a single-cell suspension of a different cell type, which results in layered spheroids [54]. An example for this type of model used spheroids to investigate anti-EGFR therapy on peripheral blood mononuclear cell (PBMC) migration. PBMC infiltration into head and neck squamous cell carcinoma spheroids was increased upon EGFR inhibition, which was less evident in heterotypic fibroblast-containing MCTS [55]. Increased recruiting into the tumor cells upon EGFR inhibition was due to

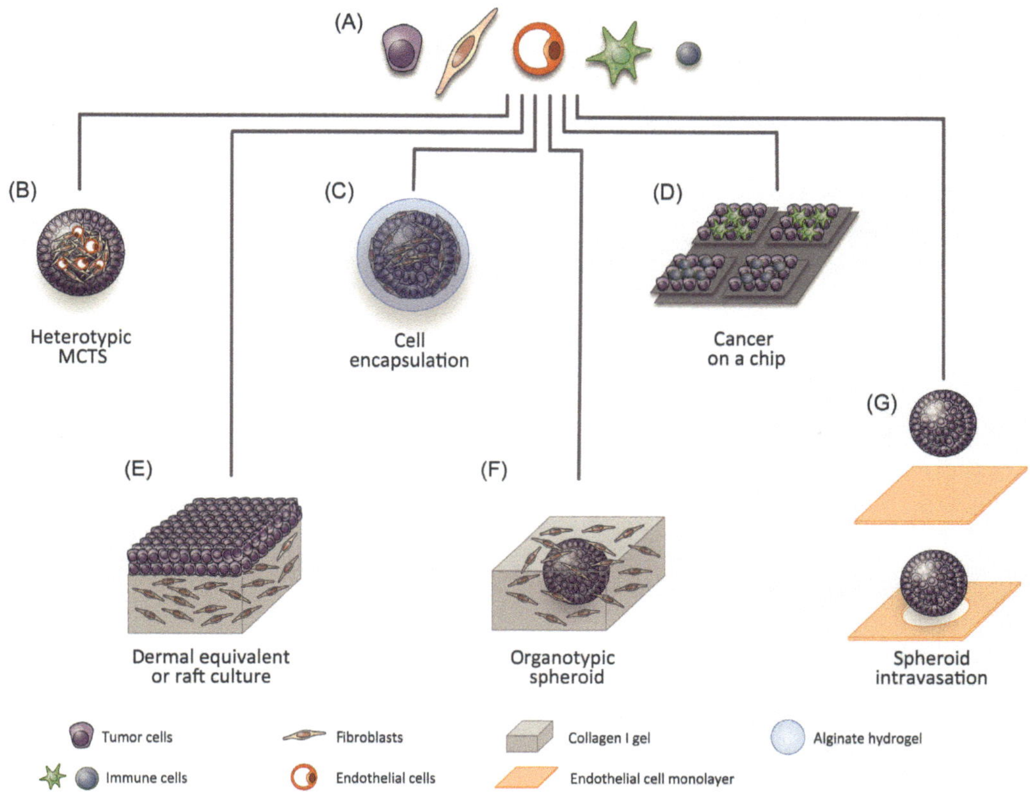

FIGURE 4.2

Heterotypic *in vitro* cancer models. (A) Heterogeneity can be achieved by mixing tumor cells and other stromal cells in different assays. (B) Heterotypic multicellular tumor spheroids (MCTS) are generated by using a tumor cell/stromal cell mixture before aggregation. (C) Cancer cells and stromal cells can also be embedded in alginate hydrogel capsules. (D) Cancer on a chip models make use of small (mostly silicone) devices, which support cell growth on the basis of microfluidics. (E) Dermal equivalent or raft cultures are generated by seeding stromal cells (e.g., fibroblasts) into extracellular matrix (like collagen I) and subsequent addition of cancer cells on top of the gel. These organotypic assays can be cultivated at the air–liquid interface for several days to weeks. (F) Alternatively tumor spheroids are embedded in collagen gels in the presence of stromal cells (e.g., fibroblasts) and tumor cell growth and invasion can be monitored over time. (G) The circular chemorepellent-induced defects assay mimics the formation of gaps in endothelial cell (EC) layers when cancer cell clusters intravasate. For this a confluent layer of labeled EC is incubated with MCTS and the formation of gaps in the EC layer is monitored over time by live microscopy.

upregulation of monocyte chemoattractant protein-1 (MCP-1/CCL2), which is known to direct macrophage, T-cell, and NK-cell migration [56].

Taken together, these studies show that homotypic and heterotypic MCTS are tools for anticancer drug testing in cellular screening assays with the option for phenotypic high-content screening approaches [57].

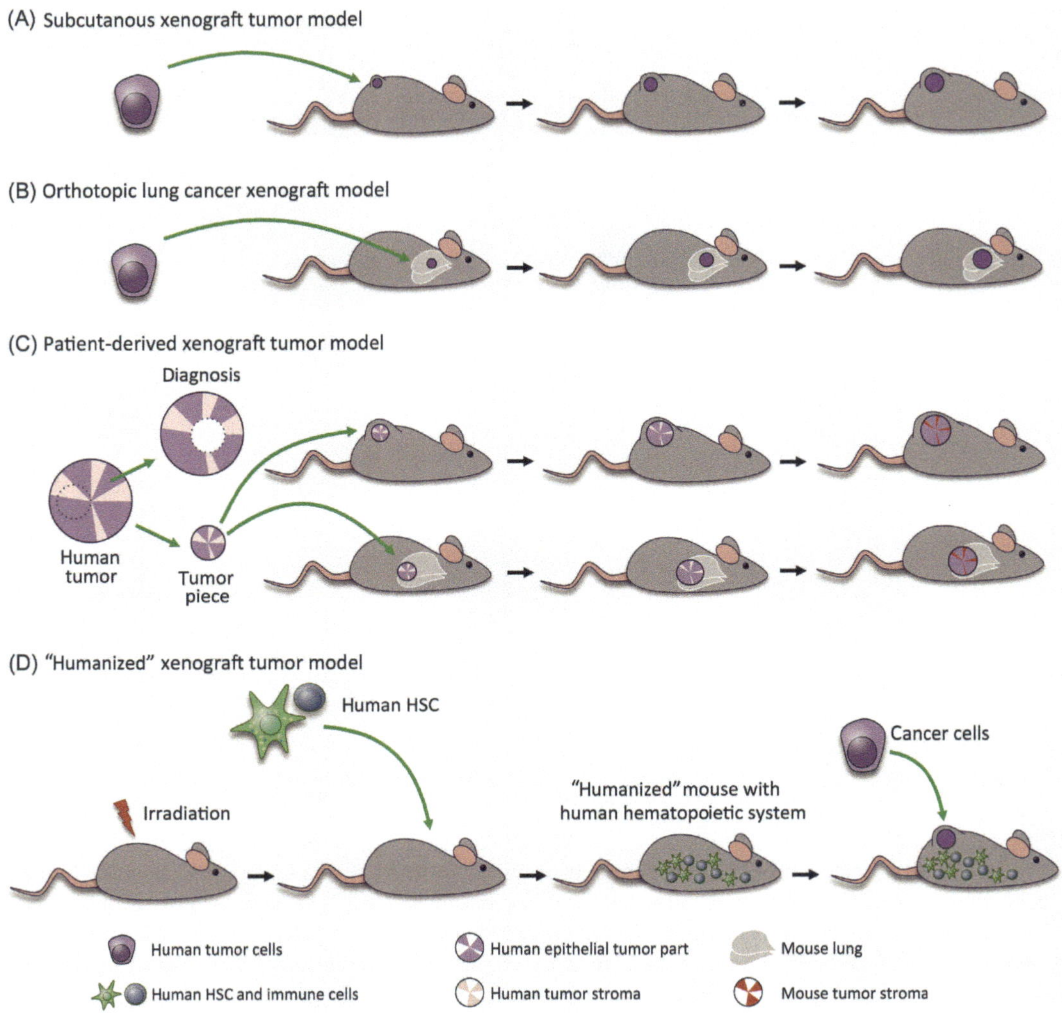

FIGURE 4.3

In vivo human cancer models. (A) Standard subcutaneous human tumor xenograft model. Tumor cells are injected beneath the skin into the flanks of immunocompromised mice and tumor growth is monitored over time. (B) Orthotopic implantation of, for example, lung cancer cells into lungs of immunocompromised mice ensure the development of lung cancers in their proper microenvironment. (C) Implantation of patient-derived tumor fragments, which are not used for diagnosis, establish so-called patient-derived xenografts in immunocompromised mice either subcutaneously or orthotopically. (D) In order to establish a human hematopoietic system with human immune cells populating the mouse, irradiated mice are transplanted with human hematopoietic stem cells (HSC) until stable engraftment is reached. Thereafter human tumor cells or PDX are implanted and tumor growth is monitored in the presence of a more or less functional human immune system.

4.2.1.2 Cell encapsulation models

Natural gelling products, such as alginate, are also well established for utilization as scaffold in 3D cultures. Here the focus lies on alginate-based systems, however, other nonanimal-derived or synthetic hydrogels or matrices are also broadly used and were excellently reviewed recently [58]. The natural polysaccharide alginate can be cross-linked by addition of Ca^{2+} resulting in immediate polymerization. This enables cell encapsulation in mini scaffolds (Figure 4.1C) [59], and subsequent bioreactor culture with high yields of encapsulated cell aggregates, as well as the production of engineered artificial tissues. Of note, easy degelling can be achieved by depletion of Ca^{2+}, which is an important aspect for rapid cell harvesting after experimentation. Animal cells do not express receptors for alginate and cannot attach to pure alginate hydrogels, thus these gels provide an inert 3D scaffold. This technique was used to generate, for example, an artificial pancreas with encapsulated insulinoma cells [60] or—using alginate composites (with protein, ceramic, polymer, etc.)—tissue-engineered bone [61,62].

If a tripeptide made of arginine–glycine–aspartic acid (Arg-Gly-Asp or RGD), which is the site of integrin binding and present in many ECM proteins, is covalently linked to alginate hydrogels, this facilitates proper cell–matrix interactions [63] required for physiological responses. Integrin binding of squamous cell carcinoma cells to RGD-peptides in 3D culture, for example, induced angiogenesis in encapsulated cocultures with endothelial cells [64]. Recently mixtures of alginate gels and matrigel have been developed allowing modulation of matrix stiffness without altering matrix composition (e.g., integrin ligand density) and structure (pore size) [65,66]. This system is well suitable for drug testing in a 3D matrix, as stiffness and pore size can be kept constant and diffusion characteristics of NMEs are controllable.

Alginate foam scaffolds, which are also commercially available (NovaMatrix 3D, FMC BioPolymer AS/NovaMatrix, Norway), enable easy cell immobilization in 3D by providing *in situ* gelation mediated by Ca^{2+} being already present in the foam [67]. These foams are also obtainable with RGD-peptide-linked alginates and the availability in a 96-well format allows cellular drug screening experiments in a 3D environment with the option to simulate heterotypic tumor–stroma interaction, by coseeding two or more cell types (Figure 4.2C). Alginate gels were already used for drug testing experiments with brain cancer [68] and nonsmall-cell lung cancer cells [69]. As reported for MCTS, cells grown in 3D in these gels displayed increased resistance to chemotherapeutics as compared to 2D.

4.2.1.3 Cancer organoid models

Recently, *in vitro* models were developed, which recapitulate the *in vivo* situation of an organ with a mixture of differentiated cells of diverse lineages, cell types at distinct differentiation stages and progenitor and stem cells in a 3D environment. These so-called "organoid" models became possible with the availability of selective adult stem cell markers and strategies to expand these stem cells *in vitro*, pioneered by the work of Hans Clevers' Lab on intestinal stem cell biology [70]. Freshly isolated adult stem cells are embedded in basement membrane extract (BME) and incubated with specific factors, start proliferating and developing into organized structures, closely resembling the composition and physiological properties of the parental organ (Figure 4.1D). The first established organoids were derived from mouse small intestinal stem cells [71]. Later, culture conditions were improved to support formation of organoids from human normal colon, intestinal adenomas, or carcinomas [72]. Of note, cocultures with colonic myofibroblasts increased the organoid-forming efficiency of colon stem cells [73]. Recently, murine and human organoids from different stages of colorectal cancer were established [74] and up to now mammary [75], trachea [76], stomach [77], prostate [78], kidney [79], pancreas [80], and liver [81] organoid cultures are available.

Importantly, organoids derived from cancers maintain the cancer cell heterogeneity of the parental tumor and cancer stem cells are reliably propagated. Cancer cell organoid models were recently summarized in Refs [82,83]. In contrast, the tissue heterogeneity and the stem cell compartment are lost during establishment of cancer cell lines. Thus, cancer cell organoids are encouraging tools for drug testing with an anticipated high clinical predictivity. Analysis of organoids in 96-well plates for high-throughput screening (HTS) attempts is technically possible [79] and might also be important cellular screening assays for phenotypic drug discovery. So far, no drug response studies in human-derived organoid cultures have been reported.

4.2.1.4 Cancer on a chip models

Progress in microfluidics research [84,85] was essential for development of novel methods for drug development. Microfluidic systems display comparable properties to the vasculature *in vivo* [86]. The use of silicone rubber to construct microfluidic systems has widened its use for cell biology studies [87] and led to the engineering of complex physiological models (organ-on-a-chip) [88], such as human liver [89,90], lung [91], or kidney [92]. A combination of these modular systems [93] is a potent tool to imitate organ systems and to study, for example, metastatic dissemination [94,95]. In addition, these chips represent promising models for pharmacokinetics (PK) and dynamics (PD) [96]. On-chip 3D *in vitro* tumor models (Figure 4.2D) were used to investigate the interaction of cancer cells with stromal cells such as fibroblasts [97], blood vessels [98], and immune cells [99]. Tumor-on-chip models better recapitulate physiological parameters than simple 2D cell cultures as drug delivery via the vascular system can be mimicked using microfluidic devices and drug uptake is better modeled in three-dimensional tumor models [100]. Furthermore, metabolic features of organisms were faithfully reproduced as colon carcinoma cells cocultured with hepatoma cells led to metabolic conversion of Tegafur to 5-fluoro uracil (5-FU) [101]. Importantly, the possibility to precisely control metabolite levels and turn over in cell culture opens up new possibilities for compound testing, for example by establishing a drug gradient to determine effective dosages and combination therapies [102,103]. Recently, immunocaptured prostate tumor cells [104] were analyzed and subjected to taxol treatment [105]. These reports demonstrate that microfluidic models can also provide a noninvasive technique to enrich patient-derived tumor cells and may be relevant in precision medicine. An excellent review of microfluidics devices was recently published [93].

4.2.1.5 Heterotypic 3D coculture models in ECMs

The ECM can modulate drug sensitivity in human cancers *in vitro* [106,107] and *in vivo* [108]. Hence, *in vitro* culture models, in which cells are embedded in ECM, better recapitulate the complexity of human tumors and their drug response behavior. Gels composed of rat or bovine collagen I or BME derived from a mouse sarcoma are frequently used to grow cells inside [109–111] or on top [22,112,113] of the matrix.

In order to establish morphological and functional properties of the corresponding tissue *in vivo* and to reconstruct cancer cell–stroma interaction, two or more cell types are cocultured combined with an ECM. These organotypic assays have already demonstrated the influence of stromal cells on tissue architecture and on the malignancy of cancer cells [12,109,114–118]. Here, the influence of stromal fibroblasts on cancer cell invasion into the ECM is well documented [22,109,112,118,119]. In order to increase the complexity of the system and being more physiologic additional cell types, like endothelial [120,121] or immune cells [122,123] can be incorporated in the model.

The most widely used organotypic model is the *dermal equivalent model* or *raft culture model* [124]. Here, fibroblasts are seeded into an ECM (collagen I or fibrin gels), which they contract and remodel for several days [22,119,125]. Subsequently cancer cells are added and attach on top of the gel surface (Figure 4.2E). After the tumor cells have reached confluence, the gels are further cultivated for up to several weeks at an air–liquid interphase [113,115,125], or remain submerged [117]. As standard chemotherapy is concerned, fibroblasts were shown to mediate resistance of melanoma cells to doxorubicin treatment in these models [115]. In cervical cancer and squamous cell carcinoma models the impact of targeted therapy using a MEK/ERK inhibitor was analyzed [125] and other tumor entities such as pancreas carcinoma [117,126] are also well recapitulated. Of note, analysis of raft cultures employ predominantly perpendicular sectioning of formaldehyde-fixed, paraffin-embedded (FFPE) specimens and subsequent immunohistochemistry (IHC) or immunofluorescence staining. This gives exactly the same readout as routine or specific histopathological evaluation of patient tumor material. Hence, a direct comparison of patient tumor material with the experimental model, using the same analysis approach, is feasible. However, a major disadvantage of this system is that most evaluations are limited to an endpoint analysis.

In the *organotypic spheroid model* two different approaches are described. Either two (or more) cell types are combined into a heterotypic spheroid [111] followed by seeding into an ECM gel prior to polymerization. With this setup a liver carcinoma model showed that the drug resistance of the cancer cells increased upon addition of fibroblasts during spheroid aggregation and/or addition of a collagen matrix [111]. In the second approach, cancer MCTS are resuspended in the liquid ECM together with single-cell suspensions of stromal cells before gelling (Figure 4.2F). In this setup cancer cells can potentially leave the MCTS and invade the fibroblast-containing scaffold. PI3K inhibition decreased cell proliferation and viability of colon carcinoma cells but not fibroblasts, as demonstrated by measuring spheroid growth and IHC staining for activated caspase 3 and phospho-H3 after FFPE [54]. In a hepatocellular cancer (HCC) model, induction invasion of tumor cell spheroids by myofibroblast coculture could be reduced by inhibition of TGF-beta or PDGFR signaling, suggesting that hepatic tumor–stroma crosstalk in HCC is dependent on the TGF-beta/PDGF axis.

The *organotypic spheroid* models are superior to the raft culture models when it comes to measurement of drug response. At the beginning of the experiment, spheroids are embedded at defined sizes into the ECM with the increase in size over culture time being a measure of tumor growth [54]. Therefore, using live imaging, the growth response upon inhibitor treatment can be determined in a time series experiment. This provides the opportunity for the experimenter to terminate the experiment at the optimal time point and process the gels for endpoint analysis. Alternatively, inhibition of tumor cell invasion starting from the spheroids might be the desired readout for drug action, which can also be determined in time course experiments. An exact distinction between invading tumor cells and fibroblasts or other stromal cell types is warranted with stably labeled cancer cells using fluorescence proteins (Dolznig et al., unpublished). So far, however, cultures of three or more cell types were sporadically described. One study showed the cocultivation of tumor cells, fibroblasts, and macrophages. However this was a "proof of principle" feasibility study and did not provide any functional interference [123].

Taken together, organotypic models represent a powerful platform for drug response studies and to analyze specific cancer-targeting agents in comparison to general cytotoxic ones (affecting also the stromal cells). A downscaling in size of these assays for a 96-well plate setting for high-throughput cellular screening approaches seems to be technically feasible.

4.2.1.6 In vitro models for intravasation/extravasation of tumor cells

As metastasis accounts for more than 90% of cancer deaths, there is an urgent need for specific drugs targeting the metastasis process or growth of already-established metastatic foci. Metastatic dissemination involves multiple processes such as breakdown of the basal membrane, tissue invasion, transmigration through blood or lymph vessels and distribution all over the body, and proliferation of a new colony with the help of the target organ microenvironment. Many of the distinct steps of metastasis are not only cell-autonomous processes, but involve stromal crosstalk. Currently there is no drug available that specifically targets a distinct step in metastatic dissemination because these remain largely elusive and robust test systems are rare. Consequently, physiologically relevant models have to be developed to provide test platforms for potential novel drugs.

The intravasation of cancer cells into the blood vessels and/or lymphatic vasculature is an early and obligatory step of metastasis. A 3D test system faithfully resembling the *in vivo* situation of breast cancer metastasis in the process of intra- and extravasation was recently developed [127]. This circular chemorepellent-induced defects assay, in which tumor spheroids induce the retraction of lymph endothelial cells (LEC) in a confluent monolayer at the spheroid–LEC contact points to generate a gap (Figure 4.2G), was used to identify molecular mechanisms and to study potential intervention strategies. Strikingly, the invasion of breast cancer cell emboli in humans *in vivo* employed the same molecular mechanisms. Using this assay for breast and colon cancer models, pathways involved in this process were described [127–129] and some natural [130,131] and synthetic compounds [132,133] interfering with intravasation have been identified.

4.2.2 EXPLANT CULTURES (TUMOR SLICES)

Explant cultures of human tumors were adopted from well-established protocols for cultivation of sections of animal embryos [134]. For human cancer, native resected tumor material is cut into 200- to 500-μm thick sections using a vibratome and these tumor slices (TSs) are further cultivated in appropriate growth medium. Culture of the slices can either be submerged or at the liquid-air interface on a supporting mesh. The TS explants were shown to preserve cell viability, tissue architecture, and global gene expression to a certain extent for 5 days [135,136]. Colon and lung TS were investigated for PI3K or MDM2 inhibitor responses [135]; human breast TS was analyzed for response to rapamycin [137] or taxol treatment [136]. These studies reported selective pathway inhibition, decreased proliferation, and/or viability upon treatment and underscore the utility of the TS to potentially predict drug response in precision medicine. However, it remains to be shown whether explant cultures will indeed be predictive for the therapeutic response in patients. In spite of manifold advantages of these model systems, there is one critical disadvantage: the viability of the cells severely declines after 4–5 days, and the TSs disintegrate. Hence, it remains to be proven that during the early and relatively short culture period, when the drug response is measured, the results are not already overlaid by a massive stress response preceding tissue disintegration, which might ultimately lead to confusing results.

4.2.3 REMODELING HUMAN CANCERS *IN VIVO*

In this chapter we only focus on human tumor models in mice and do not discuss genetically engineered mouse models, chemically induced tumor models in experimental animals, and syngeneic cancer models, which are without doubt valuable cancer models and many of these model systems are relevant for drug discovery and preclinical drug testing. Importantly, they are the only ones, in which the influence of tumor immunity can be properly studied *in vivo*. However, these models are beyond the scope of this summary of human cancer models and have recently been excellently reviewed in great detail [138].

4.2.3.1 Standard xenograft models

The discovery of how to grow human cell lines in immunocompromised mice [139] was the prerequisite to studying human cancer biology *in vivo*. Today a plethora of different immunocompromised mouse strains are available, ranging from lack of T cells in nude mice to a strain that is nonobese diabetic, severe combined immunodeficient, and lacks the common IL2 gamma chain (Gamma), abbreviated to NSG mice, lacking functional T, B, and NK cells as well as macrophages. A detailed overview of all these strains suitable for engraftment of human cancers or cancer cells is given in Ref. [138].

Conventional xenograft tumor models, which are usually established by injecting human cancer cells subcutaneously into mice (Figure 4.3A) are frequently used to determine the response to investigational anticancer compounds *in vivo*, since they are easy to handle, tumor growth is simply measurable over time, the growth rate of the tumors is generally high, providing fast results, and these models deliver highly reproducible results. However, there is increasing evidence that the results generated with subcutaneous models do not adequately predict clinical responses [16,17].

As mentioned in the introduction there is ample evidence that an organ-specific tumor microenvironment is of vital importance for tumor initiation and progression. Therefore, a more physiological response with regard to cancer architecture and histology, invasiveness and metastasis is observed in orthotopic carcinoma models [18], when human cancer cells are implanted into the organ of which the cancer was originally derived (Figure 4.3B). Obviously this is more challenging, since often surgery is needed to faithfully implant tumor material at distinct sites and tumor growth can only be monitored using *in vivo* imaging technologies. However, often more relevant results are obtained compared to subcutaneous growth. One example is that human colon cancer cell lines injected into the cecum of mice start to metastasize to the liver [140], which is also the most frequent organ where metastatic lesions in humans occur. Notably, the same cell lines are nonmetastasizing in the subcutaneous setting. Other orthotopic cancer models support this finding and therefore are increasingly used for preclinical profiling. The reader is further referred to comprehensive reviews on, for instance, orthotopic colon [141], pancreas [142], breast [143], or head and neck [144] cancer models.

4.2.3.2 Patient-derived xenograft tumor models

The establishment of cancer cell lines underlies an enormous selection pressure during cultivation on plastic surfaces without the support of stromal factors and only a small fraction of cancer cells survive this procedure and start to proliferate in 2D, most likely after having gained additional mutations [145]. Therefore, established cancer cell lines lack one crucial feature inherent to human carcinomas: the genetic, epigenetic or phenotypic heterogeneity of the cancer cells *in vivo*. This selection process can be largely avoided by directly implanting small cancer fragments subcutaneously or orthotopically into mice (Figure 4.3C) [145–147]. In these so-called patient-derived xenograft (PDX) tumor models, the initial cancer cell heterogeneity [148,149] is preserved for extended time periods [150]. A recent in-depth review of PDX tumor models and their relevance for clinical oncology is available [151]. Using PDX, the genetic variation of cancer cells and the tissue architecture, as well as the histology of the human tumor, is essentially maintained as demonstrated, for example, for colon [152], lung [153], pancreas [154,155], and ovarian [156] cancers. Of note, PDX appear to keep tumor cells epigenetically stable in contrast to cell lines [157]. However, PDX models are technical demanding, time-consuming and cost-intensive, apart from the ethical concerns of animal experiments per se. In addition, each newly established PDX has to be thoroughly characterized and compared to the initial human tumor, which is time-consuming and expensive.

Nevertheless, maintenance of the tumor characteristics makes PDX a valuable preclinical tool for drug response studies [158,159]. The predictive power of PDX tumor models for clinical response is still discussed [160–162]. On the one hand, it is possible to correlate the efficacy of treatment in the PDX tumor model with the clinical response in the corresponding patient. Using this approach a close match between the tumor response in the PDX and the patients was described for several tumor entities [152,163–165]. Alternatively, the treatment response of a collection of PDX from a certain cancer type can be related to the drug efficacy in a corresponding patient cancer collective. Here, a large number of studies are available, which compare drug responses of certain cancer subtypes in xenografted mice to compound sensitivity in patients [17,156,161,163,166–170]. A strong correlation was seen in small-cell lung cancers treated with combination chemotherapy, reporting a close match of preclinical efficacy in 75% of the cases [171] with the response rate in a prospective clinical study (76% overall response). Moreover, in another study PDX tumor models displayed a prediction rate of 90% for patient tumor sensitivity and a strong predictivity for development of resistance in 97% of cases [164,172]. In summary, considering all available data it seems that for the response of PDX tumor models to conventional chemotherapeutics a close correlation with clinical response in retrospective or prospective studies is evident.

However, as underlined recently [151], some results have to be regarded with caution. Despite very similar treatment schemes in patients and xenografts, there are striking dissimilarities in the preclinical compared to the clinical setting. The size of a tumor and its site of appearance differ significantly in patients, whereas preclinically there is a more homogeneous situation present. Moreover, patients show less response to new therapeutic approaches since the majority was pretreated with other therapies and often have relapsed [173]. Further, there are substantial differences in the definition of response between preclinical and clinical studies. In clinical studies, the response for adjuvant therapy is judged by progression-free, relapse-free, disease-free and/or metastasis-free survival, whereas in neoadjuvant therapy the direct response of tumor size to therapy can be determined. Therefore, results from PDX tumor studies are best suited to be compared to the neoadjuvant setting in patients. Finally, the definitions of response criteria are different in patients and PDX tumor models. Here, one good example is the term "partial response," which is quite diversely defined in the clinic compared to the preclinical models and therefore complicates the objective comparison [174]. Thus, all these facts imply that judgment of drug responses in the models versus the clinic might not be completely meaningful and novel concepts are needed [151]. Notably, most of the points raised here are also valid for any other *in vivo*, *in vitro*, or direct culture preclinical model, however, they have so far been more extensively addressed in the PDX models.

Despite the power of PDX as a preclinical model for human cancers, there are also several limitations. First, tumor take rates are variable, which might lead to biases in testable models. Second, the tumor–host interaction cannot be fully recapitulated due to cross-species incompatibilities and, for example, lack of cross-reactivity of paracrine factors in both directions. Third, a fully functional immune system is absent in the mice.

4.2.3.3 Xenograft models with human stroma

As pointed out above, mouse stroma to human cancer cell crosstalk and vice versa is limited to a certain extent, due to some lack of cross-species compatibility. Examples are granulocyte–macrophage colony stimulating factor, hepatocyte growth factor (HGF), or interleukin 6 (IL6). Murine HGF is not able to activate human c-Met [175] and IL6 from mouse is inactive on human cells, whereas human IL6 can activate signaling of the mouse IL6R [176]. Thus, the paracrine signaling between xenografted human cancer cells and the murine microenvironment is incomplete and highly relevant paracrine signaling loops are absent. Thus, coinjection of human stromal cells along with cancer cells is an option to restore major tumor stroma crosstalk. Indeed, coculture of human tumor cells with human stromal cells in xenografted mice increased tumor growth, angiogenesis, and progression [20,21,177,178]. A model for human breast cancer was established in mice by prepopulating the cleared mammary fat pad with human stromal fibroblasts prior to injecting human breast epithelial cells [179]. Interestingly, we are not aware of any human tumor cell/stromal cell coculture study to determine drug responses *in vivo*. However, we think that these models are promising tools for relevant drug-efficacy studies and warrant further analysis.

4.2.3.4 Xenograft models in humanized mice

Immune cells are a central component of the tumor stroma. Importantly, novel therapeutic strategies aim to exploit the immune system to specifically target tumor cells. By inducing tumor-specific immunological memory, longlasting regression and prevention of relapses might be achievable in cancer patients. However, tumor immunity is lacking in the xenograft and PDX tumor system. To cope with this disadvantage, irradiated mice are engrafted with human hematopoietic stem and progenitor cells (HSPCs) and after the human hematopoietic system has been reconstituted, the mice are subsequently inoculated with cancer cell lines or PDX (Figure 4.3D). This enables the investigation of human cancers in an *in vivo* environment in the presence of a human immune system, which is at least partially functional. Transplantation of HSPCs and peripheral T cells followed by inoculation of breast cancer cells resulted in promotion of tumor development. Functionally, the presence of CD4+ T cells was shown to promote breast cancer growth associated with a type 2 inflammation [180]. The response was dependent on dendritic cells and was partially inhibited by antagonizing IL13 [181] response. However, this field of research is currently in early development and better examination is clearly needed to judge the impact on drug response. For further advantages, potential difficulties, and graft-versus-host disease issues we refer to a recent comprehensive review [182].

4.3 ASSESSING DRUG RESPONSE IN COMPLEX MODELS

Inherent to the requirement of 3D cultures and the presence of both tumor and stromal cells to properly mimic the *in vivo* cancer situation, difficulties in assessing drug response with well-established assays arise [183]. Protocols optimized for 2D cultures are often not validated in a 3D environment where diffusion barriers are present and signals are quenched by the 3D structure. In general, only limited information is available for the suitability of standard cell viability assays in spheroid models. It seems that standard cell proliferation assays display a nonlinear correlation with increasing cell numbers or between 2D and 3D cultures, and are therefore only appropriate to provide rough estimates of proliferation rates or cell numbers [184].

A rapid, reliable, and cost-efficient method to identify antiproliferative or cytotoxic compounds is desired in cellular screening assays. In conventional 2D cultures, standard assays determining cellular metabolic activity, such as the resazurin reduction (AlamarBlue) assay, the 3-(4,5-di*m*ethyl*t*hiazol-2-yl)-2,5-diphenyl*t*etrazolium bromide assay, or quantifying intracellular ATP levels are widely used to determine drug response based on cell viability and/or cell number. Most drug response studies use these assays, despite the fact that they do not precisely reflect cellular proliferation rates as demonstrated by the comparison of the metabolic activity of different human cell types using AlamarBlue and DNA content using CyQuant/PicoGreen. Metabolic activity and cell number did not directly correlate and in fact an overestimation of cell proliferation with the metabolic assay as compared to DNA binding dyes was reported [185]. Furthermore, it was recently shown that the AlamarBlue assay could not reliably detect cytotoxic and/or cytostatic responses in MCTS [186]. This was due to tight cell–cell interactions in spheroids, which hindered resazurin (the active compound of AlamarBlue) uptake in the inner spheroid core and its subsequent reduction to resorufin, leading to reduced signals relative to the actual cell number. Treatment with cytotoxic substances disrupted tight cell–cell interactions, and despite the presence of dying cells increased the signal compared to controls, therefore the AlamarBlue assay was identified to be nonsuitable for drug response studies in MCTS. Here, endpoint methods in which cells (or spheroids) are lysed prior to measurement, such as the CellTiter-Glo assay, turned out to be most reliable.

The main question using complex cancer models is how to differentiate between the input of cancer versus stromal cells to cell viability and/or apoptosis assays. The simple indirect assays described above are nonsuited for this type of analysis. So far, in coculture models, the question of addressing the drug response in cancer and stromal cells individually remains poorly examined. In FFPE material, for example, from raft cultures, histochemical and/or IHC evaluation is employed and validated markers for mitosis (by quantifying phospho-H3 stainings), proliferation (Ki67 or PCNA positivity), or apoptosis (cleaved caspase 3 or TUNEL staining) are widely employed. In order to quantify these parameters, image analysis and cell segmentation tools are used. Generally, high-content screening is the state of the art read out for assessing individual cellular responses to drug treatment [183].

Alternatively, to measure drug efficacy specifically in cancer cells, expression of constitutively active luciferase [187] can be used as a marker for tumor cell viability in coculture with nontransfected stromal cells. More advanced is the use of triple bioluminescence imaging to quantify firefly (Fluc), gaussia (Gluc), and vargula luciferase (Vluc) expressed in three different cell types, respectively, for *in vivo* monitoring [188]. Theoretically, this approach can also be applied *in vitro* to differentially label three different cell compartments and measure the separate response of these three cell types to drug treatment.

Finally, the physiologic concentration of compounds at the cancer site and the duration of exposure has to be taken into account for *in vivo* as well as for *in vitro* studies. Extensive studies using xenograft models revealed that for proper analysis of the mode of action of a certain drug, PK as well as PD parameters have to be taken into account [189]. In line with this, clinically relevant doses of a certain NME at the site of action and proper treatment schedules can predict clinical responses [151,190]. In fact, the availability, time, and concentration of an anticancer drug reaching the tumor are of major importance; which is also very important for *in vitro* assays but is widely ignored in this area of research currently.

In summary, the employment of heterotypic cancer models for drug response analysis in 3D *in vitro* as well as *in vivo* necessitates a careful selection and verification of the readouts as well as proper pharmacologic parameters to obtain meaningful biological data.

4.4 CONCLUSIONS

Obviously only certain aspects of the complexity in human cancer development and drug response can be mimicked by any of the mentioned model systems. The availability of different drug testing models, which are constantly improved, opens the possibility for proper selection of assays tailored to the actual need and also provide novel combinations of models to obtain an accurate imitation of reality in order to better predict clinical response. As far as tumor cell heterogeneity in patients is concerned, incorporation of primary human cancer cells or material from PDX tumors is feasible into any of the above-mentioned *in vitro* models. Thus, after setup and optimization of the different models with established tumor cell lines, now more physiologic test systems can be engineered using primary cancer cells and exploited for more predictive preclinical drug testing.

Taken together, preclinical models for anticancer therapy testing have been improved considerably during the past decades, by combining three-dimensionality with tumor–stroma interactions. Emerging *in vitro* 3D coculture models represent a promising platform for evaluation of drug efficacy to supplement mouse xenograft studies, raising less ethical concerns and being relatively inexpensive. While, for example, the MCTS model is already employed for HTS, other models reviewed here still require further improvement, miniaturization, standardization, and development of suitable readouts for drug screening approaches. A careful selection of multiple preclinical *in vitro* and *in vivo* models to address different aspects of efficacy testing before advancing to clinical testing might lead to more predictable drug responses in humans.

ACKNOWLEDGMENTS

The work of HD was supported by the Herzfelder Family Foundation and the Niederösterreichische Forschungs- und Bildungsges.m.b.H (NFB). AW is recipient of a DOC fellowship from the Austrian Academy of Science.

REFERENCES

[1] Siegel R, Ma J, Zou Z, Jemal A. Cancer statistics, 2014. CA: Cancer J Clin 2014;64(1):9–29.

[2] Kola I, Landis J. Can the pharmaceutical industry reduce attrition rates? Nat Rev Drug Discov 2004;3 (8):711–15.

[3] Arrowsmith J, Miller P. Trial watch: phase II and phase III attrition rates 2011-2012. Nat Rev Drug Discov 2013;12(8):569.

[4] Bodnar AG, Ouellette M, Frolkis M, Holt SE, Chiu CP, Morin GB, et al. Extension of life-span by introduction of telomerase into normal human cells. Science 1998;279(5349):349–52.

[5] Hahn WC, Counter CM, Lundberg AS, Beijersbergen RL, Brooks MW, Weinberg RA. Creation of human tumour cells with defined genetic elements. Nature 1999;400(6743):464–8.

[6] Iyer VR, Eisen MB, Ross DT, Schuler G, Moore T, Lee JC, et al. The transcriptional program in the response of human fibroblasts to serum. Science 1999;283(5398):83–7.

[7] Bell E, Ehrlich HP, Buttle DJ, Nakatsuji T. Living tissue formed *in vitro* and accepted as skin-equivalent tissue of full thickness. Science 1981;211(4486):1052–4.

[8] Runswick SK, O'Hare MJ, Jones L, Streuli CH, Garrod DR. Desmosomal adhesion regulates epithelial morphogenesis and cell positioning. Nat Cell Biol 2001;3(9):823–30.

[9] Gudjonsson T, Ronnov-Jessen L, Villadsen R, Rank F, Bissell MJ, Petersen OW. Normal and tumor-derived myoepithelial cells differ in their ability to interact with luminal breast epithelial cells for polarity and basement membrane deposition. J Cell Sci 2002;115(Pt 1):39–50.

[10] Kunz-Schughart LA, Heyder P, Schroeder J, Knuechel R. A heterologous 3-D coculture model of breast tumor cells and fibroblasts to study tumor-associated fibroblast differentiation. Exp Cell Res 2001;266 (1):74–86.

[11] Fischbach C, Chen R, Matsumoto T, Schmelzle T, Brugge JS, Polverini PJ, et al. Engineering tumors with 3D scaffolds. Nat Methods 2007;4(10):855–60.

[12] Okawa T, Michaylira CZ, Kalabis J, Stairs DB, Nakagawa H, Andl CD, et al. The functional interplay between EGFR overexpression, hTERT activation, and p53 mutation in esophageal epithelial cells with activation of stromal fibroblasts induces tumor development, invasion, and differentiation. Genes Dev 2007;21(21):2788–803.

[13] Takagi A, Watanabe M, Ishii Y, Morita J, Hirokawa Y, Matsuzaki T, et al. Three-dimensional cellular spheroid formation provides human prostate tumor cells with tissue-like features. Anticancer Res 2007;27(1A):45–53.

[14] Mathis C, Poussin C, Weisensee D, Gebel S, Hengstermann A, Sewer A, et al. Human bronchial epithelial cells exposed *in vitro* to cigarette smoke at the air-liquid interface resemble bronchial epithelium from human smokers. Am J Physiol Lung Cell Mol Physiol 2013;304(7):L489–503.

[15] Kamb A. What's wrong with our cancer models? Nat Rev Drug Discov 2005;4(2):161–5.

[16] Staquet MJ, Byar DP, Green SB, Rozencweig M. Clinical predictivity of transplantable tumor systems in the selection of new drugs for solid tumors: rationale for a three-stage strategy. Cancer Treat Rep 1983;67(9):753–65.

[17] Voskoglou-Nomikos T, Pater JL, Seymour L. Clinical predictive value of the *in vitro* cell line, human xenograft, and mouse allograft preclinical cancer models. Clin Cancer Res 2003;9(11):4227–39.

[18] Fidler IJ. Orthotopic implantation of human colon carcinomas into nude mice provides a valuable model for the biology and therapy of metastasis. Cancer Metastasis Rev 1991;10(3):229–43.

[19] Bhowmick NA, Neilson EG, Moses HL. Stromal fibroblasts in cancer initiation and progression. Nature 2004;432(7015):332–7.

[20] Olumi AF, Grossfeld GD, Hayward SW, Carroll PR, Tlsty TD, Cunha GR. Carcinoma-associated fibroblasts direct tumor progression of initiated human prostatic epithelium. Cancer Res 1999;59 (19):5002–11.

[21] Yauch RL, Gould SE, Scales SJ, Tang T, Tian H, Ahn CP, et al. A paracrine requirement for hedgehog signalling in cancer. Nature 2008;455(7211):406–10.

[22] Gaggioli C, Hooper S, Hidalgo-Carcedo C, Grosse R, Marshall JF, Harrington K, et al. Fibroblast-led collective invasion of carcinoma cells with differing roles for RhoGTPases in leading and following cells. Nat Cell Biol 2007;9(12):1392–400.

[23] Seol HS, Kang H, Lee SI, Kim NE, Kim TI, Chun SM, et al. Development and characterization of a colon PDX model that reproduces drug responsiveness and the mutation profiles of its original tumor. Cancer Lett 2014;345(1):56–64.

[24] Chien J, Kuang R, Landen C, Shridhar V. Platinum-sensitive recurrence in ovarian cancer: the role of tumor microenvironment. Front Oncol 2013;3:251.

[25] Straussman R, Morikawa T, Shee K, Barzily-Rokni M, Qian ZR, Du J, et al. Tumour micro-environment elicits innate resistance to RAF inhibitors through HGF secretion. Nature 2012;487(7408):500–4.

[26] McMillin DW, Negri JM, Mitsiades CS. The role of tumour-stromal interactions in modifying drug response: challenges and opportunities. Nat Rev Drug Discov 2013;12(3):217–28.

[27] Baysan M, Woolard K, Bozdag S, Riddick G, Kotliarova S, Cam MC, et al. Micro-environment causes reversible changes in DNA methylation and mRNA expression profiles in patient-derived glioma stem cells. PLoS One 2014;9(4):e94045.

[28] DesRochers TM, Shamis Y, Alt-Holland A, Kudo Y, Takata T, Wang G, et al. The 3D tissue microenvironment modulates DNA methylation and E-cadherin expression in squamous cell carcinoma. Epigenetics 2012;7(1):34–46.

[29] Xu L, Deng Q, Pan Y, Peng M, Wang X, Song L, et al. Cancer-associated fibroblasts enhance the migration ability of ovarian cancer cells by increasing EZH2 expression. Int J Mol Med 2014;33(1):91–6.

[30] Kang N, Shah VH, Urrutia R. Membrane-to-nucleus signals and epigenetic mechanisms for myofibroblastic activation and desmoplastic stroma: potential therapeutic targets for liver metastasis? Mol Cancer Res 2014;13(4):604–12.

[31] Tyan SW, Hsu CH, Peng KL, Chen CC, Kuo WH, Lee EY, et al. Breast cancer cells induce stromal fibroblasts to secrete ADAMTS1 for cancer invasion through an epigenetic change. PLoS One 2012;7 (4):e35128.

[32] Mullard A. 2013 in reflection. Nat Rev Drug Discov 2014;13(1):5–7.

[33] Cook N, Jodrell DI, Tuveson DA. Predictive *in vivo* animal models and translation to clinical trials. Drug Discov Today 2012;17(5–6):253–60.

[34] Pietras K, Ostman A. Hallmarks of cancer: interactions with the tumor stroma. Exp Cell Res 2010;316 (8):1324–31.

[35] Scannell JW, Blanckley A, Boldon H, Warrington B. Diagnosing the decline in pharmaceutical R&D efficiency. Nat Rev Drug Discov 2012;11(3):191–200.

[36] Simone NL, Bonner RF, Gillespie JW, Emmert-Buck MR, Liotta LA. Laser-capture microdissection: opening the microscopic frontier to molecular analysis. Trends Genet 1998;14(7):272–6.

[37] Rupp C, Dolznig H, Puri C, Schweifer N, Sommergruber W, Kraut N, et al. Laser capture microdissection of epithelial cancers guided by antibodies against fibroblast activation protein and endosialin. Diagn Mol Pathol 2006;15(1):35–42.

[38] Rupp C, Scherzer M, Rudisch A, Unger C, Haslinger C, Schweifer N, et al. IGFBP7, a novel tumor stroma marker, with growth-promoting effects in colon cancer through a paracrine tumor-stroma interaction. Oncogene 2015;34(7):815–25.

[39] Smith MJ, Culhane AC, Donovan M, Coffey JC, Barry BD, Kelly MA, et al. Analysis of differential gene expression in colorectal cancer and stroma using fluorescence-activated cell sorting purification. Br J Cancer 2009;100(9):1452–64.

[40] Harbeck N, Schwarze S, Schuren E, Yamamoto N, Moniwa N, Schmitt M, et al. Model system for isolation of competent ovarian-carcinoma cells from fresh tumor-tissue by a magnetic separation technique (macs). Int J Oncol 1995;6(6):1249–54.

[41] Jia Z, Rahmatpanah FB, Chen X, Lernhardt W, Wang Y, Xia XQ, et al. Expression changes in the stroma of prostate cancer predict subsequent relapse. PLoS One 2012;7(8):e41371.

[42] Isella C, Terrasi A, Bellomo SE, Petti C, Galatola G, Muratore A, et al. Stromal contribution to the colorectal cancer transcriptome. Nat Genet 2015;47(4):312–19.

[43] Hirschhaeuser F, Menne H, Dittfeld C, West J, Mueller-Klieser W, Kunz-Schughart LA. Multicellular tumor spheroids: an underestimated tool is catching up again. J Biotechnol 2010;148(1):3–15.

[44] Frieboes HB, Zheng X, Sun CH, Tromberg B, Gatenby R, Cristini V. An integrated computational/experimental model of tumor invasion. Cancer Res 2006;66(3):1597–604.

[45] Yamada KM, Cukierman E. Modeling tissue morphogenesis and cancer in 3D. Cell 2007;130(4):601–10.

[46] Thoma CR, Zimmermann M, Agarkova I, Kelm JM, Krek W. 3D cell culture systems modeling tumor growth determinants in cancer target discovery. Adv Drug Deliv Rev 2014;69–70C:29–41.

[47] Weiswald LB, Bellet D, Dangles-Marie V. Spherical cancer models in tumor biology. Neoplasia 2015;17(1):1–15.

[48] Durand RE, Sutherland RM. Effects of intercellular contact on repair of radiation damage. Exp Cell Res 1972;71(1):75–80.

[49] West CM, Sutherland RM. The radiation response of a human colon adenocarcinoma grown in monolayer, as spheroids, and in nude mice. Radiat Res 1987;112(1):105–15.

[50] Desoize B, Gimonet D, Jardillier JC. Multicellular resistance: another mechanism for multidrug resistance? Bull Cancer 1998;85(9):785.

[51] Wenzel C, Riefke B, Grundemann S, Krebs A, Christian S, Prinz F, et al. 3D high-content screening for the identification of compounds that target cells in dormant tumor spheroid regions. Exp Cell Res 2014;323(1):131–43.

[52] Blangy A, Lane HA, d'Herin P, Harper M, Kress M, Nigg EA. Phosphorylation by p34cdc2 regulates spindle association of human Eg5, a kinesin-related motor essential for bipolar spindle formation *in vivo*. Cell 1995;83(7):1159–69.

[53] Thoma CR, Stroebel S, Rosch N, Calpe B, Krek W, Kelm JM. A high-throughput-compatible 3D microtissue co-culture system for phenotypic RNAi screening applications. J Biomol Screen 2013;18(10):1330–7.

[54] Dolznig H, Walzl A, Kramer N, Rosner M, Garin-Chesa P, Hengstschlager M. Organotypic spheroid cultures to study tumor-stroma interaction during cancer development. Drug Discov Today Dis Models 2011;8(2–3):113–19.

[55] Hoffmann TK, Schirlau K, Sonkoly E, Brandau S, Lang S, Pivarcsi A, et al. A novel mechanism for anti-EGFR antibody action involves chemokine-mediated leukocyte infiltration. Int J Cancer 2009;124(11):2589–96.

[56] Graves DT, Valente AJ. Monocyte chemotactic proteins from human tumor cells. Biochem Pharmacol 1991;41(3):333–7.

[57] Lee JA, Berg EL. Neoclassic drug discovery: the case for lead generation using phenotypic and functional approaches. J Biomol Screen 2013;18(10):1143–55.

[58] Hutmacher DW, Loessner D, Rizzi S, Kaplan DL, Mooney DJ, Clements JA. Can tissue engineering concepts advance tumor biology research? Trends Biotechnol 2010;28(3):125–33.

[59] Chan HF, Zhang Y, Ho YP, Chiu YL, Jung Y, Leong KW. Rapid formation of multicellular spheroids in double-emulsion droplets with controllable microenvironment. Sci Rep 2013;3:3462.
[60] Papas KK, Long Jr RC, Sambanis A, Constantinidis I. Development of a bioartificial pancreas: I. long-term propagation and basal and induced secretion from entrapped betaTC3 cell cultures. Biotechnol Bioeng 1999;66(4):219–30.
[61] Kolambkar YM, Boerckel JD, Dupont KM, Bajin M, Huebsch N, Mooney DJ, et al. Spatiotemporal delivery of bone morphogenetic protein enhances functional repair of segmental bone defects. Bone 2011;49(3):485–92.
[62] Venkatesan J, Bhatnagar I, Manivasagan P, Kang KH, Kim SK. Alginate composites for bone tissue engineering: a review. Int J Biol Macromol 2014;72:269–81.
[63] Rowley JA, Madlambayan G, Mooney DJ. Alginate hydrogels as synthetic extracellular matrix materials. Biomaterials 1999;20(1):45–53.
[64] Fischbach C, Kong HJ, Hsiong SX, Evangelista MB, Yuen W, Mooney DJ. Cancer cell angiogenic capability is regulated by 3D culture and integrin engagement. Proc Natl Acad Sci USA 2009;106 (2):399–404.
[65] Branco da Cunha C, Klumpers DD, Li WA, Koshy ST, Weaver JC, Chaudhuri O, et al. Influence of the stiffness of three-dimensional alginate/collagen-I interpenetrating networks on fibroblast biology. Biomaterials 2014;35(32):8927–36.
[66] Chaudhuri O, Koshy ST, Branco da Cunha C, Shin JW, Verbeke CS, Allison KH, et al. Extracellular matrix stiffness and composition jointly regulate the induction of malignant phenotypes in mammary epithelium. Nat Mater 2014;13(10):970–8.
[67] Andersen T, Markussen C, Dornish M, Heier-Baardson H, Melvik JE, Alsberg E, et al. *In situ* gelation for cell immobilization and culture in alginate foam scaffolds. Tissue Eng Part A 2014;20 (3–4):600–10.
[68] Lee DW, Choi YS, Seo YJ, Lee MY, Jeon SY, Ku B, et al. High-throughput screening (HTS) of anticancer drug efficacy on a micropillar/microwell chip platform. Anal Chem 2014;86(1):535–42.
[69] Godugu C, Patel AR, Desai U, Andey T, Sams A, Singh M. AlgiMatrix based 3D cell culture system as an *in-vitro* tumor model for anticancer studies. PLoS One 2013;8(1):e53708.
[70] Barker N, van Es JH, Kuipers J, Kujala P, van den Born M, Cozijnsen M, et al. Identification of stem cells in small intestine and colon by marker gene Lgr5. Nature 2007;449(7165):1003–7.
[71] Sato T, Vries RG, Snippert HJ, van de Wetering M, Barker N, Stange DE, et al. Single Lgr5 stem cells build crypt-villus structures *in vitro* without a mesenchymal niche. Nature 2009;459 (7244):262–5.
[72] Sato T, Stange DE, Ferrante M, Vries RG, Van Es JH, Van den Brink S, et al. Long-term expansion of epithelial organoids from human colon, adenoma, adenocarcinoma, and Barrett's epithelium. Gastroenterology 2011;141(5):1762–72.
[73] Hirokawa Y, Yip KH, Tan CW, Burgess AW. Colonic myofibroblast cell line stimulates colonoid formation. Am J Physiol Gastrointest Liver Physiol 2014;306(7):G547–56.
[74] Prasetyanti PR, Zimberlin CD, Bots M, Vermeulen L, Melo Fde S, Medema JP. Regulation of stem cell self-renewal and differentiation by Wnt and Notch are conserved throughout the adenoma-carcinoma sequence in the colon. Mol Cancer 2013;12(1):126.
[75] Dontu G, Abdallah WM, Foley JM, Jackson KW, Clarke MF, Kawamura MJ, et al. *In vitro* propagation and transcriptional profiling of human mammary stem/progenitor cells. Genes Dev 2003;17 (10):1253–70.
[76] Rock JR, Onaitis MW, Rawlins EL, Lu Y, Clark CP, Xue Y, et al. Basal cells as stem cells of the mouse trachea and human airway epithelium. Proc Natl Acad Sci USA 2009;106(31):12771–5.

[77] Barker N, Huch M, Kujala P, van de Wetering M, Snippert HJ, van Es JH, et al. Lgr5(+ve) stem cells drive self-renewal in the stomach and build long-lived gastric units *in vitro*. Cell Stem Cell 2010;6 (1):25–36.

[78] Lukacs RU, Goldstein AS, Lawson DA, Cheng D, Witte ON. Isolation, cultivation and characterization of adult murine prostate stem cells. Nat Protoc 2010;5(4):702–13.

[79] Astashkina AI, Mann BK, Prestwich GD, Grainger DW. A 3-D organoid kidney culture model engineered for high-throughput nephrotoxicity assays. Biomaterials 2012;33(18):4700–11.

[80] Greggio C, De Franceschi F, Figueiredo-Larsen M, Gobaa S, Ranga A, Semb H, et al. Artificial three-dimensional niches deconstruct pancreas development *in vitro*. Development 2013;140 (21):4452–62.

[81] Huch M, Dorrell C, Boj SF, van Es JH, Li VS, van de Wetering M, et al. *In vitro* expansion of single Lgr5+ liver stem cells induced by Wnt-driven regeneration. Nature 2013;494(7436):247–50.

[82] Ranga A, Gjorevski N, Lutolf MP. Drug discovery through stem cell-based organoid models. Adv Drug Deliv Rev 2014;69-70C:19–28.

[83] Sachs N, Clevers H. Organoid cultures for the analysis of cancer phenotypes. Curr Opin Genet Dev 2014;24:68–73.

[84] Whitesides GM. The origins and the future of microfluidics. Nature 2006;442(7101):368–73.

[85] Wlodkowic D, Cooper JM. Tumors on chips: oncology meets microfluidics. Curr Opin Chem Biol 2010;14(5):556–67.

[86] Takayama S, Ostuni E, LeDuc P, Naruse K, Ingber DE, Whitesides GM. Subcellular positioning of small molecules. Nature 2001;411(6841):1016.

[87] Kovarik ML, Ornoff DM, Melvin AT, Dobes NC, Wang Y, Dickinson AJ, et al. Micro total analysis systems: fundamental advances and applications in the laboratory, clinic, and field. Anal Chem 2013;85 (2):451–72.

[88] Huh D, Hamilton GA, Ingber DE. From 3D cell culture to organs-on-chips. Trends Cell Biol 2011;21 (12):745–54.

[89] Jang KJ, Mehr AP, Hamilton GA, McPartlin LA, Chung S, Suh KY, et al. Human kidney proximal tubule-on-a-chip for drug transport and nephrotoxicity assessment. Integr Biol 2013;5(9):1119–29.

[90] Lee PJ, Hung PJ, Lee LP. An artificial liver sinusoid with a microfluidic endothelial-like barrier for primary hepatocyte culture. Biotechnol Bioeng 2007;97(5):1340–6.

[91] Huh D, Matthews BD, Mammoto A, Montoya-Zavala M, Hsin HY, Ingber DE. Reconstituting organ-level lung functions on a chip. Science 2010;328(5986):1662–8.

[92] Young EW. Cells, tissues, and organs on chips: challenges and opportunities for the cancer tumor microenvironment. Integr Biol 2013;5(9):1096–109.

[93] Esch MB, Smith AS, Prot JM, Oleaga C, Hickman JJ, Shuler ML. How multi-organ microdevices can help foster drug development. Adv Drug Deliv Rev 2014;69-70C:158–69.

[94] Chaw KC, Manimaran M, Tay EH, Swaminathan S. Multi-step microfluidic device for studying cancer metastasis. Lab Chip 2007;7(8):1041–7.

[95] Kuo CT, Chiang CL, Chang CH, Liu HK, Huang GS, Huang RY, et al. Modeling of cancer metastasis and drug resistance via biomimetic nano-cilia and microfluidics. Biomaterials 2014;35(5):1562–71.

[96] Sung JH, Kam C, Shuler ML. A microfluidic device for a pharmacokinetic-pharmacodynamic (PK-PD) model on a chip. Lab Chip 2010;10(4):446–55.

[97] Ma H, Liu T, Qin J, Lin B. Characterization of the interaction between fibroblasts and tumor cells on a microfluidic co-culture device. Electrophoresis 2010;31(10):1599–605.

[98] Zervantonakis IK, Hughes-Alford SK, Charest JL, Condeelis JS, Gertler FB, Kamm RD. Three-dimensional microfluidic model for tumor cell intravasation and endothelial barrier function. Proc Natl Acad Sci USA 2012;109(34):13515–20.

[99] Mattei F, Schiavoni G, De Ninno A, Lucarini V, Sestili P, Sistigu A, et al. A multidisciplinary study using *in vivo* tumor models and microfluidic cell-on-chip approach to explore the cross-talk between cancer and immune cells. J Immunotoxicol 2014;11(4):337–46.

[100] Elliott NT, Yuan F. A microfluidic system for investigation of extravascular transport and cellular uptake of drugs in tumors. Biotechnol Bioeng 2012;109(5):1326–35.

[101] Sung JH, Shuler ML. A micro cell culture analog (microCCA) with 3-D hydrogel culture of multiple cell lines to assess metabolism-dependent cytotoxicity of anti-cancer drugs. Lab Chip 2009;9 (10):1385–94.

[102] Wang L, Liu W, Wang Y, Wang JC, Tu Q, Liu R, et al. Construction of oxygen and chemical concentration gradients in a single microfluidic device for studying tumor cell-drug interactions in a dynamic hypoxia microenvironment. Lab Chip 2013;13(4):695–705.

[103] Kim C, Bang JH, Kim YE, Lee SH, Kang JY. On-chip anticancer drug test of regular tumor spheroids formed in microwells by a distributive microchannel network. Lab Chip 2012;12(20):4135–42.

[104] Gleghorn JP, Pratt ED, Denning D, Liu H, Bander NH, Tagawa ST, et al. Capture of circulating tumor cells from whole blood of prostate cancer patients using geometrically enhanced differential immunocapture (GEDI) and a prostate-specific antibody. Lab Chip 2010;10(1):27–9.

[105] Kirby BJ, Jodari M, Loftus MS, Gakhar G, Pratt ED, Chanel-Vos C, et al. Functional characterization of circulating tumor cells with a prostate-cancer-specific microfluidic device. PLoS One 2012;7(4): e35976.

[106] Kenny PA. Three-dimensional extracellular matrix culture models of EGFR signalling and drug response. Biochem Soc Trans 2007;35(Pt 4):665–8.

[107] Lee JM, Mhawech-Fauceglia P, Lee N, Parsanian LC, Lin YG, Gayther SA, et al. A three-dimensional microenvironment alters protein expression and chemosensitivity of epithelial ovarian cancer cells *in vitro*. Lab Invest 2013;93(5):528–42.

[108] Rintoul RC, Sethi T. Extracellular matrix regulation of drug resistance in small-cell lung cancer. Clin Sci 2002;102(4):417–24.

[109] Dolznig H, Rupp C, Puri C, Haslinger C, Schweifer N, Wieser E, et al. Modeling colon adenocarcinomas *in vitro* a 3D co-culture system induces cancer-relevant pathways upon tumor cell and stromal fibroblast interaction. Am J Pathol 2011;179(1):487–501.

[110] Lee GY, Kenny PA, Lee EH, Bissell MJ. Three-dimensional culture models of normal and malignant breast epithelial cells. Nat Methods 2007;4(4):359–65.

[111] Yip D, Cho CH. A multicellular 3D heterospheroid model of liver tumor and stromal cells in collagen gel for anti-cancer drug testing. Biochem Biophys Res Commun 2013;433(3):327–32.

[112] Pickard A, Cichon AC, Barry A, Kieran D, Patel D, Hamilton P, et al. Inactivation of Rb in stromal fibroblasts promotes epithelial cell invasion. EMBO J 2012;31(14):3092–103.

[113] Timpson P, McGhee EJ, Erami Z, Nobis M, Quinn JA, Edward M, et al. Organotypic collagen I assay: a malleable platform to assess cell behaviour in a 3-dimensional context. J Vis Exp 2011;56: e3089.

[114] Weigelt B, Ghajar CM, Bissell MJ. The need for complex 3D culture models to unravel novel pathways and identify accurate biomarkers in breast cancer. Adv Drug Deliv Rev 2014;69-70C:42–51.

[115] Tiago M, de Oliveira EM, Brohem CA, Pennacchi PC, Paes RD, Haga RB, et al. Fibroblasts protect melanoma cells from the cytotoxic effects of doxorubicin. Tissue Eng Part A 2014;20 (17–18):2412–21.

[116] Maas-Szabowski N, Szabowski A, Stark HJ, Andrecht S, Kolbus A, Schorpp-Kistner M, et al. Organotypic cocultures with genetically modified mouse fibroblasts as a tool to dissect molecular mechanisms regulating keratinocyte growth and differentiation. J Invest Dermatol 2001;116 (5):816–20.

[117] Froeling FE, Mirza TA, Feakins RM, Seedhar A, Elia G, Hart IR, et al. Organotypic culture model of pancreatic cancer demonstrates that stromal cells modulate E-cadherin, beta-catenin, and Ezrin expression in tumor cells. Am J Pathol 2009;175(2):636–48.

[118] van Zijl F, Mair M, Csiszar A, Schneller D, Zulehner G, Huber H, et al. Hepatic tumor-stroma crosstalk guides epithelial to mesenchymal transition at the tumor edge. Oncogene 2009;28(45):4022–33.

[119] Nystrom ML, Thomas GJ, Stone M, Mackenzie IC, Hart IR, Marshall JF. Development of a quantitative method to analyse tumour cell invasion in organotypic culture. J Pathol 2005;205 (4):468–75.

[120] Seano G, Chiaverina G, Gagliardi PA, di Blasio L, Sessa R, Bussolino F, et al. Modeling human tumor angiogenesis in a three-dimensional culture system. Blood 2013;121(21):e129–37.

[121] Ghajar CM, Peinado H, Mori H, Matei IR, Evason KJ, Brazier H, et al. The perivascular niche regulates breast tumour dormancy. Nat Cell Biol 2013;15(7):807–17.

[122] Bechetoille N, Vachon H, Gaydon A, Boher A, Fontaine T, Schaeffer E, et al. A new organotypic model containing dermal-type macrophages. Exp Dermatol 2011;20(12):1035–7.

[123] Linde N, Gutschalk CM, Hoffmann C, Yilmaz D, Mueller MM. Integrating macrophages into organotypic co-cultures: a 3D *in vitro* model to study tumor-associated macrophages. PLoS One 2012;7(7): e40058.

[124] Mathes SH, Ruffner H, Graf-Hausner U. The use of skin models in drug development. Adv Drug Deliv Rev 2014;69–70C:81–102.

[125] Ridky TW, Chow JM, Wong DJ, Khavari PA. Invasive three-dimensional organotypic neoplasia from multiple normal human epithelia. Nat Med 2010;16(12):1450–5.

[126] Nobis M, McGhee EJ, Morton JP, Schwarz JP, Karim SA, Quinn J, et al. Intravital FLIM-FRET imaging reveals dasatinib-induced spatial control of src in pancreatic cancer. Cancer Res 2013;73 (15):4674–86.

[127] Kerjaschki D, Bago-Horvath Z, Rudas M, Sexl V, Schneckenleithner C, Wolbank S, et al. Lipoxygenase mediates invasion of intrametastatic lymphatic vessels and propagates lymph node metastasis of human mammary carcinoma xenografts in mouse. J Clin Invest 2011;121(5):2000–12.

[128] Vonach C, Viola K, Giessrigl B, Huttary N, Raab I, Kalt R, et al. NF-kappaB mediates the 12(S)-HETE-induced endothelial to mesenchymal transition of lymphendothelial cells during the intravasation of breast carcinoma cells. Br J Cancer 2011;105(2):263–71.

[129] Viola K, Kopf S, Huttary N, Vonach C, Kretschy N, Teichmann M, et al. Bay11-7082 inhibits the disintegration of the lymphendothelial barrier triggered by MCF-7 breast cancer spheroids; the role of ICAM-1 and adhesion. Br J Cancer 2013;108(3):564–9.

[130] Kopf S, Viola K, Atanasov AG, Jarukamjorn K, Rarova L, Kretschy N, et al. *In vitro* characterisation of the anti-intravasative properties of the marine product heteronemin. Arch Toxicol 2013;87 (10):1851–61.

[131] Viola K, Kopf S, Rarova L, Jarukamjorn K, Kretschy N, Teichmann M, et al. Xanthohumol attenuates tumour cell-mediated breaching of the lymphendothelial barrier and prevents intravasation and metastasis. Arch Toxicol 2013;87(7):1301–12.

[132] Kretschy N, Teichmann M, Kopf S, Atanasov AG, Saiko P, Vonach C, et al. *In vitro* inhibition of breast cancer spheroid-induced lymphendothelial defects resembling intravasation into the lymphatic vasculature by acetohexamide, isoxsuprine, nifedipin and proadifen. Br J Cancer 2013;108 (3):570–8.

[133] Teichmann M, Kretschy N, Kopf S, Jarukamjorn K, Atanasov AG, Viola K, et al. Inhibition of tumour spheroid-induced prometastatic intravasation gates in the lymph endothelial cell barrier by carbamazepine: drug testing in a 3D model. Arch Toxicol 2014;88(3):691–9.

[134] Gahwiler BH. Organotypic cultures of neural tissue. Trends Neurosci 1988;11(11):484–9.

[135] Vaira V, Fedele G, Pyne S, Fasoli E, Zadra G, Bailey D, et al. Preclinical model of organotypic culture for pharmacodynamic profiling of human tumors. Proc Natl Acad Sci USA 2010;107(18):8352–6.

[136] van der Kuip H, Murdter TE, Sonnenberg M, McClellan M, Gutzeit S, Gerteis A, et al. Short term culture of breast cancer tissues to study the activity of the anticancer drug taxol in an intact tumor environment. BMC Cancer 2006;6:86.

[137] Grosso SH, Katayama ML, Roela RA, Nonogaki S, Soares FA, Brentani H, et al. Breast cancer tissue slices as a model for evaluation of response to rapamycin. Cell Tissue Res 2013;352(3):671–84.

[138] Wartha K, Herting F, Hasmann M. Fit-for purpose use of mouse models to improve predictivity of cancer therapeutics evaluation. Pharmacol Ther 2014;142(3):351–61.

[139] Shimosato Y, Kameya T, Nagai K, Hirohashi S, Koide T, Hayashi H, et al. Transplantation of human tumors in nude mice. J Natl Cancer Inst 1976;56(6):1251–60.

[140] de Vries JE, Dinjens WN, De Bruyne GK, Verspaget HW, van der Linden EP, de Bruine AP, et al. *In vivo* and *in vitro* invasion in relation to phenotypic characteristics of human colorectal carcinoma cells. Br J Cancer 1995;71(2):271–7.

[141] Tseng W, Leong X, Engleman E. Orthotopic mouse model of colorectal cancer. J Vis Exp 2007;10:484.

[142] Tseng WW, Winer D, Kenkel JA, Choi O, Shain AH, Pollack JR, et al. Development of an orthotopic model of invasive pancreatic cancer in an immunocompetent murine host. Clin Cancer Res 2010;16 (14):3684–95.

[143] Ottewell PD, Coleman RE, Holen I. From genetic abnormality to metastases: murine models of breast cancer and their use in the development of anticancer therapies. Breast Cancer Res Treat 2006;96 (2):101–13.

[144] Sano D, Myers JN. Xenograft models of head and neck cancers. Head Neck Oncol 2009;1:32.

[145] Daniel VC, Marchionni L, Hierman JS, Rhodes JT, Devereux WL, Rudin CM, et al. A primary xenograft model of small-cell lung cancer reveals irreversible changes in gene expression imposed by culture *in vitro*. Cancer Res 2009;69(8):3364–73.

[146] Jin K, Teng L, Shen Y, He K, Xu Z, Li G. Patient-derived human tumour tissue xenografts in immunodeficient mice: a systematic review. Clin Transl Oncol 2010;12(7):473–80.

[147] Tentler JJ, Tan AC, Weekes CD, Jimeno A, Leong S, Pitts TM, et al. Patient-derived tumour xenografts as models for oncology drug development. Nat Rev Clin Oncol 2012;9(6):338–50.

[148] Chou J, Fitzgibbon MP, Mortales CL, Towlerton AM, Upton MP, Yeung RS, et al. Phenotypic and transcriptional fidelity of patient-derived colon cancer xenografts in immune-deficient mice. PLoS One 2013;8(11):e79874.

[149] Julien S, Merino-Trigo A, Lacroix L, Pocard M, Goere D, Mariani P, et al. Characterization of a large panel of patient-derived tumor xenografts representing the clinical heterogeneity of human colorectal cancer. Clin Cancer Res 2012;18(19):5314–28.

[150] Guenot D, Guerin E, Aguillon-Romain S, Pencreach E, Schneider A, Neuville A, et al. Primary tumour genetic alterations and intra-tumoral heterogeneity are maintained in xenografts of human colon cancers showing chromosome instability. J Pathol 2006;208(5):643–52.

[151] Decaudin D. Primary human tumor xenografted models ("tumorgrafts") for good management of patients with cancer. Anticancer Drugs 2011;22(9):827–41.

[152] Fichtner I, Slisow W, Gill J, Becker M, Elbe B, Hillebrand T, et al. Anticancer drug response and expression of molecular markers in early-passage xenotransplanted colon carcinomas. Eur J Cancer 2004;40(2):298–307.

[153] Nemati F, Bras-Goncalves R, Fontaine JJ, de Pinieux G, De Cremoux P, Chapelier A, et al. Preclinical assessment of cisplatin-based therapy versus docetaxel-based therapy on a panel of human non-small-cell lung cancer xenografts. Anticancer Drugs 2009;20(10):932–40.

[154] Loukopoulos P, Kanetaka K, Takamura M, Shibata T, Sakamoto M, Hirohashi S. Orthotopic transplantation models of pancreatic adenocarcinoma derived from cell lines and primary tumors and displaying varying metastatic activity. Pancreas 2004;29(3):193–203.
[155] Rubio-Viqueira B, Jimeno A, Cusatis G, Zhang X, Iacobuzio-Donahue C, Karikari C, et al. An *in vivo* platform for translational drug development in pancreatic cancer. Clin Cancer Res 2006;12 (15):4652–61.
[156] Molthoff CF, Calame JJ, Pinedo HM, Boven E. Human ovarian cancer xenografts in nude mice: characterization and analysis of antigen expression. Int J Cancer 1991;47(1):72–9.
[157] Wong NC, Bhadri VA, Maksimovic J, Parkinson-Bates M, Ng J, Craig JM, et al. Stability of gene expression and epigenetic profiles highlights the utility of patient-derived paediatric acute lymphoblastic leukaemia xenografts for investigating molecular mechanisms of drug resistance. BMC Genomics 2014;15:416.
[158] Krumbach R, Schuler J, Hofmann M, Giesemann T, Fiebig HH, Beckers T. Primary resistance to cetuximab in a panel of patient-derived tumour xenograft models: activation of MET as one mechanism for drug resistance. Eur J Cancer 2011;47(8):1231–43.
[159] Hidalgo M, Bruckheimer E, Rajeshkumar NV, Garrido-Laguna I, De Oliveira E, Rubio-Viqueira B, et al. A pilot clinical study of treatment guided by personalized tumorgrafts in patients with advanced cancer. Mol Cancer Ther 2011;10(8):1311–16.
[160] Talmadge JE, Singh RK, Fidler IJ, Raz A. Murine models to evaluate novel and conventional therapeutic strategies for cancer. Am J Pathol 2007;170(3):793–804.
[161] Johnson JI, Decker S, Zaharevitz D, Rubinstein LV, Venditti JM, Schepartz S, et al. Relationships between drug activity in NCI preclinical *in vitro* and *in vivo* models and early clinical trials. Br J Cancer 2001;84(10):1424–31.
[162] Scholz CC, Berger DP, Winterhalter BR, Henss H, Fiebig HH. Correlation of drug response in patients and in the clonogenic assay with solid human tumour xenografts. Eur J Cancer 1990;26(8):901–5.
[163] Marangoni E, Vincent-Salomon A, Auger N, Degeorges A, Assayag F, de Cremoux P, et al. A new model of patient tumor-derived breast cancer xenografts for preclinical assays. Clin Cancer Res 2007;13(13):3989–98.
[164] Fiebig HH, Schuchhardt C, Henss H, Fiedler L, Lohr GW. Comparison of tumor response in nude mice and in the patients. Behring Inst Mitt 1984;74:343–52.
[165] Harrap KR, Jones M, Siracky J, Pollard LA, Kelland LR. The establishment, characterization and calibration of human ovarian carcinoma xenografts for the evaluation of novel platinum anticancer drugs. Ann Oncol 1990;1(1):65–76.
[166] Inoue K, Fujimoto S, Ogawa M. Antitumor efficacy of seventeen anticancer drugs in human breast cancer xenograft (MX-1) transplanted in nude mice. Cancer Chemother Pharmacol 1983;10(3):182–6.
[167] Boven E, Winograd B, Berger DP, Dumont MP, Braakhuis BJ, Fodstad O, et al. Phase II preclinical drug screening in human tumor xenografts: a first European multicenter collaborative study. Cancer Res 1992;52(21):5940–7.
[168] Langdon SP, Hendriks HR, Braakhuis BJ, Pratesi G, Berger DP, Fodstad O, et al. Preclinical phase II studies in human tumor xenografts: a European multicenter follow-up study. Ann Oncol 1994;5 (5):415–22.
[169] Perez-Soler R, Kemp B, Wu QP, Mao L, Gomez J, Zeleniuch-Jacquotte A, et al. Response and determinants of sensitivity to paclitaxel in human non-small cell lung cancer tumors heterotransplanted in nude mice. Clin Cancer Res 2000;6(12):4932–8.
[170] Nemati F, Daniel C, Arvelo F, Legrier ME, Froget B, Livartowski A, et al. Clinical relevance of human cancer xenografts as a tool for preclinical assessment: example of *in-vivo* evaluation of topotecan-based

chemotherapy in a panel of human small-cell lung cancer xenografts. Anticancer Drugs 2010;21(1):25–32.

[171] Jimeno A, Solomon A, Karikari C, Garrido-Laguna I, Trusty D, Hruban R, et al. A prospective validation of a direct tumor xenograft model in pancreatic ductal adenocarcinoma (PDA). J Clin Oncol 2008;26(15):4500.

[172] Fiebig HH, Maier A, Burger AM. Clonogenic assay with established human tumour xenografts: correlation of *in vitro* to *in vivo* activity as a basis for anticancer drug discovery. Eur J Cancer 2004;40(6):802–20.

[173] Carey LA, Dees EC, Sawyer L, Gatti L, Moore DT, Collichio F, et al. The triple negative paradox: primary tumor chemosensitivity of breast cancer subtypes. Clin Cancer Res 2007;13(8):2329–34.

[174] Singh M, Lima A, Molina R, Hamilton P, Clermont AC, Devasthali V, et al. Assessing therapeutic responses in Kras mutant cancers using genetically engineered mouse models. Nat Biotechnol 2010;28(6):585–93.

[175] Jeffers M, Rong S, Vande Woude GF. Hepatocyte growth factor/scatter factor-Met signaling in tumorigenicity and invasion/metastasis. J Mol Med 1996;74(9):505–13.

[176] Fiorillo MT, Cabibbo A, Iacopetti P, Fattori E, Ciliberto G. Analysis of human/mouse interleukin-6 hybrid proteins: both amino and carboxy termini of human interleukin-6 are required for *in vitro* receptor binding. Eur J Immunol 1992;22(10):2609–15.

[177] Orimo A, Gupta PB, Sgroi DC, Arenzana-Seisdedos F, Delaunay T, Naeem R, et al. Stromal fibroblasts present in invasive human breast carcinomas promote tumor growth and angiogenesis through elevated SDF-1/CXCL12 secretion. Cell 2005;121(3):335–48.

[178] Craig M, Ying C, Loberg RD. Co-inoculation of prostate cancer cells with U937 enhances tumor growth and angiogenesis *in vivo*. J Cell Biochem 2008;103(1):1–8.

[179] Kuperwasser C, Chavarria T, Wu M, Magrane G, Gray JW, Carey L, et al. Reconstruction of functionally normal and malignant human breast tissues in mice. Proc Natl Acad Sci USA 2004;101(14):4966–71.

[180] Pedroza-Gonzalez A, Xu K, Wu TC, Aspord C, Tindle S, Marches F, et al. Thymic stromal lymphopoietin fosters human breast tumor growth by promoting type 2 inflammation. J Exp Med 2011;208(3):479–90.

[181] Aspord C, Pedroza-Gonzalez A, Gallegos M, Tindle S, Burton EC, Su D, et al. Breast cancer instructs dendritic cells to prime interleukin 13-secreting CD4+ T cells that facilitate tumor development. J Exp Med 2007;204(5):1037–47.

[182] Rongvaux A, Takizawa H, Strowig T, Willinger T, Eynon EE, Flavell RA, et al. Human hemato-lymphoid system mice: current use and future potential for medicine. Annu Rev Immunol 2013;31:635–74.

[183] Ramirez CN, Antczak C, Djaballah H. Cell viability assessment: toward content-rich platforms. Expert Opin Drug Discov 2010;5(3):223–33.

[184] Ng KW, Leong DT, Hutmacher DW. The challenge to measure cell proliferation in two and three dimensions. Tissue Eng 2005;11(1-2):182–91.

[185] Quent VM, Loessner D, Friis T, Reichert JC, Hutmacher DW. Discrepancies between metabolic activity and DNA content as tool to assess cell proliferation in cancer research. J Cell Mol Med 2010;14(4):1003–13.

[186] Walzl A, Unger C, Kramer N, Unterleuthner D, Scherzer M, Hengstschlager M, et al. The resazurin reduction assay can distinguish cytotoxic from cytostatic compounds in spheroid screening assays. J Biomol Screen 2014;19(7):1047–59.

[187] McMillin DW, Delmore J, Weisberg E, Negri JM, Geer DC, Klippel S, et al. Tumor cell-specific bioluminescence platform to identify stroma-induced changes to anticancer drug activity. Nat Med 2010;16(4):483–9.
[188] Maguire CA, Bovenberg MS, Crommentuijn MH, Niers JM, Kerami M, Teng J, et al. Triple bioluminescence imaging for *in vivo* monitoring of cellular processes. Mol Ther Nucleic Acids 2013;2:e99.
[189] Peterson JK, Houghton PJ. Integrating pharmacology and *in vivo* cancer models in preclinical and clinical drug development. Eur J Cancer 2004;40(6):837–44.
[190] Kerbel RS. Human tumor xenografts as predictive preclinical models for anticancer drug activity in humans: better than commonly perceived-but they can be improved. Cancer Biol Ther 2003;2(4 Suppl. 1):S134–9.

CHAPTER

BIOMARKER DEVELOPMENT IN TARGETING CANCER EPIGENETIC

5

Christophe Cisarovsky[1], Magdalena Benetkiewicz[2], Sandrine Faivre[3], Eric Raymond[3] and Armand de Gramont[4]

[1]Research Laboratory in Oncology, University of Lausanne, Centre Hospitalier Universitaire Vaudois, Lausanne, Switzerland [2]Groupe Coopérateur Multidisciplinaire en Oncologie, Paris, France [3]Medical Oncology, Department of Oncology, University of Lausanne, Centre Hospitalier Universitaire Vaudois, Lausanne, Switzerland [4]New Drug Evaluation Laboratory, Centre of Experimental Therapeutics and Medical Oncology, Department of Oncology, University of Lausanne, Centre Hospitalier Universitaire Vaudois, Lausanne, Switzerland

CHAPTER OUTLINE

5.1 INTRODUCTION

Since the completion of the human genome project, the search for tumor-driving mutations has reached its paramount level to drive anticancer research in the era of molecular medicine. There is currently a growing trend toward transcending the standard pathology classification to group tumors by their mutanomes. Many initiatives worldwide have endorsed this approach and started trans-organ gene-alteration-driven clinical trials. However, novel paradigms such as cancer epigenetics are nowadays challenging visions restricted only to the analyses of tumor genetic mutations. Epigenetics refers to an inheritable cellular phenotype resulting from changes in chromosomes without modifications of the DNA sequence [1]. DNA methylation and histone modification are the two main mechanisms of epigenetic modifications in cells. These modifications are essential to

G. Egger & P. Arimondo (Eds): Drug Discovery in Cancer Epigenetics. DOI: http://dx.doi.org/10.1016/B978-0-12-802208-5.00005-9

many critical cellular processes, for example, gene transcription or DNA repair with important consequences on cellular differentiation and development. Epigenetic deregulations have thus been associated with many diseases, including cancer, providing an alternative path to sequence alterations for the propagation of cancer cell tumorigenicity [2]. Links between genomic and epigenomic alterations to tumor development have been reinforced with the identification of several epigenetic regulators that are mutated in cancer cells [3]. In the process of tumor development, epigenetic modifications include silencing of tumor suppressor genes and sometimes enhancing oncogene expression [4,5]. The recent advances in deciphering the epigenomic landscape of tumors have led to better understanding of the mechanisms leading to cancer and promising new opportunities for its treatment. Targeting epigenetic factors is likely to have a global impact on tumor cells by potentially affecting hundreds of genes at the same time; in fact, while being mostly indolent to normal cells, epigenetic modulators are challenging tumor cell plasticity. Exploiting the epigenetic vulnerabilities of tumor cells has led to the development of several targeting strategies and drug development [6]. A few molecules have already reached clinical validation such as vorinostat and azacytidine in hematology, but and several new therapeutics are currently in development.

Disease characterization, drug development, improved therapeutic management, and development of biomarkers became essential tools to generate relevant information. In this chapter we outline the need for biomarkers focusing on target engagement and predictive biomarkers, as well as the challenges of biomarker evaluation methodologies, including DNA methylation, and from bench to clinic validation.

5.2 WHY DO WE NEED BIOMARKERS?

In its war against cancer, modern medicine has acknowledged that most cancers will not be cured with a single drug despite being highly targeted and that the 10 hallmarks of cancer represent as many paths for cancer cells to escape different treatments [7]. Rethinking the approach to treat a heterogeneous and ever-evolving disease has initiated a strong movement to understand tumors in the vicinity of their peculiar microenvironment, encompassing the complexity of tumorigenesis in individual patients, and the will of personalizing therapeutics. The benefits are potentially numerous for patients, practitioners, and healthcare providers. To this end it is necessary to use relevant disease-, drug-, and treatment-dependent indicators, or biomarkers, and to measure the relevance of any given therapeutic strategy. Biomarkers are measurable indicators that objectively distinguish normal from pathological states or the response to a specific therapeutic intervention [8]. Biomarkers are essential tools from early drug development to monitor target engagement and explicit drug-dependent mechanisms of action at the preclinical stage, to the later clinical development to monitor drug-related tumor response and toxicity, or to predict drug efficacy (Table 5.1). In fact, biomarkers are used for numerous purposes and can be based on very diverse parameters from radiological imaging (e.g., tumor size or density according to RECIST (Response Evaluation Criteria in Solid Tumors) or Choi criteria) and physiological indicators (e.g., prostate-specific antigen or carcinoembryonic antigen levels) to cellular (e.g., undifferentiated status of tumor cells), molecular (e.g., HER2 overexpression), and genetic (e.g., KRAS, BRAF, EGFR mutation status) factors.

Table 5.1 Biomarkers Types and Method Examples

In Vivo Biomarkers Type	Purposes	Example of Methods
Pharmacodynamic (proximal or distal)	• Drug-dependent pharmacological effects on the body • Target engagement • Drug-dependent mechanism of action	• IHC, PCR, and ELISA-based methods • Mass spectrometry assays • Thermal shift assays
Pharmacokinetic	• Body-dependent effects on drug fate • ADME (absorption, distribution, metabolization, excretion)	• Mass spectrometry assays • DCE-MRI-based methods
Prognostic	• Disease/tumor-related outcome	• Tumor imaging-based methods • IHC, PCR, and ELISA-based methods
Monitoring	• Drug-related toxicities • Drug-related tumor response	• Tumor imaging-based methods • PCR-based methods from liquid biopsies • ELISA-based methods
Predictive (companion biomarker)	• Identification of patients who are likely or not to respond to treatment	• IHC, PCR, and ELISA-based methods
Surrogate	• Prediction of outcome according to treatment response	• Tumor imaging-based methods • IHC, PCR, and ELISA-based methods

ELISA, enzyme-linked immunosorbent assay; IHC, immunohistochemistry; PCR, polymerase chain reaction; PS, performance status; DCE-MRI, dynamic contrast-enhanced magnetic resonance imaging.

Prognostic biomarkers are useful in anticipating the likely course of the disease irrespective of the treatment. The other types of biomarkers are linked to the treatments. Pharmacodynamic (PD) biomarkers are related to the action of the molecule on the body. They can be divided into proximal and distal PD biomarkers depending on the "distance" they are from the mode of action of the molecule; proximal PD biomarkers will directly or indirectly evaluate target engagement, whereas distal PD biomarkers will evaluate the biological consequences of hitting the target. Predictive biomarkers, also called companion biomarkers, are very useful in predicting treatment efficacy, while surrogate biomarkers are used to anticipate therapeutic endpoints such as progression-free and overall survival. Finally, monitoring biomarkers are useful in following the disease and/or treatment course and have important applications in daily clinical practice in the management of disease (to plan medical interventions) and its treatment (to pursue or discontinue treatment, adapt dosage, etc.).

The same biomarker can have multiple usages, such as HER2 expression in breast cancer. For instance, overexpression of HER2 may be regarded as a biomarker of poor prognosis as well as a predictive biomarker of response to the anti-HER2 therapeutics, trastuzumab, and TDM-1 (trastuzumab emtansine).

Considering the early stage of drug development, and especially the preclinical stage, the most important biomarkers will be those that are clearly strengthening the proof of mechanism of the molecule. Proximal PD biomarkers highlight target engagement, that is, they show that the new drug binds to a specific receptor and enzyme or closely influences their functions and immediate downstream biochemical consequences. At this stage, distal PD biomarkers are also necessary to evaluate the biological relevance of targeting a specific mechanism in terms of biological consequences that can be pertinent in treating tumors while sparing normal cells. These biological consequences may strongly vary (inhibiting proliferation or invasion, inducing apoptosis or necrosis, affecting cell metabolism, differentiation, etc.) and, besides tumor cells, involve other cellular components of the tumor microenvironment (angiogenesis, immune regulation, etc.). Evaluation of these disease biomarkers may necessitate advanced preclinical models (*in vivo* testing) or even early clinical evaluation to help reach the proof of principle of the new molecular entity.

For the success of personalized medicine, predictive biomarkers are perhaps the most important type of biomarker to ensure that patients receive the best therapeutic options in terms of immediate efficacy and toxicity but also in terms of global disease management and therapeutic strategy, thus improving the overall health cost burden. As an example, despite concern about the cost-effectiveness of anti-EGFR therapies, KRAS and BRAF testing substantially reduce the costs of anti-EGFR treatments by helping select the right patient population for monoclonal antibodies directed to EGFR [9,10]. Nevertheless, "breakthroughs" in clinical adoption of biomarkers are rare. While 30–50% of drugs currently in development are coupled with companion biomarkers, only a tenth of these biomarkers will reach the clinic [11], highlighting the challenge of developing companion biomarkers and the efforts to put in place to improve this transition. Indeed, biomarker development is a multistep process that would benefit from a fit-for-purpose strategy and rigorous evaluation both in method and clinical validation. Pitfalls frequently identified to explain the attrition rates in the development of biomarkers are frequently linked to the lack of knowledge of pharmaceutical and biotech institutions in clinical and regulatory requirements for biomarker development.

5.3 MONITORING TARGET ENGAGEMENT

Following *in vitro* binding and selectivity assays based on biochemical methods from purified proteins, it is necessary to develop assays and biomarkers that will assess target engagement in cells, tissues from preclinical animal models, and eventually from patients' derived tumors or surrogate tissues. PD biomarkers measure and monitor drug–target interactions and their immediate biological consequences. Proximal PD biomarkers faithfully reporting target engagements provide clues on drug efficacy when measuring its downstream biological consequences through distal PD biomarkers (e.g., apoptosis induction, growth inhibition that may be further correlated to tumor response), or drug toxicities whether there are directly linked to the main intended targets or associated with off-target activities. Both proximal and distal PD biomarkers are thus critical tools in evaluating the potential of new molecules and enhancing drug development success. However, it is not straightforward to measure target engagement in cells or tissues also because *in vivo* target engagement might be quite different from the prediction of *in vitro* experiments [12].

A number of indirect qualitative or semiquantitative methods already exist and several are emerging to evaluate target engagement in cells and tissues. Early on, the measurement of target engagement is often based on evaluation of modifications that affect the substrate or the product as exemplified by enzyme-targeted drugs (substrate phosphorylation or product generation when targeting kinases or proteases for instance) for which enzyme-linked immunosorbent assay (ELISA), radioligand, sodium dodecyl sulfate-polyacrylamide gel electrophoresis, or immunohistochemistry-based assays are conventional methods. For drugs targeting kinases, emerging methods, such as the one presented by Paweletz et al., use a differential phosphoproteomic approach to measure target engagement of Ser/Thr protein kinase by monitoring and quantifying changes in novel phosphorylated residues outside the conserved T-loop residues [13]. Other mass spectrometry approaches, such as the kinobeads matrix methodology, can evaluate drug binding and downstream signaling effects by measuring the consequences of drug exposure to a captured subproteome made of the majority of the human kinome and several hundred associated proteins [14]. Similar or alternate methods analyzing the native kinase binding profile of drugs by differential proteomic analyses may be amenable to the evaluation of target engagement in animal models and tissues from patients [15,16].

Another, and complementary, promising approach uses properties of ligands to thermally stabilize their targets to evaluate target engagement and monitor other drug-related processes in cells and tissues [17–19]. Cellular thermal shift assays are inspired from the protein thermal melting curves and use the ability of a compound to specifically shift the melting curve when bound to its target. It is therefore possible to generate specific melting curves for each target and compound to assess drug–target engagement. Moreover, by comparing the thermal profiles obtained after drug treatment of intact cells versus cell extracts, these methods could discriminate the effects directly linked to target binding from those induced by downstream signaling [18].

Radioligands or radiotracers, very precise imaging methods, such as positron emission tomography, may also be useful PD biomarkers in determining the rate and extent of radioligand delivery to tissues or quantifying drug target engagement through the design of target-specific radioligands [20]. Substrate and activity-based probes using fluorescent signal have also been successfully used to visualize and measure drug-specific biological efficacy by measuring protease activity such as caspase activation for drug-induced apoptosis [21,22].

Although it seems necessary to link *in vitro* target engagement assays to clinical biomarkers, these assays will certainly be different with the introduction of many new preanalytical variables when moving to latter phases of development and the extraordinary challenge of precisely measuring the imprecise. In the process, expensive methodologies could be cost-effective by accelerating drug development and thereby save time-related costs of clinical trials by readily selecting the right drug to the right patient. On the other hand, in routine clinical practice it seems legitimate to search for simple, cheap, and reliable methods to evaluate PD biomarkers. Despite these hurdles the development of target engagement biomarkers, from *in vitro* to *in vivo* animal models and then to clinical applications, is crucial to feed go/no-go decisions but also for individual patients to optimize drug delivery, therapeutic strategies, and eventually improve patient clinical outcomes.

5.4 PREDICTIVE BIOMARKERS

Predictive biomarkers are essential tools with regard to personalized medicine and health economics, and are crucial to improve the success rate of novel therapies. To date, most validated predictive biomarkers are based on genetic alterations (HER2 amplification, *KIT, KRAS, BRAF, EGFR* mutations, EML4-ALK translocation, COL1A1-PDGFB fusion, and BCR-ABL translocation).

These biomarkers are positively identifying tumor growth drivers (such as HER2, KIT, ALK, and BRAF) and/or negatively predicting tumor efficacy by highlighting specific drug resistance mechanisms (such as KRAS).

For instance, BRAF mutations have been identified in up to 60% of all metastatic melanomas, most of them being V600E mutations [23]. This alteration leads to oncogenesis via unregulated MAPK signaling, and offers a rational therapeutic strategy [24]. The use of vemurafenib and dabrafenib, inhibitors of MAPK signaling in patients with melanoma and the *BRAF* V600E mutation, has been associated with improved responses and survivals [25,26]. Another routinely measured predictive biomarker is HER2 in breast cancer, which is amplified in up to 25% of breast cancers and mandatory in selecting patients for treatment with anti-HER2 therapy [27]. The addition of trastuzumab to chemotherapy in patients with advanced *HER2*-positive breast tumors was found to prolong time-to-disease progression and survival, as well as to increase the response rate [28]. Lung cancers with ALK translocations rely on ALK kinase activity as their oncogenic driver, and most are highly sensitive to therapies with ALK-targeted inhibitors. The ALK inhibitor crizotinib resulted in a response rate of ~60% and a long progression-free survival of ~10 months in a lung cancer setting [29]. C-KIT is expressed in all gastrointestinal stromal tumors and activating mutations of *c-KIT* are found in the vast majority of them (80%) [30]. The c-KIT inhibitor imatinib proved effective against *c-KIT*-driven gastrointestinal tumors, producing a 50–75% objective response rate in phase II and III clinical trials [31,32]. KRAS mutation status is a strong predictive biomarker negatively predicting success of anti-EGFR therapies [33]. Several studies in metastatic colorectal cancer patients have reported that patients whose tumors have KRAS mutations do not benefit from treatment with anti-EGFR monoclonal antibodies panitumumab or cetuximab [34,35].

As shown through these examples, tumor biology is essential for the identification of biomarkers, often from preclinical experimentation to their implementation into clinical practice. However, constant scrutiny is necessary when considering biological parameters; in fact, tumor biology is often more complex than a simplistic pathway scheme, as are methods in evaluating biomarkers that are often difficult to standardize and make ultimate clinical trial validation quite challenging (see Sections 5.5 and 5.6).

Despite some feats and encouraging results, development of biomarker-based therapies confronts important hurdles. Among them is insufficient knowledge of the mechanism of action of agents and their multifaceted nature, often acting at various levels on multiple targets. For example, on the basis of trastuzumab's mechanism of action, HER2 overexpression was expected to predict benefit from trastuzumab, however this apparent candidate biomarker turns out to be ambiguous. HER2 testing for HER2 amplification either by fluorescence *in situ* hybridization or immunochemistry [36], became benchmark criteria for selecting HER2-positive tumors, thus selecting patients likely to benefit from trastuzumab therapy. However, quantitative measurements of HER2 were challenged in predicting benefits of adjuvant trastuzumab. Two trastuzumab trials, the NSABP B-31 [37] and the NCCTG N9831 [38], evaluated the predictive value of HER2 status in women with HER-positive breast cancer in an adjuvant setting. Approximately 5–10% of tumors in each trial were HER2-negative. However, the efficacy of adjuvant trastuzumab added to anthracycline-/taxane-based adjuvant chemotherapy benefited not only for HER2-positive breast cancer but also for HER2-negative breast cancer [39]. Another example is erlotinib, a small molecule anti-EGFR inhibitor. Data from anti-EGFR monoclonal antibodies suggested that anti-EGFR therapy was ineffective against *KRAS* mutant colorectal cancer [40–42]. However, results of the phase III

DREAM trial demonstrated that the addition of erlotinib to bevacizumab after bevacizumab-based induction chemotherapy resulted in statistically significant survival benefit and that benefit from erlotinib was not governed by *KRAS* tumor status [43].

Another hurdle of biomarker-based therapies is that the efficacy of targeted therapies might be related to the specific genetic background of the tumor. For example, data on efficacy and resistance to BRAF pathway inhibitor vemurafenib suggested important differences in the biology of *BRAF*-mutant tumors in different cancer types. In fact, patients with colon cancer treated with vemurafenib show only a very limited response (5%) compared with that achieved in melanoma patients (80%) for which the *BRAF V600E* mutation may be considered as predictive biomarker [44]. Corcoran et al. nicely showed that colon cancer insensitivity to BRAF inhibition was due to a rapid EGFR-dependent reactivation of the MAPK pathway [45]. Consequently, the strategy of combined BRAF/MEK inhibition was tested in *BRAF*-mutant colorectal cancer. However, the effectiveness of this concomitant inhibition using dabrafenib and trametinib was limited in colorectal cancer, unlike in the BRAF-mutant metastatic melanoma that has produced a dramatic response rate with the combination of both drugs. Additional studies suggested that the MAPK pathway has to be adequately inhibited in order to mediate the response in metastatic colorectal cancer [46–48].

As shown, predictive biomarkers are very important to drug development but subjected to various challenges. Among these challenges, methods and clinical validation are further developed in the next sections.

5.5 CHALLENGES IN METHOD VALIDATION

The implementation of biomarkers into clinical practice presents several challenges, especially related to standardization of biological, clinical, and logistical criteria across worldwide medical practice. In the process of biomarker development, robust laboratory methodologies are necessary at all analytical phases, from preanalytical (sample definition, handling, and processing) to analytical (data and quality control recording), and postanalytical (data reporting and interpretation) [49].

5.5.1 FROM BENCH TO BEDSIDE

Despite the extensive efforts in cancer biomarker discovery over the last years, only a handful of the numerous known and initially promising biomarkers have been robustly validated and translated into major clinical successes. There are several reasons for so-called "biomarker breakthroughs" failures. The most important challenges in the clinical acceptance of a biomarker, that is, in identifying its true value in a clinical setting, are mostly related to inadequate performance or to false discovery and to the lack of high-quality well-controlled clinical trial data due to study design, interpretation, and statistical deficiencies [50]. In addition, there are multiple other reasons that limit effective translation from biomarker discovery into routine clinical practice. These often originate from a lack of understanding of disease biology, absence of correlation between putative biomarkers and clinical endpoints, substantial patient and tumor heterogeneity, limited understanding of optimum analytical, diagnostic and regulatory requirements for biomarker validation, and the lack of standardized data guidelines.

Poor knowledge of the underlying mechanisms of the drug or the disease and thus criteria required for patient stratification may result in inefficient development of robust biomarkers and misconceptions of the drug potential for safety and efficacy. In fact, a single biomarker may be insufficient to predict a disease-relevant response and a single molecule may have different clinical activities than those expected, such as crizotinib, an ALK inhibitor, initially developed as a MET inhibitor or sorafenib, initially developed as a RAF kinase inhibitor and subsequently used as an antiangiogenic drug due to its activity on VEGF and PDGF receptors.

One ongoing significant hurdle in the current biomarker-driven decision-making is the fact that biomarker discoveries based on a "one-size-fits-all" strategy upon the population average are translated into clinical decisions for unique individual patients. However, finding a marker for a disease does not always mean that everyone with those characteristics can be treated with the same drug or drug combination. Thus, single biomarker approaches are increasingly regarded as being susceptible to misdirecting treatment options without a deeper understanding of the biomarker in its specific context. In fact, EGFR overexpression for anti-EGFR therapies, BRAF mutation for anti-BRAF therapies, and anti-HER2 phase III clinical trials based on a "one-size-fits-all" strategy have sometimes been disappointing [51,52]. This challenge is particularly relevant in the light of inter- and intra-patient tumor heterogeneity [53–55].

Clinical trial endpoints are not always easy to define or quantify, which adds to the limitations of assessment of putative biomarkers compared to clinical outcome measures. International initiatives, such as the Definition for the Assessment of Time-to-event Endpoints in CANcer trials (DATECAN) program, are currently ongoing to develop standardized definitions of commonly used time-to-event endpoints, enabling appropriate comparison of future clinical trials.

Although efforts have been made for the adoption of standard approaches in biomarker development processes [56–58], this continues to be the major limitation to studies investigating biomarkers. In the context of the development of clinically validated biomarkers, an early validation of analytical methods to the rigorous standards is critical to the development of robust biomarkers that can be taken forward with realistic prospects of biomarker study success. Moreover, appropriate study design, description, selection of study population, and flawless interpretation of the results have a major impact on whether a biomarker can be considered ready for clinical routine use.

Overall, the path of a biomarker from the bench to the clinic is complex and challenging, therefore every step needs to be properly planned and executed to enable successful biomarker-driven trials.

5.5.2 KEY CONCEPT: FIT-FOR-PURPOSE METHODOLOGY

In the biomarker validation process every effort should be made to ensure that a biomarker is relevant for its intended use—a principle commonly referred to as "fit-for-purpose" [59–62]. This idea is derived from the fact that depending on the intended purpose of a biomarker and the nature of the analytical methods used to generate the data, the validation objectives and levels of validation requirements will probably be different [60,63].

The fit-for-purpose method validation is an umbrella terminology for distinct stages of the validation process [64], including the first and probably the most critical preanalytical validation stage (definition of purpose and selection of the preanalytical variables, and analytical methods), exploratory purpose stage (the basic assay performance), in-study validation stage (further assessment of fitness-of-purpose and the robustness of the assay in the clinical context and the establishment of definitive acceptance criteria), and a more complete validation for advanced application stages (the formal performance of the assay, e.g., method robustness, extended specificity, and sensitivity) [60,61]. Consequently, method validation is an ongoing, iterative process, which critical steps lean on the predefined purpose of the biomarker. Typically, the objectivity of method validation should increase from the initial discovery/validation to more advanced validation and should be customized according to biomarker type, the intended use of the biomarker application (pharmacodynamics, monitoring, prognostic, predictive, surrogate, etc.), variability, and prevalence. Stringency in biomarker method validation also needs to address the type of analytical methods, the purpose of the clinical trial, the type of process-related data generated, and the intended use of the data [60]. Such a strategy allows for efficient biomarker and drug development with proper gathering of supportive information.

To aid generating a biomarker method validation plan, a biomarker assay can be of various functional categories, each requiring a distinct level of validation [61,65]. A definitive quantitative assay (e.g., mass spectrometry assays) makes use of calibrators and a regression model to calculate absolute quantitative values for unknowns (reference standard is fully characterized and representative of the biomarker); a relative quantitative assay (e.g., ELISAs) uses a response–concentration calibration with reference standards that are not fully representative of the biomarker (precision can be validated but accuracy only estimated); quasi-quantitative assay (e.g., a real-time quantitative reverse transcription PCR (qRT-PCR)-based assay) that does not employ a calibration standard; and qualitative assay (including most immunohistochemical assays) [60,61]. Qualitative assays only require proof of sufficient sensitivity and specificity to detect the target analyte that varies from drug to drug and from indication to indication, while quantitative methods additionally require an evaluation of accuracy [65]. Moreover, there are significant tradeoffs between platforms for a predictive biomarker assay, and some platforms are more technically difficult to validate than others.

In light of the high drop-out rate in biomarker development, a fit-for-purpose and robust strategy for method validation may enhance the ability to assess whether drug candidates are promising early in development, to allow a more economical and timely application of developmental resources and progress of biomarker assays, and to improve the cost-effectiveness of drugs by targeting treatment to those patients with the best balance of risk and benefit. Finally, such a strategy may ultimately lead to quick regulatory approval for the use of these biomarkers in a defined clinical context. Together, a "right" fit-for-purpose biomarkers strategy should be an essential step to high-quality drug development for each drug, and not only an anchor to clinical development.

5.6 CLINICALLY VALIDATING BIOMARKERS

A necessary, sound understanding of tumor biology and mechanisms of action of the therapeutics cannot substitute for the formal validation of biomarkers through properly designed well-controlled clinical trials and data analysis, including relevant statistical assessment.

By definition a prognostic marker identifies patients with differing risks (high/low) of a specific outcome with respect to no treatment or a standard treatment, therefore its validation is straightforward as it is associated with the disease or the patient and not with a specific therapy. In terms of statistical identification this does not require specific study design and data can be derived from small retrospective studies of uniformly treated patients with the standard drugs (or placebo if that is the standard) [66–68]. It is, however, recommended that reliable conclusions on the usefulness of a prognostic biomarker, for medical decision-making or planning of clinical trials, should more likely rest on randomized, prospective clinical trials specifically for patients in whom the optimal treatment approach is uncertain from clinicopathological parameters. In fact, studies of prognostic factor are often included within studies evaluating new therapeutic strategies. Although such trials have well-defined patient populations, study treatment, and complete follow-up information on the primary study endpoint, prognostic biomarker validation in terms of statistical association can still be quite challenging. A highly significant *P* value and large effect size (the odds, risk, or hazard ratio) do not necessarily mean that the prognostic signature or the biomarker will be of clinical utility, as it does not imply that the predictive accuracy is sufficient to justify the clinical use of the biomarker. For example, in a study testing the prognostic value of 70 genes (MammaPrint™ gene signature) [69] in breast cancer of women with 1–3 involved lymph nodes, the positive predictive accuracy of the signature was 0.63 (about two-thirds of the patients with a poor-prognosis signature would be expected to develop metastases within 5 years) and the negative predictive value was 0.9 (only one patient in 10 with a good-prognosis signature would be expected to develop metastases within 5 years). However, such a molecular signature prediction may be difficult to translate into a treatment decision since close to 40% of patients with a poor-prognosis signature will not develop metastases. Thus, a large effect size and/or a highly significant *P* value may not be a proper measure of predictive accuracy [70–72]. In addition, identification of a potential prognostic biomarker should be followed by achievement of a broad consensus on validation in a multicenter setting or in cross-validation experiments.

A predictive biomarker stratifies patients based on the likelihood of patient response (the efficacy and/or toxicity) to a particular treatment, for example, biomarker-positive and -negative subgroups. Robust statistical validation of predictive markers is complex and can present substantial challenges. Specifically, it requires data from a high-quality randomized controlled trial (RCT) that includes patients with both high and low levels of the biomarker. Although prospectively designed clinical trials are the gold standard approach to validating a predictive marker, when it is ethically or logistically impossible (large trial and long time to complete), well-designed retrospective analyses of completed RCTs, a so-called retrospective-prospective approach, may be sufficient to accumulate evidence of marker effectiveness more rapidly. For example, in advanced colorectal cancer, KRAS and NRAS were identified retrospectively as predictors of the effectiveness of the EGFR-directed monoclonal inhibitors panitumumab and cetuximab [34,35,42,73–75]. However, to yield convincing evidence, so-called retrospective–prospective study design must be predefined, that is provide analytical and statistical details (including hypotheses, sample size, cut-points, scoring system for classifying risk-groups, interpretation of statistical significance, and estimation of effect, etc.), and conducted with reference to standardized procedures and protocols. In particular, if such a retrospective validation can be demonstrated in data from different independent RCTs, this will provide strong evidence for a predictive effect [76].

A biomarker has great potential predictive value when it could be concluded that biomarker-positive patients would have better outcome than biomarker-negative patients. In this scenario, an RCT based on a marker by treatment interaction design (also known as an all-comers design or stratified design) is clearly the best option as it provides the highest (definitive) level of evidence [71,76–79]. In this design, the marker status is used as a stratification factor and randomly assigns all patients to treatments under investigation within biomarker-positive and -negative subgroups, so the efficiency of testing to detect differences between the groups is maximized. In addition, only patients with a valid marker result are allowed to be randomized, and the sample size is prospectively calculated to provide adequate power to test for a different treatment effect within each marker-based subgroup [76]. A specific limitation of interaction design is that interaction analyses require a large sample size for reliably assessing an interaction compared with testing a treatment effect in biomarker-positive or -negative patients [80]. Therefore, the number of interaction trials capable of validating a predictive biomarker is likely to be small. The selection design (enrichment or targeted design) is often used as an alternative approach [77,81,82]. This design is based on the paradigm that any treatment benefit is limited to a subgroup of patients who either express or do not express a specific biomarker, and thus only patients who are biomarker-positive enter the validation trial and are treated with a new agent. Such an approach is specifically interesting if the number of truly benefiting patients is expected to be fairly small. In this instance, successful enrichment can improve the efficiency of a trial design by increasing the power and decreasing the required sample size or duration [62,71,83]. Although trials can confirm the use of the biomarker for identifying those who will benefit from treatment [25,84–86], they cannot imply that the marker is truly predictive, as they provide no data on the treatment effectiveness, if any, on biomarker-negative patients. In such situations, patients lacking the biomarker but achieving response could be overlooked, as was shown to be the case of the anti-HER2 monoclonal agent trastuzumab for breast cancer in two RCTs that used enrichment designs [37,38].

To clinically validate a biomarker is challenging, but the expectations are high for patients and drug developers and thus deserve to be carefully planned and executed.

5.7 THE CHALLENGES OF CANCER EPIGENETICS FOR BIOMARKER DEVELOPMENT

To date, there is no validated FDA-recognized epigenetic biomarker as a companion to approved drugs. However, extensive researches are in progress and there is no doubt that epigenetic biomarkers will gain more and more focus and progressively enter routine clinical practice. Some DNA methylation-based biomarkers are already used as diagnostic and predictive biomarkers and a few are available as commercial clinical test kits. A comprehensive description of epigenetic biomarkers will be given in Chapter 17. The following section focuses on DNA methylation strategies for biomarker development.

5.7.1 STRATEGIES FOR DNA METHYLATION BIOMARKERS

DNA methylation profiling has become a major approach in biomarker development, providing a range of applications such as early cancer detection, diagnosis, assessment of tumor prognosis, therapeutic stratification, and post-therapeutic patient follow-up [87–90]. CpG methylation pattern could improve breast cancer classification or early lung cancer diagnosis [91,92]. Some predictive epigenetic biomarkers have already been introduced to the clinic, such as MGMT gene promoter methylation as a predictive biomarker for alkylating agents, like temolozomide, in the treatment of glioblastoma multiform [89,93]. Methylation of the TFAP2E gene in colorectal cancer or methylation of hMLH1 gene promoter in epithelial ovarian cancer has been associated with resistance to chemotherapy [94,95].

Typically, methods for DNA methylation analysis include three main categories: methylation-sensitive restriction enzymes digestion, affinity-based, and the most widely used bisulfite modification of genomic DNA [96,97]. Methods based on restriction enzyme digestion, such as genome-scale approaches (e.g., restriction landmark genome scanning (RLGS) [98,99], amplification of inter-methylated sites (AIMS) [100]), array hybridization techniques (e.g., methylated CpG island amplification (MCA) [101], differential methylation hybridization (DMH), HpaII tiny fragments enrichment by ligation-mediated PCR (HELP) [102], microarray-based methylation assessment of single samples (MMASS) [103]), and sequencing approaches (e.g., methylation-sensitive cut counting (MSCC) [104] or Methyl-seq [105]) have been developed. Methylation-sensitive restriction enzyme digestion followed by PCR allows accurate methylation testing. However, given that the enzymatic digestion might be incomplete, especially when low-quality DNA samples are used, the risk of encountering false-positive results is increased. Moreover, methylation of only a few CpG dinucleotide sites per locus may be assessed by this method.

Affinity-enrichment-based technologies such as methylated DNA immunoprecipitation (MeDIP) [106,107] and methylated CpG island recovery assay (MIRA) [108,109] have proven to be useful for DNA methylation analyses. These assays allow for rapid and specific genome-wide assessment of DNA methylation, but require high DNA input, substantial adjustments for varying CpG density at different DNA regions, and do not provide DNA methylation information at singlet CpG dinucleotide resolution. In addition, they have bias toward the sequences surrounding the methylated CpG islands and CpG-dense sites at different part of the genome. Therefore, repeat sequences are sometimes over-represented in affinity-based analysis. These techniques are now shifting toward analysis by next-generation sequencing (NGS) techniques, which provide comprehensive, high-throughput, and highly sensitive DNA methylation mapping with relatively low-input DNA that is free of amplification artifacts.

The majority of approaches for DNA methylation detection utilize bisulfite-treated genomic DNA as the gold standard method to provide information about the DNA methylation changes within the genome [110,111]. The advantages of this analysis include high accuracy, sensitivity, rapid detection, suitability for genome-wide high-throughput methylation analyses, single CpG resolution, and detection of strand-specific methylation patterns [112]. However, few hurdles for reliable DNA methylation detection are encountered with these assays such as significant DNA degradation (up to 96%) [113] and reduced sequence complexity due to the conversion process, incomplete denaturation before bisulfite modification, possible artifacts by incomplete bisulfite conversion, and differential PCR efficiency for methylated versus unmethylated versions of the same sequence [114]. Bisulfite-converted DNA has been adapted, for example to array hybridization (e.g., bisulfite methylation profiling (BiMP) [115], GoldenGate BeadArray [116], Illumina Infinium [117]) and sequencing approaches (e.g., reduced representation bisulfite sequencing

(RRBS) [118], bisulfite padlock probes (BSPP) [119,120], whole genome shotgun bisulfite sequencing (WGSGS) [121,122], single-molecule real-time sequencing (SMRT) [123], and NGS [124]).

An alternative approach for methylation analysis is matrix-assisted laser desorption ionization time-of-flight mass spectrometry (MALDI-TOF-MS) [125]. This approach is implemented in the EpiTYPER platform that is applied for quantitative base-specific methylation analysis of specific loci [125]. Although most often it is applied to obtain candidate gene methylation analyses, it can also be a suitable strategy for biomarker discovery.

DNA methylation biomarkers are promising and valuable biomarkers. There are several advantages of using CpG methylation as biomarkers since they are specific to tumors, stable over time from various sampling sources (blood, fresh tissues, or paraffin-embedded materials), and easy to detect on actual platforms. However, despite availability of many different methodologies for DNA methylation-based biomarker discovery, only a small fraction of these methods has been implemented into routine clinical practice [126]. The main reasons for this slow implementation rate are the lack of agreements on optimal methodology as well as the lack of uniform standards of methodology, detection sensitivity thresholds, and reference standards for use in DNA methylation detection. Due to the potential global effects of epigenetics, change in methylation of a single gene or expression of a single micro-RNA, for instance, is unlikely to be a strong biomarker except in a few cases. Integrating multiple biomarkers from different biological processes to yield the required specificity and sensitivity creates a complex dataset to be analyzed, and provides a challenge to integrate data streams originating from different technologies. Unbiased methodology and data analysis approaches are therefore essential to hold promise for highly specific epigenetic biomarkers.

5.8 CONCLUSION

There is no doubt that the development of biomarkers is an essential step in the path of true personalized medicine for the future. From the preclinical stage to final clinical development, all investigators are using biomarkers. However, as discussed above, biomarkers have different purposes, which have consequences on their own usage in terms of method development and validation as well as level of proof. Indeed, target engagement biomarkers at the preclinical stage are usually designed to answer specific, yet critical, questions. However their intended use is not to be immediately transposable to the clinical setting. Many variables, preanalytical or analytical, may be very different from the preclinical stage to the clinical situation, in which the methods for biomarker evaluation could be specifically adapted. For instance, some new drugs are aimed at epigenetic modifications in very specific cell populations, such as stem cells, which may need specific method development. In the process of drug development, associated biomarkers, whether they are PK/PD, tailoring or predictive, may necessitate that methods are fitted to their intended use and rely on highly robust and harmonized standards at all stages. To this end, there should be a bidirective connection between preclinical and clinical implementation of methods and readouts to ultimately support the cost-utility of biomarkers, that is, their implementation and adoption in development and clinical decision-making.

REFERENCES

[1] Berger SL, Kouzarides T, Shiekhattar R, Shilatifard A. An operational definition of epigenetics. Genes Dev 2009;23:781–3.

[2] Birgisdottir V, Stefansson OA, Bodvarsdottir SK, Hilmarsdottir H, Jonasson JG, Eyfjord JE. Epigenetic silencing and deletion of the BRCA1 gene in sporadic breast cancer. Breast Cancer Res 2006;8:R38.

[3] Shen H, Laird PW. Interplay between the cancer genome and epigenome. Cell 2013;153:38–55.

[4] Berdasco M, Esteller M. Aberrant epigenetic landscape in cancer: how cellular identity goes awry. Dev Cell 2010;19:698–711.

[5] Ehrlich M. DNA hypomethylation in cancer cells. Epigenomics 2009;1:239–59.

[6] Mair B, Kubicek S, Nijman SMB. Exploiting epigenetic vulnerabilities for cancer therapeutics. Trends Pharmacol Sci 2014;35:136–45.

[7] Hanahan D. Rethinking the war on cancer. Lancet 2014;383:558–63.

[8] Group BDW. Biomarkers and surrogate endpoints: preferred definitions and conceptual framework. Clin Pharm Ther 2001;69:89–95.

[9] Behl AS, Goddard KAB, Flottemesch TJ, Veenstra D, Meenan RT, Lin JS, et al. Cost-effectiveness analysis of screening for KRAS and BRAF mutations in metastatic colorectal cancer. J Natl Cancer Inst 2012;104:1785–95.

[10] Blank PR, Moch H, Szucs TD, Schwenkglenks M. KRAS and BRAF mutation analysis in metastatic colorectal cancer: a cost-effectiveness analysis from a Swiss perspective. Clin Cancer Res 2011;17:6338–46.

[11] Huriez A. Personalized medicine, introduction, business impact on the healthcare sector and regulatory aspects; 2013. Autumn Assembly of EPSA, Valentia, October 2013. Available at: http://www.epsa-online.org/doc/aa13/Symposium%20-%20Alain%20Huriez%20-%20Regulatory.pdf.

[12] Basilico C, Pennacchietti S, Vigna E, Chiriaco C, Arena S, Bardelli A, et al. Tivantinib (ARQ197) displays cytotoxic activity that is independent of its ability to bind MET. Clin Cancer Res 2013;19:2381–92.

[13] Paweletz CP, Andersen JN, Pollock R, Nagashima K, Hayashi ML, Yu SU, et al. Identification of direct target engagement biomarkers for kinase-targeted therapeutics. PLoS One 2011;6:e26459.

[14] Bantscheff M, Eberhard D, Abraham Y, Bastuck S, Boesche M, Hobson S, et al. Quantitative chemical proteomics reveals mechanisms of action of clinical ABL kinase inhibitors. Nat Biotechnol 2007;25:1035–44.

[15] Patricelli MP, Nomanbhoy TK, Wu J, Brown H, Zhou D, Zhang J, et al. *In situ* Kinase profiling reveals functionally relevant properties of native kinases. Chem Biol 2011;18:699–710.

[16] Sharma K, Weber C, Bairlein M, Greff Z, Kéri G, Cox J, et al. Proteomics strategy for quantitative protein interaction profiling in cell extracts. Nat Methods 2009;6:741–4.

[17] Molina DM, Jafari R, Ignatushchenko M, Seki T, Larsson EA, Dan C, et al. Monitoring drug target engagement in cells and tissues using the cellular thermal shift assay. Science 2013;341:84–7.

[18] Savitski MM, Reinhard FBM, Franken H, Werner T, Savitski MF, Eberhard D, et al. Tracking cancer drugs in living cells by thermal profiling of the proteome. Science 2014;346:1255784.

[19] Vedadi M, Niesen FH, Allali-Hassani A, Fedorov OY, Finerty PJ, Wasney GA, et al. Chemical screening methods to identify ligands that promote protein stability, protein crystallization, and structure determination. Proc Natl Acad Sci 2006;103:15835–40.

[20] Matthews PM, Rabiner EA, Passchier J, Gunn RN. Positron emission tomography molecular imaging for drug development. Br J Clin Pharmacol 2012;73:175–86.

[21] Edgington LE, Berger AB, Blum G, Albrow VE, Paulick MG, Lineberry N, et al. Noninvasive optical imaging of apoptosis by caspase-targeted activity-based probes. Nat Med 2009;15:967–73.

[22] Edgington LE, Verdoes M, Bogyo M. Functional imaging of proteases: recent advances in the design and application of substrate-based and activity-based probes. Curr Opin Chem Biol 2011;15:798–805.

[23] Davies H, Bignell GR, Cox C, Stephens P, Edkins S, Clegg S, et al. Mutations of the BRAF gene in human cancer. Nature 2002;417:949–54.

[24] Wan PT, Garnett MJ, Roe SM, Lee S, Niculescu-Duvaz D, Good VM, et al. Mechanism of activation of the RAF-ERK signaling pathway by oncogenic mutations of B-RAF. Cell 2004;116:855–67.

[25] Chapman PB, Hauschild A, Robert C, Haanen JB, Ascierto P, Larkin J, et al. Improved survival with vemurafenib in melanoma with BRAF V600E mutation. N Engl J Med 2011;364:2507–16.

[26] Hauschild A, Grob JJ, Demidov LV, Jouary T, Gutzmer R, Millward M, et al. Dabrafenib in BRAF-mutated metastatic melanoma: a multicentre, open-label, phase 3 randomised controlled trial. Lancet 2012;380:358–65.

[27] Ross JS, Slodkowska EA, Symmans WF, Pusztai L, Ravdin PM, Hortobagyi GN. The HER-2 receptor and breast cancer: ten years of targeted anti-HER-2 therapy and personalized medicine. Oncologist 2009;14:320–68.

[28] Slamon DJ, Leyland-Jones B, Shak S, Fuchs H, Paton V, Bajamonde A, et al. Use of chemotherapy plus a monoclonal antibody against HER2 for metastatic breast cancer that overexpresses HER2. N Engl J Med 2001;344:783–92.

[29] Kwak EL, Bang YJ, Camidge DR, Shaw AT, Solomon B, Maki RG, et al. Anaplastic lymphoma kinase inhibition in non-small-cell lung cancer. N Engl J Med 2010;363:1693–703.

[30] Debiec-Rychter M, Sciot R, Le Cesne A, Schlemmer M, Hohenberger P, van Oosterom AT, et al. KIT mutations and dose selection for imatinib in patients with advanced gastrointestinal stromal tumours. Eur J Cancer 2006;42:1093–103.

[31] Blanke CD, Demetri GD, von Mehren M, Heinrich MC, Eisenberg B, Fletcher JA, et al. Long-term results from a randomized phase II trial of standard- versus higher-dose imatinib mesylate for patients with unresectable or metastatic gastrointestinal stromal tumors expressing KIT. J Clin Oncol 2008;26:620–5.

[32] Singer S, Rubin BP, Lux ML, Chen CJ, Demetri GD, Fletcher CD, et al. Prognostic value of KIT mutation type, mitotic activity, and histologic subtype in gastrointestinal stromal tumors. J Clin Oncol 2002;20:3898–905.

[33] Ciardiello F, Tortora G. EGFR antagonists in cancer treatment. N Engl J Med 2008;358:1160–74.

[34] Amado RG, Wolf M, Peeters M, Van Cutsem E, Siena S, Freeman DJ, et al. Wild-type KRAS is required for panitumumab efficacy in patients with metastatic colorectal cancer. J Clin Oncol 2008;26:1626–34.

[35] Karapetis CS, Khambata-Ford S, Jonker DJ, O'Callaghan CJ, Tu D, Tebbutt NC, et al. K-ras mutations and benefit from cetuximab in advanced colorectal cancer. N Engl J Med 2008;359:1757–65.

[36] Perez EA, Suman VJ, Davidson NE, Martino S, Kaufman PA, Lingle WL, et al. HER2 testing by local, central, and reference laboratories in specimens from the North Central cancer treatment group N9831 intergroup adjuvant trial. J Clin Oncol 2006;24:3032–8.

[37] Paik S, Taniyama Y, Geyer Jr CE. Anthracyclines in the treatment of HER2-negative breast cancer. J Natl Cancer Inst 2008;100:2–4.

[38] Perez EA, Reinholz MM, Hillman DW, Tenner KS, Schroeder MJ, Davidson NE, et al. HER2 and chromosome 17 effect on patient outcome in the N9831 adjuvant trastuzumab trial. J Clin Oncol 2010;28:4307–15.

[39] Pogue-Geile KL, Kim C, Jeong JH, Tanaka N, Bandos H, Gavin PG, et al. Predicting degree of benefit from adjuvant trastuzumab in NSABP trial B-31. J Natl Cancer Inst 2013;105:1782–8.

[40] Lievre A, Bachet JB, Boige V, Cayre A, Le Corre D, Buc E, et al. KRAS mutations as an independent prognostic factor in patients with advanced colorectal cancer treated with cetuximab. J Clin Oncol 2008;26:374–9.

[41] Lievre A, Bachet JB, Le Corre D, Boige V, Landi B, Emile JF, et al. KRAS mutation status is predictive of response to cetuximab therapy in colorectal cancer. Cancer Res 2006;66:3992–5.
[42] Bokemeyer C, Van Cutsem E, Rougier P, Ciardiello F, Heeger S, Schlichting M, et al. Addition of cetuximab to chemotherapy as first-line treatment for KRAS wild-type metastatic colorectal cancer: pooled analysis of the CRYSTAL and OPUS randomised clinical trials. Eur J Cancer 2012;48:1466–75.
[43] Tournigand C, Chibaudel B, Samson B, Scheithauer W, Vernerey D, Mésange P, et al. Bevacizumab with or without erlotinib as maintenance therapy in patients with metastatic colorectal cancer (GERCOR DREAM; OPTIMOX3): a randomised, open-label, phase 3 trial. Lancet Oncol 2015: pii: S1470-2045 (15)00216-8. http://dx.doi.org/10.1016/S1470-2045(15)00216-8. [Epub ahead of print]
[44] Yang H, Higgins B, Kolinsky K, Packman K, Bradley WD, Lee RJ, et al. Antitumor activity of BRAF inhibitor vemurafenib in preclinical models of BRAF-mutant colorectal cancer. Cancer Res 2012;72:779–89.
[45] Corcoran RB, Ebi H, Turke AB, Coffee EM, Nishino M, Cogdill AP, et al. EGFR-mediated re-activation of MAPK signaling contributes to insensitivity of BRAF mutant colorectal cancers to RAF inhibition with vemurafenib. Cancer Discov 2012;2:227–35.
[46] Corcoran R, Falchook G, Infante J, et al. Pharmacodynamic and efficacy analysis of the BRAF inhibitor dabrafenib (GSK436) in combination with the MEK inhibitor trametinib (GSK212) in patients with BRAFV600 mutant colorectal cancer (CRC). J Clin Oncol 2013;(Suppl):31 abstr 3507
[47] Flaherty KT, Infante JR, Daud A, Gonzalez R, Kefford RF, Sosman J, et al. Combined BRAF and MEK inhibition in melanoma with BRAF V600 mutations. N Engl J Med 2012;367:1694–703.
[48] Mao M, Tian F, Mariadason JM, Tsao CC, Lemos R, Dayyani F, et al. Resistance to BRAF inhibition in BRAF-mutant colon cancer can be overcome with PI3K inhibition or demethylating agents. Clin Cancer Res 2013;19:657–67.
[49] de Gramont A, Watson S, Ellis LM, Rodon J, Tabernero J, de Gramont A, et al. Pragmatic issues in biomarker evaluation for targeted therapies in cancer. Nat Rev Clin Oncol 2015;12:197–212.
[50] Diamandis EP. The failure of protein cancer biomarkers to reach the clinic: why, and what can be done to address the problem? BMC Med 2012;10:87.
[51] Waddell T, Chau I, Cunningham D, Gonzalez D, Okines AF, Okines C, et al. Epirubicin, oxaliplatin, and capecitabine with or without panitumumab for patients with previously untreated advanced oesophagogastric cancer (REAL3): a randomised, open-label phase 3 trial. Lancet Oncol 2013;14:481–9.
[52] Lordick F, Kang YK, Chung HC, Salman P, Oh SC, Bodoky G, et al. Capecitabine and cisplatin with or without cetuximab for patients with previously untreated advanced gastric cancer (EXPAND): a randomised, open-label phase 3 trial. Lancet Oncol 2013;14:490–9.
[53] Fedele C, Tothill RW, McArthur GA. Navigating the challenge of tumor heterogeneity in cancer therapy. Cancer Discov 2014;4:146–8.
[54] Longo DL. Tumor heterogeneity and personalized medicine. N Engl J Med 2012;366:956–7.
[55] Bedard PL, Hansen AR, Ratain MJ, Siu LL. Tumour heterogeneity in the clinic. Nature 2013;501:355–64.
[56] Dancey JE, Dobbin KK, Groshen S, Jessup JM, Hruszkewycz AH, Koehler M, et al. Guidelines for the development and incorporation of biomarker studies in early clinical trials of novel agents. Clin Cancer Res 2010;16:1745–55.
[57] Meyer GJ. Guidelines for reporting information in studies of diagnostic test accuracy: the STARD initiative. J Pers Assess 2003;81:191–3.
[58] Whiting P, Rutjes AW, Reitsma JB, Bossuyt PM, Kleijnen J. The development of QUADAS: a tool for the quality assessment of studies of diagnostic accuracy included in systematic reviews. BMC Med Res Methodol 2003;3:25.
[59] Tan DS, Thomas GV, Garrett MD, Banerji U, de Bono JS, Kaye SB, et al. Biomarker-driven early clinical trials in oncology: a paradigm shift in drug development. Cancer J 2009;15:406–20.

[60] Lee JW, Devanarayan V, Barrett YC, Weiner R, Allinson J, Fountain S, et al. Fit-for-purpose method development and validation for successful biomarker measurement. Pharm Res 2006;23:312–28.
[61] Lee JW, Weiner RS, Sailstad JM, Bowsher RR, Knuth DW, O'Brien PJ, et al. Method validation and measurement of biomarkers in nonclinical and clinical samples in drug development: a conference report. Pharm Res 2005;22:499–511.
[62] Cummings J, Ward TH, Dive C. Fit-for-purpose biomarker method validation in anticancer drug development. Drug Discov Today 2010;15:816–25.
[63] Ilyin SE, Belkowski SM, Plata-Salaman CR. Biomarker discovery and validation: technologies and integrative approaches. Trends Biotechnol 2004;22:411–16.
[64] Shah VP, Midha KK, Findlay JW, Hill HM, Hulse JD, McGilveray IJ, et al. Bioanalytical method validation—a revisit with a decade of progress. Pharm Res 2000;17:1551–7.
[65] Cummings J, Ward TH, Greystoke A, Ranson M, Dive C. Biomarker method validation in anticancer drug development. Br J Pharmacol 2008;153:646–56.
[66] Hammond ME, Taube SE. Issues and barriers to development of clinically useful tumor markers: a development pathway proposal. Semin Oncol 2002;29:213–21.
[67] Hayes DF, Bast RC, Desch CE, Fritsche Jr H, Kemeny NE, Jessup JM, et al. Tumor marker utility grading system: a framework to evaluate clinical utility of tumor markers. J Natl Cancer Inst 1996;88:1456–66.
[68] Simon R, Altman DG. Statistical aspects of prognostic factor studies in oncology. Br J Cancer 1994;69:979–85.
[69] Sotiriou C, Pusztai L. Gene-expression signatures in breast cancer. N Engl J Med 2009;360:790–8.
[70] Royston P, Parmar MK, Altman DG. Visualizing length of survival in time-to-event studies: a complement to Kaplan-Meier plots. J Natl Cancer Inst 2008;100:92–7.
[71] Buyse M, Sargent DJ, Grothey A, Matheson A, de Gramont A. Biomarkers and surrogate end points—the challenge of statistical validation. Nat Rev Clin Oncol 2010;7:309–17.
[72] Pepe MS, Janes H, Longton G, Leisenring W, Newcomb P. Limitations of the odds ratio in gauging the performance of a diagnostic, prognostic, or screening marker. Am J Epidemiol 2004;159:882–90.
[73] Oliner K, Douillard J, Siena S, Tabernero J, Burkes R, Barugel M. Analysis of KRAS/NRAS and BRAF mutations in the phase III PRIME study of panitumumab and FOLFOX vs FOLFOX as first line treatment for metastatic colorectal cancer. In: ASCO Annual Meeting. Abstract 3511;2013.
[74] Peeters M, Oliner KS, Parker A, Siena S, Van Cutsem E, Huang J, et al. Massively parallel tumor multigene sequencing to evaluate response to panitumumab in a randomized phase III study of metastatic colorectal cancer. Clin Cancer Res 2013;19:1902–12.
[75] Seymour MT, Brown SR, Middleton G, Maughan T, Richman S, Gwyther S, et al. Panitumumab and irinotecan versus irinotecan alone for patients with KRAS wild-type, fluorouracil-resistant advanced colorectal cancer (PICCOLO): a prospectively stratified randomised trial. Lancet Oncol 2013;14:749–59.
[76] Mandrekar SJ, Sargent DJ. Clinical trial designs for predictive biomarker validation: theoretical considerations and practical challenges. J Clin Oncol 2009;27:4027–34.
[77] Mandrekar SJ, Sargent DJ. Clinical trial designs for predictive biomarker validation: one size does not fit all. J Biopharm Stat 2009;19:530–42.
[78] Sargent DJ, Conley BA, Allegra C, Collette L. Clinical trial designs for predictive marker validation in cancer treatment trials. J Clin Oncol 2005;23:2020–7.
[79] Freidlin B, McShane LM, Korn EL. Randomized clinical trials with biomarkers: design issues. J Natl Cancer Inst 2010;102:152–60.
[80] Lin WJ, Chen JJ. Class-imbalanced classifiers for high-dimensional data. Brief Bioinform 2013;14:13–26.
[81] Simon R. The use of genomics in clinical trial design. Clin Cancer Res 2008;14:5984–93.

[82] Maitournam A, Simon R. On the efficiency of targeted clinical trials. Stat Med 2005;24:329–39.
[83] Simon R, Maitournam A. Evaluating the efficiency of targeted designs for randomized clinical trials. Clin Cancer Res 2004;10:6759–63.
[84] Romond EH, Perez EA, Bryant J, Suman VJ, Geyer Jr CE, Davidson NE, et al. Trastuzumab plus adjuvant chemotherapy for operable HER2-positive breast cancer. N Engl J Med 2005;353:1673–84.
[85] Piccart-Gebhart MJ, Procter M, Leyland-Jones B, Goldhirsch A, Untch M, Smith I, et al. Trastuzumab after adjuvant chemotherapy in HER2-positive breast cancer. N Engl J Med 2005;353:1659–72.
[86] Joensuu H, Kellokumpu-Lehtinen PL, Bono P, Alanko T, Kataja V, Asola R, et al. Adjuvant docetaxel or vinorelbine with or without trastuzumab for breast cancer. N Engl J Med 2006;354:809–20.
[87] Rawson JB, Manno M, Mrkonjic M, Daftary D, Dicks E, Buchanan DD, et al. Promoter methylation of Wnt antagonists DKK1 and SFRP1 is associated with opposing tumor subtypes in two large populations of colorectal cancer patients. Carcinogenesis 2011;32:741–7.
[88] Muller HM, Widschwendter A, Fiegl H, Ivarsson L, Goebel G, Perkmann E, et al. DNA methylation in serum of breast cancer patients: an independent prognostic marker. Cancer Res 2003;63:7641–5.
[89] Esteller M, Garcia-Foncillas J, Andion E, Goodman SN, Hidalgo OF, Vanaclocha V, et al. Inactivation of the DNA-repair gene MGMT and the clinical response of gliomas to alkylating agents. N Engl J Med 2000;343:1350–4.
[90] Brock MV, Hooker CM, Ota-Machida E, Han Y, Guo M, Ames S, et al. DNA methylation markers and early recurrence in stage I lung cancer. N Engl J Med 2008;358:1118–28.
[91] Kim Y, Kim D-H. CpG island hypermethylation as a biomarker for the early detection of lung cancer. Methods Mol Biol 2015;1238:141–71.
[92] Stefansson OA, Moran S, Gomez A, Sayols S, Arribas-Jorba C, Sandoval J, et al. A DNA methylation-based definition of biologically distinct breast cancer subtypes. Mol Oncol 2015;9:555–68.
[93] Hegi ME, Diserens A-C, Gorlia T, Hamou M-F, de Tribolet N, Weller M, et al. MGMT gene silencing and benefit from temozolomide in glioblastoma. N Engl J Med 2005;352:997–1003.
[94] Ebert MPA, Tänzer M, Balluff B, Burgermeister E, Kretzschmar AK, Hughes DJ, et al. TFAP2E-DKK4 and chemoresistance in colorectal cancer. N Engl J Med 2012;366:44–53.
[95] Watanabe Y, Ueda H, Etoh T, Koike E, Fujinami N, Mitsuhashi A, et al. A change in promoter methylation of hMLH1 is a cause of acquired resistance to platinum-based chemotherapy in epithelial ovarian cancer. Anticancer Res 2007;27:1449–52.
[96] Noehammer C, Pulverer W, Hassler MR, Hofner M, Wielscher M, Vierlinger K, et al. Strategies for validation and testing of DNA methylation biomarkers. Epigenomics 2014;6:603–22.
[97] Laird PW. Principles and challenges of genomewide DNA methylation analysis. Nat Rev Genet 2010;11:191–203.
[98] Costello JF, Plass C, Cavenee WK. Restriction landmark genome scanning. Methods Mol Biol 2002;200:53–70.
[99] Smiraglia DJ, Plass C. The study of aberrant methylation in cancer via restriction landmark genomic scanning. Oncogene 2002;21:5414–26.
[100] Frigola J, Ribas M, Risques RA, Peinado MA. Methylome profiling of cancer cells by amplification of inter-methylated sites (AIMS). Nucleic Acids Res 2002;30:e28.
[101] Estecio MR, Yan PS, Ibrahim AE, Tellez CS, Shen L, Huang TH, et al. High-throughput methylation profiling by MCA coupled to CpG island microarray. Genome Res 2007;17:1529–36.
[102] Oda M, Glass JL, Thompson RF, Mo Y, Olivier EN, Figueroa ME, et al. High-resolution genome-wide cytosine methylation profiling with simultaneous copy number analysis and optimization for limited cell numbers. Nucleic Acids Res 2009;37:3829–39.
[103] Ibrahim AE, Thorne NP, Baird K, Barbosa-Morais NL, Tavare S, Collins VP, et al. MMASS: an optimized array-based method for assessing CpG island methylation. Nucleic Acids Res 2006;34:e136.

[104] Ball MP, Li JB, Gao Y, Lee JH, LeProust EM, Park IH, et al. Targeted and genome-scale strategies reveal gene-body methylation signatures in human cells. Nat Biotechnol 2009;27:361–8.
[105] Brunner AL, Johnson DS, Kim SW, Valouev A, Reddy TE, Neff NF, et al. Distinct DNA methylation patterns characterize differentiated human embryonic stem cells and developing human fetal liver. Genome Res 2009;19:1044–56.
[106] Mohn F, Weber M, Schubeler D, Roloff TC. Methylated DNA immunoprecipitation (MeDIP). Methods Mol Biol 2009;507:55–64.
[107] Weber M, Hellmann I, Stadler MB, Ramos L, Paabo S, Rebhan M, et al. Distribution, silencing potential and evolutionary impact of promoter DNA methylation in the human genome. Nat Genet 2007;39:457–66.
[108] Rauch T, Pfeifer GP. Methylated-CpG island recovery assay: a new technique for the rapid detection of methylated-CpG islands in cancer. Lab Invest 2005;85:1172–80.
[109] Rauch TA, Pfeifer GP. The MIRA method for DNA methylation analysis. Methods Mol Biol 2009;507:65–75.
[110] Frommer M, McDonald LE, Millar DS, Collis CM, Watt F, Grigg GW, et al. A genomic sequencing protocol that yields a positive display of 5-methylcytosine residues in individual DNA strands. Proc Natl Acad Sci USA 1992;89:1827–31.
[111] Clark SJ, Harrison J, Paul CL, Frommer M. High sensitivity mapping of methylated cytosines. Nucleic Acids Res 1994;22:2990–7.
[112] Fraga MF, Esteller M. DNA methylation: a profile of methods and applications. Biotechniques 2002;33:632, 634, 636–649.
[113] Grunau C, Clark SJ, Rosenthal A. Bisulfite genomic sequencing: systematic investigation of critical experimental parameters. Nucleic Acids Res 2001;29:E65-5.
[114] Warnecke PM, Stirzaker C, Melki JR, Millar DS, Paul CL, Clark SJ. Detection and measurement of PCR bias in quantitative methylation analysis of bisulphite-treated DNA. Nucleic Acids Res 1997;25:4422–6.
[115] Reinders J, Delucinge Vivier C, Theiler G, Chollet D, Descombes P, Paszkowski J. Genome-wide, high-resolution DNA methylation profiling using bisulfite-mediated cytosine conversion. Genome Res 2008;18:469–76.
[116] Bibikova M, Fan JB. GoldenGate assay for DNA methylation profiling. Methods Mol Biol 2009;507:149–63.
[117] Bibikova M, Le J, Barnes B, Saedinia-Melnyk S, Zhou L, Shen R, et al. Genome-wide DNA methylation profiling using infinium(R) assay. Epigenomics 2009;1:177–200.
[118] Meissner A, Gnirke A, Bell GW, Ramsahoye B, Lander ES, Jaenisch R. Reduced representation bisulfite sequencing for comparative high-resolution DNA methylation analysis. Nucleic Acids Res 2005;33:5868–77.
[119] Berman BP, Weisenberger DJ, Laird PW. Locking in on the human methylome. Nat Biotechnol 2009;27:341–2.
[120] Li JB, Gao Y, Aach J, Zhang K, Kryukov GV, Xie B, et al. Multiplex padlock targeted sequencing reveals human hypermutable CpG variations. Genome Res 2009;19:1606–15.
[121] Johnson MD, Mueller M, Game L, Aitman TJ. Single nucleotide analysis of cytosine methylation by whole-genome shotgun bisulfite sequencing. Curr Protoc Mol Biol 2012;Chapter 21:Unit 21 23.
[122] Lister R, Pelizzola M, Dowen RH, Hawkins RD, Hon G, Tonti-Filippini J, et al. Human DNA methylomes at base resolution show widespread epigenomic differences. Nature 2009; 462:315–22.
[123] Flusberg BA, Webster DR, Lee JH, Travers KJ, Olivares EC, Clark TA, et al. Direct detection of DNA methylation during single-molecule, real-time sequencing. Nat Methods 2010;7:461–5.

[124] Grada A, Weinbrecht K. Next-generation sequencing: methodology and application. J Invest Dermatol 2013;133:e11.
[125] Ehrich M, Nelson MR, Stanssens P, Zabeau M, Liloglou T, Xinarianos G, et al. Quantitative high-throughput analysis of DNA methylation patterns by base-specific cleavage and mass spectrometry. Proc Natl Acad Sci USA 2005;102:15785–90.
[126] Rodriguez-Paredes M, Esteller M. Cancer epigenetics reaches mainstream oncology. Nat Med 2011;17:330–9.

CHAPTER

NETWORKS AND CONSORTIA FOR EPIGENETIC DRUG DISCOVERY

6

Daniel J. Weisenberger[1] and Peter J. Brown[2]

[1]*USC/Norris Comprehensive Cancer Center, University of Southern California, Los Angeles, CA, USA*

[2]*Structural Genomics Consortium, University of Toronto, Toronto, ON, Canada*

CHAPTER OUTLINE

6.1 INTRODUCTION

The current era of genomic medicine began with the releases of the first human genome draft sequences in 2001 [1,2], made possible by international networks comprised of thousands of scientists in a 20-year time frame. These efforts were largely labor-intensive, by which Sanger-based

G. Egger & P. Arimondo (Eds): Drug Discovery in Cancer Epigenetics. DOI: http://dx.doi.org/10.1016/B978-0-12-802208-5.00006-0

sequencing methods were applied to large-insert clones that were then assembled into complete draft sequences. Since these reports, next-generation sequencing technologies have been developed that currently enable deep-coverage, whole-genome sequencing, and alignment to the reference genome at a small fraction of the cost and time.

These advancements have propelled interest in several aspects of molecular biology to identify: (i) genome sequence and organizational differences based on phylogeny and taxonomy; (ii) DNA sequence alterations that are risk factors for disease, as well as drivers of disease initiation and progression; (iii) gene regulatory networks and mechanisms of gene activation and repression by interrogating the RNA expression, microRNA (miRNA) expression, DNA methylation and chromatin modifications within a cell type, between cell types, as functions of development, and molecular programs in both normal tissues and those involved in disease; (iv) biomarkers for disease diagnosis, treatment response, and monitoring; and (v) novel targets for drug delivery in disease treatment.

Determining these features in a genome-wide manner in human cells is challenging for several reasons including the requirements for stratifying molecular data for large numbers of samples as functions of age, gender, race, ethnicity, environmental exposures, and family history. Moreover, genomic features of human diseases, such as human cancers, also require tissue-specific interrogations, thereby increasing experimental complexity and scope. In light of these challenges, domestic and international multi-institutional consortia have been formed to systematically generate, analyze and integrate genomic datasets from large sample collections across multiple disease and cell types (Table 6.1). Among these, Encyclopedia of DNA Elements (ENCODE), Human Epigenome Project (HEP), International Cancer Genomics Consortium (ICGC), International Human Epigenome Consortium (IHEC), NIH Roadmap Epigenomics Initiative (NREI), Stand Up to Cancer (SU2C), Structural Genomics Consortium (SGC), The Cancer Genome Atlas (TCGA), and Therapeutically Applicable Research to Generate Effective Treatments (TARGET) have been vital in identifying genomic and epigenomic alterations in normal tissues, cancer tissues, and cancer cell lines. This chapter summarizes these consortia with respect to the development of epigenetic-based therapeutics.

6.2 EPIGENETICS DEFINED

Conrad Waddington first described epigenetics in 1942 as the combination of genetics and epigenesis describing how the interactions of genes and their environment result in specific phenotypes, specifically with respect to cell fates during development and cellular differentiation (summarized in Ref. [3]). Currently, epigenetics is described as mitotically or meiotically heritable changes in phenotype that are not based on DNA sequence, essentially describing changes in gene activity and expression potential rather than changes in gene structure. As described in Chapters 1 and 2, the major contributing mechanisms of epigenetics-based gene regulation are the interplay of DNA methylation, DNA 5-hydroxymethylation, histone modifications, chromatin modifications, nucleosome positioning, transcription factor regulatory networks, and noncoding RNAs (miRNA and long noncoding RNA (lncRNA)).

Table 6.1 Genetic and Epigenetic Characterization Efforts in Multi-Institutional Consortia

Entity	Website URL	Experimental Scope	Data Types
Encyclopedia of Genomic Elements (ENCODE)	www.encodeproject.org	1600+ human cell types in 30 organs; 454 mice; and 173 *Drosophila melanogaster* cell types	Chromatin modifications, DNA methylation, gene expression (RNA, miRNA), nucleosome occupancy, transcription factor binding sites
Human Epigenome Project (HEP)	www.epigenome.org	90 genes in *MHC* locus	DNA methylation
International Cancer Genome Consortium (ICGC)	www.icgc.org	25,000 cancer genomes from 50 tumor types	DNA methylation, gene expression (RNA, miRNA), somatic DNA alterations (mutations, copy number)
International Human Epigenome Consortium (IHEC)	www.ihec-epigenomes.org	1000 epigenomes in 250 cell types	Chromatin modifications, DNA methylation, gene expression (RNA, miRNA), nucleosome occupancy, transcription factor binding, and transcription start sites
NIH Roadmap Epigenomics Initiative (NREI)	www.roadmapepigenomics.org	Epigenomic characterization of 250 nondiseased adult and fetal cell types	Chromatin modifications, DNA methylation, gene expression (RNA, miRNA), nucleosome occupancy, transcription factor, and other protein–DNA binding sites
Structural Genomics Consortium (SGC)	www.thegsc.org	Structures for 1700+ proteins; 25 chemical probes targeting epigenetic proteins	Protein crystallography, generation of antibodies, chemical probes, constructs, plasmids, and vectors
Stand Up to Cancer (SU2C)	www.standup2cancer.org	Primary tumors (AML, breast, colorectal, lung, MDS, ovarian tumors) biopsied after treatment with epigenetics-based inhibitors in phase II/III clinical trials	DNA methylation, gene expression
Therapeutically Applicable Research to Generate Effective Treatments (TARGET)	www.ocg.cancer.gov/programs/target	Genomic characterization of primary childhood ALL, AML, kidney, neuroblastoma, and osteosarcoma tissues	DNA methylation, gene expression (RNA, miRNA), somatic DNA alterations (mutations, copy number)
The Cancer Genome Atlas (TCGA)	www.cancergenome.nih.gov	11,000 primary tumors from 30 human tumor types	DNA methylation, gene expression (RNA, miRNA), somatic DNA alterations (mutations, copy number)

6.2.1 DNA METHYLATION

In mammalian cells, DNA methylation predominantly occurs in the context of 5′-CpG-3′ through the addition of a methyl-group to the C-5 position of cytosine (5-methylcytosine (5-mC)) through the efforts of DNA methyltransferases (DNMTs) which use *S*-adenosylmethionine (SAM) as a cofactor (reviewed in Ref. [4]). However, 5-mC is prone to spontaneous deamination to uracil at an order of magnitude higher rate than unmethylated cytosine. Consequently, the prevalence of CpG sites in the human genome is 20% of what is expected. While most of the human genome is CpG-depleted, there are regions of genomic DNA, termed CpG islands, that have their expected CpG content and are generally unmethylated in normal somatic tissues [4]. DNA methylation markings are stable, with successive rounds of cell division, however, DNA methylation profiles are dramatically altered in human cancers and other diseases. CpG islands, as well as the regions flanking CpG islands, termed CpG island shores, are generally thought to acquire DNA methylation (DNA hypermethylation) [5–7]. Alternatively, CpG-poor regions of the genome, including CpG island shelves and open seas, as well as repetitive elements and lamin-attachment sites are generally hypomethylated in human cancers [8–11]. Promoter DNA hypermethylation can be associated with reduced gene expression [7], while gene body DNA methylation may be associated with increased gene expression [12–14]. Methylated DNA also serves as a recruitment site for methyl-CpG binding domain proteins (MBDs) that interact with and influence chromatin structure in modulating gene activity [15].

6.2.2 DNA 5-HYDROXYMETHYLATION

5-Hydroxymethylation (5-hydroxymethylcytosine (5-hmC)) was first described as a product of 5-mC oxidation by TET1 (ten-eleven translocase) (reviewed in Refs [16,17]). Two additional TET enzymes, TET2 and TET3, were subsequently identified, with each TET enzyme functioning as 2-oxoglutarate- and iron-dependent dioxygenases that are similar in function to several known histone lysine demethylases. TET enzymes can catalyze the conversion of 5-mC to not only 5-hmC, but also the conversion of 5-hmC to 5-formylcytosine (5-fC) and 5-carboxylcytosine (5-caC) [16,17]. 5-fC and 5-caC are substrates for thymine DNA glycosylase-mediated base excision repair that ultimately results in replacement with an unmethylated cytosine. 5-hmC is present at lower levels (<1%) than 5-mC (4–5%), but 5-hmC marks are found at gene promoters, gene bodies, and enhancers in virtually every tissue type. While the function of 5-hmC as an epigenetic mark is currently uncertain, 5-hmC profiles are altered in human cancers [16,17]. Taken together, 5-hmC represents an important step in enzyme-catalyzed DNA demethylation, and potential biomarkers and source of drug targeting in cancer.

6.2.3 CHROMATIN STRUCTURE AND NUCLEOSOME POSITIONING

Chromatin structure and nucleosome positioning are closely associated with DNA methylation in gene regulation, and both display alterations in human cancers. Chromatin structure is largely determined by posttranslational modifications of specific amino acids on histone N-terminal tails, by which histone methyltransferases (HMTs), histone acetyltransferases (HATs), histone phosphorylases, and other enzymes catalyze the recognition (readers), addition (writers), and removal (erasers) of these functional groups, thereby influencing chromatin structure, and ultimately, gene activity potential (reviewed in Ref. [18]). Chromatin modifications also delineate between genes that: (i) display inducible or tissue-specific expression profiles, (ii) display constitutive expression, and (iii) are not expressed. For instance, histone lysine monomethylation (H3K4, H3K9, H3K27, H3K79, and H4K20) and acetylation (H3K9Ac, H3K14Ac, and H3K27Ac) are correlated with gene activation. Other marks associated with gene activation include histone H3 lysine 4 tri-methylation (H3K4me3), H3K36me3, and H3K79me2. Repressed regions of the genome are enriched for H3K9me2, H3K9me3, H3K27me2, and H3K27me3 marks [18–20].

The evolution and application of whole-genome sequencing technologies have also unveiled unexpected DNA sequence alterations in human cancers. One example is the identification of somatic alterations of genes coding for chromatin modifier genes, including chromatin remodelers, as well as histone readers, writers, and erasers, across multiple human cancer types (reviewed in Ref. [18]). For example, the chromatin remodeler AT-rich interactive domain-containing protein 1A (*ARID1A*) shows loss-of-function mutations across several tumor types, including those derived from brain, breast, colorectum, endometrium, lung, ovary, stomach, and pancreas. The *MLL* family of histone H3K4 methyltransferases harbor loss-of-function alterations and translocations across bladder, breast, brain, liver, lung, and prostate cancers, as well as in leukemias and lymphomas. In addition, the *EZH2* gene, which encodes a histone H3K27 methyltransferase and component of the regulatory polycomb repressive complex 2 (PRC2), shows loss-of-function and gain-of-function alterations in lymphomas, but is overexpressed in bladder, breast, colorectal, gastric, kidney, and lung cancers and lymphomas. *KDM5C/JARID1C* and *KDM6A/UTX*, which encode for histone H3K4 and H3K27 demethylases, respectively, exhibit reduction of function in bladder tumors, gliomas, and meningiomas. While the phenotypes resulting from these alterations are not completely known, these represent drugable targets for potential therapeutic applications in human cancers.

In addition to chromatin modifications, nucleosome accessibility has also received attention as a major influence in gene regulation. Active transcription requires a nucleosome-depleted region (NDR) at least one nucleosome immediately upstream of the transcription start site (TSS) to allow for transcription machinery binding. Alternatively, inactive genes do not have an NDR, and those that are epigenetically silenced in cancer generally include nucleosomes [21–24].

miRNAs are short RNA sequences of 20–22 nucleotides in length that are transcribed from either their own promoters (or from intronic gene regions) spread throughout the genome (reviewed in Refs [25,26]). Mature miRNAs form double-stranded complexes with their target mRNAs, and, as a result, signal the degradation of the mRNA–miRNA complex or inhibit translation due to imperfect pairing of these miRNA–mRNA complexes. Importantly, this allows for regulation of multiple coding mRNAs from a single miRNA, suggesting that miRNAs have tremendous influence in gene regulation. In support of this, over 2500 mature human miRNAs have been annotated (www.mirbase.org), however, the effects of each miRNA in the human genome are unclear. miRNAs and their biosynthesis pathways are frequently altered in human cancers [25,26]. Indeed, over- and underexpression of enzymes of miRNA synthesis pathways have been identified in several human cancer types. Moreover, miRNA expression is also altered in human cancers via promoter DNA methylation, binding of hormones, growth factors, and cytokines at individual miRNA promoters, as well as the functions of enhancers near the miRNA gene locus. With these roles in mind, miRNA functions represent powerful therapeutic targets for human cancer treatment, especially with the exogenous delivery of short interfering RNAs in interfering with the translation of the target mRNA.

lncRNAs are generally classified as noncoding RNA molecules greater than 200 nucleotides in length. Over 32,000 lncRNAs have been identified in the human genome, and are expressed from intragenic regions, intergenic regions, and specific regions in the human genome. When expressed, lncRNAs interact with DNA elements, RNAs, and proteins, resulting in transcriptional regulation or inhibition of mRNA function and translation [27]. In this regard, lncRNAs interact with transcription factors, chromatin regulators, mRNAs, and RNA-binding proteins. Of note, H19 and *Xist* are two well-characterized lncRNAs that interact with chromatin modifiers (CTCF (CCCTC-binding factor) and PRC2, respectively). Much like mRNA and miRNAs, lncRNAs are also deregulated in human cancers by DNA promoter hypermethylation, copy number alterations, DNA sequence variation, and transcription factor binding, all of which are critically important for tumor initiation, proliferation, and metastasis [27]. Therefore, lncRNAs represent important biomarkers, as well as therapeutic targets for cancer detection, surveillance, and treatment.

6.3 UTILITY OF EPIGENETIC DRUG INHIBITORS FOR CANCER THERAPY

The complex and multifaceted organization of the epigenome has prompted the development of therapeutic targets for disease treatment, namely for human cancers, which display substantial alterations in DNA methylation and chromatin modification profiles. As described in Chapter 7, the first-generation epigenetic inhibitors 5-azacytidine (5-Aza-CR, Vidaza) and 5-aza-2′-deoxycytidine (5-Aza-CdR, Decitabine, Dacogen) have been approved for treatment of myelodysplastic syndrome (MDS) and acute myeloid leukemia (AML) patients [28,29]. Similar to the aza-substituted cytidine analogs, zebularine (Pyrimidin-2-one β-D-ribofuranoside) was also shown to inhibit DNA methylation in both *in vitro* and *in vivo* settings after oral administration [30,31]. Finally, the DNA methylation inhibitor S110 (AzapG), a dinucleotide consisting of 5-Aza-CdR upstream of a deoxyguanosine, was shown to be an effective DNA methylation inhibitor in both *in vitro* and *in vivo* settings and showed improved stability and reduced toxicity compared to aza-substituted mononucleosides [32,33].

Small-molecule inhibitors of histone-modifying enzymes have also been increasingly used in clinical trials for their antitumor efficacies (reviewed in Refs [34,35]). These are explained in greater detail in Chapters 7–10. The histone deacetylase 1 (HDAC1) and HDAC2 inhibitor entinostat is currently in phase II clinical trials for treatment of breast cancer, nonsmall-cell lung cancer (NSCLC), and several forms of leukemia. Vorinostat (also known as SAHA (suberoylanilide hydroxamic acid)), an inhibitor of HDAC1, HDAC2, HDAC3, and HDAC6, is approved for the treatment of progressive, recurrent, and persistent T-cell lymphoma. Other HDAC inhibitors currently in clinical trials include panobinostat for the treatment of AML, MDS, and multiple myeloma patients and mocetinostat for the treatment of diffuse B-cell and follicular lymphomas.

In addition to the use of HDAC inhibitors for cancer treatment, inhibition of the lysine-specific histone demethylase LSD1 was shown to promote drug sensitivity to leukemia cells, and 3-deazaneplanocin (DZNep) was shown to inhibit H3K27me3 and H4K20me3 occupancy, as well as activation of developmental genes and those silenced in cancer [35]. Moreover, naturally occurring, small-molecule inhibitors of HATs, such as curcumin, garcinol, and anacardiac acid, were shown to cause apoptosis and downregulated gene expression levels after treatment [35].

6.4 HUMAN EPIGENOME PROJECT (HEP)

While the completion of the human genome draft sequences led to unparalleled discoveries through large collaborative efforts, it should be noted that the first epigenetics-based consortium, the Human Epigenome Project (www.epigenome.org) was organized in 1999—prior to the release of the first human genome draft sequence—with the goals of characterizing DNA methylation variations of all human genes across all major tissue types with the goals of improving the understanding of biological genomics, as well as for disease diagnosis and monitoring. This included the comprehensive analyses of DNA methylation variation at regulatory gene regions, as well as intergenic regions of the genome. The key aims are to correlate aberrantly methylated regions with disease and tissue/cell-specific DNA methylation profiles with gene expression levels in order to develop novel, personalized therapeutics and diagnostic assays for the diseased state and other biological covariates [36].

The consortium members include the Wellcome Trust Sanger Institute (Cambridge, United Kingdom), Epigenomics AG (Seattle, Washington, United States), and The Centre National de Génotypage (Evry, France). The HEP pilot project investigated the methylome of the human major histocompatibility complex (*MHC*) located on chromosome 6, a locus associated with several diseases in the human genome. In this initial report [37], HEP utilized bisulfite-DNA sequencing and mass spectrometry of PCR amplicons to determine the DNA methylation landscape of 90 genes in a 3.8-Mb region of the MHC locus of brain, breast, liver, lung, and muscle tissues. The data show a bimodal distribution, with most sequenced reads as either hypo- or hypermethylated. In addition, HEP identified extensive tissue-specific and interindividual DNA methylation, as well as promoter DNA methylation correlative with gene expression levels.

6.5 ENCYCLOPEDIA OF DNA ELEMENTS (ENCODE)

The release of the first human genome reference map provided a framework for understanding molecular features in a holistic fashion within cell or tissue types. However, this also led to substantial challenges with respect to characterizing the functions of protein-coding regions, noncoding DNA and RNA elements, and enhancers within individual cell types, especially with respect to their applications in human health and disease. The ENCODE (www.encodeproject.org) consortium was developed as a means to meet these challenges by developing technologies and generating extensive, high-resolution maps of functional and structural features of the human genome [38].

ENCODE is subdivided into *pilot*, *technology development*, and *production* phases. The goal of the *pilot phase* was the development of strategies and high-throughput experimental procedures for characterizing coding regions of the human genome. In the *technology development phase*, insufficiencies in technology, genome coverage, and informatics were identified in order to improve the detection of known genomic elements as well as identifying new ones. The *production phase* is the culmination of both the pilot and development phases in characterizing and analyzing 99% of the human genome in a wide array of tissue types and disease states.

The *pilot* phase of ENCODE involved 35 research groups from the biotechnology industry, the United States National Human Genome Research Institute (NHGRI), as well as universities and institutes in the United States, Europe, and Asia. These groups first compiled data on transcription, by characterizing the complete gene sequences based on 44 regions of the genome, including expression, chromatin structure and histone modifications, transcription factor binding sites, copy number variation, and extensive computational analyses in curating 1% (30 Mb) of the human genome [39]. ENCODE revealed that: (i) the genome is extensively transcribed, (ii) chromatin structure and histone modifications are predictive of gene activity, and (iii) chromatin structure is highly correlated with the timing of DNA replication. In addition, ENCODE identified novel non-coding transcripts and protein-coding TSS, together with accompanying chromatin modifications and transcription factor binding sites.

In the *technology development* phase, ENCODE utilized next-generation sequencing technologies to characterize genome-wide characteristics of functional/coding DNA elements in the human genome. This ambitious effort resulted in over 30 reports in *Nature* and *Genome Research* in 2012 (overview described in Ref. [40]) in which 1640 data sets involving 147 cell types were generated including characterization of mRNA and miRNA expression profiles, chromatin structure and modifications, DNaseI-hypersensitive sites, DNA methylation, transcription factors, and other regulatory elements (www.nature.com/encode/category/research-papers). All molecular data described in these reports are available on the ENCODE Portal (www.encodeproject.org).

Currently, ENCODE has compiled functional elements in 1610 human cell types including primary and immortalized cells, tissues, stem cells, and differentiated cells. These samples span 30 organ types, and are further organized by age, subcellular location, and cell cycle phase. Similar data sets are also available for mouse (454 cell types) and *Drosophila melanogaster* (173 cell types). ENCODE has published 519 reports, and a total of 820 reports have been published by the community. The ENCODE data sets provide a rich resource for not only integrated, functional epigenomic characterization across several species, but also potential drug targets for disease treatment.

6.6 NIH ROADMAP EPIGENOMICS INITIATIVE (NREI)

In 2007, the United States National Institutes of Health (NIH) Common Fund allocated funds for the development of the Roadmap Epigenomics Initiative (roadmapepigenomics.org) in order to evaluate the extent to which epigenetic aberrations contribute to health and disease [41]. This is especially true for epigenetic mechanisms of stem cell differentiation and cellular development, which become deregulated in human diseases in response to internal and external cues. With this goal in mind, the NREI aims to characterize baseline epigenomic profiles of normal cell/tissue types as a function of developmental and differentiation status, as well as environmental exposures. NREI is subcategorized into Reference Epigenome Mapping Centers (REMC) and Epigenomics Data Analysis and Coordination Centers (EDACC). The REMCs generate epigenomic data for several cell types, while the EDACCs perform data analyses and organization across the consortium. In addition, additional centers develop and integrate novel technologies for epigenomic characterization, as well as biomarker discovery and mechanistic alterations in human diseases. To date, epigenomic data generated by the Roadmap Initiative Consortium have been used in over 350 peer-reviewed reports covering all phases of the project.

NREI supports four REMCs (MIT/Harvard/Broad Institute, University of California San Francisco, Ludwig Institute for Cancer Research, and University of Washington) [41–44], who generate epigenome reference maps comprised of DNA methylation, expression, protein binding sites, nucleosome occupancy, and chromatin modifications. The data are freely available on several online data portals (reviewed in Refs [45,46]). These include the Human Epigenome Atlas data portal (www.genboree.org/epigenomeatlas/index.rhtml), the NCBI Epigenomics Gateway (http://www.ncbi.nlm.nih.gov/epigenomics), as well as UCSC visualization tracks on the Human Epigenome Browser (www.epigenomebrowser.org; genomebrowser.wustl.edu). NREI has also provided analysis tools and pipelines for mapping sequencing reads (RMAPBS, AutoGenome), as well as web browser visualization (STAR Genome Browser).

The DNA methylation platforms include whole-genome bisulfite sequencing (WGBS), reduced-representation bisulfite sequencing (RRBS), methylated DNA immunoprecipitation sequencing (MeDIP-seq), and methylated restriction enzyme sequencing (MRE-seq). The gene expression assay platforms include RNA-seq and small RNA sequencing (smRNA-seq) technologies. Nucleosome occupancy is measured by DNaseI hypersensitivity, while protein–DNA interactions are mapped using digital genomic footprinting (DGF). The most comprehensive aspect of NREI is ChIP-seq-based characterization of 30 histone modifications, including methylation and acetylation of tail residues on histones H2, H3, and H4.

A total of 128 adult and 121 fetal tissues/cell lines have been profiled for one or more of the data types listed above. Within this collection, NREI has placed a strong emphasis on profiling human embryonic stem (ES) cell lines, as well as ES-derived and induced pluripotent stem (IPS) cells, due to their promising utility in regenerative medicine treatments. Fetal tissues include those collected from brain, gastrointestinal, heart, kidney, lung, muscle, placenta, spleen and thymus, hematopoietic stem cells, and white blood cells, and the adult tissues include those collected from adipose, breast, brain, genitourinary, gastrointestinal, muscle, and stroma. RRBS-sequencing-based DNA methylation profiling data are available for nearly all cell types. In addition, H1 and H9 ES cells have been profiled for 20 histone modifications using ChIP-seq technology. The majority of ChIP-seq data are specific for H3K27me3, H3K36me3, H3K4me1, H3K4me3, H3K9Ac, H3K9me3, and H3K27Ac markings, which represent the most commonly profiled histone modifications. These markings are also characterized in ES cells, ES-derived cells, IPS cells, adult adipose, brain, breast, gastrointestinal, muscle and genitourinary tissues, and CD4+, CD8+, and CD34+ white blood cell types, among others. The wide array of epigenomic data types across hundreds of cell types on a genome-wide scale represents a powerful and unparalleled integrated data set, from which the roles of epigenetics in development, aging, and disease can be elucidated. These epigenomic data also serve as a launching point for the development of therapeutics for the restoration of normal epigenetic states for disease treatment.

The REMC has organized integrated reference epigenomes into four classes (classes 1–4) [46] based on the breadth of epigenomic characterization for each sample. Data mandatory for all four classes, and the minimal amount for class 4 epigenomes, include ChIP-seq data specific for H3K27me3, H3K36me3, H3K4me1, H3K4me3, and H3K27Ac/H3K9Ac markings, as well as sequence, array or other type of DNA methylation and gene expression data. For instance, class 1 epigenomes are required to include the core elements listed above, as well as whole-genome bisulfite sequence DNA methylation data, RNA-seq expression data, and sites of DNaseI hypersensitivity. A report from early 2014 described a total of 92 epigenomes (7, 51, 14, and 20 epigenomes in classes 1, 2, 3, and 4, respectively) [46].

Since the 2014 report, NREI continued to accumulate and analyze epigenomic data. In early 2015, NREI published a series of reports describing integrative analyses of epigenomic data sets that not only describe extensive data production and the establishment of integrated, genome-wide maps across 111 tissue types and differentiation states [47], but also provided novel insights into: (i) the contributions of epigenomics in stem cell and neural cell differentiation [48,49]; (ii) the regulation of transcription factor binding by DNA methylation in ES cell differentiation [50]; (iii) haplotype-specific chromatin modifications of enhancers and promoters [51]; (iv) tissue-specific regulation of lncRNA expression [52]; (v) revealing an immune signature in Alzheimer's disease by evaluating chromatin modifications and expression dynamics [53]; (vi) mapping of genetic and epigenetic autoimmune disease variants [54]; and (vii) harnessing the power of epigenomic features in cancer cells of origin to predict cancer mutation profiles and somatic mutation density [55].

The REMCs reported on data from over 2800 individual epigenomic data sets, encompassing over 150 billion mapped sequencing reads and greater than 3100× total genome coverage [47]. Included are over 1800 ChIP-seq-based histone modification datasets, 360 DNaseI hypersensitivity datasets to interrogate DNA/chromatin accessibility, 277 DNA methylation datasets, and 166 gene expression datasets. Integration of individual data types resulted in 111 human cells/tissues with complete epigenomes that contain all of the following data types: ChIP-seq-based H3K4me1, H3K4me3, H3K9Ac, H3K9me3, H3K36me3, H3K27Ac, and H3K27me3 data, DNaseI hypersensitivity, DNA methylation (MRE, RRBS, and WGBS datasets), and gene expression data based on RNA-seq and array platforms [47].

Analyses of these data, as well as datasets from ENCODE, established genome-wide maps of gene regulation across human tissues and during development and differentiation [47]. The Roadmap Consortium showed that specific DNA methylation and chromatin accessibility signatures are correlated with specific histone modifications, and that chromatin modifications are more predictive of gene expression levels compared to DNA methylation and chromatin accessibility profiles [47]. Tissue-specific epigenomic profiles have important roles in gene regulation, due to enrichment of regulatory domains in tissue-specific enhancers, as well as enrichment of disease-associated genetic alterations in tissue-specific promoters. The data generated by the Roadmap Initiative provide a means to develop a wide array of potential use of epigenetic inhibitors in human disease treatment.

6.7 INTERNATIONAL HUMAN EPIGENOME CONSORTIUM (IHEC)

The International Human Epigenome Consortium (www.ihec-epigenomes.org) was launched in 2010 with the ultimate goal of understanding how epigenetics has shaped human health and disease, both across generations as well as in response to environmental stimuli. IHEC plans to characterize, analyze, and release reference epigenome datasets comprised of histone modifications (ChIP-seq), DNA methylation (WGBS), gene expression (RNA-seq), and TSS for all coding genes, for 1000 epigenomes in over 250 human cell/tissue types (including model organisms), stratified across several cellular states, including differentiated, immortalized, proliferative, senescent, stem, and undifferentiated cells. Several cell types, including ES, pluripotent, adipose, cardiovascular, cord blood, fibroblasts, and skin are from NREI. IHEC reference epigenomes must include DNA methylation (WGBS), expression (RNA-seq) and ChIP-seq data for several histone modifications, including H3K4me1, H3K4me3, H3K9me3, H3K27Ac, H3K27me3, and H3K36me3 marks. Data for nucleosome positioning (DNaseI-seq) and protein–DNA binding sites are optional. These data, together with rigorous bioinformatics development, serve as a launching point for characterizing the functional aspects of epigenetic marks in human health and disease.

In addition to collaborative efforts from the Epigenomics Roadmap Initiative, the IHEC Network includes members from six countries, including the Canadian Institutes for Health Research, the European Commission, the European Institute of Oncology (Italy), the Federal Ministry of Education and Research (Germany), Japan Science and Technology Agency, and the National Institute of Health (Korea), the French National Agency of Research, UK Funders Alliance, and the Australian National Health and Medical Research Council.

IHEC collaborations include CREST, DEEP, Epigenome Platform Program, and BLUEPRINT consortia. The *Core Research for Evolutional Science and Technology* (CREST; www.crest-ihec.jp) team works closely with IHEC in characterizing complete reference epigenomes of human gut epithelia (colon, liver, stomach), vascular endothelial cells, and reproductive tissues (germ, placenta, trophoblasts, and layer cells) in the development of stem-cell-based therapeutic approaches [56].

The *Deutsches Epigenom Programm (DEEP)* (www.deutsches-epigenom-programm.de) is a German-based group with the goals of comprehensive epigenomic characterization of inflammatory and metabolic diseases. DEEP utilizes next-generation sequencing-based assays to interrogate DNA methylation, RNA expression, chromatin modifications, and DNaseI-hypersensitive sites for 70 normal and diseased cell types including adipocytes, digestive epithelial cells, hepatocytes, macrophages, monocytes, and T cells. Ultimately, DEEP aims to harmonize epigenomic data production, bioinformatics, and functional modeling for disease detection, surveillance, monitoring, and treatment.

The *Epigenomic Platform Program*, based in the Centre for Epigenome Mapping Technologies, Genome Sciences Centre at the BC Cancer Agency (www.bcgsc.ca/project/centre-for-epigenome-mapping-technologies-cemt), aims to produce sequence-level reference epigenome maps consisting of 5-mC, 5-hmC, DNA sequence, expression (mRNA and miRNA), and histone modifications (H3K4me1, H3K4me3, H3K9Ac, H3K9me3, H3K27Ac, H3K27me3, and H3K36me3 marks) in several normal and malignant tissues, including brain, breast, colorectum, thyroid, white blood cells, and induced pluripotent cells. Data are available at the Canadian Epigenetics, Environment and Health Research Consortium (CEEHRC) Platform (www.epigenomes.ca) and can be visualized with custom tracks on UCSC and Washington University in St Louis browser tracks.

BLUEPRINT (www.blueprint-epigenome.eu) is a multi-institution effort from 43 academic groups spanning 11 countries in the European Union. The ultimate goals are to: (i) compile epigenomic-based reference maps of hematopoietic cell differentiation in 50 healthy and

50 neoplastic blood cell types, based on DNA methylation (WGBS), gene expression (RNA-seq), chromatin modifications (ChIP-seq) and nucleosome accessibility (DNaseI-seq) of individual blood cell types; (ii) compile similar epigenetic maps of immune-type cells from patients with type 1 diabetes; (iii) identify biomarkers of AML and acute lymphoblastic leukemia (ALL) for potential clinical utility and patient monitoring; (iv) characterize the consequences of epigenetic variation in two blood cell types from 100 healthy individuals as well as in mouse models; (v) develop and improve protocols for epigenomic characterization, analyses, and bioinformatics pipeline development; (vi) develop new polyclonal and monoclonal antibodies for IHEC molecular characterization, including those for specific histone modifications, and (vii) use novel RNA interference screens to identify potential epigenetic-based drug targets and develop novel compounds that interfere with epigenetic regulation through the use of a drug discovery system targeting the epigenetic assembly directly from cells or tissues. With these goals in mind, over 140 reports have been published from work performed by the BLUEPRINT consortium from 2011 to 2014. Taken together, these data serve as a rich resource of molecular alterations in blood cell differentiation and blood-based malignancies in the development of novel prognostic biomarkers and therapeutics for surveillance and treatment of blood-based diseases, including cancer and diabetes.

6.8 THE CANCER GENOME ATLAS (TCGA)

The Cancer Genome Atlas (www.cancergenome.nih.gov) is a multi-institutional consortium aimed at performing comprehensive molecular profiling of 11,000 primary tumors spanning 30 cancer types, with a target accrual of 500 tumors for each tumor type. However, the inclusion of rare tumor types was complicated by smaller numbers of available tumors, therefore, TCGA targeted analyzing 50–200 tumors for these rare cancers. Molecular profiling includes mutation detection (whole-genome and whole-exome sequencing), gene expression profiling (RNA-seq and miRNA-seq), somatic copy number alterations (single nucleotide polymorphism (SNP) arrays), and DNA methylation profiling (Illumina BeadArrays). The novelty of TCGA stems from large sample numbers, centralized pathological review, selection of samples with a high fraction of tumor nuclei, genomic characterizations over multiple assay platforms using nucleic acids isolated from the same tissue samples, and data integration for high-level pathway interpretations [57]. All molecular and clinical data are available for download from the TCGA Data Portal (https://tcga-data.nci.nih.gov/tcga).

TCGA has published high-profile integrative analyses of 16 tumor types since its inception in 2006, along with additional integrated "Pan-Cancer" analyses across 12 cancer types, and expanded analyses across larger tumor groups are planned. The Pan-Cancer analyses are powered by the fact that nucleic acids from the same tissue samples are used for each individual data production platform. Common and unique tumor subgroups, signaling pathways and associations with clinical features can be used for making clinical decisions developing novel therapeutic approaches [58]. Recently, TCGA identified 11 major molecular subgroups from integrative analyses of multiplatform data from over 3500 tumors comprising the Pan-Cancer-12 tumor panel [59]. Interestingly, many subgroups were organized by tissue of origin, somatic mutations, and expression profiles. Moreover, tissue of origin and subgroup were also prognostic indicators of patient survival.

6.8.1 CIMP SUBGROUPS OF HUMAN CANCERS

A main analytical focus of TCGA DNA methylation data has involved the characterization of novel DNA methylation-based subgroups of tumors and correlating this with the other molecular and clinical data sets. Currently, there is wide interest in DNA methylation-based subgroups of human cancers and their relationship to clinical features for diagnostic and treatment purposes. Issa and colleagues first described a distinct subset of human colorectal cancers with extensive DNA hypermethylation of a subset of CpG islands that remained unmethylated in the remaining colorectal tumors [60], and are therefore distinguished from general cancer-specific DNA methylation with the classification as positive for a CpG Island Methylator Phenotype (CIMP). Colorectal CIMP-positive tumors are enriched for $BRAF^{V600E}$ mutations, microsatellite instability (MSI), located in the proximal colon, are more prevalent in patients of older age, female gender, and demonstrate improved survival, as shown previously and confirmed by TCGA [61–63]. Subsequent from the discovery of colorectal CIMP, CIMPs have been characterized in tumors derived from brain, breast, endometrium, lung (adenocarcinoma), and stomach [64–69]. However, the molecular events that drive CIMP-specific DNA methylation for each tumor type remain elusive, as no universal presentation of CIMP is present across tumor types.

Most notably, the TCGA consortium was instrumental in characterizing a CIMP in glioblastoma (grade IV glioma). TCGA showed a near complete correlation between glioblastoma multiforme (GBM) tumors with a specific heterogeneous point mutation at the isocitrate dehydrogenase 1 (*IDH1*) gene locus ($IDH1^{\mathrm{R132H}}$) and a DNA hypermethylation phenotype (G-CIMP (glioma-CIMP)) [69] in ~15% of GBM cases. All TCGA primary GBM tumors with an *IDH1* mutation ($IDH1^{MUT}$) were G-CIMP, however, a small number of G-CIMP tumors were wild type for *IDH1* ($IDH1^{WT}$). G-CIMP tumors were enriched for *TP53* alterations, reduced copy number alterations, and correlated with improved survival and younger patient age. While $\mathrm{IDH1}^{\mathrm{WT}}$ functions in the citric acid cycle by converting isocitrate to alpha-ketoglutarate (α-KG) (reviewed in Ref. [70]), $\mathrm{IDH1}^{\mathrm{MUT}}$ also catalyzes the conversion of α-KG to D-2-hydroxyglutarate (2-HG) [71–73], resulting in elevated 2-HG levels [71,74,75], and the inhibition of TET-based DNA demethylation. TET inhibition as a result of $\mathrm{IDH1}^{\mathrm{MUT}}$ supports the hypermethylated DNA landscape in G-CIMP tumors, and this landmark mechanistic discovery has great clinical promise not only for diagnostic purposes, but also using epigenetic therapies for treatment of patients with G-CIMP tumors.

TCGA also characterized two gastric cancer CIMP subgroups from the large collection of molecular data [66]. One group, *microsatellite unstable* tumors, displayed hypermutation, MSI, *MLH1* epigenetic silencing, and a colorectal-like CIMP DNA methylation profile (Gastric CIMP). Interestingly, a second subgroup, Epstein–Barr Virus (*EBV*)*-positive* gastric tumors, are present in approximately 10% of gastric cancer cases, display EBV infection, inactivating mutations of the chromatin remodeler *ARID1A*, *CDKN2A* silencing, and a novel DNA hypermethylation signature (EBV-CIMP), in which extensive DNA methylation beyond CIMP is featured. Both the Gastric-CIMP and EBV-CIMP subgroups are additional examples demonstrating strong associations of DNA methylation with clinically unique features.

6.8.2 MGMT DNA METHYLATION IN PREDICTING UTILITY OF TEMOZOLOMIDE TREATMENT FOR GBM PATIENTS

While DNA methylation-based subgroups have immediate diagnostic utility, their exploitation for treatment has not been realized. GBM is one cancer type in which DNA methylation information drives treatment decisions. Temozolomide (TMZ) is a widely used chemotherapeutic agent for treating GBM patients. TMZ methylates the *N*-7, *O*-3, and *O*-6 nucleotide positions (reviewed in Ref. [76]), which results in DNA strand nicking and subsequent mismatch repair. *O*-6-Methylguanine lesions are particularly problematic for DNA repair systems, as the repair machinery cannot successfully incorporate the correct base opposite to this type of lesion. Consequently, DNA damage accumulates and is thought to promote an apoptotic response, resulting in cell death. MGMT (*O*-6-methylguanine methyltransferase) removes methyl groups from the *O*-6 position of guanines, thereby counteracting the efficacy of TMZ. Since GBM tumors with silenced *MGMT* via promoter DNA hypermethylation are sensitive to TMZ, *MGMT* DNA methylation is a predictive biomarker in predicting TMZ treatment efficacy. *MGMT* DNA hypermethylation was identified in nearly 50% of TCGA GBM tumors, and was also more prevalent in G-CIMP tumors compared to non-G-CIMP tumors [64]. Interestingly, *MGMT* DNA methylation correlated with patient treatment response in only a subset of GBM patients, suggesting that additional molecular factors appear to influence TMZ efficacy as a therapeutic option. These findings are of critical importance for GBM patients, and suggest further development of characterizing the molecular differences between responders and nonresponders is critical for clinical drug development and efficacy.

6.9 THERAPEUTICALLY APPLICABLE RESEARCH TO GENERATE EFFECTIVE TREATMENTS (TARGET)

The US National Cancer Institute (NCI) Office of Cancer Genomics (OCG) has additional cancer genome characterization efforts for novel therapeutic utility. One effort, Therapeutically Applicable Research to Generate Effective Treatments (http://target.nci.nih.gov), is a multi-institutional consortium aimed at characterizing the genomic alterations that are responsible for the initiation and progression of pediatric cancers for the development of innovative therapeutics. The TARGET network includes clinicians and basic/translational scientists from 11 academic institutions, institutes, and children's hospitals, and leverages existing efforts from several NCI programs, including Children's Oncology Group (COG) and Cancer Therapy Evaluation Program (CTEP) for clinical trial organization and support, Strategic Partnering to Evaluate Cancer Signatures (SPECS) and Cancer Target Discovery and Development (CTD^2) in the development of cancer-specific molecular signatures for incorporation into clinical trials, and TCGA.

The need for the development of new therapeutics by TARGET stems from several trends in pediatric oncology [77]: (i) 1:300 children will develop cancer before age 20, and pediatric cancer incidence rates have increased over the last 45 years; (ii) although 5-year survival rates have improved during this time, the majority of patients suffer painful side effects and quality of life problems including heart, lung and liver damage, high risk for secondary cancers, as well as growth, development, and cognitive impairment; (iii) 5-year survival rates show substantial differences between tumor types.

The genome characterization approach used by TARGET is similar to TCGA, in which DNA sequencing, gene expression, DNA methylation, and SNP-based copy number variation data are collected and organized from nucleic acids obtained from the same tumor for streamlined data analysis and integration. TARGET samples also are accompanied by clinical covariates, and all data are available on the TARGET data matrix (http://target.nci.nih.gov/dataMatrix). TARGET is currently analyzing the cancer genomes of ALL, AML, osteosarcoma, neuroblastoma, and kidney cancers (mostly Wilms tumors), with disease-specific data production and analyses performed at specific network sites. TARGET has recently published reports of somatic mutations and copy number alterations in neuroblastoma [78], in which alterations in the chromatin remodeler *ATRX* (alpha thalassemia/mental retardation syndrome X-linked) are present in nearly 10% of cases, as well as genome-wide DNA sequence and expression alterations in ALL [79].

6.10 INTERNATIONAL CANCER GENOMICS CONSORTIUM (ICGC)

The International Cancer Genomics Consortium (www.icgc.org) is a collection of 74 research teams from 17 countries across five continents whose goal is to generate, catalog, and analyze the complete genomic, epigenomic, and transcriptomic alterations, as well as accompanying clinical data, for 50 human cancer types that are sources of human mortality across the globe. The target accrual is at least 500 primary tumors for each tumor type, for a total of 25,000 cancer genomes. ICGC has allowed collections of fewer than 500 tumors in order to provide the most comprehensive data possible for these rare tumor types.

ICGC is structured into three modules: (i) scientific centers that generate and analyze molecular data, (ii) funding centers that provide oversight, and (iii) a scientific steering committee that provides project guidelines—with a bidirectional flow of guidance, collaboration, and information between each group. With these goals in mind, several committees and working groups have been established under the ICGC Executive Committee and the International Scientific Steering Committee. These serve as the administrative groups for all facets of the project, from data production, analyses of individual and integrated data sets, data portal access, and ethics and policy guidelines.

As with TCGA and TARGET, ICGC aims to generate DNA sequence, epigenomic and gene expression data at base-pair resolution (whenever possible) on nucleic acids from the same tumor material for seamless data integration. ICGC also aims to catalog somatic sequence alterations at a 3% frequency using normal-adjacent tissues as control. All molecular and clinical data are made available on the ICGC Data Portal (https://dcc.icgc.org). The ICGC Data Portal also includes molecular data generated by collaborating consortia, including TCGA and TARGET, for a collection of over 12,000 primary tumors across 18 tumor types.

Since 2012, ICGC has actively reported advancements in cancer-associated mutation characterization, as well as technological improvements for data production and analyses of complex sequence-based data. Of note, ICGC published a meticulous analysis of the mutational landscape of over 7000 primary tumors across 30 human cancer types, in which 21 signatures were characterized.

Interestingly, strong associations were found with age at diagnosis, smoking status, *BRCA1/ BRCA2* mutations, DNA damage, and DNA mismatch repair [80]. In a separate analysis of 402 genomes from adult and pediatric cancers, ICGC demonstrated that mutations in the X-chromosome are cancer-associated, are featured across most cancer types, are generally confined to the inactive X-chromosome in women, and are higher than autosomal mutations in female cancer patients [81].

The ICGC pancreatic cancer subgroup analyzed epigenetic regulation from primary pancreatic ductal adenocarcinomas and normal-adjacent tissues using DNA methylation and gene expression profiling [82]. Interestingly, this report identified widespread DNA hypermethylation-based epigenetic silencing. Moreover, the hypermethylated loci are enriched for genes involved in WNT, TGF-β, cell adhesion, axon guidance, and integrin signaling pathways. In addition, these analyses showed suppression of the SLIT-ROBO axon-signaling pathway, DNA hypomethylation-associated *ITGA2* and *MET* overexpression, both of which correlated with poor patient survival, and implicate their potential use as biomarkers for patient outcome prediction and exploitation for directed therapy.

6.11 STRUCTURAL GENOMICS CONSORTIUM (SGC)

The Structural Genomics Consortium was created in 2004 as a public–private partnership (PPP) with the goal of solving structures of human proteins at an unprecedented rate, and making these freely available to the scientific community. The concept was based on the successful SNP Consortium and arose out of discussions among pharmaceutical industry scientists who sought to create a precompetitive project, the output of which would be a shared resource.

Established pharmaceutical companies (Pharma), including GlaxoSmithKline (GSK), as well as public sources (Wellcome Trust, Canadian Institutes of Health Research, Genome Canada, the Canadian Foundation of Innovation, and the Ontario Ministry of Research and Innovation), provided initial funding for the SGC. In the 10 years since its inception, the SGC has deposited over 1600 structures of unique human proteins in the Protein Data Bank (PDB; www.rcsb.org) and published protocols for the expression, purification, and crystallization of these proteins on their website (www.thegsc.org). The SGC has evolved considerably since its inception, which now includes collaborations with 10 pharmaceutical companies, and is performing research that pushes the precompetitive boundary further along the drug discovery pipeline.

Since 2009, the SGC has used the structures and proteins that it generates to develop chemical probes—drug-like tool compounds for preclinical target discovery research. In a unique program, this is accomplished collaboratively with industry, these chemical probes are made available to the scientific community for use without restriction. The SGC is a collaborative, worldwide network, encompassing 550 scientists and publishes its research in peer-reviewed journals at a current rate of over 75 articles per year.

The SGC is incorporated as a charity that receives funds from consortium members and then distributes those funds as grants to the academic sites that execute the research plan (primary sites at the Universities of Oxford and Toronto). Participating granting agencies also provide direct grants to the universities in their respective countries. The SGC board is responsible for the oversight of all SGC activities and includes representatives from each participating Funding Partner. Several scientific committees, comprising both independent academic experts as well as scientists from the pharmaceutical funding companies, provide scientific oversight. Both the board and the scientific committees review progress on a quarterly basis. As the Chemical Probes project involves multiple partners (both academic and Pharma), monthly joint management committee meetings involving all participants are held, enabling the sharing of progress and nomination of novel chemical probes to be reviewed by the relevant external scientific committee.

In order to meet organizational milestones, SGC was divided into three phases. In *Phase I* (2004–2007), consortium members from the University of Toronto and Oxford University focused on the structural biology of proteins involved in metabolism, signaling pathways, chromatin biology and epigenetics, parasitology, and ubiquitin biology, with the goal of depositing 390 protein structures in the first 3 years of the project. Phase I was extremely successful, as 445 structures were characterized.

In *Phase II* (2007–2011), Swedish Granting Agencies joined the consortium to fund scientists at the Karolinska Institutet, Stockholm, who made important contributions to the structural biology of proteins involved in nucleotide metabolism and poly-ADP ribosylation (macro- and PARP domains). In addition, a new initiative was launched in Oxford to solve the structures of integral membrane proteins, a challenging aspect of structural biology ideally suited to a collaborative approach. Anticipating the explosion of interest in epigenetics, the Toronto and Oxford groups initiated a collaborative project to discover small-molecule chemical probes for epigenetic proteins. This concept and organizational model proved timely, and Novartis and Pfizer joined GSK in the Pharma aspect of the consortium. Having shown that the open-access collaborative model was successful for structural biology and chemical probes, SGC expanded this concept to include recombinant antibodies for epigenetic proteins. This was funded by Life Technologies (which also handled antibody distribution) and involved key academic collaborators from the Universities of Chicago, Toronto, and Braunschweig. Over 200 recombinant antibodies were successfully prepared and robustly validated [83]. Again, the SGC exceeded its objectives for structural biology by producing 692 novel structures, exceeding the target of 660 structures.

Phase III (2011–2015) saw the expansion of the epigenetic chemical probes program to include seven new collaborators (Eli Lilly, Abbvie, Boehringer Ingelheim, JNJ, Takeda, Bayer, and Merck). As of January 2015, 25 chemical probes for epigenetic proteins have been discovered (http://www.thesgc.org/chemical-probes/epigenetics), all of which are freely available to the scientific community without restriction on their use. The PPP for chemical probes, in which Pharma partners and SGC collaborate to discover potent, selective, cell-active ligands for epigenetic targets, is described in Figure 6.1. Once a probe is declared it is shared with key academic collaborators, which allows exploration of target validation in the public domain. After publication, there are no restrictions for dissemination of knowledge to start a proprietary program for commercial purposes. Following the successful programs for structure, chemical probes, and antibodies, additional proposals are in progress for kinase inhibitors [84], target discovery projects profiling chemical probes against patient-derived disease cells [85], and open-access clinical trials [86,87].

After 10 years of operation and contributing structural information for approximately 15% of all human proteins, pioneering the concept of open-access epigenetic chemical probes and recombinant

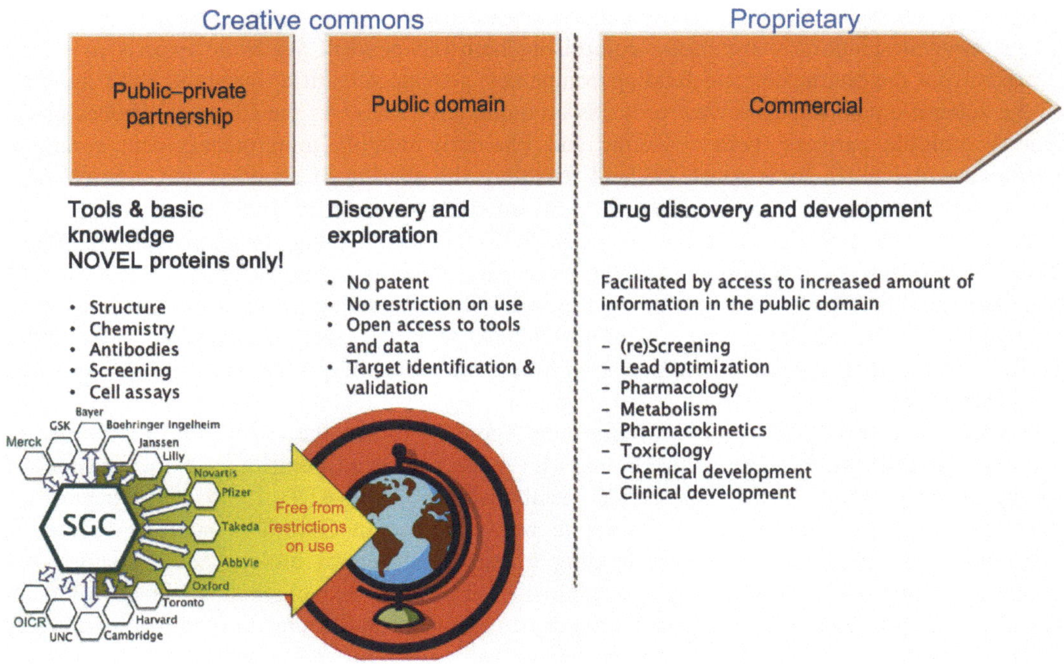

FIGURE 6.1 Operation of the Structural Genomics Consortium Chemical Probes Program.

Collaborations between the public–private partnership and the public domain ensure that human protein structures are rapidly solved and made freely available to the public. In turn, the Consortium uses the structural data to develop chemical probes for preclinical target discovery research.

antibodies, the SGC has shown that high-impact research of academic and industrial importance can be performed openly without consideration of intellectual property or personal gain. The fruits of research are shared openly with the scientific community in an effort to reduce the risk of target validation failure (usually in the clinic) and improve drug discovery efficiency. An early example of this success is JQ1, a chemical probe for the BET subfamily of bromodomains. In 2010, a GSK-enabled collaboration between SGC-Oxford and the Dana-Farber Cancer Institute demonstrated the efficacy of JQ1 in cell and animal models of the rare cancer, NUT midline carcinoma [88]. JQ1 was made available to the research community and was used extensively in cellular and animal studies by several groups demonstrating favorable effects and important insight into many cancers and models of inflammation, heart disease, and spermatogenesis [89,90]. These studies, as well as those of GSK [91], are almost certainly responsible for the launching of internal bromodomain programs in Pharma and biotech industries, leading to six clinical trials underway for BET inhibitors. Analogous discoveries have been reported for other chemical probes [92–96], and the SGC continues to use its collaborative, open-access ethos to facilitate new medicines for unmet medical needs.

6.12 STAND UP TO CANCER (SU2C) EPIGENETICS DREAM TEAM

In light of the diverse advancements and strategies in drugging the epigenome for disease treatment, their use in treating patients with solid tumors has only slowly developed. In 2008, Stand Up to Cancer (SU2C; www.standup2cancer.org) was developed from the combined efforts of the American Association for Cancer Research (AACR) and the Entertainment Industry Foundation (EIF), with the explicit requirement of using private funds to rapidly apply advancements in cancer research for use in clinical trials. In this regard, SU2C initially funded five Dream Teams, each consisting of researchers and clinicians in a multi-institutional, team-oriented format, with the goal of identifying and addressing key clinical problems in cancer treatment and expediting clinical trials with the ultimate objective of improving patient care and quality of life. One group, the Epigenetics Dream Team, is focused on developing therapeutic approaches using epigenetic-based therapies for the treatment of breast, colorectal, lung, and ovarian cancers.

The Epigenetics Dream Team showed that treatment of cell line models of human breast, colon, and lung tumors with low doses of Vidaza resulted in the reexpression of genes specific for immune response and cancer testis antigens, including interferon signaling, antigen processing and cytokines/chemokines [5,97,98]. Moreover, combinatorial treatments of HDAC inhibitors and low doses of Vidaza over several treatment cycles have been shown to also induce gene expression of genes involved in cell cycle regulation, cytoskeletal organization, and DNA damage response [99]. Indeed, clinical trials in which Vidaza and the HDAC inhibitor entinostat were administered to patients with recurrent NSCLC were overall well tolerated, and showed prolonged patient survival [100]. In addition, the Epigenetics Dream Team is currently evaluating the recently characterized DNA methylation inhibitor S110 in phase II clinical trials for treatment of ovarian, liver, and colorectal cancers, as well as AML, MDS, either alone, or in combination with other therapeutic agents.

6.13 SUMMARY AND CONCLUSIONS

In conclusion, multi-institutional consortia across the world have been instrumental in propagating genomic and epigenomic characterization of normal and diseased tissues. The findings from these consortia are unparalleled and simply not possible without collaborative, team-based approaches. These consortia have been extremely successful in not only generating molecular data in a genome-wide fashion, but also in developing bioinformatics and data processing pipelines for public use, expediting data analysis, and ultimately discovering novel genomic and epigenomic features for the development of disease surveillance and novel therapeutic strategies. These consortia have laid the foundation for future molecular characterization efforts with the goal of ending disease and improving quality of life.

ACKNOWLEDGMENTS AND DISCLOSURE

We thank the members of each consortium for their tireless efforts in genome characterization in order to improve human health and disease treatment.

Daniel J. Weisenberger is a consultant for Zymo Research Corporation, which distributes commercially available products for DNA methylation-based experiments. Zymo Research did not sponsor this material.

REFERENCES

[1] Lander ES, Linton LM, Birren B, Nusbaum C, Zody MC, Baldwin J, et al. Initial sequencing and analysis of the human genome. Nature 2001;409:860–921.

[2] Venter JC, Adams MD, Myers EW, Li PW, Mural RJ, Sutton GG, et al. The sequence of the human genome. Science 2001;291:1304–51.

[3] Bhattacharya S, Zhang Q, Andersen ME. A deterministic map of Waddington's epigenetic landscape for cell fate specification. BMC Syst Biol 2011;5:85.

[4] Jones PA. Functions of DNA methylation: islands, start sites, gene bodies and beyond. Nat Rev Genet 2012;13:484–92.

[5] Baylin SB. The cancer epigenome: its origins, contributions to tumorigenesis, and translational implications. Proc Am Thorac Soc 2012;9:64–5.

[6] Irizarry RA, Ladd-Acosta C, Wen B, Wu Z, Montano C, Onyango P, et al. The human colon cancer methylome shows similar hypo- and hypermethylation at conserved tissue-specific CpG island shores. Nat Genet 2009;41:178–86.

[7] Jones PA, Baylin SB. The fundamental role of epigenetic events in cancer. Nat Rev Genet 2002;3:415–28.

[8] Berman BP, Weisenberger DJ, Aman JF, Hinoue T, Ramjan Z, Liu Y, et al. Regions of focal DNA hypermethylation and long-range hypomethylation in colorectal cancer coincide with nuclear lamina-associated domains. Nat Genet 2012;44:40–6.

[9] Hansen KD, Timp W, Bravo HC, Sabunciyan S, Langmead B, McDonald OG, et al. Increased methylation variation in epigenetic domains across cancer types. Nat Genet 2011;43:768–75.

[10] Hon GC, Hawkins RD, Caballero OL, Lo C, Lister R, Pelizzola M, et al. Global DNA hypomethylation coupled to repressive chromatin domain formation and gene silencing in breast cancer. Genome Res 2012;22:246–58.

[11] Lister R, Pelizzola M, Dowen RH, Hawkins RD, Hon G, Tonti-Filippini J, et al. Human DNA methylomes at base resolution show widespread epigenomic differences. Nature 2009;462:315–22.

[12] Bender CM, Gonzalgo ML, Gonzales FA, Nguyen CT, Robertson KD, Jones PA. Roles of cell division and gene transcription in the methylation of CpG islands. Mol Cell Biol 1999;19:6690–8.

[13] Kulis M, Heath S, Bibikova M, Queiros AC, Navarro A, Clot G, et al. Epigenomic analysis detects widespread gene-body DNA hypomethylation in chronic lymphocytic leukemia. Nat Genet 2012;44:1236–42.

[14] Yang X, Han H, De Carvalho DD, Lay FD, Jones PA, Liang G. Gene body methylation can alter gene expression and is a therapeutic target in cancer. Cancer Cell 2014;26:577–90.

[15] Jorgensen HF, Bird A. MeCP2 and other methyl-CpG binding proteins. Ment Retard Dev Disabil Res Rev 2002;8:87–93.

[16] Pfeifer GP, Xiong W, Hahn MA, Jin SG. The role of 5-hydroxymethylcytosine in human cancer. Cell Tissue Res 2014;356:631–41.

[17] Rodger EJ, Chatterjee A, Morison IM. 5-Hydroxymethylcytosine: a potential therapeutic target in cancer. Epigenomics 2014;6:503–14.

[18] Shen H, Laird PW. Interplay between the cancer genome and epigenome. Cell 2013;153:38–55.

[19] Barski A, Cuddapah S, Cui K, Roh TY, Schones DE, Wang Z, et al. High-resolution profiling of histone methylations in the human genome. Cell 2007;129:823–37.

[20] Steger DJ, Lefterova MI, Ying L, Stonestrom AJ, Schupp M, Zhuo D, et al. DOT1L/KMT4 recruitment and H3K79 methylation are ubiquitously coupled with gene transcription in mammalian cells. Mol Cell Biol 2008;28:2825–39.

[21] Mellor J. The dynamics of chromatin remodeling at promoters. Mol Cell 2005;19:147–57.

[22] Li B, Carey M, Workman JL. The role of chromatin during transcription. Cell 2007;128:707–19.
[23] Heintzman ND, Stuart RK, Hon G, Fu Y, Ching CW, Hawkins RD, et al. Distinct and predictive chromatin signatures of transcriptional promoters and enhancers in the human genome. Nat Genet 2007;39:311–18.
[24] Schones DE, Cui K, Cuddapah S, Roh TY, Barski A, Wang Z, et al. Dynamic regulation of nucleosome positioning in the human genome. Cell 2008;132:887–98.
[25] Adams BD, Kasinski AL, Slack FJ. Aberrant regulation and function of microRNAs in cancer. Curr Biol 2014;24:R762–76.
[26] Kita Y, Vincent K, Natsugoe S, Berindan-Neagoe I, Calin GA. Epigenetically regulated microRNAs and their prospect in cancer diagnosis. Expert Rev Mol Diagn 2014;14:673–83.
[27] Yang G, Lu X, Yuan L. LncRNA: a link between RNA and cancer. Biochim Biophys Acta 2014; 1839:1097–109.
[28] Issa JP, Kantarjian HM, Kirkpatrick P. Azacitidine. Nat Rev Drug Discov 2005;4:275–6.
[29] Christman JK. 5-Azacytidine and 5-aza-2′-deoxycytidine as inhibitors of DNA methylation: mechanistic studies and their implications for cancer therapy. Oncogene 2002;21:5483–95.
[30] Cheng JC, Matsen CB, Gonzales FA, Ye W, Greer S, Marquez VE, et al. Inhibition of DNA methylation and reactivation of silenced genes by zebularine. J Natl Cancer Inst 2003;95:399–409.
[31] Cheng JC, Yoo CB, Weisenberger DJ, Chuang J, Wozniak C, Liang G, et al. Preferential response of cancer cells to zebularine. Cancer Cell 2004;6:151–8.
[32] Chuang JC, Warner SL, Vollmer D, Vankayalapati H, Redkar S, Bearss DJ, et al. S110, a 5-Aza-2′-deoxycytidine-containing dinucleotide, is an effective DNA methylation inhibitor *in vivo* and can reduce tumor growth. Mol Cancer Ther 2010;9:1443–50.
[33] Yoo CB, Jeong S, Egger G, Liang G, Phiasivongsa P, Tang C, et al. Delivery of 5-aza-2′-deoxycytidine to cells using oligodeoxynucleotides. Cancer Res 2007;67:6400–8.
[34] Arrowsmith CH, Bountra C, Fish PV, Lee K, Schapira M. Epigenetic protein families: a new frontier for drug discovery. Nat Rev Drug Discov 2012;11:384–400.
[35] Heerboth S, Lapinska K, Snyder N, Leary M, Rollinson S, Sarkar S. Use of epigenetic drugs in disease: an overview. Genet Epigenet 2014;6:9–19.
[36] Eckhardt F, Beck S, Gut IG, Berlin K. Future potential of the Human Epigenome Project. Expert Rev Mol Diagn 2004;4:609–18.
[37] Rakyan VK, Hildmann T, Novik KL, Lewin J, Tost J, Cox AV, et al. DNA methylation profiling of the human major histocompatibility complex: a pilot study for the human epigenome project. PLoS Biol 2004;2:e405.
[38] ENCODE Project Consortium. The ENCODE (ENCyclopedia Of DNA Elements) Project. Science 2004;306:636–40.
[39] ENCODE Project Consortium, Birney E, Stamatoyannopoulos JA, Dutta A, Guigo R, Gingeras TR, Margulies EH, et al. Identification and analysis of functional elements in 1% of the human genome by the ENCODE pilot project. Nature 2007;447:799–816.
[40] ENCODE Project Consortium. An integrated encyclopedia of DNA elements in the human genome. Nature 2012;489:57–74.
[41] Bernstein BE, Stamatoyannopoulos JA, Costello JF, Ren B, Milosavljevic A, Meissner A, et al. The NIH Roadmap Epigenomics Mapping Consortium. Nat Biotechnol 2010;28:1045–8.
[42] Milosavljevic A. Putting epigenome comparison into practice. Nat Biotechnol 2010;28:1053–6.
[43] Harris RA, Wang T, Coarfa C, Nagarajan RP, Hong C, Downey SL, et al. Comparison of sequencing-based methods to profile DNA methylation and identification of monoallelic epigenetic modifications. Nat Biotechnol 2010;28:1097–105.
[44] Milosavljevic A. Emerging patterns of epigenomic variation. Trends Genet 2011;27:242–50.

[45] Chadwick LH. The NIH Roadmap Epigenomics Program data resource. Epigenomics 2012;4:317–24.
[46] Satterlee JS, Beckel-Mitchener A, McAllister K, Procaccini DC, Rutter JL, Tyson FL, et al. Community resources and technologies developed through the NIH Roadmap Epigenomics Program. Methods Mol Biol 2015;1238:27–49.
[47] Roadmap Epigenomics Consortium, Kundaje A, Meuleman W, Ernst J, Bilenky M, Yen A, Heravi-Moussavi A, et al. Integrative analysis of 111 reference human epigenomes. Nature 2015;518:317–30.
[48] Dixon JR, Jung I, Selvaraj S, Shen Y, Antosiewicz-Bourget JE, Lee AY, et al. Chromatin architecture reorganization during stem cell differentiation. Nature 2015;518:331–6.
[49] Ziller MJ, Edri R, Yaffe Y, Donaghey J, Pop R, Mallard W, et al. Dissecting neural differentiation regulatory networks through epigenetic footprinting. Nature 2015;518:355–9.
[50] Tsankov AM, Gu H, Akopian V, Ziller MJ, Donaghey J, Amit I, et al. Transcription factor binding dynamics during human ES cell differentiation. Nature 2015;518:344–9.
[51] Leung D, Jung I, Rajagopal N, Schmitt A, Selvaraj S, Lee AY, et al. Integrative analysis of haplotype-resolved epigenomes across human tissues. Nature 2015;518:350–4.
[52] Amin V, Harris RA, Onuchic V, Jackson AR, Charnecki T, Paithankar S, et al. Epigenomic footprints across 111 reference epigenomes reveal tissue-specific epigenetic regulation of lincRNAs. Nat Commun 2015;6:6370.
[53] Gjoneska E, Pfenning AR, Mathys H, Quon G, Kundaje A, Tsai LH, et al. Conserved epigenomic signals in mice and humans reveal immune basis of Alzheimer's disease. Nature 2015;518:365–9.
[54] Farh KK, Marson A, Zhu J, Kleinewietfeld M, Housley WJ, Beik S, et al. Genetic and epigenetic fine mapping of causal autoimmune disease variants. Nature 2015;518:337–43.
[55] Polak P, Karlic R, Koren A, Thurman R, Sandstrom R, Lawrence MS, et al. Cell-of-origin chromatin organization shapes the mutational landscape of cancer. Nature 2015;518:360–4.
[56] Bae JB. Perspectives of international human epigenome consortium. Genomics Inform 2013;11:7–14.
[57] Weisenberger DJ. Characterizing DNA methylation alterations from The Cancer Genome Atlas. J Clin Invest 2014;124:17–23.
[58] Cancer Genome Atlas Research Network, Weinstein JN, Collisson EA, Mills GB, Shaw KR, Ozenberger BA, Ellrott K, et al. The Cancer Genome Atlas Pan-Cancer analysis project. Nat Genet 2013;45:1113–20.
[59] Hoadley KA, Yau C, Wolf DM, Cherniack AD, Tamborero D, Ng S, et al. Multiplatform analysis of 12 cancer types reveals molecular classification within and across tissues of origin. Cell 2014;158:929–44.
[60] Toyota M, Ahuja N, Ohe-Toyota M, Herman JG, Baylin SB, Issa JP. CpG island methylator phenotype in colorectal cancer. Proc Natl Acad Sci USA 1999;96:8681–6.
[61] The Cancer Genome Atlas Research Network. Comprehensive molecular characterization of human colon and rectal cancer. Nature 2012;487:330–7.
[62] Hinoue T, Weisenberger DJ, Lange CP, Shen H, Byun HM, Van Den Berg D, et al. Genome-scale analysis of aberrant DNA methylation in colorectal cancer. Genome Res 2012;22:271–82.
[63] Weisenberger DJ, Siegmund KD, Campan M, Young J, Long TI, Faasse MA, et al. CpG island methylator phenotype underlies sporadic microsatellite instability and is tightly associated with BRAF mutation in colorectal cancer. Nat Genet 2006;38:787–93.
[64] Brennan CW, Verhaak RG, McKenna A, Campos B, Noushmehr H, Salama SR, et al. The somatic genomic landscape of glioblastoma. Cell 2013;155:462–77.
[65] The Cancer Genome Atlas Research Network. Comprehensive molecular profiling of lung adenocarcinoma. Nature 2014;511:543–50.
[66] The Cancer Genome Atlas Research Network. Comprehensive molecular characterization of gastric adenocarcinoma. Nature 2014;513:202–9.
[67] The Cancer Genome Atlas Research Network. Comprehensive molecular portraits of human breast tumours. Nature 2012;490:61–70.

[68] The Cancer Genome Atlas Research Network. Integrated genomic characterization of endometrial carcinoma. Nature 2013;497:67–73.
[69] Noushmehr H, Weisenberger DJ, Diefes K, Phillips HS, Pujara K, Berman BP, et al. Identification of a CpG island methylator phenotype that defines a distinct subgroup of glioma. Cancer Cell 2010;17:510–22.
[70] Cohen AL, Holmen SL, Colman H. IDH1 and IDH2 mutations in gliomas. Curr Neurol Neurosci 2013;13:345.
[71] Dang L, White DW, Gross S, Bennett BD, Bittinger MA, Driggers EM, et al. Cancer-associated IDH1 mutations produce 2-hydroxyglutarate. Nature 2009;462:739–44.
[72] Jin G, Reitman ZJ, Duncan CG, Spasojevic I, Gooden DM, Rasheed BA, et al. Disruption of wild-type IDH1 suppresses D-2-hydroxyglutarate production in IDH1-mutated gliomas. Cancer Res 2013; 73:496–501.
[73] Ward PS, Patel J, Wise DR, Abdel-Wahab O, Bennett BD, Coller HA, et al. The common feature of leukemia-associated IDH1 and IDH2 mutations is a neomorphic enzyme activity converting alpha-ketoglutarate to 2-hydroxyglutarate. Cancer Cell 2010;17:225–34.
[74] Gross S, Cairns RA, Minden MD, Driggers EM, Bittinger MA, Jang HG, et al. Cancer-associated metabolite 2-hydroxyglutarate accumulates in acute myelogenous leukemia with isocitrate dehydrogenase 1 and 2 mutations. J Exp Med 2010;207:339–44.
[75] Sellner L, Capper D, Meyer J, Langhans CD, Hartog CM, Pfeifer H, et al. Increased levels of 2-hydroxyglutarate in AML patients with IDH1-R132H and IDH2-R140Q mutations. Eur J Haematol 2010;85:457–9.
[76] Friedman HS. Temozolomide in early stages of newly diagnosed malignant glioma and neoplastic meningitis. Semin Oncol 2000;27:35–40.
[77] Ward E, DeSantis C, Robbins A, Kohler B, Jemal A. Childhood and adolescent cancer statistics, 2014. CA Cancer J Clin 2014;64:83–103.
[78] Pugh TJ, Morozova O, Attiyeh EF, Asgharzadeh S, Wei JS, Auclair D, et al. The genetic landscape of high-risk neuroblastoma. Nat Genet 2013;45:279–84.
[79] Mullighan CG. The molecular genetic makeup of acute lymphoblastic leukemia. Hematology Am Soc Hematol Educ Program 2012;2012:389–96.
[80] Alexandrov LB, Nik-Zainal S, Wedge DC, Aparicio SA, Behjati S, Biankin AV, et al. Signatures of mutational processes in human cancer. Nature 2013;500:415–21.
[81] Jager N, Schlesner M, Jones DT, Raffel S, Mallm JP, Junge KM, et al. Hypermutation of the inactive X chromosome is a frequent event in cancer. Cell 2013;155:567–81.
[82] Nones K, Waddell N, Song S, Patch AM, Miller D, Johns A, et al. Genome-wide DNA methylation patterns in pancreatic ductal adenocarcinoma reveal epigenetic deregulation of SLIT-ROBO, ITGA2 and MET signaling. Int J Cancer 2014;135:1110–18.
[83] Ferrara F, D'Angelo S, Gaiotto T, Naranjo L, Tian H, Graslund S, et al. Recombinant renewable polyclonal antibodies. MAbs 2015;7:32–41.
[84] Knapp S, Arruda P, Blagg J, Burley S, Drewry DH, Edwards A, et al. A public-private partnership to unlock the untargeted kinome. Nat Chem Biol 2013;9:3–6.
[85] Edwards AM, Arrowsmith CH, Bountra C, Bunnage ME, Feldmann M, Knight JC, et al. Preclinical target validation using patient-derived cells. Nat Rev Drug Discov 2015;14:149–50.
[86] Edwards AM, Bountra C, Kerr DJ, Willson TM. Open access chemical and clinical probes to support drug discovery. Nat Chem Biol 2009;5:436–40.
[87] Norman T, Edwards A, Bountra C, Friend S. The precompetitive space: time to move the yardsticks. Sci Transl Med 2011;3:76cm10.
[88] Filippakopoulos P, Qi J, Picaud S, Shen Y, Smith WB, Fedorov O, et al. Selective inhibition of BET bromodomains. Nature 2010;468:1067–73.

[89] Ghurye RR, Stewart HJ, Chevassut TJ. Bromodomain inhibition by JQ1 suppresses lipopolysaccharide-stimulated interleukin-6 secretion in multiple myeloma cells. Cytokine 2015;71:415–17.
[90] Matzuk MM, McKeown MR, Filippakopoulos P, Li Q, Ma L, Agno JE, et al. Small-molecule inhibition of BRDT for male contraception. Cell 2012;150:673–84.
[91] Nicodeme E, Jeffrey KL, Schaefer U, Beinke S, Dewell S, Chung CW, et al. Suppression of inflammation by a synthetic histone mimic. Nature 2010;468:1119–23.
[92] Vedadi M, Barsyte-Lovejoy D, Liu F, Rival-Gervier S, Allali-Hassani A, Labrie V, et al. A chemical probe selectively inhibits G9a and GLP methyltransferase activity in cells. Nat Chem Biol 2011;7:566–74.
[93] Kruidenier L, Chung CW, Cheng Z, Liddle J, Che K, Joberty G, et al. A selective jumonji H3K27 demethylase inhibitor modulates the proinflammatory macrophage response. Nature 2012;488:404–8.
[94] Barsyte-Lovejoy D, Li F, Oudhoff MJ, Tatlock JH, Dong A, Zeng H, et al. (R)-PFI-2 is a potent and selective inhibitor of SETD7 methyltransferase activity in cells. Proc Natl Acad Sci USA 2014;111:12853–8.
[95] Picaud S, Da Costa D, Thanasopoulou A, Filippakopoulos P, Fish PV, Philpott M, et al. PFI-1, a highly selective protein interaction inhibitor, targeting BET bromodomains. Cancer Res 2013;73:3336–46.
[96] Yu W, Chory EJ, Wernimont AK, Tempel W, Scopton A, Federation A, et al. Catalytic site remodelling of the DOT1L methyltransferase by selective inhibitors. Nat Commun 2012;3:1288.
[97] Baylin SB. Resistance, epigenetics and the cancer ecosystem. Nat Med 2011;17:288–9.
[98] Wrangle J, Wang W, Koch A, Easwaran H, Mohammad HP, Vendetti F, et al. Alterations of immune response of Non-Small Cell Lung Cancer with Azacytidine. Oncotarget 2013;4:2067–79.
[99] Ahuja N, Easwaran H, Baylin SB. Harnessing the potential of epigenetic therapy to target solid tumors. J Clin Invest 2014;124:56–63.
[100] Juergens RA, Wrangle J, Vendetti FP, Murphy SC, Zhao M, Coleman B, et al. Combination epigenetic therapy has efficacy in patients with refractory advanced non-small cell lung cancer. Cancer Discov 2011;1:598–607.

PART

CLASSES OF EPIGENETIC DRUGS

III

CHAPTER

DNA METHYLTRANSFERASE INHIBITORS

7

Noël J.-M. Raynal[1] and Jean-Pierre J. Issa[2]

[1]Département de Pharmacologie, Université de Montréal and Sainte-Justine University Hospital Research Center, Montréal, QC, Canada [2]Fels Institute for Cancer Research and Molecular Biology, Temple University School of Medicine, Philadelphia, PA, USA

CHAPTER OUTLINE

7.1 DISCOVERY OF AZA-ANALOGS AND HISTORICAL HIGHLIGHT

In the early days of cancer chemotherapy, treatments were based on the synthesis of nucleoside analogs developed as cytotoxic agents that induced antineoplastic effects by targeting actively dividing cancer cells. In the 1960s, a group of chemists led by Dr J. Vesely at the Czechoslovak Academy of Science in Prague synthesized two cytidine analogs, 5-azacytidine (azacitidine) and 5-aza-2′-deoxycytidine (decitabine) which produced significant activity in preclinical models of leukemia [1,2]. The mechanism of action of these nucleoside analogs was not well-characterized but was considered to be mostly cytotoxic. Indeed, the knowledge of their anticancer activity was restricted to their antineoplastic effects, S-phase specificity, and incorporation into ribonucleic acid [3,4]. The *in vitro* data using these aza-analogs showing potent antileukemic activity stimulated the development of new clinical trials with azacitidine against acute lymphoblastic leukemia (ALL)

G. Egger & P. Arimondo (Eds): Drug Discovery in Cancer Epigenetics. DOI: http://dx.doi.org/10.1016/B978-0-12-802208-5.00007-2

and acute myeloid leukemia (AML). Interestingly, curative properties of these drugs were observed in mice models with leukemia, with decitabine administration of 20 mg/kg as a 15-h continuous intravenous infusion [5]. Initial clinical studies were conducted in the 1970s in children with relapsed acute leukemia. These clinical trials, using azacitidine, reported impressive activity, particularly in patients with AML resistant to cytosine arabinoside [6–8]. Complete remissions were observed in 14% of the patients. The duration of complete remissions varied between 2 and 8 months after the beginning of treatment [6–8]. In the early 1980s, clinical trials with decitabine were completed in children and adults with advanced leukemia. In this trial, decitabine produced interesting responses and complete remissions despite short duration of responses [9,10]. In these early studies, the maximum tolerated dose (MTD) was identified as around 1500–2250 mg/m^2 and was associated with delayed and prolonged myelosuppression in children that were previously treated with several rounds of cytotoxic chemotherapy [9,10]. Interestingly, these aza-analogs were administered at relatively high doses, similarly to other S-phase-specific drugs, such as cytarabine arabinoside. However, the lack of rapid responses and the lack of prolonged complete remissions reduced the interest in pursuing further the use of these new drugs. The limited knowledge regarding the mechanism of action of these aza-analogs contributed to their clinical misuse and trial failures. Indeed, the responses obtained in early clinical trials using high doses of decitabine and azacitidine caused significant cytotoxicity and produced only responses of short duration mainly in leukemia and no antitumor effects were reported against solid tumors [11].

The discovery of their mechanism of action emerged in the 1980s by Dr Peter A. Jones' laboratory showing that azanucleotides can produce cellular differentiation by inducing DNA hypomethylation [12–14]. Interestingly, cellular differentiation was observed at the lower doses which were associated with DNA demethylation [14]. The highest doses failed to induce cellular differentiation because cell viability was significantly impaired. These pioneer studies led to the emergence of the field of epigenetics as one of the most flourishing fields in cancer research and also expanded in many other fields in biology. Thus, two decades following the discovery of aza-analogs, a new era began for these molecules. Their use as classical antineoplastic drugs was beginning to be disregarded while focusing on their effects on DNA demethylation. The proof of principle of their pharmacologic activity to target DNA methylation, was confirmed rapidly by both preclinical and clinical studies from Dr Richard L. Momparler's laboratory [15,16]. Interestingly, all cancer cells regardless of their phenotype showed drug sensitivity to the antineoplastic effects of decitabine or azacitidine. Moreover, silencing of tumor suppressor genes by aberrant DNA methylation was reported in all cell types which appeared to be a promising target in cancer chemotherapy.

The growing knowledge in DNA methylation aberrations in cancer between the 1990s and now has renewed interest in decitabine and azacitidine in the clinic [17,18]. In mammals, DNA methylation is catalyzed by the addition of a methyl group on a cytosine next to a guanine of a CpG site by the DNA methyltransferase (DNMT). DNA methylation is the key component of the epigenetic machinery controlling gene expression and genome integrity [19]. The role of DNA methylation marks and its association with gene expression has been extensively studied [20]. In cancer cells, abnormal DNA methylation (local gains in promoter regions and genome-wide losses) play key roles in carcinogenesis. Silencing of hundreds of tumor suppressor genes is caused by the simple modification of cytosine into 5′-methyl-deoxycytosine in gene promoters. In recent years, it has become clear that hypermethylated promoters of tumor suppressor genes represent one of the hallmarks of human cancers which is now considered of equivalent importance to genetic mutations causing genic loss [21]. Several mechanisms playing important roles in cancer are involved in DNA methylation gain such as aging, inflammation, mutations in *DNMT*, or mutations in proteins involved directly or indirectly in DNA demethylation (such as *TET2*, *IDH1*, *IDH2*, and *WT1*) [19,22]. This epigenetic silencing of tumor suppressor genes contributes to cancer phenotype by inducing a deep reprogramming of cancer cells. Aberrant gene expression profiles will be maintained throughout cancer cell divisions. Aberrant DNA hypermethylation by silencing key tumor suppressor genes contributes to cancer cell proliferation, change in phenotype, loss of differentiation and apoptosis, and escape from the immune system [19]. The maintenance of aberrant DNA hypermethylation signature at these key genes requires the copying of DNA hypermethylation patterns throughout cancer cell divisions which are maintained by the DNMTs. Unlike genetic mutations, epigenetic silencing caused by DNA hypermethylation can be chemically reversed. Hence, pharmacologic inhibition of DNA hypermethylation became an attractive target since azanucleotides can specifically inactivate DNMT and induce DNA demethylation of dividing cancer cells [23,24].

The findings regarding DNA methylation and the activity of azanucleosides to preferentially induce terminal differentiation and DNA demethylation at low doses resulted in revisiting the concept of their mechanism of action (which is described in more details in the following sections). These drugs are considered as hypomethylating agents with a dual mechanism of action. At low doses (in the nanomolar range), azacitidine and decitabine produce anticancer effects by inhibiting DNA methylation and inducing tumor suppressor gene reactivation, which is dose-dependent [25]. However, at higher doses (in the micromolar range), these drugs also induce cytotoxic effects likely due to DNA damage by inducing numerous DNA adducts caused by the formation of covalent bonds between DNMTs and the DNA [26]. Thus, new clinical trials were designed to optimize the dosage regimens to favor the epigenetic mechanism of these drugs at low doses with prolonged exposure rather than high doses that produce cytotoxic effects in both cancer cells and normal cells. Interestingly, low doses with long exposures were more effective than higher doses in patients with myeloid malignancies which validated their clinical use as DNA demethylating drugs rather than cytotoxic drugs [27]. The clinical trials focused mainly on hematologic malignancies and in particular on myelodysplastic syndromes (MDS) and AML. So far, clinical trials in solid tumors with DNA demethylating drugs as monotherapy have not revealed promising effects in patients with advanced cancers [28].

The efficacy of azacitidine and decitabine against myelogenous malignancies led to their approval by the US Food and Drug Administration (FDA) and the European Medicines Agency (EMA). In the United States, azacitidine and decitabine were approved in 2004 and 2006, respectively, for the treatment of MDS. In Europe, azacitidine was approved in 2008 for the treatment of MDS, AML, and chronic myelomonocytic leukemia while decitabine was approved in 2012 for the treatment of AML [29,30].

7.2 THE PHARMACOLOGY OF DNMT INHIBITORS

7.2.1 CHEMISTRY OF AZANUCLEOSIDE ANALOGS

DNMT inhibitors azacitidine or decitabine are analogs of cytidine or 2′-deoxycytidine, respectively (Figure 7.1). The chemical difference between the natural bases and their corresponding

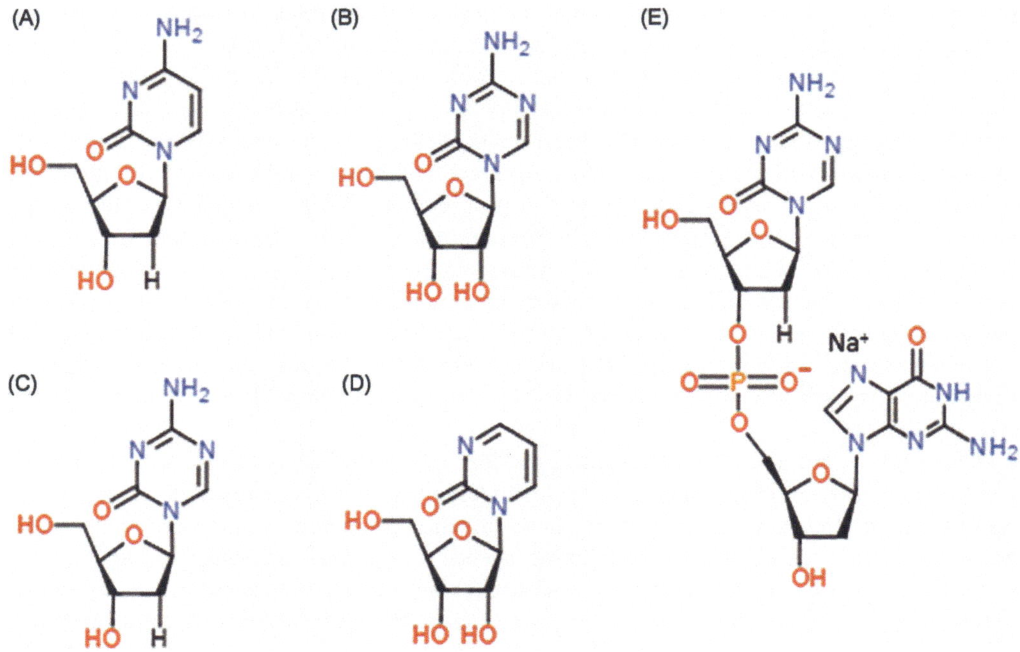

FIGURE 7.1

Chemical structure of (A) cytosine, (B) azacitidine, (C) decitabine, (D) zebularine, and (E) SGI-110.

drug analogs is that the carbon in the fifth position of the pyrimidine ring has been exchanged by a nitrogen. The molecular weights of azacitidine and decitabine are 244.20 and 228.21 g/mol, respectively. The presence of the nitrogen in the pyrimidine ring makes these drugs, while in solution, chemically unstable depending on temperature and pH. At pH 7 and 37 °C in aqueous solution, the half-lives of azacitidine and decitabine are 7 and 12 h, respectively [31]. A reduction in the temperature to 5–10 °C increases drug stability close to 98% for more than 8 h [17]. Chemical stability is an issue that has to be taken into account for drug preparation and drug administration both in the laboratory and in the clinic.

7.2.2 MOLECULAR PHARMACOLOGY OF DECITABINE

Decitabine or 5-aza-2′-deoxycytidine enters the cell by the equilibrative nucleoside transporters hENT1, hENT2, and hENT3, like the natural cytosine (Figure 7.2). Decitabine is a prodrug that has to be converted into a nucleoside triphosphate to become an active drug. Once in its triphosphate form, the active drug can be incorporated into the DNA of dividing cells by the DNA polymerase. The activation of the prodrug relies on three successive phosphorylation steps. The first phosphorylation is catalyzed by the deoxycytidine kinase (DCK) to produce the monophosphorylated form of decitabine or 5-aza-deoxycytidine-5′-monophosphate (5-aza-dCMP). Subsequently, two additional

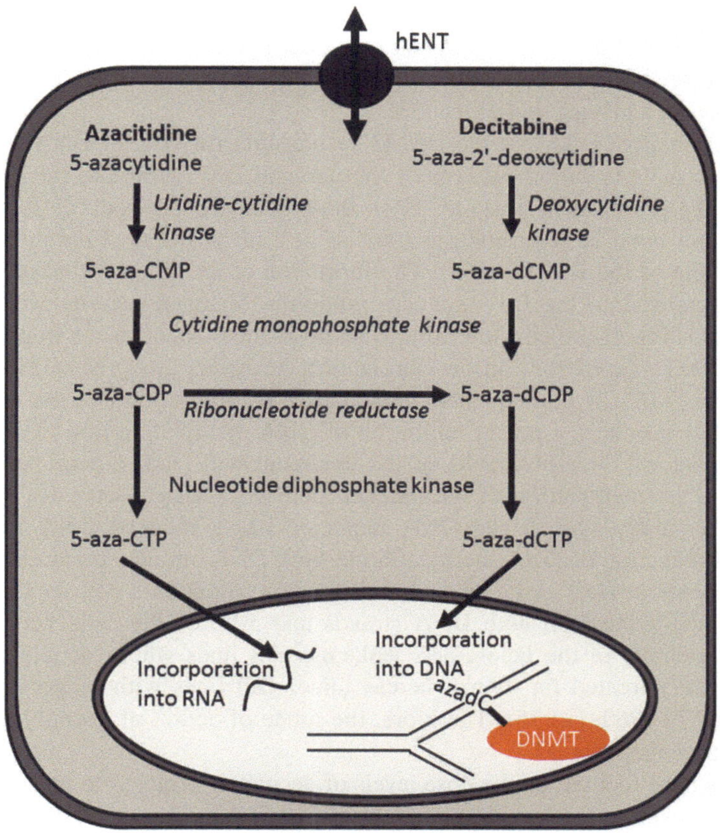

FIGURE 7.2

Molecular mechanisms of azacitidine and decitabine activation into active drugs.

phosphorylation reactions catalyzed by the deoxycytidine-5′-monophosphate (dCMP) kinase and the nucleoside diphosphokinase will produce 5-aza-deoxycytidine-5′-diphosphate (5-aza-dCDP) and 5-aza-deoxycytidine-5′-triphosphate (5-aza-dCTP), respectively [17]. These three chemical reactions are necessary to generate the active drug, 5-aza-dCTP, which can be incorporated into the DNA of dividing cells (Figure 7.2). The active form is a good substrate for the DNA polymerase that has a similar affinity for the natural dCTP to 5-aza-dCTP. Thus, decitabine is easily incorporated instead of the natural dCTP during the S-phase of the cell cycle [18,32]. Due to its pharmacologic properties, decitabine belongs to the family of S-phase-specific drugs, meaning that it is only active when incorporated in the DNA of cells in the S-phase of the cell cycle. The S-phase specificity enables selective and effective incorporation of decitabine into the DNA of rapidly dividing cancer cells, reducing, thereby, demethylation effects in normal cells which have a slower replicating rate. Unlike other S-phase-specific drugs, such as cytarabine arabinoside or gemcitabine, decitabine does not immediately block the cell cycle (even at micromolar concentrations), making it a weak cytotoxic drug. The effects of decitabine on cancer cell growth are delayed and in some cancer cell lines it becomes significant only after two or three cell divisions, that is, growth inhibition can be detected after 48–72 h posttreatment. In addition, in mammalian cells, decitabine is not known to produce mutagenic lesions [33].

During cell replication, DNMT1, the DNMT responsible for 90% of DNA methylation maintenance, copies DNA methylation patterns of the parental DNA strand to the newly synthesized DNA molecule. The presence of 5-aza-dCTP in the nascent DNA results in the formation of an irreversible covalent bond between the nitrogen in the fifth position of the modified pyrimidine and the catalytic site of the DNMT1 [34]. The formation of an adduct between the DNMT1 and decitabine incorporated into the DNA induces ubiquitin-E3 ligase activity which results in the proteasomal degradation of the enzyme [35]. In addition, decitabine seems to promote the degradation of free DNMT1 (not bound to the chromatin); however, this precise mechanism remains to be characterized [35]. The consequence of decitabine incorporation into the DNA and the degradation of DNMT1 produces a potent inhibition of DNA methylation into the DNA of daughter cells after cell division [17]. The results of the treatment will lead to hemimethylated daughter cells, where the DNA methylation of the parental cell will still conserve its DNA methylation level, whereas the newly synthesized DNA molecule has a reduced DNA methylation level. Thus, to obtain a reduction of DNA methylation in both DNA strands, the duration of decitabine treatment has to cover at least two cycles of cell division to obtain a cell population where 50% of them are hypomethylated on both DNA strands and 50% of the cells with hemimethylated DNA [17]. For example, in the laboratory, leukemic cell lines whose doubling time is around 18–24 h are typically treated for 48 h, whereas tumor cell lines with longer doubling time are usually treated for 72–96 h [25,36]. Therefore, the mode of action of decitabine and azacitidine is qualified as time-dependent.

Several studies have focused on the dose levels of decitabine required to induce DNA demethylation. Interestingly, even in the low nanomolar range, decitabine can significantly demethylate the DNA. In a dose range from low nanomolar to low micromolar concentration, decitabine induces a dose-dependent DNA demethylation effect [25]. Importantly, these low doses are compatible with cell division; that is, cancer cells survive decitabine exposure and will not slow down their growth rate, until two to three cell cycles are completed posttreatment. Apoptosis and cell differentiation are then observed in significant number of cancer cells arising several days after treatment. This phenomenon is referred to the delayed action of this epigenetic drug. However, at higher doses starting at the low micromolar range and above, decitabine produces a pronounced cytotoxic effect through an increase in DNA adducts between the DNA strand and the DNMT. Consequently, high doses of decitabine in cancer cells produce cytotoxic effects that result in stopping cell cycle progression. Even though the DNMT1 enzymes are trapped, DNA demethylation levels cannot be detected using high doses of decitabine. At high doses, cancer cells will stop their proliferation, leaving no option to synthesize demethylated DNA strands. Thus, the demethylating effect of decitabine is described by a U-shaped curve where a low-dose-dependent demethylation is observed until DNA methylation is no longer measured at higher doses [19]. DNA demethylating drugs are often described as drugs with a dual mechanism of action, being epigenetic at low doses and cytotoxic at higher doses [19]. Doses that induce cytotoxic effects are associated with unacceptable toxicity against normal cells; thereby limiting dose regimens in the clinic to doses that are considered to produce epigenetic activity [11].

Decitabine induces DNA demethylation and gene reactivation of silenced genes by DNA hypermethylation, at least, at concentrations that do not inhibiting significantly cell growth immediately after drug treatment [37–39]. Molecular studies have confirmed using single-gene approaches or whole-genome next-generation sequencing technologies that drug-induced DNA demethylation at the promoter level leads to the reactivation of silenced tumor suppressor genes in cancer cells such as *TIMP3*, *p15*, *p16*, *e-cadherin*, and *RASSF* [19,40]. Promoter DNA demethylation has been studied for several decades as a target of epigenetic therapy of cancer. It is noteworthy that decitabine-induced DNA demethylation targets all regions of the DNA, including promoters and gene bodies. Genome-wide analysis has confirmed that DNA hypermethylation found in gene bodies correlates well with gene expression [41]. The function of DNA methylation in gene bodies is believed to potentially repress alternative promoters, retrotransposon elements, and other functional elements to ensure transcription efficacy. Hypomethylation of gene bodies, induced by demethylating drugs, was shown to particularly downregulate sets of genes that are involved in metabolic processes and oncogenes regulated by c-MYC [42,43]. Interestingly, high levels of c-MYC target genes seem to rely on the DNMT3B activity, another DNMT known as a *de novo* methyltransferase. Surprisingly, DNA methylation induced by DNMT3B in gene bodies rapidly increased after decitabine treatment cessation [42]. Thus, DNMT3B could represent an interesting therapeutic target to prevent DNA methylation rebound in gene bodies of oncogenes and other c-MYC targets. This study suggests the possibility that DNMT inhibitors contribute to inhibit c-MYC-associated genes, which are overexpressed in virtually all cancers [44]. Despite a lack of demethylation specificity of DNMT inhibitors, remethylation kinetic after treatment withdrawal is dependent on a chromatin context whereby promoter hypomethylation is maintained while gene body methylation is short-lived. The effect of DNMT inhibitors of gene bodies revealed a new target for these drugs causing decrease expression of overexpressed metabolic genes or oncogenes [42].

7.2.3 MOLECULAR PHARMACOLOGY OF AZACITIDINE

Like decitabine, azacitidine is an analog of the natural cytosine where the carbon in the fifth position of the pyrimidine ring has been exchanged by a nitrogen (Figure 7.1). The only difference between azacitidine and decitabine is that the sugar moiety of azacitidine is a ribose whereas decitabine has a deoxyribose molecule. Azacitidine can cross the cytoplasmic membrane through human concentrative nucleoside transporter-1 and becomes phosphorylated by the uridine-cytidine kinase, the uridine-monophosphate-cytosine-monophosphate kinase, and finally by the nucleoside diphosphate kinase to generate the 5-azacytidine-triphosphate (Figure 7.2). The triphosphate form of azacitidine is mainly incorporated into newly synthesized RNA molecules where it disrupts RNA metabolism, RNA methylation, and protein synthesis [45]. Thus, azacitidine can cause substantial cytotoxicity in its ribonucleotide form [2,26,46]. These effects may influence both cancer and normal cells, resulting in greater *in vitro* and *in vivo* side effects. Moreover, azacitidine can be converted into a deoxyribonucleotide DNA basis by the ribonucleotide reductase. This pathway will lead to the conversion of 5-aza-CDP to 5-aza-dCDP which will then be phosphorylated into 5-aza-dCTP, the active decitabine molecule. Around 10% of azacitidine will be incorporated into the DNA and will inhibit DNA methylation by the same mechanism of action as decitabine.

In total, the hypomethylation activity of azacitidine has only about 10% the potency of decitabine since most of azacitidine (80–90%) is incorporated into RNA [18].

7.2.4 DISTRIBUTION AND METABOLISM

The pharmacokinetic parameters of decitabine were initially described in the early clinical trials in pediatric patients with acute leukemia or in adult patients with solid tumors at elevated concentrations of decitabine at steady state [9,47]. More recently, Cashen et al. developed a model based on low doses of decitabine [48]. Interestingly, all studies found that the plasmatic half-life of decitabine and azacitidine in human is around 15–25 min. The short half-lives of both azacitidine and decitabine are the results of the enzymatic degradation of the cytidine deaminase (CDA). These enzymes normally decompose natural pyrimidine but can metabolize the azanucleotide prodrugs and their mono-, di-, and triphosphate forms. These enzymes render these drugs inactive by converting them into 5-azauridine compounds. The high level of CDA in the liver and spleen cells is largely responsible for the short plasmatic half-lives of these drugs.

Azanucleotides in human have an extensive tissue distribution with a mean volume at steady state of 62.7–89.2 l/m^2 [47,48]. Interestingly, decitabine could be detected at therapeutic levels in the cerebrospinal fluids of pediatric patients with leukemia with CNS involvement [9]. The ability of decitabine to cross the blood–brain barrier was established in animal studies [49].

7.2.5 MOLECULAR MECHANISMS OF RESISTANCE TO DNMT INHIBITORS

Resistance mechanisms to DNMT inhibitors have been observed in cancer cell lines and in cancer patients. In the clinic about half of the patients respond to hypomethylating agents and the majority of responders tend to relapse within 2 years of treatment despite continuation of therapy [19,27,50,51]. Unfortunately, the outcome of patients is poor after failure of DNMT inhibitors with an overall survival time below 6 months [52–54]. Since there are no therapeutic options for patients after failure of DNMT inhibitors, understanding the mechanism of resistance is an important area of research. The mechanisms of resistance to azacitidine or decitabine are believed to be associated with either alterations in drug pharmacological pathways or epigenetic tolerance (which refers to cancer cell tolerance and survival with hypomethylated DNA).

Historically, the first studies describing drug resistance mechanisms to DNMT inhibitors focused on drug activation pathways [55,56]. Specifically, resistance was observed at the initial phosphorylation step of decitabine by the DCK. This rate-limiting step is critical for decitabine conversion from a prodrug to its active form. A reduction of expression or the genetic inactivation of DCK reduces decitabine's activation, thereby decreasing its incorporation into DNA and its demethylation activity [55,57–59]. Cells lacking DCK activity are completely resistant to decitabine or other cytosine analogs such as cytarabine arabinoside [56].

A second pharmacologic mechanism involved in decitabine resistance is its degradation by the CDA, an enzyme involved in pyrimidine turnover [17,60]. The CDA deaminates decitabine to 5-aza-deoxyuridine, thus reducing the amount of active drug leading to less epigenetic and antineoplastic activity. Azacitidine and decitabine are sensitive to CDA activity, which decomposes both prodrugs and their mono-, di-, and triphosphate forms. Interestingly, it was reported that CDA are overexpressed in several cancer cell lines, which is associated with DNMT inhibition resistance [58].

CDA are also highly expressed in hepatocytes which are responsible for the degradation of several cytosine analogs including decitabine, making this organ a sanctuary where cancer cells can survive chemotherapy.

In preclinical models of leukemia, Qin et al. observed that the half maximal (50%) inhibitory concentration (IC_{50}) of decitabine can have a 1000-fold difference between cancer cell lines [58]. The IC_{50} values correlated well with DNA demethylation of repeated elements such as long interspersed nuclear elements. Interestingly, decitabine sensitivity correlated with cytarabine sensitivity. Since both drugs share the same pharmacological activation pathway, the data confirm the possibility of a pharmacologic mechanism of resistance. Moreover, resistant cancer cells show a pattern of low expression of DCK and hENT and high expression of CDA [58]. In addition, decitabine resistance was also found to be associated with homozygous mutations of DCK, thereby blocking decitabine activation. Therefore, these resistance mechanisms are preferentially selected by cancer cells to reduce or block DNMT inhibitors' incorporation into the DNA by decreasing DCK expression and increasing their degradation by CDA. Importantly, the presence of these specific alterations in gene expression of DCK and CDA was confirmed in a patient sample in clinical trials in MDS [59]. Indeed, Qin et al. demonstrated that a gene expression ratio of CDA/DCK was threefold higher in nonresponders as compared to responders, suggesting that this mechanism could be involved in primary resistance.

Targeting pharmacologic resistance to decitabine using chemotherapeutic approaches has been explored by several investigators. Several strategies to circumvent the DCK-associated resistance mechanism were tested both in preclinical models and clinical trials. In the case of decitabine resistance, one possibility is to switch to azacitidine therapy. Indeed, azacitidine does not rely on DCK for pharmacologic activation. Pharmacological activity of azacitidine after decitabine failure has been reported several times [6,30]. An additional strategy to overcome DCK-associated resistance is to use 3-deazauridine (3-DU). The active phosphorylated form of 3-DU inhibits CTP synthetase and thus reduces the intracellular level of CTP and dCTP. Cancer cells that are DCK-deficient are dependent on the activity of CTP synthetase as the only other cellular pathway able to generate CTP and dCTP necessary for DNA synthesis. Thus, pharmacologic inhibition of CTP synthetase in DCK-deficient cells generates a cytotoxic effect targeting specifically these cancer cells since dCTP is essential for DNA synthesis. In preclinical studies and pilot clinical trials, 3-DU was shown to overcome DCK-associated resistance to decitabine [57]. In addition, the effect of 3-DU by reducing the pool of dCTP, increases the incorporation of decitabine into the DNA which results in a synergistic antineoplastic activity [55,57].

Pharmacologic approaches are also currently tested to reduce CDA activity to avoid DNMT inhibitor degradation. CDA inhibition, by tetrahydrouridine or zebularine, produces synergistic effects in combination with decitabine and continues to be under investigation [61–63]. Interestingly, the combination of tetrahydrouridine was reported to synergize with decitabine at very low doses. The inhibition of CDA activity in the liver increases the half-life of decitabine in this sanctuary organ for cancer cells. The combination of tetrahydrouridine and decitabine was shown to significantly reduce the presence of liver metastasis [62,63].

Other mechanisms may also contribute to resistance to DNMT inhibitors. Secondary resistance to demethylating drugs was also found to occur despite DNA demethylation suggesting the possibility of the existence of nonpharmacologic mechanisms of drug resistance [59,64]. Indeed, clinical samples showed similar levels of DNA hypomethylation when isolated at relapse than during responses to therapy [59,65]. To support this notion, genetically modified cancer cell lines lacking DNMT3b and hypomorphic for DNMT1 can survive with hypomethylated DNA [66]. Cancer cell tolerance to this hypomethylated state is not clearly understood. However, it is likely that cancer cells can use alternative mechanisms to repress key tumor suppressor genes by using repressive chromatin marks, thereby overcoming the DNA demethylation effect. Interestingly, several studies have highlighted the importance of the chromatin state in association with gene reactivation in cancer. It was demonstrated that drug-induced DNA demethylation was not sufficient to trigger gene reactivation and that chromatin repressive marks present in the promoter regions of demethylated genes can efficiently maintain gene silencing [36,67]. Thus, cancer cells have the possibility to use an alternative epigenetic mechanism using chromatin to repress the demethylated genes and thereby become resistant to DNMT inhibitors.

In conclusion, several mechanisms have been identified to contribute to DNMT inhibitor resistance. These data provide the rationale to use DNMT inhibitors in combination with other drugs to overcome drug resistance by using CDA inhibitors or targeting chromatin repressor complexes [19].

7.3 THE CLINICAL TRIALS IN HEMATOLOGICAL MALIGNANCIES LEADING TO APPROVAL OF DNMT INHIBITORS

Historically, the early clinical trials with azacitidine and decitabine were performed prior to understanding their epigenetic mechanism of action. These drugs were used in trials mainly in leukemia at high cytotoxic doses [6,9]. Early clinical trials with decitabine described first complete remissions with high dosage and long infusions (36 h of continuous infusion, producing plasma levels around 1.75 μM at steady state). Even though some patients went into complete remission after the first cycle, most responders had extensive leukemic blast cells in their bone marrow at day 21, indicating a delayed action of the decitabine. However, in these studies which included both ALL and AML patients, the remission rate was about 22% but the duration of responses was generally short [10]. Although, complete remissions were reported mainly in children, the results of these trials were associated with hematopoietic toxicity, which reduced the interest in these drugs. Interestingly, drug resistance was detected in two patients due to DCK reduction and increase in CDA activity [68].

The discovery of the mechanism of action related to DNMT inhibition and the possibility to reverse DNA hypermethylation in cancer have stimulated several investigators to pursue additional clinical trials [69]. The use of azacitidine and decitabine was resurrected by pioneering clinical studies that were designed around the concept that low drug dosage and long exposure could increase the therapeutic activity of azanucleotides [27,45,50,51,70]. Indeed, these drugs showed more potent activity at doses below 10% of their MTD [11].

The focus of clinical trials with decitabine shifted toward myeloid leukemia after the results showing activity in advanced patients with MDS resulting in hematological and complete responses [71,72]. Several dose regimens were explored using low doses. In one pioneer study, decitabine was administered as an intravenous infusion of 45–50 mg/m^2/day for 3 days in high-risk MDS patients showing a response rate in half of the patients and four complete hematologic responses [71]. Subsequent studies focused on the evaluation of different drug schedules in elderly patients with high-risk MDS. Firstly, a dose regimen assessed the activity of a continuous infusion for

3 days every 6 weeks of decitabine at 50 mg/m^2/day [72]. Another study in MDS tested decitabine infusion at 15 mg/m^2 IV over 3 h every 8 h for 3 days every 6 weeks [73]. Results of both studies demonstrated overall response rates in 50% of the patients and complete remission rates of 20–29%. However, median survival in patients who achieved complete remissions was superior using the 15 mg/m^2 study (19 months) as comparted to the 50 mg/m^2 study. These results were confirmed in a retrospective analysis of 177 elderly patients with intermediate/high-risk MDS which showed the superiority of low-dose versus high-dose regimens [74].

Clinical trials with MDS patients comparing decitabine at 15 mg/m^2 IV over 3 h every 8 h for 3 consecutive days (135 mg/m^2/course) every 6 weeks to best supportive care showed that high-risk patients experienced a longer median survival than patients treated with best supportive care (12 vs 6.8 months, respectively; $P = 0.03$) [50]. The results of this study led to decitabine FDA approval in the United States. The major toxicity associated with decitabine was febrile neutropenia and grade 3 or 4 nonhematologic toxicities consisted of hyperbilirubinemia and pneumonia. The efficacy of decitabine in MDS was confirmed in a European study using a similar dose-schedule showing a higher overall response rate and a longer progression-free survival as compared to best supportive care [75]. The median duration of the hematological responses was about 9–10 months [50,75].

Interestingly, other schedules and routes of decitabine administration were explored. A group at the University of Texas MD Anderson Cancer Center explored the impact of a dose-intense schedule and its association with DNA demethylation [76]. Three low-dose decitabine dose-intense regimens were tested based on the rationale that repeated low-dose exposures target more cancer cells in S-phase. These schedules consisted of 100 mg/m^2/course which were administered every 4 weeks as (i) a 10 mg/m^2 intravenous infusion over 1 h daily for 10 days or (ii) 20 mg/m^2 intravenous infusion over 1 h daily for 5 days, or (iii) 20 mg/m^2 subcutaneous injection daily for 5 days. The 5-day schedule using intravenous infusion produced the best complete remission rate (39%) and demethylation effects [76]. The results of this new dose-schedule were confirmed in a phase II, multicentric single-arm clinical trial [77]. Even though decitabine trials in MDS confirmed response rates and the delayed progression to AML (40% complete response, over 70% total response), these data did not show an improvement in survival time [50,75]. It is noteworthy that these studies used a limited number of treatment cycles [76,78]. However, when decitabine was administered until disease progression or death, the treatment resulted in a very high response rate and survival rates were superior to the survival of patients treated with cytotoxic chemotherapy. Decitabine administered at a dose of 20 mg/m^2 by a 1-h intravenous infusion for 5 consecutive days of a 4-week cycle has been approved by the United States for use in adult patients with MDS [29].

A single-agent decitabine was also tested against refractory or relapsed acute leukemia. An early phase I study performed by Rivard et al. explored escalating high doses (plasmatic concentration at steady state estimated at 1 μM) and long infusions (8–44 h) in children with relapsed and refractory acute leukemia [9,10]. Complete remissions were achieved in 4 of 21 ALL patients and 2 of 6 AML patients and antileukemic effects (including partial remissions) were observed in 86% of the patients. The major toxicity of decitabine was myelosuppression. Another study by Petti et al. reported a 25% complete remission rate in poor-prognosis AML patients with decitabine administered as an intravenous infusion at 90–120 mg/m^2 over 4 h, three times a day, for 3 days, every 4–6 weeks [79]. More recently, Issa et al. demonstrated that lower doses of decitabine ranging from 5 to 20 mg/m^2 given as intravenous infusion, over 1 h daily, for 5 days a week, showed some efficacy in AML and in CML [27]. This approach inspired other investigators to explore low-dose regimens in patients with relapsed or refractory AML and older *de novo* AML patients. Low-dose decitabine (20 mg/m^2 given as intravenous infusion, daily for 10 days, every 4 weeks) was tested in elderly AML patients producing complete responses in 47% of the patients. The median overall survival was 13.8 months and the median disease-free survival for patients with complete response was 11.5 months [80]. Cashen et al. also tested low-dose decitabine in previously untreated AML patients over 60 years old who were not eligible to cytotoxic chemotherapy [81]. In this study, complete responses were observed in 24% of the patients including patients with poor-risk cytogenetics. Toxicity profiles were similar to what was reported in MDS studies such as myelosuppression, febrile neutropenia, and fatigue. Decitabine (20 mg/m^2 in 1-h intravenous infusion for 5 consecutive days of a 4-week cycle) has been approved by the EMA and the FDA in adult patients aged ≥ 65 years with *de novo* or secondary AML who are not candidates for standard induction therapy [29,82].

Using a similar approach to decitabine, azacitidine was tested in multiple international randomized trials at low doses for several cycles of administration (Table 7.1). Phase III studies in MDS with azacitidine (administered until progression or death) produced responses ranging from 30% to 60%, which resulted in improved survival rates as compared to supportive care or cytotoxic chemotherapy (low dose ara-C) [51,70]. Azacitidine was approved in the United States for the treatment of MDS and in Europe for patients with intermediate-2 and high-risk MDS who are not eligible for hematopoietic stem cell transplantation, chronic myelomonocytic leukemia with 10–29% blasts and AML with 20–30% blasts and multilineage dysplasia [29]. In the clinic, azacitidine is administered at 75 mg/m^2 daily for 7 days, every month. A delay in starting the next cycle or a dose reduction may be necessary if a patient needs to recover from the drug-induced hematopoietic toxicity. Azacitidine treatment for a minimum of six cycles is recommended before evaluating the effects of treatment since more than 90% of responders achieved initial response after six cycles [83]. Treatment continuation after initial response was shown to improve the response in 48% of patients and the median duration of the hematological response was about 13 months [51,70,84].

The effects of demethylating drugs differ from cytotoxic therapies. Azacitidine and decitabine typically do not induce responses after the first cycle, which is often observed with chemotherapy. It is common to see patients with no evidence of therapeutic activity for 3 months before the achievement of complete remission with continuation of therapy [11,89]. Responses to hypomethylating drugs are mainly seen in patients who were not previously treated and after several cycles of chemotherapy [11]. Except reversible hematopoietic toxicity, the side effects are quite different compared to cytotoxic chemotherapy. Usual chemotherapy side effects such as mucositis, hair loss, diarrhea, and renal failure are quite rare with demethylating agents [11].

Typical clinical failures to azacitidine or decitabine therapies can be divided into three groups: (i) one-third of patients progress to AML; (ii) one-third of patients do not respond and still suffer from progressive MDS; (iii) one-third of patients decide to stop further therapy or die [30]. Due to poor performance status, most of these patients are not eligible for intensive chemotherapy or allogeneic stem cell transplant after epigenetic therapy failure. Thus, the majority of these patients will need low-intensity therapies which have not yet been identified. Further investigations are clearly needed to clarify the mechanisms of primary or secondary resistance to DNMT inhibitors and to identify the therapeutic option after DNMT inhibitor failure.

Table 7.1 Selected Clinical Trials with DNA Methylation Inhibitors

DNMT Inhibitor	Dose/Schedule	Disease	Results	Comments	References
Azacitidine	75 mg/m^2, SQ, daily ×7, every 4 week	MDS	CR 7%, PR 16%, HI 37%	Improved PFS vs supportive care	[70]
Azacitidine	75 mg/m^2, SQ, daily ×7, every 4 week	MDS	CR 17%, PR 12%, HI 20%	Improved survival vs conventional therapy	[51]
Decitabine	15 mg/m^2, IV over 3 h every 8 h for 9 doses, every 6 week	MDS	CR 24%, PR 10%, HI 14%	Well tolerated in patients older than 75 years with bad prognostic	[74]
Decitabine	15 mg/m^2, IV over 3 h every 8 h for 9 doses, every 6 week	MDS	CR 9%, PR 8%, HI 13%	Improved PFS vs supportive care	[50]
Decitabine	5–20 mg/m^2, IV over 1 h daily ×10	Refractory or relapsed leukemia	CR 18%, PR 6%, HI 8%	Low doses induce greater responses	[27,,85]
Decitabine	20 mg/m^2, IV over 1 h daily ×5, every 4 week	MDS	CR 34%, PR 1%, HI 37%	Improved survival vs conventional therapy	[76,78]
Decitabine	20–40 mg/m^2/day 3-day continuous infusion (28-day cycle)	Metastatic solid tumors	PD 14/14	CR 0%	[28]
Decitabine	2 mg/m^2/day 7-day continuous infusion (28-day cycle)	Solid tumors (ovarian, renal, breast, colon)	SD 2/9, PD 7/9	22% of response (including SD)	[86]
Decitabine	20–30 mg/m^2/day 3-day continuous infusion (35-day cycle)	Solid tumors (lung, esophagus, and pleura)	SD 13/28, PD 15/28	46% of response (including SD)	[87]
Decitabine	2.5–20 mg/m^2/day 1-h infusion for 5 or 10 days	Solid tumors and lymphomas	PR 1/28, SD 17/28	64% of response (including SD)	[88]

CR, complete remission; PR, partial remission; HI, hematological improvement; PFS, progression-free survival; SD, stable disease; SQ, subcutaneous; IV, intravenous.

7.4 BIOMARKERS OF CLINICAL OUTCOME WITH DNMT INHIBITOR THERAPY

Measurements of DNA demethylation induced by azacitidine and decitabine were studied as biomarkers of clinical response. Drug-induced DNA demethylation was observed in patients in several studies and most importantly in patients treated with low doses, thereby validating the concept of low-dose therapy [85,90]. Some studies with decitabine showed a correlation between DNA demethylation in multiple genes and clinical responses [76,91]. Reduction in DNA methylation after decitabine therapy in leukemia patients was dose-dependent and peaked 10–15 days after the initiation of therapy. Interestingly, DNA hypermethylation levels were restored to baseline at 4–6 weeks posttreatment [26,89,92,93]. Classical examples of tumor suppressor gene demethylation and associated gene reactivation were described for *p15* and *ER* [76,85,94–96]. Interestingly, tumor suppressor gene demethylation was not identified in all studies [92,94,96]. The correlation between the levels of DNA demethylation and clinical responses to decitabine has not been clearly established [38,92]. The demethylation effect was not confirmed with azacitidine probably because this drug is 10 times less potent at inducing DNA demethylation than decitabine [46,97]. Intriguingly, in chronic myelogenous leukemia, there is an inverse relation between DNA demethylation and clinical responses [19].

Correlative studies are always difficult to analyze since several factors may modify the results, such as elimination of demethylated clones, the limited number of tumor cells in patient samples, and the difference in drug activity when comparing decitabine and azacitidine [65]. Changes in cancer cell population with possible clonal replacement by resistant cells to drug-induced hypomethylation while on therapy bring more complexity to the analysis. Moreover, one can speculate that if the treatment is effective, most of the hypomethylated cells will die and will be eliminated; making the detection of DNA hypomethylation almost impossible with the current technologies.

7.5 LACK OF EFFICACY OF DNMT INHIBITORS IN SOLID TUMORS

The clinical use of azacytidine and decitabine has mainly been focused on hematological malignancies. With the emergence of *in vitro* and *in vivo* data showing that decitabine and azacitidine have antitumor activity, new clinical trials tested their activities against solid tumors. In early clinical trials, decitabine failed to produce significant antitumor activity in adult patients with melanoma, colorectal cancer, head and neck cancers, and renal carcinomas. Only one partial remission was seen in a patient with malignant melanoma [98]. Using higher doses of decitabine (660 mg/m^2 per cycle every 5–7 weeks), Momparler et al. reported a completed remission in one out of nine chemonaive stage IV metastatic nonsmall-cell lung cancer patients. This patient remained in complete remission for more than 7 years. The treatment produced a high plasmatic level of 3 μM at steady state during an 8-h infusion [99,100]. Even at this dose, the toxicity profile was reversible and in the range of what is observed using lower doses of decitabine [100].

Several studies have used decitabine at different doses in solid tumors but reported only rare clinical responses. Using a 72-h continuous intravenous infusion of a 28-day cycle of decitabine at 20–40 mg/m^2/day, no complete remission was achieved but the treatment was well-tolerated [28]. Significantly lower doses of decitabine (2 mg/m^2/day) administered as a 7-day continuous infusion did not produce complete remissions even though hematopoietic toxicity was observed, suggesting even at these levels the drug had some activity on replicating cells [86]. In summary, clinical trials using decitabine against solid tumors did not show promising responses even though occasional remissions were observed [28,87,99]. Interestingly, all these trials in solid tumors have tested a wide range of different dose-schedules ranging from low epigenetic doses to high cytotoxic doses (low nanomolar to low micromolar), and short to long infusions (8–72 h) in either refractory or newly diagnosed patients [101].

A large amount of preclinical data using cancer cell lines and *in vivo* models of solid tumors has demonstrated the efficacy of decitabine and azacitidine and does not support the lack of clinical responses. Either azacitidine or decitabine can efficiently demethylate, induce gene reactivation, induce tumor cell differentiation, and produce cancer cell reprogramming at nontoxic doses [39,102]. The lack of efficacy of DNMT inhibitors in the clinic against solid tumors remains to be understood. Several hypotheses have emerged. It may be caused by various parameters such as tumor cell biology, drug penetration in tumor masses, or a slower replicating cell cycle in solid tumors. Since synergistic responses combining DNMT inhibitors with other epigenetic drugs, cytotoxic drugs, or immunotherapy were observed in preclinical models, new clinical trials are testing these approaches in solid tumors. Complete remissions have been reported in nonsmall-cell lung cancer patients with azacitidine in combination with HDAC inhibitor [103]. Combinations with platinum compounds and decitabine have resulted in promising responses, which suggested that DNMT inhibition could be used as a priming therapy to sensitize tumor cells to other chemotherapeutic drugs (see also Chapter 15) [101,104,105].

7.6 SECOND GENERATION OF DNMT INHIBITORS

The first generation of azanucleotides, developed in the 1960s, have poor chemical stability in solution and are sensitive to enzymatic decomposition by CDA that are highly expressed in cancer cells and hepatocytes [38]. The second generation of demethylating drugs were designed to circumvent these issues. Zebularine was originally synthesized and evaluated as an inhibitor of CDA to prevent deamination of nucleoside analogs (Figure 7.1) [38]. However, zebularine has a weak DNMT inhibition activity. It is phosphorylated by the uridine-cytidine kinase and converted by ribonucleotide reductase into 2′-deoxyzebularine-5′-diphosphate. Then, 2′-deoxyzebularine-5′-diphosphate is converted to 2′-deoxyzebularine-5′-triphosphate and incorporated into the DNA of dividing cells. Based on its chemical stability, zebularine can be administered orally and was shown to induce antitumor activity against T-cell lymphoma [38]. Zebularine also produced a synergistic interaction with decitabine by inhibiting CDA activity [106]. Unfortunately, zebularine produced toxicity in primates during its preclinical assessment which blocked its progression into clinical trials [107]. Other nucleoside analogs were synthesized such as 5-fluoro-2′-deoxycyditine or other compounds with different molecular structures (procainamide, hydralazine, epigallocatechin gallate) did not induce significant DNA demethylation and their efficacy was much lower than the azanucleoside analogs [37,108].

Since azanucleotides have a very short half-life in the plasma and in cells due to CDA, several strategies have been developed to reduce their degradation. Recently, it was reported that short oligonucleotides containing decitabine can effectively inhibit DNA methylation in living cells. One of these compounds, called guadecitabine (2′-deoxy-5-azacytidylyl-(3′→5′)-2′-deoxyguanosine sodium salt or SGI-110) is a dinucleotide composed by a decitabine and a guanosine (Figure 7.1). The two nucleotides are linked by a phosphodiester bond. Interestingly, this dinucleotide has an improved aqueous stability and is protected from deamination by CDA as compared to decitabine.

The mechanism of action of SGI-110 seems to require incorporation of the decitabine group into DNA presumably after degradation of the phosphodiester bond by phosphodiesterases. The addition of a deoxyguanosine to the decitabine was shown to maintain the activity of decitabine to reduce DNA methylation [109–111]. Guadecitabine was recently introduced in the clinic demonstrating clinical efficacy and DNA demethylation activity [30].

7.7 CONCLUSIONS AND PERSPECTIVES

The introduction and the approval of DNMT inhibitors in the clinic provided the proof of principle for the epigenetic therapy of cancer. These drugs have been a major breakthrough, especially for the treatment of MDS, and they changed the natural history of the disease, leading to a reduction in the need for transfusions and delayed progression into AML. These drugs are also approved for the treatment of AML for patients who are not eligible for standard induction therapy. However, clinical results have shown that these drugs are not curative. Subgroups of patients are not responsive to the treatment and the majority of the patients relapse during therapy [112]. There are an increasing number of patients whose diseases are refractory to DNMT inhibitors because of resistance mechanisms or changes in the biology of their disease. The question of emergence of drug resistance in responders poses a great clinical challenge. In addition, the reasons for the failure to translate the preclinical data against solid tumors in patients are still unclear. Several studies support a dual mechanism of resistance involving alterations in drug metabolism (increased deamination and decreased phosphorylation) and also an epigenetic tolerance to hypomethylated DNA [55,57–59,62]. Therapeutic approaches have been studied where decitabine is combined with drugs that target the mechanism of drug resistance such as CDA inhibitors or using second generation of DNMT inhibitors. Combinations with other epigenetic drugs will likely synergize and reduce the ability of cancer cells to proliferate and resist to epigenetic reprogramming.

The issue of optimal dosage in solid tumors needs to be carefully addressed. Based on the experience against MDS and AML, many different approaches had to be tested before the identification of the optimal dose-schedule leading to the most interesting clinical results. Against solid tumors, phase I/II studies have been designed to treat refractory patients to several lines of chemotherapy, which is an approach that has failed in the past for DNMT inhibitors since these drugs need to be administered for at least two to six cycles prior to achieving complete remission [11]. The selection of appropriate patients will be a key determinant in this aspect. Optimizing dose-schedules of decitabine to the maximum level of DNA hypomethylation rather than the MTD will further improve clinical results.

REFERENCES

[1] Sorm F, Vesely J. Effect of 5-aza-2′-deoxycytidine against leukemic and hemopoietic tissues in AKR mice. Neoplasma 1968;15(4):339–43.

[2] Sorm F, Piskala A, Cihak A, Vesely J. 5-Azacytidine, a new, highly effective cancerostatic. Experientia 1964;20(4):202–3.

[3] Li LH, Olin EJ, Buskirk HH, Reineke LM. Cytotoxicity and mode of action of 5-azacytidine on L1210 leukemia. Cancer Res 1970;30(11):2760–9.
[4] Li LH, Olin EJ, Fraser TJ, Bhuyan BK. Phase specificity of 5-azacytidine against mammalian cells in tissue culture. Cancer Res 1970;30(11):2770–5.
[5] Momparler RL. Epigenetic therapy of cancer with 5-aza-2′-deoxycytidine (decitabine). Semin Oncol 2005;32(5):443–51.
[6] Karon M, Sieger L, Leimbrock S, Finklestein JZ, Nesbit ME, Swaney JJ. 5-Azacytidine: a new active agent for the treatment of acute leukemia. Blood 1973;42(3):359–65.
[7] Hrodek O, Vesely J. 5-Azacytidine in childhood leukemia. Neoplasma 1971;18(5):493–503.
[8] McCredie KB, Bodey GP, Burgess MA, Gutterman JU, Rodriguez V, Sullivan MP, et al. Treatment of acute leukemia with 5-azacytidine (NSC-102816). Cancer Chemother Rep 1973;57(3):319–23.
[9] Rivard GE, Momparler RL, Demers J, Benoit P, Raymond R, Lin K, et al. Phase I study on 5-aza-2′-deoxycytidine in children with acute leukemia. Leuk Res 1981;5(6):453–62.
[10] Momparler RL, Rivard GE, Gyger M. Clinical trial on 5-aza-2′-deoxycytidine in patients with acute leukemia. Pharmacol Ther 1985;30(3):277–86.
[11] Issa JP, Kantarjian HM. Targeting DNA methylation. Clin Cancer Res 2009;15(12):3938–46.
[12] Constantinides PG, Jones PA, Gevers W. Functional striated muscle cells from non-myoblast precursors following 5-azacytidine treatment. Nature 1977;267(5609):364–6.
[13] Taylor SM, Jones PA. Multiple new phenotypes induced in 10T1/2 and 3T3 cells treated with 5-azacytidine. Cell 1979;17(4):771–9.
[14] Jones PA, Taylor SM. Cellular differentiation, cytidine analogs and DNA methylation. Cell 1980;20(1):85–93.
[15] Wilson VL, Jones PA, Momparler RL. Inhibition of DNA methylation in L1210 leukemic cells by 5-aza-2′-deoxycytidine as a possible mechanism of chemotherapeutic action. Cancer Res 1983;43(8):3493–6.
[16] Momparler R, Bouchard J, Onetto N, Rivard GE. 5-aza-2′-deoxycytidine therapy in patients with acute leukemia inhibits DNA methylation. Leuk Res 1984;8(2):181–5.
[17] Momparler RL. Pharmacology of 5-Aza-2′-deoxycytidine (decitabine). Semin Hematol 2005;42(3 Suppl. 2):S9–16.
[18] Momparler RL. Cancer epigenetics. Oncogene 2003;22(42):6479–83.
[19] Taby R, Issa JP. Cancer epigenetics. CA Cancer J Clin 2010;60(6):376–92.
[20] Jones PA. Functions of DNA methylation: islands, start sites, gene bodies and beyond. Nat Genet 2012;13(7):484–92.
[21] Jones PA, Baylin SB. The fundamental role of epigenetic events in cancer. Nat Genet 2002;3(6):415–28.
[22] Wang Y, Xiao M, Chen X, Chen L, Xu Y, Lv L, et al. WT1 recruits TET2 to regulate its target gene expression and suppress leukemia cell proliferation. Mol Cell 2015;57(4):662–73.
[23] Jones PA. At the tipping point for epigenetic therapies in cancer. J Clin Invest 2014;124(1):14–16.
[24] You JS, Jones PA. Cancer genetics and epigenetics: two sides of the same coin? Cancer Cell 2012;22(1):9–20.
[25] Lemaire M, Chabot GG, Raynal NJ, Momparler LF, Hurtubise A, Bernstein ML, et al. Importance of dose-schedule of 5-aza-2′-deoxycytidine for epigenetic therapy of cancer. BMC Cancer 2008;8:128.
[26] Jabbour E, Issa JP, Garcia-Manero G, Kantarjian H. Evolution of decitabine development: accomplishments, ongoing investigations, and future strategies. Cancer 2008;112(11):2341–51.
[27] Issa JP, Garcia-Manero G, Giles FJ, Mannari R, Thomas D, Faderl S, et al. Phase 1 study of low-dose prolonged exposure schedules of the hypomethylating agent 5-aza-2′-deoxycytidine (decitabine) in hematopoietic malignancies. Blood 2004;103(5):1635–40.
[28] Aparicio A, Eads CA, Leong LA, Laird PW, Newman EM, Synold TW, et al. Phase I trial of continuous infusion 5-aza-2′-deoxycytidine. Cancer Chemother Pharmacol 2003;51(3):231–9.

[29] Derissen EJ, Beijnen JH, Schellens JH. Concise drug review: azacitidine and decitabine. Oncologist 2013;18(5):619–24.
[30] Estey EH. Epigenetics in clinical practice: the examples of azacitidine and decitabine in myelodysplasia and acute myeloid leukemia. Leukemia 2013;27(9):1803–12.
[31] Flisikowska T, Thorey IS, Offner S, Ros F, Lifke V, Zeitler B, et al. Efficient immunoglobulin gene disruption and targeted replacement in rabbit using zinc finger nucleases. PLoS One 2011;6(6):e21045.
[32] Bouchard J, Momparler RL. Incorporation of 5-Aza-2′-deoxycytidine-5′-triphosphate into DNA. Interactions with mammalian DNA polymerase alpha and DNA methylase. Mol Pharmacol 1983;24(1):109–14.
[33] Momparler RL, Samson J, Momparler LF, Rivard GE. Cell cycle effects and cellular pharmacology of 5-aza-2′-deoxycytidine. Cancer Chemother Pharmacol 1984;13(3):191–4.
[34] Schermelleh L, Spada F, Easwaran HP, Zolghadr K, Margot JB, Cardoso MC, et al. Trapped in action: direct visualization of DNA methyltransferase activity in living cells. Nat Methods 2005;2(10):751–6.
[35] Patel K, Dickson J, Din S, Macleod K, Jodrell D, Ramsahoye B. Targeting of 5-aza-2′-deoxycytidine residues by chromatin-associated DNMT1 induces proteasomal degradation of the free enzyme. Nucleic Acids Res 2010;38(13):4313–24.
[36] Raynal NJ, Si J, Taby RF, Gharibyan V, Ahmed S, Jelinek J, et al. DNA methylation does not stably lock gene expression but instead serves as a molecular mark for gene silencing memory. Cancer Res 2012;72(5):1170–81.
[37] Stresemann C, Brueckner B, Musch T, Stopper H, Lyko F. Functional diversity of DNA methyltransferase inhibitors in human cancer cell lines. Cancer Res 2006;66(5):2794–800.
[38] Rius M, Lyko F. Epigenetic cancer therapy: rationales, targets and drugs. Oncogene 2012; 31(39):4257–65.
[39] Tsai HC, Li H, Van Neste L, Cai Y, Robert C, Rassool FV, et al. Transient low doses of DNA-demethylating agents exert durable antitumor effects on hematological and epithelial tumor cells. Cancer Cell 2012;21(3):430–46.
[40] Baylin SB, Jones PA. A decade of exploring the cancer epigenome—biological and translational implications. Nat Rev Cancer 2011;11(10):726–34.
[41] Maunakea AK, Nagarajan RP, Bilenky M, Ballinger TJ, D'Souza C, Fouse SD, et al. Conserved role of intragenic DNA methylation in regulating alternative promoters. Nature 2010;466(7303):253–7.
[42] Yang X, Han H, De Carvalho DD, Lay FD, Jones PA, Liang G. Gene body methylation can alter gene expression and is a therapeutic target in cancer. Cancer Cell 2014;26(4):577–90.
[43] Momparler RL, Dore BT, Momparler LF. Effect of 5-aza-2′-deoxycytidine and retinoic acid on differentiation and c-myc expression in HL-60 myeloid leukemic cells. Cancer Lett 1990;54(1–2):21–8.
[44] Kasinathan S, Henikoff S. 5-Aza-CdR delivers a gene body blow. Cancer Cell 2014;26(4):449–51.
[45] Issa JP, Kantarjian H. Azacitidine. Nat Rev Drug Discov 2005;Suppl. S6–7.
[46] Momparler RL, Momparler LF, Samson J. Comparison of the antileukemic activity of 5-AZA-2′-deoxycytidine, 1-beta-D-arabinofuranosylcytosine and 5-azacytidine against L1210 leukemia. Leuk Res 1984;8(6):1043–9.
[47] van Groeningen CJ, Leyva A, O'Brien AM, Gall HE, Pinedo HM. Phase I and pharmacokinetic study of 5-aza-2′-deoxycytidine (NSC 127716) in cancer patients. Cancer Res 1986;46(9):4831–6.
[48] Cashen AF, Shah AK, Todt L, Fisher N, Dipersio J. Pharmacokinetics of decitabine administered as a 3-h infusion to patients with acute myeloid leukemia (AML) or myelodysplastic syndrome (MDS). Cancer Chemother Pharmacol 2008;61(5):759–66.
[49] Chabot GG, Rivard GE, Momparler RL. Plasma and cerebrospinal fluid pharmacokinetics of 5-Aza-2′-deoxycytidine in rabbits and dogs. Cancer Res 1983;43(2):592–7.
[50] Kantarjian H, Issa JP, Rosenfeld CS, Bennett JM, Albitar M, DiPersio J, et al. Decitabine improves patient outcomes in myelodysplastic syndromes: results of a phase III randomized study. Cancer 2006;106(8):1794–803.

[51] Fenaux P, Mufti GJ, Hellstrom-Lindberg E, Santini V, Finelli C, Giagounidis A, et al. Efficacy of azacitidine compared with that of conventional care regimens in the treatment of higher-risk myelodysplastic syndromes: a randomised, open-label, phase III study. Lancet Oncol 2009;10(3):223–32.

[52] Prebet T, Gore SD, Esterni B, Gardin C, Itzykson R, Thepot S, et al. Outcome of high-risk myelodysplastic syndrome after azacitidine treatment failure. J Clin Oncol 2011;29(24):3322–7.

[53] Bryan J, Kantarjian H, Garcia-Manero G, Jabbour E. Pharmacokinetic evaluation of decitabine for the treatment of leukemia. Expert Opin Drug Metab Toxicol 2011;7(5):661–72.

[54] Prebet T, Gore SD, Thepot S, Esterni B, Quesnel B, Beyne Rauzy O, et al. Outcome of acute myeloid leukaemia following myelodysplastic syndrome after azacitidine treatment failure. Br J Haematol 2012;157(6):764–6.

[55] Momparler RLM, Vesely J, Momparler LF, Rivard GE. Synergistic action of 5-Aza-2′-deoxycytidine and 3-Deazauridine on L1210 leukemic cells and EMT_6- tumor cells. Cancer Res 1979;39:3822–7.

[56] Momparler RL, Momparler LF. Chemotherapy of L1210 and L1210/ARA-C leukemia with 5-aza-2′-deoxycytidine and 3-deazauridine. Cancer Chemother Pharmacol 1989;25(1):51–4.

[57] Raynal NJ, Momparler LF, Rivard GE, Momparler RL. 3-Deazauridine enhances the antileukemic action of 5-aza-2′-deoxycytidine and targets drug-resistance due to deficiency in deoxycytidine kinase. Leuk Res 2011;35(1):110–18.

[58] Qin T, Jelinek J, Si J, Shu J, Issa JP. Mechanisms of resistance to 5-aza-2′-deoxycytidine in human cancer cell lines. Blood 2009;113(3):659–67.

[59] Qin T, Castoro R, El Ahdab S, Jelinek J, Wang X, Si J, et al. Mechanisms of resistance to decitabine in the myelodysplastic syndrome. PLoS One 2011;6(8):e23372.

[60] Eliopoulos N, Cournoyer D, Momparler RL. Drug resistance to 5-aza-2′-deoxycytidine, 2′,2′-difluorodeoxycytidine, and cytosine arabinoside conferred by retroviral-mediated transfer of human cytidine deaminase cDNA into murine cells. Cancer Chemother Pharmacol 1998;42(5):373–8.

[61] Lemaire M, Momparler LF, Bernstein ML, Marquez VE, Momparler RL. Enhancement of antineoplastic action of 5-aza-2′-deoxycytidine by zebularine on L1210 leukemia. Anticancer Drugs 2005;16(3):301–8.

[62] Ebrahem Q, Mahfouz RZ, Ng KP, Saunthararajah Y. High cytidine deaminase expression in the liver provides sanctuary for cancer cells from decitabine treatment effects. Oncotarget 2012;3(10):1137–45.

[63] Lavelle D, Vaitkus K, Ling Y, Ruiz MA, Mahfouz R, Ng KP, et al. Effects of tetrahydrouridine on pharmacokinetics and pharmacodynamics of oral decitabine. Blood 2012;119(5):1240–7.

[64] Issa JP, Gharibyan V, Cortes J, Jelinek J, Morris G, Verstovsek S, et al. Phase II study of low-dose decitabine in patients with chronic myelogenous leukemia resistant to imatinib mesylate. J Clin Oncol 2005;23(17):3948–56.

[65] Issa JP. The myelodysplastic syndrome as a prototypical epigenetic disease. Blood 2013; 121(19):3811–17.

[66] Paz MF, Wei S, Cigudosa JC, Rodriguez-Perales S, Peinado MA, Huang TH, et al. Genetic unmasking of epigenetically silenced tumor suppressor genes in colon cancer cells deficient in DNA methyltransferases. Hum Mol Genet 2003;12(17):2209–19.

[67] Si J, Boumber YA, Shu J, Qin T, Ahmed S, He R, et al. Chromatin remodeling is required for gene reactivation after decitabine-mediated DNA hypomethylation. Cancer Res 2010;70(17):6968–77.

[68] Onetto N, Momparler RL, Momparler LF, Gyger M. *In vitro* biochemical tests to evaluate the response to therapy of acute leukemia with cytosine arabinoside or 5-AZA-2′-deoxycytidine. Semin Oncol 1987;14(2 Suppl. 1):231–7.

[69] Kelly TK, De Carvalho DD, Jones PA. Epigenetic modifications as therapeutic targets. Nat Biotechnol 2010;28(10):1069–78.

[70] Silverman LR, Demakos EP, Peterson BL, Kornblith AB, Holland JC, Odchimar-Reissig R, et al. Randomized controlled trial of azacitidine in patients with the myelodysplastic syndrome: a study of the cancer and leukemia group B. J Clin Oncol 2002;20(10):2429–40.

[71] Zagonel V, Lo Re G, Marotta G, Babare R, Sardeo G, Gattei V, et al. 5-Aza-2′-deoxycytidine (Decitabine) induces trilineage response in unfavourable myelodysplastic syndromes. Leukemia 1993;7(Suppl. 1):30–5.
[72] Wijermans PW, Krulder JW, Huijgens PC, Neve P. Continuous infusion of low-dose 5-Aza-2′-deoxycytidine in elderly patients with high-risk myelodysplastic syndrome. Leukemia 1997;11(1):1–5.
[73] Wijermans P, Lubbert M, Verhoef G, Bosly A, Ravoet C, Andre M, et al. Low-dose 5-aza-2′-deoxycytidine, a DNA hypomethylating agent, for the treatment of high-risk myelodysplastic syndrome: a multicenter phase II study in elderly patients. J Clin Oncol 2000;18(5):956–62.
[74] Wijermans PW, Lubbert M, Verhoef G, Klimek V, Bosly A. An epigenetic approach to the treatment of advanced MDS; the experience with the DNA demethylating agent 5-aza-2′-deoxycytidine (decitabine) in 177 patients. Ann Hematol 2005;84(Suppl. 1):9–17.
[75] Lubbert M, Suciu S, Baila L, Ruter BH, Platzbecker U, Giagounidis A, et al. Low-dose decitabine versus best supportive care in elderly patients with intermediate- or high-risk myelodysplastic syndrome (MDS) ineligible for intensive chemotherapy: final results of the randomized phase III study of the European organisation for research and treatment of Cancer leukemia group and the German MDS study group. J Clin Oncol 2011;29(15):1987–96.
[76] Kantarjian H, Oki Y, Garcia-Manero G, Huang X, O'Brien S, Cortes J, et al. Results of a randomized study of 3 schedules of low-dose decitabine in higher-risk myelodysplastic syndrome and chronic myelomonocytic leukemia. Blood 2007;109(1):52–7.
[77] Steensma DP, Baer MR, Slack JL, Buckstein R, Godley LA, Garcia-Manero G, et al. Multicenter study of decitabine administered daily for 5 days every 4 weeks to adults with myelodysplastic syndromes: the alternative dosing for outpatient treatment (ADOPT) trial. J Clin Oncol 2009;27(23):3842–8.
[78] Kantarjian HM, O'Brien S, Huang X, Garcia-Manero G, Ravandi F, Cortes J, et al. Survival advantage with decitabine versus intensive chemotherapy in patients with higher risk myelodysplastic syndrome: comparison with historical experience. Cancer 2007;109(6):1133–7.
[79] Petti MC, Mandelli F, Zagonel V, De Gregoris C, Merola MC, Latagliata R, et al. Pilot study of 5-aza-2′-deoxycytidine (Decitabine) in the treatment of poor prognosis acute myelogenous leukemia patients: preliminary results. Leukemia 1993;7(Suppl. 1):36–41.
[80] Blum W, Garzon R, Klisovic RB, Schwind S, Walker A, Geyer S, et al. Clinical response and miR-29b predictive significance in older AML patients treated with a 10-day schedule of decitabine. Proc Natl Acad Sci USA 2010;107(16):7473–8.
[81] Cashen AF, Schiller GJ, O'Donnell MR, DiPersio JF. Multicenter, phase II study of decitabine for the first-line treatment of older patients with acute myeloid leukemia. J Clin Oncol 2010;28(4):556–61.
[82] Curran MP. Decitabine: a review of its use in older patients with acute myeloid leukaemia. Drugs Aging 2013;30(6):447–58.
[83] Silverman LR, Fenaux P, Mufti GJ, Santini V, Hellstrom-Lindberg E, Gattermann N, et al. Continued azacitidine therapy beyond time of first response improves quality of response in patients with higher-risk myelodysplastic syndromes. Cancer 2011;117(12):2697–702.
[84] Silverman LR, McKenzie DR, Peterson BL, Holland JF, Backstrom JT, Beach CL, et al. Further analysis of trials with azacitidine in patients with myelodysplastic syndrome: studies 8421, 8921, and 9221 by the Cancer and leukemia group B. J Clin Oncol 2006;24(24):3895–903.
[85] Yang AS, Doshi KD, Choi SW, Mason JB, Mannari RK, Gharybian V, et al. DNA methylation changes after 5-aza-2′-deoxycytidine therapy in patients with leukemia. Cancer Res 2006;66(10):5495–503.
[86] Samlowski WE, Leachman SA, Wade M, Cassidy P, Porter-Gill P, Busby L, et al. Evaluation of a 7-day continuous intravenous infusion of decitabine: inhibition of promoter-specific and global genomic DNA methylation. J Clin Oncol 2005;23(17):3897–905.
[87] Schrump DS, Fischette MR, Nguyen DM, Zhao M, Li X, Kunst TF, et al. Phase I study of decitabine-mediated gene expression in patients with cancers involving the lungs, esophagus, or pleura. Clin Cancer Res 2006;12(19):5777–85.

[88] Stewart DJ, Issa JP, Kurzrock R, Nunez MI, Jelinek J, Hong D, et al. Decitabine effect on tumor global DNA methylation and other parameters in a phase I trial in refractory solid tumors and lymphomas. Clin Cancer Res 2009;15(11):3881–8.

[89] Oki Y, Jelinek J, Shen L, Kantarjian HM, Issa JP. Induction of hypomethylation and molecular response after decitabine therapy in patients with chronic myelomonocytic leukemia. Blood 2008;111(4):2382–4.

[90] Stresemann C, Bokelmann I, Mahlknecht U, Lyko F. Azacytidine causes complex DNA methylation responses in myeloid leukemia. Mol Cancer Ther 2008;7(9):2998–3005.

[91] Shen L, Kantarjian H, Guo Y, Lin E, Shan J, Huang X, et al. DNA methylation predicts survival and response to therapy in patients with myelodysplastic syndromes. J Clin Oncol 2010;28(4):605–13.

[92] Oki Y, Aoki E, Issa JP. Decitabine—bedside to bench. Crit Rev Oncol Hematol 2007;61(2):140–52.

[93] Soriano AO, Yang H, Faderl S, Estrov Z, Giles F, Ravandi F, et al. Safety and clinical activity of the combination of 5-azacytidine, valproic acid, and all-trans retinoic acid in acute myeloid leukemia and myelodysplastic syndrome. Blood 2007;110(7):2302–8.

[94] Gore SD, Baylin S, Sugar E, Carraway H, Miller CB, Carducci M, et al. Combined DNA methyltransferase and histone deacetylase inhibition in the treatment of myeloid neoplasms. Cancer Res 2006; 66(12):6361–9.

[95] Daskalakis M, Nguyen TT, Nguyen C, Guldberg P, Kohler G, Wijermans P, et al. Demethylation of a hypermethylated P15/INK4B gene in patients with myelodysplastic syndrome by 5-Aza-2′-deoxycytidine (decitabine) treatment. Blood 2002;100(8):2957–64.

[96] Blum W, Klisovic RB, Hackanson B, Liu Z, Liu S, Devine H, et al. Phase I study of decitabine alone or in combination with valproic acid in acute myeloid leukemia. J Clin Oncol 2007;25(25):3884–91.

[97] Fandy TE, Herman JG, Kerns P, Jiemjit A, Sugar EA, Choi SH, et al. Early epigenetic changes and DNA damage do not predict clinical response in an overlapping schedule of 5-azacytidine and entinostat in patients with myeloid malignancies. Blood 2009;114(13):2764–73.

[98] Abele R, Clavel M, Dodion P, Bruntsch U, Gundersen S, Smyth J, et al. The EORTC early clinical trials cooperative group experience with 5-aza-2′-deoxycytidine (NSC 127716) in patients with colorectal, head and neck, renal carcinomas and malignant melanomas. Eur J Cancer Clin Oncol 1987;23 (12):1921–4.

[99] Momparler RL, Bouffard DY, Momparler LF, Dionne J, Belanger K, Ayoub J. Pilot phase I-II study on 5-aza-2′-deoxycytidine (Decitabine) in patients with metastatic lung cancer. Anticancer Drugs 1997;8(4):358–68.

[100] Momparler RL, Ayoub J. Potential of 5-aza-2′-deoxycytidine (Decitabine) a potent inhibitor of DNA methylation for therapy of advanced non-small cell lung cancer. Lung cancer 2001;34(Suppl. 4): S111–15.

[101] Nie J, Liu L, Li X, Han W. Decitabine, a new star in epigenetic therapy: the clinical application and biological mechanism in solid tumors. Cancer Lett 2014;354(1):12–20.

[102] Azad N, Zahnow CA, Rudin CM, Baylin SB. The future of epigenetic therapy in solid tumours—lessons from the past. Nat Rev Clin Oncol 2013;10(5):256–66.

[103] Juergens RA, Wrangle J, Vendetti FP, Murphy SC, Zhao M, Coleman B, et al. Combination epigenetic therapy has efficacy in patients with refractory advanced non-small cell lung cancer. Cancer Discov 2011;1(7):598–607.

[104] Glasspool RM, Brown R, Gore ME, Rustin GJ, McNeish IA, Wilson RH, et al. A randomised, phase II trial of the DNA-hypomethylating agent 5-aza-2′-deoxycytidine (decitabine) in combination with carboplatin vs carboplatin alone in patients with recurrent, partially platinum-sensitive ovarian cancer. Br J Cancer 2014;110(8):1923–9.

[105] Appleton K, Mackay HJ, Judson I, Plumb JA, McCormick C, Strathdee G, et al. Phase I and pharmacodynamic trial of the DNA methyltransferase inhibitor decitabine and carboplatin in solid tumors. J Clin Oncol 2007;25(29):4603–9.

[106] Lemaire M, Momparler LF, Raynal NJ, Bernstein ML, Momparler RL. Inhibition of cytidine deaminase by zebularine enhances the antineoplastic action of 5-aza-2′-deoxycytidine. Cancer Chemother Pharmacol 2009;63(3):411–16.

[107] Johnson WD, Harder JB, Naylor J, McCormick DL, Detrisac CJ, Glaze ER, et al. A pharmacokinetic/pharmacodynamic approach to evaluating the safety of zebularine in non-human primates. Cancer Res 2006;66:309.

[108] Chuang JC, Yoo CB, Kwan JM, Li TW, Liang G, Yang AS, et al. Comparison of biological effects of non-nucleoside DNA methylation inhibitors versus 5-aza-2′-deoxycytidine. Mol Cancer Ther 2005; 4(10):1515–20.

[109] Yoo CB, Jeong S, Egger G, Liang G, Phiasivongsa P, Tang C, et al. Delivery of 5-aza-2′-deoxycytidine to cells using oligodeoxynucleotides. Cancer Res 2007;67(13):6400–8.

[110] Chuang JC, Warner SL, Vollmer D, Vankayalapati H, Redkar S, Bearss DJ, et al. S110, a 5-Aza-2′-deoxycytidine-containing dinucleotide, is an effective DNA methylation inhibitor *in vivo* and can reduce tumor growth. Mol Cancer Ther 2010;9(5):1443–50.

[111] Srivastava P, Paluch BE, Matsuzaki J, James SR, Collamat-Lai G, Karbach J, et al. Immunomodulatory action of SGI-110, a hypomethylating agent, in acute myeloid leukemia cells and xenografts. Leuk Res 2014;38(11):1332–41.

[112] Kadia TM, Jabbour E, Kantarjian H. Failure of hypomethylating agent-based therapy in myelodysplastic syndromes. Semin Oncol 2011;38(5):682–92.

CHAPTER

SMALL MOLECULE INHIBITORS OF HISTONE DEACETYLASES AND ACETYLTRANSFERASES AS POTENTIAL THERAPEUTICS IN ONCOLOGY

8

Thea van den Bosch, Niek G.J. Leus, Tirza Timmerman and Frank J. Dekker

Pharmaceutical Gene Modulation, Groningen Research Institute of Pharmacy, University of Groningen, Groningen, The Netherlands

CHAPTER OUTLINE

8.1 INTRODUCTION

Cancer is caused by uncontrolled cell proliferation and a failure to undergo apoptosis. This process is regulated by differential expression of hundreds of genes. Activation of pro- or antiapoptotic gene expression encompasses a wide variety of coregulatory mechanisms acting at different levels of the transcription activation process. Among these mechanisms the regulation of gene expression by posttranslational modifications of histone proteins has gained significant attention over recent years. Exploration of histone-modifying enzymes is currently a major topic in drug discovery programs in academia and pharmaceutical companies. Nevertheless, it proves to be difficult to

G. Egger & P. Arimondo (Eds): Drug Discovery in Cancer Epigenetics. DOI: http://dx.doi.org/10.1016/B978-0-12-802208-5.00008-4

define specific posttranslational modifications and specific histone-modifying enzymes as molecular targets for drug discovery in particular diseases such as cancer. The increasing number of posttranslational histone modifications, as well as the diversity in their types and functions that are being discovered, illustrates the potential complexity on this level [1].

The most investigated type of histone modification is lysine acetylation, which is regulated by histone acetyltransferases (HATs) and histone deacetylases (HDACs) [2,3]. These enzymes influence the way in which DNA is wrapped around histones and packaged into chromatin, a highly organized and dynamic protein–DNA complex [4].

HATs enable the transfer of an acetyl group from acetyl CoA to a lysine residue, which results in neutralization of positively charged lysine residues. Because of this neutralization, the interaction between histones and DNA is weakened. Acetylation is generally considered to enhance transcription, since it enables easier access of RNA polymerase and transcription factors to the promoter regions. HDACs, on the other hand, deacetylate lysine residues, which restores their positive charge. This strengthens the histone–DNA interaction, and therefore, HDACs are considered to be transcriptional repressors [5]. In general, increased levels of histone acetylation (hyperacetylation) are associated with increased transcriptional activity, whereas decreased levels of acetylation (hypoacetylation) are associated with repression of gene expression (Figure 8.1).

Steady-state levels of acetylation of the histones and also nonhistone proteins result from the balance between opposing activities of HATs and HDACs [6]. When the equilibrium between HATs and HDACs is altered, dysregulation of (disease-associated) genes could contribute to diseases like cancer and chronic inflammation. Inhibitors of these enzymes have the potential to attenuate pathological responses implicated in these diseases. Inhibitors of HATs (HATi) have remained relatively unexplored, leaving opportunities for development. In contrast, lots of effort has been put into the development of HDAC inhibitors (HDACi). Indeed, HDAC activity is upregulated in many types of cancer, and HDACi are successfully used as drugs for the treatment of cancer. Currently, four HDACi have been approved for treatment of hematological cancer types. In addition, a number are being tested in clinical trials, and efforts are focused on new HDACi discovery and development.

The mechanisms through which HDACi exert their cytotoxic effects are also of interest. It is generally assumed that the main mechanism is inhibition of histone deacetylation, however, other mechanisms have been indicated, involving inhibition of deacetylation of nonhistone proteins. It is important to understand such specific mechanisms of action. On this level, the transcription factor nuclear factor κB (NF-κB) is one that has been implicated. Indeed, cancer is in some cases driven by well-defined stimulus-regulated transcription factors such as NF-κB [7,8]. The NF-κB pathway is relatively well investigated and several regulatory mechanisms have been described to control the NF-κB response to specific stimuli. Intriguingly, it has been reported that acetylation of the NF-κB transcription factor plays a crucial role in regulating the intensity, length, and specificity of its responses [9,10]. NF-κB can be (de)acetylated at various lysine residues resulting in NF-κB activation or deactivation [11]. This raises the idea that HATi and HDACi also influence such signaling pathways.

In the field of HATi and HDACi we highlight the latest developments, preclinical studies, and promising new agents. Also, we describe the role of acetylation and deacetylation of the nonhistone protein NF-κB in relation to its role in carcinogenesis, and how it may be targeted by HDACi. Finally, as an alternative area of application, we will discuss the potential of HDACi for the treatment of inflammatory diseases.

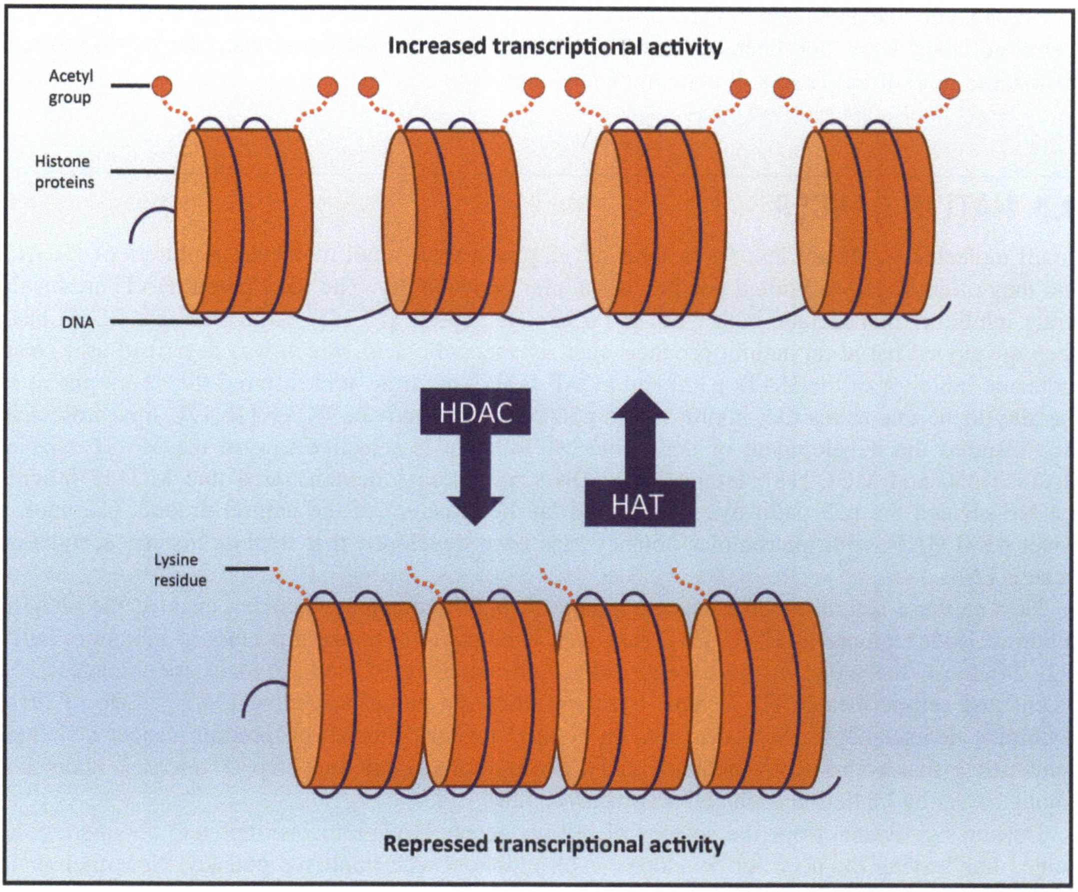

FIGURE 8.1

Schematic representation of the conformational transition of chromatin by histone-modifying enzymes. Histone modifications by (de)acetylation represent an epigenetic mechanism regulating chromatin remodeling, thereby influencing inflammatory gene expression. The addition of acetyl groups on lysine residues (acetylation), by HATs, reduces the affinity between histones and DNA, thereby loosening the chromatin (euchromatin) and activating transcription factors like NF-κB. In contrast, deacetylation induced by HDACs leads to a more compact status of chromatin (heterochromatin) and hence to repression of inflammatory gene expression.

8.2 CLASSES OF HATs

HATs are classified based on their cellular localization in nuclear HATs (type A) and cytoplasmic HATs (type B) [12]. Whereas little is known about cytoplasmic HATs, several nuclear HATs have been identified and divided into three major families based on their primary structure homology: the general control nonderepressible 5 (GCN5)-related *N*-acetyltransferase family, represented by GCN5 and PCAF (p300/CBP associated factor); the p300/CBP family, including p300 and CBP (CREB binding protein); the MYST family, which includes Tip60 (TAT interacting protein 60) and monocytic leukemia zinc finger protein. Although other nuclear HAT families have been identified, they have not been extensively studied. This demonstrates that, despite significant efforts, the field of HATs in still not completely explored.

8.3 HATi IN CANCER

Small molecule HATi are less far in their development than small molecule inhibitors of HDACs and they often display a limited potency or cellular permeability. The most potent HATi are bisubstrate inhibitors that include both CoA and a histone peptide [13,14]. As an alternate, HATi have been developed based on natural products such as anacardic acid, which was described as a small molecule inhibitor of the HATs p300 and PCAF [15]. Anacardic acid inspired the development of the alkylidene malonates that inhibit HATs p300/CBP and activate PCAF [16,17]. Anacardic acid also inspired the development of derivative MG149 that is selective toward the MYST type of HATs Tip60 and MOF [18]. Interestingly, DNA microarrays demonstrated that MG149 inhibits the NF-κB and the p53 pathways [19]. Based on the structure of the natural product Garcinol, a novel p300 HATi with micromolar potency, has been developed that inhibits histone acetylation in cells [20].

Next to these natural-product-derived inhibitors, high-throughput screening enabled the identification of isothiazolones as HATi [21]. However, attempts to optimize this class of inhibitors failed [22–24] most likely due to the exceptionally high reactivity of isothiazolones for thiolates [25]. Potent and cell-permeable HATi have been identified by virtual screening [26,27]. One of these inhibitors, denoted C646, has been used in cellular models. Studies on prostate cancer cell lines demonstrate that both siRNA-mediated and C646-mediated inhibition of p300 increase apoptosis, among others by inhibition of the NF-κB pathway [28].

Despite significant efforts the potency of cell-permeable HATi remains limited to the micromolar range, thus leaving the need for development of inhibitors with improved potency. Nevertheless, in the past few years HATi have been described that are better characterized with respect to HAT inhibition as well as their cellular potency. Despite difficulties in the development of potent and selective HATi it is reasonable to expect that the development of HATi and the identification of their applications will reach a more mature level in the coming years.

8.4 CLASSES OF HDACs

Mammalian HDAC enzymes can be classified into four main groups, based on their homology with orthologs identified in yeast [29]. Class I HDACs, including HDAC1, 2, 3, and 8 are predominantly found within the nucleus, due to the presence of a nuclear localization sequence and the absence of a nuclear export signal sequence within HDAC1, 2, and 8. However, HDAC3 has both a nuclear import and export signal, which enables localization in both the cytoplasm and the nucleus [29,30]. Class I HDACs have a ubiquitous tissue distribution [29]. Class II HDACs are subdivided into two groups, IIA (HDAC4, 5, 7, and 9) and IIB (HDAC6 and 10), and are predominantly found in the cytoplasm [31]. Class II HDACs are able to shuttle between the cytoplasm and the nucleus, and have a more tissue-specific distribution than class I HDACs [29]. Sharing similarities with both class I and class II HDACs, there is HDAC11, which is the only member of class IV. Class III HDACs are also called sirtuins (SIRT1–SIRT7) and are found in the cytoplasm. These HDACs act via different mechanisms than class I and class II and require the cofactor NAD^+ for their activity [32].

8.5 HDACi IN CANCER

Based on their chemical structure HDACi can be grouped into four main classes, that is, hydroxamic acids, 2-amino-benzamides, cyclic peptides, and short-chain fatty acids. From these, the first two classes have been investigated most intensively and will be focused on here (see Table 8.1).

The hydroxamic acid type of HDACi generally lacks selectivity among the Zn^{2+}-dependent HDACs [44] and in particular for class I HDACs. Three such HDACi of the hydroxamic acid type, vorinostat, belinostat, and panobinostat, obtained FDA approval for treatment of cancer [45]. Vorinostat (suberoylanilide hydroxamic acid, SAHA) obtained FDA approval for treatment of cutaneous T-cell lymphoma (CTCL) in patients with progressive, persistent, or recurrent symptoms [46] and applications in other hematological malignancies have also been described [47]. The HDACi belinostat has been granted FDA approved for use against peripheral T-cell lymphoma (PTCL) [48]. Recently, panobinostat was also FDA-approved for the treatment of relapsed multiple myeloma, in combination with bortezomib and dexamethasone [49].

Next to these hydroxamic acids, the natural-product-derived HDACi romidepsin is also FDA-approved for application in relapsed or refractory CTCL and PTCL [45,50]. These FDA-approved HDACi lack selectivity among class I HDACs, which is generally considered to be a

Table 8.1 Inhibition Activities (IC_{50} (nM)) of Small Molecule Inhibitors on Class I HDACs

	HDACi	HDAC1	HDAC2	HDAC3	HDAC8	Refs
Hydroxamic acids	TSA	2–300	3–32	4–200	100–456	[33–35]
	SAHA (vorinostat)	21–148	144–418	6–509	38–1700	[33,36–38]
	LBH589 (panobinostat)	3–4	3–5	4–6	32–248	[33,38]
	ITF2357 (givinostat)	28–198	56–325	21–157	854[a]	[33,39]
	Belinostat	41	125	30	216	[33]
	LAQ824 (dacinostat)	2–5	5 ± 1[a]	4–12	140–162	[33,36]
	KBH-A42	47[a]	32[a]	231[a]	222[a]	[40,41]
	ITF3056	2033[a]	>3000[a]	>3000[a]	285[a]	[39]
2-Amino-benzamides	MS-275 (entinostat)	181–300	306–1155	499–8000	2700–>10,000	[33,35–37]
	MGCD0103 (mocetinostat)	34–98	22–34	22–988	>10,000	[33,36,42]
	MI192	4800[a]	30[a]	16[a]	>10,000[a]	[43]

[a] *IC_{50} determined in one study.*

prerequisite for application of HDACi [51]. Nevertheless, the clinical successes of these inhibitors demonstrate the utility of their mechanisms of action for treatment of these types of cancer. Combination therapy of vorinostat, belinostat, or romidepsin with other agents has also been explored, with successful outcomes in some cases leading to clinical trials. For vorinostat and belinostat, examples of combination therapies include those with agents such as cisplatin, carboplatin, paclitaxel, and others [51]. Hence this may lead to important applications of these HDACi in other cancer types as well.

Unfortunately, however, application of HDACi in cancer types other than CTCL, PTCL, and multiple myeloma has not been successful so far. HDACi failed in particular in the treatment of solid tumors, which is often attributed to poor permeability of these inhibitors in the tumor. With this, however, it should be noted that other factors, such as an alternative epigenetic programming in these tumors, might also limit the success of HDACi [51].

One way that has been explored to increase clinical success is the generation of HDACi, which are more selective for certain HDAC subtypes. Examples of these include the hydroxamic-acid-derived HDACi KBH-A42 and ITF3056, which display some level of selectivity for either HDAC1 and 3 or HDAC8 (Table 8.1). Another interesting example is the hydroxamic acid derivative ACY1215, which is a selective inhibitor of HDAC6 (not shown in Table 8.1).

ACY1215 has been studied in combination with bortezomib for the treatment of multiple myeloma [52], similar to panobinostat, which was recently granted FDA approval (however, here it concerns panobinostat in combination with bortezomib as well as dexamethasone). Treatment of multiple myeloma with an HDACi in combination with bortezomib, a proteasome inhibitor, has been studied extensively. Bortezomib causes toxic aggregation of proteins (multiple myeloma cells are dependent on a high protein turnover rate). The rationale for combination treatment of bortezomib with an HDACi is that HDAC6 is thought to facilitate the removal of such misfolded proteins, thereby promoting survival and growth of myeloma cells. Inhibiting HDAC6 by applying an HDACi, next to treatment with bortezomib, is therefore thought to result in synergistic cytotoxic effects [53]. Looking at panobinostat (which is a pan-HDACi), however, many unwanted side effects are associated with its treatment [49]. In this respect, ACY1215 is a promising alternative candidate since selective inhibition of HDAC6 only is expected to give rise to fewer side effects. ACY1215 displayed synergistic cytotoxic effects in combination with bortezomib, even without additional treatment with dexamethasone, with a promising toxicity profile in mice [52]. ACY1215 is now in clinical trials, being evaluated in similar combination therapies [54].

Next to these hydroxamic acids, 2-aminobenzamide-type HDACi have been developed, which are generally more selective for HDAC1, 2, and 3. There is even subtype selectivity among these enzymes (Table 8.1). Importantly, the 2-aminobenzamide type of HDACi entinostat is under investigation in clinical trials [55]. The toxicity of this compound appears to be limited and promising effects are observed with entinostat treatment either alone or in combination with other agents. Nevertheless, the clinical efficacy until now was not sufficient to guarantee clinical approval. Currently, studies are mainly aimed at combinations of entinostat with other agents in order to increase its efficacy. MGCD0103 (mocetinostat) is another 2-aminobenzamide HDACi, which has been studied in clinical trials. There have been some indications that it displays clinical effect but a less favorable toxicity profile compared to entinostat, however, studies are still ongoing [51,56].

8.6 MECHANISMS OF ACTION OF HDACi

Histones have been recognized as the main targets of zinc-dependent HDACs. The inhibition of histone deacetylation is loosening the interaction between histones and DNA and thus upregulates transcription of proapoptotic genes that are aberrantly silenced in cancer cells. A common view is therefore that HDACi exert many of their clinical effects through modulation of histone acetylation. It has been shown that HDACi selectively induce proapoptotic gene transcriptional responses in tumor cells [57]. As a consequence, HDAC inhibition leads to p53-dependent apoptosis [58]. However, it should be noted that in some cases it has been observed that HDACi that inhibit histone deacetylation in tumor cells are not clinically effective [59,60]. This suggests that inhibition of histone deacetylation is not the only factor that plays a role in the clinical effect of HDACi.

Direct roles of HDACs in carcinogenesis have been described. For example, it has been shown that HDAC1 and HDAC2 affect extracellular signal regulated kinase 1/2 in mouse hepatocytes [61]. In this study, HDAC1 proved to act proapoptotic and HDAC2 antiapoptotic. Furthermore, roles in therapeutic resistance have been described such as for ductal adenocarcinoma in which HDAC2 proved to play a role in resistance against therapeutic agents that induce DNA damage [62]. Effects on tumor cell migration and adhesion have also been described for HDAC3 [63].

Next to this, other mechanisms through which HDACi cause cytotoxic effects are at play, involving inhibition of deacetylation of nonhistone proteins [64]. Important examples include transcription factors, which are among other factors regulated by HDACs, such as p53. In the case of p53, it has been described to be subject to regulation by HDAC2 and SIRT HDACs [65,66]. When p53 is acetylated, it is able to cause cell cycle arrest or senescence. Deacetylation renders it inactive, and this is one way in which carcinogenic cells may circumvent this pathway. HDACi treatment may then inhibit the deacetylation of p53, thereby forcing carcinogenic cells into arrest or senescence. In one study, it was demonstrated that the cytotoxic effects of HDACi entinostat partially depend upon p53, and that HDACi vorinostat can cause p53-dependent and -independent apoptosis [67]. It is important to understand such specific targets and mechanisms of action. Such information will assist the further development of HDACi and give rise to opportunities for rational design of new inhibitors. In some cases this may, for instance, also allow for combination therapy options to be exploited more easily.

One other transcription factor for which a role in carcinogenesis has been thoroughly established is NF-κB. Many studies point to a pro-oncogenic role for NF-κB (although this is not always the case) [68–71]. Inhibiting NF-κB activity may therefore be a strategy to drive carcinogenic cells into apoptosis. In this perspective the NF-κB signaling pathway is an interesting example because acetylation and deacetylation of the p65 subunit of the NF-κB transcription factor have been described to play a crucial role in the regulation of gene transcription via this pathway. Inhibition of deacetylation of specific NF-κB lysines could result in inhibition of NF-κB (pro-oncogenic) activity. It is therefore one example of a nonhistone protein, which may be affected by HDACi treatment, leading to cytotoxic effects in carcinogenic cells.

Indeed, several studies show that inhibition of deacetylases extends NF-κB transcriptional activity [33,72], whereas other studies demonstrate that inhibition of deacetylation decreases NF-κB transcriptional activity [73,74]. This indicates that the resulting effect of HDAC inhibition depends on the selectivity for specific NF-κB acetylation sites, which suggests that selective inhibition of HDACs might result in specific effects on gene transcription. It is encouraging that recent literature demonstrates the development of HDACi with improved selectivity profiles. This also provides chances for further elucidation of the role of HDACs in carcinogenesis, reduces risks of unwanted side effects, and thus offers a perspective toward drug discovery. In the following sections we will highlight further how (specific) HDACi treatment might affect NF-κB activity.

8.7 ROLE OF ACETYLATION AND DEACETYLATION IN NF-κB MEDIATED SIGNALING

The NF-κB transcription factors are a family of inducible transcription factors that play a central role in the expression of various cytokines, chemokines, and adhesion molecules, which are involved in cell survival and inflammation [75]. NF-κB transcription factors exist in homo- or heterodimeric complexes consisting of different members of the Rel family of proteins. The most prevalent and best studied of these complexes is the p50–p65 heterodimer. In quiescent cells, the p50–p65 complex is present in the cytoplasm in an inactive form, bound to inhibitory proteins known as IκBs. The NF-κB pathway can be activated via two different routes; the canonical pathway or the alternative pathway. The canonical pathway is activated by inducers such as inflammatory cytokines, that is, tumor necrosis factor α (TNF-α), interleukin (IL)-1; bacterial products (lipopolysaccharide, LPS), or oxidative stress (H_2O_2), resulting in phosphorylation, ubiquitylation, and finally degradation of IκBα. Subsequently, the p50–p65 heterodimer is released and translocates into the nucleus, followed by specific upregulation of gene expression [70]. The alternative pathway is activated by lymphotoxin β, CD40 ligand, B-cell activating factor, and receptor activator of NF-κB ligand, resulting in activation of RelB/p52 complexes [76]. Activation of the alternative pathway regulates genes required for lymph-organogenesis and B-cell activation. The canonical pathway plays an important role in chronic inflammatory diseases like inflammatory bowel disease, rheumatoid arthritis, chronic obstructive pulmonary disease, and asthma [77,78].

Acetylation of specific lysine residues in the p65 NF-κB transcription factor plays a key role in the regulation of its transcriptional capacity, its DNA binding ability, and duration of its action [79]. Seven lysine residues of p65 have been described to be subjected to acetylation, respectively, 122, 123, 218, 221, 310, 314, and 315. Acetylation of lysines 122 and 123 decreases DNA binding, whereas acetylation at lysines 218 and 221 increases binding to κB enhancers. Acetylation at lysine 310 is required for full transcriptional activity of p65, however, it does not affect the DNA binding or its assembly with IκBα [80]. Acetylation at lysine 314 and 315 does not affect the general transcriptional activity of the NF-κB complex, but rather increases promoter selectivity [81].

These examples demonstrate that direct acetylation or deacetylation of specific lysine residues of NF-κB plays a crucial role in the regulation of NF-κB-mediated gene expression, which raises the idea that apoptosis in cancer and inflammatory responses can be regulated by modulating NF-κB acetylation levels with HAT and HDACi. However, it is not completely clear which specific HATs or HDACs need to be targeted. Consequently, a challenge lies ahead in determining whether specific

HATs or HDACs target specific lysine acetylation in the NF-κB pathway. To this end, the development of highly potent molecules that selectively target specific HATs or HDACs is important.

8.8 HDAC–NF-κB INTERACTIONS IN NF-κB SIGNALING

HDAC–NF-κB interactions in signaling have not been fully understood, however, several studies have shed more light on the role of HDAC isoenzymes in the NF-κB pathway. Regarding class I HDACs, it has been demonstrated that HDAC1 negatively regulated NF-κB-mediated gene transcription through a direct association with p65, whereas HDAC2 does not interact with NF-κB directly but regulates NF-κB activity via its interaction with HDAC1 [82]. Inhibition of these HDACs with trichostatin A (TSA) resulted in increased levels of an integrated NF-κB-dependent reporter gene. Additionally, in this study, increasing levels of the IL-8 gene were observed in TNF-α-stimulated HeLa cells subjected to TSA. It has been shown that both HDAC1 and HDAC2 regulate NF-κB-mediated IL-8 expression, which is likely due to the increased acetylation of the IL-8 promoter [82]. HDAC1 has been shown to be a major nuclear HDAC controlling IL-1β-dependent inflammatory responses in intestinal epithelial cells by prolonging nuclear maintenance of NF-κB [83]. Finally, in contrast, it has been demonstrated that inhibition of HDAC1 (using siRNA) led to a reduction of LPS-induced gene expression of inducible nitric oxide synthase (iNOS) and IL-6 in bone-marrow-derived macrophages, but did not reduce the levels of LPS-induced IL-1β or TNFα [84]. Taken together, these studies imply that HDAC1 specifically increases the gene expression of some NF-κB target genes.

A number of studies have shown that HDAC3 deacetylates the NF-κB p65 subunit, leading to repression of its transcriptional activity [80]. An early study on the NF-κB–HDAC interaction demonstrated that HDAC3 plays an essential role in the deacetylation of p65 (promoting effective binding to IκBα), thereby negatively regulating TNFα-induced luciferase activity [10], which was abolished in the presence of TSA. Similarly, it has been demonstrated that HDAC3-dependent NF-κB deacetylation promotes interaction with the inhibitor protein IκBα resulting in nuclear export of NF-κB and hence repression of gene expression [85]. The positive regulatory role of HDAC3 involves binding to the NF-κB p65 subunit and concomitant deacetylation at various lysines. Interestingly, by using mutagenesis it was demonstrated that acetylation at lysine 314 and 315, in addition to acetylation of lysine 122 and 123, negatively regulates NF-κB activity. Moreover, it was shown that HDAC3 is involved in the removal of the inhibitory NF-κB p65 acetylation at lysine 122, 123, 314, and 315 [86]. Also, a recent study demonstrated that a decrease in nuclear accumulation of NF-κB in MDA-MB-231 cells was accompanied by a decrease in HDAC3 [87], though the role of HDAC3 in the reduction of NF-κB accumulation was not studied in detail. A schematic representation of the regulation of NF-κB activity by acetylation and the role of HDAC3 is shown in Figure 8.2.

The inconsistencies in these studies, with regard to NF-κB activity, can most likely be explained by the employment of different cell types in combination with the lack of selectivity of the applied HDACi (if applicable) since most HDACi target zinc-dependent HDACs nonselectively (Table 8.1). Moreover, this hampers the elucidation of the relevance of a specific HDAC in a

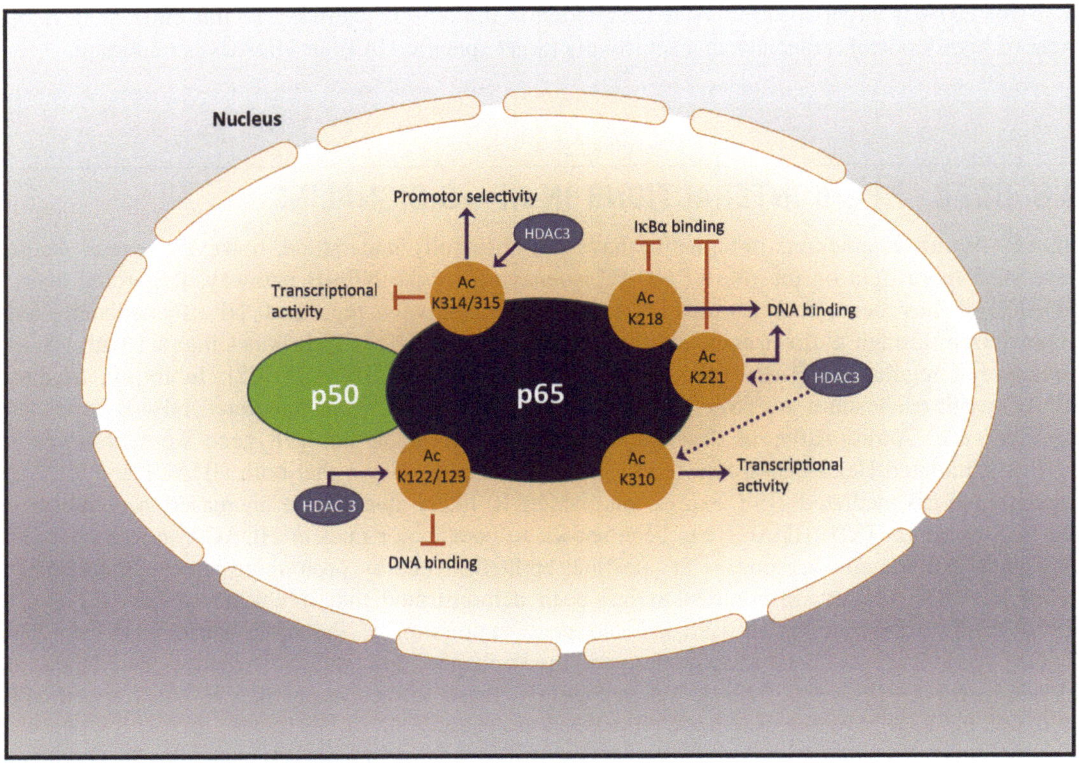

FIGURE 8.2

Regulation of NF-κB activity by lysine acetylation. Acetylation of specific lysine residues of NF-κB p65 regulate its transcriptional activity, its DNA binding ability, and duration of its action. HDAC3 deacetylates NF-κB p65 on lysines 122, 123, 314, and 315, thereby positively regulating inflammatory gene expression. A tentative role for HDAC3 in deacetylating lysine 221 and 310 has been described.

specific pathway and/or disease model. Consequently, novel HDACi with improved selectivity need to be developed to attenuate NF-κB-mediated responses in cell proliferation.

8.9 ALTERNATIVE AREA OF APPLICATION: HDACi AS ANTI-INFLAMMATORY DRUGS FOR THE TREATMENT OF (NF-κB-MEDIATED) INFLAMMATION

After their introduction in cancer therapy, HDACi are gaining attention for application in other diseases as well. Their potential applications range from neurodegenerative (Alzheimer's and Huntington's disease) to inflammatory diseases such as asthma, rheumatoid arthritis, and also viral infections [88–91]. Of particular interest are their anti-inflammatory properties, which are observed at 10–100-fold lower doses than those used in treating malignancies [92]. For example,

50% growth inhibition of tumor cells (A549) requires concentrations up to 5 μM for the HDACi vorinostat, whereas 50% inhibition of IL-1β secretion in primary peripheral blood mononuclear cells (PBMCs) requires concentrations as low as 50 nM [93]. Similar data have been reported for ITF2357 [74]. Several HDACi including TSA, phenylbutyrate, SAHA, and ITF2357 have demonstrated anti-inflammatory effects both *in vitro* and *in vivo* [94–97].

For example, it has been demonstrated that pan-HDACi such as TSA and vorinostat delay NF-κB nuclear translocation, thereby reducing gene expression upon TNF-α stimulation [98]. Also, TSA repressed iNOS expression in IL-1β/LPS/interferon γ (IFNγ)-stimulated murine macrophages (RAW264.7 cells) [99]. Furthermore, inhibition of class I/II HDACs by TSA and nicotinamide effectively blocked the TNF-α and IL-6 production by macrophages in patients with rheumatoid arthritis [100], and suppressed IL-8 production in rheumatoid arthritis synovial explants, which was not observed in other tissues. Furthermore, by blocking class I and class II HDACs with TSA and nicotinamide it was shown that these HDACs are essential in the pathogenesis of rheumatoid arthritis by promoting inflammation, suggesting a potential role for HDACi for the treatment of this disease.

Also, cytokine array analysis revealed that HDAC3 inhibition by TSA resulted in decreased monocyte chemoattractant protein 1 secretion [101]. Recently it has also been demonstrated that TSA suppresses induction of inflammatory cytokines in an acute-on-chronic liver failure (ACLF) model in rats through regulating the acetylation of NF-κB [102]. Serum TNFα, IFNγ, IL-10, and IL-18, which are greatly upregulated in patients suffering from ACLF, were significantly reduced in the TSA-treated group compared to a control group. Moreover, a decrease in the expression of HDAC1–3 was observed suggesting that class I HDACs may promote the inflammatory response observed in ACLF. In contrast, however, in OP9 preadipocytes it was shown that HDAC inhibition by TSA enhanced the expression of inflammatory proteins and NF-κB-dependent transcriptional activity, which might be caused by the increase in the acetylation of NF-κB p65 at lysine 310 and duration of the nuclear translocation of NF-κB [103]. The latter effect was presumably due to acetylation of p65 at lysine 218 or 221, thereby attenuating NF-κB interaction with IκBα.

It has to be taken into consideration that TSA is a pan-HDACi, so other HDACs might be involved in the studies discussed above as well. In fact, the majority of HDACi are broad acting or commonly known as "pan" HDACi that target multiple ZN^{2+}-dependent HDAC enzymes within the two main classes (class I and class II). For most inhibitors, it is not precisely known which particular HDAC isoenzymes they inhibit (and to what extent), making it difficult to elucidate their mechanism(s) of action. However, papers have been published on the selectivity of small molecule inhibitors of HDACs and their inhibitory activity of class I HDACs [33,34,36–38].

More selective HDACi have also been studied in relation to inflammation. For example, the benzamide type HDACi MS-275 (Table 8.1), with a moderate selectivity for HDAC 1 [35], demonstrated anti-inflammatory effects in several inflammation models, for example, a rat experimental autoimmune prostatitis [104] and mouse models of arthritis [105]. MS-275 also demonstrated anti-inflammatory effects in human rheumatoid arthritis synovial fibroblastic E11 cells [106]. Moreover, it was shown that MS-275 inhibited LPS-induced NF-κB nuclear accumulation (up to ~75%), and increased the association between NF-κB and the HAT p300. Less acetylated NF-κB was observed in the nucleus when cells were treated with MS-275, which confirms the inhibitory effect of MS-275 on NF-κB nuclear accumulation. In addition, IL-6 and IL-18 were reduced in a concentration-dependent manner in E11 cells [106]. Moreover, in THP-1 cells (human monocytic cell line), MS-275 suppressed LPS-induced IL-1β, IL-6, IL-18, and TNF-α expression [106]. Also, KBH-A42, a δ-lactam-based HDACi[40], inhibited cytokine production in RAW264.7 macrophage cells and in a LPS-induced mouse endotoxemia model [107]. The effect of KBH-A42 on enzyme activity of various HDAC isoenzymes demonstrated HDAC1−2 selectivity, which was corroborated by docking studies [41]. However, the effect of KBH-A42 on NF-κB is not precisely known. The effect of MGCD0103 (a potent HDAC1−3 selective inhibitor [42]) on NF-κB is not known, however, it would be interesting to study the potential of MGCD0103 to attenuate inflammatory responses.

Also, a recent study demonstrated that MI192 (a HDAC2−3 selective inhibitor [43]) reduced the TNF-α production and dose-dependently suppressed IL-6 production in primary PBMCs isolated from rheumatoid arthritis patients [108]. A few HDAC8 selective inhibitors have been developed to this point, which mostly display antiproliferative effects toward several human lung cancer cell lines (A549, H1299, and CL1-5) [109,110]. Recently it has been shown that inhibition of HDAC8 (using inhibitor ITF3056) reduces gene expression and production of proinflammatory cytokines *in vitro* and *in vivo* [39], indicating that selective inhibitors for HDAC8 could be effective in suppressing (NF-κB-mediated) inflammatory responses. Yet, to our knowledge, ITF3056 is the only inhibitor (with a moderate specificity for HDAC8), which has been reported for application in inflammation.

8.10 CONCLUDING REMARKS

Cell proliferation and apoptosis are processes in which gene expression plays a key role. Lysine acetylation of histones and nonhistone proteins play important roles in the regulation of gene expression. Protein lysine residues are acetylated by HATs and deacetylated by HDACs. Despite significant efforts, small molecule HATi have often limited potency or cellular permeability, thus leaving room for innovations in this area. In contrast, HDACi are much further developed and three inhibitors obtained clinical approval for treatment of hematological cancers. All four clinically approved HDACi bind to Zn^{2+}-dependent HDACs and their selectivity for HDAC isoenzymes is limited. Current development is aimed at improving the selectivity of HDACi and broadening their applications to other types of cancer and to other diseases. Generally, HDACi are expected to increase gene transcription of proapoptotic genes by increasing histone acetylation and thus enabling transcription of aberrantly silenced genes. Nevertheless, this mechanism does not always seem to work because tumor cells in which histone deacetylation is inhibited do not always undergo apoptosis. Several alternative mechanisms have been described to explain this phenomenon.

Apart from histone acetylation it becomes increasingly clear that HATs and HDACs regulate also posttranslational modifications of nonhistone proteins such as, for example, the NF-κB transcription factor, which is involved in cell proliferation and resistance for apoptosis. The importance of lysine acetylation in the regulation of the NF-κB pathway raises the idea that HATi and HDACi also influence such signaling pathways. Small molecules interfering with the activity of histone-modifying enzymes like HDACs have great potential to regulate the NF-κB signaling pathway. It should, however, be noted that the regulatory machinery of NF-κB-mediated cancer and/or inflammation is complex and therefore it is not always clear which HDAC is involved for which lysine deacetylation. Therefore, it often remains unclear which HDAC isoenzyme should be targeted by HDACi. As a result, a major challenge lies ahead in determining which HDAC isoenzymes target which specific lysine acetylation(s) of NF-κB. Finally, the key issues in attenuating NF-κB-mediated signaling are the selectivity of the applied small molecule HDACi, duration of the response and the subset of genes that is activated. Shedding more light on this is a key issue to broaden the application of HDACi in oncology, and other applications such as inflammatory diseases.

ACKNOWLEDGMENTS

We acknowledge financial support by the European Research Council with an ERC starting grant (309782) to F.J.D. We acknowledge COST action "epigenetics from bench to bedside TD0905" for building a European network on epigenetics.

REFERENCES

[1] Choudhary C, Weinert BT, Nishida Y, Verdin E, Mann M. The growing landscape of lysine acetylation links metabolism and cell signalling. Nat Rev Mol Cell Biol 2014;15:536–50.
[2] Amit I, Garber M, Chevrier N, Leite AP, Donner Y, Eisenhaure T, et al. Unbiased reconstruction of a mammalian transcriptional network mediating pathogen responses. Science 2009;326:257–63.
[3] Dekker FJ, van den Bosch T, Martin NI. Small molecule inhibitors of histone acetyltransferases and deacetylases are potential drugs for inflammatory diseases. Drug Discov Today 2014;19:654–60.
[4] Kouzarides T. Chromatin modifications and their function. Cell 2007;128:693–705.
[5] Cairns BR. The logic of chromatin architecture and remodelling at promoters. Nature 2009;461:193–8.
[6] Strahl BD, Allis CD. The language of covalent histone modifications. Nature 2000;403:41–5.
[7] Yu Y, Wan Y, Huang C. The biological functions of NF-kappaB1 (p50) and its potential as an anti-cancer target. Curr Cancer Drug Targets 2009;9:566–71.
[8] Ben-Neriah Y, Karin M. Inflammation meets cancer, with NF-kappaB as the matchmaker. Nat Immunol 2011;12:715–23.
[9] Chen L, Mu Y, Greene WC. Acetylation of RelA at discrete sites regulates distinct nuclear functions of NF-kappaB. EMBO J 2002;21:6539–48.
[10] Chen L, Fischle W, Verdin E, Greene WC. Duration of nuclear NF-kB action regulated by reversible acetylation. Science 2001;293:1653–7.
[11] Ito K. Impact of post-translational modifications of proteins on the inflammatory process. Biochem Soc Trans 2007;35:281–3.
[12] Peterson CL, Laniel M. Histones and histone modifications. Curr Biol 2007;14:R546–51.
[13] Lau OD, Kundu TK, Soccio RE, Ait-Si-Ali S, Khalil EM, Vassilev A, et al. HATs off: selective synthetic inhibitors of the histone acetyltransferases p300 and PCAF. Mol Cell 2000;5:589–95.
[14] Wu J, Xie N, Wu Z, Zhang Y, Zheng YG. Bisubstrate inhibitors of the MYST HATs Esa1 and Tip60. Bioorg Med Chem 2009;17:1381–6.

[15] Balasubramanyam K, Swaminathan V, Ranganathan A, Kundu TK. Small molecule modulators of histone acetyltransferase p300. J Biol Chem 2003;278:19134–40.
[16] Sbardella G, Castellano S, Vicidomini C, Rotili D, Nebbioso A, Miceli M, et al. Identification of long alkylidenemalonates as novel small molecule modulators of histone acetyltransferases. Bioorg Med Chem Lett 2008;18:2788–92.
[17] Castellano S, Milite C, Feoli A, Viviano M, Mai A, Novellino E, et al. Identification of structural features of 2-alkylidene-1,3-dicarbonyl derivatives that induce inhibition and/or activation of histone acetyltransferases KAT3B/p300 and KAT2B/PCAF. ChemMedChem 2015;10:144–57.
[18] Ghizzoni M, Wu J, Gao T, Haisma HJ, Dekker FJ, George Zheng Y. 6-Alkylsalicylates are selective Tip60 inhibitors and target the acetyl-CoA binding site. Eur J Med Chem 2012;47:337–44.
[19] Legartova S, Stixova L, Strnad H, Kozubek S, Martinet N, Dekker FJ, et al. Basic nuclear processes affected by histone acetyltransferases and histone deacetylase inhibitors. Epigenomics 2013;5:379–96.
[20] Milite C, Feoli A, Sasaki K, La Pietra V, Balzano AL, Marinelli L, et al. A novel cell-permeable, selective, and noncompetitive inhibitor of KAT3 histone acetyltransferases from a combined molecular pruning/classical isosterism approach. J Med Chem 2015;58:2779–98.
[21] Stimson L, Rowlands MG, Newbatt YM, Smith NF, Raynaud FI, Rogers P, et al. Isothiazolones as inhibitors of PCAF and p300 histone acetyltransferase activity. Mol Cancer Ther 2005;4:1521–32.
[22] Dekker FJ, Ghizzoni M, Van der Meer N, Wisastra R, Haisma HJ. Inhibition of the PCAF histone acetyl transferase and cell proliferation by isothiazolones. Bioorg Med Chem 2009;17:460–6.
[23] Ghizzoni M, Haisma HJ, Dekker FJ. Reactivity of isothiazolones and isothiazolone-1-oxides in the inhibition of the PCAF histone acetyltransferase. Eur J Med Chem 2009;44:4855–61.
[24] Gorsuch S, Bavetsias V, Rowlands MG, Aherne GW, Workman P, Jarman M, et al. Synthesis of isothiazol-3-one derivatives as inhibitors of histone acetyltransferases (HATs). Bioorg Med Chem 2009;17:467–74.
[25] Wisastra R, Ghizzoni M, Maarsingh H, Minnaard AJ, Haisma HJ, Dekker FJ. Isothiazolones; thiol-reactive inhibitors of cysteine protease cathepsin B and histone acetyltransferase PCAF. Org Biomol Chem 2011;9:1817–22.
[26] Bowers EM, Yan G, Mukherjee C, Orry A, Wang L, Holbert MA, et al. Virtual ligand screening of the p300/CBP Histone acetyltransferase: identification of a selective small molecule inhibitor. Chem Biol 2010;17:471–82.
[27] Zeng FQ, Peng SM, Li L, Mu LB, Zhang ZH, Zhang ZY, et al. Structure-based identification of drug-like inhibitors of p300 histone acetyltransferase. Yao Xue Xue Bao 2013;48:700–8.
[28] Santer FR, Hoschele PP, Oh SJ, Erb HH, Bouchal J, Cavarretta IT, et al. Inhibition of the acetyltransferases p300 and CBP reveals a targetable function for p300 in the survival and invasion pathways of prostate cancer cell lines. Mol Cancer Ther 2011;10:1644–55.
[29] Yang XJ, Seto E. The Rpd3/Hda1 family of lysine deacetylases: from bacteria and yeast to mice and men. Nat Rev Mol Cell Biol 2008;9:206–18.
[30] de Ruijter AJ, van Gennip AH, Caron HN, Kemp S, van Kuilenburg AB. Histone deacetylases (HDACs): characterization of the classical HDAC family. Biochem J 2003;370:737–49.
[31] Parra M, Verdin E. Regulatory signal transduction pathways for class IIa histone deacetylases. Curr Opin Pharmacol 2010;10:454–60.
[32] Dali-Youcef N, Lagouge M, Froelich S, Koehl C, Schoonjans K, Auwerx J. Sirtuins: the ‘magnificent seven’, function, metabolism and longevity. Ann Med 2007;39:335–45.
[33] Khan N, Jeffers M, Kumar S, Hackett C, Boldog F, Khramtsov N, et al. Determination of the class and isoform selectivity of small-molecule histone deacetylase inhibitors. Biochem J 2008;409:581–9.
[34] Marek L, Hamacher A, Hansen FK, Kuna K, Gohlke H, Kassack MU, et al. Histone deacetylase (HDAC) inhibitors with a novel connecting unit linker region reveal a selectivity profile for HDAC4 and HDAC5 with improved activity against chemoresistant cancer cells. J Med Chem 2013;56:427–36.

[35] Hu E, Dul E, Sung CM, Chen Z, Kirkpatrick R, Zhang GF, et al. Identification of novel isoform-selective inhibitors within class I histone deacetylases. J Pharmacol Exp Ther 2003;307:720–8.

[36] Beckers T, Burkhardt C, Wieland H, Gimmnich P, Ciossek T, Maier T, et al. Distinct pharmacological properties of second generation HDAC inhibitors with the benzamide or hydroxamate head group. Int J Cancer 2007;121:1138–48.

[37] Ning ZQ, Li ZB, Newman MJ, Shan S, Wang XH, Pan DS, et al. Chidamide (CS055/HBI-8000): a new histone deacetylase inhibitor of the benzamide class with antitumor activity and the ability to enhance immune cell-mediated tumor cell cytotoxicity. Cancer Chemother Pharmacol 2012;69:901–9.

[38] Wei DG, Chiang V, Fyne E, Balakrishnan M, Barnes T, Graupe M, et al. Histone deacetylase inhibitor romidepsin induces HIV expression in CD4 T cells from patients on suppressive antiretroviral therapy at concentrations achieved by clinical dosing. PLoS Pathog 2014;10:e1004071.

[39] Li S, Fossati G, Marchetti C, Modena D, Pozzi P, Reznikov LL, et al. Specific inhibition of histone deacetylase 8 reduces gene expression and production of proinflammatory cytokines *in vitro* and *in vivo*. J Biol Chem 2015;290:2368–78.

[40] Kim HM, Lee K, Park BW, Ryu DK, Kim K, Lee CW, et al. Synthesis, enzymatic inhibition, and cancer cell growth inhibition of novel delta-lactam-based histone deacetylase (HDAC) inhibitors. Bioorg Med Chem Lett 2006;16:4068–70.

[41] Kang MR, Kang JS, Han SB, Kim JH, Kim DM, Lee K, et al. A novel delta-lactam-based histone deacetylase inhibitor, KBH-A42, induces cell cycle arrest and apoptosis in colon cancer cells. Biochem Pharmacol 2009;78:486–94.

[42] Marson CM, Matthews CJ, Yiannaki E, Atkinson SJ, Soden PE, Shukla L, et al. Discovery of potent, isoform-selective inhibitors of histone deacetylase containing chiral heterocyclic capping groups and a N-(2-aminophenyl)benzamide binding unit. J Med Chem 2013;56:6156–74.

[43] Boissinot M, Inman M, Hempshall A, James SR, Gill JH, Selby P, et al. Induction of differentiation and apoptosis in leukaemic cell lines by the novel benzamide family histone deacetylase 2 and 3 inhibitor MI-192. Leuk Res 2012;36:1304–10.

[44] Bolden JE, Peart MJ, Johnstone RW. Anticancer activities of histone deacetylase inhibitors. Nat Rev Drug Discov 2006;5:769–84.

[45] Chun P. Histone deacetylase inhibitors in hematological malignancies and solid tumors. Arch Pharm Res 2015;38:933–49.

[46] Mann BS, Johnson JR, Cohen MH, Justice R, Pazdur R. FDA approval summary: vorinostat for treatment of advanced primary cutaneous T-cell lymphoma. Oncologist 2007;12:1247–52.

[47] O'Connor OA, Heaney ML, Schwartz L, Richardson S, Willim R, MacGregor-Cortelli B, et al. Clinical experience with intravenous and oral formulations of the novel histone deacetylase inhibitor suberoylanilide hydroxamic acid in patients with advanced hematologic malignancies. J Clin Oncol 2006;24:166–73.

[48] McDermott J, Jimeno A. Belinostat for the treatment of peripheral T-cell lymphomas. Drugs Today (Barc) 2014;50:337–45.

[49] Garnock-Jones KP. Panobinostat: first global approval. Drugs 2015;75:695–704.

[50] Coiffier B, Pro B, Prince HM, Foss F, Sokol L, Greenwood M, et al. Results from a pivotal, open-label, phase II study of romidepsin in relapsed or refractory peripheral T-cell lymphoma after prior systemic therapy. J Clin Oncol 2012;30:631–6.

[51] Mottamal M, Zheng S, Huang TL, Wang G. Histone deacetylase inhibitors in clinical studies as templates for new anticancer agents. Molecules 2015;20:3898–941.

[52] Santo L, Hideshima T, Kung AL, Tseng JC, Tamang D, Yang M, et al. Preclinical activity, pharmacodynamic, and pharmacokinetic properties of a selective HDAC6 inhibitor, ACY-1215, in combination with bortezomib in multiple myeloma. Blood 2012;119:2579–89.

[53] Kaufman JL, Fabre C, Lonial S, Richardson PG. Histone deacetylase inhibitors in multiple myeloma: rationale and evidence for their use in combination therapy. Clin Lymphoma Myeloma Leuk 2013;13:370–6.
[54] Ashjian E, Redic K. Multiple myeloma: updates for pharmacists in the treatment of relapsed and refractory disease. J Oncol Pharm Pract 2015, epub ahead of print.
[55] Knipstein J, Gore L. Entinostat for treatment of solid tumors and hematologic malignancies. Expert Opin Investig Drugs 2011;20:1455–67.
[56] Boumber Y, Younes A, Garcia-Manero G. Mocetinostat (MGCD0103): a review of an isotype-specific histone deacetylase inhibitor. Expert Opin Investig Drugs 2011;20:823–9.
[57] Bolden JE, Shi W, Jankowski K, Kan CY, Cluse L, Martin BP, et al. HDAC inhibitors induce tumor-cell-selective pro-apoptotic transcriptional responses. Cell Death Dis 2013;4:e519.
[58] Henderson C, Mizzau M, Paroni G, Maestro R, Schneider C, Brancolini C. Role of caspases, Bid, and p53 in the apoptotic response triggered by histone deacetylase inhibitors trichostatin-A (TSA) and suberoylanilide hydroxamic acid (SAHA). J Biol Chem 2003;278:12579–89.
[59] Siu LL, Pili R, Duran I, Messersmith WA, Chen EX, Sullivan R, et al. Phase I study of MGCD0103 given as a three-times-per-week oral dose in patients with advanced solid tumors. J Clin Oncol 2008;26:1940–7.
[60] Schrump DS, Fischette MR, Nguyen DM, Zhao M, Li X, Kunst TF, et al. Clinical and molecular responses in lung cancer patients receiving Romidepsin. Clin Cancer Res 2008;14:188–98.
[61] Lei WW, Zhang KH, Pan XC, Wang DM, Hu Y, Yang YN, et al. Histone deacetylase 1 and 2 differentially regulate apoptosis by opposing effects on extracellular signal-regulated kinase 1/2. Cell Death Dis 2010;1:e44.
[62] Fritsche P, Seidler B, Schuler S, Schnieke A, Gottlicher M, Schmid RM, et al. HDAC2 mediates therapeutic resistance of pancreatic cancer cells via the BH3-only protein NOXA. Gut 2009;58:1399–409.
[63] Hayashi A, Horiuchi A, Kikuchi N, Hayashi T, Fuseya C, Suzuki A, et al. Type-specific roles of histone deacetylase (HDAC) overexpression in ovarian carcinoma: HDAC1 enhances cell proliferation and HDAC3 stimulates cell migration with downregulation of E-cadherin. Int J Cancer 2010;127:1332–46.
[64] Ververis K, Hiong A, Karagiannis TC, Licciardi PV. Histone deacetylase inhibitors (HDACIs): multitargeted anticancer agents. Biologics 2013;7:47–60.
[65] Wagner T, Brand P, Heinzel T, Kramer OH. Histone deacetylase 2 controls p53 and is a critical factor in tumorigenesis. Biochim Biophys Acta 2014;1846:524–38.
[66] Olmos Y, Brosens JJ, Lam EW. Interplay between SIRT proteins and tumour suppressor transcription factors in chemotherapeutic resistance of cancer. Drug Resist Updat 2011;14:35–44.
[67] Sonnemann J, Marx C, Becker S, Wittig S, Palani CD, Kramer OH, et al. P53-dependent and P53-independent anticancer effects of different histone deacetylase inhibitors. Br J Cancer 2014;110:656–67.
[68] Bradford JW, Baldwin AS. IKK/nuclear factor-kappaB and oncogenesis: roles in tumor-initiating cells and in the tumor microenvironment. Adv Cancer Res 2014;121:125–45.
[69] Pal S, Bhattacharjee A, Ali A, Mandal NC, Mandal SC, Pal M. Chronic inflammation and cancer: potential chemoprevention through nuclear factor kappa B and p53 mutual antagonism. J Inflamm (Lond) 2014;11:23-9255-11-23.
[70] Hoesel B, Schmid JA. The complexity of NF-kappaB signaling in inflammation and cancer. Mol Cancer 2013;12:86-4598-12-86.
[71] Jing H, Lee S. NF-kappaB in cellular senescence and cancer treatment. Mol Cells 2014;37:189–95.
[72] Rosato RR, Kolla SS, Hock SK, Almenara JA, Patel A, Amin S, et al. Histone deacetylase inhibitors activate NF-kappaB in human leukemia cells through an ATM/NEMO-related pathway. J Biol Chem 2010;285:10064–77.
[73] Faraco G, Pittelli M, Cavone L, Fossati S, Porcu M, Mascagni P, et al. Histone deacetylase (HDAC) inhibitors reduce the glial inflammatory response *in vitro* and *in vivo*. Neurobiol Dis 2009;36:269–79.

[74] Leoni F, Fossati G, Lewis EC, Lee JK, Porro G, Pagani P, et al. The histone deacetylase inhibitor ITF2357 reduces production of pro-inflammatory cytokines *in vitro* and systemic inflammation *in vivo*. Mol Med 2005;11:1–15.

[75] Chen LF, Greene WC. Shaping the nuclear action of NF-kappaB. Nat Rev Mol Cell Biol 2004;5:392–401.

[76] Bonizzi G, Karin M. The two NF-kappaB activation pathways and their role in innate and adaptive immunity. Trends Immunol 2004;25:280–8.

[77] Lawrence T. The nuclear factor NF-kappaB pathway in inflammation. Cold Spring Harb Perspect Biol 2009;1:a001651.

[78] Wisastra R, Dekker FJ. Inflammation, cancer and oxidative lipoxygenase activity are intimately linked. Cancers (Basel) 2014;6:1500–21.

[79] Ghizzoni M, Haisma HJ, Maarsingh H, Dekker FJ. Histone acetyltransferases are crucial regulators in NF-kappaB mediated inflammation. Drug Discov Today 2011;16:504–11.

[80] Huang B, Yang XD, Lamb A, Chen LF. Posttranslational modifications of NF-kappaB: another layer of regulation for NF-kappaB signaling pathway. Cell Signal 2010;22:1282–90.

[81] Buerki C, Rothgiesser KM, Valovka T, Owen HR, Rehrauer H, Fey M, et al. Functional relevance of novel p300-mediated lysine 314 and 315 acetylation of RelA/p65. Nucleic Acids Res 2008;36:1665–80.

[82] Ashburner BP, Westerheide SD, Baldwin Jr AS. The p65 (RelA) subunit of NF-kappaB interacts with the histone deacetylase (HDAC) corepressors HDAC1 and HDAC2 to negatively regulate gene expression. Mol Cell Biol 2001;21:7065–77.

[83] Gonneaud A, Gagne JM, Turgeon N, Asselin C. The histone deacetylase Hdac1 regulates inflammatory signalling in intestinal epithelial cells. J Inflamm (Lond) 2014;11:43-014-0043-2.

[84] Serrat N, Sebastian C, Pereira-Lopes S, Valverde-Estrella L, Lloberas J, Celada A. The response of secondary genes to lipopolysaccharides in macrophages depends on histone deacetylase and phosphorylation of C/EBPbeta. J Immunol 2014;192:418–26.

[85] Chen LF, Williams SA, Mu Y, Nakano H, Duerr JM, Buckbinder L, et al. NF-kappaB RelA phosphorylation regulates RelA acetylation. Mol Cell Biol 2005;25:7966–75.

[86] Ziesche E, Kettner-Buhrow D, Weber A, Wittwer T, Jurida L, Soelch J, et al. The coactivator role of histone deacetylase 3 in IL-1-signaling involves deacetylation of p65 NF-kappaB. Nucleic Acids Res 2013;41:90–109.

[87] Mohapatra DK, Reddy DS, Ramaiah MJ, Ghosh S, Pothula V, Lunavath S, et al. Rugulactone derivatives act as inhibitors of NF-kappaB activation and modulates the transcription of NF-kappaB dependent genes in MDA-MB-231cells. Bioorg Med Chem Lett 2014;24:1389–96.

[88] Fischer A. Targeting histone-modifications in Alzheimer's disease. What is the evidence that this is a promising therapeutic avenue? Neuropharmacology 2014;80:95–102.

[89] Royce SG, Karagiannis TC. Histone deacetylases and their inhibitors: new implications for asthma and chronic respiratory conditions. Curr Opin Allergy Clin Immunol 2014;14:44–8.

[90] Grabiec AM, Tak PP, Reedquist KA. Function of histone deacetylase inhibitors in inflammation. Crit Rev Immunol 2011;31:233–63.

[91] Wightman F, Ellenberg P, Churchill M, Lewin SR. HDAC inhibitors in HIV. Immunol Cell Biol 2012;90:47–54.

[92] Dinarello CA, Fossati G, Mascagni P. Histone deacetylase inhibitors for treating a spectrum of diseases not related to cancer. Mol Med 2011;17:333–52.

[93] Leoni F, Zaliani A, Bertolini G, Porro G, Pagani P, Pozzi P, et al. The antitumor histone deacetylase inhibitor suberoylanilide hydroxamic acid exhibits antiinflammatory properties via suppression of cytokines. Proc Natl Acad Sci USA 2002;99:2995–3000.

[94] Halili MA, Andrews MR, Labzin LI, Schroder K, Matthias G, Cao C, et al. Differential effects of selective HDAC inhibitors on macrophage inflammatory responses to the Toll-like receptor 4 agonist LPS. J Leukoc Biol 2010;87:1103–14.

[95] Nasu Y, Nishida K, Miyazawa S, Komiyama T, Kadota Y, Abe N, et al. Trichostatin A, a histone deacetylase inhibitor, suppresses synovial inflammation and subsequent cartilage destruction in a collagen antibody-induced arthritis mouse model. Osteoarthr Cartil 2008;16:723–32.
[96] Grabiec AM, Korchynskyi O, Tak PP, Reedquist KA. Histone deacetylase inhibitors suppress rheumatoid arthritis fibroblast-like synoviocyte and macrophage IL-6 production by accelerating mRNA decay. Ann Rheum Dis 2012;71:424–31.
[97] Furlan A, Monzani V, Reznikov LL, Leoni F, Fossati G, Modena D, et al. Pharmacokinetics, safety and inducible cytokine responses during a phase 1 trial of the oral histone deacetylase inhibitor ITF2357 (givinostat). Mol Med 2011;17:353–62.
[98] Imre G, Gekeler V, Leja A, Beckers T, Boehm M. Histone deacetylase inhibitors suppress the inducibility of nuclear factor-kappaB by tumor necrosis factor-alpha receptor-1 down-regulation. Cancer Res 2006;66:5409–18.
[99] Yu Z, Zhang W, Kone BC. Histone deacetylases augment cytokine induction of the iNOS gene. J Am Soc Nephrol 2002;13:2009–17.
[100] Grabiec AM, Krausz S, de Jager W, Burakowski T, Groot D, Sanders ME, et al. Histone deacetylase inhibitors suppress inflammatory activation of rheumatoid arthritis patient synovial macrophages and tissue. J Immunol 2010;184:2718–28.
[101] Kim Y, Kim K, Park D, Lee E, Lee H, Lee YS, et al. Histone deacetylase 3 mediates allergic skin inflammation by regulating expression of MCP1 protein. J Biol Chem 2012;287:25844–59.
[102] Zhang Q, Yang F, Li X, Wang LW, Chu XG, Zhang H, et al. Trichostatin a protects against experimental acute-on-chronic liver failure in rats through regulating the acetylation of nuclear factor-kappaB. Inflammation 2015;38:1364–73.
[103] Sato T, Kotake D, Hiratsuka M, Hirasawa N. Enhancement of inflammatory protein expression and nuclear factor Kappab (NF-Kappab) activity by trichostatin A (TSA) in OP9 preadipocytes. PLoS One 2013;8:e59702.
[104] Zhang ZY, Schluesener HJ. HDAC inhibitor MS-275 attenuates the inflammatory reaction in rat experimental autoimmune prostatitis. Prostate 2012;72:90–9.
[105] Lin HS, Hu CY, Chan HY, Liew YY, Huang HP, Lepescheux L, et al. Anti-rheumatic activities of histone deacetylase (HDAC) inhibitors *in vivo* in collagen-induced arthritis in rodents. Br J Pharmacol 2007;150:862–72.
[106] Choo QY, Ho PC, Tanaka Y, Lin HS. Histone deacetylase inhibitors MS-275 and SAHA induced growth arrest and suppressed lipopolysaccharide-stimulated NF-kappaB p65 nuclear accumulation in human rheumatoid arthritis synovial fibroblastic E11 cells. Rheumatology (Oxford) 2010;49:1447–60.
[107] Choi Y, Park SK, Kim HM, Kang JS, Yoon YD, Han SB, et al. Histone deacetylase inhibitor KBH-A42 inhibits cytokine production in RAW 264.7 macrophage cells and *in vivo* endotoxemia model. Exp Mol Med 2008;40:574–81.
[108] Gillespie J, Savic S, Wong C, Hempshall A, Inman M, Emery P, et al. Histone deacetylases are dysregulated in rheumatoid arthritis and a novel histone deacetylase 3-selective inhibitor reduces interleukin-6 production by peripheral blood mononuclear cells from rheumatoid arthritis patients. Arthritis Rheum 2012;64:418–22.
[109] Tang W, Luo T, Greenberg EF, Bradner JE, Schreiber SL. Discovery of histone deacetylase 8 selective inhibitors. Bioorg Med Chem Lett 2011;21:2601–5.
[110] Huang WJ, Wang YC, Chao SW, Yang CY, Chen LC, Lin MH, et al. Synthesis and biological evaluation of ortho-aryl N-hydroxycinnamides as potent histone deacetylase (HDAC) 8 isoform-selective inhibitors. ChemMedChem 2012;7:1815–24.

CHAPTER

TARGETING HISTONE METHYLATION

9

The Development of Selective Methyltransferase and Demethylase Inhibitors

Marco P. Licciardello[1] **and Stefan Kubicek**[1,2]

[1]CeMM Research Center for Molecular Medicine of the Austrian Academy of Sciences, Vienna, Austria [2]Christian Doppler Laboratory for Chemical Epigenetics and Antiinfectives, CeMM Research Center for Molecular Medicine of the Austrian Academy of Sciences, Vienna, Austria

CHAPTER OUTLINE

G. Egger & P. Arimondo (Eds): Drug Discovery in Cancer Epigenetics. DOI: http://dx.doi.org/10.1016/B978-0-12-802208-5.00009-6

9.1 GLOBAL AND LOCUS-SPECIFIC CONTROL OF HISTONE LYSINE METHYLATION

Methylation on lysine and arginine residues is controlled by the interplay of methyltransferases and demethylases. These enzymes are remarkable in their high degree of substrate and product specificity, a feature that discriminates them from many other epigenetic regulators. Typically, histone methyltransferases and demethylases not only discriminate their substrate based on its sequence context, but also tightly regulate the degree of methylation of their substrates and products. Thereby, these enzymes control mono-, di-, and trimethylation states on the epsilon amino group of lysines and mono-, symmetric di-, and asymmetric dimethylation of arginines.

At least 11 lysines and arginines in histones can be methylated, and the different biological functions conferred by these modifications require tight control of their respective methylation states. While most chromatin modifiers are ubiquitously expressed, some epigenetic enzymes exert their activities only in specific tissues. Furthermore, binding partners can change the activity and specificity of chromatin modifiers. The catalytic activity of these enzymes relies on the presence and amount of certain cofactors. Many of these cofactors are key metabolites in central biochemical pathways, and thereby link chromatin to cellular metabolism. All of the above mechanisms directly affect the catalytic activities of methyltransferases and demethylases, and thereby change the global levels of the modifications they catalyze. However, global modification levels are not important for transcriptional outcomes, rather locus-specific modifications at the histones contacting the enhancers, promoters, and gene bodies. Since chromatin-modifying enzymes typically do not harbor DNA-binding domains themselves, they are normally recruited to certain loci through their interaction with specific transcription factors. In theory, all of the mechanisms described above—expression, activation, and recruitment of methyltransferases and demethylases—provide druggable targets for the development of small-molecule modulators.

9.1.1 REGULATION OF HISTONE METHYLATION BY EXPRESSION OF THE MODIFIERS

One important level of control is through the specific expression of the different enzymes that catalyze the formation and removal of histone methylation. Methyltransferases are grouped in a large family of around 59 proteins [1]. Of these, 50 harbor a SET domain characteristic of histonemethyltransferases. In addition, the enzyme DOT1L has been shown to methylate histone H3 lysine 79, despite being more closely related to the family of 8 arginine methyltransferases. Two families of histone lysine demethylases exist: LSD1 and LSD2 are closely related to other amine oxidases, including MAO-A/B. By mechanism, these enzymes can only remove mono- and di- but not trimethylation. In contrast, the family of approximately 23 active Jumonji C (JmjC) domain proteins can affect all three lysine methylation states by hydroxylation. In contrast to lysine methylation, the mechanism for removal of arginine methylation is less clear. While JMJD6 has been claimed to be a direct arginine demethylase [2], crystallographic evidence rather points to a role of this JmjC domain protein in the hydroxylation of the lysine side chains [3]. PAD4-dependent demethylimination [4] provides an alternative mechanism to remove arginine methylation, but generates citrulline instead of unmodified arginine. Furthermore, the efficiency of this reaction with methylated substrates is unclear.

The majority of chromatin modifiers are robustly and ubiquitously expressed across human tissues at the RNA level. However, notable exceptions exist, and some enzymes are restricted to certain tissues. For example, PRDM7, 9, 14, and KDM4D are selectively expressed in testes and SMYD1 is highly enriched in heart and skeletal muscle (Figure 9.1).

Further refining this expression map from the gene to the transcript level additionally increases the resolution and can result in tissue-specific functions through isoform-specific protein complexes.

The expression of methyltransferases and demethylases can also be modulated by a number of external stimuli. For example, the methyltransferase SETDB2 is strongly activated following influenza infection, and contributes to the pathology of bacterial superinfections [6]. Indeed, some infectious agents encode their own methyltransferases and demethylases to affect the host environment [7,8], express proteins that bind and modulate endogenous enzymes [9], or provide novel substrates to divert host methyltransferase activities [10]. Other cell-extrinsic changes in addition to infectious agents have also been shown to modulate the expression of histone methyltransferases and demethylases. For example, nutrient starvation increases the expression of KDM7A [11], and several demethylases are upregulated in hypoxic conditions [12–15].

9.1.2 ENZYMATIC REACTION MECHANISMS LINK HISTONE METHYLATION TO METABOLITE BALANCE

Epigenetics is often described as linking the genome to the environment. This happens on several levels, the most direct being the use of cellular metabolites as cofactors by histone-modifying enzymes (Figure 9.2).

Like DNA methyltransferases, all histone methyltransferases use S-adenosylmethionine (SAM) as a cofactor. In the course of the reaction, the activated S-methyl group is transferred to the substrate lysine or arginine, and the cofactor is converted to S-adenosylhomocysteine (SAH). Interestingly, a low ratio of SAM/SAH has been observed in cancer cells and shown to contribute to their proliferative potential [16]. SAM is generated from methionine and ATP by the action of the enzyme methionine adenosyltransferase. The byproduct of the methylation reaction SAH is normally converted to homocysteine and adenosine, while homocysteine is remethylated to methionine in a

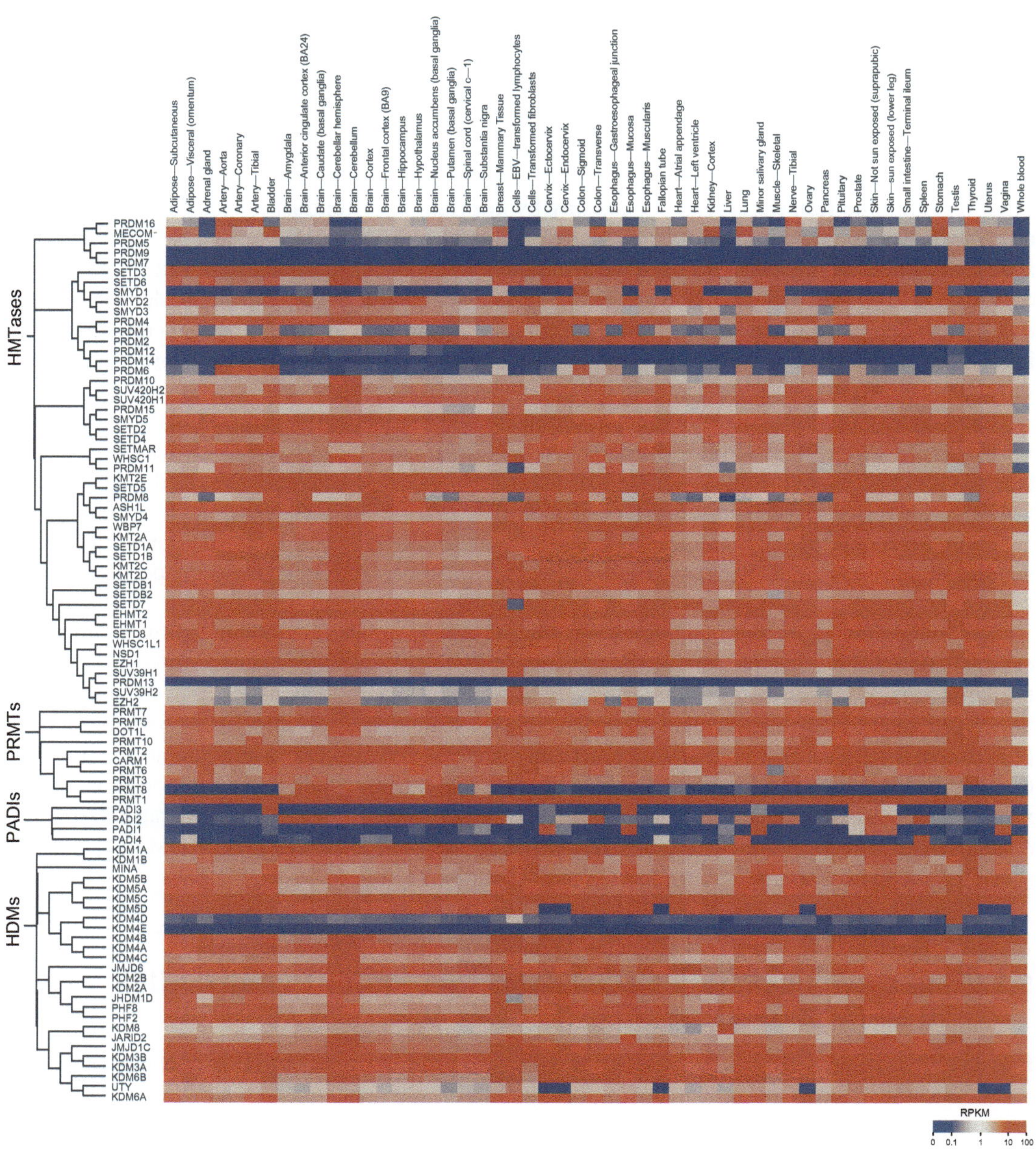

FIGURE 9.1

Tissue-specific expression of histone methyltransferases and demethylases. Enzymes are clustered by their phylogenetic conservation (*y*-axis). The heatmap is color-coded by the expression RPKM (reads per kilobase per million mapped reads) values across the different human tissues [5].

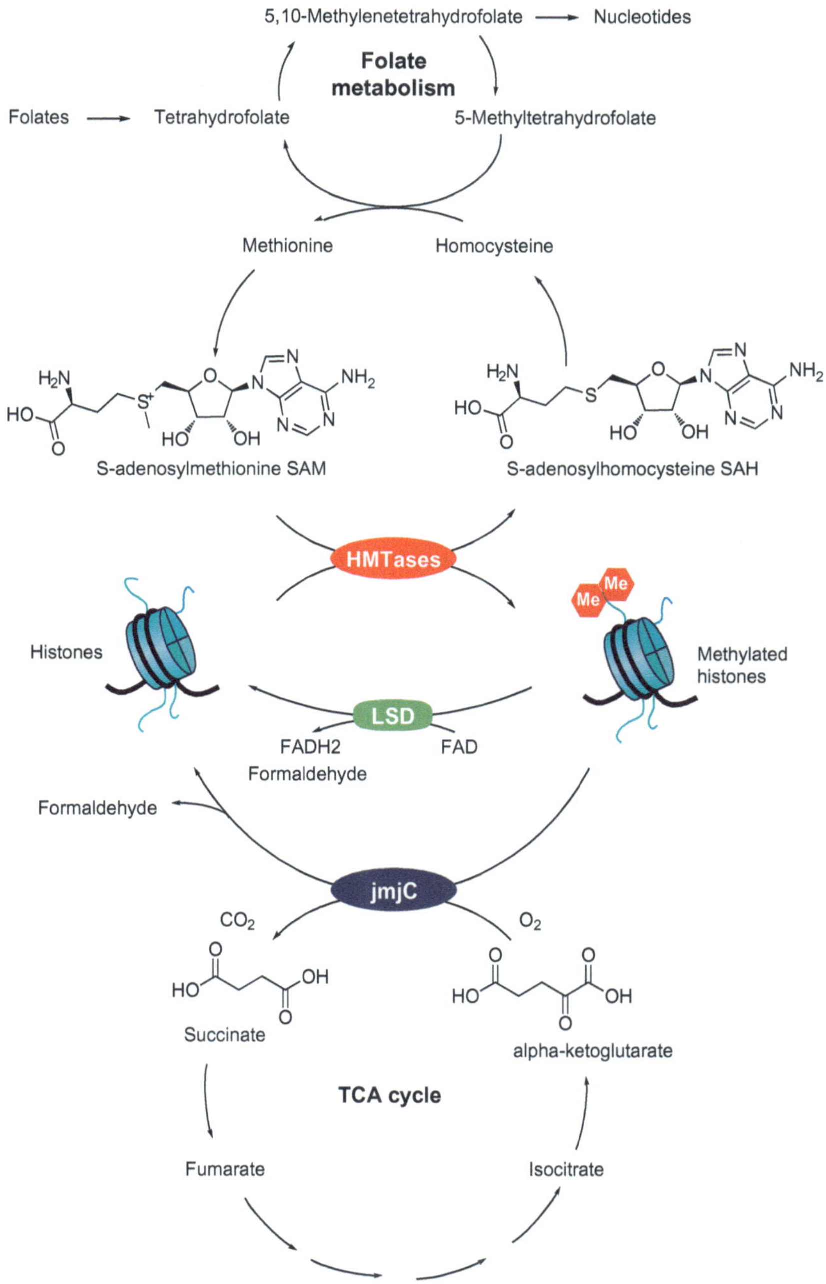

FIGURE 9.2

Chromatin modifiers integrate metabolic pathways. Histone methyltransferases require S-adenosylmethionine while demethylases use either an FAD-dependent amine oxidation mechanism or an iron, alpha-ketoglutarate, and oxygen-dependent hydroxylation. In both cases, the methyl group is lost as formaldehyde.

5-methyltetrahydrofolate-dependent reaction. Thus, the methyl group ultimately originates from folates, the dietary source of methyl donors. A methyl-deficient diet has been shown to cause liver carcinogenesis [17], but it remains unclear to what extent this effect is due to changes in DNA and histone methylation or more general methionine metabolism.

All histone demethylases oxidize the methyl group to formaldehyde, thereby linking the histone demethylation to intracellular redox chemistry. While LSD1 and LSD2 use FAD as a cofactor, interestingly these enzymes additionally contain a folate-binding site and appear to capture the produced formaldehyde as 5,10-methylene-THF, thereby feeding it back to the methylation cycle [18]. In contrast, the JmjC family uses iron, oxygen, and 2-oxoglutarate as cofactors. As such, these enzymes sense cellular oxygen levels, and their own transcription and enzymatic function is regulated accordingly [19]. The other essential cofactor, the key TCA cycle metabolite 2-oxoglutarate, is converted to succinate in the demethylase reaction. Similar to altered SAM/SAH ratios in cancer, changes in the levels of cofactors and byproducts can profoundly affect the activity of demethylases, change chromatin structure and contribute to tumorigenesis. Succinate is normally converted to fumarate by succinate dehydrogenase (SDH). Interestingly, SDH mutations observed in familial cases of paraganglioma, pheochromocytoma, and gastrointestinal stromal tumor cause alterations in the epigenome [20,21]. Even more dramatically, mutations in isocitrate dehydrogenases (IDH1 and IDH2) induce the enzymes to convert isocitrate to the cancer-specific 2-hydroxyglutarate instead of 2-oxoglutarate [22]. These mutations result in a dramatic accumulation of 2-hydroxyglutarate inhibiting histone demethylase activity and resulting in altered chromatin modifications. For this reason, 2-hydroxyglutarate has been referred to as an oncometabolite. Mutant IDH1 and IDH2 have been recognized as targets for cancer therapy, and specific inhibitors are currently being evaluated in the clinic [23,24].

9.1.3 PROTEIN–PROTEIN INTERACTIONS CONTROL THE RECRUITMENT AND ACTIVITIES OF HISTONE-METHYLATING ENZYMES

Histone methyltransferases and demethylases do not act in isolation, but in close association with other protein, RNA, and DNA factors in the chromatin environment. These factors change their overall activity, their specificity for products and substrates, and their genomic locations.

Often, functional protein complexes are required for methyltransferases and demethylases to exert their catalytic activities. The best-characterized example of an enzyme requiring interacting proteins is the H3K27 methyltransferase EZH2. This enzyme is part of the Polycomb repressive complex 2 (PRC2). EZH2 activity requires interaction with its binding partners EED and SUZ12. [25,26]. Similarly, the H3K4 methylation system requires not only the methyltransferase MLL, but also its interaction partners WDR5, RBBP5, ASH2L, and DPY30 for full activity [27]. In the absence of its interaction partners, MLL is a weak methyltransferase capable of only generating H3K4me1 [27–29]. In complex with ASH2L, WDR5, and RBBP5 there is a dramatic increase in overall methyltransferase activity, and dimethylated products are predominantly formed. The histone demethylase LSD1 was first described as a H3K4me1/2-specific enzyme [30]. However, at the promoters of androgen receptor target genes, the specificity is altered to H3K9me1/2 [31]. In part, this specificity switch is caused by phosphorylation of T6 on the histone tail [32]. The same substrate specificity for H3K9 methylation is also observed in a neuron-specific splice isoform of

LSD1 [33]. It remains to be seen how splice variants, LSD1 interaction partners, and histone substrate modifications synergize in controlling the enzyme's specificity.

In addition to structural complex components modulating the activities and specificities of methyltransferases, interaction partners are also essential for controlling the genomic location of these enzymes. Several methyltransferases and demethylases directly bind sequence-specific transcription factors, and in other cases the chromatin association is mediated indirectly through interaction partners of the enzymes.

9.2 ASSAYING HISTONE METHYLATION

An important factor in translating the knowledge on the biology of histone methylation pathways into the development of epigenetic drugs is the ability to faithfully model these processes in biologically relevant assays compatible with high-throughput screening (HTS). Screening technologies are usually grouped in two main categories: biochemical assays make use of purified enzymes and substrates, while phenotypic, cell-based formats provide more physiologically relevant information. Biochemical assays avoid many complications intrinsic to cellular systems, including compound uptake and degradation, and therefore are preferred for optimizing hit compounds in medicinal chemistry campaigns. However, these formats also have a number of shortcomings, including their reliance on purified proteins not necessarily resembling intracellular complex composition and the risk of assay artifacts. Therefore, in our opinion the best approach to rapidly define a set of relevant compounds is developing cellular and biochemical screens in parallel, addressing large compound libraries with the more effective of the two technologies and immediately following up on hit compounds with the complementary format.

9.2.1 BIOCHEMICAL ASSAYS ALLOW RAPID DETECTION OF ENZYMATIC ACTIVITIES

A number of biochemical assay formats have been used for identifying small-molecule modulators of histone methyltransferase and demethylase activities. These include binding assays, technologies that directly quantify the substrates and products of these proteins, and coupled assays that typically convert cofactor levels to quantitative readouts.

9.2.1.1 Binding assays for methyltransferases and demethylases

The most universal format to identify hit compounds for any protein target is the binding assay. Classical biophysical binding assays, including surface plasmon resonance and isothermal calorimetry, allow the exact determination of binding kinetics. However, the limited throughput of these methods makes them more suitable for follow-up studies. The same is true for a demethylase binding assay developed for JMJD2E using non-denaturing mass spectrometry [34]. The most HTS-friendly binding assay relies on the stabilization of proteins by small molecules against thermal unfolding and is known as thermal shift, thermofluor, or differential scanning fluorimetry [35,36]. Typically, in these assays a dye like Sypro orange is used that changes its fluorescent properties when bound to hydrophobic amino acid stretches. Such sequences are normally buried in the interior of the protein and only become surface-exposed upon unfolding. While binding approaches are universal, they disregard the inherent activity of enzymes, and are therefore rarely used for histone methyltransferases and demethylases. Prior knowledge on the active sites and endogenous ligands can provide a way to tune binding assays to these enzyme classes. In an interesting twist to thermofluor technology, the fluorescent property changes of the endogenous cofactor FAD upon binding to LSD1 have been used in an assay called ThermoFAD [37]. In the protein-bound state, the inherent fluorescence of the flavin cofactor is quenched by the interaction with the protein. Only upon thermal unfolding of LSD1 is a strong increase in fluorescence detected, with strong correlation to the Sypro orange assay. Therefore, this assay is theoretically suitable not only to measure the proper folding of LSD1 or identify general binders, but also to screen specifically for compounds that directly compete with FAD and therefore result in FAD fluorescence at much lower temperatures. A different approach has been developed for the family of JmjC demethylases [38]. In this case, a synthetic fluorescein-labeled ligand binds the active site of JHDM1A and allows monitoring of the binding process by fluorescence polarization. Both endogenous cofactors and known demethylase inhibitors have been shown to compete off this molecule in an assay that allows for binding quantification. For methyltransferases, label-free ligand replacement has recently been developed and shown to be suitable for the identification of potent inhibitors of DOT1L [39]. In a similar approach, a biotin-labeled inhibitor was used in homogeneous time-resolved fluorescence assays with G9a [40].

9.2.1.2 Direct detection of modified substrates or products

Small molecules identified as protein binders in the assays described above do not necessarily change the function of their target proteins, and if they do they may affect their protein–protein interactions, intracellular localization, stability, folding, or enzymatic activity. Based on experience, the most "druggable" of these potential mechanisms of action is inhibiting enzymatic activities, and often active-site binding inhibitors are desired. To find such molecules, it is advisable to directly measure the turnover of the reaction by quantifying the amounts of newly generated products, the remaining substrates, or their ratio.

The original assay format for detecting methyltransferase activity used radioactively labeled SAM [41] and monitored the incorporation of the radioactive label into substrate peptides, histones, and nucleosomes by gel-based methods or direct scintillation counting. The high sensitivity of this assay format has also resulted in the implementation of scintillation proximity assays for lysine and arginine methyltransferases in HTS [42,43]. While histone demethylase activity cannot directly be assayed with scintillation proximity assays, the unique substrate and product requirements have allowed monitoring of demethylase activity by coupling to the appropriate methyltransferases [44].

The vast majority of methyltransferase and demethylase assays use antibody-based detection methods. For these approaches, thoroughly validated antibodies that clearly discriminate between substrate and product methylation states are a prerequisite. Unfortunately, the limited characterization and lot-related variability of commercial products often precludes the establishment of highly robust assays. In order to identify efficient and reliable antibodies, we have characterized a panel of more than 150 monoclonal products for their binding specificity to a library of more than 100 differentially modified histone peptides (E. Schirghuber and S. Kubicek, unpublished). In line with similar initiatives [45–49], we observe that at best 30% of histone modification antibodies show selective binding to the mark they were developed against. For biochemical screening, some off-target specificities are typically permissible, as both substrate and products are normally well-known. However, for cross-assay comparison it is important to extensively characterize each new antibody lot for its efficacy to quantitatively discriminate the different methylation states.

With highly specific antibodies at our disposal, either the substrate or the product can be quantified. Because many of the chromatin modifiers have rather low activities, often it is preferable to detect the product formed as this allows better dynamic range in the detection system. Numerous high-throughput screens have been described that use different immunological assay formats, including enzyme-linked immunosorbent assay (ELISA) [50,51], dissociation-enhanced lanthanide fluorescent immunoassay (DELFIA) [52–54], AlphaScreen [55,56], or homogeneous time-resolved fluorescence (HTRF) [53].

While most assays use antibodies against the differentially methylated substrates and products, it is also possible to quantify the cosubstrates and products. This strategy has the advantage of a broad applicability to all members of an enzyme class. For methyltransferses, such an assay has been developed in FRET format by competing away a labeled SAH from an antibody by the SAH generated in the course of the reaction (Epigeneous Assay, CisBio).

Radioactive and antibody-based methods have the advantage of allowing switching peptide, histone, and nucleosome substrates without much optimization. Implementing such changes in mass spectrometry-based methods is typically more challenging, as it requires extensive adjustment of the sample preparation process. In return, mass spectrometry provides many distinct advantages, most importantly the high resistance to assay artifacts. Typically the abundance of both substrate and product is determined with high accuracy, allowing the sensitive identification of even weak inhibitors or activators. A number of different mass spectrometric techniques have been applied, with RapidFire [57–59] and SAMDI [60] approaches suitable for screening of mid-sized compound libraries, and more conventional HPLC-ESI-MS [34] and MALDI-TOF-MS [61] for lower-throughput applications. While most of these technologies focus on the detection of modified histone peptides, a universal HPLC-MS assay quantifying the byproduct SAH has been described [62]. Similarly, succinate has been detected by mass spectrometry as an indicator of JmjC demethylase activity [63].

9.2.1.3 Indirect detection methods

Indirect detection methods typically use a coupled reaction mechanism whereby products of the original reaction are further converted by additional enzymes to allow absorbance, fluorescence, or luminescence readouts. These methods have the advantage of often permitting continuous activity monitoring over time and are usually relatively cheap, as they do not require specific antibodies. Their disadvantage is the high propensity to assay artifacts and false-positive hits, as all modulators of the coupling enzymes or quenchers of the final signal will appear as hits.

In the case of histone methyltransferases, several approaches have been described that couple the production of SAH to a fluorescent readout. In one example, SAH is converted to adenosine monophosphate (AMP) by the action of the enzymes SAH hydrolase and adenosine kinase. This makes the assay amenable to detection systems well established for targeting kinases, for example, detection of AMP in a fluorescence polarization setup [64] or luminescent detection of the remaining adenosine kinase substrate ATP [65].

For demethylases, the generation of formaldehyde is often used in coupled assays. With the original identification of the LSD1 demethylase, two such coupled assays were described [30]: in formaldehyde dehydrogenase assays, the enzyme is used to oxidize formaldehyde to formic acid, coupled to the reduction of NAD+ and the measurement of the fluorescence by the resulting NADH. Alternatively, in a non-enzymatic reaction, formaldehyde can also react with dimedone to generate the product formaldemethone, which is detected by mass spectrometry.

9.2.1.4 Profiling of hit compounds against enzyme panels

The high interest in methyltransferases and demethylases in the pharmaceutical industry has encouraged commercial suppliers to provide assaying services to profile hit compounds biochemically against panels of methyltransferases and demethylases. In terms of completeness, these panels are still way behind those available for kinases, but at least 20 methyltransferases and 15 demethylases can currently be profiled in this way (BPS Bioscience, DiscovRx). Furthermore, the Structural Genomics Consortium [66,67] has established binding and enzymatic assays for a wide variety of histone methyltransferases and demethylases.

9.2.2 CELL-BASED ASSAYS FOR HISTONE METHYLATION PATHWAYS

Cell-based assays test compounds in an environment they also encounter in vivo and are therefore ideal to probe for biological relevance. In particular, the influence of interactors on the structure and function of chromatin-modifying enzymes is often poorly represented in biochemical assays with recombinant proteins. Furthermore, the cellular environment ensures the concomitant expression of the relevant closely related proteins, which will compete for small-molecule binding and allow the detection of off-targets. The only disadvantages of cellular assays are the dependence of intracellular concentrations on compound uptake and export, and the higher risk for artefacts due to compound degradation. In general, cell-based screens can be categorized into antibody-based, reporter-based, and mass spectrometric methods.

9.2.2.1 Antibody-based cellular assays

Highly selective antibodies that retain their specificity in a cellular environment have been used in various different assays. For small sets of compounds, Western blotting for selective marks has been the method of choice. In high-throughput applications, these are usually converted to the sandwich formats of ELISA and AlphaScreen. In an interesting approach it has been shown that high-binding plates can be used to directly immobilize histones allowing antibody-based detection of modifications on these highly abundant proteins without the need to precoat plates with pan-histone capture antibodies [68]. In addition, highly specific antibodies can be used in immunofluorescence staining protocols on permeabilized cells, as has been implemented for H3K27me3 regulated by the counteracting activities of EZH2 [69] and JMJD3 [70].

Global approaches to cellular histone methylation as described above yield fast results for some highly expressed, nonredundant, highly active enzymes. However, for most methyltransferases and demethylases the redundancy of activities within subfamilies, together with the low contribution of most enzymes to total modification levels preclude the use of global epigenetic changes for the quantification of endogenous activities. Overexpression systems are an alternative, but carry some of the risks intrinsic to biochemical assays.

The mapping of enzyme-binding sites and histone modification patterns on the genome of many cell types in the International Human Epigenome and ENCODE Consortia provides a way to directly address activities of histone-methylating enzymes at their most relevant target genes. Chromatin immunoprecipitation (ChIP) is the method of choice, and can now routinely be performed on 96-well plates. Further technology improvements and decreases in sequencing costs might make it possible to use ChIPs for screening of small preselected compound libraries in the future.

9.2.2.2 Reporter-based cellular assays

Cellular reporters for histone methylation allow the rapid and cost-effective identification of small molecules affecting these pathways. These assays are often developed using a specific methylation-dependent promoter driving the expression of a luciferase construct. However, the interplay of many chromatin factors for transcription control make such formats prone to identifying compounds acting by alternative mechanisms. In a more specific approach, a FRET pair was generated in the fusion of a histone tail to a chromo-domain, labeled by CFP and YFP on both termini [71]. Only when the histone peptide becomes methylated does the construct fold back on itself and generate a fluorescent signal. By using the chromodomains of HP1 and Polycomb, this system has been optimized for the discovery of K9 and K27 methyltransferase inhibitors.

9.2.2.3 MS-based cellular assays

Mass spectrometry has long been the method of choice to identify and confirm novel histone modifications, as well as to quantify these marks on histones from cellular extracts [72,73]. Typically, these modifications are detected on small peptides generated by proteolytic digestion of histones extracted from cellular lysates. Furthermore, top-down methods have been developed that allow the detection of the co-occurrence of marks at distant sites on the same histone protein [74]. Both these methods are ideal for in-depth analysis of specificity of small sets of inhibitors identified with other technologies, but cannot easily be up-scaled to even mid-throughput compound screening. In one of the most large-scale histone proteomics efforts so far, 115 cancer cell lines were characterized regarding their global histone modification patterns [75].

9.3 TEN WAYS TO TARGET HISTONE METHYLATION

The biochemistry of histone methyltransferases and demethylases enables multiple modes of targeting their enzymatic function. Several of these inhibitory mechanisms have been confirmed in cocrystal structures (Figure 9.3).

9.3.1 METHYLTRANSFERASES CAN BE INHIBITED BY PREVENTING SAH TURNOVER

Methyltransferases are dependent on SAM as a methyl-donor and are inhibited by the byproduct SAH. In many cancer cells, alterations in the SAM/SAH ratio have been observed.

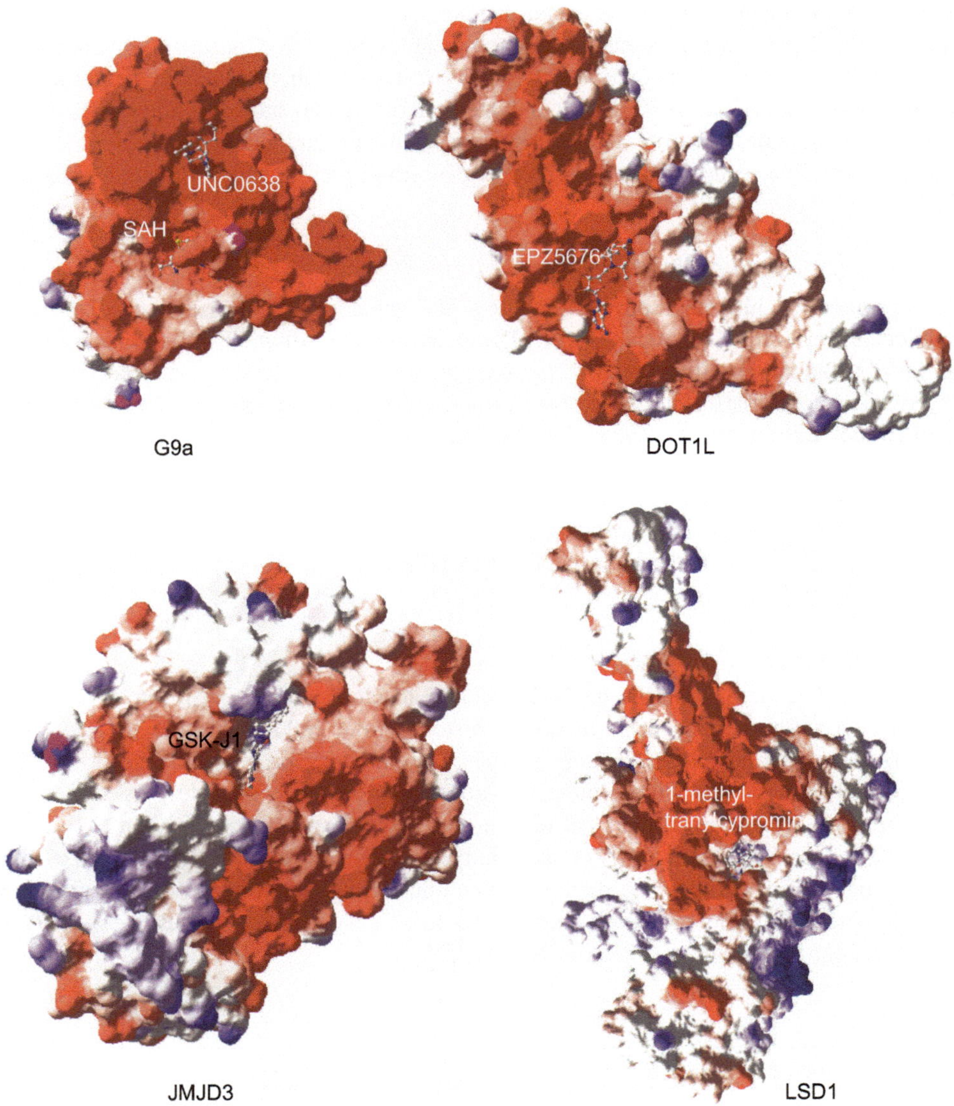

FIGURE 9.3

Representative crystal structures of histone methyltransferases and demethylases bound to small-molecule inhibitors. For histone methyltransferases, both the peptide-binding pocket (as shown for G9a, 3RJW) and the SAM-binding pocket (as for DOT1L, 4HRA) have been targeted successfully. Demethylases are represented by JMJD3 bound to GSK-J1 (4ASK) and LSD with a tranylcypromine derivative (4UVB).

This could be mimicked in an interesting biological experiment, by overexpressing nicotinamide N-methyltransferase and thereby generating a methyl sink and diminishing the cellular SAM pool [16].

The first attempts to inhibit methyltransferases were initiated before the discovery of histone methyltransferases in 2000 [41] and focused on small-molecule-modifying enzymes like catechol O-metyltransferase (COMT). These efforts aimed at manipulating the levels of the cofactor SAM and its reaction product SAH [76]. For example, it was soon recognized that SAH itself blocks methyltransferase activity by product inhibition [77]. Other compounds exploiting the same mechanism were the SAH analog sinefungin and various related molecules developed by medicinal chemistry [76,78–80].

Another way to influence the balance between SAM and SAH is to prevent the degradation of SAH by SAH hydrolase. This enzyme is inhibited by 3-deazaadenosine analogs [81,82], adenosine dialdehyde, 3-deaza-(+/−)-aristeromycin [83], and neoplanocin A [84]. The analog 3-deazaneplanocin A (DZNep) was first isolated as an SAH hydrolase inhibitor in 1986 [85], but recently experienced a comeback. This was largely based on the claim that the histone methyltransferase EZH2 is inhibited with some selectivity by this compound and that DZNep treatment results in the degradation of this H3K27 methylating enzyme [86]. However, this specificity claim was quickly refuted as it was shown that the compound is a pan-methyltransferase inhibitor inducing a general increase in the levels of SAH [87]. Nevertheless, DZNep is unfortunately still widely employed by the scientific community in studies of EZH2 function, even though now much more specific compounds have been made available.

Another compound described as a specific methyltransferase inhibitor is chaetocin, found in a screen of a natural compound library for SU(VAR)3-9 targeting agents [88]. However, the structure of this compound is not very attractive for further development, the dependence of its activity on its disulfide bonds has made the specificity disputed [89,90], and the compound targets other proteins, including thioredoxin reductase [91].

Dietary folates are essential for replenishing the pool of C1 donors and SAM. Therefore, it is not surprising that a folate-deficient diet results in DNA and histone methylation changes [92–94]. In rodent models, a methyl-deficient diet alone can cause liver carcinogenesis [95–98]. Interestingly, the folate pathway is also targeted in cancer with antifolates like methotrexate. Currently, little is known as to what extent such antifolates interfere with the balance between SAM and SAH and whether changes in histone and DNA methylation contribute to the efficacy of these drugs.

9.3.2 TARGETING THE PEPTIDE-BINDING POCKET OF METHYLTRANSFERASES

The structure of SET domain methyltransferases consists of a knot-like fold that contains binding sites for the cofactor SAM and the substrate peptide on opposite surfaces [99–101]. During the catalytic reaction the methyl is threaded through a hydrophobic hole connecting the two binding sites. Therefore, it is possible to target either of the two pockets with small molecules.

The first specific inhibitor of a histone methyltransferase was BIX-01294 that we identified by HTS as a compound binding the peptide-binding pocket of H3K9 methylating enzymes G9a and GLP [54]. The compound acts as an uncompetitive inhibitor and preferentially binds the

SAM-bound enzyme. A crystal structure of BIX-01294 bound to GLP has been obtained [102] and resulted in the development of more potent analogs, including UNC0638 [103].

Targeting the peptide-binding pocket of methyltransferases has the advantage of more easily achieving selectivity for the different enzyme subfamilies with unique substrate selectivity. At the same time it bears the risk of inhibiting other enzymes targeting the same stretch of histone sequence. Along these lines, an analog of BIX-01294 has been shown to inhibit the demethylases targeting the same amino acid [104].

9.3.3 THE SAM-BINDING POCKET AS UNIVERSAL TARGET OF ALL METHYLTRANSFERASES

The majority of inhibitors developed against methyltransferases target the SAM-binding pocket. This strategy is similar to the kinase field, where ATP-competitive inhibitors are the agents of choice. SAM competitive compounds bear the risk of having unknown side effects across other subfamilies of histone methyltransferases. However, the more limited field of only 59 histone methyltransferases compared to more than 600 human kinases makes target specificity testing easier. Importantly, other small-molecule methyltransferases, SAM-radical enzymes, and SAM-binding proteins might also be inhibited and need to be excluded as off-targets. Some compounds that have been developed to bind the SAM-binding pocket closely resemble the endogenous cofactor. This is especially true for the clinical DOT1L inhibitor EPZ-5676 [105]. This compound inhibits DOT1L in the picomolar range with more than 30,000-fold selectivity over other methyltransferases. However, the close resemblance of this compound to SAM also comes at the cost of limited oral bioavailabilty and relatively short plasma half-life [106]. Ultimately, these features might have contributed to the discontinuation of clinical trials with this compound for lack of efficiency as a single agent. Therefore, for other methyltransferases, typically structures less closely resembling the cofactor were identified from screening campaigns and then optimized by medicinal chemistry. A list of the most potent inhibitors is provided in Figure 9.4.

9.3.4 ALLOSTERIC AND INDIRECT INHIBITION OF METHYLTRANSFERASES

Allosteric inhibition of methyltransferases utilizing binding pockets other than the cofactor and substrate pockets is a rarely used strategy to target this enzyme class. The best example is an inhibitor of arginine methyltransferase PRMT3 [107,108]. Another way to inhibit methyltransferases is indirectly through preventing their binding to interaction partners that are required for their catalytic activity or recruitment to certain chromatin regions. For example, the compound OICR-9429 binds WDR5 and thus prevents the interaction with the SET/MLL methyltransferase complex [109]. Similarly, the targeted disruption of the interaction between EZH2 and EED is a way to inhibit the methyltransferase activity of the PRC2 complex [110].

9.3.5 IRREVERSIBLE INHIBITORS OF LSD1/2 COVALENTLY BIND THE COFACTOR FAD

The first attempts to target the amine oxidase class of demethylases followed the strategies employed for the inhibition of other amine oxidases, including the well-known monoamine

UNC0638
G9a/
GLP

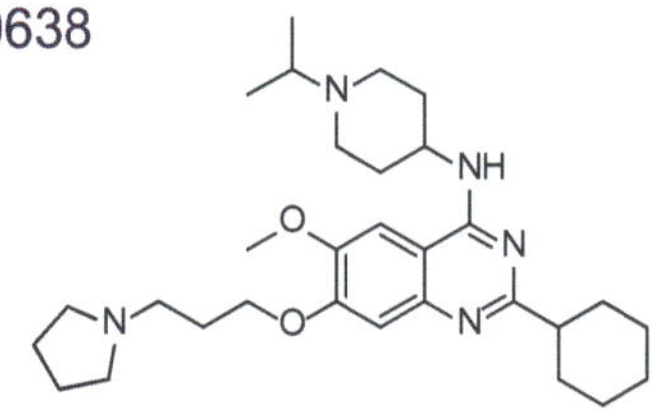

LLY-507
SMYD2

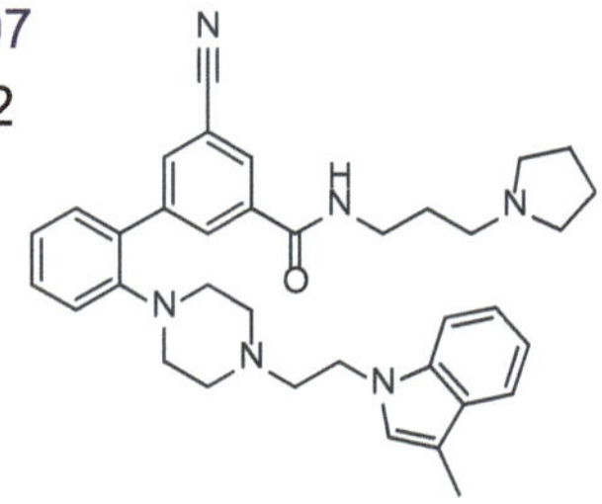

EPZ-6438
EZH2

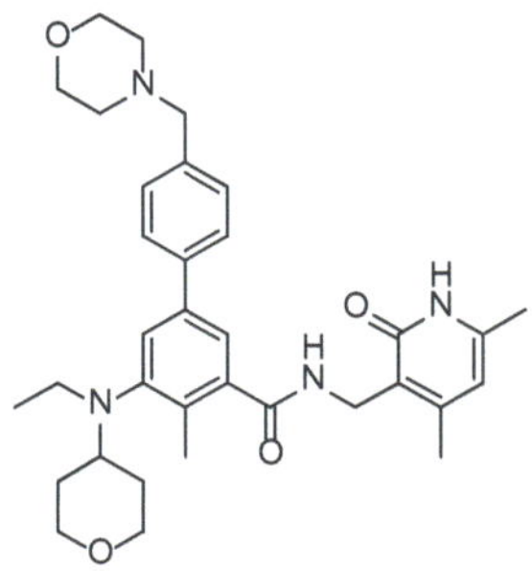

EPZ-5676
DOT1L

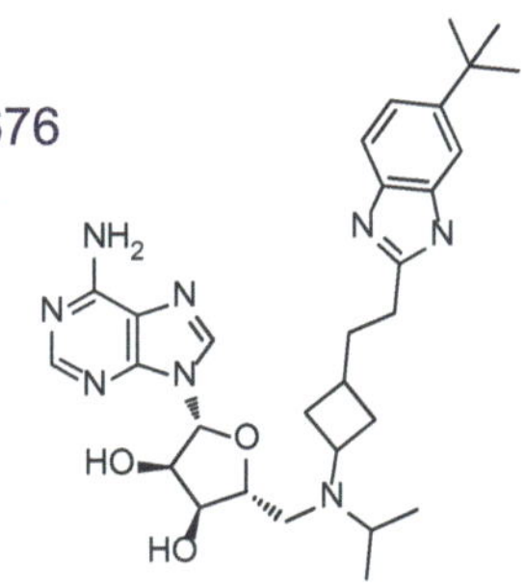

EPZ015666 (GSK3235025)
PRMT5

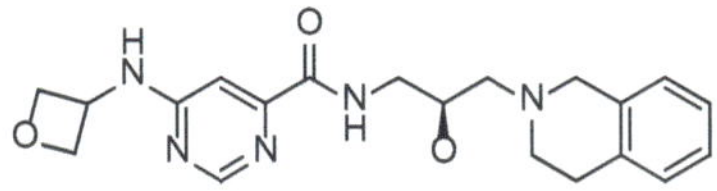

Methylstat
pan-jmjC

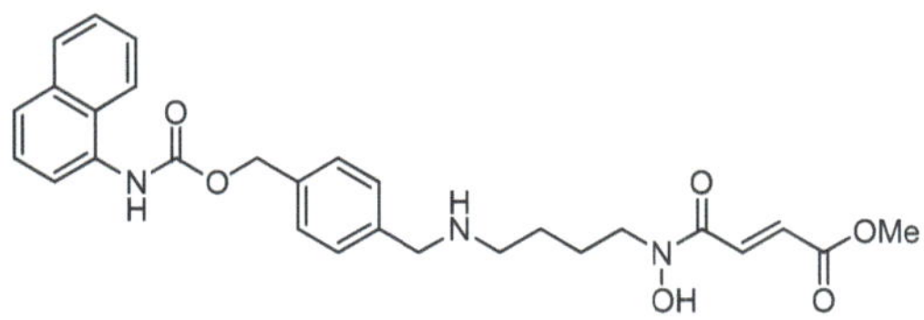

KDM5-C49
JARID

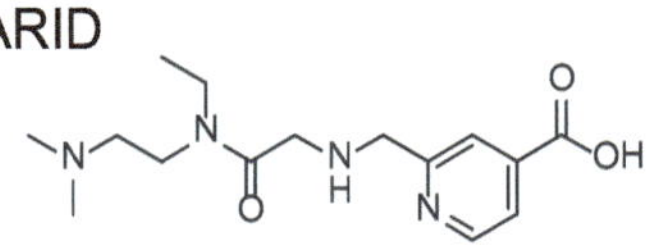

GSK-J1
JMJD3/
UTX/
JARID

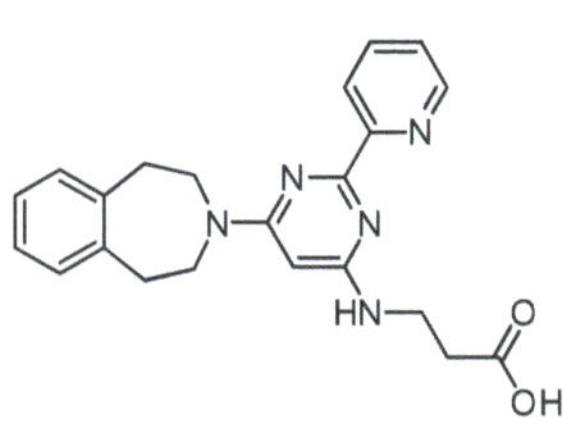

ORY-1001
LSD1

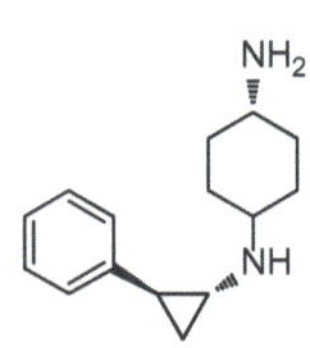

FIGURE 9.4

Chemical probes for histone methyltransferases and demethylases. For each enzyme, the best-characterized available chemical probes are indicated.

oxidases MAO-A and MAO-B. Small-molecule inhibitors of these enzymes are targeted drugs approved for neurological diseases and indeed the clinical MAO inhibitor tranylcypromine also inhibits LSD1 [111]. The mode of inhibition is conserved and the irreversible covalent binding to LSD1 has been confirmed by crystallography [112]. At the active site of LSD1 the compound reacts through a radical mechanism with the enzyme's cofactor FAD resulting in a ring opening of tranylcypromine's cyclopropyl ring and the generation of a tetracyclic addition product of FAD and the inhibitor. This reaction mechanism has been further exploited in the search for more selective and potent LSD1 inhibitors [113–119]. Two of the most potent and selective compounds of this class, GSK2879552 [120] and ORY-1001 [121], have now entered clinical trials.

9.3.6 THE EMERGENCE OF THE FIRST SUBSTRATE COMPETITIVE LSD1 INHIBITORS

While the optimization of covalent irreversible inhibitors of LSD1 has been relatively straightforward, the identification of potent reversible inhibitors proved more challenging. Both virtual and physical screening approaches as well as structure-guided design and fragment-based approaches have been attempted. Ultimately, these efforts have proven successful, and GSK-354/GSK690 and HCI2509/SP2509 [122–124] are highly specific and inhibit LSD1 at concentrations below 100 nM.

9.3.7 METABOLIC TARGETING OF JMJC DEMETHYLASES

JmjC demethylases are dependent on the cofactors Fe^{2+}, oxygen, and 2-oxoglutarate and these cofactors directly link the enzymatic activity to the metabolic state of the cell. In the course of the demethylation reaction the 2-oxoglutarate is converted to succinate. Succinate is then further processed by SDH. Inhibiting this enzyme, for example with 3-nitropropionic acid [125,126], results in the intracellular accumulation of succinate and changes in the histone methylation patterns [127]. In addition to 2-oxoglutarate, JmjC demethylases also require iron in the 2+ oxidation state. Therefore, these enzymes can be inhibited by other metal ions that compete with the iron, for example nickel [128,129]. Finally, JmjC demethylases require oxygen for their function and are inhibited in hypoxic conditions [129–132].

9.3.8 POTENT SMALL-MOLECULE INHIBITORS BINDING THE ACTIVE SITES OF JMJC DEMETHYLASES

The first compounds employed as inhibitors of JmjC demethylases closely resemble the cofactor 2-oxoglutarate. These include the compounds N-oxalylglyine and 2,4-pyridine dicarboxylate that were previously known as inhibitors of 2-oxoglutarate-dependent hydroxylases and also act on histone demethylases [133]. Further medicinal chemistry of this compound class has resulted in more specific compounds. Methylstat is a compound that covalently links a histone lysine mimetic to a 2-oxoglutarate analog and thereby acts as a pan-JmjC demethylase inhibitor with specificity over other 2-oxoglutarate-dependent hydroxylases [134]. Other modifications to the N-oxalylglycine and 2,4-pyridine dicarboxylate cores have resulted in compounds with selectivity for certain demethylase classes. Daminozine and compound 9 [135] act preferentially on KDM2/7 and KDM5-C49/70

(WO 2014053491), and the close analog KDOAM25 targets KDM5. The first JmjC domain inhibitor scaffold derived from unbiased screening was 8-hydroxyquinolines. This core structure was first identified from screening more than 200,000 compounds for KDM4 inhibitors. From the pan-demethylase inhibitor IOX-1, structural refinements led to the KDM4-specific variants SD70 [136] and ML324 [137]. Another class of demethylase inhibitors is represented by GSKJ1/4 that was first described as a specific inhibitor of the H3K27me3 demethylases JMJD3 and UTX [138] and later found to also inhibit the JARID subfamily [139].

9.3.9 PRMTs AND PADs

The first PRMT inhibitors developed from screening or targeted synthesis were compounds with low activity and specificity [140–142]. Further efforts and optimization have yielded several compounds with a high degree of selectivity. MD023 is a chemical probe for type I PRMTs 1, 3, 4, 6, and 8, which catalyze asymmetric dimethylation of arginine residues. Furthermore, a PRMT5 selective compound is available [143].

Arginine methylation can be removed by PAD4 via dimethylimination, converting arginine to citrulline [4]. This conversion does not require the presence of methylation, and indeed methylation is not beneficial for the reaction in vitro [144]. The irreversible inhibitors F-amidine and Cl-amidine and their analogs [145–149] as well as peptidomimetics [150] have been proposed as PAD4 inhibitors. Recently, GSK developed a 50 nM reversible inhibitor, GSK484, which binds to the low-calcium form of the enzyme [151].

9.3.10 METHYLATION MODULATORS OF THE FUTURE

The most advanced compounds modulating histone methylation have been developed in a very similar fashion, based on biochemical assays and medicinal chemistry. Therefore, they typically are active site binders that globally affect the activity of a particular enzyme on all its target sites. A number of complementary mechanisms can be envisioned. Inhibitors might specifically target the expression, translation, stability or nuclear import of selected enzymes. Furthermore, they could exert specific effects on a subset of complexes or only at defined genomic locations. Already, histone modifications can be altered at will in a locus-specific way with genome-editing techniques [152–154].

The growing popularity of cell-based assays, including targeted reporter assays and more genome-scale approaches, such as ChIP-Seq and Chem-Seq [155], might make it possible to identify small molecules that specifically change methylation patterns only at defined genomic loci. Fine-tuning compound effects to precisely activate and repress certain genes will enable a wide range of applications from cellular reprogramming to specific repression of oncogenes.

9.4 THE HOTTEST THERAPEUTIC TARGETS IN HISTONE METHYLATION

For the most exciting targets, significant effort both in industry and academia went into the development of specific probes and lead candidates. The most advanced compounds for selected enzyme classes are indicated in Figure 9.4.

9.4.1 EZH2 INHIBITORS PROMISE APPLICATIONS IN LYMPHOMA AND BEYOND

EZH2 was soon identified by the pharmaceutical industry as the most promising target to develop inhibitors for applications in oncology. This enzyme, together with the weakly active closely related EZH1 protein, is responsible for all H3K27 methylation in mammalian cells. The interest in targeting the protein therapeutically first emerged when its overexpression was detected in several malignancies, including prostate [156] and breast cancer [157]. Activity around the protein soared when activating mutations in the catalytic SET domain were identified in follicular and diffuse large B-cell lymphomas [158,159]. These mutations result in a dramatic increase in global H3K27me3 levels and a decrease in mono- and dimethylation at this residue.

Early approaches to validate EZH2 as a therapeutic target relied on the poorly characterized effects of 3-deazaneplanocin A (DZNep) [86]. This molecule inhibits SAH hydrolase and thereby generates pleotropic effects through product inhibition of all SAM-dependent methyltransferases. Apparently, EZH2 is most sensitive to this inhibition and DZNep results in a degradation of the EZH2 complex on the protein level. Nevertheless, the nature of this molecule makes it impossible to pinpoint EZH2-specific effects and it is worrying to see its continued use in preclinical studies despite the availability of much better compounds.

The discovery of compounds directly and specifically inhibiting the EZH2 methyltransferase function proved not to be straightforward. A particular complication in developing histone methyltransferase assays with this protein is its absolute dependence on binding partners EED and SUZ12 for enzymatic activity [25]. Furthermore, the protein prefers histone octamers and nucleosomes over peptide substrates. Ultimately, these technical challenges could be overcome by several groups and medicinal chemistry efforts resulted in the generation of chemical probes and clinical candidates targeting either the mutated protein specifically or both mutated and wild-type EZH2 [160–162].

Currently, three molecules are in clinical trials: tazemetostat/EPZ-6438, GSK2816126, and CPI-1205. These compounds are being tested in diffuse large B-cell and follicular lymphomas, and the Epizyme compound has already entered phase II testing for these indications. At the same time, additional malignancies have been identified as potentially benefitting from EZH2 inhibition. These include the susceptibility of the large group of solid tumors with mutations in the SWI/SNF (BAF) chromatin remodeling complex [163,164], MLL leukemias, MYC-driven prostate cancer, embryonal rhabdomyosarcoma, and other solid tumors.

9.4.2 DOT1L IS ESSENTIAL IN MLL-FUSION LEUKEMIAS

DOT1L is a particular enzyme in that it is the only histone lysine methyltransferase not harboring a SET domain. Moreover, it is the only enzyme known so far that is capable of methylating histone H3K79 [165]. Interestingly, this enzyme is phylogenetically much more closely related to protein arginine than lysine methyltransferases. The protein interacts with AF10, a common fusion partner in MLL-rearranged leukemias [166]. In these chromosomal translocations, the catalytic SET domain of MLL is lost, and thereby fusion proteins lose their capability to catalyze the formation of H3K4me3. Rather, they recruit other chromatin-modifying activities and, in the case of DOT1L, generate a H3K79 methylated chromatin environment at their target genes [167]. These leukemias are dependent on DOT1L-mediated H3K79 methylation [168] and a small-molecule DOT1L inhibitor selectively kills MLL-fusion leukemias [169]. Accordingly, a clinical phase I study was started with EPZ-5676 [105], a picomolar selective inhibitor of DOT1L, in patients suffering from MLL rearranged leukemias. While this first human trial is exciting, the chemical structure and in vivo pharmacokinetic parameters [106] might have contributed to the decision to stop enrolling patients for insufficient efficacy as monotherapy. Certainly alternative compounds targeting this protein are worth developing.

9.4.3 LSD1 INHIBITORS

The third enzyme involved in histone methylation for which inhibitors have entered the clinic is the demethylase LSD1. This was the first enzyme for which histone lysine demethylase activity has been shown [30]. With the closely related protein LSD2, it forms a special family of demethylases that use an FAD-dependent amine oxidase mechanism to remove mono- and dimethylation. This reaction mechanism is shared with amine oxidases in the brain, particularly the well-known monoamine oxidases MAO-A and MAO-B, targets of antidepressives. Therefore, it is not surprising that among the first compounds used to inhibit LSD1 was the approved drug tranylcypromine, which inhibits LSD1 in the low micromolar range [111]. This and related cyclopropylamines act as irreversible inhibitors by undergoing a ring opening of the cyclopropyl followed by covalent attachment to the flavin prosthetic group [112,170].

Targeting LSD1 is interesting for its contribution to androgen receptor signaling in prostate cancer [31] and its role in breast cancer metastasis [171]. Other applications for LSD1 inhibitors have been proposed in virology, where LSD1 inhibition reduces herpes virus infection [172].

Specific inhibitors for LSD1 were developed and tested in cell line panels for toxicity. There, particularly small-cell lung cancers (SCLC) and acute myeloid leukemias (AML) responded [120]. Only 30% of SCLC lines showed sensitivity but responders were characterized by high levels of MYC and a DNA methylation signature at 45 genomic loci. Currently GSK2879552 and ORY-1001 are tested in phase I studies in small-cell lung carcinoma and acute myeloid leukemia [120].

9.4.4 METHYLATION DRUGS OF THE FUTURE

In the world of arginine methyltransferases PRMT5 has emerged as the most desirable enzyme for pharmacological inhibition in cancer. This enzyme has also been dubbed the "guardian of the germline" [173], as its knockout prevents the maturation of primordial germ cells. Overexpression of PRMT5 has been detected in multiple cancers, including lymphomas, glioblastomas, and lung and ovarian cancer. The enzyme has both nuclear functions, where it generates H2A/H4R3me2, and cytoplasmic functions on substrates like the Sm proteins involved in splicing. Whether one of these functions or their combined effect is responsible for the tumorigenic potential is currently unclear. A selective inhibitor of the enzyme has been developed: EPZ015666 (GSK3235025) inhibits the enzyme in the low-nanomolar range and kills mantle cell lymphoma in xenografts [143].

Of the JmjC family, the JARID/KDM5 subfamily appears most attractive for therapeutic targeting. Particularly KDM5A/JARID1A/RBP2 and KDM5B/JARID1B/PLU1 are often found overexpressed in different solid tumors, including lung [174] and breast cancer [175]. Interestingly, chromosomal translocations were detected in acute megakaryoblastic leukemia that fuse KDM5A to NUP98 and thus activate the demethylase [176]. Loss of KDM5A decreases tumor formation and increases survival in a mouse model of neuroendocrine tumor formation driven by loss of Rb or Men1 [177]. An important mechanism by which KDM5 demethylases contribute to cancer pathogenicity is via the induction of drug resistance. Following treatment with different targeted cancer drugs, multiple cell lines respond by massive cell death but $<1\%$ drug-tolerant persisters survive [178]. These cells are characterized by global epigenetic changes driven by the upregulation of KDM5 demethylases. Similarly, a slow-growing subpopulation of KDM5B-positive melanoma cells [179] appears essential for the maintenance of tumor growth. It will be interesting to see whether specific KDM5 inhibitors can prevent the emergence of this reversible drug-tolerant state in a more specific way compared to HDAC inhibitors.

Several attempts have been taken to obtain KDM5-specific inhibitors. The first specific JmjC demethylase inhibitors that were developed to target the KDM6 family [138] also inhibit KDM5 enzymes [139]. Another compound, PBIT, was identified from a high-throughput screen to target KDM5B [180]. However, the low potency of this compound, together with the limited characterization for specificity, prevents its usage as a chemical probe. Probably the best currently available KDM5 inhibitors were developed by EpiTherapeutics (PCT WO 2014053491). These compounds act at submicromolar concentrations to inhibit H3K4 demethylation in cellular assays. A cocrystal structure with KDM5B has been solved and is publically accessible (5A3T).

The most interesting SET domain methyltransferases, for which no compounds have entered clinical trials, are SMYD3 and the NSD family. SMYD3 acts as a proto-oncogene in prostate cancer through its methyltransferase activity [181]. Furthermore, it is also involved in carcinogenesis of colorectal cancers, gastric cancer, and hepatocellular carcinoma [182]. SMYD3 methylates several residues in the histone tails [183] but the mostly cytoplasmic enzyme also has nonhistone substrates, including VEGF [184] and MAP3K2 [185], thus contributing to pancreatic and lung cancer. First inhibitors against SMYD3 have been described [186] but their potency and limited characterization prevent their use as chemical probes.

The family of NSD1/2/3 enzymes catalyzes H3K36 methylation. Particularly NSD1 has received a lot of interest, as the protein is also named MMSET for its up-regulation and translocation in multiple myeloma [187]. It is also found up-regulated in other cancers, including neuroblastoma [188] and forms a highly regulated network with EZH2 to control transcriptional activation, repression, and carcinogenesis [189,190]. NSD1 and NSD3 are found in translocations to NUP98 in AML [191–193] and NSD3 translocations occur in midline carcinomas [194]. In contrast, genetic loss of the enzymes causes Sotos syndrome (NSD1) [195] and Wolf–Hirschhorn syndrome (NSD2) [196]. A submicromolar inhibitor of NSD2 was described and might be the starting point for further optimization [197].

While the current focus on exploiting methyltransferases and demethylases is in oncology applications, several other therapeutic areas show a strong potential for targeting histone-methylating pathways. This is based on the role of these enzymes in infection and inflammation, in cellular reprogramming for regenerative medicine, and in neurology. These data promise a bright future for specific inhibitors of histone methyltransferases and demethylases and highlight the need to develop specific chemical probe compounds for all enzymes in this class.

REFERENCES

[1] Copeland RA, Solomon ME, Richon VM. Protein methyltransferases as a target class for drug discovery. Nat Rev Drug Discov 2009;8:724–32.

[2] Chang B, Chen Y, Zhao Y, Bruick RK. JMJD6 is a histone arginine demethylase. Science 2007;318:444–7.

[3] Mantri M, Krojer T, Bagg EA, Webby CJ, Butler DS, Kochan G, et al. Crystal structure of the 2-oxoglutarate- and Fe(II)-dependent lysyl hydroxylase JMJD6. J Mol Biol 2010;401:211–22.

[4] Wang Y, Wysocka J, Sayegh J, Lee YH, Perlin JR, Leonelli L, et al. Human PAD4 regulates histone arginine methylation levels via demethylimination. Science 2004;306:279–83.

[5] Mele M, Ferreira PG, Reverter F, DeLuca DS, Monlong J, Sammeth M, et al. Human genomics. The human transcriptome across tissues and individuals. Science 2015;348:660–5.

[6] Schliehe C, Flynn EK, Vilagos B, Richson U, Swaminathan S, Bosnjak B, et al. The methyltransferase Setdb2 mediates virus-induced susceptibility to bacterial superinfection. Nat Immunol 2015;16:67–74.

[7] Murata M, Azuma Y, Miura K, Rahman MA, Matsutani M, Aoyama M, et al. Chlamydial SET domain protein functions as a histone methyltransferase. Microbiology 2007;153:585–92.

[8] Alvarez-Venegas R. Bacterial SET domain proteins and their role in eukaryotic chromatin modification. Front Genet 2014;5:65.

[9] Eberle CA, Zayas M, Stukalov A, Pichlmair A, Alvisi G, Muller AC, et al. The lysine methyltransferase SMYD3 interacts with hepatitis C virus NS5A and is a negative regulator of viral particle production. Virology 2014;462–463:34–41.

[10] Marazzi I, Ho JS, Kim J, Manicassamy B, Dewell S, Albrecht RA, et al. Suppression of the antiviral response by an influenza histone mimic. Nature 2012;483:428–33.

[11] Osawa T, Muramatsu M, Wang F, Tsuchida R, Kodama T, Minami T, et al. Increased expression of histone demethylase JHDM1D under nutrient starvation suppresses tumor growth via down-regulating angiogenesis. Proc Natl Acad Sci USA 2011;108:20725–9.

[12] Beyer S, Kristensen MM, Jensen KS, Johansen JV, Staller P. The histone demethylases JMJD1A and JMJD2B are transcriptional targets of hypoxia-inducible factor HIF. J Biol Chem 2008;283: 36542–52.

[13] Pollard PJ, Loenarz C, Mole DR, McDonough MA, Gleadle JM, Schofield CJ, et al. Regulation of Jumonji-domain-containing histone demethylases by hypoxia-inducible factor (HIF)-1alpha. Biochem J 2008;416:387–94.

[14] Wellmann S, Bettkober M, Zelmer A, Seeger K, Faigle M, Eltzschig HK, et al. Hypoxia upregulates the histone demethylase JMJD1A via HIF-1. Biochem Biophys Res Commun 2008;372:892–7.

[15] Xia X, Lemieux ME, Li W, Carroll JS, Brown M, Liu XS, et al. Integrative analysis of HIF binding and transactivation reveals its role in maintaining histone methylation homeostasis. Proc Natl Acad Sci USA 2009;106:4260–5.

[16] Ulanovskaya OA, Zuhl AM, Cravatt BF. NNMT promotes epigenetic remodeling in cancer by creating a metabolic methylation sink. Nat Chem Biol 2013;9:300–6.

[17] Henning SM, Swendseid ME, Coulson WF. Male rats fed methyl- and folate-deficient diets with or without niacin develop hepatic carcinomas associated with decreased tissue NAD concentrations and altered poly(ADP-ribose) polymerase activity. J Nutr 1997;127:30–6.

[18] Luka Z, Pakhomova S, Loukachevitch LV, Calcutt MW, Newcomer ME, Wagner C. Crystal structure of the histone lysine specific demethylase LSD1 complexed with tetrahydrofolate. Protein Sci 2014;23:993–8.

[19] Shmakova A, Batie M, Druker J, Rocha S. Chromatin and oxygen sensing in the context of JmjC histone demethylases. Biochem J 2014;462:385–95.

[20] Killian JK, Kim SY, Miettinen M, Smith C, Merino M, Tsokos M, et al. Succinate dehydrogenase mutation underlies global epigenomic divergence in gastrointestinal stromal tumor. Cancer Discov 2013;3:648–57.

[21] Letouze E, Martinelli C, Loriot C, Burnichon N, Abermil N, Ottolenghi C, et al. SDH mutations establish a hypermethylator phenotype in paraganglioma. Cancer Cell 2013;23:739–52.

[22] Dang L, White DW, Gross S, Bennett BD, Bittinger MA, Driggers EM, et al. Cancer-associated IDH1 mutations produce 2-hydroxyglutarate. Nature 2009;462:739–44.

[23] Rohle D, Popovici-Muller J, Palaskas N, Turcan S, Grommes C, Campos C, et al. An inhibitor of mutant IDH1 delays growth and promotes differentiation of glioma cells. Science 2013;340:626–30.

[24] Wang F, Travins J, DeLaBarre B, Penard-Lacronique V, Schalm S, Hansen E, et al. Targeted inhibition of mutant IDH2 in leukemia cells induces cellular differentiation. Science 2013;340:622–6.

[25] Cao R, Zhang Y. SUZ12 is required for both the histone methyltransferase activity and the silencing function of the EED-EZH2 complex. Mol Cell 2004;15:57–67.

[26] Kuzmichev A, Jenuwein T, Tempst P, Reinberg D. Different EZH2-containing complexes target methylation of histone H1 or nucleosomal histone H3. Mol Cell 2004;14:183–93.

[27] Patel A, Dharmarajan V, Vought VE, Cosgrove MS. On the mechanism of multiple lysine methylation by the human mixed lineage leukemia protein-1 (MLL1) core complex. J Biol Chem 2009;284:24242–56.

[28] Cao F, Chen Y, Cierpicki T, Liu Y, Basrur V, Lei M, et al. An Ash2L/RbBP5 heterodimer stimulates the MLL1 methyltransferase activity through coordinated substrate interactions with the MLL1 SET domain. PLoS One 2010;5:e14102.

[29] Dou Y, Milne TA, Ruthenburg AJ, Lee S, Lee JW, Verdine GL, et al. Regulation of MLL1 H3K4 methyltransferase activity by its core components. Nat Struct Mol Biol 2006;13:713–19.

[30] Shi Y, Lan F, Matson C, Mulligan P, Whetstine JR, Cole PA, et al. Histone demethylation mediated by the nuclear amine oxidase homolog LSD1. Cell 2004;119:941–53.

[31] Metzger E, Wissmann M, Yin N, Muller JM, Schneider R, Peters AH, et al. LSD1 demethylates repressive histone marks to promote androgen-receptor-dependent transcription. Nature 2005;437:436–9.

[32] Wissmann M, Yin N, Muller JM, Greschik H, Fodor BD, Jenuwein T, et al. Cooperative demethylation by JMJD2C and LSD1 promotes androgen receptor-dependent gene expression. Nat Cell Biol 2007;9:347–53.

[33] Laurent B, Ruitu L, Murn J, Hempel K, Ferrao R, Xiang Y, et al. A specific LSD1/KDM1A isoform regulates neuronal differentiation through H3K9 demethylation. Mol Cell 2015;57:957–70.

[34] Rose NR, Woon EC, Kingham GL, King ON, Mecinovic J, Clifton IJ, et al. Selective inhibitors of the JMJD2 histone demethylases: combined nondenaturing mass spectrometric screening and crystallographic approaches. J Med Chem 2010;53:1810–18.

[35] Simeonov A. Recent developments in the use of differential scanning fluorometry in protein and small molecule discovery and characterization. Expert Opin Drug Discov 2013;8:1071–82.

[36] Ericsson UB, Hallberg BM, Detitta GT, Dekker N, Nordlund P. Thermofluor-based high-throughput stability optimization of proteins for structural studies. Anal Biochem 2006;357:289–98.

[37] Forneris F, Orru R, Bonivento D, Chiarelli LR, Mattevi A. ThermoFAD, a Thermofluor-adapted flavin ad hoc detection system for protein folding and ligand binding. Febs J 2009;276:2833–40.

[38] Xu W, Podoll JD, Dong X, Tumber A, Oppermann U, Wang X. Quantitative analysis of histone demethylase probes using fluorescence polarization. J Med Chem 2013;56:5198–202.

[39] Yi JS, Federation AJ, Qi J, Dhe-Paganon S, Hadler M, Xu X, et al. Structure-Guided DOT1L probe optimization by label-free ligand displacement. ACS Chem Biol 2015;10:667–74.

[40] Lin Y, Fan H, Frederiksen M, Zhao K, Jiang L, Wang Z, et al. Detecting S-adenosyl-L-methionine-induced conformational change of a histone methyltransferase using a homogeneous time-resolved fluorescence-based binding assay. Anal Biochem 2012;423:171–7.

[41] Rea S, Eisenhaber F, O'Carroll D, Strahl BD, Sun ZW, Schmid M, et al. Regulation of chromatin structure by site-specific histone H3 methyltransferases. Nature 2000;406:593–9.

[42] Wu J, Xie N, Feng Y, Zheng YG. Scintillation proximity assay of arginine methylation. J Biomol Screen 2012;17:237–44.
[43] Ibanez G, Shum D, Blum G, Bhinder B, Radu C, Antczak C, et al. A high throughput scintillation proximity imaging assay for protein methyltransferases. Comb Chem High Throughput Screen 2012;15:359–71.
[44] Yu W, Eram MS, Hajian T, Szykowska A, Burgess-Brown N, Vedadi M, et al. A scintillation proximity assay for histone demethylases. Anal Biochem 2014;463:54–60.
[45] Nishikori S, Hattori T, Fuchs SM, Yasui N, Wojcik J, Koide A, et al. Broad ranges of affinity and specificity of anti-histone antibodies revealed by a quantitative peptide immunoprecipitation assay. J Mol Biol 2012;424:391–9.
[46] Perez-Burgos L, Peters AH, Opravil S, Kauer M, Mechtler K, Jenuwein T. Generation and characterization of methyl-lysine histone antibodies. Methods Enzymol 2004;376:234–54.
[47] Rothbart SB, Dickson BM, Raab JR, Grzybowski AT, Krajewski K, Guo AH, et al. An interactive database for the assessment of histone antibody specificity. Mol Cell 2015;59:502–11.
[48] Bock I, Dhayalan A, Kudithipudi S, Brandt O, Rathert P, Jeltsch A. Detailed specificity analysis of antibodies binding to modified histone tails with peptide arrays. Epigenetics 2011;6:256–63.
[49] Egelhofer TA, Minoda A, Klugman S, Lee K, Kolasinska-Zwierz P, Alekseyenko AA, et al. An assessment of histone-modification antibody quality. Nat Struct Mol Biol 2011;18:91–3.
[50] Nielsen AL, Kristensen LH, Stephansen KB, Kristensen JB, Helgstrand C, Lees M, et al. Identification of catechols as histone-lysine demethylase inhibitors. FEBS Lett 2012;586:1190–4.
[51] Woo J, Kim HY, Byun BJ, Chae CH, Lee JY, Ryu SY, et al. Biological evaluation of tanshindiols as EZH2 histone methyltransferase inhibitors. Bioorg Med Chem Lett 2014;24:2486–92.
[52] Schmitt ML, Ladwein KI, Carlino L, Schulz-Fincke J, Willmann D, Metzger E, et al. Heterogeneous antibody-based activity assay for lysine specific demethylase 1 (LSD1) on a histone peptide substrate. J Biomol Screen 2014;19:973–8.
[53] Wang C, Caron M, Burdick D, Kang Z, Auld D, Hill WA, et al. A sensitive, homogeneous, and high-throughput assay for lysine-specific histone demethylases at the H3K4 site. Assay Drug Dev Technol 2012;10:179–86.
[54] Kubicek S, O'Sullivan RJ, August EM, Hickey ER, Zhang Q, Teodoro ML, et al. Reversal of H3K9me2 by a small-molecule inhibitor for the G9a histone methyltransferase. Mol Cell 2007;25:473–81.
[55] Quinn AM, Allali-Hassani A, Vedadi M, Simeonov A. A chemiluminescence-based method for identification of histone lysine methyltransferase inhibitors. Mol Biosyst 2010;6:782–8.
[56] Gauthier N, Caron M, Pedro L, Arcand M, Blouin J, Labonte A, et al. Development of homogeneous nonradioactive methyltransferase and demethylase assays targeting histone H3 lysine 4. J Biomol Screen 2012;17:49–58.
[57] Plant M, Dineen T, Cheng A, Long AM, Chen H, Morgenstern KA. Screening for lysine-specific demethylase-1 inhibitors using a label-free high-throughput mass spectrometry assay. Anal Biochem 2011;419:217–27.
[58] Hutchinson SE, Leveridge MV, Heathcote ML, Francis P, Williams L, Gee M, et al. Enabling lead discovery for histone lysine demethylases by high-throughput RapidFire mass spectrometry. J Biomol Screen 2012;17:39–48.
[59] Leveridge M, Buxton R, Argyrou A, Francis P, Leavens B, West A, et al. Demonstrating enhanced throughput of RapidFire mass spectrometry through multiplexing using the JmjD2d demethylase as a model system. J Biomol Screen 2014;19:278–86.
[60] Wigle TJ, Swinger KK, Campbell JE, Scholle MD, Sherrill J, Admirand EA, et al. A high-throughput mass spectrometry assay coupled with redox activity testing reduces artifacts and false positives in lysine demethylase screening. J Biomol Screen 2015;20(6):810–20.

[61] King ON, Li XS, Sakurai M, Kawamura A, Rose NR, Ng SS, et al. Quantitative high-throughput screening identifies 8-hydroxyquinolines as cell-active histone demethylase inhibitors. PLoS One 2010;5:e15535.
[62] Li S, Gu XJ, Hao Q, Fan H, Li L, Zhou S, et al. A liquid chromatography/mass spectrometry-based generic detection method for biochemical assay and hit discovery of histone methyltransferases. Anal Biochem 2013;443:214–21.
[63] Tsukada Y, Fang J, Erdjument-Bromage H, Warren ME, Borchers CH, Tempst P, et al. Histone demethylation by a family of JmjC domain-containing proteins. Nature 2006;439:811–16.
[64] Klink TA, Staeben M, Twesten K, Kopp AL, Kumar M, Dunn RS, et al. Development and validation of a generic fluorescent methyltransferase activity assay based on the transcreener AMP/GMP assay. J Biomol Screen 2012;17:59–70.
[65] Drake KM, Watson VG, Kisielewski A, Glynn R, Napper AD. A sensitive luminescent assay for the histone methyltransferase NSD1 and other SAM-dependent enzymes. Assay Drug Dev Technol 2014;12:258–71.
[66] Arrowsmith CH, Audia JE, Austin C, Baell J, Bennett J, Blagg J, et al. The promise and peril of chemical probes. Nat Chem Biol 2015;11:536–41.
[67] Arrowsmith CH, Bountra C, Fish PV, Lee K, Schapira M. Epigenetic protein families: a new frontier for drug discovery. Nat Rev Drug Discov 2012;11:384–400.
[68] Xie W, Ames RS, Li H. A cell-based high-throughput screening assay to measure cellular histone h3 lys27 trimethylation with a modified dissociation-enhanced lanthanide fluorescent immunoassay. J Biomol Screen 2012;17:99–107.
[69] Luense S, Denner P, Fernandez-Montalvan A, Hartung I, Husemann M, Stresemann C, et al. Quantification of histone H3 Lys27 trimethylation (H3K27me3) by high-throughput microscopy enables cellular large-scale screening for small-molecule EZH2 inhibitors. J Biomol Screen 2015;20:190–201.
[70] Mulji A, Haslam C, Brown F, Randle R, Karamshi B, Smith J, et al. Configuration of a high-content imaging platform for hit identification and pharmacological assessment of JMJD3 demethylase enzyme inhibitors. J Biomol Screen 2012;17:108–20.
[71] Lin CW, Jao CY, Ting AY. Genetically encoded fluorescent reporters of histone methylation in living cells. J Am Chem Soc 2004;126:5982–3.
[72] Peters AH, Kubicek S, Mechtler K, O'Sullivan RJ, Derijck AA, Perez-Burgos L, et al. Partitioning and plasticity of repressive histone methylation states in mammalian chromatin. Mol Cell 2003;12:1577–89.
[73] Garcia BA, Barber CM, Hake SB, Ptak C, Turner FB, Busby SA, et al. Modifications of human histone H3 variants during mitosis. Biochemistry 2005;44:13202–13.
[74] Britton LM, Gonzales-Cope M, Zee BM, Garcia BA. Breaking the histone code with quantitative mass spectrometry. Expert Rev Proteomics 2011;8:631–43.
[75] Jaffe JD, Wang Y, Chan HM, Zhang J, Huether R, Kryukov GV, et al. Global chromatin profiling reveals NSD2 mutations in pediatric acute lymphoblastic leukemia. Nat Genet 2013;45:1386–91.
[76] Borchardt RT, Wu YS. Potential inhibitors of S-adenosylmethionine-dependent methyltransferases. 1. Modification of the amino acid portion of S-adenosylhomocysteine. J Med Chem 1974;17:862–8.
[77] Deguchi T, Barchas J. Inhibition of transmethylations of biogenic amines by S-adenosylhomocysteine. Enhancement of transmethylation by adenosylhomocysteinase. J Biol Chem 1971;246:3175–81.
[78] Berry DR, Abbott BJ. Incorporation of 14C-labeled compounds into sinefungin (A9145), a nucleoside antifungal antibiotic. J Antibiot (Tokyo) 1978;31:185–91.
[79] Coward JK, Slisz EP. Analogs of S-adenosylhomocysteine as potential inhibitors of biological transmethylation. Specificity of the S-adenosylhomocysteine binding site. J Med Chem 1973;16:460–3.
[80] Coward JK, Sweet WD. Analogs of S-adenosylhomocysteine as potential inhibitors of biological transmethylation. Synthesis and biological activity of homocysteine derivatives bridged to adenine. J Med Chem 1972;15:381–4.

[81] Medzihradsky JL, Zimmerman TP, Wolberg G, Elion GB. Immunosuppressive effects of the S-adenosylhomocysteine hydrolase inhibitor, 3-deazaadenosine. J Immunopharmacol 1982;4:29–41.
[82] Houston DM, Dolence EK, Keller BT, Patel-Thombre U, Borchardt RT. Potential inhibitors of S-adenosylmethionine-dependent methyltransferases. 8. Molecular dissections of carbocyclic 3-deazaadenosine as inhibitors of S-adenosylhomocysteine hydrolase. J Med Chem 1985;28:467–71.
[83] Aarbakke J, Miura GA, Prytz PS, Bessesen A, Slordal L, Gordon RK, et al. Induction of HL-60 cell differentiation by 3-deaza-(+/ −)-aristeromycin, an inhibitor of S-adenosylhomocysteine hydrolase. Cancer Res 1986;46:5469–72.
[84] Borchardt RT, Keller BT, Patel-Thombre U. Neplanocin A. A potent inhibitor of S-adenosylhomocysteine hydrolase and of vaccinia virus multiplication in mouse L929 cells. J Biol Chem 1984;259:4353–8.
[85] Glazer RI, Knode MC, Tseng CK, Haines DR, Marquez VE. 3-Deazaneplanocin A: a new inhibitor of S-adenosylhomocysteine synthesis and its effects in human colon carcinoma cells. Biochem Pharmacol 1986;35:4523–7.
[86] Tan J, Yang X, Zhuang L, Jiang X, Chen W, Lee PL, et al. Pharmacologic disruption of Polycomb-repressive complex 2-mediated gene repression selectively induces apoptosis in cancer cells. Genes Dev 2007;21:1050–63.
[87] Miranda TB, Cortez CC, Yoo CB, Liang G, Abe M, Kelly TK, et al. DZNep is a global histone methylation inhibitor that reactivates developmental genes not silenced by DNA methylation. Mol Cancer Ther 2009;8:1579–88.
[88] Greiner D, Bonaldi T, Eskeland R, Roemer E, Imhof A. Identification of a specific inhibitor of the histone methyltransferase SU(VAR)3-9. Nat Chem Biol 2005;1:143–5.
[89] Cherblanc FL, Chapman KL, Brown R, Fuchter MJ. Chaetocin is a nonspecific inhibitor of histone lysine methyltransferases. Nat Chem Biol 2013;9:136–7.
[90] Greiner D, Bonaldi T, Eskeland R, Roemer E, Imhof A. Reply to "Chaetocin is a nonspecific inhibitor of histone lysine methyltransferases". Nat Chem Biol 2013;9:137.
[91] Tibodeau JD, Benson LM, Isham CR, Owen WG, Bible KC. The anticancer agent chaetocin is a competitive substrate and inhibitor of thioredoxin reductase. Antioxid Redox Signal 2009;11:1097–106.
[92] Zhou HR, Zhang FF, Ma ZY, Huang HW, Jiang L, Cai T, et al. Folate polyglutamylation is involved in chromatin silencing by maintaining global DNA methylation and histone H3K9 dimethylation in Arabidopsis. Plant Cell 2013;25:2545–59.
[93] Dobosy JR, Fu VX, Desotelle JA, Srinivasan R, Kenowski ML, Almassi N, et al. A methyl-deficient diet modifies histone methylation and alters Igf2 and H19 repression in the prostate. Prostate 2008;68:1187–95.
[94] Lambrot R, Xu C, Saint-Phar S, Chountalos G, Cohen T, Paquet M, et al. Low paternal dietary folate alters the mouse sperm epigenome and is associated with negative pregnancy outcomes. Nat Commun 2013;4:2889.
[95] Mikol YB, Hoover KL, Creasia D, Poirier LA. Hepatocarcinogenesis in rats fed methyl-deficient, amino acid-defined diets. Carcinogenesis 1983;4:1619–29.
[96] Copeland DH, Salmon WD. The occurrence of neoplasms in the liver, lungs, and other tissues of rats as a result of prolonged choline deficiency. Am J Pathol 1946;22:1059–79.
[97] Ghoshal AK, Farber E. The induction of liver cancer by dietary deficiency of choline and methionine without added carcinogens. Carcinogenesis 1984;5:1367–70.
[98] James SJ, Pogribny IP, Pogribna M, Miller BJ, Jernigan S, Melnyk S. Mechanisms of DNA damage, DNA hypomethylation, and tumor progression in the folate/methyl-deficient rat model of hepatocarcinogenesis. J Nutr 2003;133:3740S–7S.
[99] Qian C, Zhou MM. SET domain protein lysine methyltransferases: structure, specificity and catalysis. Cell Mol Life Sci 2006;63:2755–63.

[100] Cheng X, Collins RE, Zhang X. Structural and sequence motifs of protein (histone) methylation enzymes. Annu Rev Biophys Biomol Struct 2005;34:267–94.

[101] Dillon SC, Zhang X, Trievel RC, Cheng X. The SET-domain protein superfamily: protein lysine methyltransferases. Genome Biol 2005;6:227.

[102] Chang Y, Zhang X, Horton JR, Upadhyay AK, Spannhoff A, Liu J, et al. Structural basis for G9a-like protein lysine methyltransferase inhibition by BIX-01294. Nat Struct Mol Biol 2009;16:312–17.

[103] Vedadi M, Barsyte-Lovejoy D, Liu F, Rival-Gervier S, Allali-Hassani A, Labrie V, et al. A chemical probe selectively inhibits G9a and GLP methyltransferase activity in cells. Nat Chem Biol 2011;7:566–74.

[104] Upadhyay AK, Rotili D, Han JW, Hu R, Chang Y, Labella D, et al. An analog of BIX-01294 selectively inhibits a family of histone H3 lysine 9 Jumonji demethylases. J Mol Biol 2012;416:319–27.

[105] Daigle SR, Olhava EJ, Therkelsen CA, Basavapathruni A, Jin L, Boriack-Sjodin PA, et al. Potent inhibition of DOT1L as treatment of MLL-fusion leukemia. Blood 2013;122:1017–25.

[106] Basavapathruni A, Olhava EJ, Daigle SR, Therkelsen CA, Jin L, Boriack-Sjodin PA, et al. Nonclinical pharmacokinetics and metabolism of EPZ-5676, a novel DOT1L histone methyltransferase inhibitor. Biopharm Drug Dispos 2014;35:237–52.

[107] Siarheyeva A, Senisterra G, Allali-Hassani A, Dong A, Dobrovetsky E, Wasney GA, et al. An allosteric inhibitor of protein arginine methyltransferase 3. Structure 2012;20:1425–35.

[108] Kaniskan HU, Szewczyk MM, Yu Z, Eram MS, Yang X, Schmidt K, et al. A potent, selective and cell-active allosteric inhibitor of protein arginine methyltransferase 3 (PRMT3). Angew Chem Int Ed Engl 2015;54:5166–70.

[109] Grebien F, Vedadi M, Getlik M, Giambruno R, Grover A, Avellino R, et al. Pharmacological targeting of the Wdr5-MLL interaction in C/EBPalpha N-terminal leukemia. Nat Chem Biol 2015;11:571–8.

[110] Kim W, Bird GH, Neff T, Guo G, Kerenyi MA, Walensky LD, et al. Targeted disruption of the EZH2-EED complex inhibits EZH2-dependent cancer. Nat Chem Biol 2013;9:643–50.

[111] Lee MG, Wynder C, Schmidt DM, McCafferty DG, Shiekhattar R. Histone H3 lysine 4 demethylation is a target of nonselective antidepressive medications. Chem Biol 2006;13:563–7.

[112] Yang M, Culhane JC, Szewczuk LM, Jalili P, Ball HL, Machius M, et al. Structural basis for the inhibition of the LSD1 histone demethylase by the antidepressant trans-2-phenylcyclopropylamine. Biochemistry 2007;46:8058–65.

[113] Malcomson T, Yelekci K, Borrello MT, Ganesan A, Semina E, De Kimpe N, et al. Cis-cyclopropylamines as mechanism-based inhibitors of monoamine oxidases. Febs J 2015;282(16):3190–8.

[114] Mimasu S, Umezawa N, Sato S, Higuchi T, Umehara T, Yokoyama S. Structurally designed trans-2-phenylcyclopropylamine derivatives potently inhibit histone demethylase LSD1/KDM1. Biochemistry 2010;49:6494–503.

[115] Binda C, Valente S, Romanenghi M, Pilotto S, Cirilli R, Karytinos A, et al. Biochemical, structural, and biological evaluation of tranylcypromine derivatives as inhibitors of histone demethylases LSD1 and LSD2. J Am Chem Soc 2010;132:6827–33.

[116] Valente S, Rodriguez V, Mercurio C, Vianello P, Saponara B, Cirilli R, et al. Pure enantiomers of benzoylamino-tranylcypromine: LSD1 inhibition, gene modulation in human leukemia cells and effects on clonogenic potential of murine promyelocytic blasts. Eur J Med Chem 2015;94:163–74.

[117] Vianello P, Botrugno OA, Cappa A, Ciossani G, Dessanti P, Mai A, et al. Synthesis, biological activity and mechanistic insights of 1-substituted cyclopropylamine derivatives: a novel class of irreversible inhibitors of histone demethylase KDM1A. Eur J Med Chem 2014;86:352–63.

[118] Gooden DM, Schmidt DM, Pollock JA, Kabadi AM, McCafferty DG. Facile synthesis of substituted trans-2-arylcyclopropylamine inhibitors of the human histone demethylase LSD1 and monoamine oxidases A and B. Bioorg Med Chem Lett 2008;18:3047–51.

[119] Benelkebir H, Hodgkinson C, Duriez PJ, Hayden AL, Bulleid RA, Crabb SJ, et al. Enantioselective synthesis of tranylcypromine analogues as lysine demethylase (LSD1) inhibitors. Bioorg Med Chem 2011;19:3709–16.

[120] Mohammad HP, Smitheman KN, Kamat CD, Soong D, Federowicz KE, Van Aller GS, et al. A DNA hypomethylation signature predicts antitumor activity of LSD1 inhibitors in SCLC. Cancer Cell 2015;28:57–69.

[121] Maes T, Carceller E, Salas J, Ortega A, Buesa C. Advances in the development of histone lysine demethylase inhibitors. Curr Opin Pharmacol 2015;23:52–60.

[122] Sankar S, Theisen ER, Bearss J, Mulvihill T, Hoffman LM, Sorna V, et al. Reversible LSD1 inhibition interferes with global EWS/ETS transcriptional activity and impedes Ewing sarcoma tumor growth. Clin Cancer Res 2014;20:4584–97.

[123] Sorna V, Theisen ER, Stephens B, Warner SL, Bearss DJ, Vankayalapati H, et al. High-throughput virtual screening identifies novel N′-(1-phenylethylidene)-benzohydrazides as potent, specific, and reversible LSD1 inhibitors. J Med Chem 2013;56:9496–508.

[124] Fiskus W, Sharma S, Shah B, Portier BP, Devaraj SG, Liu K, et al. Highly effective combination of LSD1 (KDM1A) antagonist and pan-histone deacetylase inhibitor against human AML cells. Leukemia 2014;28:2155–64.

[125] Scallet AC, Haley RL, Scallet DM, Duhart HM, Binienda ZK. 3-Nitropropionic acid inhibition of succinate dehydrogenase (complex II) activity in cultured Chinese hamster ovary cells: antagonism by L-carnitine. Ann N Y Acad Sci 2003;993:305–12. [discussion 345-309]

[126] Xiao M, Yang H, Xu W, Ma S, Lin H, Zhu H, et al. Inhibition of alpha-KG-dependent histone and DNA demethylases by fumarate and succinate that are accumulated in mutations of FH and SDH tumor suppressors. Genes Dev 2012;26:1326–38.

[127] Cervera AM, Bayley JP, Devilee P, McCreath KJ. Inhibition of succinate dehydrogenase dysregulates histone modification in mammalian cells. Mol Cancer 2009;8:89.

[128] Chen H, Giri NC, Zhang R, Yamane K, Zhang Y, Maroney M, et al. Nickel ions inhibit histone demethylase JMJD1A and DNA repair enzyme ABH2 by replacing the ferrous iron in the catalytic centers. J Biol Chem 2010;285:7374–83.

[129] Chen H, Kluz T, Zhang R, Costa M. Hypoxia and nickel inhibit histone demethylase JMJD1A and repress Spry2 expression in human bronchial epithelial BEAS-2B cells. Carcinogenesis 2010;31:2136–44.

[130] Tausendschon M, Dehne N, Brune B. Hypoxia causes epigenetic gene regulation in macrophages by attenuating Jumonji histone demethylase activity. Cytokine 2011;53:256–62.

[131] Guo X, Tian Z, Wang X, Pan S, Huang W, Shen Y, et al. Regulation of histone demethylase KDM6B by hypoxia-inducible factor-2alpha. Acta Biochim Biophys Sin (Shanghai) 2015;47:106–13.

[132] Hancock RL, Dunne K, Walport LJ, Flashman E, Kawamura A. Epigenetic regulation by histone demethylases in hypoxia. Epigenomics 2015;1–21.

[133] Rose NR, Ng SS, Mecinovic J, Lienard BM, Bello SH, Sun Z, et al. Inhibitor scaffolds for 2-oxoglutarate-dependent histone lysine demethylases. J Med Chem 2008;51:7053–6.

[134] Luo X, Liu Y, Kubicek S, Myllyharju J, Tumber A, Ng S, et al. A selective inhibitor and probe of the cellular functions of Jumonji C domain-containing histone demethylases. J Am Chem Soc 2011;133:9451–6.

[135] Suzuki T, Ozasa H, Itoh Y, Zhan P, Sawada H, Mino K, et al. Identification of the KDM2/7 histone lysine demethylase subfamily inhibitor and its antiproliferative activity. J Med Chem 2013;56:7222–31.

[136] Jin C, Yang L, Xie M, Lin C, Merkurjev D, Yang JC, et al. Chem-seq permits identification of genomic targets of drugs against androgen receptor regulation selected by functional phenotypic screens. Proc Natl Acad Sci USA 2014;111:9235–40.

[137] Rai G, Kawamura A, Tumber A, Liang Y, Vogel JL, Arbuckle JH, et al. Discovery of ML324, a JMJD2 demethylase inhibitor with demonstrated antiviral activity. 2010.
[138] Kruidenier L, Chung CW, Cheng Z, Liddle J, Che K, Joberty G, et al. A selective jumonji H3K27 demethylase inhibitor modulates the proinflammatory macrophage response. Nature 2012;488:404–8.
[139] Heinemann B, Nielsen JM, Hudlebusch HR, Lees MJ, Larsen DV, Boesen T, et al. Inhibition of demethylases by GSK-J1/J4. Nature 2014;514:E1–2.
[140] Spannhoff A, Heinke R, Bauer I, Trojer P, Metzger E, Gust R, et al. Target-based approach to inhibitors of histone arginine methyltransferases. J Med Chem 2007;50:2319–25.
[141] Spannhoff A, Machmur R, Heinke R, Trojer P, Bauer I, Brosch G, et al. A novel arginine methyltransferase inhibitor with cellular activity. Bioorg Med Chem Lett 2007;17:4150–3.
[142] Cheng D, Yadav N, King RW, Swanson MS, Weinstein EJ, Bedford MT. Small molecule regulators of protein arginine methyltransferases. J Biol Chem 2004;279:23892–9.
[143] Chan-Penebre E, Kuplast KG, Majer CR, Boriack-Sjodin PA, Wigle TJ, Johnston LD, et al. A selective inhibitor of PRMT5 with *in vivo* and *in vitro* potency in MCL models. Nat Chem Biol 2015;11:432–7.
[144] Raijmakers R, Zendman AJ, Egberts WV, Vossenaar ER, Raats J, Soede-Huijbregts C, et al. Methylation of arginine residues interferes with citrullination by peptidylarginine deiminases *in vitro*. J Mol Biol 2007;367:1118–29.
[145] Luo Y, Knuckley B, Lee YH, Stallcup MR, Thompson PR. A fluoroacetamidine-based inactivator of protein arginine deiminase 4: design, synthesis, and *in vitro* and *in vivo* evaluation. J Am Chem Soc 2006;128:1092–3.
[146] Subramanian V, Knight JS, Parelkar S, Anguish L, Coonrod SA, Kaplan MJ, et al. Design, synthesis, and biological evaluation of tetrazole analogs of Cl-amidine as protein arginine deiminase inhibitors. J Med Chem 2015;58:1337–44.
[147] Dreyton CJ, Jones JE, Knuckley BA, Subramanian V, Anderson ED, Brown SJ, et al. Optimization and characterization of a pan protein arginine deiminase (PAD) inhibitor. Probe Reports from the NIH Molecular Libraries Program 2010.
[148] Jones JE, Slack JL, Fang P, Zhang X, Subramanian V, Causey CP, et al. Synthesis and screening of a haloacetamidine containing library to identify PAD4 selective inhibitors. ACS Chem Biol 2012;7:160–5.
[149] Knuckley B, Causey CP, Pellechia PJ, Cook PF, Thompson PR. Haloacetamidine-based inactivators of protein arginine deiminase 4 (PAD4): evidence that general acid catalysis promotes efficient inactivation. Chembiochem 2010;11:161–5.
[150] Trabocchi A, Pala N, Krimmelbein I, Menchi G, Guarna A, Sechi M, et al. Peptidomimetics as protein arginine deiminase 4 (PAD4) inhibitors. J Enzyme Inhib Med Chem 2015;30:466–71.
[151] Lewis HD, Liddle J, Coote JE, Atkinson SJ, Barker MD, Bax BD, et al. Inhibition of PAD4 activity is sufficient to disrupt mouse and human NET formation. Nat Chem Biol 2015;11:189–91.
[152] Mendenhall EM, Williamson KE, Reyon D, Zou JY, Ram O, Joung JK, et al. Locus-specific editing of histone modifications at endogenous enhancers. Nat Biotechnol 2013;31:1133–6.
[153] Hilton IB, D'Ippolito AM, Vockley CM, Thakore PI, Crawford GE, Reddy TE, et al. Epigenome editing by a CRISPR-Cas9-based acetyltransferase activates genes from promoters and enhancers. Nat Biotechnol 2015;33:510–17.
[154] Kearns NA, Pham H, Tabak B, Genga RM, Silverstein NJ, Garber M, et al. Functional annotation of native enhancers with a Cas9-histone demethylase fusion. Nat Methods 2015;12:401–3.
[155] Anders L, Guenther MG, Qi J, Fan ZP, Marineau JJ, Rahl PB, et al. Genome-wide localization of small molecules. Nat Biotechnol 2014;32:92–6.
[156] Varambally S, Dhanasekaran SM, Zhou M, Barrette TR, Kumar-Sinha C, Sanda MG, et al. The polycomb group protein EZH2 is involved in progression of prostate cancer. Nature 2002;419:624–9.
[157] Kleer CG, Cao Q, Varambally S, Shen R, Ota I, Tomlins SA, et al. EZH2 is a marker of aggressive breast cancer and promotes neoplastic transformation of breast epithelial cells. Proc Natl Acad Sci USA 2003;100:11606–11.

[158] McCabe MT, Graves AP, Ganji G, Diaz E, Halsey WS, Jiang Y, et al. Mutation of A677 in histone methyltransferase EZH2 in human B-cell lymphoma promotes hypertrimethylation of histone H3 on lysine 27 (H3K27). Proc Natl Acad Sci USA 2012;109:2989–94.

[159] Morin RD, Johnson NA, Severson TM, Mungall AJ, An J, Goya R, et al. Somatic mutations altering EZH2 (Tyr641) in follicular and diffuse large B-cell lymphomas of germinal-center origin. Nat Genet 2010;42:181–5.

[160] McCabe MT, Ott HM, Ganji G, Korenchuk S, Thompson C, Van Aller GS, et al. EZH2 inhibition as a therapeutic strategy for lymphoma with EZH2-activating mutations. Nature 2012;492:108–12.

[161] Qi W, Chan H, Teng L, Li L, Chuai S, Zhang R, et al. Selective inhibition of EZH2 by a small molecule inhibitor blocks tumor cells proliferation. Proc Natl Acad Sci U S A 2012;109:21360–5.

[162] Knutson SK, Wigle TJ, Warholic NM, Sneeringer CJ, Allain CJ, Klaus CR, et al. A selective inhibitor of EZH2 blocks H3K27 methylation and kills mutant lymphoma cells. Nat Chem Biol 2012;8:890–6.

[163] Bitler BG, Aird KM, Garipov A, Li H, Amatangelo M, Kossenkov AV, et al. Synthetic lethality by targeting EZH2 methyltransferase activity in ARID1A-mutated cancers. Nat Med 2015;21:231–8.

[164] Fillmore CM, Xu C, Desai PT, Berry JM, Rowbotham SP, Lin YJ, et al. EZH2 inhibition sensitizes BRG1 and EGFR mutant lung tumours to TopoII inhibitors. Nature 2015;520:239–42.

[165] Feng Q, Wang H, Ng HH, Erdjument-Bromage H, Tempst P, Struhl K, et al. Methylation of H3-lysine 79 is mediated by a new family of HMTases without a SET domain. Curr Biol 2002;12:1052–8.

[166] Okada Y, Feng Q, Lin Y, Jiang Q, Li Y, Coffield VM, et al. hDOT1L links histone methylation to leukemogenesis. Cell 2005;121:167–78.

[167] Krivtsov AV, Feng Z, Lemieux ME, Faber J, Vempati S, Sinha AU, et al. H3K79 methylation profiles define murine and human MLL-AF4 leukemias. Cancer Cell 2008;14:355–68.

[168] Bernt KM, Zhu N, Sinha AU, Vempati S, Faber J, Krivtsov AV, et al. MLL-rearranged leukemia is dependent on aberrant H3K79 methylation by DOT1L. Cancer Cell 2011;20:66–78.

[169] Daigle SR, Olhava EJ, Therkelsen CA, Majer CR, Sneeringer CJ, Song J, et al. Selective killing of mixed lineage leukemia cells by a potent small-molecule DOT1L inhibitor. Cancer Cell 2011;20:53–65.

[170] Schmidt DM, McCafferty DG. trans-2-Phenylcyclopropylamine is a mechanism-based inactivator of the histone demethylase LSD1. Biochemistry 2007;46:4408–16.

[171] Wang Y, Zhang H, Chen Y, Sun Y, Yang F, Yu W, et al. LSD1 is a subunit of the NuRD complex and targets the metastasis programs in breast cancer. Cell 2009;138:660–72.

[172] Hill JM, Quenelle DC, Cardin RD, Vogel JL, Clement C, Bravo FJ, et al. Inhibition of LSD1 reduces herpesvirus infection, shedding, and recurrence by promoting epigenetic suppression of viral genomes. Sci Transl Med 2014;6:265ra169.

[173] Berrens RV, Reik W. Prmt5: a guardian of the germline protects future generations. Embo J 2015;34:689–90.

[174] Teng YC, Lee CF, Li YS, Chen YR, Hsiao PW, Chan MY, et al. Histone demethylase RBP2 promotes lung tumorigenesis and cancer metastasis. Cancer Res 2013;73:4711–21.

[175] Yamamoto S, Wu Z, Russnes HG, Takagi S, Peluffo G, Vaske C, et al. JARID1B is a luminal lineage-driving oncogene in breast cancer. Cancer Cell 2014;25:762–77.

[176] de Rooij JD, Hollink IH, Arentsen-Peters ST, van Galen JF, Berna Beverloo H, Baruchel A, et al. NUP98/JARID1A is a novel recurrent abnormality in pediatric acute megakaryoblastic leukemia with a distinct HOX gene expression pattern. Leukemia 2013;27:2280–8.

[177] Lin W, Cao J, Liu J, Beshiri ML, Fujiwara Y, Francis J, et al. Loss of the retinoblastoma binding protein 2 (RBP2) histone demethylase suppresses tumorigenesis in mice lacking Rb1 or Men1. Proc Natl Acad Sci USA 2011;108:13379–86.

[178] Sharma SV, Lee DY, Li B, Quinlan MP, Takahashi F, Maheswaran S, et al. A chromatin-mediated reversible drug-tolerant state in cancer cell subpopulations. Cell 2010;141:69–80.

[179] Roesch A, Fukunaga-Kalabis M, Schmidt EC, Zabierowski SE, Brafford PA, Vultur A, et al. A temporarily distinct subpopulation of slow-cycling melanoma cells is required for continuous tumor growth. Cell 2010;141:583–94.

[180] Sayegh J, Cao J, Zou MR, Morales A, Blair LP, Norcia M, et al. Identification of small molecule inhibitors of Jumonji AT-rich interactive domain 1B (JARID1B) histone demethylase by a sensitive high throughput screen. J Biol Chem 2013;288:9408–17.

[181] Vieira FQ, Costa-Pinheiro P, Almeida-Rios D, Graca I, Monteiro-Reis S, Simoes-Sousa S, et al. SMYD3 contributes to a more aggressive phenotype of prostate cancer and targets Cyclin D2 through H4K20me3. Oncotarget 2015;6:13644–57.

[182] Hamamoto R, Furukawa Y, Morita M, Iimura Y, Silva FP, Li M, et al. SMYD3 encodes a histone methyltransferase involved in the proliferation of cancer cells. Nat Cell Biol 2004;6:731–40.

[183] Silva FP, Hamamoto R, Kunizaki M, Tsuge M, Nakamura Y, Furukawa Y. Enhanced methyltransferase activity of SMYD3 by the cleavage of its N-terminal region in human cancer cells. Oncogene 2008;27:2686–92.

[184] Kunizaki M, Hamamoto R, Silva FP, Yamaguchi K, Nagayasu T, Shibuya M, et al. The lysine 831 of vascular endothelial growth factor receptor 1 is a novel target of methylation by SMYD3. Cancer Res 2007;67:10759–65.

[185] Mazur PK, Reynoird N, Khatri P, Jansen PW, Wilkinson AW, Liu S, et al. SMYD3 links lysine methylation of MAP3K2 to Ras-driven cancer. Nature 2014;510:283–7.

[186] Peserico A, Germani A, Sanese P, Barbosa AJ, Di Virgilio V, Fittipaldi R, et al. A SMYD3 small-molecule inhibitor impairing cancer cell growth. J Cell Physiol 2015;230:2447–60.

[187] Chesi M, Nardini E, Lim RS, Smith KD, Kuehl WM, Bergsagel PL. The t(4;14) translocation in myeloma dysregulates both FGFR3 and a novel gene, MMSET, resulting in IgH/MMSET hybrid transcripts. Blood 1998;92:3025–34.

[188] Hudlebusch HR, Skotte J, Santoni-Rugiu E, Zimling ZG, Lees MJ, Simon R, et al. MMSET is highly expressed and associated with aggressiveness in neuroblastoma. Cancer Res 2011;71:4226–35.

[189] Popovic R, Martinez-Garcia E, Giannopoulou EG, Zhang Q, Ezponda T, Shah MY, et al. Histone methyltransferase MMSET/NSD2 alters EZH2 binding and reprograms the myeloma epigenome through global and focal changes in H3K36 and H3K27 methylation. PLoS Genet 2014;10:e1004566.

[190] Asangani IA, Ateeq B, Cao Q, Dodson L, Pandhi M, Kunju LP, et al. Characterization of the EZH2-MMSET histone methyltransferase regulatory axis in cancer. Mol Cell 2013;49:80–93.

[191] Rosati R, La Starza R, Veronese A, Aventin A, Schwienbacher C, Vallespi T, et al. NUP98 is fused to the NSD3 gene in acute myeloid leukemia associated with t(8;11)(p11.2;p15). Blood 2002;99: 3857–60.

[192] Wang GG, Cai L, Pasillas MP, Kamps MP. NUP98-NSD1 links H3K36 methylation to Hox-A gene activation and leukaemogenesis. Nat Cell Biol 2007;9:804–12.

[193] Jaju RJ, Fidler C, Haas OA, Strickson AJ, Watkins F, Clark K, et al. A novel gene, NSD1, is fused to NUP98 in the t(5;11)(q35;p15.5) in de novo childhood acute myeloid leukemia. Blood 2001;98: 1264–7.

[194] French CA, Rahman S, Walsh EM, Kuhnle S, Grayson AR, Lemieux ME, et al. NSD3-NUT fusion oncoprotein in NUT midline carcinoma: implications for a novel oncogenic mechanism. Cancer Discov 2014;4:928–41.

[195] Kurotaki N, Imaizumi K, Harada N, Masuno M, Kondoh T, Nagai T, et al. Haploinsufficiency of NSD1 causes Sotos syndrome. Nat Genet 2002;30:365–6.

[196] Nimura K, Ura K, Shiratori H, Ikawa M, Okabe M, Schwartz RJ, et al. A histone H3 lysine 36 trimethyltransferase links Nkx2-5 to Wolf-Hirschhorn syndrome. Nature 2009;460:287–91.

[197] di Luccio E. Inhibition of nuclear receptor binding SET Domain 2/Multiple myeloma SET domain by LEM-06 implication for epigenetic cancer therapies. J Cancer Prev 2015;20:113–20.

CHAPTER

BROMODOMAINS AS ANTICANCER TARGETS

10

Panagis Filippakopoulos[1,2] and Stefan Knapp[1,3]

[1]*Structural Genomics Consortium, Nuffield Department of Clinical Medicine, University of Oxford, Oxford, UK*

[2]*Ludwig Institute for Cancer Research, Nuffield Department of Clinical Medicine, University of Oxford, Oxford, UK*

[3]*Institute for Pharmaceutical Chemistry, Johann Wolfgang Goethe-University, Frankfurt am Main, Germany*

CHAPTER OUTLINE

10.1 INTRODUCTION

In the last two decades, a complex array of chemical modifications has been described on chromatin, which dynamically regulates chromatin structure and gene expression in a hereditary fashion. These regulatory mechanisms are often described as the "epigenome" [1], referring to modifications such as methylation or hydroxymethylation regulating gene expression on the DNA level [2] as well as the complex set of reversible post translational modifications (PTMs) found on histone proteins [3]. The intricate combination of PTMs constitutes the "histone code," a cellular language involving proteins that introduce (writers), remove (erasers), or recognize (readers) these chemical modifications [4]. Lysine acetylation is one of the most prominent PTMs and affects chromatin structure. Acetylation of the amino group on lysines results in neutralization of the positive charge of this residue, which is very abundant on histones, resulting in weaker histone–DNA interactions

G. Egger & P. Arimondo (Eds): Drug Discovery in Cancer Epigenetics. DOI: http://dx.doi.org/10.1016/B978-0-12-802208-5.00010-2

and decompaction of chromatin. This so-called *open chromatin structure* leads to transcriptional activation. However, lysine acetylation also creates binding sites for bromodomains (BRDs), acetyl-lysine-dependent epigenetic reader domains, which recruit transcriptional regulators to acetylate chromatin sites, which may lead to locus-specific transcriptional enhancement or repression. In cancer, deregulation of acetylation levels results in aberrant expression of growth- as well as survival-promoting genes, but also in downregulation of tumor suppressor genes.

10.2 ACETYLATION AND TRANSCRIPTION

Acetylation of the lysine side-chain is one of the most frequent PTMs found proteome-wide [5,6], controlled by enzymes that write it (lysine acetyltransferases) or erase it (lysine deacetylases). Enzymes that are known to recognize histones as substrates are also called histone acetyltransferases (HATs) and histone deacetylases (HDACs). However, most HATs and HDACs are now known to recognize a large number of sites on both histone and nonhistone proteins. The nuclear HATs (class A) acetylate histone and nonhistone proteins, while the cytoplasmic (class B) HATs are less characterized and are thought to acetylate newly synthesized histones, playing a role in DNA repair and chromatin assembly [7,8]. Class A HATs include GCN5L2/PCAF [9], CREBBP/EP300 [10], the transcription initiation factor TFIID (TAF) family [11] as well as the MYST family [12], consisting of the five human HATs hMOF, TIP60, HBO1 (HAT bound to Orc1, also named MYST2), monocytic leukemia zinc finger protein (MOZ), and MORF. Lysine acetylation affects chromatin condensation and is thought to be associated with transcriptional activation via formation of euchromatin, while acetylation of nonhistone proteins may result in alteration of protein–protein interactions and DNA binding as well as subcellular localization and protein stability [13].

Enzymes that remove acetylation from histones (HDACs) are grouped into four classes based on their homology to the corresponding yeast proteins. Classes I, II, and IV are usually referred to as "classical" HDACs and comprise HDAC1, HDAC2, HDAC3, and HDAC8 (class I), HDAC4, HDAC5, HDAC6, HDAC7, HDAC9, and HDAC10 (class II), and HDAC11 (class IV), sharing a common mechanism of action, involving binding of a Zn^{2+} ion at their active site, necessary for catalytic function. Classical HDACs differ in their tissue-specific expression as well as subcellular localization, with class I HDACs mainly acting in the nucleus, while class IIa HDACs are expressed in a tissue-specific fashion and shuttle between the nucleus and the cytoplasm [14]. Class III HDACs comprise the sirtuins SIRT1, SIRT2, SIRT3, SIRT4, SIRT5, SIRT6, and SIRT7, which require NAD^+ as a cofactor [15].

HDAC involvement in carcinogenesis has been reported in diverse tumor settings involving a range of different mechanisms, most common of which is deregulation of their expression [15,16]. HDAC overexpression usually leads to upregulation of genes involved in development and proliferation [17] as well as downregulation of tumor suppressor genes [18]. Mutations that lead to truncated proteins have also been identified. For instance, HDAC2 is truncated due to a frameshift mutation in sporadic carcinomas with microsatellite instability and in tumors arising in individuals with hereditary nonpolyposis colorectal cancer syndrome [19]. There is also recent evidence that HDACs may negatively regulate microRNAs, which would otherwise function as tumor suppressors, resulting in tumorigenesis [20].

Given their role in regulating gene transcription in cancer, it is not surprising that HATs and HDACs have been targeted for the development of novel pharmaceuticals. While efforts to generate potent inhibitors for HATs have not been very successful over the years, with only a handful of weak inhibitors reported so far [21], HDACs have proven to be much more amenable to chemical inhibition, leading to the development of clinically successful HDAC inhibitors, particularly in oncology [15,22–24]. Most of the developed inhibitors target class I HDACs nonselectively and the precise mechanism of their action is not well understood. However, it is widely accepted that disruption of interactions with other proteins and inhibition of enzymatic activity play a role in the observed phenotypes [16,25,26]. It is noteworthy that two inhibitors, suberoylanilide hydroxamic acid (vorinostat, Zolinza™) and depsipeptide (romidepsin, Istodax™) have been approved by the US Food and Drug Administration [27].

The readout of the acetylated lysine side-chain is principally mediated by BRDs, evolutionary conserved protein–protein interaction modules named after the "brahma" gene [28] where the first BRD core motif was identified [29]; more recently YEATES domains have also been shown to bind to acetylated lysine motifs mainly on histone proteins [30].

There are 61 BRDs that are present in 42 diverse proteins in human, including HAT and HAT-associated proteins such as GCN5L2/PCAF [9], CREBBP/EP300 [10] nuclear scaffolding proteins such as the polybromo PB1 [31] and bromodomain-9 (BRD9) [32], ATP-dependent chromatin-remodeling complexes such as BAZ1B [33], helicases such as SMARCA [34], transcriptional coactivators such as the TAF [35] and tripartite motif (TRIM)/transcription intermediary factor [36], transcriptional regulators such as the bromo and extraterminal (BET) proteins [37], methyltransferases such as ASH1L [38], and the nuclear body proteins SP100/SP110/SP140 [39–41]. BRD-containing proteins have been involved in the development of a number of diseases including inflammatory and neurological diseases as well as diverse tumors [42].

10.3 ARCHITECTURE OF BRDs

The BRD module is comprised of four alpha helices (αZ, αA, αB, and αC) linked by diverse loop segments of variable length (ZA- and BC-loops). The acetylated lysine is recognized by a conserved asparagine within a cavity formed in the core of the helical bundle which is hydrophobic and can accommodate the neutralized lysine side-chain, while the regions flanking the main recognition site are highly charged, contributing to binding and selectivity for diverse acetylated peptide sequences (Figure 10.1A). A number of BRDs lack this asparagine docking residue and contain a threonine (e.g., PHIPA) or a tyrosine (e.g., ASH1L) at this position, however there is no structural or biochemical evidence that these unusual modules recognize acetylated peptides. Based on their structural similarities, the human family of BRD modules was recently classified in eight distinct families [43] (Figure 10.1B).

10.4 LINEAR SEQUENCE MOTIFS RECOGNIZED BY BRDs

BRD modules are found in multidomain proteins, very often harboring additional reader domains. The combination of reader modules, including methyl-lysine reader modules such as the frequently

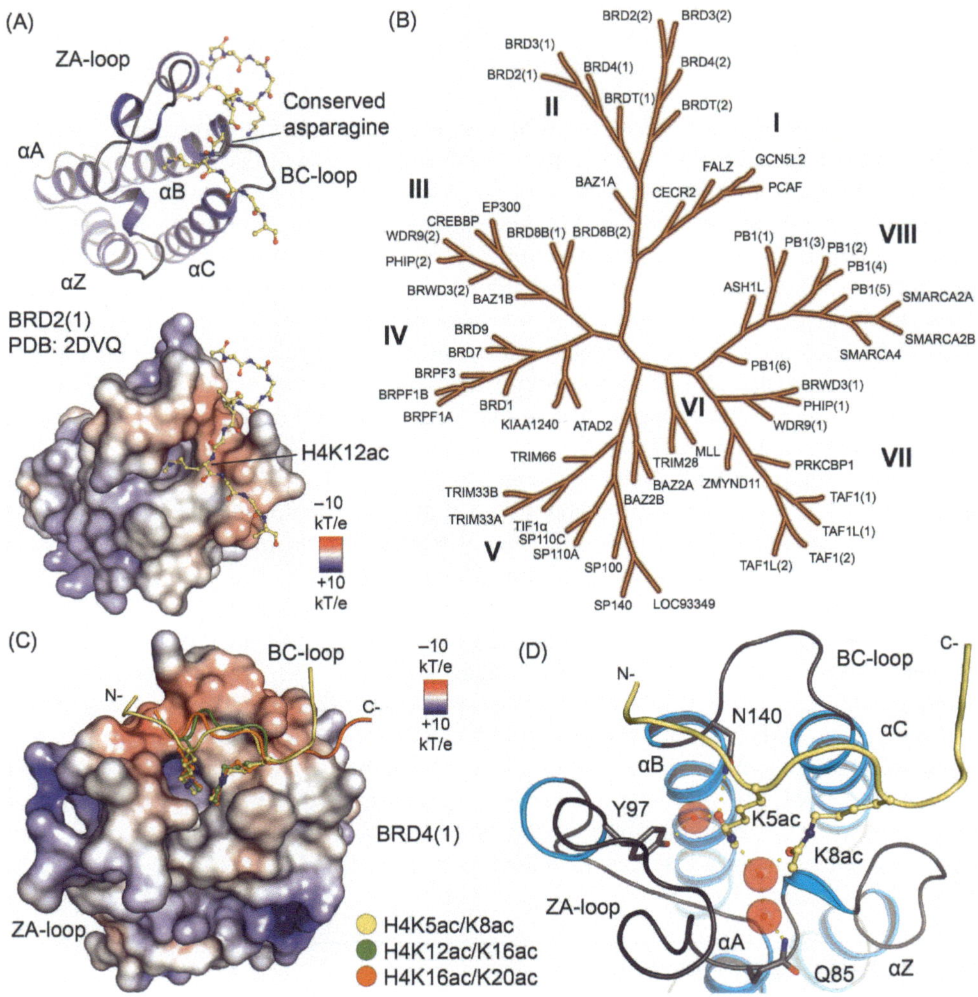

FIGURE 10.1

BRD architecture, structural phylogeny, and peptide recognition. (A) BRD fold. The conserved helical bundle (helices αZ, αA, αB, αC) is held together by diverse loop regions (AZ- and BC-loops) presenting a hydrophobic pocket where acetylated peptides insert and dock onto a conserved asparagine residue (N156 in BRD2(1) taken from PDB ID: 2DVQ). The region surrounding the central Kac recognition cavity is highly charged, driving interactions with the peptide backbone (lower panel). (B) Structural analysis allowed clustering of the human BRD modules into eight distinct subfamilies, shown in roman numerals. (C) Histone H4 peptides crystallized with the first domain of BRD4 exhibited a conserved binding mode whereby two acetylated lysines insert into the central BRD cavity occupying the bulk of the recognition site. An overlay of the H4K5ac/K8ac (PDB ID 3UVW), H4K12ac/K16ac (PDB ID: 3UVX), and H4K16ac/K20ac (PDB ID 3UVY) is shown in surface representation, colored by the electrostatic potential as indicated in the inset. All peptides orient in the same direction within the cavity aligning their N-terminus next to the beginning of the BC-loop. (D) While the first acetyl lysine (K5ac in the case of the H4K5ac/K8ac diacetylated peptide from PDB ID: 3UVW) directly engages the protein via the conserved asparagine (N140 in BRD4(1)) as well as Y97 via a water-mediated bridge, the second acetylated lysine (K8ac in the case of the H4K5ac/K8ac diacetylated peptide from PDB ID: 3UVW) binds to the first lysine via a network of water-mediated bridges which further stabilizes peptide binding by interactions with Q85.

found plant-homeodomains (PHD) [44], bromo adjacent homology (BAH) domains [45], proline-tryptophan-tryptophan-proline domains [46], or other protein–protein interaction modules such as SAND domains (named after Sp100, AIRE-1, NucP41/75, DEAF-1) [47] and kinase-inducible domain (KIX) [48], suggests that the reading process of chromatin is complex. Thus multiple histone PTMs may be read by these protein modules, which may be present on different histones or different nucleosomes altogether. This complexity poses a significant challenge in our understanding of the epigenetic code. However, recent advances in peptide array technologies have allowed the systematic parsing of a large number of histone modifications, shedding light on the potential linear motifs recognized by epigenetic readers. A recent large-scale comprehensive study identified sequences spanning acetylated lysine regions on histones that can be recognized by members of the human BRD family. Large-scale structural analysis also sheds light in the diversity of BRD modules, particularly on the distribution of charge on their recognition surfaces, suggesting that many BRD modules may not recognize histone tails and are probably recruited to other nuclear or non nuclear acetylated protein complexes [43]. Intriguingly, human BRDs potently interact with multiply acetylated peptides as well as with peptides that contain additional PTMs in addition to acetylation, suggesting a reading process governed by specific PTM combinations rather than single marks.

The first example revealing the structural mechanisms of such recognition processes has been established by the structure of a diacetylated peptide simultaneously binding to the acetyl-lysine binding site in the testis-specific isoform of the BET murine protein BRDT [49]. Peptide array data and in-solution studies showed that this is a feature shared by all N-terminal BET proteins (Figure 10.1C) [43]. High-resolution cocrystal structures of the first BRD of BRD4 with a number of diacetylated histone H4 peptides revealed a conserved mode of interaction whereby the N-terminal acetyl lysine engaged the conserved asparagine (N140 in BRD4(1)) directly, while the second acetylated lysine packed inside the BRD recognition cavity and participated in a water-mediated network of interactions that stabilized the peptide complex, engaging both the protein backbone as well as the first acetylated lysine, occupying the entire binding cavity of the BRD (Figure 10.1D). Although there have not been any structures reported of BRD/peptide complexes that carry acetylation as well as other PTMs so far, it is likely that differences in peptide properties attributed to, for example, the charge carried by other PTMs (e.g., the negative charge present in phosphorylated residues) will significantly influence the reading process of acetylated lysine epitopes. This observation suggests that BRD recognition of acetylated and other PTM-containing sequences is an integrated process and the interpreted signals may be generated by diverse signaling pathways, in order to switch a specific transcriptional program on or off.

10.5 FUNCTIONAL ROLES OF BRD-CONTAINING PROTEINS

The combination of different reader modules and the recognition of specific mark combinations facilitate a highly specific assembly of regulatory complexes at distinct sites. The HATs CREBBP/EP300, for example, exhibit a multidomain structure combining several reading modules (PHD, KIX) in addition to a BRD [10]. CREBBP and EP300 share a high degree of sequence similarity acting as transcriptional coactivators controlling cellular key processes such as neuronal plasticity, genomic stability, memory formation, development, energy homeostasis, and cell growth. Homozygous knockout of CREBBP in mice is lethal (in utero) with signs of defective blood vessel formation in the central nervous system, developmental retardation, and delays in both primitive and definitive hematopoiesis [50], while homozygous deletion of EP300 results in lethality between days 9 and 11.5 of gestation, with defects in heart development, cell proliferation, and neurulation [51]. The highly related BRD-containing HATs PCAF and GCN5 also play a role as transcriptional coactivators [9]. Homozygous deletion of PCAF has a mild phenotypic response, resulting in short-term memory deficits and an exaggerated response to acute stress and conditioned fear, associated with increased plasma corticosterone levels [52]. In contrast, genetic knockout of *Gcn5* in mice is lethal during embryogenesis due to severe growth retardation, failure in the development of dorsal mesoderm lineages and anterior neural tube closure [53,54]. The diverse phenotypes observed in knockout studies for both of the highly related BRD-containing HAT families suggest that each family member has distinct biological functions.

The methyltransferase ASH1L harbors a BRD as well as methyl-lysine binding modules of the PHD and BAH family [55]. As a member of the trithorax group of transcriptional activators, ASH1L activates in *Drosophila* ultrabithorax expression, while its mammalian homologs have been associated with actively transcribed genes. The mixed lineage leukemia protein MLL is another multidomain methyltransferase containing a BRD, which plays an essential role in regulating the expression of genes regulating self-renewal of hematopoietic stem cells and leukemia-initiating cells in cancer cells containing MLL fusions by regulating *HOX* gene expression [56–58].

The evolutionary conserved AAA ATPase nuclear coregulator cancer-associated protein ATAD2 is a transcriptional coregulator of c-Myc, estrogen (ER), and androgen receptors and a required element for the recruitment of E2F transcription factors to their target sites. ATAD2 associates with acetylated chromatin via its BRD during late mitosis, regulating the expression of genes responsible for cell-cycle progression [59–61].

Dual BRD proteins of the BET subfamily play an important role in cell-cycle progression by regulating growth-promoting genes. The four members of this family (BRD2, BRD3, BRD4, and the testis-specific BRDT) share a conserved domain architecture including two tandem N-terminal BRDs, and an extraterminal protein–protein interaction domain. BRD2 regulates expression of diverse genes, including cyclin D1, by associating with the switch/sucrose nonfermenting (SWI/SNF) complex as well as with E2F transcription factors [62,63]. An interesting phenotype has been observed in mice bearing a mutation in the *Brd2* promoter region, which gives rise to reduced levels of BRD2, resulting in extreme obesity without glucose intolerance [64]. BRD4 and BRDT mediate transcriptional elongation by recruiting the positive transcription elongation factor b (a complex of the kinase CDK9 and its activator cyclin T) to acetylated transcriptional start sites via their BRDs [65,66]. In contrast to most other BRD-containing proteins, BETs remain associates with hypoacetylated condensed chromatin during mitosis, suggesting a role in epigenetic memory and gene bookmarking [67,68]. BET protein levels are critical for the control of cell cycle; for instance overexpression of BRD4 leads to cell-cycle arrest, while genetic knockdown of BRD4 in cultured human cells significantly reduces cell growth [69,70]. Depletion of BRD4 (or BRD2) is lethal in mice, while *Brd4*$^{+/-}$ mice exhibit severe developmental defects [71–73]. BRDT is an essential component for spermatogenesis as exhibited by specific deletion of its first BRD, which results in abnormal spermatids and sterility in mice [74], while genome-wide association studies linked polymorphism found in BRDT to sterility in European men [75].

The largest subunit of the TFIID transcription factor, the TAF1 RNA polymerase II (TATA box binding protein [TBP]-associated factor, 250 kDa formerly called TAFII250) also contains tandem BRDs. These C-terminal tandem modules recognize and bind to diacetylated histone H4 (acetylated K5/K12 or K8/K16) as well as diacetylated p53 (acetylated at K373/K382) at the *p21* promoter [35,76]. TAF1 modulates the rate of transcriptional initiation by binding to transcriptional start sites while interacting with other transcriptional regulators [11], activating and regulating several essential biological processes including cell cycle and apoptosis, DNA damage response, and myogenesis [77–80]. The testis-specific homolog of TAF1, TAF1L, is X-linked and may act as a functional substitute for TAF1 during male meiosis, binding to the TBP and can functionally substitute for TAF1 in a temperature-sensitive hamster cell line [81].

WD repeat proteins also contain tandem BRD modules and are involved in many diverse processes, including apoptosis, signal transduction, gene regulation, and cell-cycle progression [82,83]. The WD repeat protein BRWD1/WDR9 is broadly expressed, with expression levels that are dynamic during mouse development; it plays a role in spermatogenesis and the oocyte-transition and associates with the SWI/SNF complex [84,85]. Interestingly the function of BRWD3 has been genetically linked in *Drosophila* to the JAK/STAT pathway, however little is known about its function in human [86].

SWI/SNF complexes exhibit chromatin-remodeling activity, thus contributing to transcriptional activation or repression of genes. Components of SWI/SNF complexes were first identified in screens for mutants that resulted in defects in mating type switching in yeast or which were unable to grow on sucrose [87–89]. Later, gene expression studies demonstrated that SWI/SNF function as a transcriptional regulators affecting 5% of all genes in yeast [90]. The mammalian SWI/SNF complexes also play a key role controlling cell differentiation and proliferation as well as components of the embryonic stem cell core pluripotency transcriptional network [91,92], while mediating critical interactions between nuclear receptors and hormones [34,93–95]. SWI/SNF complexes contain a core subunit, which is responsible for altering chromatin structure in an ATP-dependent fashion, resulting in an open chromatin conformation, which allows interactions with transcription factors [96]. In human there are two such ATPase components, which are mutually exclusive in SWI/SNF complexes, SMARCA4 (named after the *Drosophila* homolog Brama *Brg1* gene) and the related SMARCA2 (or BRM). SMARCA4 has been shown to have chromatin-remodeling activity which is important for the traversal of the nucleosome by RNA polymerase II [97]. It has also been reported to associate with Rb proteins inducing cell-cycle arrest and HDAC-dependent transcriptional repression, leading to transcriptional downregulation of cell-cycle regulatory genes [98]. The polybromo and Brg1-associated factor (PBAF) SWI/SNF complex contains the polybromo protein (PB1 or BAF180) required for ligand-dependent trans-activation by nuclear hormone receptors, which contain six BRD modules together with two BAH and a homeobox DNA-binding domain and play a key role in cell-cycle regulation and senescence [99–101]. PBAF complexes mediate the activation of vitamin D receptor-dependent transcription. Mice lacking PB1 exhibit defects in heart development due to impaired epithelial-to-mesenchymal transition and arrested maturation of epicardium as a result of the downregulation of the growth factors TGF, FGF, and VEGF [102,103].

Chromatin-remodeling complexes of the imitation-SWI (ISWI) family also contain BRDs. ISWI complexes are key regulators of chromatin structure, heterochromatin replication, and transcription; they assemble into seven or more different complexes containing a central core ATPase of the two SNF2 like mammalian homologs *Snf2L* and *Snf2h* of yeast ISWI. The nucleosome remodeling factor ISWI complex contains the BRD PHD finger transcription factor, a protein which was identified to be highly expressed in patients with Alzheimer's disease as "fetal Alz-50 reactive clone 1," as well as in fetal brain in patients with neurodegenerative diseases (Fetal Alzheimer Antigen, FALZ) [104,105]. The ATP-utilizing Chromatin remodeling and assembly Factor/Williams syndrome transcription factor ISWI complex contains BAZ1, a protein of the BRD adjacent zinc finger family, represented by four related genes in human (*BAZ1A*, *BAZ1B*, *BAZ2A*, and *BAZ2B*) sharing a similar domain organization including a PHD-BRD cassette. BAZ1A was first identified in HeLa nuclear extracts as a factor associating with SNF2H, forming a complex with ATP-dependent chromatin-remodeling activity required for the DNA-replication machinery to penetrate condensed chromatin structures [106]. SNF2H/BAZ1A is particularly enriched in replicating pericentromeric heterochromatin and RNAi knockdown of BAZ1A impairs replication of condensed chromatin [107,108]. BAZ2A is a key subunit of the nucleolar remodeling complex which mediates transcriptional silencing of noncoding RNA [109,110].

Members of the TRIM family of transcriptional regulators contain BRD modules, together with a RING finger and one or two B-boxes (zinc binding motifs) as well as an associated coiled-coil region [111]. TRIM24 contains an N-terminal TRIM domain, a nuclear receptor (LxxLL) interaction motif, and a C-terminal PHD BRD, which help associate with chromatin and allow for ligand-dependent activation of the androgen receptor and the retinoic acid receptor as well as for interactions with thyroid, vitamin D_3, and ER receptors [112–115].

TRIM28 associates with the heterochromatin-associated factors HP1 (HP1α, HP1β, HP1γ) in order to promote silencing of euchromatic genes and is also a corepressor for Krüppel-associated box-domain-containing zinc finger proteins, which play an important role in early embryogenesis [116–119]. Importantly, TRIM28 recruitment to centromeres is required for the induction of the parietal and visceral endoderm differentiation pathways [120].

The ubiquitin ligase TRIM33 targets SMAD4, affecting canonical TGFβ signaling involving SMAD4 complexes with receptor-phosphorylated SMAD2 and SMAD3 and its depletion in human cell lines inhibits SMAD4-dependent cell proliferation, while mice deficient of *Trim33* die in utero, suggesting an important role in development [121–123].

The TRIM66 member of this subfamily is mainly expressed in testes where it associates with heterochromatin-associated factors and functions as a transcriptional silencer, although it does not participate in any nuclear receptor complexes [124].

The nuclear body SP family of BRDs (SP100, SP110, and SP140) contains a SAND DNA-binding domain followed by a tandem PHD-BRD cassette and regulates diverse cellular functions, such as DNA repair and genome stability, chromatin organization, and transcription [125–130]. Nuclear bodies have been associated with the pathogenesis of acute promyelocytic leukemia (PML), are autoantibody targets in patients with primary biliary cirrhosis and are also involved in viral response [131–133]. A comprehensive list of BRD-containing proteins and their biological role(s) is given in Table 10.1.

Table 10.1 BRD-Containing Proteins, Their Function and Role(s) in Cancer

Protein	Function	Localization[a]	Ref.	Role(s) in Cancer	Ref.
ASH1L	Methyltransferase	N, C	[38]	Mutations or copy number changes detected in lung cancer	[134]
ATAD2	Transcriptional regulator	N	[60]	Overexpressed in prostate, breast, and lung cancer and high expression levels correlate with poor prognosis in some cancers. Coactivator of E2F transcription factors. Controls the expression of B-MYB, the histone methyltransferase EZH2 as well as other growth-promoting proteins. Involved in ER-induced cell proliferation and cell-cycle progression of breast cancer cells	[59,61,135–137]
ATAD2B (KIAA1240)	Transcriptional regulator	N	[137]	Highly expressed in gastrointestinal and breast carcinomas	[137]
BAZ1A	Chromatin-remodeling factor	N	[138]	None identified yet	
BAZ1B	Chromatin-remodeling factor, transcriptional regulator	N	[33]	Aberrant expression in colon cancer	[139]
BAZ2A	Transcriptional repressor	N, C	[140]	Chromosomal rearrangements in pediatric pre-B acute lymphoblastic leukemia (ALL)	[141]
BAZ2B	Unknown	N, C	[104]	None identified yet	
BRD1	Transcriptional regulator, scaffold protein	N, C	[142]	Chromosomal rearrangements detected in leukemia with PAX5 transcription factors	[143]
BRD2	Transcriptional regulator	N	[63]	Interacts with chromatin-binding domain in Kaposi's sarcoma-associated herpesvirus latency-associated nuclear antigen 1. BRD2 expression is downregulated in nasopharyngeal carcinoma cells	[144,145]
BRD3	Transcriptional regulator	N	[63]	A chromosomal aberration involving BRD3 and nuclear protein in testis (NUT) giving rise to a malignant subtype of midline carcinoma	[146]
BRD4	Transcriptional regulator	N	[147]	Translocation t(15;9)(q14;q34) with NUT giving rise to a BRD4-NUT fusion protein and development of carcinomas. BRD4 controls expression of growth-promoting genes such as Myc, aurora kinase, and BCL2. Required for expression of target genes of the super elongation complex in MLL. Drives expression of FOSL1 in lung carcinoma. BRD4 is required for the replication of tumor viruses	[148–152]
BRD7	Transcriptional regulator	N	[153]	Tumor suppressor required for efficient transcription of a subset of p53 target genes	[154]
BRD8B	Transcriptional regulator	N	[155]	BRD8 regulates cellular survival and sensitivity to spindle poisons in colon cancer	[156]

(Continued)

Table 10.1 BRD-Containing Proteins, Their Function and Role(s) in Cancer *Continued*

Protein	Function	Localization[a]	Ref.	Role(s) in Cancer	Ref.
BRD9	SWI/SNF component	N, C	[32]	DNA copy number changes in lung cancer	[157]
BRDT	Chromatin-remodeling factor	N	[49]	Overexpression detected in nonsmall-cell lung cancer (NSCLC) and squamous cell carcinomas of the head and neck as well as of the esophagus	[158]
BRPF1	Transcriptional activator	N, C	[159]	Component of the monocytic leukemia zinc finger protein MOZ/MORF complex, regulates HOX gene expression	[142,159]
BRPF3	Transcriptional regulator	N	[160]	Component of the monocytic leukemia zinc finger protein MOZ/MORF complex	[142]
BRWD3	JAK-STAT signaling	N, C	[86]	Chromosomal aberration leading to the disruption of the BRWD3 locus has been detected in B-cell chronic lymphocytic leukemia	[83]
CECR2	Chromatin-remodeling factor	N	[161]	DNA damage response protein required for DSB repair	[162]
CREBBP	Histone acetyltransferase	N	[10]	Chromosomal aberrations and mutations detected in acute leukemia. Inhibitor of RUNX1-mediated transcription. Required for self-renewal of leukemic stem cells. Cofactor of androgen receptor in prostate cancer	[163,164]
EP300	Histone acetyltransferase	N	[10]	Mutated, identified in epithelial cancers. Chromosomal translocation detected in leukemia	[165,166]
FALZ	Transcription factor	N	[167]	Overexpression is highly predictive of brain metastasis in early and advanced lung cancer	[168]
GCN5L2	Histone acetyltransferase	N	[9]	None identified yet	
LOC93349 (SP140L)	Unknown	U	NA	None identified yet	
MLL	Histone methyltransferase	N	[169]	Multiple chromosomal translocations give rise to acute lymphoid leukemia and acute myeloid leukemia (AML)	[170]
PB1	Chromatin-remodeling factor	N, C	[31]	Tumor suppressor and regulator of p21 expression. Truncations and mutants in PB1 are a cause of renal cell carcinoma (RCC) and other cancer types	[171,172]
PCAF	Histone acetyltransferase	N		Inhibitor of the mitogenic activity of the adenoviral oncoprotein E1A. Acetylates b-catenin enhancing its stability. Overexpressed in some pediatric tumors	[9,173,174]
PHIP	Insulin signaling	N	[175]	Overexpression is an adverse predictor of survival in melanoma patients. Regulates PI3K/AKT signaling	[176,177]
PRKCBP1	Transcriptional regulator	N	[178]	T-cell lymphoma antigen. High mutation frequency in patients with colorectal cancers. Fusion to the RelA gene in acute erythroid leukemia and to CEP250 in breast cancer patients	[179–182]

SMARCA2	Chromatin-remodeling factor, splicing regulator	N	[183]	Loss of function mutations and downregulation of this tumor suppressor have been detected in multiple tumor types	[184–186]
SMARCA4	Chromatin-remodeling factor	N	[187]	Germline loss of function gives rise to rhabdoid tumor predisposition syndrome type 2 (RTPS2). Mutation detected in many diverse cancers	[188]
SP100	Transcriptional regulator	N, C	[39]	SP bromodomain proteins are part of PML nuclear bodies. SP100 has tumor suppressor function that is lost in some cancer types. SP100 inhibits ETS1 DNA binding and regulates ETS1 transcriptional activity. SP140 has been identified as a susceptibility gene for chronic myeloid leukemia (CML)	[39,189–191]
SP110	Transcriptional regulator	N	[40]		
SP140	Transcriptional regulator	N, C	[41]		
TAF1	Transcription initiation	N	[11]	None identified yet	
TAF1L	Transcription initiation	N	[81]	None identified yet	
TRIM24	Transcriptional regulator	N, C	[192]	Overexpression of TRIM24 correlates with tumor progression in lung cancer and survival of breast cancer patients but promotes development of hepatocellular carcinoma by formation of a complex with TRIM33 and TRIM28. TRIM24 expression increases gradually in CML throughout the accelerated phase leading to the blast phase. TRIM33 is overexpressed in colon cancer	[192–197]
TRIM28	Transcriptional regulator	N	[198]		
TRIM33	Control of transcription elongation	N	[199]		
TRIM66	Transcriptional repressor	N	[124]		
WDR9	Chromatin-remodeling factor	N	[85]	None identified yet	
ZMYND11	Transcriptional repressor	N	[200]	Copy number alteration in hematologic cancers; linked to BRCA2 pathway; mRNA predicts clinical outcome in triple negative breast tumor patients	[201–203]

[a]*Nuclear or cytoplasmic.*

10.6 DEREGULATION OF BRD FUNCTION(S) IN CANCER

BRD proteins have been found to be overexpressed in diverse tumors and their expression levels have been linked to tumor progression, sometimes predicting clinical outcome.

Overexpression of TRIM24 was found to correlate with survival of breast cancer patients [193], while in NSCLC high expression levels were found in tumor cells when compared to normal lung epithelium [194] and knockdown of TRIM24 inhibited cell-cycle progression and induced apoptosis. A link to CML has also been established, with TRIM24 expression increasing gradually through the accelerated phase to the blast phase [195], suggesting that TRIM24 expression is intrinsic to CML progression. In the same study it has also been observed that TRIM24 is overexpressed in the CD34-positive compartment, suggesting that the signaling pathways governing CML progression and CML stem cell homeostasis may be related. Intriguingly, a tumor suppressor role has been reported for TRIM24 in liver, where knockout mice were found to be predisposed to the development of both spontaneous and chemically induced hepatocellular carcinoma [196].

ATAD2 is overexpressed in a number of tumors, particularly in gastrointestinal tumors, large B-cell lymphoma, hepatocellular carcinoma, and breast and lung cancers [204–207], correlating with poor prognosis in breast and lung cancer patients, while high expression levels have been linked to distant recurrence [208]. Genetic knockouts showed that ATAD2 is essential for proliferation and survival of tumor cells and for promoting cell proliferation, survival, and cell migration, in agreement with its key function regulating growth-promoting transcription factors.

The testis-specific BET BRDT has been also found overexpressed in digestive tract tumors [209], as well as NSCLC, where almost half of the biopsies examined showed high expression of this protein [158].

10.7 BRDs AS REGULATORS OF TUMOR-PROMOTING GENES

BRD-containing proteins play key roles as regulators of transcriptional programs that lead to cell proliferation. Members of the BET subfamily regulate the expression of the mitotic kinase *Aurora B*, *NF-κB*, and *c-MYC*, as well as antiapoptotic genes such as *BCL2*. Small molecule inhibition of BET proteins first established a link to *c-MYC* repression in diverse tumors [148,149, 210–223].

The selective downregulation of the transcription of growth-promoting and antiapoptotic genes as well as the highly context-dependent effects of BET inhibition on transcription in diverse tumor cells is difficult to explain by displacement of BET proteins from gene promoters. Recent ChIP-seq data showed that the BET family member BRD4 and the mediator complex co-occupy a large number of enhancer regions as well as a small set of exceptionally large superenhancers that increase expression of key transcription factors, such as the *Myc* oncogene, in cancer, while inhibition of BRD4 bromodomains by BET inhibitors resulted in preferential displacement of BRD4 from superenhancers. Intriguingly, superenhancers seem to be responsible for the increased expression of key oncogenic drivers in diverse tumors providing a strategy for the selective targeting of oncogenes aberrantly expressed in cancer by employing BET inhibitors [224–226].

10.8 BRD-CONTAINING PROTEINS AS TUMOR SUPPRESSORS

SMARCA4, a central component of the SWI/SNF chromatin-remodeling complex, has been identified as a tumor suppressor gene [227,228]. Several rhabdoid tumors were found to carry inactivating mutations, while *SMARCA4* expression is silenced in many human tumor cell lines and tumor tissue. It is noteworthy that the frequency of SMARCA4 loss varies significantly across different tumor types and it is particularly frequent in lung cancer [229]. In addition, *SMARCA4* expression does not seem to follow *c-MYC* amplification, suggesting that the two factors may play a common role in lung cancer. Indeed, SMARCA4 was shown to inhibit c-MYC-mediated transcription in HeLa cells [230], while its loss correlates with the aggressiveness of lung cancers [184] and its inactivation is often accompanied by loss of the related gene product SMARCA2 [231], suggesting that the combined loss of the ATPase activity of both proteins may lead to the observed aggressive phenotype of these tumors. Cells that express a mutant *SMARCA4* gene lacking ATPase activity increase in both overall volume and nuclear size and also in area of attachment [232]. However, the two proteins are not functionally redundant, despite having a high degree of homology and several overlapping functions [229]. SMARCA4 has also been implicated as a tumor suppressor in melanoma, binding to the melanoma susceptibility gene *p16INK4a* [233]. Apart from SMARCAs, other components of the SWI/SNF complex are frequently mutated, silenced or truncated in tumors, including the polybromo protein PB1, a component of the PBAF subclass of SWI/SNF complexes, found to be mutated in a high percentage of RCC [171]. Genetic knockdown of more than 60% of PB1 resulted in a significant increase in proliferation in renal carcinoma cell lines, supporting PB1's role as a tumor suppressor. The expression of the zinc finger and MYND-containing protein 11 (ZMYND11) was also found to correlate with survival of 120 triple negative breast cancer patients suggesting it has a tumor suppressor role [201].

10.9 ONCOGENIC FUSIONS INVOLVING BRDs

BRDs have been identified in highly oncogenic fusion products, with NUT midline carcinoma (NMC) being the best-characterized cancer with a chromosomal translocation, involving the *NUT* gene, of unknown function located on chromosome 15q14, which is found fused to *BRD4* on chromosome 19p13.1 (or sometimes to *BRD3* on chromosome 9q34.2), creating a *BRD4-NUT* fusion gene whose product is driven by the *BRD4* promoter and resulting in aggressive carcinomas that have no organ or tissue of origin [234]. The fusion gene product contains the N-terminus of BRD4 (or BRD3) including both BRDs and essentially the entire coding region of NUT [235]. NMCs are rare, highly aggressive and poorly differentiated squamous cell carcinomas, arising in the midline of the body, head, neck, and mediastinum, although recently BET-NUT oncoproteins were also found in cancers with different tissue origin [150]. NUT is largely unstructured and contains an acidic region responsible for binding to EP300 and under normal conditions it is known to shuttle between the nucleus and the cytoplasm. Its fusion to BET members is localized in the nucleus, suggesting that the BRD portion of the BET part of the fusion is responsible for tethering the fusion product to chromatin, resulting in a knock-on effect on transcription. Genetic knockdown of BRD4-NUT by siRNA in NMC cells or chemical inhibition of its interaction with chromatin result in terminal squamous differentiation and in G1 growth arrest, supporting the notion that the BET-NUT fusion maintains an undifferentiated state of NMC cells [235,236]. Data suggest that the BRD4-NUT fusion acts to recruit the HAT EP300, resulting in local hyperacetylation, which facilitates the recruitment of additional BRD4, leading to transcriptional stimulation of proproliferative and antidifferentiation genes, such as *Myc*, while maintaining their expression [237,238].

Chromosomal aberrations have also been identified in leukemia involving the *CREBBP/EP300* locus [239]. The first translocation characterized was the t(8,16)(p11,p13) translocation, which fuses the gene encoding MOZ to the amino terminus of CREBBP, giving rise to the M4/M5 subtype of AML [240,241]. Although this translocation may result in activation of the acetyltransferase function in the fusion protein it is rare, accounting for only 0.4% of patients with AML. A number of MOZ fusions with EP300 have also been described [242,243]. Fusions of the mixed lineage leukemia (MLL) gene products with *Crebbp/Ep300* have also been reported in relapsed MLL after treatment with topoisomerase inhibitors [244,245]. Interestingly, most of the rearrangements involving *Crebbp* target the same 13-kb genomic interval, suggesting that this region contains an unstable genomic element accounting for the higher frequency of *Crebbp* rearrangements compared to *Ep300* [246]. Chromosomal translocations involving the BRD of BRD1 and the transcription factor and master regulator of B-cell development PAX5 have also been identified in ALL, but their role in tumorigenesis is not well understood [143]. Protein kinase C-binding protein 1 (PRKCBP1 or ZMYND8) has also been found to participate in chromosomal translocation together with RelA in acute erythroid leukemia patients, as well as with CEP250 in breast cancer patients [179,180].

10.10 INHIBITION OF BRDs BY SMALL MOLECULES

The hydrophobic nature of the acetyl-lysine recognition cavity found on BRDs has been an attractive target for the development of cell-permeable acetyl-lysine competitive drug-like molecules. Early efforts resulted in fragment-like weak binders, such as NP1 targeting the interaction of the BRD of PCAF with the HIV-1 TAT protein [247] or MS2126 and MS7972 targeting the BRD of CREBBP [248]. The most potent of these initial weak binders, MS7972, effectively suppressed recruitment of CREBBP to acetylated p53 via its BRD, modulating p53 target gene expression, including the cell-cycle inhibitor *p21*. A common feature shared by these early binders is that they do not directly engage the conserved asparagine moiety found in most BRDs in an acetyl-lysine mimetic mode. The evolution of these compounds led to ischemin, an azobenzene scaffold exhibiting low μM affinity toward the BRD of CREBBP, which efficiently protected rat cardiomyocytes against myocardial damage by reducing p53-induced apoptosis instigated by doxorubicin treatment [249]. The first potent inhibitors were disclosed in the patent literature in 2009 and were identified as antitumor agents targeting BET BRDs [250]. This discovery led to the rapid development of pan-BET inhibitors, pioneered by the triazolo-thienodiazepine JQ1 [236] and the triazolo-benzodiazepine I-BET762 [251,252], which adopt an acetyl-lysine mimetic binding mode completely blocking binding of histone peptides to BET BRDs. These initial diazepine derivatives contained a stereo center making synthesis and separation challenging, a problem that was later overcome by replacing the asymmetric carbon with a nitrogen in the equivalent triazolo-benzotriazepine pan-BET inhibitiors [253]. Substitution of the methyl-triazolo acetyl-lysine mimetic with the isosteric methyl-isoxazole allowed for the development of a number of isoxazole-based inhibitors, which exhibited good ligand efficiency and potencies [254,255], leading to I-BET151, which exhibited improved pharmacokinetic properties while blocking clonogenic growth of MLL-fusion-driven leukemia cells by altering transcriptional programs that are responsible for cell-cycle progression and apoptosis [149]. Interestingly, screening of kinase inhibitors identified several kinase tool compounds and clinically approved inhibitors that exhibited low nano-molar affinity for BRD4 and other BET BRDs, such as the PLK inhibitor BI-2536, the JAK inhibitor fedratinib (also known as TG-101348) [256], and the clinically used CDK inhibitor dinaciclib [257]. Other kinase inhibitors were found to be much weaker but still able to bind to BET proteins, such as the PI3K inhibitor LY294002 [258], the CDK1/2/5/9 inhibitor dinaciclib, and the Syk/BTK inhibitor fostamatinib [259], suggesting that kinases and BRDs, albeit diverse structurally, share binding site characteristics, hinting on the possibility of rationally designing equipotent dual inhibitors. A number of fragment-like templates targeting BET BRDs have also been developed and are summarized elsewhere [260]. Structural coverage of the BRD family allowed for the assessment of the druggability of the central acetyl-lysine binding cavity which exhibits a good level of enclosure and its composition suggested that targeting BRDs other than BETs should be possible [261]. Indeed, potent and selective inhibitors have now been disclosed for the BRDs of CREBBP [262,263], BAZ2A/B [264], and BRPF1 [265], while weaker inhibitor classes have also been developed for other BRDs [266–268]. These compounds have allowed for the validation of the function of BRD-containing proteins, shedding light on their biological role. The most advanced pan-BET inhibitors are shown in Figure 10.2.

10.11 BETTER TOOLS FOR CLINICAL APPLICATIONS

Small molecule inhibitors targeting the BET class of BRDs entered clinical validation in 2012 seeking to attenuate BET function in leukemia and NMCs [260]. Registered clinical trials that aim to target BRDs of the BET subfamily in diverse tumor settings are summarized in Table 10.2.

Initial reports from the running trials are very promising, however concerns have risen not only because of the different tissue-specific programs that BET proteins control, but also because of their involvement in cellular health and stem cell fate. While BRD2 regulates differentiation in neurones and adipose tissue [64,269], BRD3 facilitates the recruitment of GATA1 affecting hematopoiesis [270], BRDT affects spermatogenesis [66,271], and BRD4 controls transcriptional activation in nuclear compartments [272], there is recent evidence that *in vivo* silencing of BET proteins may have severe effects in multiple tissues. BRD4 silencing, for example, exhibited reversible epidermal hyperplasia, alopecia, and decreased cellular diversity and stem cell depletion in the small intestine in mice. Additionally, BRD4 silencing in intestines resulted in impaired regeneration following irradiation, suggesting that BRD4 suppression and cytotoxic therapies may induce undesirable synergistic effects [273]. BETs may also have overlapping roles, which are currently not entirely understood. For example, while BRD3 has been shown to have a role in hematopoiesis by participating in GATA1-dependent transcription, recent evidence suggests that although BRD2 and BRD4 bind to a smaller number of GATA1-related genes their depletion

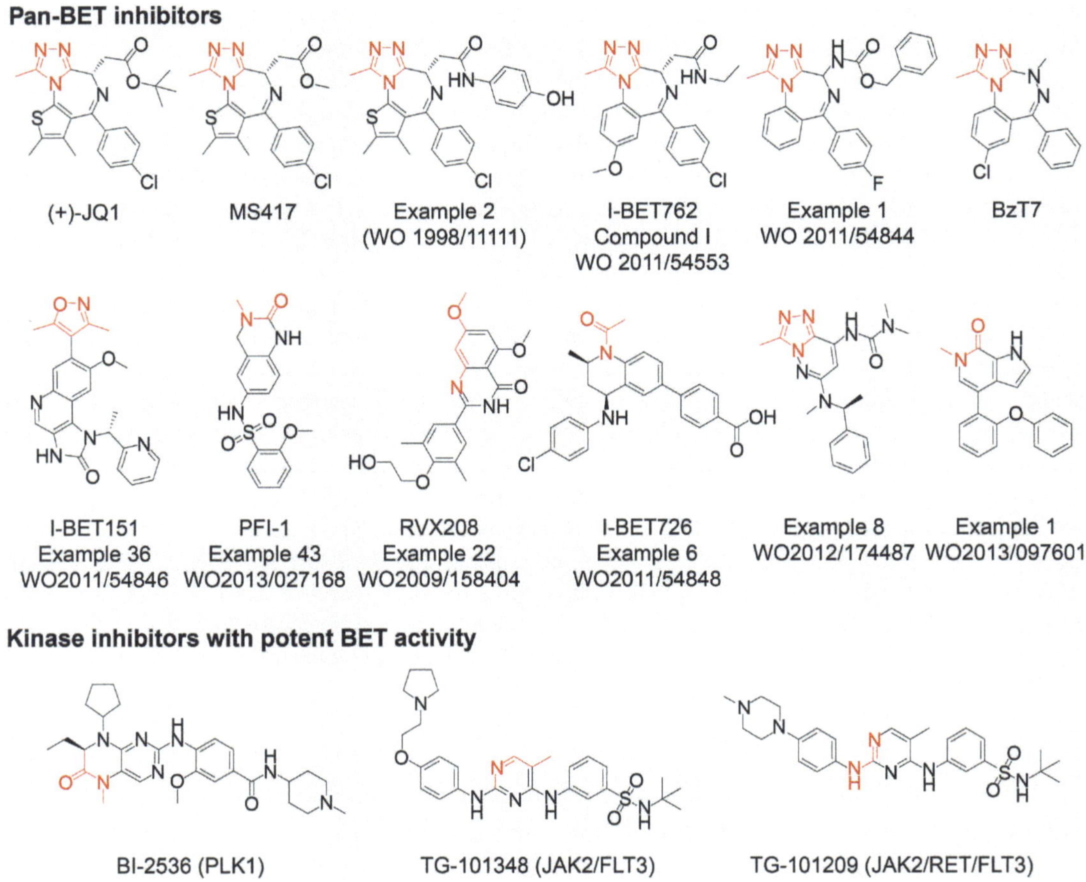

FIGURE 10.2

Advanced BET inhibitors. Different chemotypes optimized for BET BRDs are shown. All compounds exhibit high potency and selectivity for the BET subfamily of BRDs without distinguishing between proteins. Kinase inhibitors that have been identified as potent BET ligands are shown on the lower panel.

blunted erythroid gene activation, while depletion of BRD3 only affected erythroid transcription in the context of BRD2 deficiency [274].

The multidomain architecture of BRD-containing proteins, as well as the presence of several reader modules within these proteins, poses another dimension of complication in targeting the protein–protein interactions, which they participate in. For instance, selective compounds targeting the BRD of CREBBP were not able to displace the full-length protein from chromatin, suggesting that the contribution of other domains plays an important role in its participation to larger complexes [262]. This poses a challenge but at the same time offers an opportunity to target multiple domains involved in complex formation thus attenuating the transcriptional output of the system explored.

Table 10.2 Registered Clinical Trials Seeking to Target BET Proteins in Diverse Tumor Settings (www.clinicaltrials.org)

Trial	Phase	Company	Start Date	Indications	Compound	Compound Structure
NCT01587703	I	GlaxoSmithKline	03/2012	Midline carcinoma	GSK525762	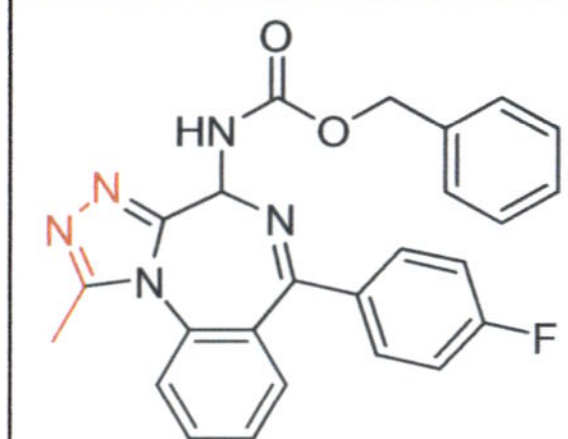 Example 1—WO 2011/54844
NCT01943851	I	GlaxoSmithKline	05/2014	Cancer		
NCT01949883	I	Constellation	09/2013	Lymphoma	CPI-0610	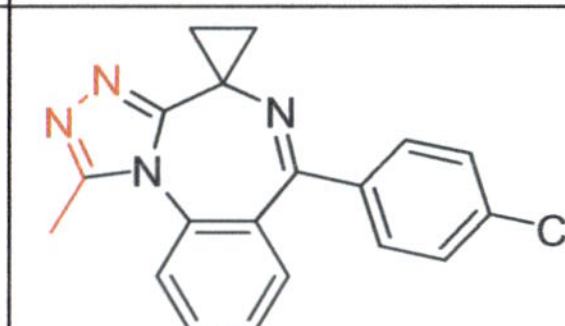 Example 71—WO 2010/075456
NCT02158858	I	Constellation	06/2014	AML, Myelodysplastic syndrome (MDS), myelodysplastic/myeloproliferative neoplasms		
NCT02157636	I	Constellation	07/2014	Multiple myeloma		
NCT01713582	I/II	OncoEthix	12/2012	AML; Other hematological malignacies	OXT015 (in-license from Tanabe Mitsubishi)	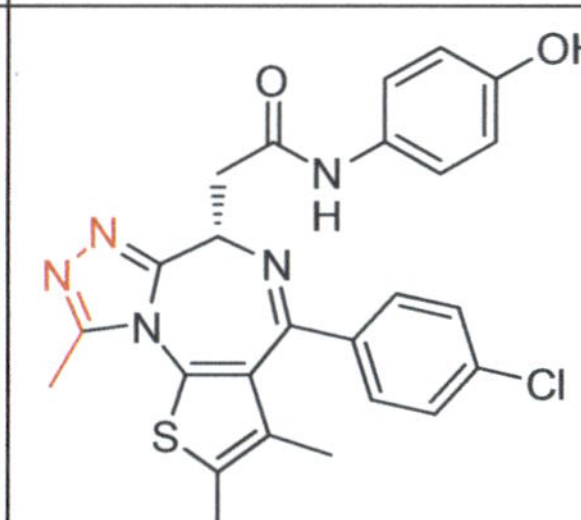 Example 1—WO 2009/84693
NCT02259114	I	OncoEthix	11/2014	NUT midline carcinoma; Triple negative breast cancer; Nonsmall-cell lung cancer with rearranged anaplastic lymphoma kinase (ALK) gene/fusion protein or KRAS mutation; Castrate-resistant prostate cancer; Pancreatic ductal adenocarcinoma		
NCT02296476	I/II	OncoEthix	11/2014	Glioblastoma multiforme		
NCT02303782	I/II	OncoEthix	01/2015	AML	OXT015 Vidaza (azacitidine)	

(Continued)

Table 10.2 Registered Clinical Trials Seeking to Target BET Proteins in Diverse Tumor Settings (www.clinicaltrials.org) ***Continued***

Trial	Phase	Company	Start Date	Indications	Compound	Compound Structure
NCT01987362	I	Tensha Therapeutics	10/2013	Solid tumors	TEN-010	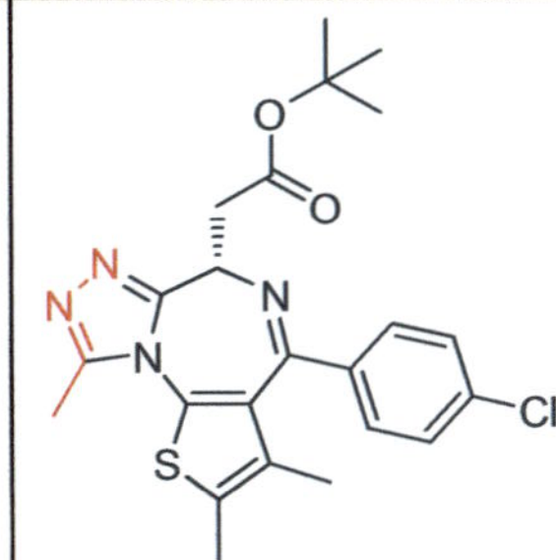 JQ1 or equivalent
NCT02308761	I	Tensha Therapeutics	10/2014	AML MDS		
NCT02369029	I	Bayer	02/2015	Neoplasms	BAY1238097	Not disclosed

Many studies have assessed the effect of BET inhibition on multiple survival pathways, leading to a rationale supporting the combination of BET inhibitors with other agents exhibiting an effect on different diseases. These efforts have successfully shown that sensitization to agents such as dexamethazone in cellular models of ALL, temozolomide (a DNA alkylation/methylation agent) in glioblastoma, the anthracycline daunorubicin in *DNMT3A*-mutated leukemia cells, the FLT3 inhibitor ponatinib in cultured and primary CD34(+) human AML blast progenitor cells, the ER degrader fulvestrant in *in vivo* breast tumor models, rapamycin in human osteosarcoma cells *in vitro* and *in vivo*, or rapamycin in pancreatic neuroendocrine tumors [215,217,275–279] is possible. These index studies have provided the early proof of concept that will hopefully lead to better clinical outcome and in enhanced targeting of different diseases. Notably, a recently registered phase I clinical trial (clinicaltrials.gov identifier: NCT02303782) has been seeking to treat AML patients by combining the triazolo-thienodiazepine OXT015 with the DNA demethylating agent azacitidine (Vidaza). In addition, the similarities that have been implied between BRDs of the BET subfamily and diverse kinases, based on the high affinity that several kinase-class inhibitors have exhibited for BET BRDs, offer a unique opportunity to explore poly-pharmacology, combining within one scaffold the effects elicited by these diverse drug targets, but the tractability of such an effort remains to be seen [256].

10.12 CONCLUSION

Protein–protein interaction modules of the BRD class play diverse roles in chromatin biology, controlling in part the transcription of key genes that affect survival. BRD-containing proteins are deregulated in disease leading to aberrant expression of proinflammatory cytokines and oncogenes. The availability of pan-BET inhibitors has allowed for the validation of this druggable class of proteins, unraveling the important biology, which depends on them. With several clinical trials currently on track and numerous new optimized scaffolds being generated and tested at large by the scientific community, our understanding of the biological significance of BRD-containing proteins is going to be augmented, leading to insight of BRD involvement in diverse diseases. The additional chemical tools that slowly start to emerge, selectively targeting different subclasses of BRD-containing proteins will allow for wider validation of these important drug targets. These efforts will inevitably lead to improved compounds that will be further developed in order to yield safe and effective novel drugs.

ACKNOWLEDGMENTS

We apologize to all authors whose important contributions could not be acknowledged due to space limitations. The authors are grateful for support for their research received from the SGC, a registered public-private partnership that receives funds from AbbVie, Bayer, Boehringer Ingelheim, Genome Canada through Ontario Genomics Institute Grant OGI-055, GlaxoSmithKline, Janssen, Lilly Canada, the Novartis Research Foundation, the Ontario Ministry of Economic Development and Innovation, Pfizer, Takeda, and the Wellcome Trust (Grant # 092809/Z/10/Z). P.F. is supported by a Wellcome Trust Career Development Fellowship (Grant # 095751/Z/11/Z).

REFERENCES

[1] Bernstein BE, Meissner A, Lander ES. The mammalian epigenome. Cell 2007;128(4):669–81.

[2] Kohli RM, Zhang Y. TET enzymes, TDG and the dynamics of DNA demethylation. Nature 2013;502 (7472):472–9.

[3] Tessarz P, Kouzarides T. Histone core modifications regulating nucleosome structure and dynamics. Nat Rev Mol Cell Biol 2014;15(11):703–8.

[4] Strahl BD, Allis CD. The language of covalent histone modifications. Nature 2000;403(6765):41–5.

[5] Choudhary C, Kumar C, Gnad F, et al. Lysine acetylation targets protein complexes and co-regulates major cellular functions. Science 2009;325(5942):834–40.

[6] Zhao S, Xu W, Jiang W, et al. Regulation of cellular metabolism by protein lysine acetylation. Science 2010;327(5968):1000–4.

[7] Marmorstein R, Trievel RC. Histone modifying enzymes: structures, mechanisms, and specificities. Biochim Biophys Acta 2009;1789(1):58–68.

[8] Yang X, Yu W, Shi L, et al. HAT4, a Golgi apparatus-anchored B-type histone acetyltransferase, acetylates free histone H4 and facilitates chromatin assembly. Mol Cell 2011;44(1):39–50.

[9] Yang XJ, Ogryzko VV, Nishikawa J, Howard BH, Nakatani Y. A p300/CBP-associated factor that competes with the adenoviral oncoprotein E1A. Nature 1996;382(6589):319–24.

[10] Kalkhoven E. CBP and p300: HATs for different occasions. Biochem Pharmacol 2004;68(6):1145–55.

[11] Wassarman DA, Sauer F. TAF(II)250: a transcription toolbox. J Cell Sci 2001;114(Pt 16):2895–902.

[12] Avvakumov N, Cote J. The MYST family of histone acetyltransferases and their intimate links to cancer. Oncogene 2007;26(37):5395–407.

[13] Kouzarides T. Acetylation: a regulatory modification to rival phosphorylation? EMBO J 2000;19(6): 1176–9.

[14] Parra M, Verdin E. Regulatory signal transduction pathways for class IIa histone deacetylases. Curr Opin Pharmacol 2010;10(4):454–60.

[15] Barneda-Zahonero B, Parra M. Histone deacetylases and cancer. Mol Oncol 2012;6:579–89.

[16] Lee JH, Choy ML, Marks PA. Mechanisms of resistance to histone deacetylase inhibitors. Adv Cancer Res 2012;116:39–86.

[17] Hagelkruys A, Sawicka A, Rennmayr M, Seiser C. The biology of HDAC in cancer: the nuclear and epigenetic components. Handb Exp Pharmacol 2011;206:13–37.

[18] Magnaghi-Jaulin L, Groisman R, Naguibneva I, et al. Retinoblastoma protein represses transcription by recruiting a histone deacetylase. Nature 1998;391(6667):601–5.

[19] Ropero S, Fraga MF, Ballestar E, et al. A truncating mutation of HDAC2 in human cancers confers resistance to histone deacetylase inhibition. Nat Genet 2006;38(5):566–9.

[20] Buurman R, Gurlevik E, Schaffer V, et al. Histone deacetylases activate hepatocyte growth factor signaling by repressing microRNA-449 in hepatocellular carcinoma cells. Gastroenterology 2012;143 (3):811–20.

[21] Tohyama S, Tomura A, Ikeda N, et al. Discovery and characterization of NK13650s, naturally occurring p300-selective histone acetyltransferase inhibitors. J Org Chem 2012;77(20):9044–52.

[22] Bolden JE, Peart MJ, Johnstone RW. Anticancer activities of histone deacetylase inhibitors. Nat Rev Drug Discov 2006;5(9):769–84.

[23] Garber K. HDAC inhibitors overcome first hurdle. Nat Biotechnol 2007;25(1):17–19.

[24] McConkey DJ, White M, Yan W. HDAC inhibitor modulation of proteotoxicity as a therapeutic approach in cancer. Adv Cancer Res 2012;116:131–63.

[25] Khan O, La Thangue NB. HDAC inhibitors in cancer biology: emerging mechanisms and clinical applications. Immunol Cell Biol 2012;90(1):85–94.

[26] Licciardi PV, Ververis K, Tang ML, El-Osta A, Karagiannis TC. Immunomodulatory effects of histone deacetylase inhibitors. Curr Mol Med 2013;13(4):640–7.
[27] Monneret C. Histone deacetylase inhibitors. Eur J Med Chem 2005;40(1):1–13.
[28] Tamkun JW, Deuring R, Scott MP, et al. Brahma: a regulator of Drosophila homeotic genes structurally related to the yeast transcriptional activator SNF2/SWI2. Cell 1992;68(3):561–72.
[29] Haynes SR, Dollard C, Winston F, Beck S, Trowsdale J, Dawid IB. The bromodomain: a conserved sequence found in human, Drosophila and yeast proteins. Nucleic Acids Res 1992;20(10):2603.
[30] Li Y, Wen H, Xi Y, et al. AF9 YEATS domain links histone acetylation to DOT1L-mediated H3K79 methylation. Cell 2014;159(3):558–71.
[31] Xue Y, Canman JC, Lee CS, et al. The human SWI/SNF-B chromatin-remodeling complex is related to yeast rsc and localizes at kinetochores of mitotic chromosomes. Proc Natl Acad Sci USA 2000;97 (24):13015–20.
[32] Kadoch C, Hargreaves DC, Hodges C, et al. Proteomic and bioinformatic analysis of mammalian SWI/ SNF complexes identifies extensive roles in human malignancy. Nat Genet 2013;45(6):592–601.
[33] Cavellan E, Asp P, Percipalle P, Farrants AK. The WSTF-SNF2h chromatin remodeling complex interacts with several nuclear proteins in transcription. J Biol Chem 2006;281(24):16264–71.
[34] Trotter KW, Archer TK. The BRG1 transcriptional coregulator. Nucl Recept Signal 2008;6:e004.
[35] Jacobson RH, Ladurner AG, King DS, Tjian R. Structure and function of a human TAFII250 double bromodomain module. Science 2000;288(5470):1422–5.
[36] Venturini L, You J, Stadler M, et al. TIF1gamma, a novel member of the transcriptional intermediary factor 1 family. Oncogene 1999;18(5):1209–17.
[37] Wu SY, Chiang CM. The double bromodomain-containing chromatin adaptor Brd4 and transcriptional regulation. J Biol Chem 2007;282(18):13141–5.
[38] Gregory GD, Vakoc CR, Rozovskaia T, et al. Mammalian ASH1L is a histone methyltransferase that occupies the transcribed region of active genes. Mol Cell Biol 2007;27(24):8466–79.
[39] Yordy JS, Moussa O, Pei H, Chaussabel D, Li R, Watson DK. SP100 inhibits ETS1 activity in primary endothelial cells. Oncogene 2005;24(5):916–31.
[40] Bloch DB, Nakajima A, Gulick T, et al. Sp110 localizes to the PML-Sp100 nuclear body and may function as a nuclear hormone receptor transcriptional coactivator. Mol Cell Biol 2000;20(16):6138–46.
[41] Zong RT, Das C, Tucker PW. Regulation of matrix attachment region-dependent, lymphocyte-restricted transcription through differential localization within promyelocytic leukemia nuclear bodies. EMBO J 2000;19(15):4123–33.
[42] Muller S, Filippakopoulos P, Knapp S. Bromodomains as therapeutic targets. Expert Rev Mol Med 2011;13:e29.
[43] Filippakopoulos P, Picaud S, Mangos M, et al. Histone recognition and large-scale structural analysis of the human bromodomain family. Cell 2012;149(1):214–31.
[44] Aasland R, Gibson TJ, Stewart AF. The PHD finger: implications for chromatin-mediated transcriptional regulation. Trends Biochem Sci 1995;20(2):56–9.
[45] Chambers AL, Pearl LH, Oliver AW, Downs JA. The BAH domain of Rsc2 is a histone H3 binding domain. Nucleic Acids Res 2013;41(19):9168–82.
[46] Stec I, Wright TJ, van Ommen GJ, et al. WHSC1, a 90 kb SET domain-containing gene, expressed in early development and homologous to a Drosophila dysmorphy gene maps in the Wolf-Hirschhorn syndrome critical region and is fused to IgH in t(4;14) multiple myeloma. Hum Mol Genet 1998;7 (7):1071–82.
[47] Gibson TJ, Ramu C, Gemund C, Aasland R. The APECED polyglandular autoimmune syndrome protein, AIRE-1, contains the SAND domain and is probably a transcription factor. Trends Biochem Sci 1998;23(7):242–4.

[48] Radhakrishnan I, Perez-Alvarado GC, Parker D, Dyson HJ, Montminy MR, Wright PE. Solution structure of the KIX domain of CBP bound to the transactivation domain of CREB: a model for activator:coactivator interactions. Cell 1997;91(6):741–52.

[49] Moriniere J, Rousseaux S, Steuerwald U, et al. Cooperative binding of two acetylation marks on a histone tail by a single bromodomain. Nature 2009;461(7264):664–8.

[50] Tanaka Y, Naruse I, Hongo T, et al. Extensive brain hemorrhage and embryonic lethality in a mouse null mutant of CREB-binding protein. Mech Dev 2000;95(1–2):133–45.

[51] Yao TP, Oh SP, Fuchs M, et al. Gene dosage-dependent embryonic development and proliferation defects in mice lacking the transcriptional integrator p300. Cell 1998;93(3):361–72.

[52] Maurice T, Duclot F, Meunier J, et al. Altered memory capacities and response to stress in p300/CBP-associated factor (PCAF) histone acetylase knockout mice. Neuropsychopharmacology 2008;33(7):1584–602.

[53] Xu W, Edmondson DG, Evrard YA, Wakamiya M, Behringer RR, Roth SY. Loss of Gcn5l2 leads to increased apoptosis and mesodermal defects during mouse development. Nat Genet 2000;26(2):229–32.

[54] Bu P, Evrard YA, Lozano G, Dent SY. Loss of Gcn5 acetyltransferase activity leads to neural tube closure defects and exencephaly in mouse embryos. Mol Cell Biol 2007;27(9):3405–16.

[55] Nakamura T, Blechman J, Tada S, et al. huASH1 protein, a putative transcription factor encoded by a human homologue of the Drosophila ash1 gene, localizes to both nuclei and cell–cell tight junctions. Proc Natl Acad Sci USA 2000;97(13):7284–9.

[56] Yu BD, Hess JL, Horning SE, Brown GA, Korsmeyer SJ. Altered Hox expression and segmental identity in Mll-mutant mice. Nature 1995;378(6556):505–8.

[57] McMahon KA, Hiew SY, Hadjur S, et al. Mll has a critical role in fetal and adult hematopoietic stem cell self-renewal. Cell Stem Cell 2007;1(3):338–45.

[58] Mohan M, Lin C, Guest E, Shilatifard A. Licensed to elongate: a molecular mechanism for MLL-based leukaemogenesis. Nat Rev Cancer 2010;10(10):721–8.

[59] Zou JX, Revenko AS, Li LB, Gemo AT, Chen HW. ANCCA, an estrogen-regulated AAA+ ATPase coactivator for ERalpha, is required for coregulator occupancy and chromatin modification. Proc Natl Acad Sci USA 2007;104(46):18067–72.

[60] Ciro M, Prosperini E, Quarto M, et al. ATAD2 is a novel cofactor for MYC, overexpressed and amplified in aggressive tumors. Cancer Res 2009;69(21):8491–8.

[61] Revenko AS, Kalashnikova EV, Gemo AT, Zou JX, Chen HW. Chromatin loading of E2F-MLL complex by cancer-associated coregulator ANCCA via reading a specific histone mark. Mol Cell Biol 2010;30(22):5260–72.

[62] Denis GV, McComb ME, Faller DV, Sinha A, Romesser PB, Costello CE. Identification of transcription complexes that contain the double bromodomain protein Brd2 and chromatin remodeling machines. J Proteome Res 2006;5(3):502–11.

[63] LeRoy G, Rickards B, Flint SJ. The double bromodomain proteins Brd2 and Brd3 couple histone acetylation to transcription. Mol Cell 2008;30(1):51–60.

[64] Wang F, Liu H, Blanton WP, Belkina A, Lebrasseur NK, Denis GV. Brd2 disruption in mice causes severe obesity without Type 2 diabetes. Biochem J 2010;425(1):71–83.

[65] Dey A, Ellenberg J, Farina A, et al. A bromodomain protein, MCAP, associates with mitotic chromosomes and affects G(2)-to-M transition. Mol Cell Biol 2000;20(17):6537–49.

[66] Gaucher J, Boussouar F, Montellier E, et al. Bromodomain-dependent stage-specific male genome programming by Brdt. EMBO J 2012;31(19):3809–20.

[67] Dey A, Chitsaz F, Abbasi A, Misteli T, Ozato K. The double bromodomain protein Brd4 binds to acetylated chromatin during interphase and mitosis. Proc Natl Acad Sci USA 2003;100(15):8758–63.

[68] Kanno T, Kanno Y, Siegel RM, Jang MK, Lenardo MJ, Ozato K. Selective recognition of acetylated histones by bromodomain proteins visualized in living cells. Mol Cell 2004;13(1):33–43.

[69] Maruyama T, Farina A, Dey A, et al. A mammalian bromodomain protein, Brd4, interacts with replication factor C and inhibits progression to S phase. Mol Cell Biol 2002;22(18):6509–20.

[70] Wu SY, Lee AY, Hou SY, et al. Brd4 links chromatin targeting to HPV transcriptional silencing. Genes Dev 2006;20(17):2383–96.

[71] Houzelstein D, Bullock SL, Lynch DE, Grigorieva EF, Wilson VA, Beddington RS. Growth and early postimplantation defects in mice deficient for the bromodomain-containing protein Brd4. Mol Cell Biol 2002;22(11):3794–802.

[72] Gyuris A, Donovan DJ, Seymour KA, et al. The chromatin-targeting protein Brd2 is required for neural tube closure and embryogenesis. Biochim Biophys Acta 2009;1789(5):413–21.

[73] Shang E, Wang X, Wen D, Greenberg DA, Wolgemuth DJ. Double bromodomain-containing gene Brd2 is essential for embryonic development in mouse. Dev Dyn 2009;238(4):908–17.

[74] Shang E, Nickerson HD, Wen D, Wang X, Wolgemuth DJ. The first bromodomain of Brdt, a testis-specific member of the BET sub-family of double-bromodomain-containing proteins, is essential for male germ cell differentiation. Development 2007;134(19):3507–15.

[75] Aston KI, Krausz C, Laface I, Ruiz-Castane E, Carrell DT. Evaluation of 172 candidate polymorphisms for association with oligozoospermia or azoospermia in a large cohort of men of European descent. Hum Reprod 2010;25(6):1383–97.

[76] Li AG, Piluso LG, Cai X, Gadd BJ, Ladurner AG, Liu X. An acetylation switch in p53 mediates holo-TFIID recruitment. Mol Cell 2007;28(3):408–21.

[77] Lin CY, Tuan J, Scalia P, Bui T, Comai L. The cell cycle regulatory factor TAF1 stimulates ribosomal DNA transcription by binding to the activator UBF. Curr Biol 2002;12(24):2142–6.

[78] Buchmann AM, Skaar JR, DeCaprio JA. Activation of a DNA damage checkpoint response in a TAF1-defective cell line. Mol Cell Biol 2004;24(12):5332–9.

[79] Deato MD, Tjian R. Switching of the core transcription machinery during myogenesis. Genes Dev 2007;21(17):2137–49.

[80] Kimura J, Nguyen ST, Liu H, Taira N, Miki Y, Yoshida K. A functional genome-wide RNAi screen identifies TAF1 as a regulator for apoptosis in response to genotoxic stress. Nucleic Acids Res 2008;36(16):5250–9.

[81] Wang PJ, Page DC. Functional substitution for TAF(II)250 by a retroposed homolog that is expressed in human spermatogenesis. Hum Mol Genet 2002;11(19):2341–6.

[82] Ramos VC, Vidal-Taboada J, Bergonon S, et al. Characterisation and expression analysis of the WDR9 gene, located in the Down critical region-2 of the human chromosome 21. Biochim Biophys Acta 2002;1577(3):377–83.

[83] Kalla C, Nentwich H, Schlotter M, et al. Translocation t(X;11)(q13;q23) in B-cell chronic lymphocytic leukemia disrupts two novel genes. Genes Chromosomes Cancer 2005;42(2):128–43.

[84] Philipps DL, Wigglesworth K, Hartford SA, et al. The dual bromodomain and WD repeat-containing mouse protein BRWD1 is required for normal spermiogenesis and the oocyte-embryo transition. Dev Biol 2008;317(1):72–82.

[85] Huang H, Rambaldi I, Daniels E, Featherstone M. Expression of the Wdr9 gene and protein products during mouse development. Dev Dyn 2003;227(4):608–14.

[86] Muller P, Kuttenkeuler D, Gesellchen V, Zeidler MP, Boutros M. Identification of JAK/STAT signalling components by genome-wide RNA interference. Nature 2005;436(7052):871–5.

[87] Neigeborn L, Carlson M. Genes affecting the regulation of SUC2 gene expression by glucose repression in *Saccharomyces cerevisiae*. Genetics 1984;108(4):845–58.

[88] Stern M, Jensen R, Herskowitz I. Five SWI genes are required for expression of the HO gene in yeast. J Mol Biol 1984;178(4):853–68.

[89] Winston F, Carlson M. Yeast SNF/SWI transcriptional activators and the SPT/SIN chromatin connection. Trends Genet 1992;8(11):387–91.

[90] Sudarsanam P, Iyer VR, Brown PO, Winston F. Whole-genome expression analysis of snf/swi mutants of Saccharomyces cerevisiae. Proc Natl Acad Sci USA 2000;97(7):3364–9.

[91] Ho L, Jothi R, Ronan JL, Cui K, Zhao K, Crabtree GR. An embryonic stem cell chromatin remodeling complex, esBAF, is an essential component of the core pluripotency transcriptional network. Proc Natl Acad Sci USA 2009;106(13):5187–91.

[92] Singhal N, Graumann J, Wu G, et al. Chromatin-remodeling components of the BAF complex facilitate reprogramming. Cell 2010;141(6):943–55.

[93] Link KA, Burd CJ, Williams E, et al. BAF57 governs androgen receptor action and androgen-dependent proliferation through SWI/SNF. Mol Cell Biol 2005;25(6):2200–15.

[94] Aoyagi S, Trotter KW, Archer TK. ATP-dependent chromatin remodeling complexes and their role in nuclear receptor-dependent transcription *in vivo*. Vitam Horm 2005;70:281–307.

[95] Debril MB, Gelman L, Fayard E, Annicotte JS, Rocchi S, Auwerx J. Transcription factors and nuclear receptors interact with the SWI/SNF complex through the BAF60c subunit. J Biol Chem 2004;279 (16):16677–86.

[96] Peterson CL, Dingwall A, Scott MP. Five SWI/SNF gene products are components of a large multisubunit complex required for transcriptional enhancement. Proc Natl Acad Sci USA 1994;91(8):2905–8.

[97] Subtil-Rodriguez A, Reyes JC. BRG1 helps RNA polymerase II to overcome a nucleosomal barrier during elongation, *in vivo*. EMBO Rep 2010;11(10):751–7.

[98] Pal S, Yun R, Datta A, et al. mSin3A/histone deacetylase 2- and PRMT5-containing Brg1 complex is involved in transcriptional repression of the Myc target gene cad. Mol Cell Biol 2003;23 (21):7475–87.

[99] Lemon B, Inouye C, King DS, Tjian R. Selectivity of chromatin-remodelling cofactors for ligand-activated transcription. Nature 2001;414(6866):924–8.

[100] Ryme J, Asp P, Bohm S, Cavellan E, Farrants AK. Variations in the composition of mammalian SWI/SNF chromatin remodelling complexes. J Cell Biochem 2009;108(3):565–76.

[101] Burrows AE, Smogorzewska A, Elledge SJ. Polybromo-associated BRG1-associated factor components BRD7 and BAF180 are critical regulators of p53 required for induction of replicative senescence. Proc Natl Acad Sci USA 2010;107(32):14280–5.

[102] Wang Z, Zhai W, Richardson JA, et al. Polybromo protein BAF180 functions in mammalian cardiac chamber maturation. Genes Dev 2004;18(24):3106–16.

[103] Huang X, Gao X, Diaz-Trelles R, Ruiz-Lozano P, Wang Z. Coronary development is regulated by ATP-dependent SWI/SNF chromatin remodeling component BAF180. Dev Biol 2008;319(2):258–66.

[104] Jones MH, Hamana N, Nezu J, Shimane M. A novel family of bromodomain genes. Genomics 2000;63 (1):40–5.

[105] Bowser R, Giambrone A, Davies P. FAC1, a novel gene identified with the monoclonal antibody Alz50, is developmentally regulated in human brain. Dev Neurosci 1995;17(1):20–37.

[106] Bochar DA, Savard J, Wang W, et al. A family of chromatin remodeling factors related to Williams syndrome transcription factor. Proc Natl Acad Sci USA 2000;97(3):1038–43.

[107] Poot RA, Dellaire G, Hulsmann BB, et al. HuCHRAC, a human ISWI chromatin remodelling complex contains hACF1 and two novel histone-fold proteins. EMBO J 2000;19(13):3377–87.

[108] Collins N, Poot RA, Kukimoto I, Garcia-Jimenez C, Dellaire G, Varga-Weisz PD. An ACF1-ISWI chromatin-remodeling complex is required for DNA replication through heterochromatin. Nat Genet 2002;32(4):627–32.

[109] Strohner R, Nemeth A, Jansa P, et al. NoRC—a novel member of mammalian ISWI-containing chromatin remodeling machines. EMBO J 2001;20(17):4892–900.
[110] Zhou Y, Grummt I. The PHD finger/bromodomain of NoRC interacts with acetylated histone H4K16 and is sufficient for rDNA silencing. Curr Biol 2005;15(15):1434–8.
[111] Meroni G, Diez-Roux G. TRIM/RBCC, a novel class of "single protein RING finger" E3 ubiquitin ligases. Bioessays 2005;27(11):1147–57.
[112] Le Douarin B, Nielsen AL, Garnier JM, et al. A possible involvement of TIF1 alpha and TIF1 beta in the epigenetic control of transcription by nuclear receptors. EMBO J 1996;15(23):6701–15.
[113] Le Douarin B, You J, Nielsen AL, Chambon P, Losson R. TIF1alpha: a possible link between KRAB zinc finger proteins and nuclear receptors. J Steroid Biochem Mol Biol 1998;65(1–6):43–50.
[114] Remboutsika E, Lutz Y, Gansmuller A, Vonesch JL, Losson R, Chambon P. The putative nuclear receptor mediator TIF1alpha is tightly associated with euchromatin. J Cell Sci 1999;112(Pt 11):1671–83.
[115] Peng H, Feldman I, Rauscher III FJ. Hetero-oligomerization among the TIF family of RBCC/TRIM domain-containing nuclear cofactors: a potential mechanism for regulating the switch between coactivation and corepression. J Mol Biol 2002;320(3):629–44.
[116] Friedman JR, Fredericks WJ, Jensen DE, et al. KAP-1, a novel corepressor for the highly conserved KRAB repression domain. Genes Dev 1996;10(16):2067–78.
[117] Cammas F, Oulad-Abdelghani M, Vonesch JL, Huss-Garcia Y, Chambon P, Losson R. Cell differentiation induces TIF1beta association with centromeric heterochromatin via an HP1 interaction. J Cell Sci 2002;115(Pt 17):3439–48.
[118] Li Y, Kirschmann DA, Wallrath LL. Does heterochromatin protein 1 always follow code? Proc Natl Acad Sci USA 2002;99(Suppl. 4):16462–9.
[119] Bartova E, Pachernik J, Kozubik A, Kozubek S. Differentiation-specific association of HP1alpha and HP1beta with chromocentres is correlated with clustering of TIF1beta at these sites. Histochem Cell Biol 2007;127(4):375–88.
[120] Cammas F, Herzog M, Lerouge T, Chambon P, Losson R. Association of the transcriptional corepressor TIF1beta with heterochromatin protein 1 (HP1): an essential role for progression through differentiation. Genes Dev 2004;18(17):2147–60.
[121] Dupont S, Zacchigna L, Cordenonsi M, et al. Germ-layer specification and control of cell growth by Ectodermin, a Smad4 ubiquitin ligase. Cell 2005;121(1):87–99.
[122] He W, Dorn DC, Erdjument-Bromage H, Tempst P, Moore MA, Massague J. Hematopoiesis controlled by distinct TIF1gamma and Smad4 branches of the TGFbeta pathway. Cell 2006;125(5):929–41.
[123] Kim J, Kaartinen V. Generation of mice with a conditional allele for Trim33. Genesis 2008;46 (6):329–33.
[124] Khetchoumian K, Teletin M, Mark M, et al. TIF1delta, a novel HP1-interacting member of the transcriptional intermediary factor 1 (TIF1) family expressed by elongating spermatids. J Biol Chem 2004;279(46):48329–41.
[125] Zhong S, Hu P, Ye TZ, Stan R, Ellis NA, Pandolfi PP. A role for PML and the nuclear body in genomic stability. Oncogene 1999;18(56):7941–7.
[126] Boisvert FM, Hendzel MJ, Bazett-Jones DP. Promyelocytic leukemia (PML) nuclear bodies are protein structures that do not accumulate RNA. J Cell Biol 2000;148(2):283–92.
[127] Wu WS, Vallian S, Seto E, et al. The growth suppressor PML represses transcription by functionally and physically interacting with histone deacetylases. Mol Cell Biol 2001;21(7):2259–68.
[128] Boisvert FM, Kruhlak MJ, Box AK, Hendzel MJ, Bazett-Jones DP. The transcription coactivator CBP is a dynamic component of the promyelocytic leukemia nuclear body. J Cell Biol 2001;152 (5):1099–106.

[129] Wang J, Shiels C, Sasieni P, et al. Promyelocytic leukemia nuclear bodies associate with transcriptionally active genomic regions. J Cell Biol 2004;164(4):515–26.
[130] Dellaire G, Ching RW, Ahmed K, et al. Promyelocytic leukemia nuclear bodies behave as DNA damage sensors whose response to DNA double-strand breaks is regulated by NBS1 and the kinases ATM, Chk2, and ATR. J Cell Biol 2006;175(1):55–66.
[131] Ishov AM, Maul GG. The periphery of nuclear domain 10 (ND10) as site of DNA virus deposition. J Cell Biol 1996;134(4):815–26.
[132] Bernardi R, Pandolfi PP. Structure, dynamics and functions of promyelocytic leukaemia nuclear bodies. Nat Rev Mol Cell Biol 2007;8(12):1006–16.
[133] Stinton LM, Swain M, Myers RP, Shaheen AA, Fritzler MJ. Autoantibodies to GW bodies and other autoantigens in primary biliary cirrhosis. Clin Exp Immunol 2011;163(2):147–56.
[134] Liu J, Lee W, Jiang Z, et al. Genome and transcriptome sequencing of lung cancers reveal diverse mutational and splicing events. Genome Res 2012;22:2315–27.
[135] Fouret R, Laffaire J, Hofman P, et al. A comparative and integrative approach identifies ATPase family, AAA domain containing 2 as a likely driver of cell proliferation in lung adenocarcinoma. Clin Cancer Res 2012;18(20):5606–16.
[136] Kalashnikova EV, Revenko AS, Gemo AT, et al. ANCCA/ATAD2 overexpression identifies breast cancer patients with poor prognosis, acting to drive proliferation and survival of triple-negative cells through control of B-Myb and EZH2. Cancer Res 2010;70(22):9402–12.
[137] Leachman NT, Brellier F, Ferralli J, Chiquet-Ehrismann R, Tucker RP. ATAD2B is a phylogenetically conserved nuclear protein expressed during neuronal differentiation and tumorigenesis. Dev Growth Differ 2010;52(9):747–55.
[138] Racki LR, Yang JG, Naber N, et al. The chromatin remodeller ACF acts as a dimeric motor to space nucleosomes. Nature 2009;462(7276):1016–21.
[139] Hennig EE, Mikula M, Rubel T, Dadlez M, Ostrowski J. Comparative kinome analysis to identify putative colon tumor biomarkers. J Mol Med (Berl) 2012;90(4):447–56.
[140] Zhou Y, Schmitz KM, Mayer C, Yuan X, Akhtar A, Grummt I. Reversible acetylation of the chromatin remodelling complex NoRC is required for non-coding RNA-dependent silencing. Nat Cell Biol 2009;11(8):1010–16.
[141] Panagopoulos I, Strombeck B, Isaksson M, Heldrup J, Olofsson T, Johansson B. Fusion of ETV6 with an intronic sequence of the BAZ2A gene in a paediatric pre-B acute lymphoblastic leukaemia with a cryptic chromosome 12 rearrangement. Br J Haematol 2006;133(3):270–5.
[142] Ullah M, Pelletier N, Xiao L, et al. Molecular architecture of quartet MOZ/MORF histone acetyltransferase complexes. Mol Cell Biol 2008;28(22):6828–43.
[143] Nebral K, Denk D, Attarbaschi A, et al. Incidence and diversity of PAX5 fusion genes in childhood acute lymphoblastic leukemia. Leukemia 2009;23(1):134–43.
[144] Viejo-Borbolla A, Ottinger M, Bruning E, et al. Brd2/RING3 interacts with a chromatin-binding domain in the Kaposi's Sarcoma-associated herpesvirus latency-associated nuclear antigen 1 (LANA-1) that is required for multiple functions of LANA-1. J Virol 2005;79(21):13618–29.
[145] Zhou M, Peng C, Nie XM, et al. Expression of BRD7-interacting proteins, BRD2 and BRD3, in nasopharyngeal carcinoma tissues. Ai Zheng 2003;22(2):123–7.
[146] French CA, Ramirez CL, Kolmakova J, et al. BRD-NUT oncoproteins: a family of closely related nuclear proteins that block epithelial differentiation and maintain the growth of carcinoma cells. Oncogene 2008;27(15):2237–42.
[147] Yang Z, He N, Zhou Q. Brd4 recruits P-TEFb to chromosomes at late mitosis to promote G1 gene expression and cell cycle progression. Mol Cell Biol 2008;28(3):967–76.

[148] Delmore JE, Issa GC, Lemieux ME, et al. BET bromodomain inhibition as a therapeutic strategy to target c-Myc. Cell 2011;146(6):904–17.
[149] Dawson MA, Prinjha RK, Dittmann A, et al. Inhibition of BET recruitment to chromatin as an effective treatment for MLL-fusion leukaemia. Nature 2011;478(7370):529–33.
[150] French CA. Pathogenesis of NUT midline carcinoma. Annu Rev Pathol 2012;7:247–65.
[151] Lockwood WW, Zejnullahu K, Bradner JE, Varmus H. Sensitivity of human lung adenocarcinoma cell lines to targeted inhibition of BET epigenetic signaling proteins. Proc Natl Acad Sci USA 2012;109:19408–13.
[152] You J, Srinivasan V, Denis GV, et al. Kaposi's sarcoma-associated herpesvirus latency-associated nuclear antigen interacts with bromodomain protein Brd4 on host mitotic chromosomes. J Virol 2006;80(18):8909–19.
[153] Kaeser MD, Aslanian A, Dong MQ, Yates III JR, Emerson BM. BRD7, a novel PBAF-specific SWI/SNF subunit, is required for target gene activation and repression in embryonic stem cells. J Biol Chem 2008;283(47):32254–63.
[154] Drost J, Mantovani F, Tocco F, et al. BRD7 is a candidate tumour suppressor gene required for p53 function. Nat Cell Biol 2010;12(4):380–9.
[155] Cai Y, Jin J, Tomomori-Sato C, et al. Identification of new subunits of the multiprotein mammalian TRRAP/TIP60-containing histone acetyltransferase complex. J Biol Chem 2003;278(44):42733–6.
[156] Yamada HY, Rao CV. BRD8 is a potential chemosensitizing target for spindle poisons in colorectal cancer therapy. Int J Oncol 2009;35(5):1101–9.
[157] Kang JU, Koo SH, Kwon KC, Park JW, Kim JM. Gain at chromosomal region 5p15.33, containing TERT, is the most frequent genetic event in early stages of non-small cell lung cancer. Cancer Genet Cytogenet 2008;182(1):1–11.
[158] Grunwald C, Koslowski M, Arsiray T, et al. Expression of multiple epigenetically regulated cancer/germline genes in nonsmall cell lung cancer. Int J Cancer 2006;118(10):2522–8.
[159] Laue K, Daujat S, Crump JG, et al. The multidomain protein Brpf1 binds histones and is required for Hox gene expression and segmental identity. Development 2008;135(11):1935–46.
[160] Doyon Y, Cayrou C, Ullah M, et al. ING tumor suppressor proteins are critical regulators of chromatin acetylation required for genome expression and perpetuation. Mol Cell 2006;21(1):51–64.
[161] Fairbridge NA, Dawe CE, Niri FH, Kooistra MK, King-Jones K, McDermid HE. Cecr2 mutations causing exencephaly trigger misregulation of mesenchymal/ectodermal transcription factors. Birth Defects Res A Clin Mol Teratol 2010;88(8):619–25.
[162] Lee SK, Park EJ, Lee HS, Lee YS, Kwon J. Genome-wide screen of human bromodomain-containing proteins identifies Cecr2 as a novel DNA damage response protein. Mol Cells 2012;34(1):85–91.
[163] Mullighan CG, Zhang J, Kasper LH, et al. CREBBP mutations in relapsed acute lymphoblastic leukaemia. Nature 2011;471(7337):235–9.
[164] Gong AY, Eischeid AN, Xiao J, et al. miR-17-5p targets the p300/CBP-associated factor and modulates androgen receptor transcriptional activity in cultured prostate cancer cells. BMC Cancer 2012;12(1):492.
[165] Gayther SA, Batley SJ, Linger L, et al. Mutations truncating the EP300 acetylase in human cancers. Nat Genet 2000;24(3):300–3.
[166] Chaffanet M, Gressin L, Preudhomme C, Soenen-Cornu V, Birnbaum D, Pebusque MJ. MOZ is fused to p300 in an acute monocytic leukemia with t(8;22). Genes Chromosomes Cancer 2000;28(2):138–44.
[167] Li H, Ilin S, Wang W, et al. Molecular basis for site-specific read-out of histone H3K4me3 by the BPTF PHD finger of NURF. Nature 2006;442(7098):91–5.
[168] Grinberg-Rashi H, Ofek E, Perelman M, et al. The expression of three genes in primary non-small cell lung cancer is associated with metastatic spread to the brain. Clin Cancer Res 2009;15(5):1755–61.

[169] Dou Y, Milne TA, Tackett AJ, et al. Physical association and coordinate function of the H3 K4 methyltransferase MLL1 and the H4 K16 acetyltransferase MOF. Cell 2005;121(6):873–85.
[170] Krivtsov AV, Armstrong SA. MLL translocations, histone modifications and leukaemia stem-cell development. Nat Rev Cancer 2007;7(11):823–33.
[171] Varela I, Tarpey P, Raine K, et al. Exome sequencing identifies frequent mutation of the SWI/SNF complex gene PBRM1 in renal carcinoma. Nature 2011;469(7331):539–42.
[172] Xia W, Nagase S, Montia AG, et al. BAF180 is a critical regulator of p21 induction and a tumor suppressor mutated in breast cancer. Cancer Res 2008;68(6):1667–74.
[173] Ge X, Jin Q, Zhang F, Yan T, Zhai Q. PCAF acetylates {beta}-catenin and improves its stability. Mol Biol Cell 2009;20(1):419–27.
[174] Armas-Pineda C, Arenas-Huertero F, Perezpenia-Diazconti M, et al. Expression of PCAF, p300 and Gcn5 and more highly acetylated histone H4 in pediatric tumors. J Exp Clin Cancer Res 2007;26 (2):269–76.
[175] Podcheko A, Northcott P, Bikopoulos G, et al. Identification of a WD40 repeat-containing isoform of PHIP as a novel regulator of beta-cell growth and survival. Mol Cell Biol 2007;27(18):6484–96.
[176] De Semir D, Nosrati M, Bezrookove V, et al. Pleckstrin homology domain-interacting protein (PHIP) as a marker and mediator of melanoma metastasis. Proc Natl Acad Sci USA 2012;109(18):7067–72.
[177] Tokuda E, Fujita N, Oh-hara T, et al. Casein kinase 2-interacting protein-1, a novel Akt pleckstrin homology domain-interacting protein, down-regulates PI3K/Akt signaling and suppresses tumor growth *in vivo*. Cancer Res 2007;67(20):9666–76.
[178] Fossey SC, Kuroda S, Price JA, Pendleton JK, Freedman BI, Bowden DW. Identification and characterization of PRKCBP1, a candidate RACK-like protein. Mamm Genome 2000;11(10):919–25.
[179] Panagopoulos I, Micci F, Thorsen J, et al. Fusion of ZMYND8 and RELA genes in acute erythroid leukemia. PloS One 2013;8(5):e63663.
[180] Wada Y, Matsuura M, Sugawara M, et al. Development of detection method for novel fusion gene using GeneChip exon array. J Clin Bioinform 2014;4(1):3.
[181] Eichmuller S, Usener D, Dummer R, Stein A, Thiel D, Schadendorf D. Serological detection of cutaneous T-cell lymphoma-associated antigens. Proc Natl Acad Sci USA 2001;98(2):629–34.
[182] Park J, Betel D, Gryfe R, et al. Mutation profiling of mismatch repair-deficient colorectal cncers using an in silico genome scan to identify coding microsatellites. Cancer Res 2002;62(5):1284–8.
[183] Harikrishnan KN, Chow MZ, Baker EK, et al. Brahma links the SWI/SNF chromatin-remodeling complex with MeCP2-dependent transcriptional silencing. Nat Genet 2005;37(3):254–64.
[184] Reisman D, Glaros S, Thompson EA. The SWI/SNF complex and cancer. Oncogene 2009;28 (14):1653–68.
[185] Shain AH, Giacomini CP, Matsukuma K, et al. Convergent structural alterations define SWItch/ Sucrose NonFermentable (SWI/SNF) chromatin remodeler as a central tumor suppressive complex in pancreatic cancer. Proc Natl Acad Sci USA 2012;109(5):E252–9.
[186] Liu G, Gramling S, Munoz D, et al. Two novel BRM insertion promoter sequence variants are associated with loss of BRM expression and lung cancer risk. Oncogene 2011;30(29):3295–304.
[187] Rada-Iglesias A, Bajpai R, Swigut T, Brugmann SA, Flynn RA, Wysocka J. A unique chromatin signature uncovers early developmental enhancers in humans. Nature 2011;470(7333):279–83.
[188] Schneppenheim R, Fruhwald MC, Gesk S, et al. Germline nonsense mutation and somatic inactivation of SMARCA4/BRG1 in a family with rhabdoid tumor predisposition syndrome. Am J Hum Genet 2010;86(2):279–84.
[189] Negorev DG, Vladimirova OV, Kossenkov AV, et al. Sp100 as a potent tumor suppressor: accelerated senescence and rapid malignant transformation of human fibroblasts through modulation of an embryonic stem cell program. Cancer Res 2010;70(23):9991–10001.

[190] Nicewonger J, Suck G, Bloch D, Swaminathan S. Epstein-Barr virus (EBV) SM protein induces and recruits cellular Sp110b to stabilize mRNAs and enhance EBV lytic gene expression. J Virol 2004;78 (17):9412–22.
[191] Di Bernardo MC, Crowther-Swanepoel D, Broderick P, et al. A genome-wide association study identifies six susceptibility loci for chronic lymphocytic leukemia. Nat Genet 2008;40(10):1204–10.
[192] Tsai WW, Wang Z, Yiu TT, et al. TRIM24 links a non-canonical histone signature to breast cancer. Nature 2010;468(7326):927–32.
[193] Chambon M, Orsetti B, Berthe ML, et al. Prognostic significance of TRIM24/TIF-1alpha gene expression in breast cancer. Am J Pathol 2011;178(4):1461–9.
[194] Li H, Sun L, Tang Z, et al. Overexpression of TRIM24 correlates with tumor progression in non-small cell lung cancer. PloS One 2012;7(5):e37657.
[195] Quintas-Cardama A, Qiu YH, Post SM, et al. Reverse phase protein array profiling reveals distinct proteomic signatures associated with chronic myeloid leukemia progression and with chronic phase in the CD34-positive compartment. Cancer 2012;118(21):5283–92.
[196] Herquel B, Ouararhni K, Khetchoumian K, et al. Transcription cofactors TRIM24, TRIM28, and TRIM33 associate to form regulatory complexes that suppress murine hepatocellular carcinoma. Proc Natl Acad Sci USA 2011;108(20):8212–17.
[197] Khetchoumian K, Teletin M, Tisserand J, et al. Loss of Trim24 (Tif1alpha) gene function confers oncogenic activity to retinoic acid receptor alpha. Nat Genet 2007;39(12):1500–6.
[198] Rowe HM, Jakobsson J, Mesnard D, et al. KAP1 controls endogenous retroviruses in embryonic stem cells. Nature 2010;463(7278):237–40.
[199] Bai X, Kim J, Yang Z, et al. TIF1gamma controls erythroid cell fate by regulating transcription elongation. Cell 2010;142(1):133–43.
[200] Masselink H, Bernards R. The adenovirus E1A binding protein BS69 is a corepressor of transcription through recruitment of N-CoR. Oncogene 2000;19(12):1538–46.
[201] Wen H, Li Y, Xi Y, et al. ZMYND11 links histone H3.3K36me3 to transcription elongation and tumour suppression. Nature 2014;508(7495):263–8.
[202] Yang H, Zhang C, Zhao X, et al. Analysis of copy number variations of BS69 in multiple types of hematological malignancies. Ann Hematol 2010;89(10):959–64.
[203] Hughes-Davies L, Huntsman D, Ruas M, et al. EMSY links the BRCA2 pathway to sporadic breast and ovarian cancer. Cell 2003;115(5):523–35.
[204] Chen X, Cheung ST, So S, et al. Gene expression patterns in human liver cancers. Mol Biol Cell 2002;13(6):1929–39.
[205] Alizadeh AA, Eisen MB, Davis RE, et al. Distinct types of diffuse large B-cell lymphoma identified by gene expression profiling. Nature 2000;403(6769):503–11.
[206] van't Veer LJ, Dai H, van de Vijver MJ, et al. Gene expression profiling predicts clinical outcome of breast cancer. Nature 2002;415(6871):530–6.
[207] Ma XJ, Salunga R, Tuggle JT, et al. Gene expression profiles of human breast cancer progression. Proc Natl Acad Sci USA 2003;100(10):5974–9.
[208] Caron C, Lestrat C, Marsal S, et al. Functional characterization of ATAD2 as a new cancer/testis factor and a predictor of poor prognosis in breast and lung cancers. Oncogene 2010;29(37):5171–81.
[209] Ding WM, Yang L, Zheng Y, Sun HY, Liang H, Wang JJ. Preparation of anti-BRDT-NY polyclonal antibody and expression of BRDT-NY protein in digestive tract tumors. Xi Bao Yu Fen Zi Mian Yi Xue Za Zhi 2012;28(8):847–50.
[210] Chaidos A, Caputo V, Gouvedenou K, et al. Potent antimyeloma activity of the novel bromodomain inhibitors I-BET151 and I-BET762. Blood 2014;123(5):697–705.

[211] Henssen A, Thor T, Odersky A, et al. BET bromodomain protein inhibition is a therapeutic option for medulloblastoma. Oncotarget 2013;4(11):2080–95.
[212] Mertz JA, Conery AR, Bryant BM, et al. Targeting MYC dependence in cancer by inhibiting BET bromodomains. Proc Natl Acad Sci USA 2011;108(40):16669–74.
[213] Zuber J, Shi J, Wang E, et al. RNAi screen identifies Brd4 as a therapeutic target in acute myeloid leukaemia. Nature 2011;478(7370):524–8.
[214] Herrmann H, Blatt K, Shi J, et al. Small-molecule inhibition of BRD4 as a new potent approach to eliminate leukemic stem- and progenitor cells in acute myeloid leukemia AML. Oncotarget 2012;3(12):1588–99.
[215] Fiskus W, Sharma S, Qi J, et al. Highly active combination of BRD4 antagonist and histone deacetylase inhibitor against human acute myeloid leukemia (AML) cells. Mol Cancer Ther 2014;13(5):1142–54.
[216] Ott CJ, Kopp N, Bird L, et al. BET bromodomain inhibition targets both c-Myc and IL7R in high-risk acute lymphoblastic leukemia. Blood 2012;120(14):2843–52.
[217] Da Costa D, Agathanggelou A, Perry T, et al. BET inhibition as a single or combined therapeutic approach in primary paediatric B-precursor acute lymphoblastic leukaemia. Blood Cancer J 2013;3:e126.
[218] Roderick JE, Tesell J, Shultz LD, et al. c-Myc inhibition prevents leukemia initiation in mice and impairs the growth of relapsed and induction failure pediatric T-ALL cells. Blood 2014;123(7):1040–50.
[219] Cheng Z, Gong Y, Ma Y, et al. Inhibition of BET bromodomain targets genetically diverse glioblastoma. Clin Cancer Res 2013;19(7):1748–59.
[220] Segura MF, Fontanals-Cirera B, Gaziel-Sovran A, et al. BRD4 sustains melanoma proliferation and represents a new target for epigenetic therapy. Cancer Res 2013;73(20):6264–76.
[221] Wyce A, Degenhardt Y, Bai Y, et al. Inhibition of BET bromodomain proteins as a therapeutic approach in prostate cancer. Oncotarget 2013;4(12):2419–29.
[222] Venkataraman S, Alimova I, Balakrishnan I, et al. Inhibition of BRD4 attenuates tumor cell self-renewal and suppresses stem cell signaling in MYC driven medulloblastoma. Oncotarget 2014;5(9):2355–71.
[223] Shao Q, Kannan A, Lin Z, Stack Jr BC, Suen JY, Gao L. BET protein inhibitor JQ1 attenuates Myc-amplified MCC tumor growth *in vivo*. Cancer Res 2014;74(23):7090–102.
[224] Loven J, Hoke HA, Lin CY, et al. Selective inhibition of tumor oncogenes by disruption of super-enhancers. Cell 2013;153(2):320–34.
[225] Hnisz D, Abraham BJ, Lee TI, et al. Super-enhancers in the control of cell identity and disease. Cell 2013;155(4):934–47.
[226] Mansour MR, Abraham BJ, Anders L, et al. Oncogene regulation. An oncogenic super-enhancer formed through somatic mutation of a noncoding intergenic element. Science 2014;346(6215):1373–7.
[227] Wong AK, Shanahan F, Chen Y, et al. BRG1, a component of the SWI-SNF complex, is mutated in multiple human tumor cell lines. Cancer Res 2000;60(21):6171–7.
[228] Muchardt C, Yaniv M. When the SWI/SNF complex remodels...the cell cycle. Oncogene 2001;20(24):3067–75.
[229] Medina PP, Romero OA, Kohno T, et al. Frequent BRG1/SMARCA4-inactivating mutations in human lung cancer cell lines. Hum Mutat 2008;29(5):617–22.
[230] Cheng SW, Davies KP, Yung E, Beltran RJ, Yu J, Kalpana GV. c-MYC interacts with INI1/hSNF5 and requires the SWI/SNF complex for transactivation function. Nat Genet 1999;22(1):102–5.
[231] Decristofaro MF, Betz BL, Rorie CJ, Reisman DN, Wang W, Weissman BE. Characterization of SWI/SNF protein expression in human breast cancer cell lines and other malignancies. J Cell Physiol 2001;186(1):136–45.

[232] Hill DA, Chiosea S, Jamaluddin S, et al. Inducible changes in cell size and attachment area due to expression of a mutant SWI/SNF chromatin remodeling enzyme. J Cell Sci 2004;117(Pt 24):5847–54.
[233] Becker TM, Haferkamp S, Dijkstra MK, et al. The chromatin remodelling factor BRG1 is a novel binding partner of the tumor suppressor p16INK4a. Mol Cancer 2009;8:4.
[234] French C. NUT midline carcinoma. Nat Rev Cancer 2014;14(3):149–50.
[235] French CA. Demystified molecular pathology of NUT midline carcinomas. J Clin Pathol 2010;63 (6):492–6.
[236] Filippakopoulos P, Qi J, Picaud S, et al. Selective inhibition of BET bromodomains. Nature 2010;468 (7327):1067–73.
[237] Grayson AR, Walsh EM, Cameron MJ, et al. MYC, a downstream target of BRD-NUT, is necessary and sufficient for the blockade of differentiation in NUT midline carcinoma. Oncogene 2014;33 (13):1736–42.
[238] Wang R, You J. Mechanistic analysis of the role of bromodomain-containing protein 4 (BRD4) in BRD4-NUT oncoprotein-induced transcriptional activation. J Biol Chem 2015;290(5):2744–58.
[239] Iyer NG, Ozdag H, Caldas C. p300/CBP and cancer. Oncogene 2004;23(24):4225–31.
[240] Borrow J, Stanton Jr VP, Andresen JM, et al. The translocation t(8;16)(p11;p13) of acute myeloid leukaemia fuses a putative acetyltransferase to the CREB-binding protein. Nat Genet 1996;14(1):33–41.
[241] Panagopoulos I, Fioretos T, Isaksson M, et al. Fusion of the MORF and CBP genes in acute myeloid leukemia with the t(10;16)(q22;p13). Hum Mol Genet 2001;10(4):395–404.
[242] Lai JL, Jouet JP, Bauters F, Deminatti M. Chronic myelogenous leukemia with translocation (8;22): report of a new case. Cancer Genet Cytogenet 1985;17(4):365–6.
[243] Kitabayashi I, Aikawa Y, Yokoyama A, et al. Fusion of MOZ and p300 histone acetyltransferases in acute monocytic leukemia with a t(8;22)(p11;q13) chromosome translocation. Leukemia 2001;15 (1):89–94.
[244] Rowley JD, Reshmi S, Sobulo O, et al. All patients with the T(11;16)(q23;p13.3) that involves MLL and CBP have treatment-related hematologic disorders. Blood 1997;90(2):535–41.
[245] Satake N, Ishida Y, Otoh Y, et al. Novel MLL-CBP fusion transcript in therapy-related chronic myelomonocytic leukemia with a t(11;16)(q23;p13) chromosome translocation. Genes Chromosomes Cancer 1997;20(1):60–3.
[246] Giles RH, Dauwerse JG, Higgins C, et al. Detection of CBP rearrangements in acute myelogenous leukemia with t(8;16). Leukemia 1997;11(12):2087–96.
[247] Zeng L, Li J, Muller M, et al. Selective small molecules blocking HIV-1 Tat and coactivator PCAF association. J Am Chem Soc 2005;127(8):2376–7.
[248] Sachchidanand LRS, Yan S, et al. Target structure-based discovery of small molecules that block human p53 and CREB binding protein association. Chem Biol 2006;13(1):81–90.
[249] Borah JC, Mujtaba S, Karakikes I, et al. A small molecule binding to the coactivator CREB-binding protein blocks apoptosis in cardiomyocytes. Chem Biol 2011;18(4):531–41.
[250] Miyoshi S, Ooike S, Iwata K, Hikawa H, Sugahara K. Inventors; Mitsubishi Tanabe Pharma Corporation, Japan, assignee. Antitumor agent patent WO 2009/084693 A1. 2009.
[251] Nicodeme E, Jeffrey KL, Schaefer U, et al. Suppression of inflammation by a synthetic histone mimic. Nature 2010;468(7327):1119–23.
[252] Chung CW, Coste H, White JH, et al. Discovery and characterization of small molecule inhibitors of the BET family bromodomains. J Med Chem 2011;54(11):3827–38.
[253] Filippakopoulos P, Picaud S, Fedorov O, et al. Benzodiazepines and benzotriazepines as protein interaction inhibitors targeting bromodomains of the BET family. Bioorg Med Chem 2012;20(6):1878–86.
[254] Hewings DS, Wang M, Philpott M, et al. 3,5-Dimethylisoxazoles act as acetyl-lysine-mimetic bromodomain ligands. J Med Chem 2011;54(19):6761–70.

[255] Bamborough P, Diallo H, Goodacre JD, et al. Fragment-based discovery of bromodomain inhibitors part 2: optimization of phenylisoxazole sulfonamides. J Med Chem 2012;55(2):587–96.
[256] Ciceri P, Muller S, O'Mahony A, et al. Dual kinase-bromodomain inhibitors for rationally designed polypharmacology. Nat Chem Biol 2014;10(4):305–12.
[257] Martin MP, Olesen SH, Georg GI, Schonbrunn E. Cyclin-dependent kinase inhibitor dinaciclib interacts with the acetyl-lysine recognition site of bromodomains. ACS Chem Biol 2013;8(11):2360–5.
[258] Dittmann A, Werner T, Chung CW, et al. The commonly used PI3-kinase probe LY294002 is an inhibitor of BET bromodomains. ACS Chem Biol 2014;9(2):495–502.
[259] Ember SW, Zhu JY, Olesen SH, et al. Acetyl-lysine binding site of bromodomain-containing protein 4 (BRD4) interacts with diverse kinase inhibitors. ACS Chem Biol 2014;9:1160–71.
[260] Filippakopoulos P, Knapp S. Targeting bromodomains: epigenetic readers of lysine acetylation. Nat Rev Drug Discov 2014;13(5):337–56.
[261] Vidler LR, Brown N, Knapp S, Hoelder S. Druggability analysis and structural classification of bromodomain acetyl-lysine binding sites. J Med Chem 2012;55(17):7346–59.
[262] Hay DA, Fedorov O, Martin S, et al. Discovery and optimization of small-molecule ligands for the CBP/p300 bromodomains. J Am Chem Soc 2014;136(26):9308–19.
[263] Rooney TP, Filippakopoulos P, Fedorov O, et al. A series of potent CREBBP bromodomain ligands reveals an induced-fit pocket stabilized by a cation–pi interaction. Angew Chem Int Ed Engl 2014;53 (24):6126–30.
[264] Drouin L, McGrath S, Vidler LR, et al. Structure enabled design of BAZ2-ICR, a chemical probe targeting the bromodomains of BAZ2A and BAZ2B. J Med Chem 2015;58(5):2553–9.
[265] Demont EH, Bamborough P, Chung CW, et al. 1,3-Dimethyl benzimidazolones are potent, selective inhibitors of the BRPF1 bromodomain. ACS Med Chem Lett 2014;5(11):1190–5.
[266] Ferguson FM, Fedorov O, Chaikuad A, et al. Targeting low-druggability bromodomains: fragment based screening and inhibitor design against the BAZ2B bromodomain. J Med Chem 2013;56 (24):10183–7.
[267] Fedorov O, Lingard H, Wells C, et al. [1,2,4]Triazolo[4,3-a]phthalazines: inhibitors of diverse bromodomains. J Med Chem 2014;57(2):462–76.
[268] Picaud S, Strocchia M, Terracciano S, et al. 9H-purine scaffold reveals induced-fit pocket plasticity of the BRD9 bromodomain. J Med Chem 2015;40:468–79.
[269] Tsume M, Kimura-Yoshida C, Mochida K, et al. Brd2 is required for cell cycle exit and neuronal differentiation through the E2F1 pathway in mouse neuroepithelial cells. Biochem Biophys Res Commun 2012;425(4):762–8.
[270] Lamonica JM, Deng W, Kadauke S, et al. Bromodomain protein Brd3 associates with acetylated GATA1 to promote its chromatin occupancy at erythroid target genes. Proc Natl Acad Sci USA 2011;108(22):E159–68.
[271] Matzuk MM, McKeown MR, Filippakopoulos P, et al. Small-molecule inhibition of BRDT for male contraception. Cell 2012;150(4):673–84.
[272] Jang MK, Mochizuki K, Zhou M, Jeong HS, Brady JN, Ozato K. The bromodomain protein Brd4 is a positive regulatory component of P-TEFb and stimulates RNA polymerase II-dependent transcription. Mol Cell 2005;19(4):523–34.
[273] Bolden JE, Tasdemir N, Dow LE, et al. Inducible *in vivo* silencing of Brd4 identifies potential toxicities of sustained BET protein inhibition. Cell Rep 2014;8(6):1919–29.
[274] Stonestrom AJ, Hsu SC, Jahn KS, et al. Functions of BET proteins in erythroid gene expression. Blood 2015;30:2825–34.
[275] Pastori C, Daniel M, Penas C, et al. BET bromodomain proteins are required for glioblastoma cell proliferation. Epigenetics 2014;9(4):611–20.

[276] Stewart HJ, Horne GA, Bastow S, Chevassut TJ. BRD4 associates with p53 in DNMT3A-mutated leukemia cells and is implicated in apoptosis by the bromodomain inhibitor JQ1. Cancer Med 2013;2 (6):826–35.
[277] Feng Q, Zhang Z, Shea MJ, et al. An epigenomic approach to therapy for tamoxifen-resistant breast cancer. Cell Res 2014;24(7):809–19.
[278] Lee DH, Qi J, Bradner JE, et al. Synergistic effect of JQ1 and rapamycin for treatment of human osteosarcoma. Int J Cancer 2015;136(9):2055–64.
[279] Wong C, Laddha SV, Tang L, et al. The bromodomain and extra-terminal inhibitor CPI203 enhances the antiproliferative effects of rapamycin on human neuroendocrine tumors. Cell Death Dis 2014;5: e1450.

CHAPTER

EPIGENETIC READERS INTERPRETING THE LYSINE METHYLOME—BIOLOGICAL ROLES AND DRUG DISCOVERY

11

Tobias Wagner[1], Dina Robaa[2], Wolfgang Sippl[2] and Manfred Jung[1]

[1]Institute of Pharmaceutical Sciences, Albert-Ludwigs-Universität Freiburg, Freiburg, Germany

[2]Institut für Pharmazie, Martin-Luther-Universität Halle-Wittenberg, Halle, Germany

CHAPTER OUTLINE

11.1 INTRODUCTION

Chromatin exists as an assembly of nucleosomes which are composed of basic and thus positively charged histone proteins with negatively charged DNA wrapped around them. In order to enable a high level of chromatin compaction on the one hand and the possibility for controlled access to the DNA for transcription and replication processes on the other hand, chromatin switches between silent and active states in a highly dynamic manner. Gene expression is regulated in a hierarchical fashion by various control mechanisms. Changes in chromatin regulation are critical for fundamental cellular processes, including self-renewal, differentiation, and proliferation. Misregulation may lead to abnormal gene expression patterns which might eventually contribute to human diseases like cancers.

G. Egger & P. Arimondo (Eds): Drug Discovery in Cancer Epigenetics. DOI: http://dx.doi.org/10.1016/B978-0-12-802208-5.00011-4

Proteins and enzymes that are involved in chromatin-modifying processes constitute interesting targets for drug discovery and may enable new therapeutic strategies for the treatment of diseases. On an epigenetic level, chromatin regulation is mainly defined by DNA methylation and subsequent oxidation of methylated cytosine residues within DNA on one side and by a variety of specific posttranslational modifications (PTMs) of histone proteins on the other side [1].

Enzyme families, like histone lysine methyltransferases (HMTs or KMTs), that add certain posttranslational modifications to distinct amino acid side chains within their substrate proteins are called "writers." In some cases, like lysine acetylation, the PTMs themselves have an intrinsic effect on the chromatin regulation process. Lysine acetylation removes the positive charge of the ammonium group of the unmodified lysine residue under physiological conditions leading to a weaker interaction of DNA with the histone proteins which eventually results in a more accessible chromatin state. "Erasers," on the other hand, are enzymes that are capable of removing PMTs from their specific substrates which contribute the dynamics of the epigenome. Finally, there are proteins or protein complexes bearing highly conserved binding domains that are able to recognize specific PTM sites or PTM patterns on client proteins, including histones. These so-called "reader" proteins can recruit further binding partners like linker or auxiliary proteins, transcription factors (TCF), oncoproteins or tumor suppressors, and other elements of the transcription machinery to their targets or they are effector enzymes themselves [2]. The creation of posttranslational methylation marks does not alter the charge of the amino acid side chain, as opposed to lysine acetylation, and there is only a small change in size and lipophilicity from the nonmethylated to the mono-, di-, and trimethylated state of lysines [3]. Therefore, not surprisingly, proteins that are not only capable of recognizing the methylation marks but that can also distinguish between different methylation states are responsible for a major part of the biological outcome of PTMs of histones and thus complete the dynamic process of epigenetic regulation [4]. Some proteins comprise an assembly of various reader domains in one protein and are thus able to bind more than one PTM at the same time. Hence, those binding proteins are often regarded as the readers or interpreters of the proposed epigenetic code [5]. A scheme of the epigenetic writer–eraser–reader interplay is outlined in Figure 11.1.

Increasing evidence on the involvement of reader proteins in the development and progression of different diseases and malignancies has been gained in the last few years. The data underline the promising potential of epigenetic reader proteins as drug targets. Hence, to further elucidate the function of the epigenetic reader family, small-molecule ligands that inhibit the interactions between readers and their native binding partners are of great value.

So far, compared to writers and erasers, the readers and their respective effector domains have been less intensively pursued as therapeutic targets, possibly also due to the fact that targeting protein–protein interactions with small molecules is often very difficult. However, recent drug discovery efforts resulted in potent small-molecule ligands of bromodomains which makes a case for histone readers as druggable targets [6–11]. Bromodomain-containing proteins are able to bind to specific acetylated lysine residues (KAc) within histone proteins [2]. By blocking the KAc-binding site of bromodomains the recognition of acetylated histone tails is prevented, which in turn changes the process of chromatin remodeling. The compounds named I-BET and JQ1 are among the first reported small-molecule inhibitors that selectively target the so-called BET bromodomains. Both compounds were identified during phenotypic cellular screens. JQ1 was also found to significantly inhibit tumor growth in two xenograft models of rare, rapidly fatal NUT midline

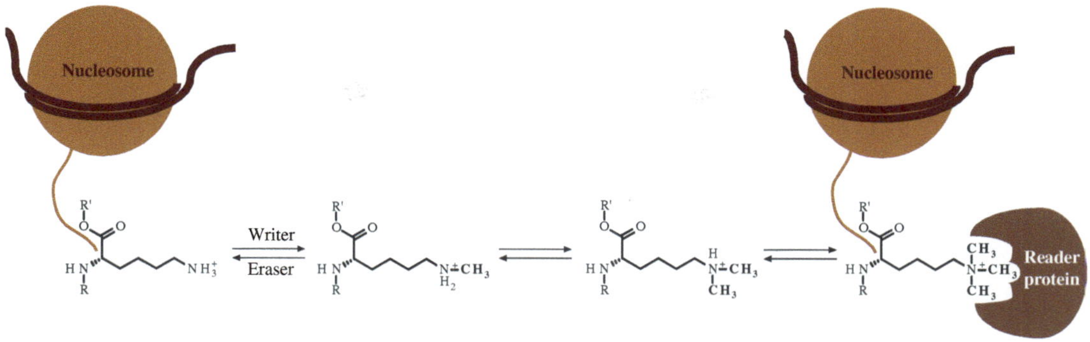

FIGURE 11.1

Schematic illustration of a nucleosome with a histone tail protruding from the center. A representative lysine residue of the amino acid chain is depicted which is a substrate for writer enzymes like histone lysine methyltransferases (KMTs). KMTs can either mono-, di-, or trimethylatespecific lysine residues within histones in a site-dependent mode. In turn, eraser enzymes like lysine demethylases are capable of removing distinct methylation marks. This epigenetic interplay is completed by reader proteins that are able to recognize specific modification patterns (exemplarily described here for trimethyl-lysines) via conserved binding domains which are discussed in this chapter.

carcinoma in mice [9]. Currently (2015), there are at least six BET bromodomain inhibitors in clinical trials targeting different types of cancers, like acute leukemia, NUT-midline carcinoma, glioblastoma, and also other diseases like atherosclerosis and type 2 diabetes (OTX015, GSK525762, Bay1238097, TEN-010, CPI-0610, RVX-20). This fact highlights the potential of targeting epigenetic reader proteins as pharmacological targets but there are still many open questions on the outcome of the inhibition of methyl-lysine readers.

11.2 METHYL-LYSINE READERS

Besides the well-known acetyl-lysine binding family of bromodomains, a large family of reader proteins is represented by the methyl-lysine binders. They all contain a specific subdomain that interacts with methylated lysine residues. The methyl-lysine binders are categorized into three families by their respective binding domains, namely the plant homeodomain (PHD) zinc finger proteins, the WD40 repeat domain-containing proteins, and the Royal family of reader proteins. The latter is further subdivided into Tudor-, chromodomain-, Pro-Trp-Trp-Pro (PWWP)-, and malignant brain tumor (MBT) repeat domain-containing proteins. For many methyl-lysine binders crystal structures have been resolved enabling a better understanding of the structural aspects involved in the interaction between methyl-lysines and the respective reader domains. The structural features of histone readers have in fact been thoroughly reviewed in recent years [12–14]. A common characteristic of all known methyl-lysine binding domains is the so-called aromatic cage (see Figure 11.2).

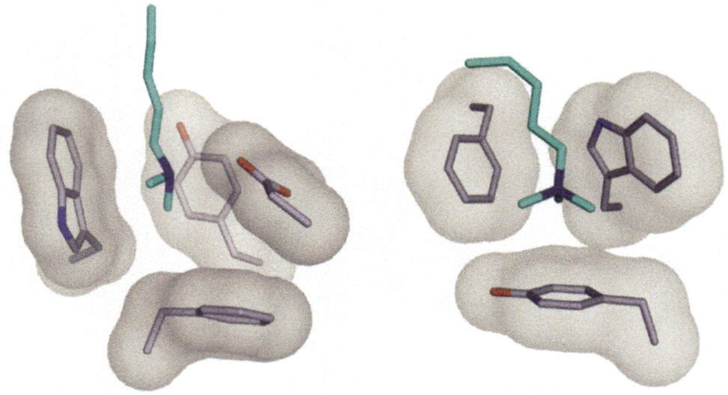

FIGURE 11.2

Comparison of the aromatic cavity of different methyl-lysine readers. Methyl-lysine residues are shown as cyan sticks, while residues forming the lysine pocket are depicted as blue-white sticks.

It is formed by two to four electron-rich aromatic amino acid residues and seems to be responsible for the specific recognition of certain methylation states of distinct lysine residues. Among the reader domains there is a remarkable selectivity for certain methylation states at their respective target site. The interaction between the methyl-lysine site and the reading domain occurs via cation–π interactions and to a lesser extent through hydrophobic van der Waals contacts [15]. Furthermore, hydrogen bonding, electrostatic contacts with the methyl ammonium moiety, and steric exclusion are an issue when the target has a lower methylation state [2]. It was also discovered that mono- and dimethyl-lysines are preferentially bound by proteins in which one wall of the aromatic cage is substituted by a negatively charged acidic amino acid residue [16].

11.3 THE ROYAL FAMILY OF EPIGENETIC READER PROTEINS

The first and most prominent family of reader proteins that is discussed here is the so-called Royal family of epigenetic reader proteins. Through sequence analysis it was observed that the members of this family show structural similarity. As was initially described, this family is further subdivided by their respective reader domains. All members of the Royal family have a similar structural fold in common. It is characterized by a barrel-like structure made up of three to five antiparallel β-sheets. Two of those β-sheets are connected by a 3_{10}- helix [12].

11.3.1 THE MBT REPEAT DOMAIN

The MBT repeat domain represents a member of the Royal family of epigenetic readers and MBT repeat domain-containing proteins mainly bind to targets that comprise mono- and dimethyl-lysines. The MBT repeat often exists as tandem repeats comprising a structural motif of about 100 amino acids. One of the first MBT repeat-containing proteins that was characterized is

L3MBTL1 and thus it is a well-explored representative of this family. *L3mbt* was initially identified as a *Drosophila* tumor suppressor gene [17]. It was shown that a depletion of L3MBTL1 promotes the erythroid differentiation of human hematopoietic progenitor cells [18]. Furthermore, L3MBTL1 represses the expression of E2F-regulated genes, such as the oncogenic and growth-related *c-myc* gene, by binding to chromatin and is therefore often depicted as "chromatin lock" [19]. L3MBTL1 is also an example for epigenetic readers that do not exclusively target histone proteins: L3MBTL1 binds the monomethylated nonhistone protein p53 (p53K382me1). The recruitment of L3MBTL1 to p53K382me leads to a repression of p53-mediated transactivation of target genes [20].

Taken together, *l3mbtl1* exhibits the typical characteristics of a tumor suppressor gene. Considering drug discovery strategies in the field of cancer, L3MBTL1 thus so far seems to be an inappropriate drug target. Rather, the fact that, like L3MBTL1, a number of epigenetic readers represent tumor suppressors raises the general question of target selectivity of potential drug candidates. Nevertheless, small-molecule probes exhibiting high affinity toward potential tumor suppressor-like readers are of great value to further investigate the biological role and cellular effects of the respective proteins.

In this respect, the way was paved by the identification of the first low-micromolar small-molecule binders of L3MBTL1 in the Frye lab in 2010 [21]. Kireev and coworkers combined their screening strategy for new L3MBTL1 ligands with a preceding virtual screening using two complementary approaches. First, they were searching for compounds containing Kme1 and Kme2 side chains. In another approach they made use of a sequential application of pharmacophore and docking techniques in order to potentially reveal more structurally remote compounds that might antagonize the L3MBTL1–histone peptide interactions. Commercially available hits were verified *in vitro* against a selection of four MBT-containing proteins. This led to a set of compounds that bind to L3MBTL1 in the low-micromolar range (see Figure 11.3).

One year later, also in the Frye lab, Gao and coworkers introduced a set of small-molecule compounds, which the authors refer to as "biophysical" probes, that were designed to mimic the shape of a single lysine residue and thus reproduce the binding affinities of cognate histone peptides (see Figure 11.4).

The active compounds are *para*-bromophenyl sulfonamides with variously substituted lysines and demonstrate the same affinity profile toward L3MBTL1 as native peptide ligands, which was validated by K_d measurements using isothermal calorimetry (ITC).With the help of the probes they addressed the question of how adding a single methyl group to Kme0 or removing it from Kme3 results in a huge gain in affinity to a reader protein like L3MBTL1, while adding a methyl moiety to Kme1 does not affect the affinity at all. The preference of L3MBTL1 for mono- and dimethyl over trimethyl-lysine could be explained by calculating the binding free energy using the free energy perturbation method. A trimethyl-lysine group placed in the aromatic cavity exhibited little polar and nonpolar contributions, compared to mono- and dimethyl-lysine. A monomethyl-lysine group showed the highest nonpolar contributions, whereas a dimethyl-lysine group exhibited the highest polar contributions. It was furthermore shown that besides forming favorable interactions, the L3MBTL1 binding pocket does not fit optimally for both methylation states (Kme1 and Kme2) from an energetic view. The fact that L3MBTL1 discriminates between mono- and dimethyl-lysine residues as well as nonmethylated and trimethylated lysine residues has most probably evolved as a "compromise" [22].

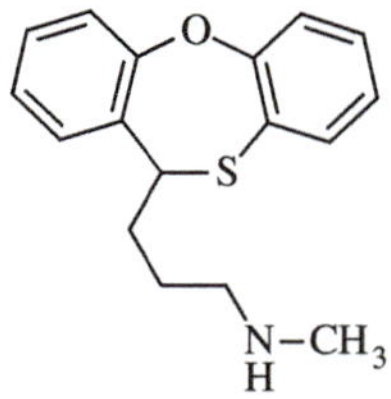

Kireev_2
IC_{50} 14 ± 0.69 μM

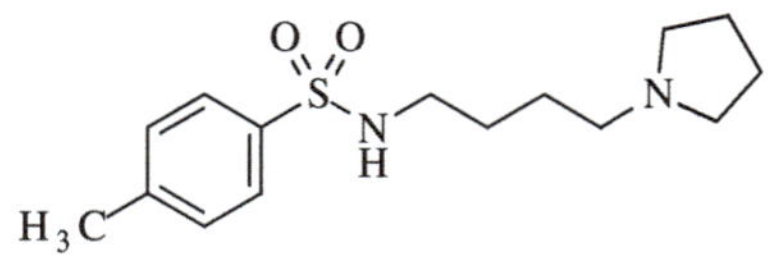

Kireev_13
IC_{50} 17 ± 0.13 μM

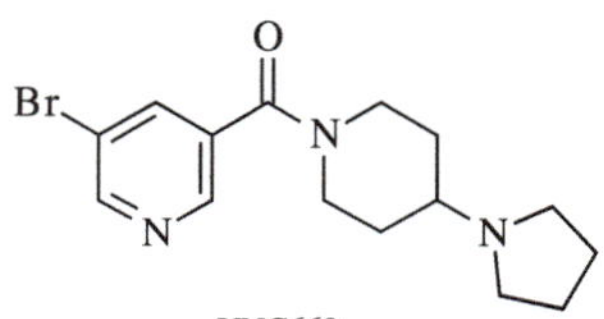

UNC669
K_d 5 ± 1 μM

FIGURE 11.3

Structures of small-molecule ligands of the L3MBTL1-MBT domain.

In 2011, Herold and coworkers successfully identified UNC669 as another low-micromolar small-molecule L3MBTL1 antagonist (see Figure 11.3). It was designed based on the crystal structure of histone-derived peptide–MBT complexes. UNC669 antagonizes the binding of a peptide fragment of the native histone-binding partner (H4K20me1) to L3MBTL1 with a fivefold stronger binding affinity. UNC669 also showed sixfold selectivity for L3MBTL1 over its close homolog L3MBTL3 and tenfold over L3MBTL4. ITC measurements revealed a K_d value of 5 μM for the binding of UNC669 to L3MBTL1, which is several times stronger than the native histone peptides. The results of structure–activity relation (SAR) studies indicate that MBT domains can accommodate larger amines, such as pyrrolidines, but that the size of the binding pocket is too small for methyl-substituted pyrimidines. Herold and coworkers were also able to solve the first cocrystal structure of a small molecule bound to L3MBTL1 which provides a deeper understanding of the binding requirements for further ligand design. The cocrystal structure shows that the pyrrolidine moiety of UNC669 mimics the binding of the methyl-lysine residue in the second domain of L3MBTL1 (see Figure 11.5) [23].

L3MBTL3, which is, as already mentioned above, a close homolog of L3MBTL1, preferably binds to dimethylated lysine residues in a rather promiscuous manner. It shows the highest affinities toward H4K20me2, H3K4me2, and H3K9me2 [24]. L3MBTL3 misregulation has been linked to the development of diseases like cancer. Like *l3mbtl1*, *l3mbtl3* was shown to be a tumor suppressor gene that blocks transcription from chromatin.

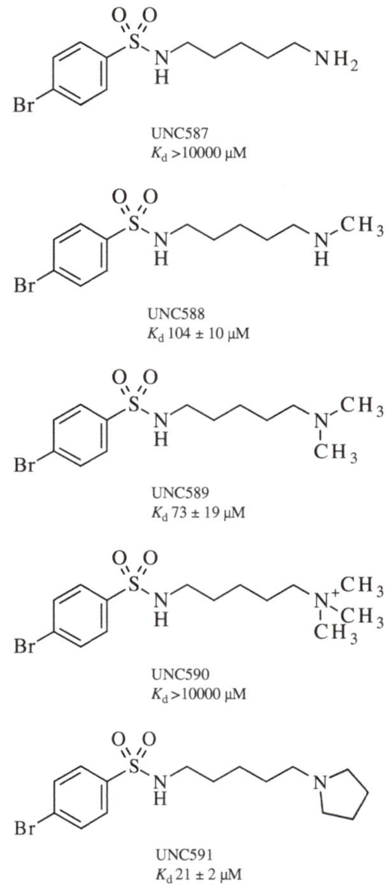

FIGURE 11.4

Structures of small-molecule "biophysical" probes for the L3MBTL1-MBT domain.

By using high-resolution single nucleotide polymorphism (SNP) genotyping to identify regions of genomic gain and loss in the genomes of 212 medulloblastomas (malignant pediatric brain tumors) it was found that *l3mbtl3* and *l3mbt2* are among the genes that show focal homozygous and hemizygous deletions. Reexpression of *l3mbtl3* after knockout was shown to attenuate the malignant phenotype of a medulloblastoma cell line which supports a critical role for histone lysine methylation (particularly at position H3K9) and L3MBTL3 in the pathogenesis of medulloblastoma [25].

Also, for L3NBTL3 promising findings have been made with regard to small-molecule tool compounds that will help to further elucidate the cellular role of L3MBTL3 and its implementation in malignancies. James and coworkers identified the symmetrical compound UNC1215 as a strong ligand of L3MBTL3 with a K_d value of 120 nM, which displays in fact a 40 times stronger affinity to L3MBTL3 than UNC669 exhibits toward L3MBTL1 (see Figure 11.6). UNC1215 shows a 50-fold selectivity for L3MBTL3 over other members of the MBT family and more than 200 additional reader domains [26].

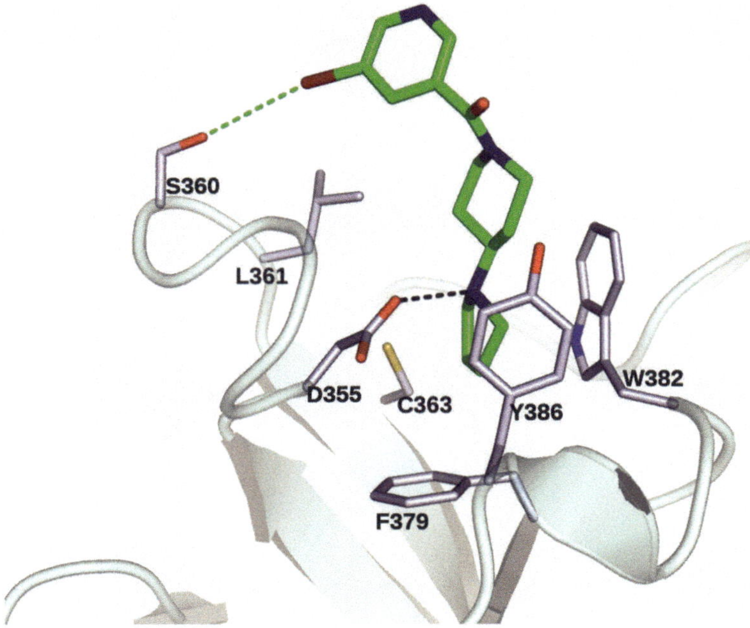

FIGURE 11.5

Crystal structure of L3MBTL1 in complex with UNC669. UNC699 is shown as green sticks and the side chains of the protein residues involved in the interaction are shown as white sticks.

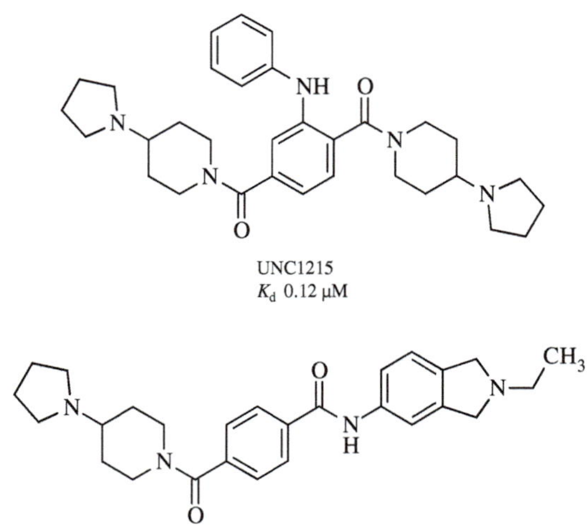

FIGURE 11.6

Structures of small-molecule ligands of the L3MBTL3-MBT domain.

In 2009, Guo and coworkers already predicted that some MBT members, such as L3MBTL3, could bind methylated histones with more than one repeat domain [27]. Indeed, the crystal structure of the complex composed of UNC1215 bound to L3MBTL3 revealed an interesting polyvalent 2:2 binding mode: each ligand molecule interacts with the first MBT domain of one chain and the second MBT domain of the other chain of the dimer. UNC1215 contains two pyrrolidine residues, both of which are embedded in the aromatic cage of the two involved MBT domains (see Figure 11.7; PDB: 4FL6) [26].

James and coworkers established a fluorescence recovery after photobleaching (FRAP) assay in order to investigate the effects of UNC1215 at a cellular level. For this purpose, human embryonic kidney 293 (HEK293) cells were transfected with a GFP fusion protein of the three MBT domains of L3MBTL3 (GFP-3MBT). Recovery time of a photobleached area in GFP-3MBT-expressing cells is reduced upon treatment with UNC1215 in a dose-dependent manner. This indicates that UNC1215 competes with cellular factors for binding of L3MBTL3.

Through traditional pulldown assays it was revealed that a biotinylated analog of UNC1215 was able to affinity purify L3MBTL3 from G-401 cell lysates. G-401 cells were shown to have fivefold higher levels of endogenous L3MBTL3 compared to other cell lines, including HEK293. Additionally, soluble UNC1215 also demonstrated potent binding to endogenous L3MBTL3 by effectively competing with biotin-UNC1215.

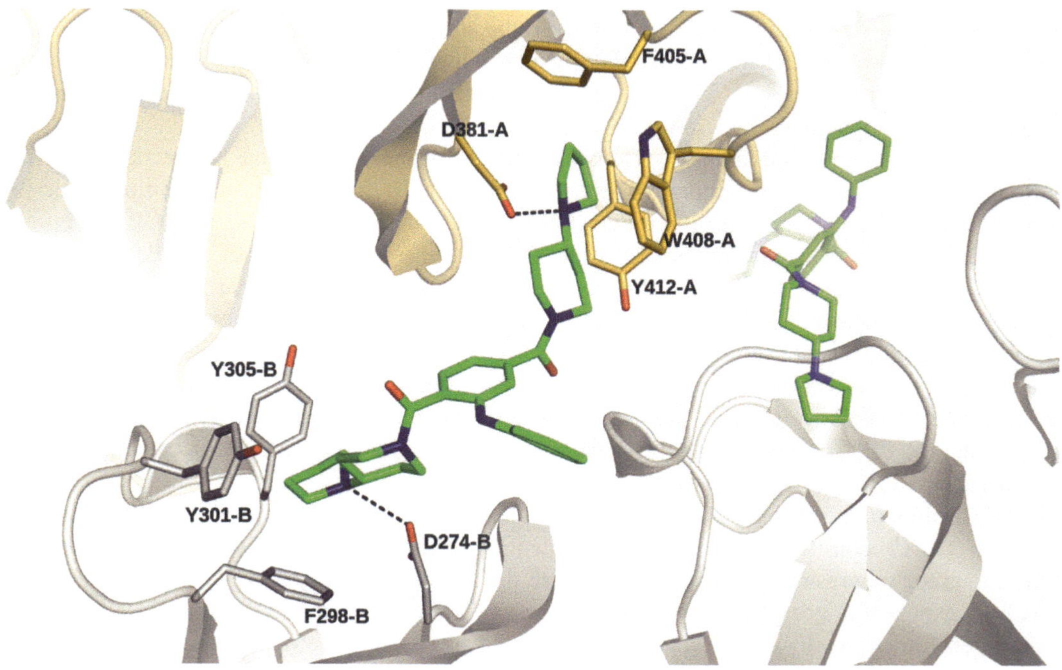

FIGURE 11.7

Unique binding mode of L3MBTL3 with UNC1215 showing a 2:2 polyvalent interaction. Chains A and B of L3MBTL3 are shown in white and yellow; respectively. UNC1215 is depicted as green sticks.

With the knowledge of the requirements for potent L3MBTL3 binding originating from the SAR with UNC1215, it was possible to improve the compound's structure in terms of selectivity against other MBT domain-containing proteins while maintaining in vitro and cellular potency which led to the discovery of UNC1679 (see Figure 11.6) [28].

11.3.2 THE CHROMODOMAIN

The name of this Royal family member is derived from "chromatin organization modifier" which already implicates its role as a histone reader protein. It is made up of 40–50 amino acids that form a three-stranded β-barrel structure with an α-helix connecting the second and third strand as well as two α-helices following the third β-strand [29].

In the two best-known subfamilies of chromodomain-containing proteins, namely HP1 and Polycomb, the histone target resides in a surface groove where the aromatic cage is made up by three aromatic residues often supplemented by one or more acidic side chains that interact with the methyl-lysine residue. Typically, two distinct recognition modes can be observed in methyl-lysine reader–ligand complexes: the cavity insertion- and the surface groove-recognition mode. The first mode generally occurs in complexes containing ligands with a lower methylation state (Kme0 and Kme1) and the latter mode is commonly seen in complexes with binding partners of a higher methylation state (Kme2 and Kme3). In the cavity insertion recognition mode, the methyl ammonium moiety of the methyl-lysine residue is buried within a deep protein cleft which potentially provides the pocket with a size-sensitive selection filter. In the surface-groove recognition mode, the binding pockets are both wider and more accessible, which allows the methyl-lysine side chain to remain along a protein surface groove. In many cases this leads to markedly but less rigid discriminations between different lysine methylation states [2].

In the case of chromodomains, various degrees of lysine methylation are recognized with a preference for trimethyl- over dimethyl-lysine residues due to their surface recognition mode [30]. The chromobox homolog (CBX) proteins CBX2, CBX4, CBX6, CBX7, and CBX8 represent a major class of human chromodomain-containing proteins and are part of the Polycomb group of proteins. They were found to be the paralogs of the *Drosophila* protein Polycomb [31]. The Polycomb group proteins form multiprotein polycomb repressive complexes (PRC) of which two main complexes are present in mammals (PRC1 and PRC2). Members of the PRC are frequently prone to dysregulation. Representing transcriptional repressors, Polycomb group genes are implemented in development, stem cell maintenance and in cancer progression [32]. Each of the five human CBX proteins recognizes and binds to H3K27me3, a mark that is created by another member of the PRC (the lysine methyltransferase EZH2).

The chromobox homolog 7 (CBX7) belongs to the best-studied representatives of the Polycomb paralogs. CBX7 constitutes an interesting potential drug target, as it is involved in silencing multiple tumor suppressors and it is implicated in different malignancies [33].

Cbx7 expression goes hand in hand with proliferation in various stem cell- and stemlike cancer cell lines [34–40]. An involvement of CBX7 in cancer development is further underlined by the observation that the depletion or knockdown of CBX7 in prostate cancer cell lines induces a senescent phenotype with reduced cell proliferation [36,37]. Moreover, *cbx7* expression was shown to be significantly increased in clinical samples of hormone-resistant prostate cancer relative to hormone-dependent prostate cancers [33]. Also, in ovarian clear cell adenocarcinoma *cbx7* expression was linked to poor prognosis [40]. Additional malignancies, where CBX7 seems to play a central role, are lymphomas [39] and T-cell leukemias [41]. However, a controversial role of CBX7 in regulation of cell growth and tumorigenesis is discussed as CBX7 was also found to exhibit tumor suppressor properties, for example, in lung adenomas and other carcinomas [42].

Nevertheless, the above-mentioned facts highlight the benefit potentially arising from further exploring CBX7 as a drug target. A first approach to find inhibitors of CBX7 resulted in the peptide-derived compound 64 by Simhadri and coworkers after earlier virtual-screening-based efforts failed to yield small-molecule lead compounds. In their studies, small trimethyl-lysine-containing peptides, derived from a truncated (Ac-FALKme$_3$S-NH$_2$) methylated SETDB1 sequence which was shown to have a somewhat higher *in vitro* affinity for CBX7 compared to the H3K27me3 sequence, were used as basic scaffolds. From this starting point (IC_{50} 12 μM) potent ligands were developed for the disruption of the CBX–H3K27me3 complex. Analogs of the initial peptide were synthesized by varying different amino acids of the scaffold (Leu, Ala, Ser, Phe) while keeping the trimethylated lysine residue. For example, it was found that substituting the C-terminal serine by simple H-bond-donating substituents like aminobenzimidazole, propanediols, or a pendent ammonium group increases the binding potency. Finally, a selected set of compounds that comprise all the favorable modifications to the FALKme3S-scaffold previously gained by SAR studies was synthesized, which revealed ligands with affinities around 200 nM for CBX7 and tenfold selectivity over CBX8 and even 400-fold selectivity over CBX1, representing the first reported inhibitors of any chromodomain (see Figure 11.8) [33].

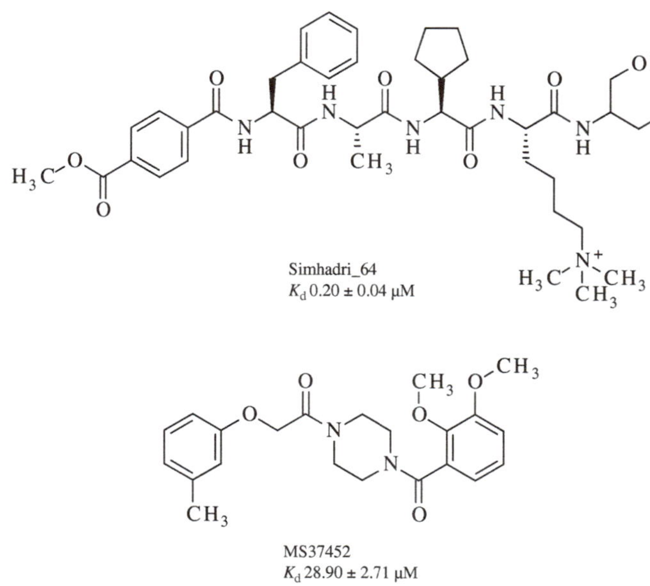

FIGURE 11.8

Peptide-like compound (Simhadri_64) and a small-molecule screening hit (MS37452) representing ligands of the chromodomain of CBX7.

Only recently, the first set of small-molecule inhibitors of the CBX7–H3K27me3 interaction has been reported by Ren et al. [43]. Like in the previously described study, they used a SETDB1-derived peptide (SETDB1K1170me3) for their screening approach, which exhibits a higher affinity toward the CBX7 chromodomain compared to the histone-derived peptide containing H3K4me3. Subsequent virtual screening of the FDA-approved drugs library, which comprises 2560 molecules and a L1 library consisting of 100160 compounds selected from commercial sources, revealed 16 initial hits from the FDA library and 40 from the L1 library. These hits were further validated *in vitro* by using a fluorescence anisotropy-based binding assay with a fluorescein isothiocyanate (FITC)-labeled SETDB1-K1170me3-derived peptide as an assay probe. By using 2D ^{1}H-^{15}N HSQC spectral analysis and the fluorescence anisotropy binding assay, five initial hits (sennoside A, suramin, aurin tricarboxylic acid, trypan blue, and Evans blue) from the FDA database and one hit (MS37452) from the L1 library were verified as ligands of the chromodomain of CBX7 (see Figure 11.8). NMR titration revealed a K_d value of 29 μM for MS37452 to CBX7 and with the help of the fluorescence anisotropy binding assay K_i values of 43 and 55 μM for MS37452 were determined for the disruption of CBX7-H3K27me3 and -H3K9me3, respectively. Next, a cocrystal structure of the CBX7–MS37452 complex was solved and by point mutation experiments it was confirmed that the amino acid residues that form the aromatic cage are essential for the CBX7–MS37452 interaction. Selectivity evaluation of MS37452 toward other chromodomain containing proteins revealed a threefold weaker affinity for CBX4 and even a 10-fold lower binding potency for CBX2, CBX6, and CBX8 than for CBX7. Moreover, MS37452 shows almost no binding to the chromodomains of heterochromatin protein 1 (HP1) β, γ, and α. As opposed to the rather surface groove-exposed binding mode of CBX7 to MS37452, the MBT domain containing and previously discussed L3MBTL1 and L3MBTL3 domains contain a rather deep and narrow binding pocket for lysine residues of a lower methylation state, which is occupied by the amine group of the small-molecule ligand UNC1215. This explains the major difference in affinity between these two methyl-lysine reader domain ligands. The lead compound, MS37452, was further characterized on a cellular level where it induced transcriptional depression of *p16/CDKN2A* by disrupting CBX7-CHD binding to H3K27me3 at the *INK4A/ARF* locus in PC3 prostate cancer cells, which was determined by chromatin immunoprecipitation (ChIP) [43].

11.3.3 THE TUDOR DOMAIN

The next group of proteins that is discussed is the structurally diverse family of Tudor domain-containing proteins. A typical feature of the Tudor domains is their wide range of recognized motifs. This is emphasized by their ability to not only interact with all different states of lysine methylation but also with methylated arginine residues [44–46]. A representative example of the Tudor domain family is the p53-binding protein 1 (53BP1) which, as its name implies, binds to the tumor suppressor p53 methylated at lysine 382 (p53K382me2) and to histone H4K20me2 [47]. As H4K20me2 is predominant at DNA damage sites, 53BP1 is capable of attracting p53 to its target. Upon DNA double-strand breaks (DSB) a specific histone code is produced which can be recognized by 53BP1 not only by binding to H4K20me2 but also to a ubiquitinylation mark on histone H2A at lysine 15 (H2AK15ub) in a bivalent mode [48]. This emphasizes that 53BP1 presumably plays an important role in DNA DSB repair by nonhomologous end joining.

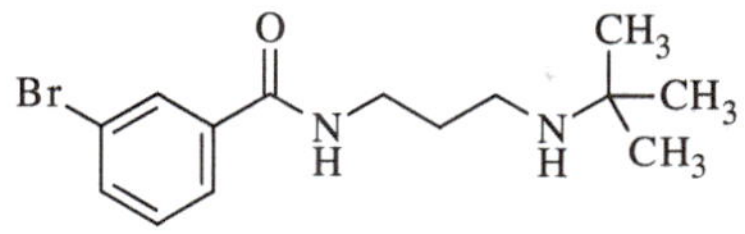

UNC2170
K_d 22 ± 2.5 μM

FIGURE 11.9

Structure of a fragment-like ligand of the Tudor domain of 53BP1.

Encouragingly, first attempts to find new small-molecule ligands of 53BP1 have been successful. Perfetti and coworkers identified UNC2170 as a micromolar fragment-like ligand of 53BP1 through a cross-screening approach (see Figure 11.9).

UNC2170 exhibits at least 17-fold selectivity for 53BP1 as compared to other methyl-lysine binding proteins that were tested. In structural studies they were able to show that the *tert*-butyl amine moiety of UNC2170 anchors the compound in the methyl-lysine binding pocket of 53BP1. They seem to be competitive to cognate peptides of native methyl-lysine ligands. UNC2170 also shows cellular activity by suppressing class switch recombination, a process which requires a functional 53BP1 Tudordomain [49].

Another interesting example of Tudor domain-containing proteins is the methyl-lysine demethylase JMJD2A which hence is both a demethylating eraser enzyme and a methyl-lysine reader protein. JMJD2A catalyzes the demethylation reaction of H3K9me2/3 and H3K4me2/3. Besides two tandem Tudor domains, JMJD2A also contains two PHD finger domains. In cocrystal structures it was shown that JMJD2A associates with H3K4me3 and also with H4K20me3 [50]. In Table 11.1, cancer forms are listed, where JMJD2A was shown to be overexpressed or dysregulated, which emphasizes the need for small-molecule probes that would help to evaluate the JMJD2A-Tudor domain as a drug target, besides inhibitors of its demethylating function [131] (see Table 11.1).

The Polycomb PRC2 complex (discussed above) not only comprises chromodomain-containing proteins (see CBX family) but also further Tudor domain-containing subunits, namely PHF1, MTF2, and PHF19. These three subunits bind to H3K36me3 and have the shared purpose of contributing to the recruitment of the PRC2 complex to its target genes for histone methylation [68]. It was observed that fusion proteins of PHF1 (MEAF6–PHF1, JAZF1–PHF1, and EPC1–PHF1) are involved in endometrial stromal sarcoma [132,133].

The Tudor-like containing protein Spindlin1 recognizes H3K4me3 exclusively by the second of three domains [71]. Initially, Spindlin1 was found to be associated with the meiotic spindle in mice after which the protein was named [134]. Later, in 2004, the human homolog *spindlin1* was discovered as an overexpressed gene fragment in ovarian cancer cells. Furthermore, *spindlin1* was shown to be highly expressed in various types of malignant tumor tissues, including nonsmall-cell lung cancers, ovarian tumors, and some hepatic carcinomas [135]. In another study several observations were made indicating that the *spindlin1* gene may influence tumorigenesis. For example, *spindlin1*-overexpressing cells undergo a complete morphological change, show an increased cell growth as well as altered cell cycle phases [136]. More precisely, the abnormal expression of *spindlin1* causes cell-cycle delay in metaphase and leads to chromosome instability [137,138].

Table 11.1 Implementation in Cancer Development, Progression and Histone-Binding Partners of Selected Reader Proteins: Proteins with a Reported Relevance as Oncoprotein or Reader Genes Shown to be Upregulated in Tumor Samples or Cancer-Derived Cell Lines are Listed and Categorized After Their Respective Binding Domain

Reader Domain	Protein	Cancer Involvement	Histone Ligands
Chromodomain	CBX7[a]	Ovarian clear cell adenocarcinoma [40]	H3K9me3 [30]
		Follicular lymphoma [39]	H3K27me3 [30]
		Prostate cancer [36]	
		Gastric cancer [51]	
	HP1α (CBX5)[a]	Lung cancer [52]	H3K9me2/3 [53]
	HP1β (CBX1)[a]	Prostate cancer [54,55]	H3K9me3 [56]
	HP1γ (CBX3)[a]	Prostate cancer [54,57]	H3K9me2/3 [58]
		Nonsmall-cell lung cancer [58,59]	
		Osteosarcoma [60]	
Tudor	53BP1	Breast cancer [61]	H3K36me2 [62]
		Ovarian carcinoma [63]	H4K20me2 [47]
	JMJD2A (TDRD14A, KDM4A)	Gastric cancer [64]	H3K4me3 [50]
		Endometrial carcinoma [65]	H4K20me3 [50]
		Breast cancer [66]	
	PHF1	Endometrial stromal sarcoma [67]	H3K36me3 [68,69]
	Spindlin1	Ovarian cancer [70]	H3K4me3 [71]
	Lamin B receptor (TDRD18)	Breast cancer [72]	H4K20me2 [73]
		Papillary thyroid carcinoma [74]	
	UHRF1 (ICBP90, Np95)	See PHD domain	H3K9me3 [75–77]
PWWP	WHSC1 (NSD2, MMSET)	See PHD finger	H3K36me2/3 [78]
	PSIP1 (p75, LEDGF)	Pediatric acute myeloid leukemia [79]	H3K36me3 [80–82]
		Acute myelogenic leukemia [83]	
		Prostate cancer [84]	
		Colon cancer [84]	
		Thyroid cancer [84]	
	HDGF	Hepatocellular carcinoma [85]	
		Nonsmall-cell lung cancer [86]	
		Pancreatic cancer [87]	
		Esophageal carcinoma [88,89]	
		Nasopharyngeal carcinoma [90]	
		Ewing's sarcoma [91]	
		Gallbladder cancer [92]	
	DNMT3A	Hepatocellular carcinoma [93]	H3K36me3 [94]
		Gastric cancer [95]	

Table 11.1 Implementation in Cancer Development, Progression and Histone-Binding Partners of Selected Reader Proteins: Proteins with a Reported Relevance as Oncoprotein or Reader Genes Shown to be Upregulated in Tumor Samples or Cancer-Derived Cell Lines are Listed and Categorized After Their Respective Binding Domain *Continued*

Reader Domain	Protein	Cancer Involvement	Histone Ligands
PHD finger	WHSC1 (NSD2, MMSET)	Serous ovarian carcinoma [96]	H3unmod [97]
		Hepatocellular carcinoma [98]	
		Multiple myeloma [99]	
		Glioblastoma [100]	
		Urinary bladder cancer [101]	
		Colon carcinoma [101]	
	JARID1A (RBBP2, KDM5A)	Nonsmall-cell lung cancer [102]	H3K4me3 [103]
		Gastric cancer [104]	
		Hepatocellular carcinoma [105]	
		Lung cancer [106]	
		Breast cancer [107]	
	Pygo PHD(via β-catenin)	Bone cancers [108]	H3K4me2/3 [109]
		Ovarian cancer [110]	
		Prostate cancer [111]	
	UHRF1 (ICBP90, Np95)	Gallbladder cancer [112]	H3unmod [75,113]
		Breast cancer [114]	
		Colorectal cancer [115]	
		Bladder cancer [116]	
		Hepatocellular carcinoma [117]	
		Lung cancer [118]	
		Laryngeal squamous cell carcinoma [119]	
	PHF8	Nonsmall-cell lung cancer [120]	H3K4me3 [121]
		Esophageal squamous cell carcinoma [120]	
		Prostate cancer [122]	
WD40 repeat	WDR5	Mixed-lineage leukemia [123]	MLL-"Win"-motif MLL1-R3765 [124]
		Bladder cancer [125]	H3unmod/H3R2 [126–128]
		Prostate cancer [129]	
		Breast cancer [130]	

[a]*Notably, for these reader proteins also tumor suppressor properties are reported indicating a controversial role in the regulation of cell growth and tumorigenesis.*

An upregulation of *spindlin1* expression in clinical tumors was shown by Wang and coworkers in 2012. They assume that Spindlin1 promotes cancer cell growth via WNT/TCF-4 signaling activation [70]. Interestingly, it was recently found that the affinity of the binding of Spindlin1 to H3K4me3 is further increased by an asymmetric methylated arginine in position 8 (H3R8me2a), which is recognized by the first Tudor-like domain. Furthermore, *in vivo* functional experiments give hints that Spindlin1 activates the signaling downstream of Wnt/β-catenin from protein arginine methyltransferase 2 (PRMT2), which creates H3R8me2a marks, and the MLL complex, which is responsible for the generation of trimethylated H3K4 residues. Together, this results in a specific "K4me3-R8me2a" pattern which can be recognized by Spindlin1. The involvement of Spindlin1 in Wnt/β-catenin signaling is further underlined by the observation that mutagenesis of the Spindlin1 reader pocket weakens the activation of Wnt target genes [139]. Wnt/β-catenin signaling plays a role in a wide range of cellular processes, including proliferation, differentiation, tissue homeostasis, tumorigenesis, apoptosis, and cell survival. Targeting this pathway could open new strategies for the treatment of a variety of diseases [140,141].

Up to now only one small-molecule ligand (UNC2170) of a Tudor-domain-containing protein has been reported. The development of lead structures for these targets is challenging due to the rather low druggability of the Tudor domains [142]. The difficulty in finding specific ligands with high affinity is also explained by the obvious promiscuity within the Tudor domain family regarding their native ligands as well as the tolerated range of methylation states. However, through their involvement in the development and progression of cancer and other malignancies, a lot of Tudor-domain-containing proteins display promising targets for the development of potential drug candidates and new therapeutic concepts in the treatment of diseases.

11.3.4 THE PWWP DOMAIN

The last member of the Royal family presented here is the PWWP domain. So far, no success in finding small-molecule ligands for these proteins has been reported. The PWWP domain was first described as the protein–protein interaction domain of the WHSC1 protein. The PWWP domain, like other members of the Royal family, contains a five-stranded β-barrel core structure with an insertion motif of variable length connecting the second and third strand and a C-terminal α-helical structure [12,78]. Interestingly, PWWP domains were first identified as DNA-binding domains and only recently were found to also bind histones [78,143]. As interaction studies have shown a very weak binding affinity of PWWP domains toward methyl-lysines on histone tails, a dual binding mechanism of PWWP domains toward histones and nucleic acids was assumed [78]. This hypothesis was supported by the observation that the PWWP domain-containing protein PDP1 (PWWP domain protein 1) is capable of simultaneously binding both double-stranded DNA and methyl-lysine histones [143]. Along with the fact that PWWP domains were found to bind lysine residues with a higher degree of methylation (Kme3) they show some typical features of this kind of reader proteins. For example, the binding pocket is open and rather surface-exposed which makes it possible to incorporate the bulkier Kme3 moiety (see surface-recognition mode, discussed in Section 11.3.2). This further leads to less selectivity toward the degree of lysine methylation. The aromatic cage of PWWP domains is made up of three conserved aromatic residues. Few interactions between the protein and its histone ligand are observed apart from the interaction with the methyl-lysine residue [12].

PSIP1 (PC4 and SFRS1 interacting protein 1) is a PWWP-domain-containing protein that exists in two isoforms (p52 and p75). The p75 isoform is also described as LEDGF (lens epithelium-derived growth factor) and was found to play roles in different diseases. One of its functions is transcriptional coactivation occurring in lentiviral integration. The capability of PSIP1 to bind chromatin via its PWWP domain is essential for the recruitment of human immunodeficiency virus (HIV-1) to active transcription units on chromatin [144]. PSIP1 shows antiapoptotic activity [145] and it was reported to be an oncogenic protein [144] which controls a caspase-independent lysosomal cell death pathway. Furthermore, it was found that PSIP1 interacts with MeCP2 (methyl-CpG binding protein) and consequently enhances its transcriptional activity. In this way, PSIP1 acts as a transcriptional coactivator that promotes resistance to oxidative stress- and chemotherapy-induced death of cancer cells [146].

Through its PWWP domain PSIP1 binds H3K36me3, a mark that is predominantly associated with actively transcribed chromatin [144]. Importantly, PSIP1 affinity to H3K36me-containing nucleosomes was shown to be 1000-fold higher compared to its affinity to the H3K36me3 peptide fragment. This supports the above-mentioned dual binding mode theory of PWWP-domain-containing proteins and the importance of nucleosomal DNA for the binding potency of this class of reader proteins [80].

Another example that highlights the usefulness of small-molecule ligands of PWWP domains for a potential use as drug candidates is the PWWP-domain-containing histone lysine methyltransferase WHSC1 (Wolf-Hirschhorn syndrome candidate 1) which combines an epigenetic writer and reader protein. WHSC1 was shown to play a major role in multiple myeloma. An overexpression of *whsc1* in multiple myeloma cells was observed upon a chromosomal translocation (t(4;14)) leading to an increased proliferation of multiple myeloma cells along with global changes in chromatin. This goes hand in hand with an unfavorable prognosis of multiple myeloma, a disease which is characterized by the malignant proliferation of terminally differentiated B cells [147]. In addition to its two PWWP domains, WHSC1 also contains five PHD finger domains [97]. The lysine methyltransferase activity of WHSC1 is implemented in the formation of H3K36me2 marks which correspond with activated transcription of typical multiple myeloma oncoproteins such as c-Myc [147]. The expression of various genes involved in the regulation of cell-cycle progression is further affected by dysregulation of WHSC1 [148]. Apart from multiple myeloma, *whsc1* expression is recurrently found in multiple different cancers compared to the respective equivalent normal tissues. In some carcinomas, high *whsc1* expression is even linked with tumor aggressiveness [101].

11.4 THE PHD FINGER FAMILY OF EPIGENETIC READER PROTEINS

The PHD fingers are regarded as a separate family from the Royal family of epigenetic reader proteins. However, similar to the Tudor domains, the PHD fingers are able to interpret different methylation states, especially on H3K4, as well as unmodified H3 tails. Here again, an involvement of dysregulated PHD fingers in various human diseases, including immune- and neurological disorders and cancers, is evident (see Table 11.1) [149]. The PHD finger is made up of about 65 amino acid residues and it is folded into two antiparallel β-sheets followed by two small α-helices.

Characteristically, the structure of the PHD finger contains a conserved zinc-coordinating Cys4-His-Cys3 motif [150]. The aromatic cage of PHD fingers is built from 2–4 aromatic residues with cysteine residues being frequently encountered as part of the aromatic cage.

Besides its epigenetic eraser functionality as lysine-specific demethylase, JARID1A also has the ability to bind H3K4me3 via its three PHD finger domains. A fusion protein of the C-terminus of JARID1A, including its third PHD finger, and nucleoporin protein 98 (NUP98) is implicated in acute myeloid leukemia (AML) [151]. This fusion protein (NUP98/JARID1A) was further shown to be a recurrent genetic abnormality in pediatric acute megakaryoblastic leukemia (AMKL) [152]. Another fusion protein of the C-terminal PHD finger of JARID1A and PHF23 constitutes a potent oncoprotein involved in the arrest of hematopoietic differentiation and induced AML in murine models. During these processes the recognition of H3K4me2/3 by the PHD finger was shown to be essential [103].

As for other previously discussed classes of binding proteins, first attempts toward the discovery of novel small-molecule ligands of PHD finger domains revealed promising results. With the help of a new Halo-Tag-based method Wagner and coworkers performed a small-molecule screen of the NIH Clinical Collection 1, which contains 446 compounds, all of which have undergone phase I – III clinical trials and address a broad range of therapeutic indications, against the third PHD finger of JARID1A (JARID1A-PHD3). They found the approved drugs disulfiram, amiodarone, and tegaserod as initial hits that showed significant inhibition of JARID1A-PHD3 binding to a H3K4me3-derived peptide. These hits were further validated by affinity pull down and fluorescence polarization (FP) assays. In a next step, a possible improvement of the amiodarone structure was evaluated by a fragment-based dose dependence study, which exposed the importance of an alkylated amine chain for the binding potency of the compounds. The rationale deduced from the previously gained results led to the identification of inhibitors with a 10-fold higher potency than amiodarone, including di-N-desethylamiodarone (IC_{50}: 26 μM) and trimethylamiodarone (WAG-003, IC_{50}: 30 μM; see Figure 11.10) [153].

A useful starting point for the development of new PHD-finger ligand lead structures was reported by Miller and coworkers in 2014.

The potent oncogene *β-catenin* is a key effector of Wnt signaling and activating mutations in the *β-catenin* gene have been proven in many types of cancer. Although this fact makes β-catenin a potential drug target it seems rather challenging to successfully target this protein directly with small molecules. The reason for this is that β-catenin is attracted to DNA by a set of adaptor proteins and TCF and that the interface of β-catenin with these TCF includes the major part of its structured domain, namely the Armadillo Repeat domain (ARD). The size of this domain is extensive and it also comprises the interface for the negative regulators of β-catenin (e.g., APC and Axin) whose binding sites even overlap with that of TCF. These facts highlight that β-catenin itself does not constitute a proper drug target. In turn, looking at the mechanism, how β-catenin is further targeted to DNA, reveals some attractive possibilities of interruption by small molecules. First, BCL9 adaptor proteins are needed which interact via a short α-helical domain (HD2) with a separate binding site at the N-terminus of the ARD of β-catenin. BCL9 binds to β-catenin simultaneously with TCF. Next, BCL9 adaptors bind to the rear of the so-called Pygo PHD finger by a separate domain (HD1). This interaction induces a slight allosteric change of the PHD finger, which in turn facilitates the binding of the protein complex to the methylated histone H3 tail (H3K4me) through the frontal surface of the PHD finger. Thus, oncogenic β-catenin could be inhibited by

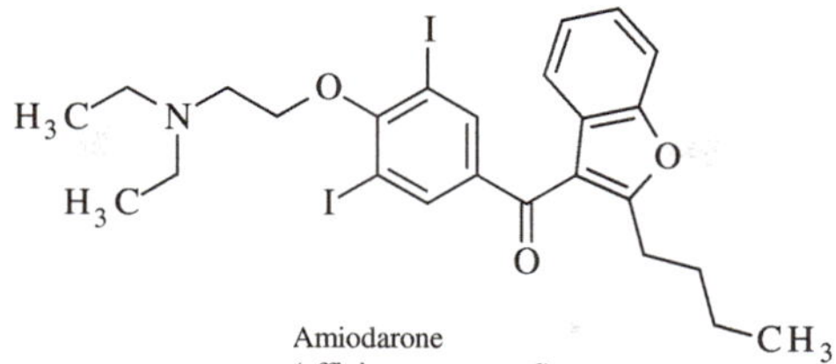

Amiodarone
(affinity not reported)

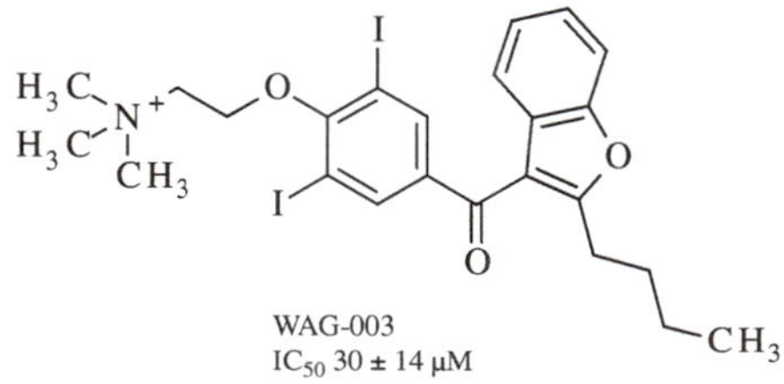

WAG-003
IC_{50} 30 ± 14 μM

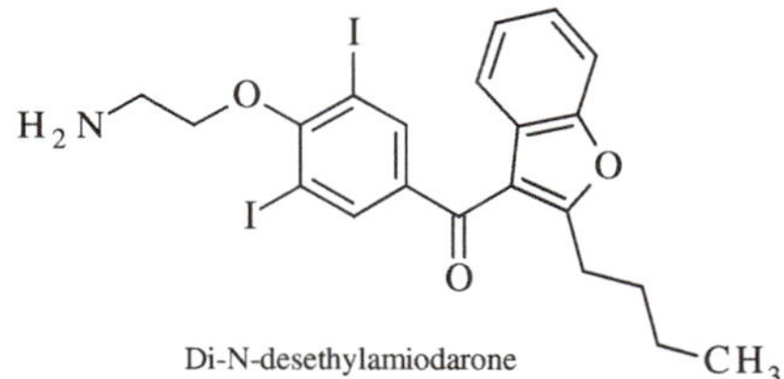

Di-N-desethylamiodarone
IC_{50} 26 ± 15 μM

FIGURE 11.10

Structures of small-molecule ligands of the JARID1A-PHD3 domain.

targeting these three unique and fairly small protein–protein interfaces of the Pygo-BCL9 complex. Indeed, Miller and coworkers have been successful with identifying the first compounds to address these targets [154]. As the HD1-interacting surface of the Pygo PHD finger is hydrophobic and unstable in aqueous solution, they decided to use the Pygo PHD finger in complex with HD1 for small-molecule (up to 500 Da) ligand screening. They performed an *in silico* aided screen for chemical fragments that bind to the PHD-HD1 complex by two-dimensional nuclear magnetic resonance (NMR). By further investigating relationships SAR of the initial hit fragments, they were able to identify two sets of privileged substructures which exhibit tight binding to the distal part of the K4me pocket and to a highly conserved narrow cleft at the rear of the PHD neighboring its HD1-binding surface. Eventually, one benzimidazole-containing scaffold was found which competes with the native ligand (the methylated H3 tail) of the PHD finger and docks into the distal part of the K4me pocket(see CF16 in Figure 11.11) [154]. In general, fragment-like compound screenings can potentially reveal useful starting points for the further development of chemical lead structures by combining different favorable structural elements. In Table 11.1, further interesting PHD finger-containing proteins are listed that might display auspicious targets for drug development.

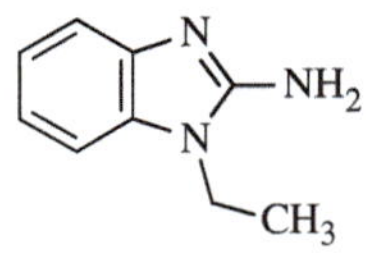

CF16

FIGURE 11.11

Structure of a chemical fragment ligand of the Pygo-PHD domain.

11.5 THE WD40 REPEAT DOMAIN FAMILY

Like the Royal family and the PHD fingers, the WD40 repeat domain-containing proteins are classified as an independent family of epigenetic readers. The WD40 repeat domain is among the most abundant domains in eukaryotic genomes and some are also involved in chromatin complexes. WD40 repeat domains are often part of proteins that mediate various protein–protein interactions. For instance, WD40-repeat-containing proteins are involved in scaffolding and the cooperative assembly and regulation of dynamic multisubunit complexes. WD40 repeats are made up of 40–60 amino acid residues flanked by a conserved Gly-His sequence near its N-terminus and another conserved Trp-Asp sequence at its C-terminus. WD40 domains typically show a low level of sequence conservation and a high degree of functional diversity making them rather promiscuous binding partners that undergo manifold types of protein–protein interactions [155].

WDR5 (WD40 repeat protein 5) is a core component of the Set1/MLL complexes where it is needed for assembly and activity of the complexes as lysine methyltransferases. The complexes play major roles in development and they are excessively expressed in many cancers. The key function of WDR5 in this complex was initially found to be its direct association with histone H3K4me2 which guides the methyltransferase complexes to its target, namely histone H3K4, where it creates di- and trimethylation marks [156]. However, structural studies later showed that WDR5 is able to bind histone H3 mainly through an arginine residue (H3R2) and that both, unmodified and methylated H3K4 peptide fragments, are equally recognized [126,127,157]. Furthermore, the crystal structures of WDR5 bound to unmodified, mono-, di-, and trimethylated H3K4 peptides exhibit the absence of the common methyl-lysine recognition motif, that is, the aromatic cage. A hydrogen bond with a glutamic acid residue (Glu322) at the surface of the protein is the sole interaction with the lysine residue of the binding partner [127,128]. Looking at the structure of WDR5 in more detail reveals an arginine binding pocket that accepts unmodified and dimethylated arginine residues (see Figure 11.12).

Regarding histone effects, the symmetric dimethylation of H3R2 (H3R2me2s) attracts WDR5 as opposed to the asymmetric dimethylation at the same site (H3R2me2a) which excludes WDR5 binding [158,159]. This implies that WDR5 is rather an arginine reader than a methyl-lysine reader. This is even more underlined by the observation that WDR5 binds several arginine-containing peptides, with some of them even lacking a lysine residue (e.g., SET1A, SET1B, and MLL1-4) [124,160].

WDR5 is therefore able to recognize either histone H3 or its catalytic partner SETD1A/B or MLL1-4 which contain the so-called Win motif, representing a minimum binding sequence

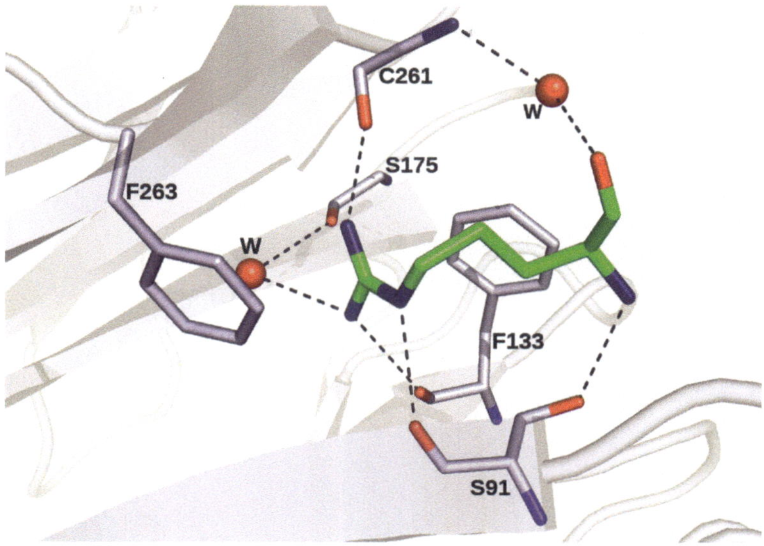

FIGURE 11.12

WDR5 recognition of arginine residues. The arginine residue is shown as green sticks and the protein residues involved in the interaction as white sticks. Water molecules are depicted as red spheres.

of WDR5. The Set1/MLL methyltransferase complexes are prone to genetic aberrations leading to various genetic diseases including cancers. Up to 80% of infant acute leukemia and about 5–10% of adult AML and acute lymphoid leukemia (ALL) cases arise from fusion proteins that are generated by chromosomal rearrangements of MLL (mixed-lineage leukemia protein; KMT2) with over 60 different translocation partners [161]. The WDR5–MLL1 interaction within the Set1/MLL complexes is essential for their assembly and enzymatic activity. Therefore the identification of inhibitors of this interaction might serve as a potential starting point for the development of new therapeutic strategies of leukemia [160,162]. In 2013, encouraging findings were made by Senisterra and coworkers who identified the first small-molecule inhibitor of the WDR5–MLL1 interaction by screening a library of 16,000 compounds. The most promising hit (WDR5-0101) revealed a K_d value of 12 μM (see Figure 11.13).

Further screening of a library of 6 million commercially available compounds led to 119 molecules with similarities to WDR5-0101. Among these compounds, WDR5-0103 exhibits the highest binding affinity with a K_d value of 450 nM (see Figure 11.13) [163].

The observation that the arginine residue contributes strongly to the binding affinity in the WDR5–peptide complex was taken up by Karatas and coworkers in order to design a series of peptidomimetic inhibitors based on the accessible WDR5 complex structures [164]. Starting from a minimum binding motif of WDR5 (-CO-ARA-NH-) three peptidomimetic compounds were identified (MM-101, MM-102, and MM-103) that bind to WDR5 with an estimated K_i of <1 nM (see Figure 11.13) [164]. Next, in the same lab, Cao and coworkers synthesized a cyclic derivative (MM-401) of MM-101 which revealed an over threefold higher affinity to WDR5 compared to MM-101 (see Figure 11.13). In a competitive FP experiment MM-401 showed an IC_{50} value of

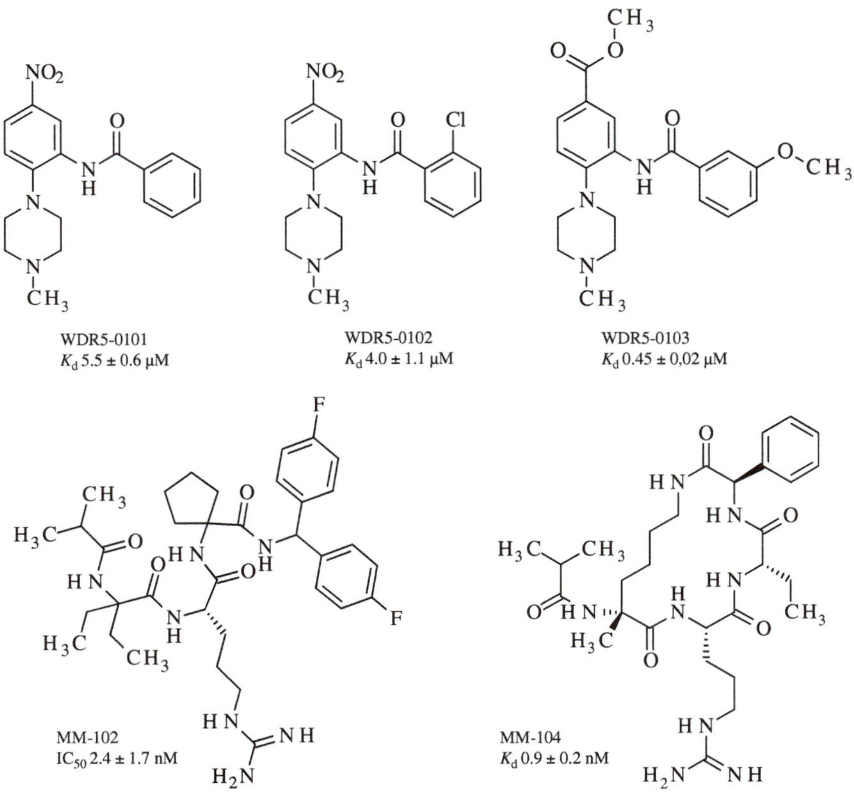

FIGURE 11.13

Structures of small-molecule inhibitors of the WDR5–MML interaction.

0.9 nM in disrupting the interaction of WDR5 with MLL1. MM-401 was further demonstrated to specifically inhibit the enzymatic activity of MLL1 by blocking the MLL1–WDR5 interaction and thus the assembly of the complex assembly. MM-401 is cytotoxic to MLL-associated leukemic cells in a MLL1-mediated H3K4 methylation-dependent manner by inducing cell-cycle arrest, apoptosis, and myeloid differentiation, whereas normal cells stay unaffected, making MM-401 an interesting candidate to be further evaluated as a potential therapeutic drug [123].

11.6 COMPUTATIONAL METHODS

The number of structural data for methyl-lysine readers has increased over the last few years. The PDB contains many crystal structures of various reader proteins in the apo form or in complex with histone-derived peptides or small-molecule ligands. These data enable scientists to improve their drug discovery strategies with the help of computational methods.

An interesting study was reported by Santiago and coworkers, who examined the general druggability of the methyl-lysine binding sites of a variety of reader domains, for which peptide-bound or inhibitor-bound structures were solved, by calculating the Dscore. As a promising result, their data show that druggable members are found for each reader fold, although methyl-lysine readers as a whole seem less druggable than for example the BET bromodomains, that recognize acetylated lysine residues. They assumed that the druggability among the methyl-lysine reader families is highly variable and that the PWWP and WD40 domains represent the most promising target classes, whereas for other members of the Royal family, a varying degree of druggability is predicted. Better druggability of the binding pockets could be envisaged if ligand binding induces conformational changes of the protein or if secondary binding pockets are present. The tendency of reader proteins to be part of multiprotein complexes, which may change their topology, also potentially affects the druggability of the binding sites, as mentioned, for example, in the case of the Pygo PHD protein [142].

The number of small-molecule ligands of methyl-lysine reader domains published so far is still fairly low. As already summarized in this chapter, only small-molecule ligands of L3MBTL1, L3MBTL3, CBX7, 53BP1, JARID1A, and Pygo PHD have been reported. In order to overcome this status, computational methods could help to identify new small-molecule inhibitors of methyl-lysine readers aided by the availability of increasing structural information on these proteins. As a matter of fact, the above-mentioned approaches of Kireev and coworkers regarding the identification of novel L3MBTL1 inhibitors was guided by *in silico* approaches. Two screening methods were employed on a large database of commercially available compounds. First, a substructure search for mono- and dimethyl-lysine moiety containing compounds was conducted. A three-point pharmacophore was built based on the interactions between L3MBTL1 and its histone peptide ligand as seen in the crystal structure. The hits that were acquired from database screening were docked into L3MBTL1 afterwards. As already mentioned, several hit compounds obtained from virtual screening actually revealed *in vitro* affinity in the micromolar range [21].

Another recent study introduces an approach to target histone readers via a pharmacophore search that was implemented to screen almost 98,000 protein–ligand complexes from the protein database (PDB) for all proteins that embody an aromatic cage. The aromatic cage of a set of representative members of the Royal family was simulated by various four-point pharmacophores which were designed out of either three aromatic amino acid residues plus an acidic feature (Asp or Glu) or four aromatic amino acid side chains. A set of ligands was extracted from the so-obtained protein–ligand complexes from which, in turn, different chemotypes were derived that are able to address the aromatic cage, including quaternary ammonium compounds and sulfonium derivatives, primary, secondary, and tertiary amines as well as imines and aromatic rings. However, these hits have not yet been verified by *in vitro* data but, as this approach revealed some substructures that are known to address aromatic cages, this work could aid in designing new inhibitors of methyl-lysine readers in the future [165].

11.7 CONCLUSION

Taken together, the field of epigenetic readers offers great possibilities and potential for drug development aimed at the treatment of various diseases, especially cancer. The well-defined structure of the aromatic cage as the major binding site of methyl-lysine readers enables feasible *in silico* and *in vitro* screening possibilities compared to otherwise rather challenging discovery of protein–protein interaction inhibitors. Increasing knowledge is gained about the structural requirements of small-molecule ligands that might serve as starting points for the development of drug candidates targeting methyl-lysine readers. Selective high-quality probes will aid strongly in assessing the druggability and utility of methyl-lysine binding proteins in drug discovery. Initial small-molecule ligands of a still-small number of reader proteins have been reported, but however, for many potential targets no small-molecule ligands are known yet, and in some cases only rather weak inhibitors have been found so far.

Furthermore, as pointed out in the text and in Table 11.1, besides representing important oncoproteins, epigenetic reader proteins are also thought to be tumor suppressor proteins in some cases. On the one hand, this raises the question of selectivity of small-molecule inhibitors regarding their intended target. On the other hand, selective small-molecule ligands of tumor suppressors are of great importance and represent useful tools for further exploring the functions of the respective proteins as well as the signaling pathways they are involved in. In cases where proteins comprise different properties, like combining writer or eraser function with reader abilities, targeting the respective reader domain should also be considered as an alternative to the so-far preferred aim of influencing the enzymatic reader or eraser process. It will be interesting to address the question if inhibitors of the reader–histone interaction will phenocopy the inhibition of the enzymatic activity. This strategy would potentially preserve enzymatic nonhistone activities of the respective protein which could help in avoiding possible side effects that might occur upon the treatment of a specific phenotype.

ACKNOWLEDGMENT AND FUNDING

MJ thanks the Deutsche Forschungsgemeinschaft for funding within CRC992 (Medical Epigenetics, MEDEP).

REFERENCES

[1] Daujat S, Weiss T, Mohn F, Lange UC, Ziegler-Birling C, Zeissler U, et al. H3K64 trimethylation marks heterochromatin and is dynamically remodeled during developmental reprogramming. Nat Struct Mol Biol 2009;16(7):777–81.

[2] Taverna SD, Li H, Ruthenburg AJ, Allis CD, Patel DJ. How chromatin-binding modules interpret histone modifications: lessons from professional pocket pickers. Nat Struct Mol Biol 2007;14(11):1025–40.

[3] Bissinger E-M, Heinke R, Sippl W, Jung M. Targeting epigenetic modifiers: inhibitors of histone methyltransferases. MedChemComm 2010;1(2):114.

[4] Lu Q, Quinn AM, Patel MP, Semus SF, Graves AP, Bandyopadhyay D, et al. Perspectives on the discovery of small-molecule modulators for epigenetic processes. J Biomol Screen 2012;17(5):555–71.

[5] Jenuwein T, Allis CD. Translating the histone code. Science 2001;293(5532):1074–80.

[6] Zeng L, Li J, Muller M, Yan S, Mujtaba S, Pan C, et al. Selective small molecules blocking HIV-1 Tat and coactivator PCAF association. J Am Chem Soc 2005;127(8):2376–7.

[7] Sachchidanand, Resnick-Silverman L, Yan S, Mutjaba S, Liu WJ, Zeng L, et al. Target structure-based discovery of small molecules that block human p53 and CREB binding protein association. Chem Biol 2006;13(1):81–90.
[8] Chung CW, Coste H, White JH, Mirguet O, Wilde J, Gosmini RL, et al. Discovery and characterization of small molecule inhibitors of the BET family bromodomains. J Med Chem 2011;54(11):3827–38.
[9] Filippakopoulos P, Qi J, Picaud S, Shen Y, Smith WB, Fedorov O, et al. Selective inhibition of BET bromodomains. Nature 2010;468(7327):1067–73.
[10] Borah JC, Mujtaba S, Karakikes I, Zeng L, Muller M, Patel J, et al. A small molecule binding to the co-activator CREB-binding protein blocks apoptosis in cardiomyocytes. Chem Biol 2011; 18(4):531–41.
[11] Lucas X, Wohlwend D, Hügle M, Schmidtkunz K, Gerhardt S, Schüle R, et al. 4-Acyl pyrroles: mimicking acetylated lysines in histone code reading. Angew Chem Int Ed 2013;52(52):14055–9.
[12] Wagner T, Robaa D, Sippl W, Jung M. Mind the methyl: methyllysine binding proteins in epigenetic regulation. ChemMedChem 2014;9(3):466–83.
[13] Yun M, Wu J, Workman JL, Li B. Readers of histone modifications. Cell Res 2011;21(4):564–78.
[14] Yap KL, Zhou M-M. Keeping it in the family: diverse histone recognition by conserved structural folds. Crit Rev Biochem Mol Biol 2010;45(6):488–505.
[15] Herold JM, Ingerman LA, Gao C, Frye SV. Drug discovery toward antagonists of Methyl-Lysine binding proteins. Curr Chem Genomics 2011;5(Suppl. 1):51–61.
[16] Musselman CA, Lalonde M-E, Côté J, Kutateladze TG. Perceiving the epigenetic landscape through histone readers. Nat Struct Mol Biol 2012;19(12):1218–27.
[17] Wismar J, Löffler T, Habtemichael N, Vef O, Geißen M, Zirwes R, et al. The Drosophila melanogaster tumor suppressor gene lethal(3)malignant brain tumor encodes a proline-rich protein with a novel zinc finger. Mech Dev 1995;53(1):141–54.
[18] Perna F, Gurvich N, Hoya-Arias R, Abdel-Wahab O, Levine RL, Asai T, et al. Depletion of L3MBTL1 promotes the erythroid differentiation of human hematopoietic progenitor cells: possible role in 20q-polycythemia vera. Blood 2010;116(15):2812–21.
[19] Trojer P, Li G, Sims RJ, Vaquero A, Kalakonda N, Boccuni P, et al. L3MBTL1, a Histone-Methylation-Dependent chromatin lock. Cell 2007;129(5):915–28.
[20] West LE, Roy S, Lachmi-Weiner K, Hayashi R, Shi X, Appella E, et al. The MBT repeats of L3MBTL1 link SET8-mediated p53 methylation at lysine 382 to target gene repression. J Biol Chem 2010; 285(48):37725–32.
[21] Kireev D, Wigle TJ, Norris-Drouin J, Herold JM, Janzen WP, Frye SV. Identification of Non-Peptide Malignant Brain Tumor (MBT) repeat antagonists by virtual screening of commercially available compounds. J Med Chem 2010;53(21):7625–31.
[22] Gao C, Herold JM, Kireev D, Wigle T, Norris JL, Frye S. Biophysical probes reveal a "Compromise" nature of the methyl-lysine binding pocket in L3MBTL1. J Am Chem Soc 2011; 133(14):5357–62.
[23] Herold JM, Wigle TJ, Norris JL, Lam R, Korboukh VK, Gao C, et al. Small-molecule ligands of methyl-lysine binding proteins. J Med Chem 2011;54(7):2504–11.
[24] Nady N, Krichevsky L, Zhong N, Duan S, Tempel W, Amaya MF, et al. Histone recognition by human malignant brain tumor domains. J Mol Biol 2012;423(5):702–18.
[25] Northcott PA, Nakahara Y, Wu X, Feuk L, Ellison DW, Croul S, et al. Multiple recurrent genetic events converge on control of histone lysine methylation in medulloblastoma. Nat Genet 2009;41(4):465–72.
[26] James LI, Barsyte-Lovejoy D, Zhong N, Krichevsky L, Korboukh VK, Herold JM, et al. Discovery of a chemical probe for the L3MBTL3 methyllysine reader domain. Nat Chem Biol 2013;9(3):184–91.
[27] Guo Y, Nady N, Qi C, Allali-Hassani A, Zhu H, Pan P, et al. Methylation-state-specific recognition of histones by the MBT repeat protein L3MBTL2. Nucl Acids Res 2009;37(7):2204–10.

[28] James LI, Korboukh VK, Krichevsky L, Baughman BM, Herold JM, Norris JL, et al. Small-molecule ligands of methyl-lysine binding proteins: optimization of selectivity for L3MBTL3. J Med Chem 2013;56(18):7358–71.
[29] Yap KL, Zhou M-M. Structure and mechanisms of lysine methylation recognition by the chromodomain in gene transcription. Biochemistry 2011;50(12):1966–80.
[30] Kaustov L, Ouyang H, Amaya M, Lemak A, Nady N, Duan S, et al. Recognition and specificity determinants of the human Cbx chromodomains. J Biol Chem 2010;286(1):521–9.
[31] Whitcomb SJ, Basu A, Allis CD, Bernstein E. Polycomb group proteins: an evolutionary perspective. Trends Genet 2007;23(10):494–502.
[32] Gieni RS, Hendzel MJ. Polycomb group protein gene silencing, non-coding RNA, stem cells, and cancer. This paper is one of a selection of papers published in this special issue, entitled the 30th annual international asilomar chromatin and chromosomes conference, and has undergone the journal's usual peer review process. Biochem Cell Biol 2009;87(5):711–46.
[33] Simhadri C, Daze KD, Douglas SF, Quon TT, Dev A, Gignac MC, et al. Chromodomain antagonists that target the polycomb-group methyllysine reader protein chromobox homolog 7 (CBX7). J Med Chem 2014;57(7):2874–83.
[34] Morey L, Pascual G, Cozzuto L, Roma G, Wutz A, Benitah Salvador A, et al. Nonoverlapping functions of the polycomb group Cbx family of proteins in embryonic stem cells. Cell Stem Cell 2012;10 (1):47–62.
[35] Gil J, Bernard D, Martinez D, Beach D. Polycomb CBX7 has a unifying role in cellular lifespan. Nat Cell Biol 2004;6(1):67–72.
[36] Bernard D, Martinez-Leal JF, Rizzo S, Martinez D, Hudson D, Visakorpi T, et al. CBX7 controls the growth of normal and tumor-derived prostate cells by repressing the Ink4a//Arf locus. Oncogene 2005;24(36):5543–51.
[37] Yap KL, Li S, Muñoz-Cabello AM, Raguz S, Zeng L, Mujtaba S, et al. Molecular interplay of the non-coding RNA ANRIL and methylated histone H3 Lysine 27 by polycomb CBX7 in transcriptional silencing of INK4a. Mol Cell 2010;38(5):662–74.
[38] van den Boom V, Rozenveld-Geugien M, Bonardi F, Malanga D, van Gosliga D, Heijink AM, et al. Nonredundant and locus-specific gene repression functions of PRC1 paralog family members in human hematopoietic stem/progenitor cells. Blood 2013;121(13):2452–61.
[39] Scott CL, Gil J, Hernando E, Teruya-Feldstein J, Narita M, Martinez D, et al. Role of the chromobox protein CBX7 in lymphomagenesis. Proc Natl Acad Sci USA 2007;104(13):5389–94.
[40] Shinjo K, Yamashita Y, Yamamoto E, Akatsuka S, Uno N, Kamiya A, et al. Expression of chromobox homolog 7 (CBX7) is associated with poor prognosis in ovarian clear cell adenocarcinoma via TRAIL-induced apoptotic pathway regulation. Int J Cancer 2014;135(2):308–18.
[41] Klauke K, Radulović V, Broekhuis M, Weersing E, Zwart E, Olthof S, et al. Polycomb Cbx family members mediate the balance between haematopoietic stem cell self-renewal and differentiation. Nat Cell Biol 2013;15(4):353–62.
[42] Forzati F, Federico A, Pallante P, Abbate A, Esposito F, Malapelle U, et al. CBX7 is a tumor suppressor in mice and humans. J Clin Invest 2012;122(2):612–23.
[43] Ren C, Morohashi K, Plotnikov AN, Jakoncic J, Smith SG, Li J, et al. Small-molecule modulators of methyl-lysine binding for the CBX7 chromodomain. Chem Biol 2015;22(2):161–8.
[44] Liu K, Guo Y, Liu H, Bian C, Lam R, Liu Y, et al. Crystal structure of TDRD3 and methyl-arginine binding characterization of TDRD3, SMN and SPF30. PLoS One 2012;7(2):e30375.
[45] Côté J, Richard S. Tudor domains bind symmetrical dimethylated arginines. J Biol Chem 2005; 280(31):28476–83.
[46] Saxe JP, Chen M, Zhao H, Lin H. Tdrkh is essential for spermatogenesis and participates in primary piRNA biogenesis in the germline. EMBO J 2013;32(13):1869–85.

[47] Botuyan MV, Lee J, Ward IM, Kim J-E, Thompson JR, Chen J, et al. Structural basis for the methylation state-specific recognition of histone H4-K20 by 53BP1 and Crb2 in DNA repair. Cell 2006;127(7):1361–73.
[48] Fradet-Turcotte A, Canny MD, Escribano-Díaz C, Orthwein A, Leung CCY, Huang H, et al. 53BP1 is a reader of the DNA-damage-induced H2A Lys 15 ubiquitin mark. Nature 2013;499(7456):50–4.
[49] Perfetti MT, Baughman BM, Dickson BM, Mu Y, Cui G, Mader P, et al. Identification of a fragment-like small molecule ligand for the methyl-lysine binding protein, 53BP1. ACS Chem Biol 2015;10(4):1072–81.
[50] Lee J, Thompson JR, Botuyan MV, Mer G. Distinct binding modes specify the recognition of methylated histones H3K4 and H4K20 by JMJD2A-tudor. Nat Struct Mol Biol 2008;15(1):109–11.
[51] Zhang XW, Zhang L, Qin W, Yao XH, Zheng LZ, Liu X, et al. Oncogenic role of the chromobox protein CBX7 in gastric cancer. J Exp Clin Cancer Res 2010;29:114.
[52] Yu YH, Chiou GY, Huang PI, Lo WL, Wang CY, Lu KH, et al. Network biology of tumor stem-like cells identified a regulatory role of CBX5 in lung cancer. Sci Rep 2012;2:584.
[53] Vad-Nielsen J, Nielsen AL. Beyond the histone tale: HP1alpha deregulation in breast cancer epigenetics. Cancer Biol Ther 2015;16(2):189–200.
[54] Itsumi M, Shiota M, Yokomizo A, Kashiwagi E, Takeuchi A, Tatsugami K, et al. Human heterochromatin protein 1 isoforms regulate androgen receptor signaling in prostate cancer. J Mol Endocrinol 2013;50(3):401–9.
[55] Shiota M, Song Y, Yokomizo A, Tada Y, Kuroiwa K, Eto M, et al. Human heterochromatin protein 1 isoform HP1beta enhances androgen receptor activity and is implicated in prostate cancer growth. Endocr Relat Cancer 2010;17(2):455–67.
[56] Lachner M, O'Carroll D, Rea S, Mechtler K, Jenuwein T. Methylation of histone H3 lysine 9 creates a binding site for HP1 proteins. Nature 2001;410(6824):116–20.
[57] Slezak J, Truong M, Huang W, Jarrard D. HP1gamma expression is elevated in prostate cancer and is superior to Gleason score as a predictor of biochemical recurrence after radical prostatectomy. BMC Cancer 2013;13:148.
[58] Zhou J, Bi H, Zhan P, Chang C, Xu C, Huang X, et al. Overexpression of HP1gamma is associated with poor prognosis in non-small cell lung cancer cell through promoting cell survival. Tumour Biol 2014;35(10):9777–85.
[59] Han SS, Kim WJ, Hong Y, Hong SH, Lee SJ, Ryu DR, et al. RNA sequencing identifies novel markers of non-small cell lung cancer. Lung Cancer 2014;84(3):229–35.
[60] Saini V, Hose CD, Monks A, Nagashima K, Han B, Newton DL, et al. Identification of CBX3 and ABCA5 as putative biomarkers for tumor stem cells in osteosarcoma. PLoS One 2012;7(8):e41401.
[61] Li X, Kong X, Wang Y, Yang Q. 53BP1 is a novel regulator of angiogenesis in breast cancer. Cancer Sci 2013;104(11):1420–6.
[62] Tong Q, Cui G, Botuyan MV, Rothbart SB, Hayashi R, Musselman CA, et al. Structural plasticity of methyllysine recognition by the tandem tudor domain of 53BP1. Structure 2015;23(2):312–21.
[63] Pennington KP, Wickramanayake A, Norquist BM, Pennil CC, Garcia RL, Agnew KJ, et al. 53BP1 expression in sporadic and inherited ovarian carcinoma: relationship to genetic status and clinical outcomes. Gynecol Oncol 2013;128(3):493–9.
[64] Hu CE, Liu YC, Zhang HD, Huang GJ. JMJD2A predicts prognosis and regulates cell growth in human gastric cancer. Biochem Biophys Res Commun 2014;449(1):1–7.
[65] Wang HL, Liu MM, Ma X, Fang L, Zhang ZF, Song TF, et al. Expression and effects of JMJD2A histone demethylase in endometrial carcinoma. Asian Pac J Cancer Prev 2014;15(7):3051–6.
[66] Berry WL, Shin S, Lightfoot SA, Janknecht R. Oncogenic features of the JMJD2A histone demethylase in breast cancer. Int J Oncol 2012;41(5):1701–6.
[67] Micci F, Gorunova L, Gatius S, Matias-Guiu X, Davidson B, Heim S, et al. MEAF6/PHF1 is a recurrent gene fusion in endometrial stromal sarcoma. Cancer Lett 2014;347(1):75–8.

[68] Qin S, Guo Y, Xu C, Bian C, Fu M, Gong S, et al. Tudor domains of the PRC2 components PHF1 and PHF19 selectively bind to histone H3K36me3. Biochem Biophys Res Commun 2013;430(2):547–53.

[69] Musselman CA, Gibson MD, Hartwick EW, North JA, Gatchalian J, Poirier MG, et al. Binding of PHF1 Tudor to H3K36me3 enhances nucleosome accessibility. Nat Commun 2013;4:2969.

[70] Wang J-X, Zeng Q, Chen L, Du J-C, Yan X-L, Yuan H-F, et al. SPINDLIN1 promotes cancer cell proliferation through activation of WNT/TCF-4 signaling. Mol Cancer Res 2012;10(3):326–35.

[71] Yang N, Wang W, Wang Y, Wang M, Zhao Q, Rao Z, et al. Distinct mode of methylated lysine-4 of histone H3 recognition by tandem tudor-like domains of Spindlin1. Proc Natl Acad Sci USA 2012;109(44):17954–9.

[72] Wazir U, Ahmed MH, Bridger JM, Harvey A, Jiang WG, Sharma AK, et al. The clinicopathological significance of lamin A/C, lamin B1 and lamin B receptor mRNA expression in human breast cancer. Cell Mol Biol Lett 2013;18(4):595–611.

[73] Hirano Y, Hizume K, Kimura H, Takeyasu K, Haraguchi T, Hiraoka Y. Lamin B receptor recognizes specific modifications of histone H4 in heterochromatin formation. J Biol Chem 2012;287 (51):42654–63.

[74] Recupero D, Annaratone L, Maletta F, Bussolati G. Nuclear shape in papillary thyroid carcinoma: a role for lamin B receptor? Rom J Morphol Embryol 2010;51(4):615–20.

[75] Gelato KA, Tauber M, Ong MS, Winter S, Hiragami-Hamada K, Sindlinger J, et al. Accessibility of different histone H3-binding domains of UHRF1 is allosterically regulated by phosphatidylinositol 5-phosphate. Mol Cell 2014;54(6):905–19.

[76] Rottach A, Frauer C, Pichler G, Bonapace IM, Spada F, Leonhardt H. The multi-domain protein Np95 connects DNA methylation and histone modification. Nucl Acids Res 2010;38(6):1796–804.

[77] Nady N, Lemak A, Walker JR, Avvakumov GV, Kareta MS, Achour M, et al. Recognition of multivalent histone states associated with heterochromatin by UHRF1 protein. J Biol Chem 2011;286 (27):24300–11.

[78] Wu H, Zeng H, Lam R, Tempel W, Amaya MF, Xu C, et al. Structural and histone binding ability characterizations of human PWWP domains. PLoS One 2011;6(6):e18919.

[79] Morerio C, Acquila M, Rosanda C, Rapella A, Tassano E, Micalizzi C, et al. t(9;11)(p22;p15) with NUP98-LEDGF fusion gene in pediatric acute myeloid leukemia. Leuk Res 2005;29(4):467–70.

[80] van Nuland R, van Schaik FM, Simonis M, van Heesch S, Cuppen E, Boelens R, et al. Nucleosomal DNA binding drives the recognition of H3K36-methylated nucleosomes by the PSIP1-PWWP domain. Epigenet Chromatin 2013;6(1):12.

[81] Cermakova K, Tesina P, Demeulemeester J, El Ashkar S, Mereau H, Schwaller J, et al. Validation and structural characterization of the LEDGF/p75-MLL interface as a new target for the treatment of MLL-dependent leukemia. Cancer Res 2014;74(18):5139–51.

[82] Eidahl JO, Crowe BL, North JA, McKee CJ, Shkriabai N, Feng L, et al. Structural basis for high-affinity binding of LEDGF PWWP to mononucleosomes. Nucl Acids Res 2013;41(6):3924–36.

[83] Huang TS, Myklebust LM, Kjarland E, Gjertsen BT, Pendino F, Bruserud O, et al. LEDGF/p75 has increased expression in blasts from chemotherapy-resistant human acute myelogenic leukemia patients and protects leukemia cells from apoptosis *in vitro*. Mol Cancer 2007;6:31.

[84] Basu A, Rojas H, Banerjee H, Cabrera IB, Perez KY, De Leon M, et al. Expression of the stress response oncoprotein LEDGF/p75 in human cancer: a study of 21 tumor types. PLoS One 2012;7(1): e30132.

[85] Hu TH, Huang CC, Liu LF, Lin PR, Liu SY, Chang HW, et al. Expression of hepatoma-derived growth factor in hepatocellular carcinoma. Cancer 2003;98(7):1444–56.

[86] Iwasaki T, Nakagawa K, Nakamura H, Takada Y, Matsui K, Kawahara K. Hepatoma-derived growth factor as a prognostic marker in completely resected non-small-cell lung cancer. Oncol Rep 2005;13 (6):1075–80.

[87] Uyama H, Tomita Y, Nakamura H, Nakamori S, Zhang B, Hoshida Y, et al. Hepatoma-derived growth factor is a novel prognostic factor for patients with pancreatic cancer. Clin Cancer Res 2006;12(20):6043–8.

[88] Yamamoto S, Tomita Y, Hoshida Y, Morii E, Yasuda T, Doki Y, et al. Expression level of Hepatoma-derived growth factor correlates with tumor recurrence of esophageal carcinoma. Ann Surg Oncol 2007;14(7):2141–9.

[89] Bao CH, Wang XT, Ma W, Wang NN, Un Nesa E, Wang JB, et al. Irradiated fibroblasts promote epithelial-mesenchymal transition and HDGF expression of esophageal squamous cell carcinoma. Biochem Biophys Res Commun, 2015.

[90] Wang S, Fang W. Increased expression of hepatoma-derived growth factor correlates with poor prognosis in human nasopharyngeal carcinoma. Histopathology 2011;58(2):217–24.

[91] Yang Y, Li H, Zhang F, Shi H, Zhen T, Dai S, et al. Clinical and biological significance of hepatoma-derived growth factor in Ewing's sarcoma. J Pathol 2013;231(3):323–34.

[92] Tao F, Ye MF, Sun AJ, Lv JQ, Xu GG, Jing YM, et al. Prognostic significance of nuclear hepatoma-derived growth factor expression in gallbladder cancer. World J Gastroenterol 2014;20(28):9564–9.

[93] Shikauchi Y, Saiura A, Kubo T, Niwa Y, Yamamoto J, Murase Y, et al. SALL3 interacts with DNMT3A and shows the ability to inhibit CpG island methylation in hepatocellular carcinoma. Mol Cell Biol 2009;29(7):1944–58.

[94] Dhayalan A, Rajavelu A, Rathert P, Tamas R, Jurkowska RZ, Ragozin S, et al. The Dnmt3a PWWP domain reads histone 3 Lysine 36 trimethylation and guides DNA methylation. J Biol Chem 2010;285 (34):26114–20.

[95] Ding WJ, Fang JY, Chen XY, Peng YS. The expression and clinical significance of DNA methyltransferase proteins in human gastric cancer. Dig Dis Sci 2008;53(8):2083–9.

[96] Yang S, Zhang Y, Meng F, Liu Y, Xia B, Xiao M, et al. Overexpression of multiple myeloma SET domain (MMSET) is associated with advanced tumor aggressiveness and poor prognosis in serous ovarian carcinoma. Biomarkers 2013;18(3):257–63.

[97] He C, Li F, Zhang J, Wu J, Shi Y. The methyltransferase NSD3 has chromatin-binding Motifs, PHD5-C5HCH, that are distinct from other NSD (Nuclear Receptor SET Domain) family members in their histone H3 recognition. J Biol Chem 2013;288(7):4692–703.

[98] Zhou P, Wu LL, Wu KM, Jiang W, Li JD, Zhou LD, et al. Overexpression of MMSET is correlation with poor prognosis in hepatocellular carcinoma. Pathol Oncol Res 2013;19(2):303–9.

[99] Keats JJ, Maxwell CA, Taylor BJ, Hendzel MJ, Chesi M, Bergsagel PL, et al. Overexpression of transcripts originating from the MMSET locus characterizes all t(4;14)(p16;q32)-positive multiple myeloma patients. Blood 2005;105(10):4060–9.

[100] Li J, Yin C, Okamoto H, Mushlin H, Balgley BM, Lee CS, et al. Identification of a novel proliferation-related protein, WHSC1 4a, in human gliomas. Neuro Oncol 2008;10(1):45–51.

[101] Hudlebusch HR, Santoni-Rugiu E, Simon R, Ralfkiaer E, Rossing HH, Johansen JV, et al. The histone methyltransferase and putative oncoprotein MMSET is overexpressed in a large variety of human tumors. Clin Cancer Res 2011;17(9):2919–33.

[102] Qi L, Zhu F, Li SH, Si LB, Hu LK, Tian H. Retinoblastoma binding protein 2 (RBP2) promotes HIF-1alpha-VEGF-induced angiogenesis of non-small cell lung cancer via the Akt pathway. PLoS One 2014;9(8):e106032.

[103] Wang GG, Song J, Wang Z, Dormann HL, Casadio F, Li H, et al. Haematopoietic malignancies caused by dysregulation of a chromatin-binding PHD finger. Nature 2009;459(7248):847–51.

[104] Li L, Wang L, Song P, Geng X, Liang X, Zhou M, et al. Critical role of histone demethylase RBP2 in human gastric cancer angiogenesis. Mol Cancer 2014;13:81.

[105] Liang X, Zeng J, Wang L, Fang M, Wang Q, Zhao M, et al. Histone demethylase retinoblastoma binding protein 2 is overexpressed in hepatocellular carcinoma and negatively regulated by hsa-miR-212. PLoS One 2013;8(7):e69784.

[106] Teng YC, Lee CF, Li YS, Chen YR, Hsiao PW, Chan MY, et al. Histone demethylase RBP2 promotes lung tumorigenesis and cancer metastasis. Cancer Res 2013;73(15):4711–21.
[107] Hou J, Wu J, Dombkowski A, Zhang K, Holowatyj A, Boerner JL, et al. Genomic amplification and a role in drug-resistance for the KDM5A histone demethylase in breast cancer. Am J Transl Res 2012;4(3):247–56.
[108] Tian J, He H, Lei G. Wnt/β-catenin pathway in bone cancers. Tumor Biol 2014;35(10):9439–45.
[109] Fiedler M, Sanchez-Barrena MJ, Nekrasov M, Mieszczanek J, Rybin V, Muller J, et al. Decoding of methylated histone H3 tail by the Pygo-BCL9 Wnt signaling complex. Mol Cell 2008;30(4):507–18.
[110] Arend RC, Londono-Joshi AI, Straughn Jr JM, Buchsbaum DJ. The Wnt/beta-catenin pathway in ovarian cancer: a review. Gynecol Oncol 2013;131(3):772–9.
[111] Kypta RM, Waxman J. Wnt/beta-catenin signalling in prostate cancer. Nat Rev Urol 2012; 9(8):418–28.
[112] Qin Y, Wang J, Gong W, Zhang M, Tang Z, Zhang J, et al. UHRF1 depletion suppresses growth of gallbladder cancer cells through induction of apoptosis and cell cycle arrest. Oncol Rep 2014;31 (6):2635–43.
[113] Lallous N, Legrand P, McEwen AG, Ramon-Maiques S, Samama JP, Birck C. The PHD finger of human UHRF1 reveals a new subgroup of unmethylated histone H3 tail readers. PLoS One 2011;6(11): e27599.
[114] Geng Y, Gao Y, Ju H, Yan F. Diagnostic and prognostic value of plasma and tissue ubiquitin-like, containing PHD and RING finger domains 1 in breast cancer patients. Cancer Sci 2013;104(2):194–9.
[115] Kofunato Y, Kumamoto K, Saitou K, Hayase S, Okayama H, Miyamoto K, et al. UHRF1 expression is upregulated and associated with cellular proliferation in colorectal cancer. Oncol Rep 2012;28 (6):1997–2002.
[116] Unoki M, Kelly JD, Neal DE, Ponder BA, Nakamura Y, Hamamoto R. UHRF1 is a novel molecular marker for diagnosis and the prognosis of bladder cancer. Br J Cancer 2009;101(1):98–105.
[117] Mudbhary R, Hoshida Y, Chernyavskaya Y, Jacob V, Villanueva A, Fiel MI, et al. UHRF1 overexpression drives DNA hypomethylation and hepatocellular carcinoma. Cancer Cell 2014;25(2):196–209.
[118] Unoki M, Daigo Y, Koinuma J, Tsuchiya E, Hamamoto R, Nakamura Y. UHRF1 is a novel diagnostic marker of lung cancer. Br J Cancer 2010;103(2):217–22.
[119] Pi JT, Lin Y, Quan Q, Chen LL, Jiang LZ, Chi W, et al. Overexpression of UHRF1 is significantly associated with poor prognosis in laryngeal squamous cell carcinoma. Med Oncol 2013;30(4):613.
[120] Shen Y, Pan X, Zhao H. The histone demethylase PHF8 is an oncogenic protein in human non-small cell lung cancer. Biochem Biophys Res Commun 2014;451(1):119–25.
[121] Horton JR, Upadhyay AK, Qi HH, Zhang X, Shi Y, Cheng X. Enzymatic and structural insights for substrate specificity of a family of jumonji histone lysine demethylases. Nat Struct Mol Biol 2010;17(1):38–43.
[122] Bjorkman M, Ostling P, Harma V, Virtanen J, Mpindi JP, Rantala J, et al. Systematic knockdown of epigenetic enzymes identifies a novel histone demethylase PHF8 overexpressed in prostate cancer with an impact on cell proliferation, migration and invasion. Oncogene 2012;31(29):3444–56.
[123] Cao F, Townsend EC, Karatas H, Xu J, Li L, Lee S, et al. Targeting MLL1 H3K4 methyltransferase activity in mixed-lineage leukemia. Mol Cell 2014;53(2):247–61.
[124] Patel A, Dharmarajan V, Cosgrove MS. Structure of WDR5 bound to mixed lineage leukemia Protein-1 peptide. J Biol Chem 2008;283(47):32158–61.
[125] Chen X, Xie W, Gu P, Cai Q, Wang B, Xie Y, et al. Upregulated WDR5 promotes proliferation, self-renewal and chemoresistance in bladder cancer via mediating H3K4 trimethylation. Sci Rep 2015;5:8293.
[126] Couture J-F, Collazo E, Trievel RC. Molecular recognition of histone H3 by the WD40 protein WDR5. Nat Struct Mol Biol 2006;13(8):698–703.

[127] Schuetz A, Allali-Hassani A, Martín F, Loppnau P, Vedadi M, Bochkarev A, et al. Structural basis for molecular recognition and presentation of histone H3 By WDR5. EMBO J 2006;25(18):4245–52.
[128] Ruthenburg AJ, Wang W, Graybosch DM, Li H, Allis CD, Patel DJ, et al. Histone H3 recognition and presentation by the WDR5 module of the MLL1 complex. Nat Struct Mol Biol 2006;13(8):704–12.
[129] Kim JY, Banerjee T, Vinckevicius A, Luo Q, Parker JB, Baker MR, et al. A role for WDR5 in integrating threonine 11 phosphorylation to lysine 4 methylation on histone H3 during androgen signaling and in prostate cancer. Mol Cell 2014;54(4):613–25.
[130] Mungamuri SK, Murk W, Grumolato L, Bernstein E, Aaronson SA. Chromatin modifications sequentially enhance ErbB2 expression in ErbB2-positive breast cancers. Cell Rep 2013;5(2):302–13.
[131] Hoffmann I, Roatsch M, Schmitt ML, Carlino L, Pippel M, Sippl W, et al. The role of histone demethylases in cancer therapy. Mol Oncol 2012;6(6):683–703.
[132] Panagopoulos I, Micci F, Thorsen J, Gorunova L, Eibak AM, Bjerkehagen B, et al. Novel fusion of MYST/Esa1-associated factor 6 and PHF1 in endometrial stromal sarcoma. PLoS One 2012;7(6):e39354.
[133] Micci F, Panagopoulos I, Bjerkehagen B, Heim S. Consistent rearrangement of chromosomal band 6p21 with generation of fusion genes JAZF1/PHF1 and EPC1/PHF1 in endometrial stromal sarcoma. Cancer Res 2006;66(1):107–12.
[134] Oh B, Hwang SY, Solter D, Knowles BB. Spindlin, a major maternal transcript expressed in the mouse during the transition from oocyte to embryo. Development 1997;124:493–503.
[135] Yue W, Sun LY, Li CH, Zhang LX, Pei XT. Screening and identification of ovarian carcinomas related genes. Ai Zheng 2004;23:141–5.
[136] Gao Y, Yue W, Zhang P, Li L, Xie X, Yuan H, et al. Spindlin1, a novel nuclear protein with a role in the transformation of NIH3T3 cells. Biochem Biophys Res Commun 2005;335(2):343–50.
[137] Zhao Q, Qin L, Jiang F, Wu B, Yue W, Xu F, et al. Structure of human Spindlin1: tandem Tudor-Like domains for cell cycle regulation. J Biol Chem 2006;282(1):647–56.
[138] Zhang P, Cong B, Yuan H, Chen L, Lv Y, Bai C, et al. Overexpression of spindlin1 induces metaphase arrest and chromosomal instability. J Cell Physiol 2008;217(2):400–8.
[139] Su X, Zhu G, Ding X, Lee SY, Dou Y, Zhu B, et al. Molecular basis underlying histone H3 lysine–arginine methylation pattern readout by Spin/Ssty repeats of Spindlin1. Genes Dev 2014;28(6):622–36.
[140] MacDonald BT, Tamai K, He X. Wnt/β-Catenin signaling: components, mechanisms, and diseases. Dev Cell 2009;17(1):9–26.
[141] Franz H, Greschik H, Willmann D, Ozretic L, Jilg CA, Wardelmann E, et al. The histone code reader SPIN1 controls RET signaling in liposarcoma. Oncotarget 2015;6(7):4773–89.
[142] Santiago C, Nguyen K, Schapira M. Druggability of methyl-lysine binding sites. J Comput Aided Mol Des 2011;25(12):1171–8.
[143] Qiu Y, Zhang W, Zhao C, Wang Y, Wang W, Zhang J, et al. Solution structure of the Pdp1 PWWP domain reveals its unique binding sites for methylated H4K20 and DNA. Biochem J 2012;442(3):527–38.
[144] Pradeepa MM, Sutherland HG, Ule J, Grimes GR, Bickmore WA, Reik W. Psip1/Ledgf p52 binds methylated histone H3K36 and splicing factors and contributes to the regulation of alternative splicing. PLoS Genet 2012;8(5):e1002717.
[145] Shun M-C, Botbol Y, Li X, Di Nunzio F, Daigle JE, Yan N, et al. Identification and characterization of PWWP domain residues critical for LEDGF/p75 chromatin binding and human immunodeficiency virus type 1 infectivity. J Virol 2008;82(23):11555–67.
[146] Leoh LS, van Heertum B, Rijck J d, Filippova M, Rios-Colon L, Basu A, et al. The stress oncoprotein LEDGF/p75 interacts with the methyl CpG binding protein MeCP2 and influences its transcriptional activity. Mol Cancer Res 2012;10(3):378–91.
[147] Min D-J, Ezponda T, Kim MK, Will CM, Martinez-Garcia E, Popovic R, et al. MMSET stimulates myeloma cell growth through microRNA-mediated modulation of c-MYC. Leukemia 2012;27(3):686–94.

[148] Brito JLR, Walker B, Jenner M, Dickens NJ, Brown NJM, Ross FM, et al. MMSET deregulation affects cell cycle progression and adhesion regulons in t(4;14) myeloma plasma cells. Haematologica 2009;94(1):78–86.
[149] Baker LA, Allis CD, Wang GG. PHD fingers in human diseases: disorders arising from misinterpreting epigenetic marks. Mutat Res-Fund Mol M 2008;647(1-2):3–12.
[150] Sanchez R, Zhou M-M. The PHD finger: a versatile epigenome reader. Trends Biochem Sci 2011;36:364–72.
[151] van Zutven LJCM, Önen E, Velthuizen SCJM, van Drunen E, von Bergh ARM, van den Heuvel-Eibrink MM, et al. Identification ofNUP98 abnormalities in acute leukemia: JARID1A (12p13) as a new partner gene. Genes Chromosomes Cancer 2006;45(5):437–46.
[152] de Rooij JDE, Hollink I, Arentsen-Peters S, van Galen J, Beverloo B, Baruchel A, et al. NUP98/JARID1A is a novel recurrent abnormality in pediatric acute megakaryoblastic leukemia with a distinct HOX gene expression pattern. Leukemia 2013;27(12):2280–8.
[153] Wagner EK, Nath N, Flemming R, Feltenberger JB, Denu JM. Identification and characterization of small molecule inhibitors of a plant homeodomain finger. Biochemistry 2012;51(41):8293–306.
[154] Miller TCR, Rutherford TJ, Birchall K, Chugh J, Fiedler M, Bienz M. Competitive binding of a benzimidazole to the histone-binding pocket of the pygo PHD finger. ACS Chem Biol 2014;9(12):2864–74.
[155] Stirnimann CU, Petsalaki E, Russell RB, Müller CW. WD40 proteins propel cellular networks. Trends Biochem Sci 2010;35(10):565–74.
[156] Wysocka J, Swigut T, Milne TA, Dou Y, Zhang X, Burlingame AL, et al. WDR5 associates with histone H3 methylated at K4 and is essential for H3 K4 methylation and vertebrate development. Cell 2005;121(6):859–72.
[157] Han Z, Guo L, Wang H, Shen Y, Deng XW, Chai J. Structural basis for the specific recognition of methylated histone H3 lysine 4 by the WD-40 protein WDR5. Mol Cell 2006;22(1):137–44.
[158] Iberg AN, Espejo A, Cheng D, Kim D, Michaud-Levesque J, Richard S, et al. Arginine methylation of the histone H3 tail impedes effector binding. J Biol Chem 2007;283(6):3006–10.
[159] Migliori V, Müller J, Phalke S, Low D, Bezzi M, Mok WC, et al. Symmetric dimethylation of H3R2 is a newly identified histone mark that supports euchromatin maintenance. Nat Struct Mol Biol 2012;19(2):136–44.
[160] Zhang P, Lee H, Brunzelle JS, Couture J-F. The plasticity of WDR5 peptide-binding cleft enables the binding of the SET1 family of histone methyltransferases. Nucl Acids Res 2012;40(9):4237–46.
[161] Muntean AG, Tan J, Sitwala K, Huang Y, Bronstein J, Connelly JA, et al. The PAF complex synergizes with MLL fusion proteins at HOX Loci to promote leukemogenesis. Cancer Cell 2010;17(6):609–21.
[162] Dharmarajan V, Lee J-H, Patel A, Skalnik DG, Cosgrove MS. Structural basis for WDR5 interaction (Win) motif recognition in human SET1 family histone methyltransferases. J Biol Chem 2012;287(33):27275–89.
[163] Senisterra G, Wu H, Allali-Hassani A, Wasney GA, Barsyte-Lovejoy D, Dombrovski L, et al. Small-molecule inhibition of MLL activity by disruption of its interaction with WDR5. Biochem J 2013;449(1):151–9.
[164] Karatas H, Townsend EC, Cao F, Chen Y, Bernard D, Liu L, et al. High-affinity, small-molecule peptidomimetic inhibitors of MLL1/WDR5 protein–protein interaction. J Am Chem Soc 2013;135(2):669–82.
[165] Campagna-Slater V, Schapira M. Finding inspiration in the protein data bank to chemically antagonize readers of the histone code. Mol Inform 2010;29(4):322–31.

CHAPTER

NONCODING RNAs: CLINICAL AND THERAPEUTIC APPLICATIONS

12

Yannick Delpu[1], Dorian Larrieu[2], Marion Gayral[2], Dina Arvanitis[3], Marlène Dufresne[2], Pierre Cordelier[2] and Jérôme Torrisani[2]

[1]Skirball Institute of Biomolecular Medicine, New York University Langone Medical Center, New York, NY, USA
[2]INSERM, UMR1037, Cancer Research Center of Toulouse, University of Toulouse-Paul Sabatier, Toulouse, France
[3]INSERM, UMR1048, Institute of Cardiovascular and Metabolic Diseases, University-Paul Sabatier, Toulouse, France

CHAPTER OUTLINE

G. Egger & P. Arimondo (Eds): Drug Discovery in Cancer Epigenetics. DOI: http://dx.doi.org/10.1016/B978-0-12-802208-5.00012-6

12.1 INTRODUCTION

Since the last decade, improvements in high-throughput sequencing apparatuses and bioinformatic softwares have allowed for better quantitative analysis of noncoding RNAs (ncRNAs). While long neglected, the interest in ncRNAs by scientific communities is rising given that they represent 60% of total RNA. Their functions, which in some cases involve complex molecular mechanisms, are progressively being understood. Though one class of ncRNAs can act using different actions, most display the ability to control gene expression without affecting the DNA sequence. Therefore they constitute a major class of epigenetic regulators. Moreover, ncRNA expression profiles were demonstrated as valuable biomarkers in various human pathologies, and this has intensified the interest in ncRNAs by pharmaceutical industries and biotechnology companies. Thus the targeting of ncRNA species has become a promising therapeutic approach for the treatment of human pathologies such as cancer.

In this chapter we will provide an overview of the biogenesis and mode of action of the different classes of ncRNAs. Data concerning the modulation of ncRNA expression in human diseases will be presented, as well as their potential as biomarkers. Finally, our attention will be focused on the therapeutic strategies developed to target ncRNAs in human disorders.

12.2 DIFFERENT CLASSES OF NONCODING RNAs: THEIR BIOGENESIS AND MODE OF ACTIONS

Several classes of ncRNAs have been identified within the last two decades (Figure 12.1). One general feature is that they are not translated into proteins. Apart from this, considerable variation exists in the generation and functioning of ncRNAs. In addition, the size of ncRNAs varies from approximately 20 nucleotides to several hundred nucleotides. To allow for an integrative review in this chapter, the different ncRNAs will be divided into two classes depending on their average size: the small and long ncRNAs.

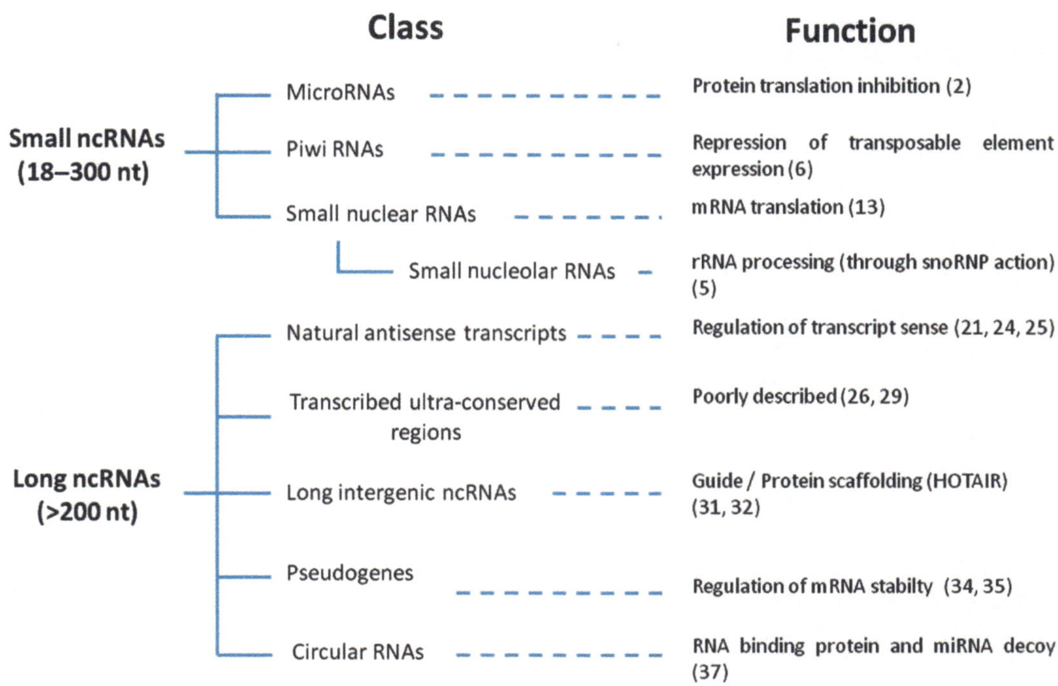

FIGURE 12.1

Classification and function of the different noncoding RNAs. ncRNAs, noncoding RNAs; rRNAs, ribosomal RNAs; snoRNP, small nucleolar ribonucleoprotein; miRNAs, micro RNAs.

12.2.1 SMALL NONCODING RNAs

12.2.1.1 MicroRNAs

MicroRNAs (miRNAs) are one of the most studied among the different categories of small ncRNAs. They are 18–22-nucleotide-long RNAs. Up to now, 1996 miRNAs have been sequenced with high confidence (miRBase 21; http://www.mirbase.org/). While scattered throughout the genome, some are grouped into clusters [1]. Canonically, they are transcribed by RNA polymerase II to produce primary miRNAs (pri-miRNAs) [2]. These pri-miRNAs are then processed in the nucleus by a complex DiGeorge syndrome critical region 8 (Drosha-DGCR8) to form precursor miRNAs (pre-miRNAs) that are exported to the cytoplasm by the exportin 5 system. In the cytoplasm they are further processed by the Dicer-TRBP (TAR RNA-binding protein 2) complex to generate the 18–22-nt double-strand mature miRNA. Incorporated to an Argonaute 2 containing RNA-induced silencing complex (RISC), the guide strand of miRNA duplexes binds, by complementary association to messenger RNAs (mRNAs), predominantly at the 3′-untranslated region (3′-UTR). miRNAs exert their function in the cytoplasm. In most cases they bind to their target mRNA in an imperfect manner. Nucleotides from positions 2 to 7 of miRNAs are crucial mRNA recognition sites that are termed "miRNA seed" sequences. miRNAs subsequently inhibit mRNA translation by increasing their degradation or decreasing their stability [2]. In plants, the miRNA–mRNA association can be perfectly complementary and is shown to result in mRNA degradation. Their small size and imperfect binding confer to each miRNA the faculty to target numerous mRNAs. Inversely, a single mRNA can be targeted by several miRNAs depending on the length and the structure of their 3′-UTR. This imperfect miRNA/mRNA association makes the identification of miRNA targets difficult. An additional complexity exists given that miRNA expression is tightly regulated at the transcriptional and posttranscriptional levels [3,4].

12.2.1.2 PIWI-interacting RNAs

PIWI-interacting RNAs (piRNA) range between 26 and 31 nucleotides long. They are usually intragenic or organized in clusters. Over 20,000 piRNAs are expressed [5]. They are transcribed as single-strand precursor RNAs from genomic regions containing transposable elements and repetitive sequences. They mature in the cytoplasm and associate with PIWI proteins, a subfamily of Argonaute proteins including PIWIL1/HIWI, PIWIL2/HILI, PIWIL3, and PIWIL4/HIWI2. A second maturation pathway consists of an auto-amplification loop, also called a "ping-pong" loop, which involves sense and antisense intermediates [5]. These riboprotein complexes are mainly responsible for the repression and mobility of transposable element expression. The repression involves degradation of transposable element transcripts by the slicer activity of PIWI proteins which are targeted by complementary recognition of piRNAs [6]. Transcriptional repression by piRNAs can also be mediated in *Drosophila* by heterochromatin silencing. In mammals, it appears that the activity of piRNAs in transposon silencing is most important during embryonic development. For example, piRNAs mediate a methylation wave responsible for silencing potentially active retrotransposons that evade the first wave of DNA methylation [7]. During genomic imprinting, it is proposed that piRNAs associate with target RNA to direct the sequence-specific methylation of *Rasgrf1* imprinted gene in paternal germ line [8].

12.2.1.3 Small nucleolar RNAs

Small nucleolar RNAs (snoRNAs) are 60–300-nucleotide-long RNAs located in the nucleolus or in Cajal bodies. They constitute one of the most abundant classes of ncRNAs [9]. Predominantly intronic, 300 different snoRNA sequences are located in the human genome. They are classified into two categories, those containing boxes C and D; and, those containing boxes H and ACA. snoRNAs are generated after splicing, debranching, and trimming of mRNA introns. Subsequently, mature snoRNAs associate with proteins to form small nucleolar ribonucleoproteins (snoRNPs). These complexes are exported into the nucleolus to participate in rRNA processing [5]. Indeed, the main function of snoRNPs takes place in the nucleolus for 2′-O-methylation [10] and pseudouridylation of rRNAs [11]. These chemical modifications ensure better folding and stability of rRNAs [12]. In the nucleus, snoRNAs can also participate in the regulation of alternative splicing.

12.2.1.4 Small nuclear RNAs

Small nuclear RNAs (snRNAs) were the first family of ncRNAs identified, and are shown to participate in protein synthesis. snRNAs are approximately 150-nt-long ncRNAs. They are transcribed by RNA polymerase II or III before maturation in the nucleus [13]. snRNAs are components of the spliceosome where they participate in pre-mRNA processing. They are also involved in the regulation of RNA polymerase II or other transcription factors and in the maintenance of telomeres.

The most abundant RNAs in the cell, ribosomal RNAs (rRNAs) are generated after transcription by RNA polymerase III of 300–400 tandem repeats organized in five clusters [14]. They are components of the large and small ribosomal subunits for the translation of mRNA into proteins.

Transfer-RNAs (tRNAs) are also involved in protein translation as carriers of amino acids to the nascent polypeptide chain. They are transcribed as pre-tRNA by RNA polymerase III from around 500 genes in human genome. They mature in the nucleus before being exported to the cytoplasm by the Los1/Xpo-t system [15].

12.2.2 LONG NONCODING RNAs

lncRNAs represent 60% of the ncRNAs within the human genome and a very heterogeneous subclass of ncRNAs [16]. They are mainly transcribed by RNA polymerase II, spliced and polyadenylated [17]. Compared to mRNAs, lncRNAs contain fewer exons and are shorter in length [18]. Translation of lncRNAs is not likely since the ATG start codon and open reading frame (ORF) are usually in a poor context, leading to the activation of nonsense-mediated decay (NMD). Therefore the biological function of lncRNAs relies essentially on the structure of the transcribed RNA molecule. Their localization is generally nuclear even though certain lncRNAs can be mutually present in the nucleus and the cytoplasm. Due to their extended length, lncRNAs can fold into complex three-dimensional structures. lncRNAs also have sequence-specific interactions with DNA and RNA to form duplex or triplex structures creating complex regulatory networks. They can act as molecular guides for chromatin organization, and transcriptional and posttranscriptional regulation. lncRNAs can be subdivided into five main categories depending of their mode of biogenesis, functions, or locations on the genome.

12.2.2.1 Natural antisense transcripts

Natural antisense transcripts (NATs) are endogenous RNAs that partially or completely overlap transcripts to influence, negatively or positively, the expression of the sense gene. It is important to separate *cis* and *trans* NATs, as *cis* NATs originate from the opposite strand of the same locus of the target gene. NATs are intragenic lncRNAs that can overlap introns, exons, promoters, enhancers, and UTRs of the sense gene partner. It was shown that 20–40% of coding genes display overlapping antisense partners [19,20]. A large fraction of antisense transcripts do not encode for proteins, suggesting that NATs have important regulatory functions in controlling the expression of sense transcripts. On the other hand, *trans* NATs are transcribed from a distinct genomic region remote from their target genes [19]. NATs can exert their function using different modes of actions which can help to classify the multitude of NATs. Discordant NAT regulation is associated with transcriptional repression of sense transcript. This regulation involves different mechanisms leading to epigenetic modifications of promoter-associated histones [21]. The targeting of this discordant regulation is of particular interest for targeted gene up-regulation in therapeutic applications (see below). For instance, the antisense noncoding RNA in the *INK4* locus (ANRIL) participates in epigenetic transcriptional repression of *INK4b/ARF/INK4a* locus by recruiting and tethering chromobox homolog 7 protein (CBX7), a component of polycomb repressive complex (PRC1) [22]. Via the physical interaction with the CCCTC-binding factor (CTCF) the NAT WD repeat, which contains antisense sequences to TP53 (Wrap53), regulates the expression of *p53* at the transcriptional level [23].

In contrast, the concordant regulation by NATs takes place in the cytoplasm and consists of RNA duplex formation with the sense transcript to modify stability or translation rate. For instance, the NAT of Zinc Finger E-Box Binding Homeobox 2 (ZEB2) gene induces ZEB2 mRNA translation by repressing the splicing of the ribosome entry site [24]. Furthermore, NATs such as beta-secretase-1 antisense (BACE1-AS) can prevent gene repression by masking the binding sites of miRNAs [25].

12.2.2.2 Transcribed-ultraconserved regions

Ultraconserved regions (UCRs) are genomic regions with a size >200 bp that display 100% homology between human, rat, and mouse genomes. There exist 481 UCRs, excluding rRNAs, that are widely distributed throughout the genome (except Chr 21 and Y) and often found in clusters [26]. They were classified into five different groups: intergenic, intronic, exonic, partly exonic, and exon-containing [27].

The majority of these regions (93%) are transcribed into transcribed-UCR (T-UCR) at least in one normal tissue and form a new class of lncRNAs [28]. Due to their high degree of conservation, T-UCRs may play fundamental roles in mammals. To date it has been identified that the UCR expression profile makes it possible to distinguish human cancers. Nevertheless, the function of T-UCRs is only starting to be described, even if their conservation suggests they exert essential functions in cell physiology [26]. Interactions with miRNAs have been described as having a biological significance [28] For example, neuroblastoma differentiation in response to all-trans-retinoic acid (ATRA), is marked by altered expression levels in a significant number of T-UCRs. Although the exact mechanism is not yet understood, the authors proposed that the T-UC.300A is connected with proliferation, invasion, and the inhibition of differentiation of NB cells before ATRA treatment [29].

12.2.2.3 Long intergenic ncRNAs

Long intergenic ncRNAs (lincRNAs) are defined as lncRNAs transcribed from the genomic region located between protein-coding genes and therefore do not overlap with exonic sequences. Over 1000 lincRNAs have been identified in the human genome [5]. Like other lncRNAs, lincRNAs are mostly located in the nucleus and sometimes in nuclear compartments such as the nucleolus [30].

One of most studied lincRNAs is HOTAIR. HOTAIR is an epigenetic regulator and plays an important role in cancer [31]. Approximately 2.2 kb, HOTAIR was shown to regulate gene expression by mediating the modulation of chromatin structures. It functions as a scaffold for histone modification complexes to coordinately interact with histone modifiers polycomb repressive complex 2 (PRC2) and lysine specific demethylase (LSD1). In turn, HOTAIR guides the proteins to specific regions of the genome and regulates gene expression through histone tail methylation [32].

12.2.2.4 Noncoding pseudogenes

Pseudogenes are inheritable genetic elements that are similar to functional genes but are non-functional as they do not encode for proteins. Their biogenesis results from the duplication of a parental gene, or the retrotransposition of an mRNA sequence into different genomic loci. The inability of pseudogenes to produce functional proteins is often the consequence of subsequent genetic alterations (frameshift mutations, creation of stop codon). There exist roughly 10,000 pseudogenes in mammalian genomes. Although they are not able to produce functional proteins, many pseudogenes (approximately 20%) are transcribed into RNAs that comprise another category of lncRNAs [33]. Several studies demonstrated the biological activity of pseudogenes in controlling parental gene expression by producing natural siRNAs [34] or antisense transcripts, and by competing with miRNA binding sites on mRNA targets [35].

12.2.2.5 Circular RNAs

Circular RNAs (circRNAs) were first discovered in plants. They arise from the covalent linkage of RNA molecules extremities by *trans*-splicing [36]. In animals, circRNAs originate from a covalent linkage produced by the spliceosome of 5′ and 3′ splicing sites of exons. Long considered as a rare event, the technological progress in genome sequencing revealed the existence of thousands of circRNAs in human and mouse tissues [37]. Only recently have we begun to uncover the biological functions of circRNAs. The function of circRNAs is believed to involve competition with other RNAs for the interaction with RNA-binding protein and miRNAs, therefore, circRNAs form a new class of posttranscriptional regulators. For example, the circRNA CDR1 (cerebellar degeneration-related protein 1) harbors dozens of conserved miR-7 seed matches. Biological function of this circDNA was proven in a zebra fish model in which CDR1 as transcript impairs cerebral development, at least in part, by interacting with miR-7 [37].

12.2.2.6 Other classes of ncRNAs

In addition to the ncRNAs described above, there exist other types of ncRNAs. To date the biogenesis and functions of these ncRNAs are unclear even if they are believed to integrate in the regulation of gene transcription. As a result none have been proposed as a therapeutic target. Promoter-associated small RNAs (PASRs) are associated with the epigenetic silencing of the human ubiquitine C promoter [38]. Transcription start site-associated RNAs (TSS-RNAs) [39] are 20–90-nt small ncRNAs for which transcription is initiated at −250 to +50 nt of active promoter regions. Promoter upstream transcripts (PROMPTs) are generally initiated in the opposite direction, closely upstream of the TSS of their associated genes [40]. Transcription initiation RNAs (tiRNAs) are tiny RNA transcribed from the same strand in close vicinity to TSS (−60 to 120 nt) [39].

12.3 ALTERED ncRNA EXPRESSION AND BIOLOGICAL CONSEQUENCE IN HUMAN PATHOLOGIES

Like mRNAs, the expression of ncRNAs is profoundly modified in human pathologies. The altered expression profiles of a number of ncRNAs are often implicated at the origin of human pathologies such as cancer. Genetic and epigenetic alterations are commonly responsible for the deregulation of ncRNA expression. In this section, we will describe the most emblematic examples of ncRNA alterations and their biological consequences in human disorders.

12.3.1 SMALL ncRNAs

12.3.1.1 MicroRNAs

In many cases, altered profiles of miRNA expression can distinguish pathological organs from normal tissues (see below). In cancer, they can be classified into two categories: oncogenic miRNAs when they repress tumor suppressor gene mRNAs, or tumor suppressor miRNAs when they target mRNAs that encode oncogenic proteins; however, certain miRNAs can be classified in both categories depending on the cellular context. Often retrieved in cancer, these alterations originate from multiple mechanisms. DNA hypermethylation, an epigenetic alteration, is responsible for silencing numerous miRNA promoter activity [41]. Gene amplification or genomic deletion can also lead to miRNA over- or downexpression. For instance, miR-151 located on Chr8q24 is frequently amplified in hepatocellular carcinoma (HCC) [42]. By direct targeting of RhoGDI A (GDP dissociation inhibitor A), a putative metastasis suppressor in HCC, miR-151 overexpression leads to the activation of Rac1, Cdc42, and RhoGTPases. In contrast, miR-15 and miR-16 located on Chr13q14 are deleted in more than half of B-cell chronic lymphocytic leukemia (B-CLL) [43]. This downregulation inversely correlates with increased expression of arginyl-tRNA synthetase (RARS).

One of the most studied miRNAs is miR-21. MiR-21 expression is increased in multiple human pathologies such as cancer [44], heart failure [45], or autoimmune disease [46]. By regulating the translation of various proteins with important functions, overexpression of miR-21 leads to severe biological consequences in human pathologies. Increased expression of miR-21 in lupus CD4+ T cells results in a downregulation of the DNA methyltransferase 1 (DNMT1) [46]. Highly expressed in HCC patients, miR-21 contribute to HCC growth by modulation of phosphatase and tensin homolog (PTEN) and PTEN-dependent pathways [47].

12.3.1.2 Piwi RNAs

Like other small ncRNAs, piRNA expression is altered in human pathologies and particularly in cancers. Global DNA methylation and focal DNA hypermethylation observed in cancers are two of the causes responsible for deregulated piRNA expression. For instance, piR-017061 is differentially expressed between pancreatic cancer tissues and normal tissues [48]. Differentially expressed piRNAs have been identified in breast cancer (BC) as compared to normal tissues. Some of them exert transcriptional and posttranscriptional gene-regulatory actions in cancer cells [49]. Altered expression of piRNAs directly participates in tumor initiation and progression. For instance, upregulation of piR-823 has been involved in multiple myeloma (MM) tumorigenesis. Targeted depletion of piR-823 results in a marked reduction of DNA methyltransferases 3A and 3B expression leading to the re expression of methylation-silenced tumor suppressor gene, *p16(INK4A)* [50]. Thus, piRNA-targeted therapeutic strategies can be considered for MM.

12.3.1.3 Small nucleolar RNAs

Emerging evidence indicates that snoRNAs may play an important function in tumorigenesis in the lung [51,52]. A recent study identified and validated 16 snoRNAs for which the expression is changed more than threefold in nonsmall-cell lung cancer compared to normal tissues [53]. Alteration of snoRNAs expression is observed not only in cancer but in a myriad of other disease states. Prader–Willi syndrome (PWS), a disorder of genomic imprinting results from the loss of paternal copies of the HBII-85C/D box snoRNA cluster [54]. Another example includes an RNA reverse transcriptase related to the vertebrate telomerase. Associated with a box H/ACA, this complex ensures the maintenance of telomere length by adding DNA repeats to the extremities on chromosomes [52]. This component of the telomerase is defective in congenital dyskeratosis where patients display shorter telomeres [55].

12.3.1.4 Small nuclear RNAs

Increased protein synthesis is a distinguishing feature of cancer cells. It is therefore not surprising to observe deregulated expression of snRNAs. Enlargement of the nucleoli (in which rRNAs are transcribed) has been long considered a hallmark of malignant cells. Epigenetic and transcriptional mechanisms are responsible for deregulated rRNA synthesis [56]. Hypomethylation of rRNA promoters increases rRNA transcription and production of ribosome in HCC [57].

12.3.2 LONG ncRNAs

12.3.2.1 Natural antisense transcripts

NAT aberrant expression is common in a vast number of human disorders. Fragile X Mental Retardation 1 antisense (FMR1-AS) is involved in Fragile X syndrome [58]. Huntington antisense (HTT-AS) expression is reduced in Huntington's disease, consequently leading to the deregulation of HHT protein [59]. The altered expression of the same NAT can be found in several other pathologies. For instance, ANRIL is upregulated in HCC, lung cancer, and esophageal squamous cell carcinoma (ESCC) [56]. Elevated expression of ANRIL is also associated with noncancerous pathology, such as atherosclerosis [60]. When overexpressed in human pathology, ANRIL promotes cell growth by repression of the *INK4b-ARF-INK4a* gene cluster. This NAT is transcriptionally upregulated by the transcription factor E2F1 in an ATM-dependent manner following DNA damage [57]. Another example of NAT is beta-secretase-1 AS (BACE1-AS) that controls BACE1, a crucial enzyme in Alzheimer's disease [25].

12.3.2.2 Transcribed-ultraconserved regions

Transcribed-ultraconserved regions are aberrant in cancers [61]. Croce's group was the first to investigate the expression profile of T-UCRs in several tumor types (including leukemia, colorectal, and HCC) [28]. These studies revealed differential expression profiles of T-UCRs in cancer samples compared to normal matching tissues (see below). They further showed that UCRs are located in fragile genomic regions altered during the malignant process. The T-UCRs, such as uc.58, uc.202, uc.207, and uc.223, are up-regulated in Barrett's esophagus carcinoma, whereas uc.214+ is downregulated [62]. Altered expression of T-UCRs may be a result of negative miRNA regulation [28]. DNA hypermethylation of promoters is a common hallmark of many types of cancer. This altered epigenetic modification is also responsible for T-UCR silencing, such as Uc.160+, Uc.283+A and Uc.346+ in several cancers [63]. Despite the few findings mentioned above, the biological consequences of altered T-UCR expression are unclear.

12.3.2.3 Long intergenic ncRNAs

Numerous lincRNAs such as ANRIL, HOTAIR, PCAT-1, PRNCR1, PCGEM1, CCAT2, and MALAT1 have been shown to correlate with human cancers and other pathologies [5]. The linc p21-AS, transcribed in the opposite direction to the *p21* gene, functions as a downstream regulator of p53 transcriptional response [64]. Indeed, a physical interaction between p21-AS and the protein hnRNP-K is required for the regulation of p53-mediated apoptosis. Interestingly, it was shown that XIST lincRNA is not only required for X-chromosome inactivation but also for cancer repression in a mouse model [65]. The metastasis-associated lung adenocarcinoma transcript 1 (MALAT-1) is an 8.7-kb transcript overexpressed in a number of human cancers. It interacts with the DNA and RNA-binding protein PSF to release it from target genes, thus inducing transcription of the proto-oncogene GAGE6 [66]. HOTAIR is one of the most-studied lncRNAs. HOTAIR is a *trans*-acting lncRNA demonstrating different target loci [31]. It interacts with the repressive complex PRC2, and partakes in PCR2 location on the genome. Altered expression of HOTAIR is observed in several types of cancers where it contributes to epigenetic modifications and suppression of multiple tumor suppressor and metastasis-promoting genes.

12.3.2.4 Pseudogenes

Pseudogene "transcription" meta-analysis from a RNA-Seq resource of 293 samples (cancer and normal tissues) revealed a subset of cancer-specific pseudogenes that can have a significant role in tumor progression [67]. The POU domain class 5 transcription factor 1B (POU5F1B) is a processed pseudogene, highly homologous to octamer-binding transcription factor 4 (OCT4). POU5F1B is located on Chr8q24 and frequently amplified in gastric cancer (GC) [68]. The inhibition of POU5F1B expression using shRNA strategy confirmed its role in promoting cancer cell growth *in vitro* and tumor growth *in vivo*. In contrast, the PTENP1 pseudogene locus is selectively lost in human cancer. The 5′-most region of the *PTENP*1 3′ UTR is ~95% similar in sequence to the 3′-UTR of *PTEN*. PTENP1 is biologically active as it can regulate cellular levels of PTEN and exert a growth-suppression role [35].

12.3.2.5 Circular RNAs

Recent researches have revealed that circRNAs can function as miRNA sponges, regulators of splicing, and transcription. Emerging evidence indicate that circRNAs play important roles in atherosclerotic vascular disease (ASVD) risk, neurological disorders, prion diseases, and cancer. CircRNAs exhibit aberrant expression in CRC and ovarian cancer [69,70]. The well-known CiRS-7 circRNA acts as a sponge inhibitor of miR-7, which modulates the expression of several oncogenes [71]. Expressed in the brain, ciRS-7 expression has not yet been described in cancer samples. Recent studies identified novel circRNA products emanating from the *ANRIL* locus. These circRNAs are involved in ANRIL regulation and ASVD [72].

12.4 UTILITY OF ncRNA BIOMARKERS FOR HUMAN DISEASES

As described earlier, ncRNA expression is modified in a number of human diseases, particularly in cancer. These modifications are often specific to the disease, cancer type, or stage. Therefore, these altered profiles have proven their utility as helpful markers for the diagnosis, the prognosis, or the prediction of human diseases. In this section, we will describe, using different examples, the utility of ncRNAs as biological markers in human disorders.

12.4.1 DIAGNOSTIC MARKERS

Among small ncRNAs, miRNAs are valuable as diagnostic markers in several studies. Due to their relative stability in biological fluids and tissue biopsies, miRNAs can be easily detected in clinical practice using automated, fast and standardized approaches [73]. Detection of cell-free miRNAs in biological fluids has proven its clinical relevance and diagnostic efficiency [74]. Mir-378 and miR-451 detection in serum revealed the presence of renal cell cancer [75]. Recently, a miR-Test detecting a signature of miRNAs has been developed for early detection of lung cancer in high-risk individuals [76]. miRNA expression profiles are shown to facilitate tumor classification according to tissue of origin [77]. piRNAs, which are also aberrantly expressed in cancers, have shown their utility in the detection of circulating tumor cells in peripheral blood of patients with GC [78]. Detection of U2 snRNA fragments has also drawn attention as potential diagnostic markers of pancreatic cancer and CRC [79].

Although less stable in biological fluid, lncRNAs have demonstrated an undeniable interest as diagnostic markers in clinical practices. The prostate-specific lincRNA, PCA3, is detected in urine and remains one of the most specific markers for the detection of prostate cancer. PCA3 detection kits are now available for clinical use [80]. The uc.73 and uc.388 T-UCRs were demonstrated to potentially be a diagnostic marker in CRC patients [81].

12.4.2 PROGNOSTIC MARKERS

The number of studies proving the interest of miRNAs as prognostic marker has been constantly increasing since the last decade. A myriad of examples are available in the literature [82]. A recent study proposed the expression of the oncomiR miR-197 as a novel prognostic indicator for non-small-cell lung cancer (NSCLC) patients [83]. Another recent work proved the power of piRNA expression as a prognostic marker in clear cell renal cell carcinomas (ccRCC). Both piR-30924 and pi-57125 are independent prognostic predictors [84]. In the same type of cancer, an independent study revealed that the upregulation of three piRNAs is highly associated with metastasis, late clinical stage, and survival [53].

lncRNAs can also display potential interest as a prognostic marker. A meta-analysis revealed that the HOTAIR lncRNA abundance is associated with poor prognosis in different types of cancers in Asia and in Western countries. Moreover, HOTAIR expression is increased in pancreatic cancer and associated with more aggressive tumors [85]. In addition, the mitochondrial lncRNA LIPCAR levels in plasma were predictive of future cardiovascular deaths in patients with heart failure independent of other predictors [86]. In high-risk neuroblastoma patients, a T-UCR signature (15 up-regulated T-UCRs) allows the discrimination of short- and long-surviving patients with high sensitivity and specificity [87]. A recent work found that the *POU5F1B* pseudogene amplification (described earlier) is associated with a poor prognosis in GC patients [68].

12.4.3 PREDICTIVE MARKERS OF THERAPEUTIC RESPONSE

The therapeutic response in patients diagnosed with the same pathology and treated with the same compound is often highly variable. This variability is usually due to different parameters not limited to disease stage and age. The molecular heterogeneity of the diseased organ also contributes to the overall therapeutic response. Expression of ncRNAs has been associated with a differential response to diverse therapeutic treatments.

Among small ncRNAs, the participation of miRNA expression in therapeutic response has largely been demonstrated in a number of studies. For instance, in metastatic renal cell carcinoma (MRCC), the expression of miR-942 is the most accurate predictor of sunitinib treatment. By indirectly provoking the upregulation of matrix metallopeptidase 9 (MMP-9) and vascular endothelial growth factor (VEGF), miR-942 expression promotes sunitinib resistance [88]. In rheumatoid arthritis (RA) pathology, antitumor necrosis factor alpha (anti-TNF-α) treatment has greatly improved the medical management of patients. However, this treatment is ineffective in a fraction of the patients. A specific miRNA plasma signature (miR-23 and miR-223) was proposed as a potential bioindicator of anti-TNF therapeutic response [89].

lncRNAs display great potential in the prediction of therapeutic treatment [16]. Maternally expressed gene (MEG3) is an imprinted gene located on Chr14q32 that encodes a lncRNA. This lncRNA participates in the regulation of cisplatin resistance of lung adenocarcinoma cells through the control of p53 and B-cell lymphoma-extra large (Bcl-xl) expression [90]. Cancer stem cells (CSC) are the tumorigenic cell population that sustain tumor growth and resist conventional chemotherapies. Low expression of the lncRNA Xist (X-inactive specific transcript) predicts drug response to histone deacetylase (HDAC) inhibitor of patient-derived xenography of breast cancer with a significant reduction of CSC population [91]. More recently, a lncRNA score (48 different lncRNAs) was derived from a cohort of patients with cytogenetically normal acute myeloid leukemia (AML). This lncRNA score correlates strongly with treatment response and survival [92].

12.5 ncRNA TARGETING STRATEGIES FOR TREATMENT OF HUMAN DISEASES

Small and long ncRNA expression is deregulated in various human diseases. Some are specifically upregulated or downregulated depending on the tissue, the developmental stage, or the disease. In addition to their biological implication in the genesis of human disease, ncRNAs are logically seen as attractive therapeutic targets. Many strategies have been developed to restore the expression of downregulated ncRNAs (Figure 12.2). These approaches are thoroughly described in Ling's review [93]. Thus, in this chapter, we will focus our interest on strategies that target overexpressed ncRNAs. We will illustrate those strategies with preclinical or clinical studies when available.

12.5.1 ANTISENSE OLIGONUCLEOTIDES

Antisense oligonucleotides (ASOs) are short single-strand DNA or RNA molecules (between 8 and 50 nt) that are complementary to selected RNAs [89,93]. They can target both small and long ncRNAs. Strategies were developed to combine the inhibition of ncRNA activity with high binding affinity and resistance to nuclease. Although these ASOs proved their efficacy *in vitro*, chemical modifications were required to protect them from nuclease degradation *in vivo*. One of the first described modifications was the 2′-O-methyl modification [94]. This chemical modification significantly increased the resistance of ASO to nuclease and hybridization to ssRNA. However, these

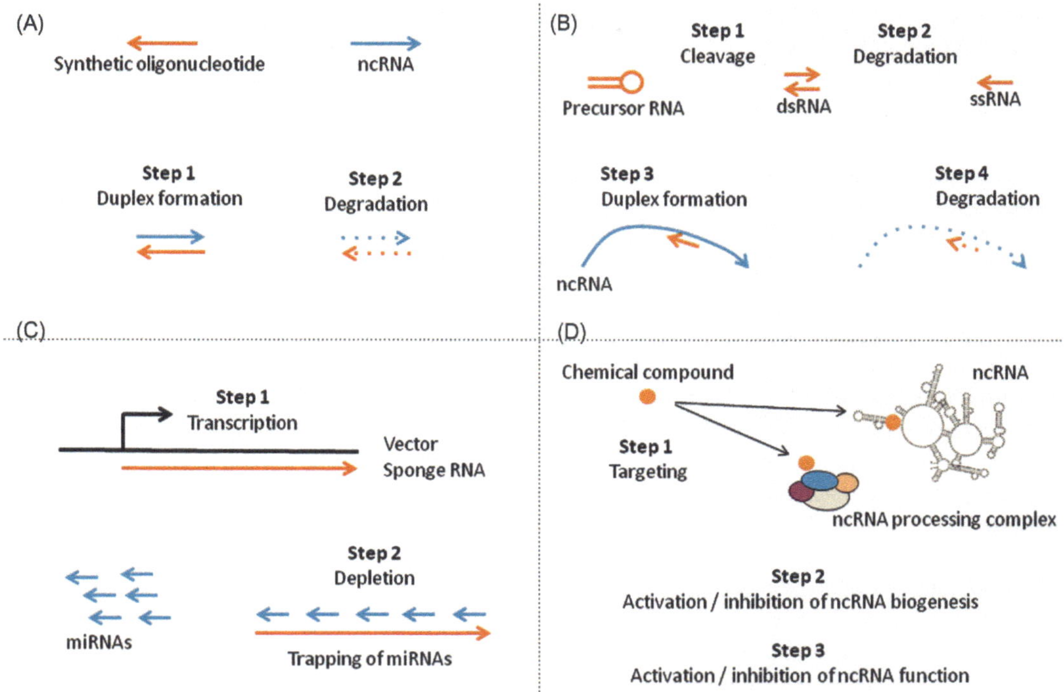

FIGURE 12.2

Strategies for the targeting of overexpressed noncoding RNAs. (A) Antisense oligonucleotides. (B) Exogenous small interfering RNA. (C) miRNA sponge vectors. (D) Small molecules. ncRNA, noncoding RNA; dsRNA, double stranded RNA; ssRNA, single stranded RNA; miRNA, micro RNA.

modified ASOs were still sensitive to serum exonuclease and therefore not ideal for clinical application. ASO targeting of lncRNAs has shown potential interest in preclinical neurological disease models. Angelman syndrome is a single-gene disorder characterized by intellectual disability and developmental delay. It is caused by maternal deficiency of the imprinted gene *UBE3A*, encoding an E3 ubiquitine ligase. All patients carry at least one copy of paternal *UBE3A*, which is intact but silenced by a lncRNA, UBE3A antisense transcript (UBE3A-AS). In a recent study, the authors developed a therapeutic intervention by reducing Ube3a-ATS with ASOs. Partial restoration of UBE3A protein in an Angelman syndrome mouse model improved certain cognitive deficits associated with the disease [95]. In addition, in a preclinical mouse model of lung cancer, the targeting of the lncRNA MALAT-1 with ASO nucleotides was shown to prevent metastasis [96].

In 2005, antagomir ASOs, which are single-stranded $\approx$23 nt-long RNA molecules that complementarily bind miRNAs, were engineered. They combined 2′-O-methyl modifications, selected phosphorothioate modifications at both ends, thereby protecting them from exonucleases, and a 3′ cholesterol tail to improve *in vivo* distribution [97]. Other ASO chemical modifications at the ribose 2′ position were developed, such as the 2′-O-methyoxyethyl (2′-MOE) [98] and 2′-fluoro (2′-F) [99]. One application of an ASO was shown in studies aimed at targeting miR-135b, which promotes lung cancer cell invasion and metastasis *in vivo*. Treatment with miR-135b antagomir in mouse orthotopic models of lung tumor delayed tumor growth and metastasis [100]. In noncancerous disease, antagomirs have also shown promising results in preclinical mouse models. Indeed, antagomir targeting of miR-34 protects pilocarpine-induced status epilepticus [101].

Locked nucleic acid (LNA) is another type of ASO modification. This modification links the oxygen at the position 2′ to the carbon at position 4′ by a methylene bridge and confers a better binding affinity and nuclease resistance. Elmén et al. provided evidence for the efficacy of LNA-modified ASO in antagonizing liver expressed miR-122 in nonhuman primates [102]. One of the most advanced human clinical approaches targeting small ncRNA is the antiviral therapy using a LNA anti-miR directed against mir-122. The clinical use of Miravirsen or SPC3649 (Santaris Pharma) aims to repress hepatitis C virus (HCV) replication. The rationale for these clinical studies stems from Sarnow's group study in which they demonstrate that miR-122 is highly expressed in human liver and binds to the 5′ noncoding region of the HCV genome to aid viral RNA folding or RNA sequestration in replication complexes. More importantly, they showed that sequestering miR-122 results in marked loss of autonomously replicating HCV RNAs [103]. The phase II study revealed that Miravirsen treatment leads to a dose-dependent reduction in HCV RNA levels that lasts after the end of the treatment (NCT01200420) [104].

Tiny LNA anti-miRs are eight-nucleotide LNA-modified ASOs [105]. They target the 5′-seed region of miRNAs and therefore interact with miRNAs sharing the same seed. Therapeutic effects of tiny LNA anti-miRs have been shown in preclinical models. The silencing of miR-21 *in vivo* ameliorates autoimmune splenomegaly in lupus mice using tiny-LNA [106].

LNA molecules are also able to target lncRNAs. Recently, a role for the lncRNA Breast Cancer Anti-Estrogen Resistance 4 (BCAR4) was reported in breast cancer metastasis [107]. Inactivation of lncBCAR4 reduces the number of lung metastasis nodules with no consequence on primary tumor size, tumor cell proliferation, or apoptosis. To evaluate the therapeutic potential of BCAR4 targeting, *in vivo* optimized LNAs against lncRNA BCAR4 were developed. Two individual systemic LNA treatments significantly reduced lung metastases in NOD/SCID mouse models.

As described earlier, discordant negative regulation of specific genes by NAT can be of particular interest from a therapeutic point of view. As NATs are often upregulated in human pathologies, their targeting leads to a specific reexpression of abnormally repressed genes. Single-strand oligonucleotide strategies to degrade NATs or block their interaction with sense partner genes were already developed and named AntagoNATs [108]. The success of this approach was brought about by the inhibition of brain-derived neurotrophic factor antisense (BDNF-AS) that leads to an upregulation of BDNF mRNA, an increased of BDNF protein expression and neuronal differentiation *in vitro* and *in vivo*.

12.5.2 EXOGENOUS SMALL INTERFERING RNAs

Small interfering RNAs (siRNAs) are generated from cytoplasmic long double-stranded RNAs (dsRNAs) that are cleaved in the cytoplasm by Dicer into functional siRNAs. They are recognized by the Ago2/RISC complex that interacts to degrade one strand. The other strand guides the inhibitory complex to perfect or closely perfect complementary mRNA. Targeted mRNAs or ncRNAs are subsequently degraded through cleavage by the RISC complex. Due to this almost perfect association with targeted mRNAs or ncRNAs, the number of mRNAs targeted by one siRNA is smaller than for miRNAs.

siRNA approaches in most clinical studies target mRNAs. One example includes targeting the Ephrin A2 (EphA2) gene using neutral liposomal siRNA delivery (phase I study/NCT01591356), or K-RAS^{G12D} gene targeting siG12D LODER, a miniature biodegradable capsule that encompasses the drug, designed and produced by Silenseed Ltd (phase II study/NCT01676259). Several siRNA strategies targeting ncRNAs have showed promising preclinical results. For instance, depletion of the lincRNA HOTAIR by siRNA inhibits tumor growth in a pancreatic cancer xenograft model [85]. Moreover, in an HCC mouse model, inhibition of HOTAIR expression by siRNA suppresses tumorigenicity *in vivo* by promoting the expression of miR-218 [109]. NATs can also be targeted by siRNA strategies. Indeed, the depletion of ANRIL NAT by siRNA disrupts the protein suppressor 2 of zeste (SUZ2) binding to the p15(INK4B) locus, increases p15(INK4B) protein expression, and inhibits cell proliferation [110].

12.5.3 miRNA SPONGE VECTORS

Sponge RNA strategies were developed to explicitly deplete cells of a specific miRNA or a miRNA family harboring the same seed sequence. This approach was inspired by natural molecular mechanisms that lead to inhibition of miRNA activity in human cells. The HCV RNA functions as a natural miR-122 "sponge" RNA in liver cells. By sequestering miR-122, HCV RNA expression leads to global derepression of host miR-122 mRNA targets [111]. Another example of the natural miRNA sponge is the circRNA ciRS-7. Highly expressed in human and mouse brain, this circRNA traps miR-7. It strongly suppresses miR-7 activity resulting in increased levels of its mRNA targets [112]. Equally, the imprinted H19 lncRNA acts as a Let-7 molecular sponge in human cells [113].

Artificial sponge RNAs were engineered to contain multiple binding sites that are complementary of mature miRNA seed sequence. They are usually transcribed from strong viral promoter (Simian Virus 40, CytoMegalo Virus). The expression of sponge RNAs is made possible by either transient plasmid transfection or stable expression using viral transduction. Artificial sponge RNAs have been predominately used for *in vitro* and preclinical studies. For instance, 4T1 cells infected by a retrovirus expressing miR-9 tandem repeated binding site sponge RNA were injected into the mammary fat pads of Balb/c mice. After 4 weeks, the animals display a 50% decrease of lung metastasis when compared to a sponge control RNA [114].

12.5.4 SMALL MOLECULES

Previous known small molecules that specifically inhibit RNA molecules have raised the idea of exploring synthetic small molecules to target lncRNAs. The secondary structures of RNA, of which lncRNAs are abundant, create numerous potential binding pockets for small molecules. The screening of small organic molecules followed by a structure–activity relationship analysis of an initial hit produced the azobenzene (compound 2) as a specific and efficient inhibitor of miR-21 [115]. Another study identified a small molecule modifier of the liver-specific miR-122 [116]. Interestingly, they showed that small molecules can be either activators or inhibitors of miRNA functions. Moreover, one of the miR-122 inhibitors was found to block HCV replication in liver cells. In 2013, Tan et al. identified a cyclobutene-type compound as a specific inhibitor of miR-1, which is closely related to cardiac development and disease [117]. Besides miR-specific small-molecule inhibition, enoxacin, a fluoroquinolone, was identified as a miRNA processing activator by binding to the miRNA biosynthesis protein TAR RNA-binding protein (TRBP). This application leads to an upregulation of 24 different mature miRNAs and mediates antitumoral effects *in vitro* and *in vivo* [118].

Despite the exhaustive research efforts, we are still far from designing potent small molecules specifically targeting miRNAs with a complete understanding of their mechanisms. The limitations of such strategies are discussed in an elegant review by Zhang et al. [119].

12.6 CONCLUSION

Over the last decade tremendous efforts have been made to identify and annotate new classes ncRNAs, and, to decipher their multiple, intricate functions. These RNA species are recognized as crucial actors in human disorders including cancer. Their expression patterns have proved to be powerful biomarkers in clinical trials. One main challenge now is to extend their use into clinical practice. Understanding the regulation and how to target specific ncRNAs paves the way for numerous companies for the development of innovative therapies. Methodologies to improve *in vivo* delivery, off-target residual effects, and efficacy are and will continue to be important future challenges. Dedicated efforts are required to generate and implement ncRNA targeting strategies to become highly effective and safe for clinical trials.

REFERENCES

[1] Bartel DP. MicroRNAs: genomics, biogenesis, mechanism, and function. Cell 2004;116(2):281–97.

[2] Eulalio A, Huntzinger E, Izaurralde E. Getting to the root of miRNA-mediated gene silencing. Cell 2008;132(1):9–14.

[3] Ha M, Kim VN. Regulation of microRNA biogenesis. Nat Rev Mol Cell Biol 2014;15(8):509–24.

[4] Shenoy A, Blelloch RH. Regulation of microRNA function in somatic stem cell proliferation and differentiation. Nat Rev Mol Cell Biol 2014;15(9):565–76.

[5] Esteller M. Noncoding RNAs in human disease. Nat Rev Genet 2011;12(12):861–74.

[6] Gunawardane LS, Saito K, Nishida KM, Miyoshi K, Kawamura Y, Nagami T, et al. A slicer-mediated mechanism for repeat-associated siRNA 5′ end formation in Drosophila. Science 2007;315(5818): 1587–90.

[7] Molaro A, Falciatori I, Hodges E, Aravin AA, Marran K, Rafii S, et al. Two waves of *de novo* methylation during mouse germ cell development. Genes Dev 2014;28(14):1544–9.

[8] Watanabe T, Tomizawa S, Mitsuya K, Totoki Y, Yamamoto Y, Kuramochi-Miyagawa S, et al. Role for piRNAs and noncoding RNA in *de novo* DNA methylation of the imprinted mouse Rasgrf1 locus. Science 2011;332(6031):848–52.

[9] Kiss T. Small nucleolar RNAs: an abundant group of noncoding RNAs with diverse cellular functions. Cell 2002;109(2):145–8.

[10] Kiss-László Z, Henry Y, Bachellerie JP, Caizergues-Ferrer M, Kiss T. Site-specific ribose methylation of preribosomal RNA: a novel function for small nucleolar RNAs. Cell 1996;85(7):1077–88.
[11] Ni J, Tien AL, Fournier MJ. Small nucleolar RNAs direct site-specific synthesis of pseudouridine in ribosomal RNA. Cell 1997;89(4):565–73.
[12] King TH, Liu B, McCully RR, Fournier MJ. Ribosome structure and activity are altered in cells lacking snoRNPs that form pseudouridines in the peptidyl transferase center. Mol Cell 2003;11 (2):425–35.
[13] Matera AG, Terns RM, Terns MP. Noncoding RNAs: lessons from the small nuclear and small nucleolar RNAs. Nat Rev Mol Cell Biol 2007;8(3):209–20.
[14] Granneman S, Baserga SJ. Ribosome biogenesis: of knobs and RNA processing. Exp Cell Res 2004;296 (1):43–50.
[15] Arts GJ, Fornerod M, Mattaj IW. Identification of a nuclear export receptor for tRNA. Curr Biol CB 1998;8(6):305–14.
[16] Malek E, Jagannathan S, Driscoll JJ. Correlation of long non-coding RNA expression with metastasis, drug resistance and clinical outcome in cancer. Oncotarget 2014;5(18):8027–38.
[17] Mercer TR, Dinger ME, Mattick JS. Long non-coding RNAs: insights into functions. Nat Rev Genet 2009;10(3):155–9.
[18] Derrien T, Johnson R, Bussotti G, Tanzer A, Djebali S, Tilgner H, et al. The GENCODE v7 catalog of human long noncoding RNAs: analysis of their gene structure, evolution, and expression. Genome Res 2012;22(9):1775–89.
[19] Katayama S, Tomaru Y, Kasukawa T, Waki K, Nakanishi M, Nakamura M, et al. Antisense transcription in the mammalian transcriptome. Science 2005;309(5740):1564–6.
[20] Chen J, Sun M, Kent WJ, Huang X, Xie H, Wang W, et al. Over 20% of human transcripts might form sense-antisense pairs. Nucleic Acids Res 2004;32(16):4812–20.
[21] Wahlestedt C. Targeting long non-coding RNA to therapeutically upregulate gene expression. Nat Rev Drug Discov 2013;12(6):433–46.
[22] Yap KL, Li S, Muñoz-Cabello AM, Raguz S, Zeng L, Mujtaba S, et al. Molecular interplay of the non-coding RNA ANRIL and methylated histone H3 lysine 27 by polycomb CBX7 in transcriptional silencing of INK4a. Mol Cell 2010;38(5):662–74.
[23] Saldaña-Meyer R, González-Buendía E, Guerrero G, Narendra V, Bonasio R, Recillas-Targa F, et al. CTCF regulates the human p53 gene through direct interaction with its natural antisense transcript, Wrap53. Genes Dev 2014;28(7):723–34.
[24] Beltran M, Puig I, Peña C, García JM, Alvarez AB, Peña R, et al. A natural antisense transcript regulates Zeb2/Sip1 gene expression during Snail1-induced epithelial-mesenchymal transition. Genes Dev 2008;22 (6):756–69.
[25] Faghihi MA, Zhang M, Huang J, Modarresi F, Van der Brug MP, Nalls MA, et al. Evidence for natural antisense transcript-mediated inhibition of microRNA function. Genome Biol 2010;11(5):R56.
[26] Bejerano G, Pheasant M, Makunin I, Stephen S, Kent WJ, Mattick JS, et al. Ultraconserved elements in the human genome. Science 2004;304(5675):1321–5.
[27] Mestdagh P, Fredlund E, Pattyn F, Rihani A, Van Maerken T, Vermeulen J, et al. An integrative genomics screen uncovers ncRNA T-UCR functions in neuroblastoma tumours. Oncogene 2010;29(24):3583–92.
[28] Calin GA, Liu C, Ferracin M, Hyslop T, Spizzo R, Sevignani C, et al. Ultraconserved regions encoding ncRNAs are altered in human leukemias and carcinomas. Cancer Cell 2007;12(3):215–29.
[29] Watters KM, Bryan K, Foley NH, Meehan M, Stallings RL. Expressional alterations in functional ultra-conserved non-coding RNAs in response to all-trans retinoic acid—induced differentiation in neuroblastoma cells. BMC Cancer 2013;13:184.
[30] Ulitsky I, Bartel DP. lincRNAs: genomics, evolution, and mechanisms. Cell 2013;154(1):26–46.

[31] Hajjari M, Salavaty A. HOTAIR: an oncogenic long non-coding RNA in different cancers. Cancer Biol Med 2015;12(1):1–9.
[32] Tsai M-C, Spitale RC, Chang HY. Long intergenic noncoding RNAs: new links in cancer progression. Cancer Res 2011;71(1):3–7.
[33] ENCODE Project Consortium, Birney E, Stamatoyannopoulos JA, Dutta A, Guigó R, Gingeras TR, et al. Identification and analysis of functional elements in 1% of the human genome by the ENCODE pilot project. Nature 2007;447(7146):799–816.
[34] Sasidharan R, Gerstein M. Genomics: protein fossils live on as RNA. Nature 2008;453(7196):729–31.
[35] Poliseno L, Salmena L, Zhang J, Carver B, Haveman WJ, Pandolfi PP. A coding-independent function of gene and pseudogene mRNAs regulates tumour biology. Nature 2010;465(7301):1033–8.
[36] Sanger HL, Klotz G, Riesner D, Gross HJ, Kleinschmidt AK. Viroids are single-stranded covalently closed circular RNA molecules existing as highly base-paired rod-like structures. Proc Natl Acad Sci USA 1976;73(11):3852–6.
[37] Memczak S, Jens M, Elefsinioti A, Torti F, Krueger J, Rybak A, et al. Circular RNAs are a large class of animal RNAs with regulatory potency. Nature 2013;495(7441):333–8.
[38] Hawkins PG, Santoso S, Adams C, Anest V, Morris KV. Promoter targeted small RNAs induce long-term transcriptional gene silencing in human cells. Nucleic Acids Res 2009;37(9):2984–95.
[39] Taft RJ, Glazov EA, Cloonan N, Simons C, Stephen S, Faulkner GJ, et al. Tiny RNAs associated with transcription start sites in animals. Nat Genet 2009;41(5):572–8.
[40] Ntini E, Järvelin AI, Bornholdt J, Chen Y, Boyd M, Jørgensen M, et al. Polyadenylation site-induced decay of upstream transcripts enforces promoter directionality. Nat Struct Mol Biol 2013;20(8):923–8.
[41] Lopez-Serra P, Esteller M. DNA methylation-associated silencing of tumor-suppressor microRNAs in cancer. Oncogene 2012;31(13):1609–22.
[42] Ding J, Huang S, Wu S, Zhao Y, Liang L, Yan M, et al. Gain of miR-151 on chromosome 8q24.3 facilitates tumour cell migration and spreading through downregulating RhoGDIA. Nat Cell Biol 2010;12(4):390–9.
[43] Calin GA, Dumitru CD, Shimizu M, Bichi R, Zupo S, Noch E, et al. Frequent deletions and down-regulation of micro- RNA genes miR15 and miR16 at 13q14 in chronic lymphocytic leukemia. Proc Natl Acad Sci USA 2002;99(24):15524–9.
[44] Pfeffer SR, Yang CH, Pfeffer LM. The role of miR-21 in cancer. Drug Dev Res 2015 June 16.
[45] Duygu B, Da Costa Martins PA. miR-21: a star player in cardiac hypertrophy. Cardiovasc Res 2015;105(3):235–7.
[46] Pan W, Zhu S, Yuan M, Cui H, Wang L, Luo X, et al. MicroRNA-21 and microRNA-148a contribute to DNA hypomethylation in lupus CD4 + T cells by directly and indirectly targeting DNA methyltransferase 1. J Immunol Baltim Md 1950 2010;184(12):6773–81.
[47] Meng F, Henson R, Wehbe-Janek H, Ghoshal K, Jacob ST, Patel T. MicroRNA-21 regulates expression of the PTEN tumor suppressor gene in human hepatocellular cancer. Gastroenterology 2007;133(2):647–58.
[48] Müller S, Raulefs S, Bruns P, Afonso-Grunz F, Plötner A, Thermann R, et al. Next-generation sequencing reveals novel differentially regulated mRNAs, lncRNAs, miRNAs, sdRNAs and a piRNA in pancreatic cancer. Mol Cancer 2015;14:94.
[49] Hashim A, Rizzo F, Marchese G, Ravo M, Tarallo R, Nassa G, et al. RNA sequencing identifies specific PIWI-interacting small non-coding RNA expression patterns in breast cancer. Oncotarget 2014;5(20):9901–10.
[50] Yan H, Wu Q-L, Sun C-Y, Ai L-S, Deng J, Zhang L, et al. piRNA-823 contributes to tumorigenesis by regulating de novo DNA methylation and angiogenesis in multiple myeloma. Leukemia 2015;29(1):196–206.

[51] Mei Y-P, Liao J-P, Shen J, Yu L, Liu B-L, Liu L, et al. Small nucleolar RNA 42 acts as an oncogene in lung tumorigenesis. Oncogene 2012;31(22):2794–804.
[52] Mannoor K, Shen J, Liao J, Liu Z, Jiang F. Small nucleolar RNA signatures of lung tumor-initiating cells. Mol Cancer 2014;13:104.
[53] Li Y, Wu X, Gao H, Jin JM, Li AX, Kim YS, et al. PIWI-interacting RNAs are dysregulated in renal cell carcinoma and associated with tumor metastasis and cancer specific survival. Mol Med Camb Mass 2015;21:381–8.
[54] Sahoo T, del Gaudio D, German JR, Shinawi M, Peters SU, Person RE, et al. Prader-Willi phenotype caused by paternal deficiency for the HBII-85 C/D box small nucleolar RNA cluster. Nat Genet 2008;40 (6):719–21.
[55] Mitchell JR, Wood E, Collins K. A telomerase component is defective in the human disease dyskeratosis congenita. Nature 1999;402(6761):551–5.
[56] Nguyen LXT, Raval A, Garcia JS, Mitchell BS. Regulation of ribosomal gene expression in cancer. J Cell Physiol 2015;230(6):1181–8.
[57] Ghoshal K, Majumder S, Datta J, Motiwala T, Bai S, Sharma SM, et al. Role of human ribosomal RNA (rRNA) promoter methylation and of methyl-CpG-binding protein MBD2 in the suppression of rRNA gene expression. J Biol Chem 2004;279(8):6783–93.
[58] Ladd PD, Smith LE, Rabaia NA, Moore JM, Georges SA, Hansen RS, et al. An antisense transcript spanning the CGG repeat region of FMR1 is upregulated in premutation carriers but silenced in full mutation individuals. Hum Mol Genet 2007;16(24):3174–87.
[59] Chung DW, Rudnicki DD, Yu L, Margolis RL. A natural antisense transcript at the Huntington's disease repeat locus regulates HTT expression. Hum Mol Genet 2011;20(17):3467–77.
[60] Holdt LM, Beutner F, Scholz M, Gielen S, Gäbel G, Bergert H, et al. ANRIL expression is associated with atherosclerosis risk at chromosome 9p21. Arterioscler Thromb Vasc Biol 2010;30(3):620–7.
[61] Peng JC, Shen J, Ran ZH. Transcribed ultraconserved region in human cancers. RNA Biol 2013;10 (12):1771–7.
[62] Fassan M, Dall'Olmo L, Galasso M, Braconi C, Pizzi M, Realdon S, et al. Transcribed ultraconserved noncoding RNAs (T-UCR) are involved in Barrett's esophagus carcinogenesis. Oncotarget 2014;5 (16):7162–71.
[63] Lujambio A, Portela A, Liz J, Melo SA, Rossi S, Spizzo R, et al. CpG island hypermethylation-associated silencing of non-coding RNAs transcribed from ultraconserved regions in human cancer. Oncogene 2010;29(48):6390–401.
[64] Huarte M, Guttman M, Feldser D, Garber M, Koziol MJ, Kenzelmann-Broz D, et al. A large intergenic noncoding RNA induced by p53 mediates global gene repression in the p53 response. Cell 2010;142 (3):409–19.
[65] Yildirim E, Kirby JE, Brown DE, Mercier FE, Sadreyev RI, Scadden DT, et al. Xist RNA is a potent suppressor of hematologic cancer in mice. Cell 2013;152(4):727–42.
[66] Li L, Feng T, Lian Y, Zhang G, Garen A, Song X. Role of human noncoding RNAs in the control of tumorigenesis. Proc Natl Acad Sci USA 2009;106(31):12956–61.
[67] Kalyana-Sundaram S, Kumar-Sinha C, Shankar S, Robinson DR, Wu Y-M, Cao X, et al. Expressed pseudogenes in the transcriptional landscape of human cancers. Cell 2012;149(7):1622–34.
[68] Hayashi H, Arao T, Togashi Y, Kato H, Fujita Y, De Velasco MA, et al. The OCT4 pseudogene POU5F1B is amplified and promotes an aggressive phenotype in gastric cancer. Oncogene 2015;34 (2):199–208.
[69] Qu S, Yang X, Li X, Wang J, Gao Y, Shang R, et al. Circular RNA: a new star of noncoding RNAs. Cancer Lett 2015;365(2):141–8.

[70] Bachmayr-Heyda A, Reiner AT, Auer K, Sukhbaatar N, Aust S, Bachleitner-Hofmann T, et al. Correlation of circular RNA abundance with proliferation—exemplified with colorectal and ovarian cancer, idiopathic lung fibrosis, and normal human tissues. Sci Rep 2015;5:8057.
[71] Hansen TB, Kjems J, Damgaard CK. Circular RNA and miR-7 in cancer. Cancer Res 2013;73(18):5609–12.
[72] Burd CE, Jeck WR, Liu Y, Sanoff HK, Wang Z, Sharpless NE. Expression of linear and novel circular forms of an INK4/ARF-associated non-coding RNA correlates with atherosclerosis risk. PLoS Genet 2010;6(12):e1001233.
[73] Hayes J, Peruzzi PP, Lawler S. MicroRNAs in cancer: biomarkers, functions and therapy. Trends Mol Med 2014;20(8):460–9.
[74] Schwarzenbach H, Nishida N, Calin GA, Pantel K. Clinical relevance of circulating cell-free microRNAs in cancer. Nat Rev Clin Oncol 2014;11(3):145–56.
[75] Redova M, Poprach A, Nekvindova J, Iliev R, Radova L, Lakomy R, et al. Circulating miR-378 and miR-451 in serum are potential biomarkers for renal cell carcinoma. J Transl Med 2012;10:55.
[76] Montani F, Marzi MJ, Dezi F, Dama E, Carletti RM, Bonizzi G, et al. miR-Test: a blood test for lung cancer early detection. J Natl Cancer Inst 2015;107(6):djv063.
[77] Rosenfeld N, Aharonov R, Meiri E, Rosenwald S, Spector Y, Zepeniuk M, et al. MicroRNAs accurately identify cancer tissue origin. Nat Biotechnol 2008;26(4):462–9.
[78] Cui L, Lou Y, Zhang X, Zhou H, Deng H, Song H, et al. Detection of circulating tumor cells in peripheral blood from patients with gastric cancer using piRNAs as markers. Clin Biochem 2011;44(13):1050–7.
[79] Baraniskin A, Nöpel-Dünnebacke S, Ahrens M, Jensen SG, Zöllner H, Maghnouj A, et al. Circulating U2 small nuclear RNA fragments as a novel diagnostic biomarker for pancreatic and colorectal adenocarcinoma. Int J Cancer J Int Cancer 2013;132(2):E48–57.
[80] Hessels D, Schalken JA. The use of PCA3 in the diagnosis of prostate cancer. Nat Rev Urol 2009;6(5):255–61.
[81] Sana J, Hankeova S, Svoboda M, Kiss I, Vyzula R, Slaby O. Expression levels of transcribed ultraconserved regions uc.73 and uc.388 are altered in colorectal cancer. Oncology 2012;82(2):114–18.
[82] Cortez MA, Ivan C, Zhou P, Wu X, Ivan M, Calin GA. microRNAs in cancer: from bench to bedside. Adv Cancer Res 2010;108:113–57.
[83] Mavridis K, Gueugnon F, Petit-Courty A, Courty Y, Barascu A, Guyetant S, et al. The oncomiR miR-197 is a novel prognostic indicator for non-small cell lung cancer patients. Br J Cancer 2015;112(9):1527–35.
[84] Busch J, Ralla B, Jung M, Wotschofsky Z, Trujillo-Arribas E, Schwabe P, et al. Piwi-interacting RNAs as novel prognostic markers in clear cell renal cell carcinomas. J Exp Clin Cancer Res CR 2015;34(1):61.
[85] Kim K, Jutooru I, Chadalapaka G, Johnson G, Frank J, Burghardt R, et al. HOTAIR is a negative prognostic factor and exhibits pro-oncogenic activity in pancreatic cancer. Oncogene 2013;32(13):1616–25.
[86] Kumarswamy R, Bauters C, Volkmann I, Maury F, Fetisch J, Holzmann A, et al. Circulating long noncoding RNA, LIPCAR, predicts survival in patients with heart failure. Circ Res 2014;114(10):1569–75.
[87] Scaruffi P, Stigliani S, Moretti S, Coco S, De Vecchi C, Valdora F, et al. Transcribed-ultra conserved region expression is associated with outcome in high-risk neuroblastoma. BMC Cancer 2009;9:441.
[88] Prior C, Perez-Gracia JL, Garcia-Donas J, Rodriguez-Antona C, Guruceaga E, Esteban E, et al. Identification of tissue microRNAs predictive of sunitinib activity in patients with metastatic renal cell carcinoma. PLoS One 2014;9(1):e86263.
[89] Castro-Villegas C, Pérez-Sánchez C, Escudero A, Filipescu I, Verdu M, Ruiz-Limón P, et al. Circulating miRNAs as potential biomarkers of therapy effectiveness in rheumatoid arthritis patients treated with anti-TNFα. Arthritis Res Ther 2015;17:49.

[90] Liu J, Wan L, Lu K, Sun M, Pan X, Zhang P, et al. The long noncoding RNA MEG3 contributes to cisplatin resistance of human lung adenocarcinoma. PLoS One 2015;10(5):e0114586.
[91] Salvador MA, Wicinski J, Cabaud O, Toiron Y, Finetti P, Josselin E, et al. The histone deacetylase inhibitor abexinostat induces cancer stem cells differentiation in breast cancer with low Xist expression. Clin Cancer Res Off J Am Assoc Cancer Res 2013;19(23):6520–31.
[92] Garzon R, Volinia S, Papaioannou D, Nicolet D, Kohlschmidt J, Yan PS, et al. Expression and prognostic impact of lncRNAs in acute myeloid leukemia. Proc Natl Acad Sci USA 2014;111(52):18679–84.
[93] Ling H, Fabbri M, Calin GA. MicroRNAs and other non-coding RNAs as targets for anticancer drug development. Nat Rev Drug Discov 2013;12(11):847–65.
[94] Cummins LL, Owens SR, Risen LM, Lesnik EA, Freier SM, McGee D, et al. Characterization of fully 2′-modified oligoribonucleotide hetero- and homoduplex hybridization and nuclease sensitivity. Nucleic Acids Res 1995;23(11):2019–24.
[95] Meng L, Ward AJ, Chun S, Bennett CF, Beaudet AL, Rigo F. Towards a therapy for Angelman syndrome by targeting a long non-coding RNA. Nature 2015;518(7539):409–12.
[96] Gutschner T, Hämmerle M, Eissmann M, Hsu J, Kim Y, Hung G, et al. The noncoding RNA MALAT1 is a critical regulator of the metastasis phenotype of lung cancer cells. Cancer Res 2013;73(3):1180–9.
[97] Krützfeldt J, Rajewsky N, Braich R, Rajeev KG, Tuschl T, Manoharan M, et al. Silencing of microRNAs *in vivo* with "antagomirs". Nature 2005;438(7068):685–9.
[98] Esau CC. Inhibition of microRNA with antisense oligonucleotides. Methods San Diego Calif 2008;44 (1):55–60.
[99] Davis S, Lollo B, Freier S, Esau C. Improved targeting of miRNA with antisense oligonucleotides. Nucleic Acids Res 2006;34(8):2294–304.
[100] Lin C-W, Chang Y-L, Chang Y-C, Lin J-C, Chen C-C, Pan S-H, et al. MicroRNA-135b promotes lung cancer metastasis by regulating multiple targets in the Hippo pathway and LZTS1. Nat Commun 2013;4:1877.
[101] Jimenez-Mateos EM, Engel T, Merino-Serrais P, Fernaud-Espinosa I, Rodriguez-Alvarez N, Reynolds J, et al. Antagomirs targeting microRNA-134 increase hippocampal pyramidal neuron spine volume *in vivo* and protect against pilocarpine-induced status epilepticus. Brain Struct Funct 2015;220(4):2387–99.
[102] Elmén J, Lindow M, Schütz S, Lawrence M, Petri A, Obad S, et al. LNA-mediated microRNA silencing in non-human primates. Nature 2008;452(7189):896–9.
[103] Jopling CL, Yi M, Lancaster AM, Lemon SM, Sarnow P. Modulation of hepatitis C virus RNA abundance by a liver-specific MicroRNA. Science 2005;309(5740):1577–81.
[104] Janssen HLA, Reesink HW, Lawitz EJ, Zeuzem S, Rodriguez-Torres M, Patel K, et al. Treatment of HCV infection by targeting microRNA. N Engl J Med 2013;368(18):1685–94.
[105] Obad S, Santos CO, dos, Petri A, Heidenblad M, Broom O, Ruse C, et al. Silencing of microRNA families by seed-targeting tiny LNAs. Nat Genet 2011;43(4):371–8.
[106] Garchow BG, Bartulos Encinas O, Leung YT, Tsao PY, Eisenberg RA, Caricchio R, et al. Silencing of microRNA-21 *in vivo* ameliorates autoimmune splenomegaly in lupus mice. EMBO Mol Med 2011;3 (10):605–15.
[107] Xing Z, Lin A, Li C, Liang K, Wang S, Liu Y, et al. lncRNA directs cooperative epigenetic regulation downstream of chemokine signals. Cell 2014;159(5):1110–25.
[108] Modarresi F, Faghihi MA, Lopez-Toledano MA, Fatemi RP, Magistri M, Brothers SP, et al. Inhibition of natural antisense transcripts *in vivo* results in gene-specific transcriptional upregulation. Nat Biotechnol 2012;30(5):453–9.
[109] Fu W-M, Zhu X, Wang W-M, Lu Y-F, Hu B-G, Wang H, et al. Hotair mediates hepatocarcinogenesis through suppressing MiRNA-218 expression and activating P14 and P16 signaling. J Hepatol 2015;63(4):886–95.

[110] Kotake Y, Nakagawa T, Kitagawa K, Suzuki S, Liu N, Kitagawa M, et al. Long non-coding RNA ANRIL is required for the PRC2 recruitment to and silencing of p15(INK4B) tumor suppressor gene. Oncogene 2011;30(16):1956–62.

[111] Luna JM, Scheel TKH, Danino T, Shaw KS, Mele A, Fak JJ, et al. Hepatitis C virus RNA functionally sequesters miR-122. Cell 2015;160(6):1099–110.

[112] Hansen TB, Jensen TI, Clausen BH, Bramsen JB, Finsen B, Damgaard CK, et al. Natural RNA circles function as efficient microRNA sponges. Nature 2013;495(7441):384–8.

[113] Kallen AN, Zhou X-B, Xu J, Qiao C, Ma J, Yan L, et al. The imprinted H19 lncRNA antagonizes let-7 microRNAs. Mol Cell 2013;52(1):101–12.

[114] Ma L, Young J, Prabhala H, Pan E, Mestdagh P, Muth D, et al. miR-9, a MYC/MYCN-activated microRNA, regulates E-cadherin and cancer metastasis. Nat Cell Biol 2010;12(3):247–56.

[115] Gumireddy K, Young DD, Xiong X, Hogenesch JB, Huang Q, Deiters A. Small-molecule inhibitors of microrna miR-21 function. Angew Chem Int Ed Engl 2008;47(39):7482–4.

[116] Young DD, Connelly CM, Grohmann C, Deiters A. Small molecule modifiers of microRNA miR-122 function for the treatment of hepatitis C virus infection and hepatocellular carcinoma. J Am Chem Soc 2010;132(23):7976–81.

[117] Tan S-B, Huang C, Chen X, Wu Y, Zhou M, Zhang C, et al. Small molecular inhibitors of miR-1 identified from photocycloadducts of acetylenes with 2-methoxy-1,4-naphthalenequinone. Bioorg Med Chem 2013;21(20):6124–31.

[118] Melo S, Villanueva A, Moutinho C, Davalos V, Spizzo R, Ivan C, et al. Small molecule enoxacin is a cancer-specific growth inhibitor that acts by enhancing TAR RNA-binding protein 2-mediated microRNA processing. Proc Natl Acad Sci USA 2011;108(11):4394–9.

[119] Zhang S, Chen L, Jung EJ, Calin GA. Targeting microRNAs with small molecules: from dream to reality. Clin Pharmacol Ther 2010;87(6):754–8.

PART

IV

CLINICAL IMPLICATIONS OF EPIGENETIC THERAPY

CHAPTER

CLINICAL APPLICATIONS OF EPIGENETIC DRUGS

13

Norbert Vey[1,2] and Thomas Prebet[1,3]

[1]Departement d'Hematologie et Unite d'Evaluation Therapeutique en Oncologie/Hematologie, Institut Paoli Calmettes, Marseille, France [2]Aix-Marseille Université, Marseille, France [3]Hematology Department, Smilow Cancer Center at Yale University, New Haven, CT, USA

CHAPTER OUTLINE

13.1 INTRODUCTION

As described in the previous chapters, the rationale to support the development of epigenetic drugs in cancer relies on the observation that in virtually all tumor types mutations affecting genes involved in the epigenetic regulation are found [1,2]. These somatic mutations induce downstream modification in the expression of genes that remain wild-type for DNA sequence. The phenomenon is thus reversible and normal function of these genes might be restored by the administration of epigenetic targeted drugs [3]. Interestingly, in parallel with these discoveries, several drugs target epigenetic marks. Over the last 10 years, many trials have been conducted in various hematologic malignancies and solid tumors. The success of demethylating agents in acute myeloid leukemia (AML)/myelodysplastic syndrome (MDS) and of histone deacetylase inhibitors (HDACi) in

G. Egger & P. Arimondo (Eds): Drug Discovery in Cancer Epigenetics. DOI: http://dx.doi.org/10.1016/B978-0-12-802208-5.00013-8

lymphoma that led to the registration of these agents has provided the proof of principle that targeting deregulated epigenome may translate into clinical activity. The next step is already ongoing with the introduction in the clinic of a new generation of drugs not targeting the epigenetic marks but the mutations and genetic alteration leading to epigenetic dysregulation.

In this chapter, we will review the state-of-the-art relative to the use of "epidrugs" as monotherapies in hematological malignancies and solid tumors including the most recently developed agents that specifically target them, and discuss the mechanisms explaining the lack of efficacy seen in solid tumors.

13.2 EPIGENETIC DRUGS IN HEMATOLOGICAL MALIGNANCIES

13.2.1 DEMETHYLATING AGENTS IN HEMATOLOGICAL MALIGNANCIES

Demethylating agents have been extensively evaluated in AML and MDSs but not in lymphoid disorders. Two agents are currently registered in the United States and/or the European Union: azacitidine (AZA) and decitabine. This section will thus focus on their use in AML and MDS.

13.2.1.1 Azacitidine

The first trials evaluating the potential antileukemic activity of AZA were performed in Europe at the end of the 1960s [4]. The drug was used as a potential cytotoxic agent and was associated with high toxicities. The drug began to be studied at much lower doses for its DNA methyl transferase inhibition properties in the late 1980s relying on pioneer *in vitro* studies. Several phase II studies [5] demonstrated the clinical activity of AZA in MDS patients including complete remissions but also hematological improvement without blast clearance [6]. A phase III randomized study from the cancer and leukemia group B (CALGB) network evaluated the efficacy of AZA compared to best supportive care (BSC) for patients with MDS (CALGB 9221, see Table 13.1 for details). This study [7] showed a 47% overall response rate to AZA compared to 17% in the BSC arm after four cycles of treatment. The median time to response was three cycles and the median duration of response to AZA was 15 months. Time to progression to AML was significantly prolonged in the AZA arm (19 months vs 8 months in the BSC arm). The study included a potential crossover after four cycles for nonresponding patients in the BSC arm; treatment with AZA after crossover was associated with a 35% response rate [8]. This crossover also limits the interpretation of the results on overall survival (OS) in this study; no significant difference in survival was found between the two initial arms of randomization (20 months vs 14 months respectively, $P = 0.10$). A *post hoc* landmark analysis suggested a potential benefit on survival for the AZA group as compared to patients treated with BSC with or without crossing over. Of note, this study also demonstrated that treatment with AZA was associated with a decrease in transfusion needs or transfusion dependency and improved quality of life. The more frequent toxicities were hematological toxicities, predominantly neutropenia, and thrombocytopenia. Their frequencies usually decrease with the number of cycles if the patient responds to therapy. Extra hematological toxicities include injection site skin reaction, vomiting, and fatigue. This study was the pivotal study supporting the registration by FDA of AZA for the treatment of MDS using a 75 mg/m^2/d for 7 consecutive days schedule administered subcutaneously or IV. This schedule of AZA may be difficult to deliver in clinical practice.

Table 13.1 Studies of AZA or Decitabine Montherapy

Study	Reference	Patients (*n*)	Patient Population	ORR (%)	CR+PR (%)	Median OS
CALGB 9221	Silverman JCO 2006	191	MDS 5-Aza 7 days	44	14	NE
AZA-001	FenauxLancOncol 2009	179	Higher-risk 5-Aza 7 days	49	29	24 m
E1905 (5-AZA arm)	Prebet JCO 2014	75	98 MDS/52 AML 5-Aza 10 days	46	20	18 m
French registry	Itzykson Blood 2011	282	High-risk MDS AZA 7 days	43	17	14 m
Spanish registry	Garcia ASH 2008	144	Intermediate-2 + high-risk: 59%; 5-Aza 7 days	74	33	NA
Korean registry	Lee Ann Hematol 2013	97	MDS 5-Aza 7 days	44	26	26 m
ADOPT	Steensma JCO 2009	99	MDS DAC 5 days	51	17	NA
EORTC	Lubbert JCO 2011	116	MDS high-risk DAC 3 days	34	19	10 m
Japan	Oki Cancer Sci 2012	37	MDS DAC 5 days	47	27	>24 m
DEC arm	Issa Cancer 2015	74	MDS/AML DAC 5 days	55	34	12 m

Note: 5-aza, azacitidine; DAC, decitabine; m, months; ORR, overall response rate.

Alternative dosing schedules of AZA have been tested ranging from 5 to 14 days [9–11]. The rates of responses seem similar between the recommended 7 consecutive days schedule and 5 days or 5 + 2 (5 days of AZA, 2 days off, 2 days of AZA) schedules. Nevertheless, none of these alternative schedules has been tested for impact on patient survival [11].

A second phase III study was conducted more recently, predominantly at European sites. The AZA-001 study [12] aimed to evaluate survival and compared AZA to three "conventional care regimens" (CCR) in 358 patients treated for high-risk International Prognostic Scoring System MDSs. CCR regimens consisted of BSC, low-dose cytarabine, or/and cytarabine plus anthracycline leukemia induction chemotherapy. The CCR was selected for an individual patient prior to randomization. Patients received AZA until progression and no crossover was allowed. The results showed survival to be superior in patients treated with AZA with a median OS of 24 months versus 15 months in the CCR arm. This corresponded to a doubling of the 2-year probability of survival. Time to progression to AML was also significantly increased (18 months vs 11 months). Subsequent subanalyses showed that AZA was beneficial for all three CCR randomization arms [13]; however, the impact on survival in the intensive chemotherapy arm was not statistically significant (probably due to the small number of patients in this arm, $n = 42$). Patients with RAEB-T per FAB classification, which are now classified as AML per WHO classification (i.e., AML with 20–30% blasts) benefited from AZA to the same extent as other MDS categories [12]. For this reason, the so-called oligoblastic AMLs are included in the marketing application label of AZA that is registered in Europe for the treatment of high-risk MDS and AML.

Other trials have included both patients with lower-risk and higher-risk MDS [8,9]. Few studies focused on the lower-risk subgroup [14]. The results in this subgroup of MDS showed substantial clinical activity and AZA may represent an alternative for patients resistant to erythropoiesis stimulating agents. For patients with lower-risk MDS, development of the oral formulation of AZA is appealing. This was initially tested using a standard 7-day schedule but the good tolerance profile and pharmacodynamic studies pointed out the interest of prolonged exposure that would produce prolonged DNA demethylation. The latest trials are currently using a 21-day schedule that showed acceptable tolerance and substantial rates of hematological improvement.

The evaluation of response is a critical and difficult issue in MDS. Treatment withdrawal is recommended in the absence of response, but several points must be taken into consideration before stopping the treatment. First, the kinetics of response to AZA is low: the median number of cycles before response was three and 87% of responses occurred before cycle 6 and 95% before cycle 8 [8]. The time to best response is even longer and could range between 2 and more than 17 cycles (i.e., 17 months of treatment). Second, survival benefit was not clearly linked to response. Finally, a vast majority of patients will relapse if AZA is stopped, regardless of the current response [15], indicating that treatment should be continued until progression; however, even when AZA treatment is continued, most patients relapse [16,17]. Based on this, AZA has become the standard frontline treatment for higher-risk MDS, but for virtually all patients who will become azacitine-resistant, there is currently no effective therapy, thus defining a new area of unmet medical needs.

In *de novo* AML, several retrospective studies confirmed the potential efficacy of AZA [18–21]. The response rate ranged from 25% to 50% with the complete remission rate ranging between 10% and 20%. The OS seems shorter than what was observed in MDS with median OS ranging between 8 and 14 months. More recently the results of a randomized phase III trial of AZA compared to conventional care in 488 patients was presented (AML-001 trial) [17]. The study design was similar to that of the AZA-001. The results showed an improvement of survival in the AZA group as compared to the control arm that did not reach statistical significance (median OS: 10.4 months vs 6.5 months respectively, $P = 0.08$). Using a prespecified analysis with censoring at the time of introduction of a new treatment (frequently being demethylating agents in the control arm), the difference in survival reached statistical significance (median OS of 12 months vs 7 months; $P = 0.019$). Of note, a retrospective study [22] suggested a potential benefit of AZA treatment in AML arising from myeloproliferative disorders.

13.2.1.2 Decitabine

Decitabine has followed a path of development parallel to that of AZA [23]. The first phase I [24] and phase II [25–28] trials with low-dose schedules were performed during the 1980s and the early 1990s and showed responses rates ranging from 30% to 55% in patients with MDS. The main limiting toxicities were hematological, with neutropenia and thrombocytopenia. The extra hematological toxicities included nausea, fatigue, and liver enzyme modifications.

A randomized phase III trial comparing decitabine to BSC [29] led to the approval of the drug by FDA. A total of 170 MDS patients were treated for a maximum of eight cycles using a 3-day schedule. The overall response rate was 30% in the decitabine arm, including 17% complete response (CR) as compared to 7% (0% CR) in the BSC arm. Responses were seen after a median of two cycles and median duration of response was 10 months. There was no significant difference in terms of survival or time to progression to AML. Per protocol, patients could receive up to eight cycles. As a result, a median of only two cycles was given and this may explain the difference in patient outcome as compared to AZA trials.

Another phase III study comparing decitabine to BSC has been published [30]. This EORTC study showed a better response rate in the decitabine arm as compared to BSC: 44% (13% CR) versus 2% (0% CR) respectively on 233 MDS patients. Time to response was 4 months and median duration of response was 8 months. This translated into a significantly better progression-free survival (PFS; 6 months vs 3 months) but OS was not significantly different (10 months vs 9 months). As with the US registration trial, the design of the trial (maximum of eight cycles of decitabine) and the low median number of administered cycles (underlining some problems of toxicity of the 3-day schedule) may explain the differences with the AZA study. However, as noted above, differences in the mechanisms of action of AZA and decitabine may explain different clinical effects [31]. Improvements of decitabine activity might be achieved with the use of prolonged administration as a more effective demethylation can be reached as shown in a study that compared biological effects of a 5-day schedule to a 3-day schedule [24]. Recently, encouraging activity was shown with the use of low doses of decitabine given over 10 days.

Most data regarding decitabine in the lower-risk MDS population stem from subgroup analyses of larger clinical trials [32]. Decitabine response rate and duration may be superior in lower-risk MDS. In *de novo* AML, a recently published phase II trial [33] showed a 25% response rate (including 24% of CR) using the 5-day schedule in 55 elderly patients with previously untreated AML. The response rate was homogeneous across the different cytogenetic subgroups. Median OS was 8 months and early mortality rate was 7%.

As previously mentioned, in most of the clinical trials with decitabine, patients stopped treatment after 6–8 cycles of treatment even if responding clinically. Some data are available regarding the outcome of these patients after stopping decitabine [27]. Virtually all patients progressed, with a median time to progression of 11 months. Forty-five percent of such patients achieved a second response following decitabine retreatment; survival appeared longer than in those relapsed patients treated with conventional care (27 months vs 18 months, respectively).

In AML, several studies have explored the impact of decitabine using the 3-day, 5-day, or 10-day schedules. A randomized phase III study of decitabine (5-day) versus BSC or low-dose cytarabine was published in 2012 and reanalyzed in 2013. The study population was limited to elderly AML patients not eligible for intensive chemotherapy. There was a survival benefit in favor of decitabine (7.7 vs 5 months with a borderline significance [$P = 0.10$]). *Post hoc* analysis performed 1 year later reached statistical significance with $P = 0.037$. Based on these results, the European Medical Agency (EMA) granted approval to decitabine in these settings whereas the US FDA refused to register the drug.

To date, there is no comparative trial comparing AZA and decitabine. Retrospective comparison of the registration trials is limited by the differences in the design of the trials and the differences in study populations. Nevertheless, a recent and controversial, meta-analysis [34] was performed pooling the results of the four randomized clinical trials [7,12,29,30]. This analysis confirmed the benefit of treating patients with demethylating agents as compared to BSC and confirmed a benefit in terms of survival for patients treated with AZA but not for patients treated with decitabine.

13.2.1.3 Other demethylating agents

Several drugs are currently being developed for AML and MDS. The most advanced drug is SGI-110, a decitabine derivative with prolonged activity related to a resistance to cytidine deaminase. Peak DNA demethylation evaluated by LINE-1 methylation is at day 8, consistent with what is expected for DAC. Clinical activity was observed in phase I and the drug was tested in a randomized phase II in high-risk AML patients [35] (treatment naïve and not eligible to intensive chemotherapy, or relapsed/refractory AML). Two schedules of SGI-110 were tested in a total of 90 patients. Complete remission rate (CR and CR with incomplete recovery, CRi) was 27% without difference between the two arms. Response in treatment-naïve AML was 43%. Phase III is ongoing.

13.2.2 HDACi IN HEMATOLOGICAL MALIGNANCIES

The effects of HDACi monotherapy in AML and MDS have been extensively evaluated and the results are summarized in Table 13.2 [36–42]. Most of the studies were phase I dose escalation trials conducted in heavily pretreated patients. Overall response rates ranged between 0% and 15% and were mostly limited to hematological improvement or transient blast clearance. *In vitro* studies have highlighted that resistance to HDACi could be associated with overexpression of HSP90 [43] and expression of genes implicated in the reactive oxygen species pathway [38]. Data from sodium valproate studies suggested that patients with lower-risk MDS were more likely to respond than patients with higher risk [36,44]. However, this patient population was poorly represented in most HDACi monotherapy trials. A new (third) generation of HDACi is currently being tested. These new drugs have been chemically optimized to increase their affinity to one or several HDAC. Results are still preliminary but some compounds seem to show a better efficacy than prior generations.

The most common dose-limiting toxicities [45,46] observed with HDACi were fatigue and gastrointestinal toxicities (vomiting, diarrhea). Neutropenia and thrombocytopenia were frequent in

Table 13.2 HDACi Single-Agent Clinical Trials in AML/MDS

Study	Study Design	*N*	ORR (%)	HI+MLFS (%)	CR+Cri (%)
Gore, Clin Cancer Res 2001	Phenylbutyrate Phase 1	27	15	15	0
Garcia Manero, Blood 2008	Vorinostat Phase 1	41	17	7	10
Schaefer, Hematologica 2009	Vorinostat Phase2 AML	37	0/9*	0/5*	0/5*
Gojo, Blood 2005	MS 275 Phase 1	38	0	NE	0
Byrd, Blood 2005	Depsipeptide Phase 1	10	0	0	0
Klimek, Clin Cancer Res 2008	Depsipeptide Phase1	12	8	0	8
Garcia Manero, Blood 2008	MGCD0103 Phase 1	23	13	13	0
Kuentgen, Cancer 2006	VPA ± ATRA	55	16	16	0

Note: HI, hematologic improvement; MLFS, marrow leukemia free state.
**Number of available patients.*

patients with advanced-phase AML/MDS. Initial concerns about cardiac toxicities (QT prolongation) raised following romidepsin trials have been allayed. Neurologic toxicity including dizziness and confusion has been seen with sodium valproate, sodium phenylbutyrate, and the benzamide derivatives (Entinostat, MGCD0103).

In contrast with AML/MDS, more convincing clinical activity of HDACi has been observed in lymphoid malignancies. Indeed, two HDACi have been registered in cutaneous T-cell lymphomas (CTCL). Vorinostat was the first in class HDAC registered based on the results of a phase II study in advanced CTCL in 2006. This study had included 74 patients. The tolerance profile was similar to what is observed in myeloid malignancies and an overall response rate of 30% was observed. Time to progression however, remained modest (median of 202 days) [47]. This study was supported by an additional phase II study in 33 patients [48]. More recently, romidepsin was also registered in this indication based on the results of two phase II trials [49,50]. The response rate was 34% in both studies and time to progression ranged between 8 and 9 months. Panobinostat has also been tested with very similar results but is not yet registered [51].

In myeloma, there is a strong rationale for using HDACi mostly for their "side properties" of cytoplasmic protein acetylation and targeting to proteasome. Synergy with proteasome inhibitors has been shown *in vitro* and *in vivo*. Several second-generation HDAC have been or are currently evaluated in myeloma. Vorinostat was recently evaluated in combination with bortezomib (a proteasome inhibitor) in a large phase III randomized trial [52] versus placebo for patients with relapsed/refractory myeloma (380 patients per arm of treatment). PFS was the primary endpoint and was significantly prolonged by 1 month as compared to placebo (7.6 vs 6.8 months, $P = 0.01$). Based on these results, the clinical development of vorinostat in multiple myeloma has not been pursued. Panobinostat was also tested in the same settings [53], median PFS for the experimental arm was prolonged as compared to placebo (12 vs 8.1 month, $P < 0.0001$). Several other agents are currently developed.

13.2.3 NEXT-GENERATION EPIGENETIC DRUGS IN HEMATOLOGICAL MALIGNANCIES

While the first generations of epigenetically targeted agents have provided the proof-of-concept of anticancer therapeutic effects, the development of targeted agents with far greater specificity represents the future of epigenetic therapies. This new generation of "epidrugs" has been designed to interfere with the products of mutated epigenetic genes or on aberrations disrupting epigenetic regulation and not on epigenetic marks. Interestingly, mutations in epigenetic-related genes are frequent in hematological diseases. However, the most frequently mutated genes that are TET2 and ASXL1, are not yet "drugged." Inhibitors of IDH and EZH2 are currently in clinical development but IDH1 and 2 mutations are present in only 5–15% of the cases and EZH2 mutations in less than 5% of the cases [54].

An increasing number of compounds have successfully been tested *in vitro* but only a limited number have yet entered clinical development with the first clinical trial for this class launched in 2012. Phase I studies of IDH1, IDH2, DOT1L, EZH2, and Brd2/4 inhibitors are currently accruing or have just been completed. In AML with IDH2 mutations, AG-221 (an IDH2 inhibitor) [55] is tolerable with mainly gastrointestinal side effects. AG-221 lead to significant clinical responses (20 objective responses on 32 evaluated patients) and *in vivo* evidence of cell differentiation, a feature previously observed only in patients successfully treated with retinoic acid for acute promyelocytic leukemia. Encouraging preliminary results are also observed with AG-120 in patients with advanced AML harboring IDH1 mutation [56]: in that study, 4/17 patients achieved a complete remission while toxicity was acceptable. OTX-015, a BRD2/4 inhibitor, is currently being evaluated in AML/MDS and in lymphoid malignancies. The most frequent side effects are gastrointestinal toxicity, hyperglycemia, and fatigue. In the AML group [57], 4/32 patients responded (including one complete remission). In the lymphoid malignancy group [58], 4/18 objective responses were observed essentially in patients with diffuse large B-cell lymphomas (DLBCL), whereas no sign of activity was observed in myeloma. EZ7438 is an EZH2 inhibitor used in advanced-phase lymphoma [59] (follicular and DLBCL), phase I data showed acceptable toxicity and significant response in 4 of the 10 evaluable patients. DOT1L inhibitors have shown compelling preclinical evidence of activity in MLL-driven leukemia. EP5676, a potent DOT1L inhibitor, is currently being tested in adults and children with MLL gene-rearranged leukemias. The safety profile is acceptable and clinical responses were observed in 8/34 patients including CRs [60].

13.3 EPIGENETIC THERAPIES IN SOLID TUMORS

13.3.1 DEMETHYLATING AGENTS IN SOLID TUMORS

The two main agents that underwent clinical development were AZA and decitabine but few studies reported their clinical activity in solid tumors. Initial studies were conducted in the late 1970s. The response rate was low, associated with significant toxicity and led to no further clinical development, but after the discovery of their demethylating effects, new studies were launched using lower dosages (see Table 13.3). Two main trials evaluated AZA in patients with various solid tumors. In one study [61], high doses (225 mg/m^2 × 5, then reduced to 150 mg/m^2) were used. One hundred and ninety-one patients were included, five "partial regressions" were seen (2.6% response rate). Another phase 2 study [62] published in 1977 included 177 patients treated with IV AZA over 10 days and included various solid tumors. Responses were reported in 17% of patients treated for advanced breast cancer. No responses were seen in a short study of AZA (150 mg/m^2 × 5) in patients with bone sarcomas [63]. The main side effects were myelosuppression, nausea, and vomiting.

Decitabine has been evaluated in patients with solid tumors in more recent studies. In three studies that included patients with various advanced cancers, only one partial response (PR) in a patient treated for thymoma (out of a total of 85 patients) was reported but some patients achieved disease stabilization [64–66]. In these studies, myelosuppression was dose-limiting. Pharmacodynamic studies showed that significant global demethylation was achieved, especially in epithelial tumors [66]. A recent study evaluated low doses of decitabine (7 mg/m^2 × 5) in patients with solid tumors and lymphomas [67]. Eight patients were included in the phase I single part of the study. One PR was achieved in a patient with an advanced pleural tumor. Restoration of p16, MAGE A1, and A3 expression was achieved following decitabine therapy. Interestingly, this study also showed that following decitabine administration, a restoration of the expression of CTR1

Table 13.3 Single-Agent Hypomethylating Agents Studies in Solid Tumors

Dug	Phase	No. of Patients	Indication	Dose	Responses	Comments	Reference
AZA	2	191	Various ST	225 mg/m^2 × 5	5 PR (2.6%)	–	[61]
AZA	2	177	Various ST	1.6 mg/kg/day × 10	17% OR in BC	–	[62]
Decitabine	1	19	Metastatic ST	20–40 mg/m^2 over 72 h (CIV)	0	Gene demethylation	[64]
Decitabine	1	31	Various ST, NHL	2.5–10 mg/m^2 d1–5, d8–12	1/28 PR	Demethylation	[66]
				15–20 mg/m^2 d1–5	17/28 SD	Decrease expression of CTR1	
Decitabine	1	35	Lung, esophagus, pleura	72 h CIV	0/25	MTD 60–75 mg/m^2	[65]
						Expression MAGE A3, p16, NY-ESO1	
Decitabine	1-2	8	Various ST	7 mg/m^2 × 5	1/8 PR	Expression p16, MAGE A1-3	[67]

Note: OR, overall response, PR, partial response; BC, breast cancer; CIV, continuous IV infusion; ST, solid tumors; NHL, non-Hodgkin lymphoma.

(a copper transporter involved in cellular platinum uptake), supporting a role for decitabine in the modulation of platinum resistance.

13.3.2 HDACi IN SOLID TUMORS

Aberrant regulation of HDACs has been implicated in tumorigenesis of a variety of solid tumors including colon, ovarian, prostate, colorectal, lung, or liver cancers (see Ref. [68] for a review) supporting the rationale for a potential therapeutic activity of HDACi. Although most of the clinical development and success were obtained in hematologic malignancies, extensive investigation of the different HDACi was conducted in various solid tumor models. More than 10 different HDACi were tested and the results of the most representative trials of HDACi as a single agent are summarized in Table 13.4. Taken together, the results show limited clinical activity. Only occasional objective responses were recorded: in the eight trials presented in Table 13.4 which included 247 patients, one CR [in a patient with non-Hodgkin lymphoma (NHL)] and six PRs (B-cell lymphoma, laryngeal, papillary thyroid cancer, and two metastatic mesotheliomas) were achieved for an overall response rate of only 3%. It is worth noting that most of these studies were phase I trials with heavily pretreated patients. In ovarian cancer, although the response rate was disappointing, interesting observations could be made: (i) sustained disease stabilization could be obtained [69,70],

Table 13.4 HDACi for Solid Tumors

HDACi	Phase	No. of Patients	Indication	Results	Reference
Vorinostat	2	14	Breast cancer	OR = 0/14 4 durable SD	[87]
Vorinostat	2	16	BC, NSCLC, colorectal	OR = 1/16	[88]
Vorinostat	1	76	Multiple ST	1 CR (NHL) 5 PR	[45]
Panobinostat	2	35	PC	11% PF at 24 wks PSA reduction	[89]
Panobinostat	2	21	NSCLC	2/19 tumor reduction	[90]
Pracinostat	2	38	Multiple ST	OR 0/32 SD 32%	[91]
Entinostat	2	28	Melanoma	0R = 0 SD in 25%	[92]
Quisonostat	1	19	Multiple ST	1 PR (melanoma)	[76]

(ii) proliferative forms were more sensitive to treatment [71], and (iii) in line with preclinical data [72], activity was higher in platinum-resistant ovarian cancers probably through HDACi-induced DNA damage and reduced DNA repair.

Toxicity of HDACi in patients with solid tumors was generally manageable. However, cardiac toxicity, constitutional symptoms (fatigue), thrombocytopenia, or gastrointestinal effects can be dose-limiting or impair the ability of the patients to tolerate prolonged treatments. Class I HDACi affect DNA damage and repair that lead to unacceptable toxicities following prolonged exposures including oncogenic effects depending on the context [71–73].

Second- and third-generation HDACi currently in early clinical development, such as CHR-3966 [74], chidamide [75] or the hydroxamides quisinostat [76], and abexinostat [77], were more potent in preclinical studies as compared to former HDACi, and have shown acceptable toxicity profiles. Although they have the same HDAC specificity as former HDACi, their more favorable PK and toxicity profiles might translate into improved clinical activity [68].

As for demethylating agents, clinical studies of HDACi in solid tumors have shown limited clinical activity as monotherapy. However, manageable toxicities make this therapeutic class amenable to various combination regimens.

13.3.3 NEXT-GENERATION EPIGENETIC DRUGS IN SOLID TUMORS

An increasing number of second-generation epigenetic drugs are currently entering clinical development. These include histone methyl transferase inhibitors, EZH2, DOTL1 or G9a, histone demethylase LSD1 inhibitors, or bromodomain inhibitors. However, for solid tumors, few data have been reported so far. In the phase I trial of the EZH2 inhibitor EPZ6438 conducted in patients with refractory NHL and solid tumors, in addition to responses observed in the NHL cohort, one patient with a rhabdoid tumor achieved a PR suggesting clinical activity at least in this specific solid tumor model (Ribrag et al., EORTC-NCI-AACR 2014 meeting, Barcelona, November, 2014). Clinical studies of the IDH1 and 2 inhibitors AG-120 and AG-221 in patients with solid tumors are currently ongoing.

13.4 WHY ARE EPIGENETIC THERAPIES MORE ACTIVE IN HEMATOLOGICAL MALIGNANCIES THAN IN SOLID TUMORS?

The clinical experience accumulated over the last 10 years with the use of epigenetic therapy has provided the proof-of-concept of clinical efficacy. However, the only convincing data came from their use in hematological malignancies, while studies conducted in solid tumors showed anecdotal responses that suggested but did not prove their clinical activity. The reasons for this discrepancy between hematological versus solid malignancies are unclear and could not be predicted from pre-clinical studies which showed activity in both types of tumors. However, several explanations can be discussed that relate to the trial designs, the drugs' mechanisms of action, or intrinsic biological differences between hematologic and solid malignancies.

Experience with demethylating agents in AML and MDS has shown that most responses are delayed [78] with only 50% of responders having responded by cycle 2. This is in line with the mechanism of action and stresses the importance of prolonged treatment before concluding treatment failure. In the trials of hypomethylating agents in solid tumors, the majority of patients received less than two cycles of therapy [65] and this might have negatively impacted the results of these studies. In addition, the strength of epigenetic therapies may yield long-term improvements in patient survival and cancer management without producing typical clinical responses. This paradigm change was illustrated by the effects of AZA in patients treated for high-risk MDS in whom survival benefit was not restricted to those with objective responses (CR-PR) but also in patients who achieved a disease stabilization with or without hematologic improvement [79]. It must be said that this was not reproduced in decitabine studies nor in trials evaluating HDACi.

Whether differences in tumor biology explain these differences also remains to be established. MDS, currently the prototype of epitherapy-sensitive disease, is a preleukemic condition. As mentioned above, the rate of objective responses reflecting reduction of the MDS clone is low, but an additional 30–40% of patients achieved hematologic improvements that may reflect beneficial effects of the drug on normal residual hematopoiesis. This may in part account for survival prolongation through reduction of the frequency of complications of the cytopenias [80] (infections, bleeding, or RBC transfusion dependency).

The importance of epigenetic deregulation in the pathophysiology of MDS has been clearly established and is further illustrated by the frequency of epigenetic regulating gene mutations [81]. In addition, the median number of mutations found per patient is lower than in acute leukemias and solid tumors. As for other targeted approaches, the multiplicity of the oncogenic pathways activated in solid tumors obviously reduces the chances of observing clinical responses with single-agent therapies, whatever their target. Moreover, the effects of the different HDAC classes seem to vary among tumor models [68]. For instance, although increased expression of class I and II HDACs correlated with invasion in gastric [82] and breast cancer [83], a low expression of HDAC5 and 10 correlated with a poor prognosis in lung cancer [84]. Given the variable spectrum of HDAC selectivity of the different classes of HDACi [85], disease-specific sensitivity to classes of HDACi might also explain the variability in responses between tumor models. The activity of HDACi on nonhistone proteins can also explain their variable effects since they can affect immune regulation, inflammation, cell survival, differentiation, or proliferation that are not equally important in the pathophysiology of various cancers [85]. Finally, the mechanisms of resistance to HDACi are not fully elucidated but include a variety of mechanisms such as drug efflux, epigenetic alterations, stress response, antiapoptotic, and prosurvival mechanisms [86] more likely to have been operational in the patient populations selected for the early-phase trials of epidrugs discussed here.

13.5 CONCLUSIONS

In the present chapter, we have seen that our scientific knowledge of epigenetics has been translated into clinical benefits for cancer patients. However, even if benefit could be striking for some patients there is still a long way to go. Cancer is a heterogeneous disease with different subtypes in each histologically defined entity and different clones in each individual patient's tumor. This landscape is able to change with time and with the selective pressure produced by exposure to conventional anticancer treatments. The ultimate goal will be to personalize as much as possible each patient treatment. Conversely, it means that we should be able to have ways to select the best treatment or the best combination. With the exception of mutation-targeted drugs like IDH1, IDH2, or EZH2 inhibitors, we currently do not have any strong biological or clinical correlate able to predict response to therapy. A lot of translational work was performed to try to solve this issue but, to date, none of the potential biomarker candidates has been validated. There is also a "gray area" linking the epigenetic effects of drugs and patient outcome. Even if the clinical efficacy of demethylating agents in AML/MDS or HDACi in CTCL has been demonstrated, there was no strong evidence linking mechanism of action (i.e., demethylation or deacetylation) with response to therapy. This suggests the role of off-target effects that may be predominant for the biological activity of the drug. In that case, pleiotropy or "lack of specificity" is beneficial.

The experience with the clinical use of epidrugs as single agents has provided a new standard of frontline treatment of MDS, a new therapeutic option for cutaneous lymphomas. Although negative in terms of clinical activity in many other tumor models, and specifically for solid tumors, it has allowed for the identification of tolerable doses and regimens and provided further rationale for their use in combination with other epidrugs or classes of anticancer therapies.

REFERENCES

[1] Baylin SB, Jones PA. A decade of exploring the cancer epigenome—biological and translational implications. Nat Rev Cancer 2011;11(10):726–34.

[2] You JS, Jones PA. Cancer genetics and epigenetics: two sides of the same coin? Cancer Cell 2012;22(1):9–20.

[3] Azad N, Zahnow CA, Rudin CM, Baylin SB. The future of epigenetic therapy in solid tumours–lessons from the past. Nat Rev Clin Oncol 2013;10(5):256–66.

[4] Von Hoff DD, Slavik M, Muggia FM. 5-Azacytidine. A new anticancer drug with effectiveness in acute myelogenous leukemia. Ann Intern Med 1976;85(2):237–45.

[5] Raj K, Mufti GJ. Azacytidine (Vidaza(R)) in the treatment of myelodysplastic syndromes. Ther Clin Risk Manag 2006;2(4):377–88.

[6] Silverman LR, Holland JF, Weinberg RS, Alter BP, Davis RB, Ellison RR, et al. Effects of treatment with 5-azacytidine on the *in vivo* and *in vitro* hematopoiesis in patients with myelodysplastic syndromes. Leukemia 1993;7(Suppl. 1):21–9.

[7] Silverman LR, Demakos EP, Peterson BL, Kornblith AB, Holland JC, Odchimar-Reissig R, et al. Randomized controlled trial of azacitidine in patients with the myelodysplastic syndrome: a study of the cancer and leukemia group B. J Clin Oncol 2002;20(10):2429–40.

[8] Silverman LR, McKenzie DR, Peterson BL, Holland JF, Backstrom JT, Beach CL, et al. Further analysis of trials with azacitidine in patients with myelodysplastic syndrome: studies 8421, 8921, and 9221 by the Cancer and Leukemia Group B. J Clin Oncol 2006;24(24):3895–903.

[9] Lyons RM, Cosgriff TM, Modi SS, Gersh RH, Hainsworth JD, Cohn AL, et al. Hematologic response to three alternative dosing schedules of azacitidine in patients with myelodysplastic syndromes. J Clin Oncol 2009;27(11):1850–6.

[10] Fandy TE, Herman JG, Kerns P, Jiemjit A, Sugar EA, Choi SH, et al. Early epigenetic changes and DNA damage do not predict clinical response in an overlapping schedule of 5-azacytidine and entinostat in patients with myeloid malignancies. Blood 2009;114(13):2764–73.

[11] Martin MG, Walgren RA, Procknow E, Uy GL, Stockerl-Goldstein K, Cashen AF, et al. A phase II study of 5-day intravenous azacitidine in patients with myelodysplastic syndromes. Am J Hematol 2009;84(9):560–4.

[12] Fenaux P, Mufti GJ, Hellstrom-Lindberg E, Santini V, Finelli C, Giagounidis A, et al. Efficacy of azacitidine compared with that of conventional care regimens in the treatment of higher-risk myelodysplastic syndromes: a randomised, open-label, phase III study. Lancet Oncol 2009;10(3):223–32.

[13] Fenaux P, Gattermann N, Seymour JF, Hellstrom-Lindberg E, Mufti GJ, Duehrsen U, et al. Prolonged survival with improved tolerability in higher-risk myelodysplastic syndromes: azacitidine compared with low dose ara-C. Brit J Haematol 2010;149:244–9.

[14] Musto P, Maurillo L, Spagnoli A, Gozzini A, Rivellini F, Lunghi M, et al. Azacitidine for the treatment of lower risk myelodysplastic syndromes: a retrospective study of 74 patients enrolled in an Italian named patient program. Cancer 2010;116(6):1485–94.

[15] Voso MT, Breccia M, Lunghi M, Poloni A, Niscola P, Finelli C, et al. Rapid loss of response after withdrawal of treatment with azacitidine: a case series in patients with higher-risk myelodysplastic syndromes or chronic myelomonocytic leukemia. Eur J Haematol 2013;90(4):345–8.

[16] Prebet T, Gore SD, Esterni B, Gardin C, Itzykson R, Thepot S, et al. Outcome of high-risk myelodysplastic syndrome after azacitidine treatment failure. J Clin Oncol 2011;29(24):3322–7.

[17] Seymour J, Dohner H, Butrym A, Wierzbowska A, Selleslag D, Jang JH, et al. Azacitidine (AZA) versus conventional care regimens (CCR) in older patients with newly diagnosed acute myeloid leukemia (>30% bone marrow blasts) with myelodysplasia-related changes: a subgroup analysis of the AZA-AML-001 trial. ASH Annual Meeting Abstracts 2014; 2014.

[18] Itzykson R, Thepot S, Berthon C, Delaunay J, Bouscary D, Cluzeau T, et al. Azacitidine for the treatment of relapsed and refractory AML in older patients. Leuk Res 2015;39(2):124–30.

[19] Lao Z, Yiu R, Wong GC, Ho A. Treatment of elderly patients with acute myeloid leukemia with azacitidine results in fewer hospitalization days and infective complications but similar survival compared with intensive chemotherapy. Asia-Pacific J Clin Oncol 2015;11(1):54–61.

[20] Ramos F, Thepot S, Pleyer L, Maurillo L, Itzykson R, Bargay J, et al. Azacitidine frontline therapy for unfit acute myeloid leukemia patients: clinical use and outcome prediction. Leuk Res 2015;39 (3):296–306.

[21] Pleyer L, Burgstaller S, Girschikofsky M, Linkesch W, Stauder R, Pfeilstocker M, et al. Azacitidine in 302 patients with WHO-defined acute myeloid leukemia: results from the Austrian Azacitidine Registry of the AGMT-Study Group. Ann Hematol 2014;93(11):1825–38.

[22] Thepot S, Itzykson R, Raffoux E, Roussel M, Dartigeas C, Dreyfus F, et al. Treatment of progression of myeloproliferative disorders (MPD) to MDS/AML by azacytidine (AZA): a preliminary report on 17 patients (pts). ASH Abstract Book 2008.
[23] Momparler RL, Cote S, Eliopoulos N. Pharmacological approach for optimization of the dose schedule of 5-Aza-2′-deoxycytidine (Decitabine) for the therapy of leukemia. Leukemia 1997;11(2):175–80.
[24] Issa JP, Garcia-Manero G, Giles FJ, Mannari R, Thomas D, Faderl S, et al. Phase 1 study of low-dose prolonged exposure schedules of the hypomethylating agent 5-aza-2′-deoxycytidine (decitabine) in hematopoietic malignancies. Blood 2004;103(5):1635–40.
[25] Wijermans P, Lubbert M, Verhoef G, Bosly A, Ravoet C, Andre M, et al. Low-dose 5-aza-2′-deoxycytidine, a DNA hypomethylating agent, for the treatment of high-risk myelodysplastic syndrome: a multicenter phase II study in elderly patients. J Clin Oncol 2000;18(5):956–62.
[26] Wijermans PW, Krulder JW, Huijgens PC, Neve P. Continuous infusion of low-dose 5-Aza-2′-deoxycytidine in elderly patients with high-risk myelodysplastic syndrome. Leukemia 1997;11(1):1–5.
[27] Ruter B, Wijermans PW, Lubbert M. Superiority of prolonged low-dose azanucleoside administration? Results of 5-aza-2′-deoxycytidine retreatment in high-risk myelodysplasia patients. Cancer 2006;106(8):1744–50.
[28] Steensma DP, Baer MR, Slack JL, Buckstein R, Godley LA, Garcia-Manero G, et al. Multicenter study of decitabine administered daily for 5 days every 4 weeks to adults with myelodysplastic syndromes: the alternative dosing for outpatient treatment (ADOPT) trial. J Clin Oncol 2009;27(23):3842–8.
[29] Kantarjian H, Issa JP, Rosenfeld CS, Bennett JM, Albitar M, DiPersio J, et al. Decitabine improves patient outcomes in myelodysplastic syndromes: results of a phase III randomized study. Cancer 2006;106(8):1794–803.
[30] Wijermans P, Suciu S, Baila L, Platzbecker U, Giagounidis A, Selleslag D, et al. Low dose decitabine versus best supportive care in elderly patients with intermediate or high risk MDS not eligible for intensive chemotherapy: final results of the randomized phase III study (06011) of the EORTC Leukemia and German MDS Study Groups. ASH Abstract Book 2008.
[31] Hollenbach PW, Nguyen AN, Brady H, Williams M, Ning Y, Richard N, et al. A comparison of azacitidine and decitabine activities in acute myeloid leukemia cell lines. PloS One 2010;5(2):e9001.
[32] Wijermans PW, Lubbert M, Verhoef G, Klimek V, Bosly A. An epigenetic approach to the treatment of advanced MDS; the experience with the DNA demethylating agent 5-aza-2′-deoxycytidine (decitabine) in 177 patients. Ann Hematol 2005;84(Suppl. 1):9–17.
[33] Cashen AF, Schiller GJ, O'Donnell MR, DiPersio JF. Multicenter, phase II study of decitabine for the first-line treatment of older patients with acute myeloid leukemia. J Clin Oncol 2010;28(4):556–61.
[34] Gurion R, Vidal L, Gafter-Gvili A, Belnik Y, Yeshurun M, Raanani P, et al. 5-Azacitidine prolongs overall survival in patients with myelodysplastic syndrome—a systematic review and meta-analysis. Haematologica 2010;95(2):303–10.
[35] Issa JP, Kantarjian H, Jabbour E, Kropf P, O'connel C, Stock W, et al. First clinical results of a randomized phase 2 study of SGI-110, a novel subcutaneous (SQ) hypomethylating agent (HMA), in adult patients with acute myeloid leukemia (AML). ASH Annual Meeting Abstracts 2013; 2013.
[36] Kuendgen A, Gattermann N. Valproic acid for the treatment of myeloid malignancies. Cancer 2007;110(5):943–54.
[37] Gore SD, Weng LJ, Figg WD, Zhai S, Donehower RC, Dover G, et al. Impact of prolonged infusions of the putative differentiating agent sodium phenylbutyrate on myelodysplastic syndromes and acute myeloid leukemia. Clin Cancer Res Off J Am Assoc Cancer Res 2002;8(4):963–70.
[38] Schaefer EW, Loaiza-Bonilla A, Juckett M, DiPersio JF, Roy V, Slack J, et al. A phase 2 study of vorinostat in acute myeloid leukemia. Haematologica 2009;94(10):1375–82.

[39] Gimsing P, Hansen M, Knudsen LM, Knoblauch P, Christensen IJ, Ooi CE, et al. A phase I clinical trial of the histone deacetylase inhibitor belinostat in patients with advanced hematological neoplasia. Eur J Haematol 2008;81(3):170–6.

[40] Garcia-Manero G, Assouline S, Cortes J, Estrov Z, Kantarjian H, Yang H, et al. Phase 1 study of the oral isotype specific histone deacetylase inhibitor MGCD0103 in leukemia. Blood 2008;112(4):981–9.

[41] Byrd JC, Marcucci G, Parthun MR, Xiao JJ, Klisovic RB, Moran M, et al. A phase 1 and pharmacodynamic study of depsipeptide (FK228) in chronic lymphocytic leukemia and acute myeloid leukemia. Blood 2005;105(3):959–67.

[42] Klimek VM, Fircanis S, Maslak P, Guernah I, Baum M, Wu N, et al. Tolerability, pharmacodynamics, and pharmacokinetics studies of depsipeptide (romidepsin) in patients with acute myelogenous leukemia or advanced myelodysplastic syndromes. Clin Cancer Res Off J Am Assoc Cancer Res 2008;14(3):826–32.

[43] Fiskus W, Rao R, Fernandez P, Herger B, Yang Y, Chen J, et al. Molecular and biologic characterization and drug-sensitivity of pan histone deacetylase inhibitor resistant acute myeloid leukemia cells. Blood 2008;112:2896–905.

[44] Kuendgen A, Schmid M, Schlenk R, Knipp S, Hildebrandt B, Steidl C, et al. The histone deacetylase (HDAC) inhibitor valproic acid as monotherapy or in combination with all-trans retinoic acid in patients with acute myeloid leukemia. Cancer 2006;106(1):112–19.

[45] Kelly WK, O'Connor OA, Krug LM, Chiao JH, Heaney M, Curley T, et al. Phase I study of an oral histone deacetylase inhibitor, suberoylanilide hydroxamic acid, in patients with advanced cancer. J Clin Oncol Off J Am Soc Clin Oncol 2005;23(17):3923–31.

[46] Gojo I, Jiemjit A, Trepel JB, Sparreboom A, Figg WD, Rollins S. Phase 1 and pharmacologic study of MS-275, a histone deacetylase inhibitor, in adults with refractory and relapsed acute leukemias. Blood 2007;109(7):2781–90.

[47] Olsen EA, Kim YH, Kuzel TM, Pacheco TR, Foss FM, Parker S, et al. Phase IIb multicenter trial of vorinostat in patients with persistent, progressive, or treatment refractory cutaneous T-cell lymphoma. J Clin Oncol 2007;25(21):3109–15.

[48] Duvic M, Talpur R, Ni X, Zhang C, Hazarika P, Kelly C, et al. Phase 2 trial of oral vorinostat (suberoylanilide hydroxamic acid, SAHA) for refractory cutaneous T-cell lymphoma (CTCL). Blood 2007;109 (1):31–9.

[49] Piekarz RL, Frye R, Turner M, Wright JJ, Allen SL, Kirschbaum MH, et al. Phase II multi-institutional trial of the histone deacetylase inhibitor romidepsin as monotherapy for patients with cutaneous T-cell lymphoma. J Clin Oncol 2009;27(32):5410–17.

[50] Whittaker SJ, Demierre MF, Kim EJ, Rook AH, Lerner A, Duvic M, et al. Final results from a multicenter, international, pivotal study of romidepsin in refractory cutaneous T-cell lymphoma. J Clin Oncol 2010;28(29):4485–91.

[51] Duvic M, Dummer R, Becker JC, Poulalhon N, Ortiz Romero P, Grazia Bernengo M, et al. Panobinostat activity in both bexarotene-exposed and -naive patients with refractory cutaneous T-cell lymphoma: results of a phase II trial. Eur J Cancer 2013;49(2):386–94.

[52] Dimopoulos M, Siegel DS, Lonial S, Qi J, Hajek R, Facon T, et al. Vorinostat or placebo in combination with bortezomib in patients with multiple myeloma (VANTAGE 088): a multicentre, randomised, double-blind study. Lancet Oncol 2013;14(11):1129–40.

[53] San-Miguel JF, Hungria VT, Yoon SS, Beksac M, Dimopoulos MA, Elghandour A, et al. Panobinostat plus bortezomib and dexamethasone versus placebo plus bortezomib and dexamethasone in patients with relapsed or relapsed and refractory multiple myeloma: a multicentre, randomised, double-blind phase 3 trial. Lancet Oncol 2014;15(11):1195–206.

[54] Patel JP, Gonen M, Figueroa ME, Fernandez H, Sun Z, Racevskis J, et al. Prognostic relevance of integrated genetic profiling in acute myeloid leukemia. New Engl J Med 2012;366(12):1079–89.

[55] Stein E, Altman JK, Collins R, DeAngelo D, Fathi AT, Flinn I, et al. AG-221, an oral, selective, first-in-class, potent inhibitor of the IDH2 mutant metabolic enzyme, induces durable remissions in a phase I study in patients with IDH2 mutation positive advanced hematologic malignancies. ASH Annual Meeting Abstracts 2014; 2014.

[56] Pollyea DA, Clinical safety and activity in a phase I trial of AG-120, a first in class, selective, potent inhibitor of the IDH1-mutant protein, in patients with IDH1 mutant positive advanced haematologic malignancies. In: 26th EORTC-NCI-AACR symposium on molecular targets and cancer therapeutics, Barcelona, Spain; 2014.

[57] Dombret H, Preudhomme C, Berthon C, Raffoux E, Thomas X, Vey N, et al. A phase 1 study of the BET-bromodomain inhibitor OTX015 in patients with advanced acute leukemia. asH Annual Meeting Abstracts 2014; 2014.

[58] Thieblemont C, Stathis A, Inghirami G, Karlin L, Morschhauser F, Gleeson M, et al. A phase 1 study of the BET-bromodomain inhibitor OTX015 in patients with non-leukemic hematologic malignancies. asH Annual Meeting Abstracts 2014; 2014.

[59] Study of E7438 (EZH2 histone methyl transferase [HMT] inhibitor) as a single agent in subjects with advanced solid tumors or with B cell lymphomas; 2014 [cited]. Available from: clinicaltrials.gov.

[60] Stein E, Garcia-Manero G, Rizzieri DA, Savona M, Tibes R, Altman JK, et al. The DOT1L inhibitor EPZ-5676: safety and activity in relapsed/refractory patients with MLL-rearranged leukemia. ash Annual Meeting Abstracts 2014; 2014.

[61] Quagliana JM, O'Bryan RM, Baker L, Gottlieb J, Morrison FS, Eyre HJ, et al. Phase II study of 5-azacytidine in solid tumors. Cancer Treat Rep 1977;61(1):51–4.

[62] Weiss AJ, Metter GE, Nealon TF, Keanan JP, Ramirez G, Swaiminathan A, et al. Phase II study of 5-azacytidine in solid tumors. Cancer Treat Rep 1977;61(1):55–8.

[63] Srinivasan U, Reaman GH, Poplack DG, Glaubiger DL, LeVine AS. Phase II study of 5-azacytidine in sarcomas of bone. Am J Clin Oncol 1982;5(4):411–15.

[64] Aparicio A, Eads CA, Leong LA, Laird PW, Newman EM, Synold TW, et al. Phase I trial of continuous infusion 5-aza-2′-deoxycytidine. Cancer Chemother Pharmacol 2003;51(3):231–9.

[65] Schrump DS, Fischette MR, Nguyen DM, Zhao M, Li X, Kunst TF, et al. Phase I study of decitabine-mediated gene expression in patients with cancers involving the lungs, esophagus, or pleura. Clin Cancer Res Off J Am Assoc Cancer Res 2006;12(19):5777–85.

[66] Stewart DJ, Issa JP, Kurzrock R, Nunez MI, Jelinek J, Hong D, et al. Decitabine effect on tumor global DNA methylation and other parameters in a phase I trial in refractory solid tumors and lymphomas. Clin Cancer Res Off J Am Assoc Cancer Res 2009;15(11):3881–8.

[67] Fan H, Lu X, Wang X, Liu Y, Guo B, Zhang Y, et al. Low-dose decitabine-based chemoimmunotherapy for patients with refractory advanced solid tumors: a phase I/II report. J Immunol Res 2014;2014:371087.

[68] West AC, Johnstone RW. New and emerging HDAC inhibitors for cancer treatment. J Clin Invest 2014;124 (1):30–9.

[69] Mackay HJ, Hirte H, Colgan T, Covens A, MacAlpine K, Grenci P, et al. Phase II trial of the histone deacetylase inhibitor belinostat in women with platinum resistant epithelial ovarian cancer and micropapillary (LMP) ovarian tumours. Eur J Cancer 2010;46(9):1573–9.

[70] Modesitt SC, Sill M, Hoffman JS, Bender DP. A phase II study of vorinostat in the treatment of persistent or recurrent epithelial ovarian or primary peritoneal carcinoma: a Gynecologic Oncology Group study. Gynecol Oncol 2008;109(2):182–6.

[71] Khabele D. The therapeutic potential of class I selective histone deacetylase inhibitors in ovarian cancer. Front Oncol 2014;4:111.

[72] Khabele D, Son DS, Parl AK, Goldberg GL, Augenlicht LH, Mariadason JM, et al. Drug-induced inactivation or gene silencing of class I histone deacetylases suppresses ovarian cancer cell growth: implications for therapy. Cancer Biol Ther 2007;6(5):795–801.

[73] Santoro F, Botrugno OA, Dal Zuffo R, Pallavicini I, Matthews GM, Cluse L, et al. A dual role for Hdac1: oncosuppressor in tumorigenesis, oncogene in tumor maintenance. Blood 2013;121 (17):3459–68.

[74] Banerji U, van Doorn L, Papadatos-Pastos D, Kristeleit R, Debnam P, Tall M, et al. A phase I pharmacokinetic and pharmacodynamic study of CHR-3996, an oral class I selective histone deacetylase inhibitor in refractory solid tumors. Clin Cancer Res Off J Am Assoc Cancer Res 2012;18(9):2687–94.

[75] Dong M, Ning ZQ, Xing PY, Xu JL, Cao HX, Dou GF, et al. Phase I study of chidamide (CS055/HBI-8000), a new histone deacetylase inhibitor, in patients with advanced solid tumors and lymphomas. Cancer Chemother Pharmacol 2012;69(6):1413–22.

[76] Venugopal B, Baird R, Kristeleit RS, Plummer R, Cowan R, Stewart A, et al. A phase I study of quisinostat (JNJ-26481585), an oral hydroxamate histone deacetylase inhibitor with evidence of target modulation and antitumor activity, in patients with advanced solid tumors. Clin Cancer Res Off J Am Assoc Cancer Res 2013;19(15):4262–72.

[77] Fouliard S, Robert R, Jacquet-Bescond A, du Rieu QC, Balasubramanian S, Loury D, et al. Pharmacokinetic/pharmacodynamic modelling-based optimisation of administration schedule for the histone deacetylase inhibitor abexinostat (S78454/PCI-24781) in phase I. Eur J Cancer 2013;49 (13):2791–7.

[78] Silverman LR, Fenaux P, Mufti GJ, Santini V, Hellstrom-Lindberg E, Gattermann N, et al. Continued azacitidine therapy beyond time of first response improves quality of response in patients with higher-risk myelodysplastic syndromes. Cancer 2011;117(12):2697–702.

[79] Gore SD, Fenaux P, Santini V, Bennett JM, Silverman LR, Seymour JF, et al. A multivariate analysis of the relationship between response and survival among patients with higher-risk myelodysplastic syndromes treated within azacitidine or conventional care regimens in the randomized AZA-001 trial. Haematologica 2013;98(7):1067–72.

[80] Santini V, Fenaux P, Mufti GJ, Hellstrom-Lindberg E, Silverman LR, List A, et al. Management and supportive care measures for adverse events in patients with myelodysplastic syndromes treated with azacitidine*. Eur J Haematol 2010;85(2):130–8.

[81] Papaemmanuil E, Gerstung M, Malcovati L, Tauro S, Gundem G, Van Loo P, et al. Clinical and biological implications of driver mutations in myelodysplastic syndromes. Blood 2013;122(22):3616–27.

[82] Weichert W, Roske A, Gekeler V, Beckers T, Ebert MP, Pross M, et al. Association of patterns of class I histone deacetylase expression with patient prognosis in gastric cancer: a retrospective analysis. Lancet Oncol 2008;9(2):139–48.

[83] Park SY, Jun JA, Jeong KJ, Heo HJ, Sohn JS, Lee HY, et al. Histone deacetylases 1, 6 and 8 are critical for invasion in breast cancer. Oncol Rep 2011;25(6):1677–81.

[84] Osada H, Tatematsu Y, Saito H, Yatabe Y, Mitsudomi T, Takahashi T. Reduced expression of class II histone deacetylase genes is associated with poor prognosis in lung cancer patients. Int J Cancer J 2004;112(1):26–32.

[85] Tang J, Yan H, Zhuang S. Histone deacetylases as targets for treatment of multiple diseases. Clin Sci (Lond) 2013;124(11):651–62.

[86] Rosato RR, Maggio SC, Almenara JA, Payne SG, Atadja P, Spiegel S, et al. The histone deacetylase inhibitor LAQ824 induces human leukemia cell death through a process involving XIAP down-regulation, oxidative injury, and the acid sphingomyelinase-dependent generation of ceramide. Mol Pharmacol 2006;69(1):216–25.

[87] Luu TH, Morgan RJ, Leong L, Lim D, McNamara M, Portnow J, et al. A phase II trial of vorinostat (suberoylanilide hydroxamic acid) in metastatic breast cancer: a California Cancer Consortium study. Clin Cancer Res 2008;14(21):7138–42.
[88] Vansteenkiste J, Van Cutsem E, Dumez H, Chen C, Ricker JL, Randolph SS, et al. Early phase II trial of oral vorinostat in relapsed or refractory breast, colorectal, or non-small cell lung cancer. Invest New Drugs 2008;26(5):483–8.
[89] Rathkopf DE, Picus J, Hussain A, Ellard S, Chi KN, Nydam T, et al. A phase 2 study of intravenous panobinostat in patients with castration-resistant prostate cancer. Cancer Chemother Pharmacol 2013; 72(3):537–44.
[90] de Marinis F, Atmaca A, Tiseo M, Giuffreda L, Rossi A, Gebbia V, et al. A phase II study of the histone deacetylase inhibitor panobinostat (LBH589) in pretreated patients with small-cell lung cancer. J Thorac Oncol 2013;8(8):1091–4.
[91] Razak AR, Hotte SJ, Siu LL, Chen EX, Hirte HW, Powers J, et al. Phase I clinical, pharmacokinetic and pharmacodynamic study of SB939, an oral histone deacetylase (HDAC) inhibitor, in patients with advanced solid tumours. Br J Cancer 2011;104(5):756–62.
[92] Hauschild A, Trefzer U, Garbe C, Kaehler KC, Ugurel S, Kiecker F, et al. Multicenter phase II trial of the histone deacetylase inhibitor pyridylmethyl-N-{4-[(2-aminophenyl)-carbamoyl]-benzyl}-carbamate in pretreated metastatic melanoma. Melanoma Res 2008;18(4):274–8.

CHAPTER

EFFICACY OF COMBINED EPIGENETIC THERAPIES

14

Noël J.-M. Raynal

Département de Pharmacologie, Université de Montréal and Sainte-Justine University Hospital Research Center, Montréal, QC, Canada

CHAPTER OUTLINE

14.1 SCIENTIFIC RATIONALE TO COMBINE EPIGENETIC DRUGS

14.1.1 INTRODUCTION

Epigenetic mechanisms play important roles in establishing and maintaining cellular identities throughout cell divisions. Epigenetics is defined as molecular mechanisms involved in regulating gene expression, such as DNA methylation, histone modifications, chromatin remodeling, and RNA-based gene regulation [1]. Dysregulation of epigenetic marks can result in gene expression alterations leading to carcinogenesis. Recently, genomic sequencing of cancer cells has revealed crosstalk between mutations and epigenetic alterations. Even though epigenetic and genetic alterations have

G. Egger & P. Arimondo (Eds): Drug Discovery in Cancer Epigenetics. DOI: http://dx.doi.org/10.1016/B978-0-12-802208-5.00014-X

been considered as two separate entities in cancer, sequencing thousands of cancer genomes has demonstrated that many mutations occur in genes that are involved in epigenome regulation [2]. These mutations play a causative role in the establishment of epigenetic alterations, leading to aberrant changes in DNA methylation, chromatin modifications, and nucleosome positioning.

The cancer epigenome is characterized by several layers of epigenetic alterations that cause a stable epigenetic reprogramming of normal gene expression profile into a malignant phenotype [2,3]. The sum of these epigenetic alterations is necessary to maintain unlimited cell proliferation, cell dedifferentiation, and metastasis. Silencing of tumor suppressor genes and activation of oncogenes are well-known examples of such reprogramming [1]. Several epigenetic mechanisms are involved to ensure that the malignant epigenome is copied and transmitted throughout cell divisions during carcinogenesis.

Proteins involved in epigenetic regulation can be classified into three categories based on their functions: *epigenetic writers* that deposit epigenetic marks on DNA or histones (i.e., methyltransferases or acetyltransferases); *epigenetic erasers* that remove these epigenetic marks (i.e., demethylases or deacetylases); and *epigenetic readers* that transmit epigenetic signals (i.e., methyl-binding proteins, bromodomains, chromodomains, or Tudor domains). Although epigenetic marks are somatically heritable, these modifications are chemically reversible, making them interesting targets for cancer therapy [4]. Targeting epigenetic alterations, which is referred to as epigenetic therapy, is a promising approach in cancer chemotherapy with the goal of reprograming cancer cell phenotype to induce cancer cell differentiation and ultimately cancer cell death [5]. Although epigenetic drugs have demonstrated clinical efficacy as monotherapy, there is still considerable room for improvement.

The current rationale to improve epigenetic therapy is to combine epigenetic drugs together to target several layers of epigenetic alterations simultaneously or to combine epigenetic drugs with standard treatment that will benefit from drug-induced epigenetic reprogramming. It is noteworthy that gene silencing and oncogene activation in cancer are the results of a coordinated action of dozens of epigenetic marks found in promoters, enhancers, insulators, and gene bodies. Thus, simultaneously targeting several epigenetic marks, using combined epigenetic therapies can lead to synergistic gene expression changes, cancer cell differentiation, and cancer cell death. Early preclinical studies have demonstrated that the combination of DNA methylation and histone deacetylase inhibitors (HDACi) can produce synergistic responses leading to tumor suppressor gene reactivation and anticancer effects [6]. Improvements in the knowledge of cancer cell epigenomic alterations have led to the development of several combined epigenetic therapies. The recent discovery of new epigenetic drugs targeting different epigenetic alterations in cancer cells offers multiple possibilities to target these alterations in cancer by affecting epigenetic writers, erasers, or readers. In addition, other studies have highlighted the potential of epigenetic drugs to enhance or synergize with other cancer therapies such as targeted treatment, immunotherapies, or standard chemotherapy. Altogether, epigenetic therapies are promising approaches against cancer, especially when used in combination.

14.1.2 TARGETING DNA METHYLATION

One of the first epigenetic targets to be identified to treat cancer was DNA methylation. Alterations in the content and the location of DNA methylation marks were reported in cancer cells as compared to normal cells. Neoplastic cells are characterized by a global DNA hypomethylation level and by a local DNA hypermethylation at promoter regions of silenced tumor suppressor genes. DNA methylation is catalyzed by a family of enzymes called DNA methyltransferases (DNMTs) which transfer a methyl group from S-adenosyl-L-methionine to the 5′ position of the pyrimidine ring of cytosines on a CpG sequence. DNMT1 ensures the maintenance of more than 90% of DNA methylation during cell replication. DNMT3A and DNMT3B are expressed throughout the cell cycle and play a major role in establishing DNA methylation during development [7]. DNA methylation is a well-known mechanism that precisely regulates gene expression. DNA hypermethylation in cytosines and more specifically in regions called CpG islands in gene promoters is associated with gene silencing. Enzymatic removal of cytosine methylation is a more complex phenomenon requiring a multistep oxidation of the methylation group catalyzed by the *TET* family proteins which form hydroxylmethyl-cytosine leading to progressive loss of DNA methylation [3].

Expression of hundreds of tumor suppressor genes is lost in cancer and associated with DNA hypermethylation in their promoter regions [8]. Genes encoding key biological processes involved in cell cycle control (*p16*), apoptosis (*DAPK*), DNA repair (*BRCA1/2*, *MGMT*, *MLH1*), Wnt signaling (*APC*), and suppression of cell invasion (*CDH1*, *CDH13*, *TIMP-3*) are typically methylated in cancer cells [3]. Global DNA hypomethylation is mainly found in repeated elements and gene bodies, and is believed to be associated with activation of repetitive elements causing DNA instability and oncogene activation, both of which contribute to carcinogenesis [3].

Targeting DNA hypermethylation in cancer with the goal to reduce DNA methylation levels of tumor suppressor genes and induce their reactivation became possible with the discovery of DNMT inhibitors (DNMTi) decitabine (5-aza-2′-deoxycytidine), azacitidine (5-aza-cytidine), and more recently guadecitabine (2′-deoxy-5-azacytidylyl-(3′→5′)-2′-deoxyguanosine sodium salt or SGI-110). The proof of concept and the clinical success of DNMTi were validated by the approval of decitabine and azacitidine for the treatment of myelodysplastic syndrome (MDS), acute myelogenous leukemia, and chronic myeolomonocytic leukemia [1,9]. However, clinical experience with these drugs has demonstrated (i) that their activity is limited to a small number of patients, (ii) that patients exhibit drug resistance, and (iii) that complete remissions are generally of short duration [9].

DNA methylation is considered as the epigenetic memory of the cells and hypomethylating agents have the potential to permanently reverse aberrant DNA hypermethylation in cancer cells [10]. When DNMTi treatment is optimized to target most replicating cells, drug-induced demethylation is observed in the majority of the cells [11]. However, gene reactivation is observed only in a limited number of cancer cells. Using sophisticated cancer cell line engineering, Si et al. [11] demonstrated that gene reactivation can still be blocked by chromatin-repressive marks in the majority of the cell population, even though drug-induced DNA demethylation was measured in promoter regions. Thus, obtaining full epigenetic reactivation is important to reset both DNA hypermethylation and chromatin-repressive marks on tumor suppressor genes [10]. These findings highlight the therapeutic limit of monotherapy using DNA demethylating drugs and strongly suggest that combinations targeting chromatin are absolutely required to enhance the therapeutic activity of DNMTi in the clinic.

14.1.3 TARGETING CHROMATIN MODIFICATIONS

Chromatin structure plays a key role in regulating gene expression by allowing DNA accessibility to transcriptional machinery and transcription factors [12]. The basic unit of chromatin is the nucleosome, which contains 147 bases of DNA wrapped around a histone octamer [13]. Each histone protein is subjected to post translational modifications that can either alter electrostatic charges of histones tails or provide recognition sites for epigenetic reader binding involved in chromatin remodeling and/or transcription control. Post translational modifications of histone proteins involve addition or removal of covalent groups such as methyl, acetyl, phosphoryl, ubiquityl, sumoyl, ADP ribosyl, deimine, proline, crotonyl, propionyl, butyryl, formyl, hydroxyl, and *O*-GlcNAcyl groups [4,14,15]. In cancer, global changes on histone marks are observed and affect gene transcription activity, thereby contributing to the malignant gene expression pattern [12].

Histone acetylation is an epigenetic mark associated with gene expression, whereas deacetylation of histone tails is associated with gene silencing. The addition of acetyl groups to lysine residues of histone tails neutralizes lysine positive charge, thereby reducing the electrostatic interaction in the negatively charged DNA. Thus, histone acetylation is associated with decondensed chromatin, allowing interaction between the bulky transcriptional machinery and promoter regions to favor gene expression. On the other hand, deacetylated histones are associated with chromatin condensation and gene silencing. Histone acetylation is catalyzed by histone acetyltransferases (HATs) and removed by histone deacetylases (HDACs). HATs are composed of three families of enzymes: Gcn5/PCAF, p300/CBP, and the MYST families. These enzymes are found mutated or translocated in leukemia (such as p300, CBP, and MYST4). HDACs are encoded by 18 different genes, which are overexpressed or mutated in various cancers. For example, truncation mutations in HDAC2 have been reported in epithelial, colon, gastric, and endometrial tumors [16]. Notably, global loss of H4K16 acetylation and H4K20 trimethylation have been reported in cancer cells and are associated with hypomethylated repetitive DNA sequences [4,17]. In addition, loss of H3K9me2 across large chromatin domains, an epigenetic mark involved in differentiation, is observed in many cancer cells [18].

Histone modifications play an important role in cancer due to their ability to bind and silence critical tumor suppressor genes. The therapeutic strategy aiming to inhibit HDAC activity and, consequently, increase histone acetylation and reactivate silenced tumor suppressor genes has been extensively studied. The efficacy of HDACi to reactivate silenced genes in cancer and induce anticancer effects has been validated by the US FDA approval of vorinostat and romidepsin for the treatment of cutaneous T-cell lymphoma and romidepsin and belinostat for peripheral T-cell lymphoma.

Histone methylation is also a critical epigenetic mark that is tightly associated with gene expression. Histone methylation can occur on arginine or lysine residues and can be found as mono-, di-, or trimethylation marks. Histone methyltransferases (HMTs) and demethylases (HDMs) add and remove methylation groups on histone residues, respectively. The main difference between methylation and acetylation is that histone methylation does not change the positive charge on lysine residues. Thus, the association of histone methylation and gene expression is considerably more complex than acetylation. Depending on the position of lysine residues on the histone tails, histone methylation can be associated either with gene activation (such as H3K4me3 and H3K36me3) or gene repression (H3K9me3 and H3K27me3) [3].

In cancer, histone methylation patterns are profoundly altered and its role as a driver event is confirmed by the identification of mutations in proteins directly or indirectly involved in histone methylation or demethylation [19]. A well-known example is the translocation of the H3K4 methyltransferase called mixed lineage leukemia (*MLL*). This mutation is considered as a driver event in cancer. In acute myeloid or lymphoid leukemia, oncogenic translocation of *MLL* creates an oncoprotein lacking its methyltransferase domain. The truncated protein is translocated with different fusion patterns that will reprogram histone marks and is associated with poor prognosis [20]. Surprisingly, *MLL* translocations were found to modify dimethylation patterns at H3K79 residues, which are deposited by DOTL1. The gene signature of *MLL-AF9* leukemia cells was found to be influenced by the alteration of H3K79 dimethylation patterns [21]. Other examples are found in the polycomb group of repressor proteins where *EZH2*, a HMT of H3K27, is found mutated in lymphoma and myeloid leukemia [2]. Overexpression of EZH2 has been reported in various cancers (prostate, lung, breast, and bladder), correlates with an increase in H3K27me3, and is associated with poor prognosis which is consistent with the function of *EZH2* in cancer [2,3]. Other HMTs such as SMYD3 and G9a are aberrantly expressed in various cancers [22]. Conversely, HDMs are also involved in carcinogenesis. Mutations in *LSD1* and *UTX* have been reported in several cancers. Therapeutic strategies have been developed to target histone methylation aberrations in cancer. At the moment, these drugs are in preclinical and clinical trial stages. Small-molecule inhibitors of these enzymes could offer a personalized approach for cancer treatment to precisely target epigenetic driver events in a particular cancer characterized by a specific mutation in epigenetic regulators [19,23].

14.1.4 TARGETING EPIGENETIC READERS

Targeting the readers of epigenetic signals represents a novel area for epigenetic therapy. Small-molecule inhibitors of epigenetic readers may have some potential either alone or in combination with inhibitors of epigenetic writers or erasers. Several studies have focused on the role of epigenetic readers in cancer and several strategies are being pursued to target their function in cancer cells.

Histone acetylation creates the binding site for epigenetic readers called bromodomains. These proteins that read acetyl marks of lysine residues have been associated with gene activation and especially c-MYC activation and associated oncogenic expression patterns [24,25]. Recently, selective small molecules have been designed to block bromodomain-containing proteins such as the bromo and extra-terminal (BET) proteins. Compounds such as JQ1, a thiodiazepine, block BET-bromodomains with a low affinity in nM ranges. Therapeutic strategies designed to inhibit bromodomain activity have successfully reprogrammed cancer cell expression profiles [24–26]. The efficiency of this approach was demonstrated in several cancer cell lines, suggesting the importance of bromodomains in cancer. To support this notion, a genetic alteration producing a fusion between the bromodomain of Brd4 with a nuclear protein in testis (NUT) is a driver event in the aggressive NUT midline carcinoma [26]. In addition Brd4 was also shown to play an important role in hematological malignancies. Pharmacologic inhibition of Brd4 by JQ1 produced downregulated c-Myc-dependent gene expression patterns and antiproliferative effects [27]. These findings demonstrate the potential of BET bromodomain inhibitors as anticancer agents.

Histone methylation at lysine residues serves as a signal for many epigenetic readers such as the plant homeodomains (PHD), chromodomains, and proteins of the Tudor, Agenet, PWWP, WDR5, EED, malignant brain tumor (MBT) domain, and others. In contrast to lysine acetylation, the addition of methyl groups (creating mono-, di-, or trimethylation) does not change the electrostatic charge of positively charged lysines. However, lysine methylation plays an important function in gene expression. Small molecules, such as UNC669 and UNC1215, were discovered to inhibit PHD domains and MBT-like proteins, respectively [28,29].

14.1.5 RATIONALE OF COMBINED EPIGENETIC THERAPIES IN CANCER

Epigenetic regulation of transcription is a complex mechanism that is regulated by several enzymes targeting DNA (methylation) and histone proteins (methylation, acetylation, etc.) that directly affect transcription factor recruitment and transcriptional machinery. DNA methylation represents the basic signal of epigenetic memory that is transmitted by methyl-binding proteins at the chromatin level by recruiting histone-modifying proteins such as HDACs [10]. The precise control of gene expression is controlled by the effect of several histone modifications at different sites, such as promoters, enhancers, or gene bodies, which is referred to as the "histone code" [30,31]. The occurrence and the coordination of both methylation and acetylation marks on active promoters and enhancers play a key role for gene activation. For example, H3K4me3 and H3K9ac are associated with active promoters. Active enhancer regions are typically marked by H3K27ac and H3K4me1 [32]. On the other hand, gene silencing is marked by the combination of repressive methylation marks such as H3K9me3 or H3K27me3 and corresponding loss of acetylation on these lysine residues.

Cancer cells are addicted to aberrant epigenetic alterations resulting from the sum of epigenetic abnormalities. Global DNA hypomethylation, region-specific DNA hypermethylation, global histone hypoacetylation, and other chromatin changes aberrantly reprogram cell phenotype. Tumor suppressor gene silencing and oncogene activation are the result of cumulative epigenetic alterations at the level of both DNA methylation and chromatin modifications. For example, silent genes are marked by promoter DNA hypermethylation embedded in compact heterochromatin marked by loss of H3K9ac and H3K4me3 and gain of H3K9me3 [33]. Cancer cells use different strategies to silence tumor suppressor genes and genes involved in regulation of development and differentiation. These genes, which are in a poised (or bivalent) transcriptional state in normal stem cells marked by both H3K27me3 and H3K4me3 at their promoters, are repressed in cancer cells. These genes lose the bivalent state plasticity and become permanently repressed by either gain in H3K9me3 and DNA hypermethylation marks or by gain in H3K27me3 marks [34].

Epigenetic therapies based on a single-agent approach affecting only one category of epigenetic marks have limited activity in leukemia and have been disappointing, especially against solid tumors. Simultaneously targeting several epigenetic alterations by combined approaches may enhance therapeutic efficiency [4,6]. The rationale behind this positive drug interaction is based on targeting mechanisms at the DNA methylation and the chromatin level involved in gene silencing in cancer cells. The mechanism explaining this synergistic interaction was clearly described by the work of Si et al. who demonstrated that DNMTi induce DNA demethylation in most of the cells but gene reactivation was only observed in a small subset of cancer cells exhibiting both DNA demethylation and chromatin resetting at the promoter of reactivated genes (removal of silencing marks and apparition of activating marks). Interestingly, genes that remained silent after drug-induced DNA demethylation were characterized by promoters with chromatin-repressive marks and nucleosomal density [11]. Indeed, the contribution of nucleosome occupancy and repressive chromatin marks produces a major block for tumor suppressor gene reactivation at the transcription start site [35].

Even though the concept of combined epigenetic therapies was developed and validated in preclinical models more than 25 years ago, no combinational therapies of epigenetic drugs have been approved for cancer treatments [6,36,37]. However, several dozens of clinical trials have been launched to investigate these combinations in patients (selected clinical trials mentioned in the chapter are described in Table 14.1).

Table 14.1 Selected Clinical Trials Tested Drugs Combinations Involving Epigenetic Drugs

Clinical Trial Number	Study Drug	Study Drug	Phase	Disease
NCT00867672	Decitabine	Valproic acid	II	Acute myeloblastic leukemia (AML)
NCT02273102	Tranylcypromine	Tretinoin	I	AML-MDS
NCT02261779	Tranylcypromine	Tretinoin	I/II	AML
NCT02303782	OTX015	Azacitidine	I/II	AML
NCT02190695	Decitabine	Carboplatin-arsenic trioxide	II	AML-MDS-CMML
NCT02159820	Decitabine	Paclitaxel	II/III	Ovarian cancer
NCT01928576	Azacitidine	Entinostat-CC-486	II	Nonsmall-cell lung cancer (NSCLC)
NCT01628471	Decitabine	Genistein	I/IIa	NSCLC

Note: Clinical trials were listed from ClinicalTrials.gov as of July 2015.

The rationale of using low doses of DNMTi to focus on epigenetic effects (rather than cytotoxic effects) offers the possibility of combinational approaches to improve clinical responses without reaching unacceptable toxicity. Several approaches will be discussed in this chapter describing the combination of epigenetic drugs targeting epigenetic writers (such as DNMT or HMT), epigenetic erasers (such as HDACs), epigenetic readers (such as bromodomains), and combination with other drug classes like cytotoxic chemotherapeutics or targeted therapy (Figure 14.1).

14.2 COMBINATIONS BETWEEN DNMT AND HDACi

Chromatin status is controlled by the dynamic interplay of a multitude of histone writers or erasers that mediate post translational modifications. HDACs are histone erasers involved in the removal of acetyl groups from lysine residues of histone tails, which produce heterochromatin (i.e., chromatin in a condensed state), resulting in gene silencing. In human, 18 HDACs have been identified and grouped into four classes. Classes I, II, and IV are zinc-dependant metalloproteases, while class III (sirtuins) use NAD^+ as a cofactor. The following discussion will focus on zinc-dependent HDACs since pharmacologic approaches targeting HDACs concern more specifically this type of enzymes.

Class I HDACs (1, 2, 3, and 8) are present in the nucleus of most cells, while class II (4, 5, 6, 7, 9, and 10) and class IV (11) HDACs shuttle between the nucleus and the cytoplasm, and their expression is more tissue-specific. HDAC6 and HDAC10 are preferentially found in the cytoplasm [38]. Class I enzymes play a role in cell proliferation and survival, whereas class II are involved in differentiation and development [38]. It is noteworthy that HDAC function is not restricted to histones since many other proteins can be deacetylated by HDACs at their lysine residues, thereby regulating their activity such as p53 or tubulin [1]. This diversity of functions goes beyond promoter deacetylation, which complicates their biological effects.

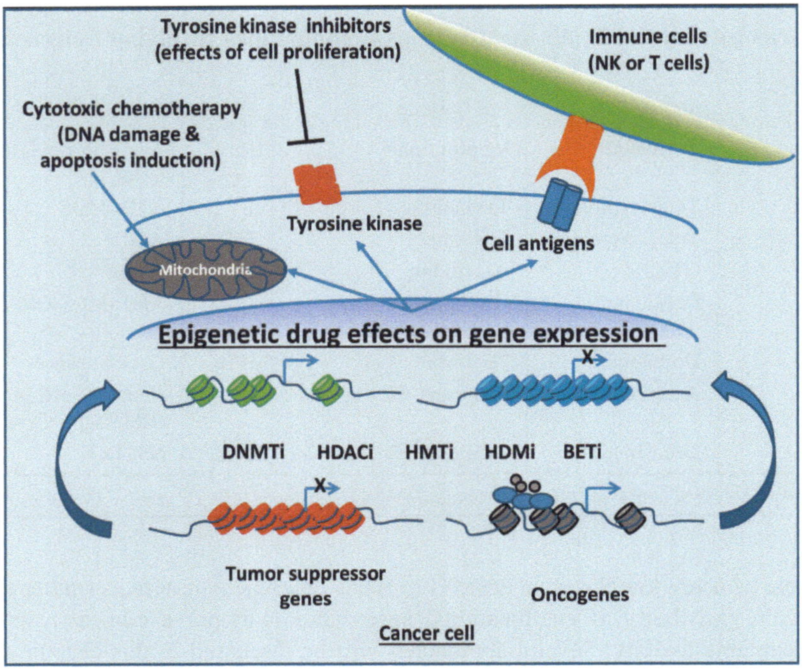

FIGURE 14.1

Drug combinations involving epigenetic drugs with other anticancer theraupeutic strategies. The combination of several epigenetic drugs such as DNMTi, HDACi, HMT inhibitor (HMTi), HDM inhibitors (HDMi), and BET inhibitors (BETi) played key roles in reprogramming the cancer cell epigenome. Their efficacy at inducing tumor suppressor gene reactivation or reducing expression of cell proliferation genes can serve as a rationale for drug combination with cytotoxic chemotherapy, targeted treatment with specific tyrosine kinase inhibitors or with immunotherapy.

Some HDACs are found overexpressed in cancer cells and expression of HDAC1, HDAC5, and HDAC7 can be used as biomarkers for tumor cells as compared to normal cells [39]. For example, HDAC1 and HDAC3 are overexpressed in nonsmall-cell lung cancer [40]. In acute promyelocytic leukemia, gene silencing is caused by PLZF-RARα fusion protein, which recruits HDAC to gene promoters [15]. In AML-ETO, HDAC1, HDAC2, and/or HDAC3 have been shown to repress AML1 target genes, and consequently blocking cell differentiation. A large amount of data support an oncogenic role of HDAC. Correlative data show that HDACs are overexpressed and associated with poor prognosis in several types of cancer (prostate, colorectal, breast, lung, liver, and stomach cancers). Overexpression of class I HDACs in tumor cells can induce cell proliferation and dedifferentiation, even though it is not clear that HDAC levels can be driver events in cancer. Interestingly, class II HDACs, like HDAC5 and HDAC10, were reported to be downregulated in lung cancer [40]. However, loss of H3K9 and H4K16 acetylation seems to be a common feature in cancer cells, suggesting the importance of targeting HDACs [15,17,41]. To support this concept, knockdowns of HDAC1, HDAC2, HDAC3, and HDAC6 in a variety of tumors (colon, breast, lung cancer, and acute promyelocytic leukemia) produce anticancer effects by blocking cell cycle and proliferation, inducing apoptosis, differentiation, senescence, and disruption of angiogenesis [39].

HDAC inhibition has been a flourishing area in the field of epigenetic therapy since the early observation where the HDACi trichostatin A produced cancer cell differentiation, gene reactivation, cell cycle arrest, and loss of clonogenicity [42]. Dozens of HDACi have been synthesized against different HDAC isoforms. Most commonly used HDACi target multiple HDACs, which complicated the understanding of their pharmacological and toxicological effects. HDACi are classified into groups based on their chemical structure, such as hydroxamic acids (trichostatin A and vorinostat), short-chain fatty acids (valproic acid, sodium butyrate, phenylbutyrate), benzamides (entinostat, mocetinostat), and cyclic peptides (apicidin, romidepsin) [39]. Three HDACi are currently approved by the FDA. Vorinostat and romidepsin are approved for the treatment of cutaneous T-cell lymphoma and romidepsin and belinostat for the treatment of peripheral T-cell lymphoma. These HDACi have weak target specificity and inhibit several HDAC isoforms. Generally, hydroxamic acids (vorinostat) are less specific, whereas benzamides (belinostat) inhibit class I HDACs [15]. Currently, more specific HDACi are being synthesized to circumvent specificity issues. Except for the case of T-cell lymphoma, the effects of HDACi in clinical trials have been generally disappointing, especially against solid tumors.

HDACi produce broad effects on cancer cells encompassing a G1-phase arrest (caused by *p21* reactivation), apoptotic effects, generation of reactive oxygen species, induction of DNA damage, and increase in immunogenicity of cancer cells through the upregulation of T-cell and NK-cell ligands [43,44]. Interestingly, HDACi have also been reported to reactivate genes silenced by DNA hypermethylation. By resetting histone acetylation, tumor suppressor genes silenced by promoter DNA hypermethylation can be re expressed for a short duration [10]. In phase I clinical trials in MDSs and AML, phenylbutyrate exhibited limited effects, which resulted at best in hematological improvements. Valproic acid seemed to be more active, particularly in MDS [45]. As single agents, other HDACi such as the benzamide MS-275 or depsipeptide showed evidence of activity in AML, which was associated with neurologic toxicity (fatigue and somnolence), infections, and gastrointestinal effects. Side effects of HDACi are predominantly gastrointestinal (nausea, vomiting) and neurologic (fatigue, somnolence, and confusion) [45]. Even though these drugs are quite well tolerated, stable disease was the best response reported in trials suggesting that combinational approaches should be investigated [40].

In a landmark article, Cameron et al. [6] demonstrated a synergistic interaction between DNMTi and HDACi by measuring significant tumor suppressor gene reactivation. Another important study demonstrated that the combination of azacitidine and the HDACi entinostat was twice as effective as azacitidine monotherapy, to reduce the growth of Kras mutant lung cancer cell lines in animal models [46]. Several types of HDACi were assessed in combination with DNMTi and demonstrated *in vitro* synergy against breast cancer, leukemia, and lymphoma cells [42,47–52]. One of the first classes was the short-chain fatty acid. Most studied compounds include valproic acid (also used in the clinic as an antiepileptic drug and mood stabilizer) and phenylbutyrate. Sequential combination of decitabine and valproic acid resulted in 90% reduction of leukemia cell viability [53]. Phenylbutyrate also produced a synergistic effect with decitabine in leukemic mice and against Calu-6 lung cancer cells [51,54]. Even though valproic acid and phenylbutyrate produced synergistic responses, the activity of these drugs is weak and usually requires millimolar concentrations to produce antineoplastic effects. Moreover, these drugs lack HDAC specificity. Hydroxamic acids such as trichostatin A (also used as an antifungal antibiotic), panobinostat, vorinostat, belinostat, selectively inhibit class I, II, and IV HDACs. The combination of decitabine with trichostatin A was also synergistic at killing MDA-MB-231 breast carcinoma cells and inducing gene reactivation of the retinoic acid receptor and estrogen receptor *alpha* gene [55]. Combination of vorinostat and decitabine results in synergistic cell cycle arrest, apoptosis, and gene reactivation *in vitro* and *in vivo* in ovarian cancer cell lines and xenografts [56]. Panobinostat, a cinnamic hydroxamic acid analog, produced synergistic responses in combination with decitabine against acute leukemia cells or diffuse large B-cell lymphoma [47,57]. Interestingly, the second generation of DNMTi, such as SGI-110, produced synergistic responses in combination with entinostat in K-ras/p53 mutant lung adenocarcinoma cell line (Calu-6) engrafted orthotopically in nude rats [58].

Although preclinical research has demonstrated the possibility of using a combination of DNMTi and HDACi, the clinical use of such a combination took some time to emerge. Pilot studies testing the combination of azacitidine and phenylbutyrate demonstrated some clinical benefit or major cytogenetic responses in patients with AML and MDS [59,60]. Other trials combining decitabine and valproic acid showed objective responses in patients with MDS and AML [61]. Decitabine was delivered at 15 mg/m^2 for 10 consecutive days with simultaneous escalating doses of valproic acid in 54 patients. The treatment produced 10 complete remissions. The median survival was 15.3 months for responders. The combination of azacitidine with valproic acid was tested in several clinical trials [62,63]. Interestingly, this combination was reported to be superior as compared to monotherapy [63]. However, other clinical trials testing the combination of azacitidine and short-chain fatty acid HDACi like phenylbutyrate, valproic acid, or entinostat did not report clinical benefits [62,64–66]. The lack of specificity and the weak activity of short-chain fatty acid HDACi could explain the weak clinical benefits when used in combination [10]. However, using more potent and selective HDACi with decitabine or azacitidine could result in better clinical outcome. The combination of decitabine and vorinostat against MDS and AML or the combination of azacitidine and entinostat against MDS, AML, and non-small cell lung cancer (NSCLC) showed interesting responses while being well tolerated [67–69]. In refractory patients with lung cancer, the combination produced 6.4 months survival, which is in a similar range of what other FDA-approved therapies can produce (such as erlotinib) [40]. However, few patients exhibited durable responses. Predictive biomarkers are being characterized to clearly identify which patients will particularly benefit from this epigenetic drug combination. Overall, the combination of DNMTi and HDACi in clinical trials has not reached the synergistic levels obtained in preclinical assays possibly due to issues in dosing schedules, HDACi selectivity, selection of most sensitive cancer indication, or disease stage [70].

Other drugs, such as retinoic acids, have been added to the combination of DNMTi and HDACi. Retinoic acids are metabolites of vitamin A and an interesting agent to use in combination with decitabine since synergistic effects were reported using colony assays in DLD-1 colon cancer cells [71]. In another study, this combination produced additive effects against HL-60 leukemia cells [72]. The clinical potential of triple combination with decitabine, valproic acid, and all-*trans* retinoic acid is investigated in a randomized phase II trial in patients over 60 years with AML who are ineligible for induction chemotherapy (ClinicalTrials.gov identifier: NCT00867672) [73].

14.3 EPIGENETIC COMBINATIONS WITH HMTi

HMTs are generally more specific toward particular histone residues as compared to HDAC as their enzymatic activity is restricted to defined lysine or arginine residues [74]. However, like HDACs, HMTs can methylate other proteins [75]. Given the importance of HMTs in a number of hematologic malignancies and solid tumors, drug discovery efforts have identified new effective HMTi, specifically to target EZH2 and DOT1L. These HMTi represent interesting epigenetic drugs that can be used in combination.

One of the first HMT targets to be extensively studied was EZH2, a component of the PCR2 complex that catalyzes H3K27 methylation. Recently, point mutations within the catalytic domain of EZH2 have been found in patients with non-Hodgkin lymphoma. This HMT is an interesting target for drug combination with DNMTi because there is a "crosstalk" between EZH2 and DNA methylation since it was demonstrated that EZH2 recruits DNMT to gene promoters [76]. Among the first drugs identified to decrease H3K27me levels was 3-deazaneplanocin-A (DZNep). However, this compound was found to be nonspecific due to its global inhibition of histone methylation marks [77]. Indeed, DZNep is a S-adenosylhomocysteine hydrolase inhibitor leading to indirect inhibition of S-adenosyl-methionine-dependent reactions, including HMTs [77]. The discovery of its nonspecificity reduced the interest of several investigators to pursue its development. Nonetheless, the combination of DZNep and decitabine produced a synergistic response using colony assays in leukemic cell lines (AML-3, HL-60, and L1210) and increased survival in mice with L1210 leukemia [78]. Interestingly, a triple combination between DZNep, decitabine, and HDACi trichostatin A showed an even greater synergy against human AML cells as demonstrated by colony assays and gene reactivation microarrays [79]. Combinatorial treatment with DZNep also demonstrated synergistic interaction with etoposide and vorinostat by measuring apoptosis induction, cell cycle inhibition, and clonogenicity reduction in rhabdoid cells [80]. In addition, another HDACi, panobinostat, demonstrated a synergistic interaction with DZNep producing increased apoptosis and reactivation of *p16*, *p21*, *p27*, and *FBX032* in AML cells but, interestingly, not in normal CD34+ bone marrow progenitor cells. The combination of DZNep and panobinostat significantly improved survival of nonobese diabetic (NOD)/severe combined immunodeficiency mice bearing HL-60 leukemia [81]. Interestingly, DZNep demonstrated synergistic activity in combination with the cytotoxic drug gemcitabine against pancreatic cancer cells (PANC-1, MIA-PaCa-2, and LPC006) [82].

High-throughput screenings have identified new EZH2 inhibitors with improved activity and selectivity, such as GSK-126 and EPZ-6438. These highly selective and bioavailable compounds have clinical potential and EPZ-6438 is in clinical trials as a single agent in patients with advanced solid tumors and with B-cell lymphomas (ClinicalTrials.gov identifier: NCT01897571). More studies are needed with the newly discovered EZH2 inhibitors in combination with other epigenetic or chemotherapeutic drugs.

In leukemia harboring the myeloid/lymphoid leukemia (*MLL*) rearrangements, epigenetic abnormalities have been associated with DOT1L, a HMT which methylates histone 3 on lysine 79 (H3K79). Indeed, the oncogenic fusion involving *MLL* loses its H3K4 methyltransferase activity but produces abnormal H3K79 methylation patterns [21,23,83]. Potent, selective DOT1L inhibitors, such as EPZ00477, selectively target *MLL*-translocated cells while not affecting non-*MLL*-translocated cells [23]. An improved version of this compound, named EPZ-5676, showed *in vivo* efficacy producing significant tumor growth inhibition, with complete eradication of tumors [84]. Based on animal study efficacy, EPZ-5676 is now advancing into phase I clinical trials (ClinicalTrials.gov identifier: NCT02141828) and administered as a 28-day continuous IV infusion of each 28-day cycle. Preclinical combinations using DOT1L inhibitors have recently been described. Using, SYC-522 a recently developed DOT1L inhibitor, *MLL*-rearranged leukemia cells were sensitized to several chemotherapeutic drugs such as mitoxantrone, etoposide, and cytarabine [85]. Recently, it has been demonstrated by genome-scale RNAi screen that the HDAC SIRT1 establishes a heterochromatin state around MLL fusion target genes after DOT1L inhibitor treatment. Interestingly, DOT1L inhibits the formation of a chromatin-repressive complex involving SIRT1 and the H3K9 methyltransferase SUV39H1 activities. Thus, DOT1L maintains an open chromatin state at *MLL* fusion target genes characterized by a gain in H3K9 acetylation and low H3K9 methylation. The combination of SIRT1 activators and DOT1L inhibitors produced enhanced antiproliferative effects in *MLL*-rearranged leukemia cells. These results could provide novel opportunities for combination therapy with DOT1L inhibitors [86].

14.4 EPIGENETIC COMBINATIONS WITH HDMi

HDMs induce lysine or arginine demethylation. These enzymes are divided into two classes: the lysine-specific demethylase (LSD) family and the Jumonji C-containing demethylases. Alterations in these enzymes have been associated with cancer [87]. Several inhibitors of LSD1 targeting its catalytic domain have been synthesized. Drugs such as tranylcypromine, geranylgeranoic acid or selective ORY-1001 have been studied in *MLL*-rearranged leukemia and neuroblastoma [88–91]. Recently, several clinical trials have been initiated using LSD1 inhibitor GSK2879552 alone against leukemia and small-cell lung cancer (ClinicalTrials.gov identifiers: NCT02177812 and NCT02034123). Interestingly, it was demonstrated that LSD1 can also demethylate DNMT1, resulting in destabilization and loss of global maintenance of DNA methylation [92]. The ability of LSD1 to affect both histone and DNA methylation makes it a promising target for epigenetic therapy. Although targeting of HDM is still in its infancy, early preclinical studies showed promising results in combination therapies. Treatment of colon cancer cells with the LSD1 inhibitor, SL11144, in combination with DNMT inhibition increased H3K4 methylation and reactivated expression of *SFRP2*, a negative regulator of Wnt signaling, in colon cancer cells resulting in significant growth inhibition of established tumors [93]. Recently, Fiskus et al. [94] reported that high expression of LSD1 in AML was associated with a block in cell differentiation and poor prognosis. Using novel LSD1 antagonist SP2509, the authors reported that treatment reduced the binding of LSD1 to the corepressor complex CoREST and increased H3K4me3 on target genes. Epigenetic changes were associated with *p21* and *p27* reactivation in AML cells. Interestingly, SP2509 treatment inhibited colony growth and induced AML cell differentiation. The combination between HDACi panobinostat and SP2509 was synergistic against AML blasts and significantly improved the survival of mice engrafted with human AML cells [94].

Other combinatorial approaches involving tranylcypromine and all-*trans* retinoic acid have been explored [95]. Interestingly, tranylcypromine sensitized nonacute promyelocytic leukemia cells to all-*trans* retinoic acid, which are usually resistant, in opposition to acute promyelocytic leukemia cells. LSD1 inhibition leads to an increase in H3K4me2 and gene reactivation specifically in myeloid-differentiation-associated genes. The combination of all-*trans* retinoic acid and tranylcypromine reduced the engraftment and produced an antileukemic effect against primary human AML cells in NOD-severe combined immunodeficient mice [95]. These data on the combination of LSD1 inhibition and differentiation effects of all-*trans* retinoic acid have paved the way for new clinical trials in AML and MDS (ClinicalTrials.gov identifiers: NCT02273102 and NCT02261779).

Selective inhibitors have also been synthesized for Jumonji C-containing demethylases. Recently, it was demonstrated that GSK-J4, a histone H3 lysine 27 (H3K27) demethylase JMJD3 inhibitor, produced a synergistic effect in combination with HDACi panobinostat in diffuse intrinsic pontine glioma, a fatal childhood cancer [96].

14.5 EPIGENETIC READER DRUG COMBINATIONS

In contrast to epigenetic writer or eraser inhibition, targeting epigenetic readers is based on the rationale of disrupting the protein interaction between epigenetic marks and proteins that translate the epigenetic signal, such as bromodomain-, chromodomain-, or Tudor domain-containing proteins. One of the most advanced therapeutic strategies is the inhibition of bromodomain-containing proteins using small molecules such as JQ1, GSK525762, I-BET762, I-BET151, or CPI-BETi [26,97]. BET domain proteins contain bromodomains that are specifically involved in lysine acetylation binding of histone tails. Members of the BET family, including BRD2, BRD3, BRD4, and BRDT, regulate transcription of a set of genes playing key roles in cancer [26,98].

Targeting acetylation readers disrupts the transmission of epigenetic signaling. One immediate result comes through the downregulation of c-MYC oncogenic signature, resulting in downregulation of its transcriptional gene targets and anticancer effects [98]. Thus, c-MYC, an oncogenic transcription factor lacking enzymatic activity and overexpressed in the vast majority of cancers, became druggable for the first time [24–26]. Treatment with BETi resulted in growth arrest and apoptosis in cell culture and induced antitumor activity in xenograft animal models of multiple myeloma, lymphoma, and acute leukemia. BETi induce a global alteration in BET and c-MYC chromatin localization, histone modifications, and RNA polymerase II distribution which result in epigenetic reprogramming of gene expression profiles. Notably, these drugs only downregulated expression of a small subset of genes. Specifically, these drugs repressed around 2% of BRD2- or BRD4-bound genes. It is noteworthy that many of these genes were transcription factors [24,27,98]. Targeting epigenetic readers seems to be a promising strategy also in neuroblastoma, where poor-prognosis stages are characterized by MYCN amplification [25]. BETi are being investigated as single agents in clinical trials using TEN-010 in AML and MDS, GSK525762 in NUT midline carcinoma, and OTX015 in recurrent glioblastoma (ClinicalTrials.gov identifiers: NCT01987362, NCT01587703, and NCT02296476).

Several studies have described enhancement or a synergistic response using bromodomain inhibitors with other epigenetic drugs. In addition, the combination with JQ1 and the SUV39H1 inhibitor chaetocin has been explored in AML cells. Chaetocin induced CD11b expression and differentiation of AML cells (HL-60 and KG-1a) at noncytotoxic concentrations. These phenotypic changes were associated with downregulation of SUV39H1 and reduction of H3K9me3. Interestingly, the combination of chaetocin with JQ1 showed synergistic cytotoxicity highlighting a new possible combination for the treatment of AML [99]. Sequential combinations with OTX015 and HDACi panobinostat or DNMTi azacitidine have been shown to produce synergic effects against KASUMI AML cells [100]. Some combinations have moved forward to clinical trials. For example, a phase Ib/II is assessing the BET-bromodomain inhibitor OTX015 with azacitidine in newly diagnosed AML patients (ClinicalTrials.gov identifier: NCT02303782).

14.6 COMBINATIONS BETWEEN EPIGENETIC DRUGS AND STANDARD CHEMOTHERAPY

Conventional cytotoxic chemotherapy regimens can rapidly induce cancer cell death. However, their clinical use and efficacy have reached a limit, since these treatments are already administered at the maximum tolerated dose. Moreover, cancer cells can develop resistance to standard chemotherapy after several rounds of treatment or drug resistance is present early in the case of refractory disease. Interestingly, a significant number of resistance mechanisms occur through epigenetic dysregulation [1]. Using epigenetic drugs at low doses in combination with standard chemotherapy can represent an approach that will reverse epigenetically induced drug-resistance mechanisms without causing unacceptable toxicity, leading to synergistic responses.

Importantly, a clear understanding of the mechanism of action of standard chemotherapy with epigenetic drugs is crucial to avoid an antagonistic drug interaction. For example, decitabine or vorinostat treatment can reverse drug-resistance mechanisms to standard chemotherapy in several cancer cells [101,102]. In contrast, DNA methylation-induced silencing of DNA repair genes, such as *MGMT*, is a good prognostic factor associated with survival benefit and sensitivity to temozolomide in pediatric glioblastoma [103]. Thus, the combination with DNA demethylating drugs may not benefit standard chemotherapy in this particular case. The key to success for combinations of epigenetic drugs with standard chemotherapy may depend on the mechanism of action of defined standard chemotherapeutic drugs and the epigenetic mechanisms involved in drug resistance.

The combination of DNMTi and standard chemotherapy has shown promising results in several preclinical investigations, some of which have reached clinical trial development. Here, we will focus on combinations involving decitabine with carboplatin, arsenic/all-*trans* retinoic acid, and clofarabine. The combination of DNMTi and platinum compound carboplatin has been studied in various cancer models such as colon (SW48) and ovarian (A2780/cp70) cancers [104]. Mechanistically, this combination is based on the reversal of *hMLH1* silencing which occurs in colon, gastric, endometrial, and ovarian cancers by promoter DNA hypermethylation. In ovarian cancer, a higher frequency of *hMLH1* promoter DNA hypermethylation is observed after chemotherapy treatment [105]. *In vitro* and *in vivo* data suggest that decitabine induction of promoter demethylation reactivates *hMLH1* expression and sensitizes tumor xenografts to several standard chemotherapeutic drugs such as cisplatin, carboplatin, temozolomide, and epirubicin [104]. Based on this rationale, phase I clinical trials have emerged in solid tumors and ovarian cancers

[101,106]. The combination was safely administered and produced one complete response and three patients had stable disease for more than 6 months [101]. Results from a phase II clinical trial in recurrent, partially platinum-sensitive ovarian cancer patients were disappointing because decitabine appeared to reduce the efficacy of carboplatin. Future investigations should explore different schedules or better patient selection [107]. Currently, several clinical trials are pursuing the study of decitabine combination with carboplatin in AML, MDS, chronic myelomonocytic leukemia, and ovarian cancer (ClinicalTrials.gov identifiers: NCT02190695 and NCT02159820).

Clofarabine is a second generation of purine analog approved for the treatment of resistant forms of pediatric acute lymphocytic leukemia. Interestingly, it also demonstrates promising efficacy against AML and MDS. Investigators studied the effect of a combination of clofarabine plus low-dose cytarabine followed by a consolidation phase with decitabine. The study population was newly diagnosed AML elderly patients (60–81 years). Dosing regimens included an induction with intravenous clofarabine at 20 mg/m^2 (days 1–5) and subcutaneous cytarabine at 20 mg twice daily on days 1–10. Patients in complete response were treated with a consolidation regimen consisting of cycles of clofarabine (days 1–3) and cytarabine (days 1–7) for 2 months followed by three cycles of decitabine (20 mg/m^2 i.v. daily on days 1–5, every month). This therapeutic strategy showed promising activity in older patients with newly diagnosed AML [108]. Final results of the study demonstrated a complete remission rate of 60% with a median overall survival of 11.1 months for all patients and 18.5 months in responding patients. This low-intensity therapy with clofarabine-cytarabine-decitabine was well tolerated and highly effective in older patients with AML [109].

Since HDACi exert pleotropic effects on cancer cells, these epigenetic drugs can cooperate with other anticancer drugs to produce potent anticancer activity. Even though the pharmacology of HDACi is relatively well-known, it still remains difficult to fully characterize the exact mechanism responsible for drug interaction activity. It is likely that the role of HDACi in combination may involve chromatin resetting and gene reactivation, but also other effects such as nonhistone acetylation, oxidative stress, and DNA damage. Another rationale to support the combination of HDACi with standard chemotherapy is to use the property of HDACi to lower cancer cells' apoptotic threshold. Indeed, synergistic responses have been reported with different HDACi when combined with activators of TRAIL and Fas pathways [44]. For example, in a randomized phase II trial with untreated NSCLC patients, the addition of HDACi vorinostat to carboplatin and paclitaxel, was well tolerated. The addition of vorinostat induced a response rate of 34% as compared to 12.5% with placebo. There was a tendancy toward increased median progression-free survival and overall survival when vorinostat was added to the chemotherapy. The addition of HDACi seemed to be a promising therapeutic strategy for treatment of NSCLC [110,111]. Valproic acid was also combined with other cytotoxic drugs such as doxorubicin and produced a 16% response rate in refractory or recurrent disease in patients with mesothelioma [112]. Combinations of vorinostat with gemcitabine, docetaxel, or with idarubin/cytarabine therapy were tested in several cancers. Particularly, the combination of vorinostat with idarubin/cytarabine therapy in AML or MDS patients produced an overall response rate of 85%, including 76% complete response and 9% in complete remission with incomplete platelet recovery [44,113]. In acute promyelocytic leukemia, a disease characterized by aberrant HDAC recruitment to gene promoters by oncogenic fusion proteins (PLZF-RARα or PML-RARα), the addition of HDACi to standard differentiation therapy with all-*trans* retinoic acid produced a synergistic effect in both preclinical and clinical studies [44]. This therapeutic strategy,

HDACi involving valproic acid and all-*trans* retinoic acid, was expanded to MDS, which resulted in prolonged second responses after valproic acid partial responses [114]. Synergistic responses were also observed with regulators of proteosomal degradation such as the Hsp90 inhibitor (17-allyl-amino-demetoxy geldanamycin) and the proteasome inhibitor bortezomib against CML and AML cell lines [44].

Surprisingly, a limited amount of data is available regarding combination of HMT, HDM, or bromodomain inhibitors with standard chemotherapy. Further studies need to characterize the potential of these epigenetic regulators in combination with current chemotherapeutic drugs.

14.7 EMERGING COMBINATIONS BETWEEN EPIGENETIC DRUGS AND TARGETED CHEMOTHERAPY KINASE INHIBITORS

Tyrosine kinase inhibition leads to signal transduction blockade in proliferation and survival pathways. Several epigenetic drugs have been tested to assess their potential in combination with compounds such as flavopiridol, a cyclin-dependent kinase inhibitor, midostaurin a multitarget protein kinase inhibitor, or MEK1/2 inhibitor PD184352. Interestingly, targeting both DNA methylation and chromatin (erasers and readers) levels seems to enhance or synergize with tyrosine kinase inhibitors. A combination of midostaurin and decitabine synergistically induces greater apoptosis of cultured FLT3-ITD expressing AML cells. In a phase I clinical trial in AML, 57% of the patients achieved stable disease while 25% had a complete hematologic response warranting additional studies exploring the sequential combination of decitabine and midostaurine [115]. Tyrosine kinase inhibitors targeting epidermal growth factor have been tested in combination with DNA demethylating drugs. Gefitinib, an orally selective epidermal growth factor receptor tyrosine kinase inhibitor with demonstrated antitumor activity in solid tumors induced a synergistic response when combined with decitabine, as evaluated by clonogenic assays or annexin V staining in colon and NSCLC [116,117]. Other epidermal growth factor receptor tyrosine kinase inhibitors, such as erlotinib, induced synergistic interaction with DNMTi azacitidine against AML cells. The combination induced cell cycle blockage, caspase-dependent apoptosis and proteasomal degradation of the antiapoptotic Bcl-2 family members, MCL-1 and BCL2L10 [118]. Interestingly, the intracellular accumulation of azacitidine was increased by erlotinib (or gefitinib), suggesting a pharmacokinetic mechanism of synergy [118]. This combination has been investigated in a phase I study in advanced solid tumor malignancies (lung, head and neck, and ovarian cancers) [119]. Erlotinib at 150 mg daily and azacitidine at 75 mg/m^2 per cycle was safe, well-tolerated and produced partial responses in lung and ovarian cancers with a median progression-free survival of 2 months.

Several studies have demonstrated the potential of HDACi to produce synergy when used in combination with tyrosine kinase inhibitors. HDACi romidepsin has been shown to enhance the activity of erlotinib in NSCLC cell lines, including wild-type and mutant cell lines for KRAS and epidermal growth factor receptor. The combination enhanced apoptosis and inhibited xenograft growth in athymic nude mice [120]. Another HDACi, LBH589 (panobinostat) demonstrated synergy with erlotinib in lung cancer cells. Mechanistically, LBH589 induced acetylation of Hsp90, depleted epidermal growth factor receptor and other survival signaling proteins leading to apoptosis in lung cancer cells harboring epidermal growth factor receptor mutations. Thus, epidermal growth factor receptor mutation status may predict treatment outcome with the combination of LBH589 and erlotinib [121]. Several combinations of HDACi and erlotinib were tested in clinical trials. Vorinostat and erlotinib in NSCLC patients were safely administered but the combination did not produce meaningful responses in epidermal growth factor receptor-mutated patients [122]. In contrast, the combination of panobinostat and erlotinib, which was tested in seven patients with nonsmall-cell lung cancer with epidermal growth factor receptor mutations, produced three partial responses and three stable diseases [123].

Combinations involving targeted therapy and bromodomain have been explored in several cancers. In osteosarcoma, a synergistic effect was reported between the BETi JQ1 and the mTOR inhibitor rapamycin, which synergistically inhibited the growth and survival of osteosarcoma cells *in vitro* and *in vivo* [124]. The BET bromodomain inhibitor OTX015 also produced synergistic effects when combined with mTOR (everolimus) and BTK (ibrutinib) inhibitors in B-cell lymphoid tumor models. These data provide the basis for clinical investigation of OTX015 in combination therapies [125].

14.8 USING EPIGENETIC DRUGS IN COMBINATION WITH IMMUNOTHERAPY

Epigenetic drugs are known to modify tumor cell phenotype, which can trigger immunologic effects and induce cancer cell elimination. Epigenetic drugs can be used to sensitize cancer cells for immunotherapy. Preclinical data using decitabine have shown that targeting DNA methylation induced the expression of CTA and HLA class I antigens [126]. Interestingly, *in vivo* modifications of cancer cell epigenome were durable post-DNMTi treatment and NY-ESO-1 and HLA class I antigens were still detectable on melanoma xenografts 1 month after decitabine treatment [126]. Importantly, epigenetic silencing by DNA hypermethylation of antigens, such as NY-ESO-1, plays a key role in cancer cell escape from T cells. Gene reactivation of these antigens can potentially allow recognition and killing by immune system cells such as engineered cytotoxic T lymphocytes [127]. HDACi trichostatin A has also been shown to enhance NK-cell activity following upregulation of NK-activating ligands on the surface of glioblastoma cells [128]. Thus, there is a strong rationale to support the combination of immunotherapy and epigenetic drugs.

The combination of decitabine with high-dose interleukin-2 has been tested in a phase I clinical trial where patients received decitabine daily for 5 days per week for 2 weeks and high-dose interleukin-2. The combination can be safely administered and may enhance the activity of interleukin-2 as demonstrated by a 31% objective response rate in melanoma patients [129]. Interestingly, mechanistic studies have demonstrated that the combination of decitabine and interleukin-12 in B6F10 mouse melanoma cells was dependent on CD4+ and CD8+ T cells [130]. Tumor mass reduction induced by decitabine seemed to involve the CD8+ and NK cells [126]. Similarly, HDACi appeared to produce anticancer effects through the cooperation of CD8+ T cells. Interestingly, anticancer activity of HDACi was abrogated in immunodeficient mice [131]. In the case of DNMTi azacitidine, transcriptome analysis showed upregulation of genes involved in innate and adaptive immunity and genes associated with immune evasion in NSCLC cell lines. Interestingly, azacitidine induced upregulation of PD-L1, a key ligand-mediator of immune tolerance which is commonly downregulated in NSLC. These data suggest that PD-1 pathway blockade in combination with azacitidine might enhance anticancer activity [132]. This new approach is being tested in a randomized phase II clinical trial with azacitidine and entinostat or oral azacitidine alone before the PD-1 blocker nivolumab in patients with recurrent metastatic NSCLC (ClinicalTrials.gov identifier: NCT01928576).

14.9 COMBINATION WITH NONTOXIC AGENTS

The anticancer activity of epigenetic drugs has also been investigated with nontoxic agents with the intention of increasing responses while limiting toxicity. As previously mentioned, the function of the host immune system is critical in the epigenetic response. Thus, avoiding additional toxicity to the hematopoietic system is important to ensure potent anticancer effects. In addition, combination with nontoxic agents can enhance or even synergize with epigenetic drugs.

Genistein, a nontoxic soybean isoflavone, is a well-known natural product with chemopreventive and anticancer activity [133]. Several studies have highlighted its potential to synergize with some cytotoxic antineoplastic drugs [134,135]. The combination of genistein and decitabine produced a synergistic response both *in vitro* and in a mouse leukemic model, which is currently under clinical trial (ClinicalTrials.gov identifier: NCT01628471) [136]. Vitamin D is another nontoxic agent that exhibits differentiation potential in human myeloid leukemia cell lines. Decitabine with vitamin D analogs produced a higher growth inhibition and cell differentiation than either drug alone in HL-60 and NB-4 myeloid leukemia cells [72].

14.10 CONCLUSION AND PERSPECTIVES

Epigenetic drug combination is a promising field of investigation that started with combinations of epigenetic drugs with one another and is now opening to standard chemotherapy, targeted drugs, and immunotherapy. With the advancement of knowledge and sequencing technologies, new and more specific inhibitors are being designed that can specifically target epigenetic erasers or readers. The understanding that cancer epigenome is a complex and coordinated network of epigenetic signals brings a strong rationale to carefully investigate epigenetic drug combinations. One of the most studied, the combination involving DNMTi and HDACi, has demonstrated the potential of this approach, which has been validated in clinical trials. However, there is still room for improvement and disturbing the cancer epigenome with other epigenetic drugs may lead to better clinical outcome and more durable responses.

The discovery that epigenetic drugs can also synergize with other chemotherapeutic drugs by lowering cancer cell apopotic threshold or reactivating specific antigens, brings an incredible opportunity to include epigenetic drugs into almost any kind of cancer treatment. Taking into consideration the diversity of cancer cell epigenetics, new drug combinations should be part of the development of personalized oncology. The development of new combinational strategies will require deep preclinical investigation to define (i) molecular targets, (ii) optimized dosing regimens, and (iii) target cancer patient population. The use of epigenetic drugs in cancer treatment is growing at an exponential rate with the study of two or more drug combinations, with the goal to provide to the vast majority of cancer patients rapid, durable, and complete remissions.

REFERENCES

[1] Issa JP, Kantarjian HM. Targeting DNA methylation. Clin Cancer Res 2009;15(12):3938–46.

[2] You JS, Jones PA. Cancer genetics and epigenetics: two sides of the same coin? Cancer Cell 2012;22(1):9–20.

[3] Baylin SB, Jones PA. A decade of exploring the cancer epigenome—biological and translational implications. Nat Rev Cancer 2011;11(10):726–34.

[4] Kelly TK, De Carvalho DD, Jones PA. Epigenetic modifications as therapeutic targets. Nat Biotechnol 2010;28(10):1069–78.

[5] Momparler RL. Cancer epigenetics. Oncogene 2003;22(42):6479–83.

[6] Cameron EE, Bachman KE, Myohanen S, Herman JG, Baylin SB. Synergy of demethylation and histone deacetylase inhibition in the re-expression of genes silenced in cancer. Nat Genet 1999;21(1):103–7.

[7] Jones PA, Liang G. Rethinking how DNA methylation patterns are maintained. Nat Rev Genet 2009;10(11):805–11.

[8] Jones PA. Functions of DNA methylation: islands, start sites, gene bodies and beyond. Nat Rev Genet 2012;13(7):484–92.

[9] Estey EH. Epigenetics in clinical practice: the examples of azacitidine and decitabine in myelodysplasia and acute myeloid leukemia. Leukemia 2013;27(9):1803–12.

[10] Raynal NJ, Si J, Taby RF, Gharibyan V, Ahmed S, Jelinek J, et al. DNA methylation does not stably lock gene expression but instead serves as a molecular mark for gene silencing memory. Cancer Res 2012;72(5):1170–81.

[11] Si J, Boumber YA, Shu J, Qin T, Ahmed S, He R, et al. Chromatin remodeling is required for gene reactivation after decitabine-mediated DNA hypomethylation. Cancer Res 2010;70(17):6968–77.

[12] Sharma S, Kelly TK, Jones PA. Epigenetics in cancer. Carcinogenesis 2010;31(1):27–36.

[13] Geutjes EJ, Bajpe PK, Bernards R. Targeting the epigenome for treatment of cancer. Oncogene 2012;31(34):3827–44.

[14] Dawson MA, Kouzarides T. Cancer epigenetics: from mechanism to therapy. Cell 2012;150(1):12–27.

[15] Falkenberg KJ, Johnstone RW. Histone deacetylases and their inhibitors in cancer, neurological diseases and immune disorders. Nat Rev Drug Discov 2014;13(9):673–91.

[16] Smith KT, Workman JL. Histone deacetylase inhibitors: anticancer compounds. Int J Biochem Cell Biol 2009;41(1):21–5.

[17] Fraga MF, Ballestar E, Villar-Garea A, Boix-Chornet M, Espada J, Schotta G, et al. Loss of acetylation at Lys16 and trimethylation at Lys20 of histone H4 is a common hallmark of human cancer. Nat Genet 2005;37(4):391–400.

[18] Wen B, Wu H, Shinkai Y, Irizarry RA, Feinberg AP. Large histone H3 lysine 9 dimethylated chromatin blocks distinguish differentiated from embryonic stem cells. Nat Genet 2009;41(2):246–50.

[19] Copeland RA, Moyer MP, Richon VM. Targeting genetic alterations in protein methyltransferases for personalized cancer therapeutics. Oncogene 2013;32(8):939–46.

[20] Krivtsov AV, Armstrong SA. MLL translocations, histone modifications and leukaemia stem-cell development. Nat Rev Cancer 2007;7(11):823–33.

[21] Bernt KM, Zhu N, Sinha AU, Vempati S, Faber J, Krivtsov AV, et al. MLL-rearranged leukemia is dependent on aberrant H3K79 methylation by DOT1L. Cancer Cell 2011;20(1):66–78.

[22] Varier RA, Timmers HT. Histone lysine methylation and demethylation pathways in cancer. Biochim Biophys Acta 2011;1815(1):75–89.
[23] Daigle SR, Olhava EJ, Therkelsen CA, Majer CR, Sneeringer CJ, Song J, et al. Selective killing of mixed lineage leukemia cells by a potent small-molecule DOT1L inhibitor. Cancer Cell 2011;20(1):53–65.
[24] Knoechel B, Roderick JE, Williamson KE, Zhu J, Lohr JG, Cotton MJ, et al. An epigenetic mechanism of resistance to targeted therapy in T cell acute lymphoblastic leukemia. Nat Genet 2014;46(4):364–70.
[25] Puissant A, Frumm SM, Alexe G, Bassil CF, Qi J, Chanthery YH, et al. Targeting MYCN in neuroblastoma by BET bromodomain inhibition. Cancer Discov 2013;3(3):308–23.
[26] Filippakopoulos P, Qi J, Picaud S, Shen Y, Smith WB, Fedorov O, et al. Selective inhibition of BET bromodomains. Nature 2010;468(7327):1067–73.
[27] Zuber J, Shi J, Wang E, Rappaport AR, Herrmann H, Sison EA, et al. RNAi screen identifies Brd4 as a therapeutic target in acute myeloid leukaemia. Nature 2011;478(7370):524–8.
[28] Wagner EK, Nath N, Flemming R, Feltenberger JB, Denu JM. Identification and characterization of small molecule inhibitors of a plant homeodomain finger. Biochemistry 2012;51(41):8293–306.
[29] James LI, Barsyte-Lovejoy D, Zhong N, Krichevsky L, Korboukh VK, Herold JM, et al. Discovery of a chemical probe for the L3MBTL3 methyllysine reader domain. Nat Chem Biol 2013;9(3):184–91.
[30] Jenuwein T, Allis CD. Translating the histone code. Science 2001;293(5532):1074–80.
[31] Hake SB, Xiao A, Allis CD. Linking the epigenetic 'language' of covalent histone modifications to cancer. Br J Cancer 2004;90(4):761–9.
[32] Mills AA. Throwing the cancer switch: reciprocal roles of polycomb and trithorax proteins. Nat Rev Cancer 2010;10(10):669–82.
[33] Jones PA, Baylin SB. The epigenomics of cancer. Cell 2007;128(4):683–92.
[34] Ohm JE, McGarvey KM, Yu X, Cheng L, Schuebel KE, Cope L, et al. A stem cell-like chromatin pattern may predispose tumor suppressor genes to DNA hypermethylation and heritable silencing. Nat Genet 2007;39(2):237–42.
[35] Lin JC, Jeong S, Liang G, Takai D, Fatemi M, Tsai YC, et al. Role of nucleosomal occupancy in the epigenetic silencing of the MLH1 CpG island. Cancer Cell 2007;12(5):432–44.
[36] Jahangeer S, Elliott RM, Henneberry RC. Beta-adrenergic receptor induction in HeLa cells: synergistic effect of 5-azacytidine and butyrate. Biochem Biophys Res Commun 1982;108(4):1434–40.
[37] Yoo CB, Jones PA. Epigenetic therapy of cancer: past, present and future. Nat Rev Drug Discov 2006;5(1):37–50.
[38] Marks PA. Histone deacetylase inhibitors: a chemical genetics approach to understanding cellular functions. Biochim Biophys Acta 2010;1799(10–12):717–25.
[39] West AC, Johnstone RW. New and emerging HDAC inhibitors for cancer treatment. J Clin Invest 2014;124(1):30–9.
[40] Vendetti FP, Rudin CM. Epigenetic therapy in non-small-cell lung cancer: targeting DNA methyltransferases and histone deacetylases. Expert Opin Biol Ther 2013;13(9):1273–85.
[41] Coolen MW, Stirzaker C, Song JZ, Statham AL, Kassir Z, Moreno CS, et al. Consolidation of the cancer genome into domains of repressive chromatin by long-range epigenetic silencing (LRES) reduces transcriptional plasticity. Nat Cell Biol 2010;12(3):235–46.
[42] Johnstone RW. Histone-deacetylase inhibitors: novel drugs for the treatment of cancer. Nat Rev Drug Discov 2002;1(4):287–99.
[43] West AC, Smyth MJ, Johnstone RW. The anticancer effects of HDAC inhibitors require the immune system. Oncoimmunology 2014;3(1):e27414.
[44] Bolden JE, Peart MJ, Johnstone RW. Anticancer activities of histone deacetylase inhibitors. Nat Rev Drug Discov 2006;5(9):769–84.
[45] Griffiths EA, Gore SD. DNA methyltransferase and histone deacetylase inhibitors in the treatment of myelodysplastic syndromes. Semin Hematol 2008;45(1):23–30.

[46] Belinsky SA, Grimes MJ, Picchi MA, Mitchell HD, Stidley CA, Tesfaigzi Y, et al. Combination therapy with vidaza and entinostat suppresses tumor growth and reprograms the epigenome in an orthotopic lung cancer model. Cancer Res 2011;71(2):454–62.
[47] Fiskus W, Buckley K, Rao R, Mandawat A, Yang Y, Joshi R, et al. Panobinostat treatment depletes EZH2 and DNMT1 levels and enhances decitabine mediated de-repression of JunB and loss of survival of human acute leukemia cells. Cancer Biol Ther 2009;8(10):939–50.
[48] Gagnon J, Shaker S, Primeau M, Hurtubise A, Momparler RL. Interaction of 5-aza-2′-deoxycytidine and depsipeptide on antineoplastic activity and activation of 14-3-3sigma, E-cadherin and tissue inhibitor of metalloproteinase 3 expression in human breast carcinoma cells. Anticancer Drugs 2003;14(3):193–202.
[49] Shaker S, Bernstein M, Momparler LF, Momparler RL. Preclinical evaluation of antineoplastic activity of inhibitors of DNA methylation (5-aza-2′-deoxycytidine) and histone deacetylation (trichostatin A, depsipeptide) in combination against myeloid leukemic cells. Leuk Res 2003;27(5):437–44.
[50] Primeau M, Gagnon J, Momparler RL. Synergistic antineoplastic action of DNA methylation inhibitor 5-AZA-2′-deoxycytidine and histone deacetylase inhibitor depsipeptide on human breast carcinoma cells. Int J Cancer 2003;103(2):177–84.
[51] Lemaire M, Momparler LF, Farinha NJ, Bernstein M, Momparler RL. Enhancement of antineoplastic action of 5-aza-2′-deoxycytidine by phenylbutyrate on L1210 leukemic cells. Leuk Lymphoma 2004;45(1):147–54.
[52] Shaker S, Bernstein M, Momparler RL. Antineoplastic action of 5-aza-2′-deoxycytidine (Dacogen) and depsipeptide on raji lymphoma cells. Oncol Rep 2004;11(6):1253–6.
[53] Yang H, Hoshino K, Sanchez-Gonzalez B, Kantarjian H, Garcia-Manero G. Antileukemia activity of the combination of 5-aza-2′-deoxycytidine with valproic acid. Leuk Res 2005;29(7):739–48.
[54] Boivin AJ, Momparler LF, Hurtubise A, Momparler RL. Antineoplastic action of 5-aza-2′-deoxycytidine and phenylbutyrate on human lung carcinoma cells. Anticancer Drugs 2002;13(8):869–74.
[55] Bovenzi V, Momparler RL. Antineoplastic action of 5-aza-2′-deoxycytidine and histone deacetylase inhibitor and their effect on the expression of retinoic acid receptor beta and estrogen receptor alpha genes in breast carcinoma cells. Cancer Chemother Pharmacol 2001;48(1):71–6.
[56] Chen MY, Liao WS, Lu Z, Bornmann WG, Hennessey V, Washington MN, et al. Decitabine and suberoylanilide hydroxamic acid (SAHA) inhibit growth of ovarian cancer cell lines and xenografts while inducing expression of imprinted tumor suppressor genes, apoptosis, G2/M arrest, and autophagy. Cancer 2011;117(19):4424–38.
[57] Kalac M, Scotto L, Marchi E, Amengual J, Seshan VE, Bhagat G, et al. HDAC inhibitors and decitabine are highly synergistic and associated with unique gene-expression and epigenetic profiles in models of DLBCL. Blood 2011;118(20):5506–16.
[58] Tellez CS, Grimes MJ, Picchi MA, Liu Y, March TH, Reed MD, et al. SGI-110 and entinostat therapy reduces lung tumor burden and reprograms the epigenome. Int J Cancer J 2014;135(9):2223–31.
[59] Gore SD, Baylin S, Sugar E, Carraway H, Miller CB, Carducci M, et al. Combined DNA methyltransferase and histone deacetylase inhibition in the treatment of myeloid neoplasms. Cancer Res 2006;66(12):6361–9.
[60] Maslak P, Chanel S, Camacho LH, Soignet S, Pandolfi PP, Guernah I, et al. Pilot study of combination transcriptional modulation therapy with sodium phenylbutyrate and 5-azacytidine in patients with acute myeloid leukemia or myelodysplastic syndrome. Leukemia 2006;20(2):212–17.
[61] Garcia-Manero G, Kantarjian HM, Sanchez-Gonzalez B, Yang H, Rosner G, Verstovsek S, et al. Phase 1/2 study of the combination of 5-aza-2′-deoxycytidine with valproic acid in patients with leukemia. Blood 2006;108(10):3271–9.
[62] Braiteh F, Soriano AO, Garcia-Manero G, Hong D, Johnson MM, Silva Lde P, et al. Phase I study of epigenetic modulation with 5-azacytidine and valproic acid in patients with advanced cancers. Clin Cancer Res 2008;14(19):6296–301.

[63] Voso MT, Santini V, Finelli C, Musto P, Pogliani E, Angelucci E, et al. Valproic acid at therapeutic plasma levels may increase 5-azacytidine efficacy in higher risk myelodysplastic syndromes. Clin Cancer Res 2009;15(15):5002–7.

[64] Lin J, Gilbert J, Rudek MA, Zwiebel JA, Gore S, Jiemjit A, et al. A phase I dose-finding study of 5-azacytidine in combination with sodium phenylbutyrate in patients with refractory solid tumors. Clin Cancer Res 2009;15(19):6241–9.

[65] Issa JP, Garcia-Manero G, Huang X, Cortes J, Ravandi F, Jabbour E, et al. Results of phase 2 randomized study of low-dose decitabine with or without valproic acid in patients with myelodysplastic syndrome and acute myelogenous leukemia. Cancer 2015;121(4):556–61.

[66] Prebet T, Sun Z, Figueroa ME, Ketterling R, Melnick A, Greenberg PL, et al. Prolonged administration of azacitidine with or without entinostat for myelodysplastic syndrome and acute myeloid leukemia with myelodysplasia-related changes: results of the US leukemia intergroup trial E1905. J Clin Oncol 2014;32(12):1242–8.

[67] Kirschbaum M, Gojo I, Goldberg SL, Bredeson C, Kujawski LA, Yang A, et al. A phase 1 clinical trial of vorinostat in combination with decitabine in patients with acute myeloid leukaemia or myelodysplastic syndrome. Br J Haematol 2014;167(2):185–93.

[68] Juergens RA, Wrangle J, Vendetti FP, Murphy SC, Zhao M, Coleman B, et al. Combination epigenetic therapy has efficacy in patients with refractory advanced non-small cell lung cancer. Cancer Discov 2011;1(7):598–607.

[69] Fandy TE, Herman JG, Kerns P, Jiemjit A, Sugar EA, Choi SH, et al. Early epigenetic changes and DNA damage do not predict clinical response in an overlapping schedule of 5-azacytidine and entinostat in patients with myeloid malignancies. Blood 2009;114(13):2764–73.

[70] Foulks JM, Parnell KM, Nix RN, Chau S, Swierczek K, Saunders M, et al. Epigenetic drug discovery: targeting DNA methyltransferases. J Biomol Screen 2012;17(1):2–17.

[71] Côté S, Momparler RL. Antineoplastic action of all-trans retinoic acid and 5-aza-2′-deoxycytidine on human DLD-1 colon carcinoma cells. Cell Pharmacol 1995;2:221–8.

[72] Momparler RL, Dore BT, Momparler LF. Effect of 5-aza-2′-deoxycytidine and retinoic acid on differentiation and c-myc expression in HL-60 myeloid leukemic cells. Cancer Lett 1990;54(1–2):21–8.

[73] Grishina O, Schmoor C, Dohner K, Hackanson B, Lubrich B, May AM, et al. DECIDER: prospective randomized multicenter phase II trial of low-dose decitabine (DAC) administered alone or in combination with the histone deacetylase inhibitor valproic acid (VPA) and all-trans retinoic acid (ATRA) in patients >60 years with acute myeloid leukemia who are ineligible for induction chemotherapy. BMC Cancer 2015;15:430.

[74] Lall S. Primers on chromatin. Nat Struct Mol Biol 2007;14(11):1110–15.

[75] Huang J, Berger SL. The emerging field of dynamic lysine methylation of non-histone proteins. Curr Opin Genet Dev 2008;18(2):152–8.

[76] Vire E, Brenner C, Deplus R, Blanchon L, Fraga M, Didelot C, et al. The Polycomb group protein EZH2 directly controls DNA methylation. Nature 2006;439(7078):871–4.

[77] Miranda TB, Cortez CC, Yoo CB, Liang G, Abe M, Kelly TK, et al. DZNep is a global histone methylation inhibitor that reactivates developmental genes not silenced by DNA methylation. Mol Cancer Ther 2009;8(6):1579–88.

[78] Momparler RL, Idaghdour Y, Marquez VE, Momparler LF. Synergistic antileukemic action of a combination of inhibitors of DNA methylation and histone methylation. Leuk Res 2012;36(8):1049–54.

[79] Momparler RL, Cote S, Momparler LF, Idaghdour Y. Epigenetic therapy of acute myeloid leukemia using 5-aza-2′-deoxycytidine (decitabine) in combination with inhibitors of histone methylation and deacetylation. Clin Epigenetics 2014;6(1):19.

[80] Unland R, Borchardt C, Clemens D, Kool M, Dirksen U, Fruhwald MC. Analysis of the antiproliferative effects of 3-deazaneoplanocin A in combination with standard anticancer agents in rhabdoid tumor cell lines. Anticancer Drugs 2015;26(3):301–11.

[81] Fiskus W, Wang Y, Sreekumar A, Buckley KM, Shi H, Jillella A, et al. Combined epigenetic therapy with the histone methyltransferase EZH2 inhibitor 3-deazaneplanocin A and the histone deacetylase inhibitor panobinostat against human AML cells. Blood 2009;114(13):2733–43.

[82] Avan A, Crea F, Paolicchi E, Funel N, Galvani E, Marquez VE, et al. Molecular mechanisms involved in the synergistic interaction of the EZH2 inhibitor 3-deazaneplanocin A with gemcitabine in pancreatic cancer cells. Mol Cancer Ther 2012;11(8):1735–46.

[83] Chen L, Deshpande AJ, Banka D, Bernt KM, Dias S, Buske C, et al. Abrogation of MLL-AF10 and CALM-AF10-mediated transformation through genetic inactivation or pharmacological inhibition of the H3K79 methyltransferase Dot1l. Leukemia 2013;27(4):813–22.

[84] Daigle SR, Olhava EJ, Therkelsen CA, Basavapathruni A, Jin L, Boriack-Sjodin PA, et al. Potent inhibition of DOT1L as treatment of MLL-fusion leukemia. Blood 2013;122(6):1017–25.

[85] Liu W, Deng L, Song Y, Redell M. DOT1L inhibition sensitizes MLL-rearranged AML to chemotherapy. PLoS ONE 2014;9(5):e98270.

[86] Chen CW, Koche RP, Sinha AU, Deshpande AJ, Zhu N, Eng R, et al. DOT1L inhibits SIRT1-mediated epigenetic silencing to maintain leukemic gene expression in MLL-rearranged leukemia. Nat Med 2015;21(4):335–43.

[87] Helin K, Dhanak D. Chromatin proteins and modifications as drug targets. Nature 2013;502(7472):480–8.

[88] Harris WJ, Huang X, Lynch JT, Spencer GJ, Hitchin JR, Li Y, et al. The histone demethylase KDM1A sustains the oncogenic potential of MLL-AF9 leukemia stem cells. Cancer Cell 2012;21(4):473–87.

[89] Maes T, Carceller E, Salas J, Ortega A, Buesa C. Advances in the development of histone lysine demethylase inhibitors. Curr Opin Pharmacol 2015;23:52–60.

[90] Sakane C, Okitsu T, Wada A, Sagami H, Shidoji Y. Inhibition of lysine-specific demethylase 1 by the acyclic diterpenoid geranylgeranoic acid and its derivatives. Biochem Biophys Res Commun 2014;444(1):24–9.

[91] Schulte JH, Lim S, Schramm A, Friedrichs N, Koster J, Versteeg R, et al. Lysine-specific demethylase 1 is strongly expressed in poorly differentiated neuroblastoma: implications for therapy. Cancer Res 2009;69(5):2065–71.

[92] Wang J, Hevi S, Kurash JK, Lei H, Gay F, Bajko J, et al. The lysine demethylase LSD1 (KDM1) is required for maintenance of global DNA methylation. Nat Genet 2009;41(1):125–9.

[93] Huang Y, Stewart TM, Wu Y, Baylin SB, Marton LJ, Perkins B, et al. Novel oligoamine analogues inhibit lysine-specific demethylase 1 and induce reexpression of epigenetically silenced genes. Clin Cancer Res 2009;15(23):7217–28.

[94] Fiskus W, Sharma S, Shah B, Portier BP, Devaraj SG, Liu K, et al. Highly effective combination of LSD1 (KDM1A) antagonist and pan-histone deacetylase inhibitor against human AML cells. Leukemia 2014;28(11):2155–64.

[95] Schenk T, Chen WC, Gollner S, Howell L, Jin L, Hebestreit K, et al. Inhibition of the LSD1 (KDM1A) demethylase reactivates the all-trans-retinoic acid differentiation pathway in acute myeloid leukemia. Nat Med 2012;18(4):605–11.

[96] Grasso CS, Tang Y, Truffaux N, Berlow NE, Liu L, Debily MA, et al. Functionally defined therapeutic targets in diffuse intrinsic pontine glioma. Nat Med 2015;21(6):555–9.

[97] Dawson MA, Kouzarides T, Huntly BJ. Targeting epigenetic readers in cancer. N Engl J Med 2012;367(7):647–57.

[98] Delmore JE, Issa GC, Lemieux ME, Rahl PB, Shi J, Jacobs HM, et al. BET bromodomain inhibition as a therapeutic strategy to target c-Myc. Cell 2011;146(6):904–17.

[99] Lai YS, Chen JY, Tsai HJ, Chen TY, Hung WC. The SUV39H1 inhibitor chaetocin induces differentiation and shows synergistic cytotoxicity with other epigenetic drugs in acute myeloid leukemia cells. Blood Cancer J 2015;5:e313.

[100] Coude MM, Braun T, Berrou J, Dupont M, Bertrand S, Masse A, et al. BET inhibitor OTX015 targets BRD2 and BRD4 and decreases c-MYC in acute leukemia cells. Oncotarget 2015;6:17698–712.

[101] Fang F, Balch C, Schilder J, Breen T, Zhang S, Shen C, et al. A phase 1 and pharmacodynamic study of decitabine in combination with carboplatin in patients with recurrent, platinum-resistant, epithelial ovarian cancer. Cancer 2010;116(17):4043–53.

[102] Bachmann PS, Piazza RG, Janes ME, Wong NC, Davies C, Mogavero A, et al. Epigenetic silencing of BIM in glucocorticoid poor-responsive pediatric acute lymphoblastic leukemia, and its reversal by histone deacetylase inhibition. Blood 2010;116(16):3013–22.

[103] Hegi ME, Diserens AC, Gorlia T, Hamou MF, de Tribolet N, Weller M, et al. MGMT gene silencing and benefit from temozolomide in glioblastoma. N Engl J Med 2005;352(10):997–1003.

[104] Plumb JA, Strathdee G, Sludden J, Kaye SB, Brown R. Reversal of drug resistance in human tumor xenografts by 2′-deoxy-5-azacytidine-induced demethylation of the hMLH1 gene promoter. Cancer Res 2000;60(21):6039–44.

[105] Strathdee G, MacKean MJ, Illand M, Brown R. A role for methylation of the hMLH1 promoter in loss of hMLH1 expression and drug resistance in ovarian cancer. Oncogene 1999;18(14):2335–41.

[106] Appleton K, Mackay HJ, Judson I, Plumb JA, McCormick C, Strathdee G, et al. Phase I and pharmacodynamic trial of the DNA methyltransferase inhibitor decitabine and carboplatin in solid tumors. J Clin Oncol 2007;25(29):4603–9.

[107] Glasspool RM, Brown R, Gore ME, Rustin GJ, McNeish IA, Wilson RH, et al. A randomised, phase II trial of the DNA-hypomethylating agent 5-aza-2′-deoxycytidine (decitabine) in combination with carboplatin vs carboplatin alone in patients with recurrent, partially platinum-sensitive ovarian cancer. Br J Cancer 2014;110(8):1923–9.

[108] Faderl S, Ravandi F, Huang X, Wang X, Jabbour E, Garcia-Manero G, et al. Clofarabine plus low-dose cytarabine followed by clofarabine plus low-dose cytarabine alternating with decitabine in acute myeloid leukemia frontline therapy for older patients. Cancer 2012;118(18):4471–7.

[109] Kadia TM, Faderl S, Ravandi F, Jabbour E, Garcia-Manero G, Borthakur G, et al. Final results of a phase 2 trial of clofarabine and low-dose cytarabine alternating with decitabine in older patients with newly diagnosed acute myeloid leukemia. Cancer 2015;121(14):2375–82.

[110] Ramalingam SS, Maitland ML, Frankel P, Argiris AE, Koczywas M, Gitlitz B, et al. Carboplatin and paclitaxel in combination with either vorinostat or placebo for first-line therapy of advanced non-small-cell lung cancer. J Clin Oncol 2010;28(1):56–62.

[111] Ramalingam SS, Parise RA, Ramanathan RK, Lagattuta TF, Musguire LA, Stoller RG, et al. Phase I and pharmacokinetic study of vorinostat, a histone deacetylase inhibitor, in combination with carboplatin and paclitaxel for advanced solid malignancies. Clin Cancer Res 2007;13(12):3605–10.

[112] Scherpereel A, Berghmans T, Lafitte JJ, Colinet B, Richez M, Bonduelle Y, et al. Valproate-doxorubicin: promising therapy for progressing mesothelioma. A phase II study. Eur Respir J 2011;37(1):129–35.

[113] Garcia-Manero G, Tambaro FP, Bekele NB, Yang H, Ravandi F, Jabbour E, et al. Phase II trial of vorinostat with idarubicin and cytarabine for patients with newly diagnosed acute myelogenous leukemia or myelodysplastic syndrome. J Clin Oncol 2012;30(18):2204–10.

[114] Kuendgen A, Knipp S, Fox F, Strupp C, Hildebrandt B, Steidl C, et al. Results of a phase 2 study of valproic acid alone or in combination with all-trans retinoic acid in 75 patients with myelodysplastic syndrome and relapsed or refractory acute myeloid leukemia. Ann Hematol 2005;84(Suppl. 1):61–6.

[115] Williams CB, Kambhampati S, Fiskus W, Wick J, Dutreix C, Ganguly S, et al. Preclinical and phase I results of decitabine in combination with midostaurin (PKC412) for newly diagnosed elderly or relapsed/refractory adult patients with acute myeloid leukemia. Pharmacotherapy 2013;33(12):1341–52.

[116] Lou YF, Zou ZZ, Chen PJ, Huang GB, Li B, Zheng DQ, et al. Combination of gefitinib and DNA methylation inhibitor decitabine exerts synergistic anti-cancer activity in colon cancer cells. PLoS One 2014;9(5):e97719.

[117] Li XY, Wu JZ, Cao HX, Ma R, Wu JQ, Zhong YJ, et al. Blockade of DNA methylation enhances the therapeutic effect of gefitinib in non-small cell lung cancer cells. Oncol Rep 2013;29(5):1975–82.

[118] Lainey E, Wolfromm A, Marie N, Enot D, Scoazec M, Bouteloup C, et al. Azacytidine and erlotinib exert synergistic effects against acute myeloid leukemia. Oncogene 2013;32(37):4331–42.
[119] Bauman J, Verschraegen C, Belinsky S, Muller C, Rutledge T, Fekrazad M, et al. A phase I study of 5-azacytidine and erlotinib in advanced solid tumor malignancies. Cancer Chemother Pharmacol 2012;69(2):547–54.
[120] Zhang W, Peyton M, Xie Y, Soh J, Minna JD, Gazdar AF, et al. Histone deacetylase inhibitor romidepsin enhances anti-tumor effect of erlotinib in non-small cell lung cancer (NSCLC) cell lines. J Thorac Oncol: Off Publ Int Assoc Study Lung Cancer 2009;4(2):161–6.
[121] Edwards A, Li J, Atadja P, Bhalla K, Haura EB. Effect of the histone deacetylase inhibitor LBH589 against epidermal growth factor receptor-dependent human lung cancer cells. Mol Cancer Ther 2007; 6(9):2515–24.
[122] Reguart N, Rosell R, Cardenal F, Cardona AF, Isla D, Palmero R, et al. Phase I/II trial of vorinostat (SAHA) and erlotinib for non-small cell lung cancer (NSCLC) patients with epidermal growth factor receptor (EGFR) mutations after erlotinib progression. Lung Cancer 2014;84(2):161–7.
[123] Gray JE, Haura E, Chiappori A, Tanvetyanon T, Williams CC, Pinder-Schenck M, et al. A phase I, pharmacokinetic, and pharmacodynamic study of panobinostat, an HDAC inhibitor, combined with erlotinib in patients with advanced aerodigestive tract tumors. Clin Cancer Res 2014;20(6):1644–55.
[124] Lee DH, Qi J, Bradner JE, Said JW, Doan NB, Forscher C, et al. Synergistic effect of JQ1 and rapamycin for treatment of human osteosarcoma. Int J Cancer 2015;136(9):2055–64.
[125] Boi M, Gaudio E, Bonetti P, Kwee I, Bernasconi E, Tarantelli C, et al. The BET bromodomain inhibitor OTX015 affects pathogenetic pathways in preclinical B-cell Tumor models and synergizes with targeted drugs. Clin Cancer Res 2015;21(7):1628–38.
[126] Sigalotti L, Fratta E, Coral S, Maio M. Epigenetic drugs as immunomodulators for combination therapies in solid tumors. Pharmacol Ther 2014;142(3):339–50.
[127] Wargo JA, Robbins PF, Li Y, Zhao Y, El-Gamil M, Caragacianu D, et al. Recognition of NY-ESO-1+ tumor cells by engineered lymphocytes is enhanced by improved vector design and epigenetic modulation of tumor antigen expression. Cancer Immunol Immunother 2009;58(3):383–94.
[128] Horing E, Podlech O, Silkenstedt B, Rota IA, Adamopoulou E, Naumann U. The histone deacetylase inhibitor trichostatin a promotes apoptosis and antitumor immunity in glioblastoma cells. Anticancer Res 2013;33(4):1351–60.
[129] Gollob JA, Sciambi CJ, Peterson BL, Richmond T, Thoreson M, Moran K, et al. Phase I trial of sequential low-dose 5-aza-2′-deoxycytidine plus high-dose intravenous bolus interleukin-2 in patients with melanoma or renal cell carcinoma. Clin Cancer Res 2006;12(15):4619–27.
[130] Kozar K, Kaminski R, Switaj T, Oldak T, Machaj E, Wysocki PJ, et al. Interleukin 12-based immunotherapy improves the antitumor effectiveness of a low-dose 5-Aza-2′-deoxycitidine treatment in L1210 leukemia and B16F10 melanoma models in mice. Clin Cancer Res 2003;9(8):3124–33.
[131] Christiansen AJ, West A, Banks KM, Haynes NM, Teng MW, Smyth MJ, et al. Eradication of solid tumors using histone deacetylase inhibitors combined with immune-stimulating antibodies. Proc Natl Acad Sci USA 2011;108(10):4141–6.
[132] Wrangle J, Wang W, Koch A, Easwaran H, Mohammad HP, Vendetti F, et al. Alterations of immune response of non-small cell lung cancer with azacytidine. Oncotarget 2013;4(11):2067–79.
[133] Raynal NJ, Momparler L, Charbonneau M, Momparler RL. Antileukemic activity of genistein, a major isoflavone present in soy products. J Nat Prod 2008;71(1):3–7.
[134] Sarkar FH, Li Y. Using chemopreventive agents to enhance the efficacy of cancer therapy. Cancer Res 2006;66(7):3347–50.
[135] Banerjee S, Li Y, Wang Z, Sarkar FH. Multi-targeted therapy of cancer by genistein. Cancer Lett 2008;269(2):226–42.
[136] Raynal NJ, Charbonneau M, Momparler LF, Momparler RL. Synergistic effect of 5-Aza-2′-deoxycytidine and genistein in combination against leukemia. Oncol Res 2008;17(5):223–30.

CHAPTER

PREDICTING AND MONITORING RESPONSES TO EPIGENETIC DRUGS

15

Semira Sheikh, Mina Bekheet, Heidi Olzscha and Nicholas B. La Thangue
Laboratory of Cancer Biology, Department of Oncology, University of Oxford, Oxford, UK

CHAPTER OUTLINE

G. Egger & P. Arimondo (Eds): Drug Discovery in Cancer Epigenetics. DOI: http://dx.doi.org/10.1016/B978-0-12-802208-5.00015-1

15.1 INTRODUCTION

Epigenetics is commonly defined as the genetic information that is inherited without any change in the underlying nucleotide sequence of DNA. Techniques such as whole-genome sequencing have greatly improved our understanding of the human epigenome and led to the discovery of a complex network of epigenetic modifications that are frequently disrupted in cancer cells [1,2]. The study of epigenetic changes in cancer has therefore rapidly evolved into an area that holds great promise for unraveling important mechanisms involved in carcinogenesis and the discovery of targeted cancer therapies. Further, characteristic epigenetic patterns associated with certain cancers may be developed into biomarkers with diagnostic, predictive, or prognostic value that could allow for more rational treatment approaches in the context of personalized medicine [1,3,4]. In this review we will provide an overview of important epigenetic modifications that have been identified in cancer and discuss how these might be applied in a clinical context, focusing both on developing targeted cancer therapies and on biomarkers.

15.1.1 EPIGENETIC MODIFICATIONS AND CHROMATIN CONTROL

The most commonly described epigenetic modification is DNA methylation, stemming from the original discovery of global hypomethylation in human cancers [5]. This was followed by the identification of hypermethylation of the gene promoter regions of tumor suppressor genes [6,7] and the finding that microRNA (miRNA) genes can be inactivated by DNA methylation [8]. Moreover, these changes occur within the context of other epigenetic modifications, such as the reversible acetylation and methylation of histone proteins, which influence and regulate chromatin and are sometimes collectively referred to as the "histone code" [9].

DNA methylation and histone posttranslational modifications (PTMs) together form a complex interdependent network that is involved in chromatin control, and alterations in epigenetic regulators leading to aberrant chromatin regulation have been shown to play an important role in the initiation and progression of cancer.

Chromatin is the physiological scaffold of DNA. It is a highly complex yet dynamic structure, which is made up of DNA, histone and nonhistone proteins, and exists in two functional conformational states: heterochromatin and euchromatin. Heterochromatin is usually considered a densely compacted and therefore transcriptionally inert state, whereas euchromatin is considered a decondensed, transcriptionally active state [3]. The dynamic remodeling of chromatin is thought to underlie most, if not all, DNA-mediated processes, and in particular gene transcription, DNA replication and repair, and apoptosis [1,3,10].

The functional unit of chromatin is the nucleosome. This consists of approximately 147 DNA base pairs wrapped around an octamer of core histone proteins H2A, H2B, H3, and H4, which are highly conserved. Nucleosomes are linked together by linker DNA. Histone proteins are small, highly evolutionarily conserved basic proteins which are found in the nuclei of most eukaryotic cells [11,12]. The four core histones each contain a globular domain which enables dimerization, and a flexible charged N-terminus, also referred to as the histone tail, which may be modified by a wide range of reversible PTMs. These can be divided into at least eight different classes, including acetylation of lysine, methylation of lysine or arginine, phosphorylation of serine or threonine, lysine ubiquitination, lysine sumoylation, and arginine deamination, where each modification is mediated by a specific enzyme [3,10,11].

15.1.2 EPIGENETIC READERS, WRITERS, AND ERASERS

Epigenetic modifications can alter the accessibility of chromatin based on both direct structural effects and their ability to facilitate protein interactions, for example, the recruitment and assembly of other multicomponent protein complexes or the crosstalk between different epigenetic marks, ultimately leading to a precise system that serves to fine-tune gene expression through epigenetic writers, erasers, and readers [3,9,11]: Thus, epigenetic writers (DNA methyltransferases (DNMTs) and histone acetyltransferases (HAT)) lay down epigenetic marks which can be modified or abolished by epigenetic erasers (histone demethylases (HDMs) or histone deacetylases (HDAC)), whereas epigenetic readers contain specific protein–protein interaction modules which allow them to recognize different covalent modifications on chromatin and to dock at specific regions within the genome [13].

15.1.3 TARGETED EPIGENETIC THERAPIES AND BIOMARKERS

Many of the epigenetic actors (or enzymes) are now being explored as potential therapeutic targets in drug design (Figure 15.1; Table 15.1). With the growing recognition that aberrant epigenetic modifications are involved in the initiation and progression of cancer [1,14], the possibility of reversing these changes presents an appealing opportunity for developing targeted inhibitors which can be used as epigenetic drugs. Moreover, the identification of epigenetic alterations that are specific to certain cancer types and subtypes may also allow the parallel development of biomarkers with companion diagnostic, prognostic, and therapeutic implications.

15.2 EPIGENETICS AND CANCER

15.2.1 DNA METHYLATION AND CANCER

It is becoming increasingly apparent that changes in epigenetic modifications play a significant role in both early and late tumor development [15]. In humans, DNA methylation refers to the enzymatic addition of a methyl ($-CH_3$) group onto cytosines that precede guanines (also termed dinucleotide CpGs) by DNMTs. CpG-rich regions are not evenly distributed in the genome; rather, they most frequently occur in the promoter regions of many genes. In normal cells, CpG islands in gene promoter regions are not usually methylated, whereas they are methylated outside gene regulatory regions [1,14]. Altered methylation patterns are amongst the first epigenetic phenomena described in human cancer, and have been observed frequently during the early stages of carcinogenesis [1,5]. Thus, hypomethylation is thought to allow expression of oncogenes; a decrease in the 5-methylcytosine content has been described widely in various cancers, in particular in colon cancer when compared to normal tissue [16]. Further studies suggest that there is extensive hypomethylation in genomic regions that are noncoding regions [17], and that the degree of DNA hypomethylation increases as benign cells transform into invasive cancer cells [14,18]. Interestingly,

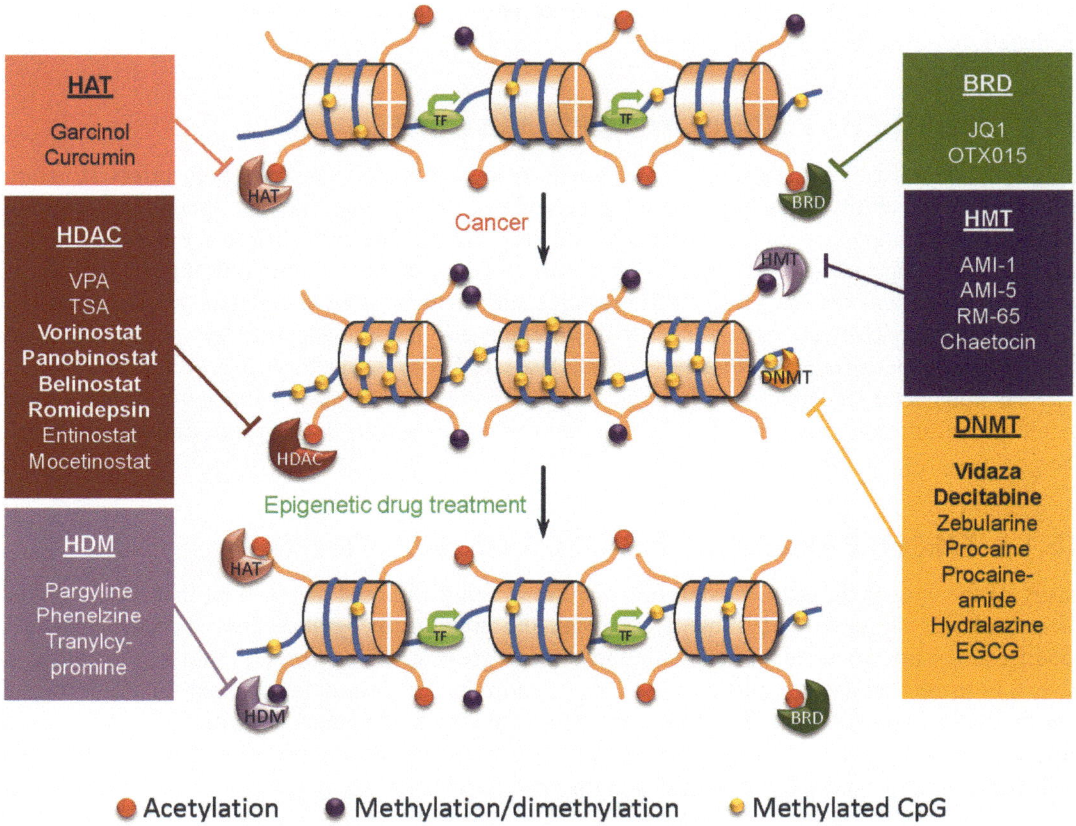

FIGURE 15.1

Aberrant epigenetic modifications in cancer and the effect of epigenetic drugs. Epigenetic modulating enzymes influence chromatin structure. Euchromatin allows TF to bind and transcribe the accessible genes (green arrows). Chromatin structure is mainly controlled by DNA methylation (yellow circles) and histone modifications such as lysine methylation (purple circles) or lysine acetylation (red circles.) DNMTs catalyze the methylation of CpG dinucleotides, resulting in the silencing of genes. Histone acetylation is catalyzed by HATs and deacetylated by HDACs. Deacetylation of chromatin leads to its compaction in a heterochromatin state with reduced transcription. Histones can also be methylated by HMTs and demethylated by HDMs. Readers of the epigenetic changes, for instance BRD-containing proteins, can mediate the information. FDA-approved epigenetic drugs are marked in bold.

it has recently been shown in melanoma that whilst a decreased 5-methylcytosine content in itself is not sufficient to distinguish a benign nevus from melanoma [19], 5-methylcytosine is converted to 5-hydroxymethylcytosine by the ten-eleven translocation (TET) family of DNA hydroxylases, and that a loss in 5-hydroxymethylcytosine in melanoma tumors was of both diagnostic and prognostic significance. The decrease in 5-hydroxymethylcytosine was shown to be related to the downregulation of isocitrate dehydrogenase 2 (IDH2) and TET enzymes, and the reintroduction of active IDH2 or TET in this context was shown to suppress melanoma growth in an animal model, highlighting a potential role of epigenetic modifiers as drivers of carcinogenesis [20].

Table 15.1 Epigenetic Modifiers as Emerging Targets for Cancer Therapy

Epigenetic Modifier	Epigenetic Protein Example	Deregulation in Cancer	Tumor Type	Example of Targeted Agents
DNMTs	DNMT1 DNMT3A DNMT3B	Mainly overexpression, some mutations	Multiple solid and hematologic (e.g., CRC, AML)	Nucleoside DNA methylation inhibitors: • Azacitidine • Decitabine • Zebularine Nonnucleoside DNA methylation inhibitors: • Procaine • Procaine-amide • Hydralazine DNMT1 antisense oligonucleotide: • MG 98
HAT	p300 CBP pCAF MOZ Tip60	Mutations, translocations, deletions	Multiple solid and hematologic (e.g., CRC, APML)	Bisubstrate inhibitors: • Lys-CoA 1 (specific p300 inhibitor) • H3-CoA-20 (pCAF inhibitor) Natural products: • Curcumin • Garcinol Synthetic inhibitors: • CTK7 • LTK-14 (selective inhibition of CBP/p300)
HDAC	HDAC 1–11 SIRT 1–7	Mainly overexpression, some mutations, deletions	Multiple solid and hematologic (e.g., glioblastoma, prostate, breast)	Hydroxamic acids: • Vorinostat • Belinostat • Panobinostat • TSA Cyclic tetrapeptides and depsipeptide: • Romidepsin Benzamides: • Entinostat • Mocetinostat Aliphatic acids: • VPA
HMT	DOT1L EZH2 BLIMP1	Amplification, overexpression, some mutations	Multiple solid and hematologic (e.g., breast, gastric, AML, lymphoma)	DOT1L enzyme/H3K79 modification: • EPZ004777 • AMI-1 • AMI-5 Set7/9 enzyme/H3K4me1/Tat/ ER modifications: • PDB 4e47 • AMI-1

(Continued)

Table 15.1 Epigenetic Modifiers as Emerging Targets for Cancer Therapy *Continued*

Epigenetic Modifier	Epigenetic Protein Example	Deregulation in Cancer	Tumor Type	Example of Targeted Agents
				• AMI-5 • RM-65 • Chaetocin EZH2 enzyme/H3K27me1/2/3 modification: • GSK343
HDM	LSD-1 JmJD3	Amplification, overexpression	Breast cancer, leukemia	Selective LSD-1 inhibitor: • GSK-LSD-1 • Pargyline • Phenelzine • Tranylcypromine Selective jumonji H3K27 inhibitor: • GSK-J1
BET/BRD proteins	BRD2 BRD3 BRD4	Mutation?	Prostate cancer, leukemia, NMC	Pan-BET/BDI: • OTX015 • I-BET 762 Selective BRD2,-3,-4 inhibitor: • I-BET 151 BRD4 inhibitor: • JQ1

Conversely, DNA hypermethylation can also be found in cancer and occurs when CpG islands in gene promoter regions become aberrantly methylated, leading to transcriptional silencing of affected genes. Hypermethylation in the promoter regions of the retinoblastoma (Rb), von Hippel Lindau (vHL) and breast-cancer susceptibility gene 1 (BRCA1) genes have all been described [6,21,22]. Transcriptional inactivation secondary to hypermethylation of CpG islands occurs due to different mechanisms, for instance by direct inhibition of the binding of transcription factors (TF) such as E2 promoter binding factor (E2F), cAMP responsive element binding protein (CREB), and nuclear factor kappa-light-chain-enhancer of activated B cells (NFκB), by cooperation with somatic genetic mutations, or by crosstalk with other chromatin-modifying elements. Thus, dysregulation of histone-modifying proteins can cause aberrant DNA methylation: Mutations in the histone lysine N-methyltransferase enhancer of zeste 2 polycomb repressive complex 2 subunit (EZH2), which is part of the polycomb repressive complex 2 (PRC2) and catalyzes the trimethylation of histone 3 lysine 27 (H3K27), have been described to lead to a gain of function that causes excessive trimethylation of H3K27 and repression of PRC2 target genes in lymphoma [23,24]. This process is thought to occur by cooperation with aberrant DNA methylation which causes gene silencing by chromatin compaction [14,25].

Of interest, studies in different types of cancers, including colon, hepatocellular, and breast cancers, have shown that different cancers exhibit different and tumor-specific methylation patterns [14,26,27]. In colorectal cancer (CRC), for instance, the existence of a subgroup of patients with a high frequency of methylated genes led to the identification of a unique CpG island methylator phenotype (CIMP) [14]. Methylation patterns therefore may be regarded as candidates for biomarker development with both diagnostic and prognostic relevance, as well as targets for epigenome-directed therapies as discussed below. An interesting question that remains is what the significance of differential methylation patterns in cancer and their role in carcinogenesis is. This is beginning to be addressed as more tools for analyzing genome-wide methylation patterns become available. One study that looked at chromosome-wide and promoter-specific methylation patterns in both normal and transformed human cells found that hypomethylation was mainly detected in gene-poor regions of transformed cells, whereas promoter hypermethylation was seen at a lower level than expected in a colon cancer cell line, giving rise to the hypothesis that promoter hypermethylation may firstly be a less common occurrence in cancer than expected, and secondly may therefore be a highly selective event [17]. Furthermore, DNA hypomethylation and hypermethylation have been proposed to arise independently of each other and may play different roles in cancer progression [28], and further mapping of the epigenome in many cancers will therefore be highly relevant in defining their underlying pathology.

15.2.2 HISTONE ACETYLATION AND CANCER

Histone acetylation, and in particular the acetylation status of histones H3 and H4, appear to play a key role in chromatin assembly, transcriptional regulation, and gene expression [3,29]. HAT acetylate proteins on their N-epsilon lysine residues, whereas HDAC provide the deacetylation event. In general, histone acetylation has been associated with an open chromatin structure and an active transcriptional state, whereas the removal of acetyl groups by HDACs has been shown to lead to compaction of nucleosomes and a closed chromatin structure, and consequently the repression of gene transcription [1,15].

Both HAT and HDAC have become implicated in the deregulation of epigenetic events during carcinogenesis in solid as well as hematological malignancies [9,10]. Aberrant histone acetylation is thought to occur due to a variety of mechanisms including chromosomal translocations, specific somatic gene mutations, and the abnormal recruitment and localization of epigenetic modifiers mediating particular acetylation events [1,11]. For instance, CREB-binding protein (CBP), a HAT and bromodomain (BRD)-containing protein that acts as a coactivator of transcription and regulates chromatin remodeling, becomes aberrantly recruited to the promoter regions of genes involved in cell division and proliferation [3]. Another example is the chromosomal translocation t(8;16) which results in a fusion of monocytic zinc finger protein (MOZ) and CBP, which enhances the intrinsic ability of MOZ to upregulate transcription of NFκB target genes, and has been identified in some types of acute leukemia [30]. The tat-interacting protein Tip60 is a HAT, which has been implicated in mediating an important acetylation-dependent switch from cell cycle arrest to apoptosis by means of regulating the acetylation status of p53. Thus, the acetylation of lysine 120 (K120) of the DNA-binding domain of p53 can be catalyzed by Tip60 and occurs rapidly after DNA damage. Mutations in K120, from lysine to arginine (K120R), as well as decreased expression of Tip60 through monoallelic loss have been described in human cancer. In cell-based studies, a K120R mutant was shown to induce a state of defective apoptosis with reduced transcription of proapoptotic target genes [3,10,31].

By contrast, HDACs are not generally somatically mutated in cancer, but rather overexpressed or aberrantly localized [10]. In a number of solid tumors, such as prostate, cervical and breast cancers, overexpression of HDAC1 and HDAC2 has been described. This finding was linked with more aggressive subtypes in patients with CRCs, and *in vitro*, shRNA knockdown of HDAC1 and HDAC2 as well as HDAC inhibition was effective at suppressing growth of cancer cells [32]. Deregulated HDAC activity has also been demonstrated in acute promyelocytic leukemia (APML), in which the retinoic acid receptor gene on chromosome 15 is fused with the promyelocytic leukemia (PML) gene on chromosome 17, leading to the characteristic t(15;17). This results in a fusion PML protein which binds HDAC with high affinity, leading to transcriptional silencing of retinoid target genes and a block in terminal myeloid differentiation of the leukemic cells [10,33–35].

15.3 EPIGENETIC THERAPIES

To date, six drugs designed to target epigenetic modifications have been approved for the treatment of hematological cancers by the US Food and Drug Administration (FDA) (Table 15.2): Two of

Table 15.2 FDA-Approved Drugs that Target the Cancer Epigenome

Drug	Mechanism of Action	Indication Approved for	Response Criteria	References
Azacytidine	DNMT inhibitor	• Myelodysplastic Syndromes	• Improvement in peripheral blood counts • Improvement in bone marrow blast percentage	[164]
Decitabine	DNMT inhibitor	• Myelodysplastic Syndromes	• Improvement in peripheral blood counts • Improvement in bone marrow blast percentage	[82]
Vorinostat	HDAC inhibitor	• Cutaneous T-cell lymphoma	• Improvement in overall % of disease manifestations by imaging, flow cytometry and Physician's Global Assessment score	[165,166]
Romidepsin	HDAC inhibitor	• Cutaneous T-cell lymphoma (in patients who have received at least one prior systemic therapy) • Peripheral T-cell lymphoma (in patients who have received at least one prior systemic therapy)	• Improvement of skin, visceral and bone marrow involvement • Flow cytometry for circulating tumour cells	[112,167]
Belinostat	HDAC inhibitor	• Peripheral T-cell lymphoma (in patients with relapsed or refractory disease)	• Improvement of nodal disease on imaging	[168,169]
Panobinostat	HDAC inhibitor	• Multiple myeloma (for patients with pre-treated disease, in combination with bortezomib and dexamethasone)	• Improvement as defined by International Myeloma Working Group Criteria	[119]

Table 15.3 FDA-Approved Drugs That Target the Cancer Epigenome

Combination Therapy	Trial Example	Type of Malignancy	Phase	NCT Number
DNMT inhibitor + conventional chemotherapy	Azacitidine + cisplatin	Solid (nonsmall-cell lung cancer and squamous cell cancer of the head and neck)	I	00901537
DNMT inhibitor + targeted antibody	Azacitidine + gemtuzumab ozogamicin	Hematological malignancy (acute leukemia)	II	00658814
DNMT inhibitor + EGFR inhibitor	Azacitidine + erlotinib	Solid (advanced)	I	00996515
DNMT inhibitor + HDI	Azacitidine + romidepsin	Solid (nonsmall-cell lung cancer)	I	01537744
HDI + conventional chemotherapy	Vorinostat + CHOP chemotherapy	Hematological (untreated T-cell lymphoma)	I/II	00787527
HDI + radiation	Vorinostat + radiotherapy	Brain metastases	I	00838929
HDI + proteasome inhibitor	Belinostat + bortezomib	Hematological (MM)	II	00431340
HDI + "IMiD"	Panobinostat + lenalidomide	Hematological (Hodgkin's lymphoma)	II	01460940

these drugs, 5-azacytidine (azacitidine, tradename Vidaza) and 5-aza-2-deoxycytidine (decitabine, tradename Dacogen), target DNMT, and have been approved for the treatment of myelodysplastic syndromes (MDS) and acute myeloid leukemia (AML) [36,37].

Vorinostat (SAHA, tradename Zolinza), romidepsin (FK228/depsipeptide, tradename Istodax), belinostat (PXD101, tradename Beleodaq), and panobinostat (LBH-589, tradename Farydaq) are HDAC inhibitors approved for the treatment of cutaneous T-cell lymphoma (CTCL), peripheral T-cell lymphoma (PTCL), and multiple myeloma (MM), respectively [38]. Several other drugs with activity against epigenetic enzymes are currently undergoing preclinical and clinical investigation, both in solid and hematological malignancies. All these compounds are also tested in combination with a variety of other anticancer agents and modalities such as radiation, and some examples are provided in Table 15.3. However, despite very promising preclinical results for some epigenetic drugs, our grasp on how they work is still rudimentary and there is a pressing need for a better understanding of how these drugs exert their activity.

15.3.1 DNMT INHIBITORS

DNMT are responsible for copying the methylation pattern of a particular parent strand of DNA to its daughter strands but may also respond to exogenous changes and add extra methyl groups [15,38]. DNMT1, the most abundant DNMT enzyme, maintains methylation of newly synthesized DNA strands, whereas DNMT3A and DNMT3B are involved in *de novo* methylation occurring in CpG islands, an event that has been linked with gene silencing.

15.3.1.1 Mechanisms of DNMT inhibitor action

Azacitidine and decitabine are both classed as cytosine nucleoside analogs; they inhibit DNMT by incorporating into nucleic acids during the S-phase of DNA replication, then forming irreversible covalent bonds with DNMT leading to DNMT trapping, and ultimately blocking its function and causing its depletion [26,36]. Because the inhibitors integrate into DNA strands during replication, more rapidly dividing cells are more susceptible to their effects [38]. *In vitro* studies of azacitidine in breast cancer cell lines have demonstrated that DNA remethylation was effectively blocked at 23 out of 26 genes that were tested and previously known to be hypermethylated [39,40]. Furthermore, select genes which display DNA hypermethylation in breast cancer cells were also shown to have RNA polymerase II stalling which could be reversed through treatment with azacitidine and also led to an increase in gene transcript levels [39].

Preclinical studies in MDS indicate that both azacitidine and decitabine cause hypomethylation as well as cytotoxic effects. Further, in mouse xenograft models derived from breast cancer cells, tumor size decreased after treatment with azacitidine, although it remains unclear whether the degree of hypomethylation achieved with drug treatment is predictive of clinical response in humans [15,36].

The reactivation of previously silenced genes such as TNF-related apoptosis-inducing ligand (TRAIL) after treatment of cancer cell lines with decitabine has led to the concept that DNMT inhibitors may be able to reprogram the microenvironment of a tumor from a chemotherapy- or hormone-refractory to a nonrefractory state, effectively leading to resensitization of a tumor [26,37].

Despite sharing structural similarities, azacitidine and decitabine have important functional differences: Decitabine is a deoxyribonucleoside which is incorporated into DNA, whereas azacitidine, on the other hand, is a ribonucleoside which is only incorporated into 10% of DNA, whereas the other 90% of the drug is incorporated into RNA. Decitabine has a more potent demethylating activity at similar dose concentrations relative to azacitidine but studies in a panel of cancer cell lines showed a poor correlation between sensitivity of cells to decitabine as compared to azacitidine [41]. Resistance of cancer cell lines to decitabine appears to be more common, and treatment with either agent has been reported to cause different gene expression profiles in various leukemia cell lines [36].

15.3.1.2 Trial evidence leading to DNMT inhibitor approval

Azacitidine and decitabine are approved for the treatment of MDS or AML in patients who are otherwise ineligible for high-dose chemotherapy. Both drugs were found to have activity in large randomized clinical trials although they have never been compared prospectively against each other [36,42]. Azacitidine was shown to improve overall survival in a cohort of patients with MDS (on azacitidine, the time to development of AML or death was 21 months compared to 12 months with best supportive care, $P = 0.007$), whereas decitabine was found to improve blood count parameters and bone marrow blast percentage without an overall survival benefit (on decitabine, time to AML or death was between 8.8 and 12.1 months compared to 10.1 and 7.8 months in two different trials, respectively, $P > 0.05$). Overall response rates (ORRs) were low, with only up to 50% of patients experiencing hematologic improvement with consequent reduction in dependency on blood products, and only up to 15% of patients achieving a complete clinical response [36,42]. Both agents are relatively well tolerated, with the main toxicities reported as being nausea, liver dysfunction, and fatigue.

It is also recognized that patients may not respond immediately to treatment with either DNMT inhibitor. The time to response, therefore, can be delayed, with the median time to response being up to 3 months. Treatment guidelines for azacitidine usually recommend a minimum number of six treatment cycles before a patient is found to be a nonresponder, and in the absence of side effects or disease progression, treatment in those patients achieving stable disease or remission, then continues indefinitely thereafter. This is based on trial evidence which shows that discontinuation of treatment can lead to a rapid loss of response and although retreatment is a possibility, the depth and durability of a subsequent response can be inferior [36].

15.3.1.3 Approaches to response prediction

There currently are no standardized biomarkers in use which allow for predicting which patients will respond to hypomethylating treatment in MDS or AML, but various approaches for response prediction have been used [43,44]: The French prognostic score for MDS evaluated a combination score of bone marrow blast percentage, abnormal karyotype and previous treatment regimens; this score has been validated in a cohort of several hundred MDS patients, and predicts a poor response to azacitidine treatment for those patients found to be in the intermediate-2 and high-risk groups [44,45]. However, when looking at a molecular level and examining the value of recurrently mutated myeloid malignancy genes such as *Tet methylcytosine dioxygenase 2 (TET2), Additional Sex Combs Like Transcriptional Regulator 1 (ASXL1), and Tumor Protein (TP53)* as potential response predictors, other groups have not been able to determine a clear pattern of mutations that was strongly associated with a lack of response to hypomethylating therapy [42]. Although the presence of some mutations, such as those leading to the loss of the tumor suppressor *TET2*, were shown to be able to predict whether patients with MDS were more likely to respond to a hypomethylating agent, there was no correlation with improved overall survival. The response rate did not differ with the agent used (i.e., either azacitidine or decitabine) [42]. It also remains unclear whether there is a definite relationship between hypomethylation and a clinical response in patients with MDS or other malignancies [36,44]. To this effect, DNA methylation profiling of single genes and combinations of genes in MDS has been carried out with conflicting results: Thus, for example, hypermethylation of $\mathrm{p15}^{\mathrm{INK4B}}$ was associated with a poor prognosis in a cohort of 50 patients but this correlation did not reach statistical significance [46]. A combination of four genes with promoter hypermethylation, $p15^{INK4B}$, *HIC1*, *CDH1*, and *ER*α, was found to be predictive of poor prognosis in patients with early-stage but not late-stage MDS: Hypermethylation of three or more genes was more frequently found in advanced MDS, and $p15^{INK4B}$ hypermethylation in particular was significantly associated with transformation to AML ($P < 0.05$) [47]. Although DNA methylation status in 10 genes in a cohort of 89 MDS patients undergoing decitabine treatment was able to identify a subgroup of patients with a specific CIMP, there was no significant association between methylation status at baseline and clinical response to treatment. However, a decrease in methylation with ongoing therapy appeared to be indicative of a better response in patients [43]. By contrast, pretreatment methylation levels of *long interspersed element (LINE-1)* were demonstrated to predict for a clinical response in elderly patients receiving azacitidine for the treatment of AML. In this small study of 40 patients, lower baseline levels of *LINE-1* methylation occurred in those patients who went on to achieve either a complete or partial remission, or hematologic improvement (22 vs 26 patients, $P = 0.06$), but subsequent treatment-related methylation changes were not found to be significant [48].

In summary, DNMT inhibitors are useful in the treatment of some hematological malignancies, but there are clear shortcomings in terms of predicting and monitoring outcome of therapy with these agents: In part, these may be related to study design, with some studies having relatively small sample sizes and including patients with very heterogeneous disease. On the other hand, the mixed response and the dynamics of methylation changes observed after treatment with hypomethylating agents may be related to the fact that these drugs could also affect tumor antigen presentation and MDS stem cell renewal [49]. Finally, there are no tools as yet in routine use that reliably monitor changes of methylation across cancer genomes, and most trials rely on more readily interpretable and accessible outcome measures such as blood counts and blast percentages, tumor size by imaging, and even sometimes quality of life [36]. At present, biomarkers for the prediction and monitoring of response to hypomethylating agents therefore remain an unmet need and further research is needed. Biomarkers that are considered promising and are currently under development include those which may be useful at baseline, including EZH2, TET2, DNMT3A, and IDH1/IDH2 levels, those that allow monitoring of response whilst on treatment, such as an increase in platelet levels after two cycles of therapy, or the methylation status of *phosphoinositide-phospholipase C beta 1*, and those that predict for relapse, such as the evaluation of leukemic stem cells by immunophenotyping.

15.3.2 HDAC INHIBITORS

HDAC inhibitors (HDI) are a class of compounds that increase acetylation of lysine residues on histone proteins as well as other, nonhistone, proteins by inhibiting the activity of HDAC enzymes. HDI are being developed for a wide range of cancer types, either as single-agent treatments or in combination with other anticancer agents as a result of their antitumor activity, which has been demonstrated *in vitro* and *in vivo*. HDI can be classified into four different groups, including hydroxamates, cyclic peptides, aliphatic acids, and benzamides [50].

15.3.2.1 The classification of HDAC

In humans, 18 HDAC enzymes have been categorized into four classes according to their protein homology to *Saccharomyces cerevisiae*, domain organization, subcellular location, and tissue specificity, and these can be further subcategorized into those that are zinc-dependent (classes I, II, and IV), and those that are zinc-independent but NAD-dependent (class III) [51]. Class I includes HDAC1, 2, 3, and 8; class II includes HDAC4, 5, 6, 7, 9, and 10; class III includes seven mammalian sirtuins (SIRT 1–7), while class IV consists of HDAC11 only [52]. In this chapter we will focus on HDAC from classes I, II, and IV since HDI inhibiting these subunits have been approved for clinical practice (Figure 15.2) [50,53].

Class I HDAC are ubiquitously expressed in various human tissues and have homology to the yeast HDAC subunit RPD3; they are mainly located in the nucleus [54]. Class II HDAC can be further subdivided into two subclasses, IIa and IIb, based on sequence homology to yeast HDAC1 (yHda1) and domain organization. The class IIa HDAC, HDAC4, 5, 7, and 9, contain a conserved C-terminal deacetylase catalytic domain homologous to yHda1, and an N-terminal domain with low similarity to any other HDAC classes. Class IIa HDAC expression is tissue-specific and the

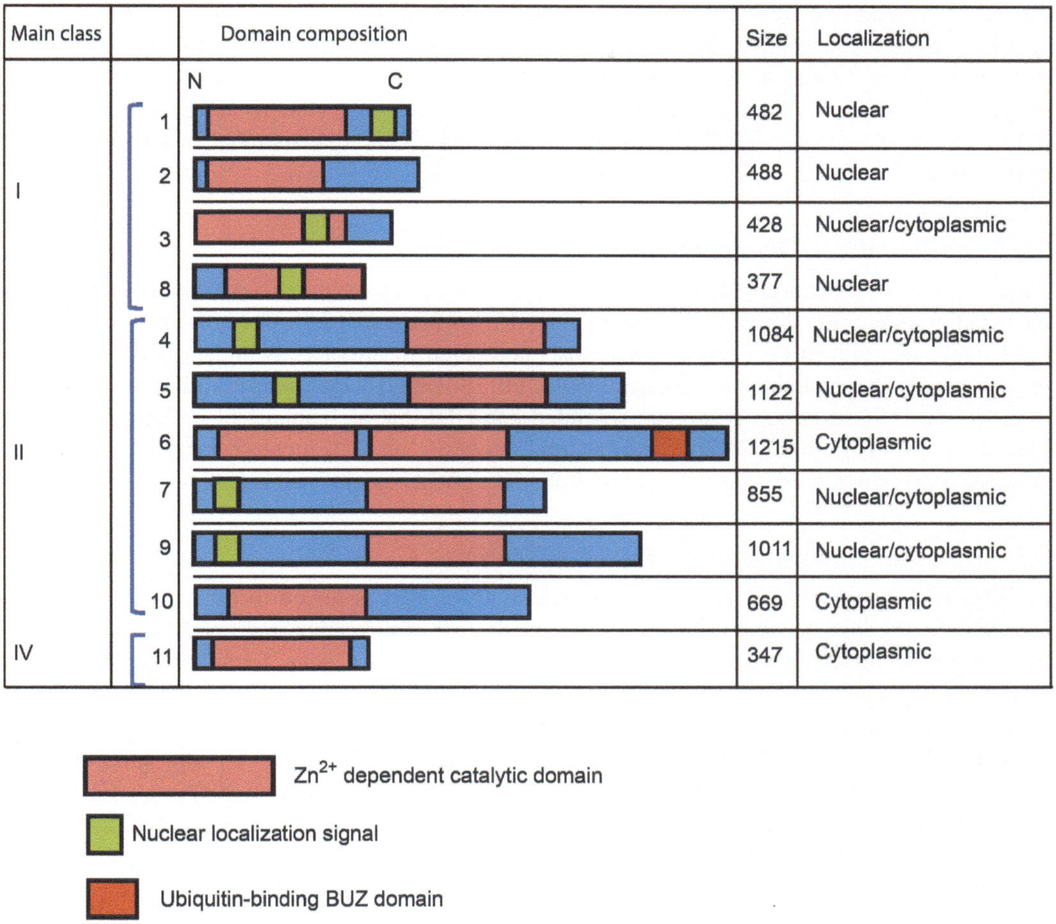

FIGURE 15.2

Main classes of HDAC.

Adapted from New et al. [9].

HDAC proteins are transported between the cytoplasm and the nucleus. On the other hand, class IIb HDAC, HDAC6 and 10, are mainly located in the cytoplasm and are characterized by the presence of two catalytic domains. HDAC11 is the sole member of class IV and is located in the nucleus; however, some studies also show that it coprecipitates cytoplasmically with HDAC6 [55].

Biochemically, HAT neutralize the positive charge of lysine residues by acetylation, leading to a relaxation of the chromatin structure, which in turn enables TF to access DNA. HDAC remove acetyl groups from the epsilon-amino groups of lysine residues on histones, which is in part responsible for transcriptional regulation. While acetylation is linked to high levels of gene transcription, deacetylation of histone tails causes transcriptional downregulation through chromatin condensation and compaction [56,57]. In normal cells, the equilibrium between HAT and HDAC activity is reflected by a balanced acetylation and deacetylation of their substrates, a feature that is disrupted in several cancers. Therefore, HDI treatment can induce diverse physiological effects, including gene expression, apoptosis and cell cycle arrest, inhibition of angiogenesis, and regulation of DNA damage repair pathways through hyperacetylation of the relevant histone and nonhistone targets [58,59]. These properties make HDI attractive candidates for the development of targeted anticancer treatments.

15.3.2.2 HDI-induced antitumor effects

A large number of structurally diverse HDI have been developed synthetically, and several of these are currently undergoing clinical trials (see Tables 15.1, 15.3, and 15.4). Of significance, fibroblasts are relatively resistant to HDI-induced apoptosis compared to tumor cells, meaning that this selective cytotoxicity can be exploited therapeutically [60]. Some HDI are specific inhibitors of a particular class of HDAC, whereas others are so-called pan-HDI that are capable of inhibiting the whole spectrum of class I, II, and IV HDACs. In general, however, all HDI increase acetylation of histone and nonhistone proteins, leading to variable biological effects as outlined below.

15.3.2.2.1 Selective alteration of gene expression

Several studies have shown that the anticancer activity of HDI is linked to their ability to regulate the genetic expression of various proteins involved in cell proliferation and apoptosis. An example is the cyclin-dependent kinase (CDK) inhibitor p21, which has been found to be significantly increased after treatment with SAHA in bladder carcinoma cells, whilst SAHA treatment in MM cells led to the marked decrease in HDAC1 in the protein complex bound to the *p21* promoter [61]. Additionally, treatment with romidepsin caused downregulation of cyclin D1 and upregulation of p21 in breast cancer cells, resulting in dephosphorylation of the tumor suppressor Rb protein and growth arrest in the early G1 phase [62]. Additionally, SAHA and romidepsin affect the gene expression of multiple Myc target genes including *cyclin, CDK, tumor necrosis factor (TNF), Bcl-2*, and caspases, ultimately resulting in apoptosis [63].

15.3.2.2.2 Activation of apoptotic pathways

HDI can induce apoptosis via both extrinsic and intrinsic pathways, and by transcription-dependent as well as transcription-independent mechanisms [64]. The extrinsic pathway of apoptosis is initiated via binding of death ligands, including TRAIL, TNF-α, and Fas ligand, to their corresponding death receptors, resulting in activation of a series of caspases and other proteases, leading to apoptosis [64,65]. Examples of this include the overexpression of TNF-ligands and -receptors on leukemic cells upon treatment with valproic acid (VPA) and the overexpression of TNF-α following the hyperacetylation of histones H3 and H4 after the treatment of leukemia cells with romidepsin [66]. On the other hand, the mitochondrial intrinsic pathway is activated in response to stimuli from inside the cell such as oxidative stress, hypoxia, or DNA damage in response to chemotherapy. Activating the mitochondrial apoptotic pathway causes the influx of several proteins, including cytochrome c, into the cytosol, which then binds Apaf-1, forming the apoptosome and activating a set of caspases similar to the extrinsic pathway [65,67]. Examples include SAHA-induced apoptosis via E2F1 recruitment and induction of Bim expression in p53-deficient colon cancer cells [68].

15.3.2.2.3 Inhibition of angiogenesis

HDIs were also found to suppress several angiogenic factors including endothelial nitric oxide synthase, hypoxia-inducible factor-1α (HIF-1α), vascular endothelial growth factor (VEGF), chemokine (C-X-C motif) receptor 4, basic fibroblast growth factor, angiopoietin, and tunica intima endothelial kinase 2. For instance, HIF-1α expression is increased in the hypoxic tumor microenvironment and the HDI trichostatin A (TSA) has been shown to inhibit hypoxia-induced angiogenesis through suppression of both HIF-1α and VEGF whilst simultaneously upregulating p53 [69]. In addition, HDI can promote HSP90 hyperacetylation by inhibition of HDAC6, leading to its decreased affinity to HIF-1α. As a result, binding to HSP70 is followed by proteasomal degradation of HIF-1α [70].

15.3.2.3 HDI and clinical outcomes

So far, four HDI have been approved for clinical use by the US FDA (Table 15.2): vorinostat was approved in October 2006 for the treatment of CTCL; romidepsin was given approval for the treatment of CTCL in November 2009, and in May 2011 for the treatment of PTCL; belinostat was approved for the treatment of patients with relapsed or refractory PTCL in July 2014, and most recently, in February 2015, panobinostat was approved as the first HDI for the treatment of MM.

Vorinostat is approved for the treatment of CTCL. The major trial supporting this was a phase II trial that recruited 74 patients with mycosis fungoides or Sézary syndrome CTCL who had failed at least two prior systemic therapies. The ORR was 29.7%, and 32% of patients had relief from pruritus symptoms. The common drug-related adverse events were diarrhea (49%), fatigue (46%), nausea (43%), and anorexia (26%) of which most occurred at grade 2 or lower [71]. More recently, two phase II studies have reported favorable responses in relapsed/refractory indolent follicular lymphoma with good safety profiles [72,73]. Despite the demonstrated effect in hematological malignancies, however, only a moderate effect of single-agent HDI has been observed in solid tumors [74–77].

Romidepsin is available to treat CTCL patients who have failed at least one prior systemic therapy. Its efficacy and safety is mainly based on a phase II trial in which 96 patients were treated with intravenous romidepsin at a dose of 14 mg/m^2/day on days 1, 8, and 15 of a 28-day cycle. The ORR was 34% and the complete response (CR) rate was 6%. The most common drug-related adverse events were nausea and fatigue [78]. Romidepsin was later also approved for the treatment of relapsed or refractory PTCL. This approval was based on the results of a multinational phase II trial conducted in 130 patients who were refractory to at least one prior systemic therapy or for whom at least one prior systemic therapy failed. The ORR was 25%, and 15% achieved CR. The most common grade 3 drug-related adverse events were thrombocytopenia (23%), neutropenia (18%), and infection (6%) [79]. As with vorinostat, several phase II clinical trials of romidepsin in solid tumors including colorectal, small-cell lung and squamous cell head and neck cancers, have not shown it to be of similar efficacy compared with its effects in hematological malignancy [80–82].

Belinostat is a hydroxamate pan-HDI that was approved for the treatment of patients with relapsed or refractory PTCL. Its approval was based on the result of a phase II trial conducted in 120 patients with relapsed/refractory PTCL. Belinostat was administered by intravenous infusion at a dose of 1000 mg/m^2 once daily on days 1–5 of a 21-day cycle. The ORR was 25.8% and median progression-free survival was 1.6 months. The most common grade 3 adverse events were thrombocytopenia (15%), neutropenia (13%), leukopenia (13%), and anemia (12%). Patients with CTCL have also shown favorable responses to belinostat, but, again, only a moderate effect was observed in solid tumors [83–85].

Panobinostat is the most recent HDI to be approved for the treatment of pretreated MM, to be used in conjunction with dexamethasone and bortezomib. The basis for the FDA approval was a phase III clinical trial in 193 patients with MM who received panobinostat in conjunction with bortezomib and dexamethasone versus bortezomib and dexamethasone alone. The HDI treatment group was shown to have a superior ORR (59% vs 41%) and improved progression-free survival (10.6 vs 5.8 months). Significant side effects were recorded as diarrhea, tiredness, and nausea. Panobinostat carries a Blackbox warning in view of severe diarrhea and fatal cardiac events that have been observed in some patients, but overall the benefits of treatment are thought to outweigh the risks [86].

15.3.2.4 New HDI in trials

Several other HDI have reached phase III clinical trials, such as, for example, panobinostat (LBH-589) for the treatment of CTCL. Other agents, such as mocetinostat (MGCD0103) and givinostat (ITF2357) are being trialed in phase II studies for the treatment of various lymphoid and myeloid malignancies. Multiple novel HDI, including more specific and targeted agents, are emerging from preclinical studies: Examples include rocilinostat (ACY-1215), a selective inhibitor of HDAC6, which is currently in phase I trials and is being evaluated for the treatment of MM in combination with bortezomib (Velcade) or lenalidomide (Revlimid), and CXD-101, a next-generation HDI with specificity for class I and IIb HDAC that is used to treat patients with relapsed and refractory tumors and uses tissue expression of a response-specific biomarker, human ultraviolet excision repair protein RAD23 homolog B (HR23B) (see section on Biomarkers), for treatment stratification (Table 15.4).

In summary, HDI have shown good anticancer activity both in preclinical and clinical trials, with relatively little effect on normal tissues and a favorable toxicity profile. Although preclinical trials show that HDI have a variety of biological effects across multiple pathways, the mechanism by which the clinical activity is mediated remains unclear. Accumulating evidence demonstrates the selective effect of HDI in altering gene expression and apoptosis. However, it has not yet been determined which inhibitor will be more therapeutically effective against specific cancer types. Positive therapeutic responses with some HDI have been shown consistently in some hematologic malignancies, but the clinical outcomes in solid tumors are disappointing when used as single agents. Therefore, many clinical trials have examined combination therapies of HDI with chemotherapeutic and biological agents in both solid and hematologic cancers (Table 15.4).

Table 15.4 Examples of HDI Currently in Preclinical or Clinical Trials
(IC50 Denotes Half Maximal Inhibitory Concentration of Compound)

Drug Name and Approval Status	Proposed Indication	Target HDAC	IC50
Approved			
Belinostat	PTCL	Pan-HDI	27 nM
Panobinostat	MM	Pan-HDI	5–20 nM
Romidepsin	CTCL, PTCL	HDAC1, HDAC2	36 and 47 nM
Vorinostat	CTCL	HDAC1, HDAC2, HDAC3, HDAC8, HDAC9	10 nM
In clinical trials			
Started phase III clinical trials			
VPA (as Mg valproate)	Cervical cancer and ovarian cancer have also been trialed in spinal muscular atrophy	HDAC1, HDAC2, HDAC3	0.7–20 mM
Started pivotal phase II clinical trials			
Belinostat (PXD101)	Relapsed ovarian cancer, and reported good results for T-cell lymphoma	Pan-HDI	27 nM
Started phase II clinical trials			
Abexinostat (PCI-24781)	Sarcoma and another for lymphoma	HDAC1, HDAC3, HDAC2, HDAC6, HDAC8	0.15–3.09 μM
Entinostat (MS-275)	Hodgkin's lymphoma, lung cancer, and breast cancer	HDAC1, HDAC2, HDAC3, HDAC9	0.5 mM
Givinostat (ITF2357)	Refractory leukemias and myelomas	HDAC2, HDAC1B, HDAC1A	7.5–16 nM
Mocetinostat (MGCD0103)	Various cancers (including follicular lymphoma, Hodgkin's lymphoma, and AML)	HDAC1, HDAC2, HDAC3, HDAC11	0.15–10 μM
Pracinostat (SB939)	Recurrent or metastatic prostate cancer	Pan-HDI	40–140 nM
Quisinostat (JNJ-26481585)	Myeloma	Modest potent to HDACs 2, 4, 10, and 11; greater than 30-fold selectivity against HDACs 3, 5, 8, and 9 and lowest potency to HDACs 6 and 7	0.11 nM
Resminostat (4SC-201)	Hodgkin's lymphoma, hepatocellular carcinoma	HDAC1, HDAC3, HDAC6, HDAC8	42.5–877 nM

(*Continued*)

Table 15.4 Examples of HDI Currently in Preclinical or Clinical Trials
(IC50 Denotes Half Maximal Inhibitory Concentration of Compound) ***Continued***

Drug Name and Approval Status	Proposed Indication	Target HDAC	IC50
Started phase I clinical trials			
4SC-202	Advanced hematological indications	Class I	1 μM
AR-42	Relapsed or treatment-resistant MM, CLL, or lymphoma		30 nM
CG200745	Solid tumors	Pan-HDI	
CXD-101	Advanced solid and hematological malignancy	Class I and class IIB HDACs	2.4–17.7 nM
Rocilinostat (ACY-1215)	MM in combination with bortezomib (Velcade, Takeda Millennium) and with lenalidomide (Revlimid, Celgene)	HDAC6	5 nM
Preclinical			
Kevetrin		Selective for HDAC2	

15.3.3 BRDs AND BRD INHIBITORS

Beyond the ability of lysine acetylation to change the charge of lysines followed by structural changes of proteins and chromatin, acetylation of proteins has to be recognized and translated into physiological function. Given the variety of acetylation events on proteins in many cellular compartments [87], binding modules are essential to mediate the different types of acetylation statuses. BRD-containing proteins are the only proteins known so far which can bind via their conserved BRD activity to monoacetylated lysine residues and act as readers of this PTM. The origin of the name "bromo" domain refers to the identification of the BRD-containing protein "Brahma" in *Drosophila melanogaster*, homolog of the SWItch/Sucrose Non-Fermentable (Swi2/Snf2) and Sth1 proteins in *S. cerevisiae* [88].

The human proteome encompasses 61 BRDs, which are present in 46 different BRD-containing proteins, and each BRD-containing protein can contain up to six different BRDs [89]. BRDs are categorized into eight groups or subfamilies according to their alignment and structural similarity [90,91]. Recently determined structures of the different BRDs or BRD-containing proteins revealed that they share evolutionary conserved folding, consisting of a left-handed four-helix bundle, named "BRD fold" [91,92]. The respective acetylation marks are recognized by a pocket consisting of an interhelical αZαA and αB-αC loops. Consequently, the amino acid residues which mediate the binding interaction between the acetylated lysine residue and the BRDs are highly conserved residues within the BRDs [93,94]. A conserved asparagine residue serves as an H-donor in a hydrogen bond between the oxygen of the acetyl group and the amide nitrogen of the asparagine [95], although in some BRDs the conserved asparagine is replaced by a different residue, for example, tyrosine or threonine [91].

Different BRD inhibitors (BDIs) have been developed. Most are based on a triazolodiazepine scaffold that mimics interactions observed with members of the bromo and extraterminal (BET) family [96,97]. However, there are a growing number of selective inhibitors under development targeting the BRDs of CBP, p300, and BAZ2B.

The 46 BRD-containing proteins are involved in a plethora of different biological functions, including histone methyltransferase (HMT) activity such as ASH1L and mixed lineage leukemia protein [98,99], the HATs and HAT-associated proteins p300/CBP, GCN5L2 (general control of amino acid synthesis protein 5-like 2) and p300/CBP-associated factor PCAF and BRD9 [92,100], the helicases SWI/SNF-related matrix associated actin-dependent regulators of chromatin subfamily A [101] or chromatin remodeling complexes including BRD adjacent to zinc finger domain protein 1B [102]. Other BRD-containing proteins are transcriptional coactivators like tripartite motif-containing proteins, TBP-associated factors [103], and nuclear scaffolding proteins, for instance polybromo 1 [104]. The BET family of BRD-containing proteins such as BRD2, BRD3, BRD4, and BRDT recognize mainly acetylated lysine residues in the histones H3 and H4 and are involved in cellular proliferation, cell cycle progression, and chromatin remodeling [105–107].

BRD-containing proteins play a role in multiple disease areas, including cancer [108], cardiovascular disease [109], inflammation [97], neurodegenerative disorders [110,111], and viral diseases including HIV/AIDS [112,113]. Given that BRD-containing proteins are involved in these different diseases, targeting these proteins could be beneficial to develop new treatment strategies. The development of BDIs specifically targeting the BRD-containing proteins CBP/p300 could also be beneficial in the treatment of neurodegenerative disorders, for instance Huntington's disease, since p300, CBP, and PCAF have been shown to interact with huntingtin aggregates [114–116]. There are several reports demonstrating the antiproliferative effects of BET inhibitors including JQ1, I-BET151, and I-BET762 in various cancer types and cell lines which are either dependent on or independent of MYC downregulation [117–120]. Combination studies of BDIs alongside drugs that are already the standard of care could demonstrate synergistic effects, for example, the combination of JQ1 with dexamethasone in cellular models of acute lymphoblastic leukemia [121], I-BET151 with temozolomide in glioblastomas [122] or with HDI [123]. Another example where BRD-containing proteins play a role in carcinogenesis is the NUT (Nuclear protein in testis) midline carcinomas (NMCs). Here, the expression of a BRD4-NUT fusion gene is enhanced by the BRD4 promotor. The resulting, predominantly nuclear, fusion protein contains either BRD4 or BRD3 and the entire NUT protein [124]. There is evidence that the BRD4-NUT protein can recruit CBP and p300, which are both BRD-containing proteins themselves, thereby enhancing their HAT activity, ultimately leading to inactive hyperacetylated chromatin domains [125]. The association of these tumors with BRD3/BRD4-NUT underpinned the rationale to initiate clinical studies with BDIs targeting the BET family. I-BET762 and TEN-010 are being explored in two phase I trials, both inhibitors belonging to the triazolobenzodiazepines [126]. Since *MYC* can contribute to the pathogenesis of many human types of cancers, the rationale behind treating with BDIs is to target *MYC* transcription by inhibiting the BRDs of putative coactivator proteins of RNA polymerase involved in transcriptional initiation and elongation [118]. The association of the *MYC* oncogene with other tumor promoters in leukemia led to a phase I clinical trial to investigate the potential and efficacy of the BDI OTX015 in acute leukemia and other hematological malignancies. This reflected its activity in B-cell tumor models and its ability to synergize with targeted drugs such as the HDI vorinostat and romidepsin [127]. The triazolothienodiazepine CPI-0610 is being tested in a phase I clinical trial for the treatment of progressive lymphoma.

In summary, BRDs are involved in numerous processes, including epigenetic modulation, and have been identified in human diseases such as cancer, inflammation, and neurological disorders. Therefore, BRDs represent attractive targets especially for epigenetic therapy, and recently developed BET BDI might contribute alone or in combination with other established anticancer drugs to the treatment of malignancies.

15.4 BIOMARKERS

15.4.1 DEFINITIONS OF BIOMARKERS

Response to epigenetic drugs can be monitored by means of biomarkers or the more complex molecular signatures. The term "biomarker" is a portmanteau of the word "biological" and "marker" and was established in 1980 [128]. Biomarkers can be defined in different ways, but what these definitions all have in common is a quantitative measurement of biological conditions [129]. The WHO also takes the effects and outcomes of treatments, interventions, or exposures including chemicals or nutrients into account, and gives the even broader definition that a biomarker includes "almost any measurement reflecting an interaction between a biological system and a potential hazard, which may be chemical, physical, or biological. The measured response may be functional and physiological, biochemical at the cellular level, or a molecular interaction" [130].

Biomarkers can be distinguished from clinical endpoints, which are defined as a characteristic or variable that reflects how a patient feels, functions, or survives. Conversely, a surrogate endpoint can be a biomarker or subset of biomarkers that is substituted for a clinical endpoint and predicts clinical benefit or lack thereof [129,131]. A molecular signature is defined as a set of biomolecules (e.g., DNA, RNA, modified chromatin, and proteins or metabolites), which form unique patterns in different tissues, disease states, and conditions. High-throughput analyses now enable the determination of molecular signatures as a set of biomarkers which have to be processed to an integrative measure [132]. Given the sometimes complex changes of biomolecules after epigenetic drug treatment, molecular signatures can be more useful in monitoring these alterations, for instance with regards to DNA methylation or protein acetylation, rather than a single biomarker.

15.4.2 CLASSIFICATION OF BIOMARKERS

Biomarkers can be classified based on different parameters. One possible classification is according to their physicochemical [133] or physiological characteristics [134]. Another system classifies biomarkers into three types: prognostic, companion diagnostic, and pharmacodynamic (PD) [4,38]. According to this classification, the aim of using a prognostic biomarker is to objectively evaluate the overall outcome of the patient. Prognostic biomarkers can be useful for the selection of patients for a therapy, but do not directly predict the response to a treatment or therapy. A companion diagnostic biomarker, sometimes also referred to as a predictive biomarker, can provide information to identify patients who are most likely to benefit from a therapeutic intervention. It can also help to identify patients likely to be at increased risk for serious adverse reactions as a result of treatment with the drug. It allows monitoring the response to treatment with the drug for the purpose of adjusting treatment (e.g., schedule, dose, or discontinuation) to achieve improved safety or efficacy. It should help to identify patients for whom the drug or treatment is likely to be successful and should help with patient stratification.

The impact of a drug on a disease can be assessed by PD biomarkers, which provide a post treatment measure of whether a pharmacological response of the target organism was obtained and to which degree [135,136]. A PD can be utilized to obtain critical data, both in preclinical models and clinical trials, to select rational combinations of targeted agents, and to optimize schedules of combination drug regimens.

15.4.3 WHAT MAKES A GOOD BIOMARKER?

The ideal biomarker should be able to give information not only on the selection of treatment, but also on responsiveness and outcome. This implies that a given biomarker can be a measure of whether the intervention alters the disease and to what extent. The ideal biomarker does not necessarily have to be directly related to an endpoint, but can be affected in parallel with the disease. Companion diagnostic biomarkers in particular should enable the stratification of patients from a diverse group into those that are likely to benefit from the treatment and those who will not, thereby taking individual variability into account [137].

Several aspects govern the development of biomarkers: The first aspect is the use of rational and reproducibly applied guidelines to decide from certain criteria which biomarkers are good candidates. Hill et al. proposed guidelines and criteria a good biomarker should fulfill. Biomarkers which conform with these guidelines are more likely to be useful and increase the likelihood that an association is causative (Table 15.2) [138]. Secondly, the use of biomarkers may allow to predict whether a given treatment has been successful, or if further treatment may be necessary, avoiding ethical problems which can be associated with measuring endpoints [139]. Thirdly, biomarkers should be straightforward to measure, and the biological material required for this easy to obtain. This is usually the case compared to endpoint analyses [139]. Similarly, measuring biomarkers should be cost-effective. The difference between conventional and molecular diagnosis of cancer is that although conventional diagnosis is based on morphological appearance of cells to distinguish between various morphological stages, molecular diagnosis relies on measurements of gene modification and expression [137,140]. With the new omics technologies, it is now possible to measure biomarkers or even complex molecular signatures in a fast and cheap way [137]. It is expected, therefore, that biomarkers will reduce the costs for molecularly targeted drugs by allowing streamlining of treatment for patients, ultimately aligning the right patient with the right drug and avoiding treatment with agents that are ineffective. In parallel with these developments, biomarkers that enable the monitoring of effects that occur on a molecular level during targeted therapy can be used to assess response and identify those that respond poorly or start to relapse on treatment.

15.4.4 BIOMARKERS IN CANCER EPIGENETICS

15.4.4.1 Examples of general and specific biomarkers reflecting epigenetic events in cancer

The specificity of a given biomarker is one of the main criteria of a useful biomarker (Table 15.5). Biomarkers in cancer can be specific to certain cancer types, or they are general biomarkers. An example of a general biomarker is HDM, an enzyme which has been demonstrated to be overexpressed or mutated in different cancer types. Different demethylases, targeting for example H3K4, H3K9, and K3K27, can be expressed at altered levels in various cancer types [141]. Examples of biomarkers that are specific for a particular type of cancer and its associated epigenetic changes are

Table 15.5 Criteria a Good Biomarker Should Fulfill: Increasing the Likelihood That an Association is Causative

Guidelines	Characteristics of Useful Biomarkers
Strength	A strong association between the biomarker and the outcome
Consistency	The association persists in different individuals
Specificity	The biomarker is associated with a specific disease
Temporality	The time courses of changes in the biomarker and outcome occur in parallel
Biological gradient (dose-responsiveness)	Increasing exposure to an intervention produces increasing effects on the biomarker
Plausibility	Credible mechanisms connect the biomarker, the pathogenesis of the disease, and the mode of action of the intervention
Coherence	The association is consistent with the natural history of the disease and the biomarker
Experimental evidence	An intervention gives results consistent with the association
Analogy	There is a similar result from which a relationship can be concluded

Modified after Refs [138,139].

as follows: In breast cancer, mutations in the *BRCA1* gene and *BRCA1* promoter hyperacetylation have been positively correlated with the disease [40,142]. Both *RASSF1A* and *PITX2* have potential as prognostic biomarkers, since they have been found to be hypermethylated in breast cancer. Tamoxifen resistance in breast cancer cells can be regulated by methylation of phosphatase and tensin homolog [143]. Epigenetic changes and regulation also play a role in CRC, and proteins which can be measured are adenomatous polyposis coli (APC) and p14 (ADP-ribosylation factor) [144]. In a cohort of more than 100 patients with CRC, a distinct CIMP was identified, underlying sporadic microsatellite instability. Another outcome of this study was that CIMP is associated with the BRAF mutation in CRC [145,146]. A very promising candidate as a biomarker in prostate cancer is the gene *GSTP1* in its hypermethylated state; in other cancer types *GSTP1* is rarely hypermethylated [147]. However, other genes such as *NKX2-5*, *DPYS*, and *NSE1* are also aberrantly methylated in prostate cancer [148]. In cervical cancer patients, promoters for genes for *MYOD1*, *CALCA*, *hTERT*, *TIMP3*, and *PGR* (progesterone receptor) were reported to be aberrantly hypermethylated. Unmethylated *MYOD1* promoter detected as circulating tumor DNA in patient serum and also in corresponding patient tissue biopsy samples predicted for a significantly better disease-free ($P = 0.04$) and overall survival ($P = 0.02$) in comparison with patients who displayed *MYOD1* promoter hypermethylation. The results suggest also that serological detection of *MYOD1* promoter hypermethylation may be of potential use as a prognostic marker for discriminating cervical cancer patients at high risk for lymph node metastasis or relapse [149].

In addition to methylation events on DNA and PTMs, miRNAs, a class of small regulatory RNA, are also known to modulate gene expression without changing the primary DNA sequence and therefore are classed as epigenetic modulators. Overexpression of certain miRNAs has been observed in chronic lymphocytic leukemia (CLL); especially two miRNA families, miR-29 and miR-223, are involved in the disease and decreased levels of these miRNAs are associated with poor prognosis [150,151].

Aberrant acetylation patterns could be observed both in solid and hematological cancers. The HAT- and BRD-containing proteins CBP and p300 can initiate carcinogenesis by causing hypoacetylation of H3K18 with subsequent activation of cell growth promotors [152,153]. Levels of HDACs have been shown to differ in specific tumor types, for example, high levels of HDACs 1 and 2 correlated with histone hypoacetylation and tumor dedifferentiation, and reduced patient survival in a study of 192 prostate carcinoma biopsies [154].

15.4.4.2 Examples of existing biomarkers that can direct epigenetic therapy

1. *Prognostic:* Diffuse large B-cell lymphoma (DLBCL) is an aggressive but extremely heterogeneous group of lymphomas, and in recent years gene expression profiling has been used to distinguish two major subgroups, termed activated B-cell-like (ABC) and germinal center B-cell-like (GCB), which respond very differently to standard chemotherapy, with ABC lymphomas proving much more challenging to cure with conventional chemotherapy regimens [155,156]. Recent evidence suggests that both the ABC and GCB subtypes harbor targetable mutations that are associated with either acetylation control such as the MYD88 L265P mutation in ABC DLBCL ([157]; Maria New, personal communication), or epigenetic regulation through mutation in *EZH2* in GCB DLBCL [158]. Gene expression profiling in conjunction with mutational analysis can therefore be used for prognostic purposes and have important implications in tailoring therapeutic strategies and identifying responsive disease.
2. *Companion diagnostic:* A promising predictive biomarker is the DNMT MGMT (O6-methylguanine-DNA-methyltransferase). Like other DNA repair genes, it can be silenced by hypermethylation and is therefore not expressed, and both the hypermethylation event as well as the protein level of MGMT can be used as a predictive biomarker for response to carmustine or temozolomide in gliomas [159,160] as well as for other alkylating agents. An example of a screen-driven approach to identify a biomarker to determine sensitivity to HDI is the protein HR23B, which is involved in DNA repair and shuttling proteins to the proteasome [161]. Initial studies revealed that HR23B governs the sensitivity of CTCL cells to HDI and provided evidence that HR23B is a predictive biomarker in CTCL. HDAC6, which is involved in autophagy, has been shown to modulate this response [162,163]. These findings are supported by a recent clinical study using the HDI belinostat for patients with unresectable hepatocellular carcinoma, where a correlation between the level of HR23B expression and a clinical response was reported. Strong HR23B expression was associated with disease stabilization upon treatment [164].
3. *Pharmacodynamic:* Measuring the acetylation of marker proteins represents an example of a PD, and has been use to guide treatment with HDI: By demonstrating a rise in acetyl-tubulin, a specific substrate of HDAC6, after treatment with a specific HDAC6 inhibitor, ACY-1215, it was possible to determine therapeutic plasma levels of this drug in phase I trials and guide dose escalation [165].

In summary, the development of biomarkers to detect epigenetic changes in cancer and monitor these during treatment, appears to be a promising strategy for delivering safe and effective care to patients in a clinical setting. Changes in the methylation and acetylation status of DNA, and indeed other PTMs, can be detected and evaluated by highly sensitive and specific technologies using minimal amounts of nucleic acid or protein. For methylation changes, these techniques include methylation-specific PCR or mass-spectrometry-based methylation detection (e.g., EpiTYPER®). With the help of mass-spectrometry-based methylation detection, for example, it is now possible to perform quantitative analysis of DNA methylation and identification of differentially methylated CpG sites in any genomic region or candidate region. With regards to acetylation status in cells and monitoring response to HDI, levels of acetylated histones can be measured either by mass spectrometry, or specific immunochemical methods, for example, with the kit EpiQuik® which allows quantification of the global acetylation status of histone H3 both in cell lines and human tissue biopsy samples. Whereas a number of nonepigenetic biomarkers such as *Human epidermal growth factor receptor 2 (HER-2), Kirsten rat sarcoma viral oncogene homolog (KRAS), or Epidermal growth factor receptor (EGFR)* have been FDA-approved and are in routine use in oncological practice, to date only a few epigenetic biomarkers are being used routinely in the context of diagnosis or treatment. One kit which uses epigenetic information as a biomarker in combination with other nonepigenetic biomarkers to detect CRC is cologuard®; besides the *KRAS* mutation and fecal immunochemical tests, the test analyzes the DNA methylation markers *NDRG4* and *BMP3* promoter regions. A composite score is built consisting of the results of the different biomarkers, and patients are considered to be cologuard-positive if they score above a certain threshold. Cologuard® was the first epigenetic biomarker kit to receive FDA approval in August 2014, and other similar devices are awaiting regulatory clearance.

15.5 CONCLUSION

Over the last decade, it has become clear that epigenetic changes are key events in the progression of many cancer types. Since epigenetic mechanisms are essential to maintain normal physiological functions of cells, a change toward an aberrant pattern can cause altered gene and protein function, ultimately leading to or contributing to malignant transformation. Other than genetic changes, which are often involved in the initiation of cancer, epigenetic changes are mostly reversible. This reversibility led to two developments in the field of cancer epigenetics, which are closely linked to each other: The emergence of epigenetic drugs, and the search for biomarkers to predict response and clinical outcome. Outcome and survival in cancer patients depend on the early diagnosis of the malignancy and its prompt management. In parallel with recent advances to discover epigenetic alterations and associated pathways, new investigation methods and omics technologies have been developed to complement already-existing methods of investigation which have their own inherent limitations. Concomitant with the discovery of reversible epigenetic changes and the development of these advanced methods, there is an increasing demand to personalize cancer medicine in order to deliver tailored treatments and limit side effects in nonresponding patients. Personalized medicine is a broad term and includes the aim of understanding the molecular basis of a disease in order to optimize and provide the most efficient medical care for each individual, depending on their unique molecular status, including their epigenetic status [137]. Personalized medicine should enable patient stratification from a diverse group into patient subgroups that are likely to benefit from the treatment and those who will not.

The ultimate aim of personalized medicine, to define the right drug for the right patient at the right time, is now a realistic possibility. This implies on the one hand the transition from broadly acting cytotoxic agents to more targeted and selectively toxic therapies, and on the other hand the quantification of outcomes. Although high-throughput technologies are able to analyze most cellular components within hours, they force data-driven studies rather than hypothesis-driven studies. Genomic, epigenomic, and proteomic data are ideally obtained in parallel from a patient at the same timepoint; the integration of these data should then give a detailed picture of all molecular changes, including protein levels, PTMs, and epigenetic changes. At least three prerequisites are necessary to obtain this overall picture of a patient: The technological modalities, including analysis and data evaluation, the expertise in interpreting the data, and references and guidelines to annotate the changes. The Human Epigenome Project (http://www.epigenome.org/) and the International Human Epigenome Consortium (http://www.ihec-epigenomes.org/) try to work out detailed information for reference purposes and compendia for utility on a daily basis. An example of an integrative omics approach is the integrative personal omics profile (iPOP) study, in which various omics methods are combined from an individual over a 14-month timespan [166]. This study revealed that longitudinal iPOP can be performed to analyze health and disease states by connecting genomic information with additional dynamic omics-derived data. These types of analyses go beyond the approach of a classical biomarker and are examples of developing complex molecular signatures.

Biomarkers and molecular signatures can help to enable the right combination of therapies. Numerous clinical studies have demonstrated the advantage of combination therapies: For instance, it is known that DNA hypermethylation compacts chromatin, hindering subsequent acetylation. This is one of the reasons for the ongoing trials of epigenetic combination therapy with HDI and DNMT inhibitors [167]. To avoid side effects, reduce resistance and lower treatment costs, the development of this dual-targeting strategy is an interesting new advancement. Dual-targeting agents, also called bivalent agents, combine two complementary chemoactive groups within a single molecular architecture. An example for a relatively new epigenetically acting drug is a dual-acting prototypical HDAC and topoisomerase II inhibitor, derived from the HDI vorinostat and the anthracycline daunorubicin [168].

Although modern technology enables us to gain a deeper understanding of epigenetic phenomena, perhaps a more holistic picture is needed to understand complex diseases such as cancer. The regulation of genetic, epigenetic and protein networks could be reflected in systems biology, including systems medicine, as we learn that organismal systems cannot be seen as a single linear chain of discrete events, but rather as an integrated network which is held in dynamic homeostasis. Being able to track epigenetic changes during the natural history of a disease, design drugs based on epigenetic molecular targets and select patients on the basis of a companion biomarker toolkit that predicts for response to a given treatment strategy will represent a major step in this direction.

REFERENCES

[1] Esteller M. Epigenetics in cancer. N Engl J Med 2008;358(11):1148–59.

[2] Hatzimichael E, Crook T. Cancer epigenetics: new therapies and new challenges. J Drug Deliv 2013;2013:529312.

[3] Ellis L, Atadja PW, Johnstone RW. Epigenetics in cancer: targeting chromatin modifications. Mol Cancer Ther 2009;8(6):1409–20.

[4] La Thangue NB, Kerr DJ. Predictive biomarkers: a paradigm shift towards personalized cancer medicine. Nat Rev Clin Oncol 2011;8(10):587–96.

[5] Feinberg AP, Vogelstein B. Hypomethylation distinguishes genes of some human cancers from their normal counterparts. Nature 1983;301(5895):89–92.

[6] Greger V, Passarge E, Hopping W, Messmer E, Horsthemke B. Epigenetic changes may contribute to the formation and spontaneous regression of retinoblastoma. Hum Genet 1989;83(2):155–8.

[7] Sakai T, Toguchida J, Ohtani N, Yandell DW, Rapaport JM, Dryja TP. Allele-specific hypermethylation of the retinoblastoma tumor-suppressor gene. Am J Hum Genet 1991;48(5):880–8.

[8] Saito Y, Jones PA. Epigenetic activation of tumor suppressor microRNAs in human cancer cells. Cell Cycle 2006;5(19):2220–2.

[9] New M, Olzscha H, La Thangue NB. HDAC inhibitor-based therapies: can we interpret the code? Mol Oncol 2012;6(6):637–56.

[10] Inche AG, La Thangue NB. Chromatin control and cancer-drug discovery: realizing the promise. Drug Discov Today 2006;11(3–4):97–109.

[11] Bannister AJ, Kouzarides T. Regulation of chromatin by histone modifications. Cell Res 2011;21 (3):381–95.

[12] Marino-Ramirez L, Jordan IK, Landsman D. Multiple independent evolutionary solutions to core histone gene regulation. Genome Biol 2006;7(12):R122.

[13] Dawson MA, Kouzarides T, Huntly BJ. Targeting epigenetic readers in cancer. N Engl J Med 2012;367 (7):647–57.

[14] Lao VV, Grady WM. Epigenetics and colorectal cancer. Nat Rev Gastroenterol Hepatol 2011;8 (12):686–700.

[15] Mummaneni P, Shord SS. Epigenetics and oncology. Pharmacotherapy 2014;34(5):495–505.

[16] Ehrlich M. DNA methylation in cancer: too much, but also too little. Oncogene 2002;21(35):5400–13.

[17] Weber M, Davies JJ, Wittig D, Oakeley EJ, Haase M, Lam WL, et al. Chromosome-wide and promoter-specific analyses identify sites of differential DNA methylation in normal and transformed human cells. Nat Genet 2005;37(8):853–62.

[18] Fraga MF, Herranz M, Espada J, Ballestar E, Paz MF, Ropero S, et al. A mouse skin multistage carcinogenesis model reflects the aberrant DNA methylation patterns of human tumors. Cancer Res 2004;64 (16):5527–34.

[19] Paz MF, Fraga MF, Avila S, Guo M, Pollan M, Herman JG. A systematic profile of DNA methylation in human cancer cell lines. Cancer Res 2003;63(5):1114–21.

[20] Lian CG, Xu Y, Ceol C, Wu F, Larson A, Dresser K, et al. Loss of 5-hydroxymethylcytosine is an epigenetic hallmark of melanoma. Cell 2012;150(6):1135–46.

[21] Esteller M, Silva JM, Dominguez G, Bonilla F, Matias-Guiu X, Lerma E, et al. Promoter hypermethylation and BRCA1 inactivation in sporadic breast and ovarian tumors. J Natl Cancer Inst 2000;92(7):564–9.

[22] Herman JG, Latif F, Weng Y, Lerman MI, Zbar B, Liu S, et al. Silencing of the VHL tumor-suppressor gene by DNA methylation in renal carcinoma. Proc Natl Acad Sci USA 1994;91(21):9700–4.

[23] McCabe MT, Ott HM, Ganji G, Korenchuk S, Thompson C, Van Aller GS, et al. EZH2 inhibition as a therapeutic strategy for lymphoma with EZH2-activating mutations. Nature 2012;492(7427):108–12.

[24] Shaknovich R, Melnick A. Epigenetics and B-cell lymphoma. Curr Opin Hematol 2011;18(4):293–9.

[25] Intlekofer AM, Younes A. Precision therapy for lymphoma—current state and future directions. Nat Rev Clin Oncol 2014;11(10):585–96.

[26] Falahi F, van Kruchten M, Martinet N, Hospers GA, Rots MG. Current and upcoming approaches to exploit the reversibility of epigenetic mutations in breast cancer. Breast Cancer Res 2014;16 (4):412.

[27] Nogueira da Costa A, Herceg Z. Detection of cancer-specific epigenomic changes in biofluids: powerful tools in biomarker discovery and application. Mol Oncol 2012;6(6):704–15.

[28] Frigola J, Sole X, Paz MF, Moreno V, Esteller M, Capella G, et al. Differential DNA hypermethylation and hypomethylation signatures in colorectal cancer. Hum Mol Genet 2005;14(2):319–26.

[29] Nebbioso A, Carafa V, Benedetti R, Altucci L. Trials with 'epigenetic' drugs: an update. Mol Oncol 2012;6(6):657–82.
[30] Katsumoto T, Yoshida N, Kitabayashi I. Roles of the histone acetyltransferase monocytic leukemia zinc finger protein in normal and malignant hematopoiesis. Cancer Sci 2008;99(8):1523–7.
[31] Sykes SM, Mellert HS, Holbert MA, Li K, Marmorstein R, Lane WS, et al. Acetylation of the p53 DNA-binding domain regulates apoptosis induction. Mol Cell 2006;24(6):841–51.
[32] Weichert W, Röske A, Niesporek S, Noske A, Buckendahl A-C, Dietel M, et al. Class I histone deacetylase expression has independent prognostic impact in human colorectal cancer: specific role of class I histone deacetylases *in vitro* and *in vivo*. Clin Cancer Res 2008;14:1669–77.
[33] Chen JD, Umesono K, Evans RM. SMRT isoforms mediate repression and anti-repression of nuclear receptor heterodimers. Proc Natl Acad Sci USA 1996;93(15):7567–71.
[34] Horlein AJ, Naar AM, Heinzel T, Torchia J, Gloss B, Kurokawa R, et al. Ligand-independent repression by the thyroid hormone receptor mediated by a nuclear receptor co-repressor. Nature 1995;377 (6548):397–404.
[35] Huang EY, Zhang J, Miska EA, Guenther MG, Kouzarides T, Lazar MA. Nuclear receptor corepressors partner with class II histone deacetylases in a Sin3-independent repression pathway. Genes Dev 2000;14 (1):45–54.
[36] Duong VH, Komrokji RS, List AF. Update on the pharmacotherapy for myelodysplastic syndromes. Expert Opin Pharmacother 2014;15(13):1811–25.
[37] Oronsky B, Oronsky N, Scicinski J, Fanger G, Lybeck M, Reid T. Rewriting the epigenetic code for tumor resensitization: a review. Transl Oncol 2014;7(5):626–31.
[38] Heerboth S, Lapinska K, Snyder N, Leary M, Rollinson S, Sarkar S. Use of epigenetic drugs in disease: an overview. Genet Epigenet 2014;6:9–19.
[39] Tao Y, Liu S, Briones V, Geiman TM, Muegge K. Treatment of breast cancer cells with DNA demethylating agents leads to a release of Pol II stalling at genes with DNA-hypermethylated regions upstream of TSS. Nucleic Acids Res 2011;39(22):9508–20.
[40] Chen Y, Toland AE, McLennan J, Fridlyand J, Crawford B, Costello JF, et al. Lack of germ-line promoter methylation in BRCA1-negative families with familial breast cancer. Genet Test 2006;10 (4):281–4.
[41] Qin T, Jelinek J, Si J, Shu J, Issa JP. Mechanisms of resistance to 5-aza-2'-deoxycytidine in human cancer cell lines. Blood 2009;113(3):659–67.
[42] Bejar R, Lord A, Stevenson K, Bar-Natan M, Perez-Ladaga A, Zaneveld J, et al. TET2 mutations predict response to hypomethylating agents in myelodysplastic syndrome patients. Blood 2014;124 (17):2705–12.
[43] Shen L, Kantarjian H, Guo Y, Lin E, Shan J, Huang X, et al. DNA methylation predicts survival and response to therapy in patients with myelodysplastic syndromes. J Clin Oncol 2010;28(4):605–13.
[44] Treppendahl MB, Kristensen LS, Gronbaek K. Predicting response to epigenetic therapy. J Clin Invest 2014;124(1):47–55.
[45] Itzykson R, Thepot S, Quesnel B, Dreyfus F, Beyne-Rauzy O, Turlure P, et al. Prognostic factors for response and overall survival in 282 patients with higher-risk myelodysplastic syndromes treated with azacitidine. Blood 2011;117(2):403–11.
[46] Tien HF, Tang JH, Tsay W, Liu MC, Lee FY, Wang CH, et al. Methylation of the p15(INK4B) gene in myelodysplastic syndrome: it can be detected early at diagnosis or during disease progression and is highly associated with leukaemic transformation. Br J Haematol 2001;112(1):148–54.
[47] Aggerholm A, Holm MS, Guldberg P, Olesen LH, Hokland P. Promoter hypermethylation of p15INK4B, HIC1, CDH1, and ER is frequent in myelodysplastic syndrome and predicts poor prognosis in early-stage patients. Eur J Haematol 2006;76(1):23–32.

[48] Cross M, Bach E, Tran T, Krahl R, Jaekel N, Niederwieser D, et al. Pretreatment long interspersed element (LINE)-1 methylation levels, not early hypomethylation under treatment, predict hematological response to azacitidine in elderly patients with acute myeloid leukemia. Onco Targets Ther 2013;6:741–8.
[49] Kantarjian H, Oki Y, Garcia-Manero G, Huang X, O'Brien S, Cortes J, et al. Results of a randomized study of 3 schedules of low-dose decitabine in higher-risk myelodysplastic syndrome and chronic myelomonocytic leukemia. Blood 2007;109(1):52–7.
[50] Marks P, Rifkind RA, Richon VM, Breslow R, Miller T, Kelly WK. Histone deacetylases and cancer: causes and therapies. Nat Rev Cancer 2001;1(3):194–202.
[51] McLaughlin F, La Thangue NB. Histone deacetylase inhibitors open new doors in cancer therapy. Biochem Pharmacol 2004;68(6):1139–44.
[52] Lucio-Eterovic AK, Cortez MA, Valera ET, Motta FJ, Queiroz RG, Machado HR, et al. Differential expression of 12 histone deacetylase (HDAC) genes in astrocytomas and normal brain tissue: class II and IV are hypoexpressed in glioblastomas. BMC Cancer 2008;8:243.
[53] Ruijter AJMDE, Gennip AHVAN, Caron HN, Kemp S, Kuilenburg ABPVAN. Histone deacetylases (HDACs): characterization of the classical HDAC family. Biochem J 2003;749:737–49.
[54] Bertrand P. Inside HDAC with HDAC inhibitors. Eur J Med Chem 2010;45.
[55] Gao L, Cueto MA, Asselbergs F, Atadja P. Cloning and functional characterization of HDAC11, a novel member of the human histone deacetylase family. J Biol Chem 2002;277:25748–55.
[56] Norton VG, Imai BS, Yau P, Bradbury EM. Histone acetylation reduces nucleosome core particle linking number change. Cell 1989;57:449–57.
[57] Ropero S, Esteller M. The role of histone deacetylases (HDACs) in human cancer. Mol Oncol 2007;1:19–25.
[58] Kim MK, Kang YJ, Kim DH, Hossain MA, Jang JY, Lee SH, et al. A novel hydroxamic acid derivative, MHY218, induces apoptosis and cell cycle arrest through downregulation of NF-κB in HCT116 human colon cancer cells. Int J Oncol 2014;44:256–64.
[59] Konstantinopoulos PA, Wilson AJ, Saskowski J, Wass E, Khabele D. Suberoylanilide hydroxamic acid (SAHA) enhances olaparib activity by targeting homologous recombination DNA repair in ovarian cancer. Gynecol Oncol 2014;133:599–606.
[60] Ungerstedt JS, Sowa Y, Xu W-S, Shao Y, Dokmanovic M, Perez G, et al. Role of thioredoxin in the response of normal and transformed cells to histone deacetylase inhibitors. Proc Natl Acad Sci USA 2005;102:673–8.
[61] Gui C, Ngo L, Xu WS, Richon VM, Marks PA. Histone deacetylase (HDAC) inhibitor activation of p21 WAF1 involves changes in promoter-associated proteins, including HDAC1. Proc Natl Acad Sci USA 2004;101:1241–6.
[62] Sandor V, Senderowicz A, Mertins S, Sackett D, Sausville E, Blagosklonny MV, et al. P21-dependent G 1 arrest with downregulation of cyclin D1 and upregulation of cyclin E by the histone deacetylase inhibitor FR901228. Br J Cancer 2000;83:817–25.
[63] Peart MJ, Smyth GK, van Laar RK, Bowtell DD, Richon VM, Marks PA, et al. Identification and functional significance of genes regulated by structurally different histone deacetylase inhibitors. Proc Natl Acad Sci USA 2005;102:3697–702.
[64] Bolden JE, Peart MJ, Johnstone RW. Anticancer activities of histone deacetylase inhibitors. Nat Rev Drug Discov 2006;5:769–84.
[65] Ashkenazi A. Targeting death and decoy receptors of the tumour-necrosis factor superfamily. Nat Rev Cancer 2002;2:420–30.
[66] Insinga A, Monestiroli S, Ronzoni S, Gelmetti V, Marchesi F, Viale A, et al. Inhibitors of histone deacetylases induce tumor-selective apoptosis through activation of the death receptor pathway. Nat Med 2005;11:71–6.

[67] Burz C, Berindan-Neagoe I, Balacescu O, Irimie A. Apoptosis in cancer: key molecular signaling pathways and therapy targets. Acta Oncol (Stockholm, Sweden) 2009;48:811–21.
[68] Zhao Y, Tan J, Zhuang L, Jiang X, Liu ET, Yu Q. Inhibitors of histone deacetylases target the Rb-E2F1 pathway for apoptosis induction through activation of proapoptotic protein Bim. Proc Natl Acad Sci USA 2005;102:16090–5.
[69] Osada H, Tatematsu Y, Saito H, Yatabe Y, Mitsudomi T, Takahashi T. Reduced expression of class II histone deacetylase genes is associated with poor prognosis in lung cancer patients. Int J Cancer 2004;112:26–32.
[70] Kong X, Lin Z, Liang D, Fath D, Sang N, Caro J. Histone deacetylase inhibitors induce VHL and ubiquitin-independent proteasomal degradation of hypoxia-inducible factor 1alpha. Mol Cell Biol 2006;26:2019–28.
[71] Olsen EA, Kim YH, Kuzel TM, Pacheco TR, Foss FM, Parker S, et al. Phase IIb multicenter trial of vorinostat in patients with persistent, progressive, or treatment refractory cutaneous T-cell lymphoma. J Clin Oncol 2007;25:3109–15.
[72] Kirschbaum M, Frankel P, Popplewell L, Zain J, Delioukina M, Pullarkat V, et al. Phase II study of vorinostat for treatment of relapsed or refractory indolent non-Hodgkin's lymphoma and mantle cell lymphoma. J Clin Oncol 2011;29:1198–203.
[73] Ogura M, Ando K, Suzuki T, Ishizawa K, Oh SY, Itoh K, et al. A multicentre phase II study of vorinostat in patients with relapsed or refractory indolent B-cell non-Hodgkin lymphoma and mantle cell lymphoma. Br J Haematol 2014;165:768–76.
[74] Blumenschein GR, Kies MS, Papadimitrakopoulou VA, Lu C, Kumar AJ, Ricker JL, et al. Phase II trial of the histone deacetylase inhibitor vorinostat (Zolinza, suberoylanilide hydroxamic acid, SAHA) in patients with recurrent and/or metastatic head and neck cancer. Invest New Drugs 2008;26:81–7.
[75] Luu TH, Morgan RJ, Leong L, Lim D, McNamara M, Portnow J, et al. A phase II trial of vorinostat (suberoylanilide hydroxamic acid) in metastatic breast cancer: a California Cancer Consortium study. Clin Cancer Res 2008;14:7138–42.
[76] Modesitt SC, Sill M, Hoffman JS, Bender DP. A phase II study of vorinostat in the treatment of persistent or recurrent epithelial ovarian or primary peritoneal carcinoma: a Gynecologic Oncology Group study. Gynecol Oncol 2008;109:182–6.
[77] Traynor A, Dubey S. Vorinostat (NSC# 701852) in patients with relapsed non-small cell lung cancer: a Wisconsin Oncology Network phase II study. J Thorac Oncol 2009;4:522–6.
[78] Whittaker SJ, Demierre M-F, Kim EJ, Rook AH, Lerner A, Duvic M, et al. Final results from a multicenter, international, pivotal study of romidepsin in refractory cutaneous T-cell lymphoma. J Clin Oncol 2010;28:4485–91.
[79] Coiffier B, Pro B, Prince HM, Foss F, Sokol L, Greenwood M, et al. Results from a pivotal, open-label, phase II study of romidepsin in relapsed or refractory peripheral T-cell lymphoma after prior systemic therapy. J Clin Oncol 2012;30:631–6.
[80] Haigentz M, Kim M, Sarta C, Lin J. Phase II trial of the histone deacetylase inhibitor romidepsin in patients with recurrent/metastatic head and neck cancer. Oral Oncol 2012;48:1644–8.
[81] Vokess EE, Otterson GA. Phase II study of the HDAC inhibitor romidepsin in relapsed small cell lung cancer (CALGB 30304). J Thorac Oncol 2010;5:1644–8.
[82] Whitehead RP, Rankin C, Hoff PMG, Gold PJ, Billingsley KG, Chapman RA, et al. Phase II trial of romidepsin (NSC-630176) in previously treated colorectal cancer patients with advanced disease: a Southwest Oncology Group study (S0336). Invest New Drugs 2009;27:469–75.
[83] Giaccone G, Rajan A, Berman A, Kelly RJ, Szabo E, Lopez-Chavez A, et al. Phase II study of belinostat in patients with recurrent or refractory advanced thymic epithelial tumors. J Clin Oncol 2011;29:2052–9.

[84] Gitlitz B, Koczywas M, Espinoza-delgado I. Phase II study of Belinostat (PXD101), a histone deacetylase inhibitor, for second line therapy of advanced malignant pleural mesothelioma. J Thorac Oncol 2012;4:97–101.
[85] Mackay HJ, Hirte H, Colgan T, Covens A, MacAlpine K, Grenci P, et al. Phase II trial of the histone deacetylase inhibitor belinostat in women with platinum resistant epithelial ovarian cancer and micropapillary (LMP) ovarian tumours. Eur J Cancer 2010;46:1573–9.
[86] San-Miguel JF, Hungria VT, Yoon SS, Beksac M, Dimopoulos MA, Elghandour A, et al. Panobinostat plus bortezomib and dexamethasone versus placebo plus bortezomib and dexamethasone in patients with relapsed or relapsed and refractory multiple myeloma: a multicentre, randomised, double-blind phase 3 trial. Lancet Oncol 2014;15(11):1195–206.
[87] Choudhary C, Kumar C, Gnad F, Nielsen ML, Rehman M, Walther TC, et al. Lysine acetylation targets protein complexes and co-regulates major cellular functions. Science 2009;325(5942):834–40.
[88] Tamkun JW, Deuring R, Scott MP, Kissinger M, Pattatucci AM, Kaufman TC, et al. Brahma: a regulator of Drosophila homeotic genes structurally related to the yeast transcriptional activator SNF2/SWI2. Cell 1992;68(3):561–72.
[89] Filippakopoulos P, Knapp S. Targeting bromodomains: epigenetic readers of lysine acetylation. Nat Rev Drug Discov 2014;13(5):337–56.
[90] Plotnikov AN, Yang S, Zhou TJ, Rusinova E, Frasca A, Zhou MM. Structural insights into acetylated-histone H4 recognition by the bromodomain-PHD finger module of human transcriptional coactivator CBP. Structure 2014;22(2):353–60.
[91] Filippakopoulos P, Picaud S, Mangos M, Keates T, Lambert JP, Barsyte-Lovejoy D, et al. Histone recognition and large-scale structural analysis of the human bromodomain family. Cell 2012;149 (1):214–31.
[92] Dhalluin C, Carlson JE, Zeng L, He C, Aggarwal AK, Zhou MM. Structure and ligand of a histone acetyltransferase bromodomain. Nature 1999;399(6735):491–6.
[93] Mujtaba S, He Y, Zeng L, Yan S, Plotnikova O, Sachchidanand, et al. Structural mechanism of the bromodomain of the coactivator CBP in p53 transcriptional activation. Mol Cell 2004;13(2):251–63.
[94] Zeng L, Zhang Q, Gerona-Navarro G, Moshkina N, Zhou MM. Structural basis of site-specific histone recognition by the bromodomains of human coactivators PCAF and CBP/p300. Structure 2008;16 (4):643–52.
[95] Owen DJ, Ornaghi P, Yang JC, Lowe N, Evans PR, Ballario P, et al. The structural basis for the recognition of acetylated histone H4 by the bromodomain of histone acetyltransferase gcn5p. EMBO J 2000;19(22):6141–9.
[96] Filippakopoulos P, Qi J, Picaud S, Shen Y, Smith WB, Fedorov O, et al. Selective inhibition of BET bromodomains. Nature 2010;468(7327):1067–73.
[97] Nicodeme E, Jeffrey KL, Schaefer U, Beinke S, Dewell S, Chung CW, et al. Suppression of inflammation by a synthetic histone mimic. Nature 2010;468(7327):1119–23.
[98] Gregory GD, Vakoc CR, Rozovskaia T, Zheng X, Patel S, Nakamura T, et al. Mammalian ASH1L is a histone methyltransferase that occupies the transcribed region of active genes. Mol Cell Biol 2007;27 (24):8466–79.
[99] Malik S, Bhaumik SR. Mixed lineage leukemia: histone H3 lysine 4 methyltransferases from yeast to human. FEBS J 2010;277(8):1805–21.
[100] Yang XJ, Ogryzko VV, Nishikawa J, Howard BH, Nakatani Y. A p300/CBP-associated factor that competes with the adenoviral oncoprotein E1A. Nature 1996;382(6589):319–24.
[101] Trotter KW, Archer TK. The BRG1 transcriptional coregulator. Nucl Recept Signal 2008;6:e004.
[102] Percipalle P, Fomproix N, Cavellan E, Voit R, Reimer G, Kruger T, et al. The chromatin remodelling complex WSTF-SNF2h interacts with nuclear myosin 1 and has a role in RNA polymerase I transcription. EMBO Rep 2006;7(5):525–30.

[103] Venturini L, You J, Stadler M, Galien R, Lallemand V, Koken MH, et al. TIF1gamma, a novel member of the transcriptional intermediary factor 1 family. Oncogene 1999;18(5):1209–17.
[104] Xue Y, Canman JC, Lee CS, Nie Z, Yang D, Moreno GT, et al. The human SWI/SNF-B chromatin-remodeling complex is related to yeast rsc and localizes at kinetochores of mitotic chromosomes. Proc Natl Acad Sci USA 2000;97(24):13015–20.
[105] Pivot-Pajot C, Caron C, Govin J, Vion A, Rousseaux S, Khochbin S. Acetylation-dependent chromatin reorganization by BRDT, a testis-specific bromodomain-containing protein. Mol Cell Biol 2003;23(15):5354–65.
[106] Gamsjaeger R, Webb SR, Lamonica JM, Billin A, Blobel GA, Mackay JP. Structural basis and specificity of acetylated transcription factor GATA1 recognition by BET family bromodomain protein Brd3. Mol Cell Biol 2011;31(13):2632–40.
[107] Lamonica JM, Deng W, Kadauke S, Campbell AE, Gamsjaeger R, Wang H, et al. Bromodomain protein Brd3 associates with acetylated GATA1 to promote its chromatin occupancy at erythroid target genes. Proc Natl Acad Sci USA 2011;108(22):E159–68.
[108] Muller S, Filippakopoulos P, Knapp S. Bromodomains as therapeutic targets. Expert Rev Mol Med 2011;13:e29.
[109] Anand P, Brown JD, Lin CY, Qi J, Zhang R, Artero PC, et al. BET bromodomains mediate transcriptional pause release in heart failure. Cell 2013;154(3):569–82.
[110] Bowser R, Giambrone A, Davies P. FAC1, a novel gene identified with the monoclonal antibody Alz50, is developmentally regulated in human brain. Dev Neurosci 1995;17(1):20–37.
[111] Jones MH, Hamana N, Shimane M. Identification and characterization of BPTF, a novel bromodomain transcription factor. Genomics 2000;63(1):35–9.
[112] Boehm D, Conrad RJ, Ott M. Bromodomain proteins in HIV infection. Viruses 2013;5(6):1571–86.
[113] Boehm D, Calvanese V, Dar RD, Xing S, Schroeder S, Martins L, et al. BET bromodomain-targeting compounds reactivate HIV from latency via a Tat-independent mechanism. Cell Cycle 2013;12(3):452–62.
[114] Steffan JS, Kazantsev A, Spasic-Boskovic O, Greenwald M, Zhu YZ, Gohler H, et al. The Huntington's disease protein interacts with p53 and CREB-binding protein and represses transcription. Proc Natl Acad Sci USA 2000;97(12):6763–8.
[115] Saura CA, Valero J. The role of CREB signaling in Alzheimer's disease and other cognitive disorders. Rev Neurosci 2011;22(2):153–69.
[116] Nucifora Jr FC, Sasaki M, Peters MF, Huang H, Cooper JK, Yamada M, et al. Interference by huntingtin and atrophin-1 with cbp-mediated transcription leading to cellular toxicity. Science 2001;291(5512):2423–8.
[117] Mertz JA, Conery AR, Bryant BM, Sandy P, Balasubramanian S, Mele DA, et al. Targeting MYC dependence in cancer by inhibiting BET bromodomains. Proc Natl Acad Sci USA 2011;108(40):16669–74.
[118] Delmore JE, Issa GC, Lemieux ME, Rahl PB, Shi J, Jacobs HM, et al. BET bromodomain inhibition as a therapeutic strategy to target c-Myc. Cell 2011;146(6):904–17.
[119] Ott CJ, Kopp N, Bird L, Paranal RM, Qi J, Bowman T, et al. BET bromodomain inhibition targets both c-Myc and IL7R in high-risk acute lymphoblastic leukemia. Blood 2012;120(14):2843–52.
[120] Lockwood WW, Zejnullahu K, Bradner JE, Varmus H. Sensitivity of human lung adenocarcinoma cell lines to targeted inhibition of BET epigenetic signaling proteins. Proc Natl Acad Sci USA 2012;109(47):19408–13.
[121] Da Costa D, Agathanggelou A, Perry T, Weston V, Petermann E, Zlatanou A, et al. BET inhibition as a single or combined therapeutic approach in primary paediatric B-precursor acute lymphoblastic leukaemia. Blood Cancer J 2013;3:e126.
[122] Pastori C, Daniel M, Penas C, Volmar CH, Johnstone AL, Brothers SP, et al. BET bromodomain proteins are required for glioblastoma cell proliferation. Epigenetics 2014;9(4):611–20.

[123] Fiskus W, Sharma S, Qi J, Valenta JA, Schaub LJ, Shah B, et al. Highly active combination of BRD4 antagonist and histone deacetylase inhibitor against human acute myelogenous leukemia cells. Mol Cancer Ther 2014;13(5):1142–54.

[124] French CA. Demystified molecular pathology of NUT midline carcinomas. J Clin Pathol 2010;63 (6):492–6.

[125] Reynoird N, Schwartz BE, Delvecchio M, Sadoul K, Meyers D, Mukherjee C, et al. Oncogenesis by sequestration of CBP/p300 in transcriptionally inactive hyperacetylated chromatin domains. EMBO J 2010;29(17):2943–52.

[126] Mirguet O, Gosmini R, Toum J, Clement CA, Barnathan M, Brusq JM, et al. Discovery of epigenetic regulator I-BET762: lead optimization to afford a clinical candidate inhibitor of the BET bromodomains. J Med Chem 2013;56(19):7501–15.

[127] Boi M, Gaudio E, Bonetti P, Kwee I, Bernasconi E, Tarantelli C, et al. The BET bromodomain inhibitor OTX015 affects pathogenetic pathways in pre-clinical B-cell tumor models and synergizes with targeted drugs. Clin Cancer Res 2015;21:1628–38.

[128] Paone JF, Waalkes TP, Baker RR, Shaper JH. Serum UDP-galactosyl transferase as a potential biomarker for breast carcinoma. J Surg Oncol 1980;15(1):59–66.

[129] Atkinson AEA. Biomarkers and surrogate endpoints: preferred definitions and conceptual framework. Clin Pharmacol Ther 2001;69(3):89–95.

[130] WHO/IPCS. Biomarkers and risk assessment: concepts and principles, environmental health criteria 155. Geneva: WHO; 1993.

[131] Food and Drug Administration Modernization Act of 1997; list of documents issued by the Food and Drug Administration that apply to medical devices regulated by the Center for Biologics Evaluation and Research. Food and Drug Administration, HHS. Notice. Fed Regist 1999; 64(79):20312–3.

[132] Berman DM, Bosenberg MW, Orwant RL, Thurberg BL, Draetta GF, Fletcher CD, et al. Investigative pathology: leading the post-genomic revolution. Lab Invest 2012;92(1):4–8.

[133] O'Connor JP, Jackson A, Asselin MC, Buckley DL, Parker GJ, Jayson GC. Quantitative imaging biomarkers in the clinical development of targeted therapeutics: current and future perspectives. Lancet Oncol 2008;9(8):766–76.

[134] Frank R, Hargreaves R. Clinical biomarkers in drug discovery and development. Nat Rev Drug Discov 2003;2(7):566–80.

[135] Sarker D, Workman P. Pharmacodynamic biomarkers for molecular cancer therapeutics. Adv Cancer Res 2007;96:213–68.

[136] Zhou Q, Gallo JM. The pharmacokinetic/pharmacodynamic pipeline: translating anticancer drug pharmacology to the clinic. AAPS J 2011;13(1):111–20.

[137] Olzscha H, New M, La Thangue NB. Personalised cancer medicine: fulfilling the promise. In: Sons JW, editor. Encyclopedia of life sciences. Chichester: John Wiley & Sons; 2013. p. 1–10.

[138] Shakir SA, Layton D. Causal association in pharmacovigilance and pharmacoepidemiology: thoughts on the application of the Austin Bradford-Hill criteria. Drug Saf 2002;25(6):467–71.

[139] Aronson JK. Biomarkers and surrogate endpoints. Br J Clin Pharmacol 2005;59(5):491–4.

[140] Raftery J. NICE and the challenge of cancer drugs. BMJ 2009;338:b67.

[141] Lim S, Metzger E, Schule R, Kirfel J, Buettner R. Epigenetic regulation of cancer growth by histone demethylases. Int J Cancer 2010;127(9):1991–8.

[142] Bean GR, Ibarra Drendall C, Goldenberg VK, Baker Jr JC, Troch MM, Paisie C, et al. Hypermethylation of the breast cancer-associated gene 1 promoter does not predict cytologic atypia or correlate with surrogate end points of breast cancer risk. Cancer Epidemiol Biomarkers Prev 2007;16 (1):50–6.

[143] Flanagan JM, Popendikyte V, Pozdniakovaite N, Sobolev M, Assadzadeh A, Schumacher A, et al. Intra- and interindividual epigenetic variation in human germ cells. Am J Hum Genet 2006;79(1):67–84.

[144] Moriyama T, Matsumoto T, Nakamura S, Jo Y, Mibu R, Yao T, et al. Hypermethylation of p14 (ARF) may be predictive of colitic cancer in patients with ulcerative colitis. Dis Colon Rectum 2007;50 (9):1384–92.
[145] Weisenberger DJ, Siegmund KD, Campan M, Young J, Long TI, Faasse MA, et al. CpG island methylator phenotype underlies sporadic microsatellite instability and is tightly associated with BRAF mutation in colorectal cancer. Nat Genet 2006;38(7):787–93.
[146] Ogino S, Nosho K, Kirkner GJ, Kawasaki T, Meyerhardt JA, Loda M, et al. CpG island methylator phenotype, microsatellite instability, BRAF mutation and clinical outcome in colon cancer. Gut 2009;58(1):90–6.
[147] Chung W, Kwabi-Addo B, Ittmann M, Jelinek J, Shen L, Yu Y, et al. Identification of novel tumor markers in prostate, colon and breast cancer by unbiased methylation profiling. PLoS One 2008;3(4): e2079.
[148] Rogers CG, Yan G, Zha S, Gonzalgo ML, Isaacs WB, Luo J, et al. Prostate cancer detection on urinalysis for alpha methylacyl coenzyme a racemase protein. J Urol 2004;172(4 Pt 1):1501–3.
[149] Widschwendter A, Muller HM, Fiegl H, Ivarsson L, Wiedemair A, Muller-Holzner E, et al. DNA methylation in serum and tumors of cervical cancer patients. Clin Cancer Res 2004;10(2):565–71.
[150] Calin GA, Cimmino A, Fabbri M, Ferracin M, Wojcik SE, Shimizu M, et al. MiR-15a and miR-16-1 cluster functions in human leukemia. Proc Natl Acad Sci USA 2008;105(13):5166–71.
[151] Sampath D, Calin GA. miRs: fine-tuning prognosis in CLL. Blood 2009;113(21):5035–6.
[152] Ferrari R, Pellegrini M, Horwitz GA, Xie W, Berk AJ, Kurdistani SK. Epigenetic reprogramming by adenovirus e1a. Science 2008;321(5892):1086–8.
[153] Rasti M, Grand RJ, Mymryk JS, Gallimore PH, Turnell AS. Recruitment of CBP/p300, TATA-binding protein, and S8 to distinct regions at the N terminus of adenovirus E1A. J Virol 2005;79(9):5594–605.
[154] Weichert W, Roske A, Gekeler V, Beckers T, Stephan C, Jung K, et al. Histone deacetylases 1, 2 and 3 are highly expressed in prostate cancer and HDAC2 expression is associated with shorter PSA relapse time after radical prostatectomy. Br J Cancer 2008;98(3):604–10.
[155] Sehn LH, Gascoyne RD. Diffuse large B-cell lymphoma: optimizing outcome in the context of clinical and biologic heterogeneity. Blood 2015;125(1):22–32.
[156] Nogai H, Dorken B, Lenz G. Pathogenesis of non-Hodgkin's lymphoma. J Clin Oncol 2011;29 (14):1803–11.
[157] Ngo VN, Young RM, Schmitz R, Jhavar S, Xiao W, Lim KH, et al. Oncogenically active MYD88 mutations in human lymphoma. Nature 2011;470(7332):115–19.
[158] Bohers E, Mareschal S, Bertrand P, Viailly PJ, Dubois S, Maingonnat C, et al. Activating somatic mutations in diffuse large B-cell lymphomas: lessons from next generation sequencing and key elements in the precision medicine era. Leuk Lymphoma 2014;Early online:1–10.
[159] Esteller M, Garcia-Foncillas J, Andion E, Goodman SN, Hidalgo OF, Vanaclocha V, et al. Inactivation of the DNA-repair gene MGMT and the clinical response of gliomas to alkylating agents. N Engl J Med 2000;343(19):1350–4.
[160] Hegi ME, Diserens AC, Gorlia T, Hamou MF, de Tribolet N, Weller M, et al. MGMT gene silencing and benefit from temozolomide in glioblastoma. N Engl J Med 2005;352(10):997–1003.
[161] Fotheringham S, Epping MT, Stimson L, Khan O, Wood V, Pezzella F, et al. Genome-wide loss-of-function screen reveals an important role for the proteasome in HDAC inhibitor-induced apoptosis. Cancer Cell 2009;15(1):57–66.
[162] Khan O, Fotheringham S, Wood V, Stimson L, Zhang C, Pezzella F, et al. HR23B is a biomarker for tumor sensitivity to HDAC inhibitor-based therapy. Proc Natl Acad Sci USA 2010;107(14):6532–7.
[163] New M, Olzscha H, Liu G, Khan O, Stimson L, McGouran J, et al. A regulatory circuit that involves HR23B and HDAC6 governs the biological response to HDAC inhibitors. Cell Death Differ 2013;20 (10):1306–16.

[164] Yeo W, Chung HC, Chan SL, Wang LZ, Lim R, Picus J, et al. Epigenetic therapy using belinostat for patients with unresectable hepatocellular carcinoma: a multicenter phase I/II study with biomarker and pharmacokinetic analysis of tumors from patients in the Mayo Phase II Consortium and the Cancer Therapeutics Research Group. J Clin Oncol 2012;30(27):3361–7.

[165] Santo L, Hideshima T, Kung AL, Tseng JC, Tamang D, Yang M, et al. Preclinical activity, pharmacodynamic, and pharmacokinetic properties of a selective HDAC6 inhibitor, ACY-1215, in combination with bortezomib in multiple myeloma. Blood 2012;119(11):2579–89.

[166] Chen R, Mias GI, Li-Pook-Than J, Jiang L, Lam HY, Miriami E, et al. Personal omics profiling reveals dynamic molecular and medical phenotypes. Cell 2012;148(6):1293–307.

[167] Gore SD, Baylin S, Sugar E, Carraway H, Miller CB, Carducci M, et al. Combined DNA methyltransferase and histone deacetylase inhibition in the treatment of myeloid neoplasms. Cancer Res 2006;66 (12):6361–9.

[168] Guerrant W, Patil V, Canzoneri JC, Oyelere AK. Dual targeting of histone deacetylase and topoisomerase II with novel bifunctional inhibitors. J Med Chem 2012;55(4):1465–77.

CHAPTER

EPISENSITIZATION: A NEW WORD FOR A NEW CONCEPT

16

Bryan T. Oronsky[1], Arnold L. Oronsky[2], Michelle Lybeck[1], Neil C. Oronsky[3], Jan J. Scicinski[1], Corey Carter[4], Gary R. Fanger[1] and Tony R. Reid[5]

[1]EpicentRx, Mountain View, CA, USA [2]InterWest Partners, Menlo Park, CA, USA [3]CFLS, San Jose, CA, USA [4]John P. Murtha Cancer Center, Walter Reed National Military Medical Center/National Cancer Institute, Bethesda, MD, USA [5]Moores Cancer Institute, University of California, San Diego, CA, USA

CHAPTER OUTLINE

G. Egger & P. Arimondo (Eds): Drug Discovery in Cancer Epigenetics. DOI: http://dx.doi.org/10.1016/B978-0-12-802208-5.00016-3

16.1 INTRODUCTION

Cancer is a complex evolution of genetic and epigenetic changes that lead to aberrant and uncontrolled cellular growth. Epigenetic changes are a key foundation to these events that silence key genes, which control cellular growth and apoptosis and contribute to transformative potential. The purpose of this chapter is to define and examine the treatment strategy of "episensitization," or resensitization by epigenetic means, which challenges the implicit and longstanding assumption in oncology that the reuse of previously tried therapies is contraindicated. Over the years several manuscripts have attempted to define the word resensitization [1], rather unsatisfactorily, given the lack of context and the use of the same root word in the definition, as "A second or subsequent sensitization, especially following desensitization." To avoid cross-purposes and confusion, we define resensitization, borrowing from and building on the above definition, as a renewed clinical benefit from chemotherapy, immunotherapy, or radiotherapy that was previously effective but to which the tumor had become desensitized or resistant. Episensitization is a hybrid term, coined by Oronsky, Scicinski, Fanger, and Reid, which follows on from clinical observations and refers to the mechanistic epigenetic underpinnings of resensitization [2].

Among the many formidable challenges in the treatment of cancer, which include inter- and intratumor heterogeneity, immune suppression and a dearth of actionable/druggable targets, the most inevitable and persistent in all forms and stages of disease is the development of acquired therapeutic resistance [3,4]. Its primary clinical manifestation is tumor progression. This chapter discusses the clinical evidence for active and passive episensitization, that is, the reversal of clinical drug resistance in terms of epigenetically mediated histone hypoacetylation and DNA hypermethylation events. Strictly speaking, an epigenetic event implies reversibility, and on the premise that cancer cells constantly sense and adapt to changes in their "bioterrain," the effect of epigenetic modulation on gene expression and the tumor microenvironment may mediate a return to a baseline state of treatment susceptibility. Accordingly, several epigenetic agents are highlighted in this chapter that, through the reactivation of silenced gene expression, "push the reset button" on acquired therapeutic resistance. These epigenetic agents are decitabine (DAC), 5-azacitidine (5-AZA), entinostat, resminostat, epigenetically targeted oncolytic viral vectors, and RRx-001. Analogous to consecutive booster immunization, active episensitization may involve repeated exposure to epigenetic agents to "prime and boost" the response of tumor cells to retried therapies.

Worldwide, cancer was responsible for an estimated 8.2 million deaths in 2012 [5] (the time lag in the collection [6], compilation and release of mortality data is 3–4 years behind the current year), primarily due to the development of acquired resistance. Episensitization is a paradigm that addresses the major challenge of drug resistance and is a highly relevant treatment strategy both as an antidote to the next-step practice of permanently switching to different drugs at each line of therapy, which rapidly exhausts the remaining available treatment options and as a blueprint to substantially prolong overall survival (OS).

"Epigenetics" is an elastic, polysemous term, possessing several different meanings in different contexts [7]. It was originally coined in the early 1940s by the British developmental biologist, Conrad Waddington [8], to refer broadly to "causal interactions between genes and their products," which manifest as the phenotype of an organism. For biologists, epigenetics is linked to phenotypic plasticity and adaptive capacity; "software" to the gene sequence's "hardware." However, to oncologists, in general, epigenetics is interpreted as reversible changes in gene expression, requiring active maintenance [9] (unlike genetic modifications) and potentially manipulatable by small-molecule DNA methyltransferase (DNMT) and histone deacetylase (HDAC) inhibitors; these changes are brought about in two main ways: (i) by CpG island methylation, thereby silencing gene expression and (ii) by histone modifications where methyl, acetyl, and other groups are added to the histones in chromatin with different effects on gene expression. For example, the acetylation of H3K27 (H3K27ac) is associated with many active genes in that region of the chromatin, while the trimethylation of histone H3 lysine 27 (H3K27me3) is associated with transcriptional silencing [10]. In summary, epigenetic modifications, such as DNA methylation and histone modification, alter chromatin structure and DNA accessibility to sequence-specific transcription factors, thereby regulating patterns of gene expression. These epigenetic modifications can contribute to the cancer progression and also may play a role in development of resistance to anticancer therapies [11].

For the purposes of this chapter, "resensitization" refers to clinical benefit in the context of previous exposure to a particular agent or combination of agents that was used effectively but where the patient become resistant while "sensitization" indicates enhanced clinical benefit in the absence of previous exposure. Episensitization is a portmanteau coinage of epigenetics and sensitization by Oronsky, Scicinski, Fanger, and Reid that refers to epigenetic reversal of gene expression patterns associated with acquired resistance to treatment [12]. A related keyword is "priming," which refers to the induction of drug sensitivity. In the context of episensitization, epigenetic agents precondition or "prime" the tumor to change its phenotype in a manner that makes it markedly more sensitive to further treatment. A "primed" state may last for months or even years (see the multiepigenetic agent, RRx-001). Under the general rubric of episensitization is the term epi-resensitization.

This chapter will serve as an introduction to the relatively underexplored strategy of tumor resensitization, requiring a lexical innovation, that is, a compound new word "epi-resensitization" most appropriate to describe resensitization by epigenetic means, and includes the definition and the main characteristics; analyzes the reasons for the failure of traditional therapies (chemotherapy, radiotherapy) and the reversible epigenetic mechanisms inherent in the microenvironment; discusses the "butterfly effect" of active and spontaneous epigenetic modulation and the therapeutic implications of treatment reintroduction; and makes the case that while episensitization occurs sporadically as "one-off" exemptions in the context of passive strategies like treatment interruptions and ultra-low doses, for example, it is only reliably and stably observed with epigenetic therapies.

16.2 THE CANCER ECOSYSTEM

Cancer has been described [13] as an ecological process whereby tumor cells "terraform" their environment, remodeling the tissue structure, releasing growth factors, cytokines, and matrix metalloproteinases and generating oxidative and metabolic stress [14], in order to out-compete the indigenous population of normal cells. This "chaotic" microenvironment rearrangement [15] sets in motion an evolving dynamic in which the cancer cell must continually finetune its genetic expression to cope with the harsh, heterologous environmental conditions anathema to normal tissue, for example, hypoxia, nutrient deprivation, poor blood flow, low pH, and increased oxidative stress. Further, effective anticancer therapies can also pressure these processes to adapt and evolve in ways that limit drug use due to changes that lead to insensitivity to drug treatment and hence patients become refractory.

To borrow from the concept of "ecogenetic feedback" in evolutionary biology [16], which describes a synergistic interaction between ecological and genetic effects, cancer cells "eco-(epi) genetically" regulate transcription to maximize the match of phenotype to the prevailing microenvironment. Thus, in a reciprocal manner, tumor cells influence the microenvironment and respond to changes in it by dynamically regulating transcription, epigenetically upregulating and downregulating the expression of particular genes [17] in real-time, which results in specific epigenetic signatures [18]. The collective behavior of these specific gene expression patterns and epigenetic marks is to optimize survivability in particular microenvironmental conditions. This built-in flexibility is known as phenotypic plasticity [19] and greatly increases the competitiveness of the cancer cell vis-à-vis normal differentiated cells, allowing the tumor to adjust its phenotype according to local environmental conditions, provided that the exaggerated costs of an adaptive plastic response are not selectively disadvantageous, since protein synthesis is energetically the most expensive process in the cell [20]. The fact that the main energy-generating pathway in eukaryotic cells, oxidative phosphorylation or OxPhos, is repressed in favor of the Warburg effect [21], a metabolic hallmark of cancer cells, whereby mitochondrial damage is circumvented by increased glycolytic flux, means that cancer cells must carefully balance their energy budgets—glycolysis is an inefficient pathway for ATP production, generating only two adenosine triphosphate (ATP) molecules per molecule of glucose, while complete oxidation of one glucose molecule by oxidative phosphorylation can generate up to 36 ATP molecules [22]; hence, spare capacity is not infinite.

Treatment of cancer with molecularly targeted agents that modulate specific genetic abnormalities like VEGF, epidermal growth factor receptor (EGFR), or BRAF rarely succeeds because their scope is too limited or "surgical": the built-in plasticity or flexibility [23,24] of the cancer epigenome facilitates rapid adaptation and the energetic costs that this adaptation entails are supportable. Consequently, *de novo* or acquired (or secondary) resistance, defined as progression after initial benefit, inevitably ensues [25], which is strongly linked to shortened OS. In contrast to these agents with selective or narrow target specificity, multidimensional epigenetic compounds act locally but "behave" globally. Through the inhibition of selective enzymes, they pleiotropically affect the expression of *multiple* genes and, in this way, not only increase energy expenditure but also globally disrupt the delicate ecological balance between the microenvironment and the tumor cells, thereby producing a consequent decrease in microenvironmental fitness.

16.3 EPIGENETIC MODULATION AND THE BUTTERFLY EFFECT

The MIT mathematician, meteorologist, and father of chaos theory Edward N. Lorenz [26] postulated the existence of a Butterfly Effect [27], whereby small initial changes, "tiny initiating events," have profound ripple effects on the outcome of the whole system. The title of a lecture that Lorenz gave in 1972 entitled "Predictability: does the flap of a butterfly's wings in Brazil set off a tornado in Texas?" proposes that the tiny disturbance from the flap of a butterfly's wings in the Amazon has the potential to propagate and amplify into a tornado halfway around the world. This concept is also illustrated in the proverb entitled "For Want of a Nail" [28]:

> *For want of a nail the shoe was lost.*
> *For want of a shoe the horse was lost.*
> *For want of a horse the rider was lost.*
> *For want of a rider the message was lost.*
> *For want of a message the battle was lost.*
> *For want of a battle the kingdom was lost.*
> *And all for the want of a horseshoe nail.*

In this proverb, the lack of a simple nail sets off a linear chain of accumulated subevents, which are concatenated together to bring about a major event, the downfall of a kingdom. Similarly, a small perturbation, the inhibition of epigenetic modulators like HDACs or DNMTs, alters the pattern of coordinated gene expression in the tumor, resulting in a multiplier effect with dramatic downstream consequences (Figure 16.1).

For example, while inhibition of HDAC enzymatic activity only affects a low percentage of protein-coding genes (on the order of 5–20%) [29], the fact that these proteins converge on pathways linked to gene expression, cell proliferation, cell migration, cell death, immune pathways, and angiogenesis creates a sum of sudden unfavorable microenvironmental changes, forcing the tumor to expend energy to adapt (expend to defend) and tipping the delicate balance between chemosensitivity and chemoresistance in favor of the former [30].

16.4 RESISTANCE IS ENERGETICALLY EXPENSIVE AND INEFFICIENT

From a resource allocation perspective, acquired resistance to cytotoxicity is energy-intensive, entailing an increased metabolic expense. Dr Robert Gatenby, a practitioner of mathematical medicine, has written extensively about the emergence of therapy resistance in cancer, and it is worth summarizing the conclusions from his evolutionary modeling studies for the purposes of this chapter. Gatenby refers to resistance as an inevitable but impermanent tradeoff, due to the high cost of

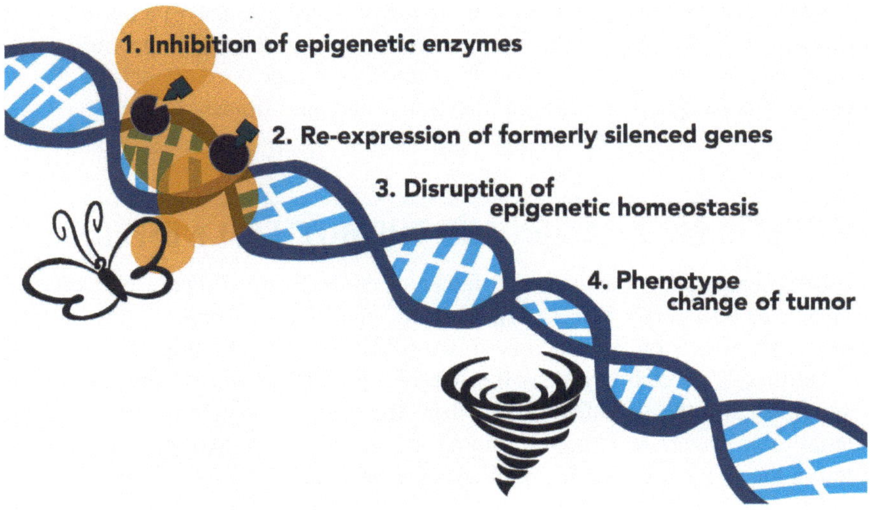

FIGURE 16.1

The butterfly effect of epigenetic inhibition. Seemingly trivial changes, in this case inhibition of epigenetic enzymes has disproportionately large consequences on the gene expression phenotype, similar to the meterological "butterfly effect," also known as sensitive dependence on initial conditions, proposed by MIT Professor Edward N. Lorenz, who described how small changes in initial weather data may result in large differences in simulated weather conditions. As in weather forecasting, seemingly trivial inputs—like the flap of a butterfly's wings in one location—can disproportionately impact later conditions a world away.

maintaining it. That is, resistance to cytotoxicity is an active process, requiring ATP expenditure to upregulate xenobiotic metabolism, DNA repair, and inactivation of cell death pathways [31]. In a case of calculus versus cancer, Gatenby demonstrates mathematically that treatment failure in oncology is, for the most part, iatrogenically mediated. In other words, that the standard one-size-fits-all continuous maximally tolerated dose or MTD more-is-better empiricism, which remains the focus of cancer treatment, paradoxically sets the stage for failure by promoting the growth of resistant clones.

It is well established that tumors, genetically and architecturally heterogeneous [32], are composed of multiple different cell types. Within that heterogeneity, which is a byproduct of genetic instability, drug-intolerant cells are initially in the minority due to asymmetric competition with drug-sensitive variants. The possession of multiple defensive mechanisms, leading to the acquisition of a refractory phenotype, is energetically expensive and inefficient [33]. Like medieval knights restricted by heavy, clunky armor, resistant subclones are outcompeted and outnumbered by nimbler, less encumbered drug-sensitive subclones (which maintain a broader overall niche) for "real estate" and resources within the tissue microenvironment. Thus, drug sensitivity is the "default" state of the cancer cell. However, the selection pressure of MTD chemo- or radiotherapy, designed to induce maximal lethality [34], shifts the "balance of power" by promoting the rapid expansion of resistant populations through the elimination of competing tumor clones. Therefore, because high-dose treatment imposes these drastic changes on the ecosystem, removing the inhibitory effect of chemotherapy-sensitive cancer cells, and thereby increasing the "fitness" of the drug-resistant variant, it paradoxically brings about the very outcome that it purports to prevent—regeneration of the malignancy. In contrast, if chemosensitive cells are not displaced, they suppress the proliferation and colonization of resistant populations. A similar situation occurs in different body niches (e.g., oral cavity, skin, vagina, stomach, urinary tract, and intestines), where the normal flora limit the ability of exogenous pathogens or indigenous opportunistic commensals [35] to gain a stable foothold and multiply. However, in the setting of antibiotic administration, for example, subsequent overcolonization due to perturbation or eradication of the normal flora may lead to superficial or invasive candidiasis and antibiotic-associated *Clostridium difficile* colitis. In the same way, resistant cancer cells opportunistically replace the sensitive cells eradicated during treatment.

For this reason, Gatenby's prescription is that less is more: on the premise that the goal of complete tumor eradication is not only, in most cases, unachievable but also incompatible with long-term survival, he proposes a "kinder, gentler" paradigm called adaptive therapy [33], a sort of "Goldilocks' approach," which preserves the viability of the drug-sensitive cells through continuous adjustment of lower-dose chemo over less frequent intervals to maintain a mathematically defined target tumor burden that is not too large, not too small, but just right.

Similar sensitization/resensitization strategies, which attempt with lower-dose regimens to achieve a stable population of chemotherapy-sensitive cells that suppress the growth of resistant clones, include treatment holidays, metronomic dosing, OPTIMOX [36,37], and COIN-like [38] intermittent treatment, and hypomethylating agents discussed below.

16.4.1 TREATMENT HOLIDAYS AND THE EPIGENETIC LANDSCAPE

Treatment holidays have several intended functions: first, to facilitate a recovery from cumulative toxicity and to control acute adverse events (e.g., nausea and diarrhea); second, to allow a return to "normal" life, however briefly; and third, to regain the previously effective treatment response after a period of continuous use.

Waddington originally proposed the phrase "epigenetic landscape" to represent the process of biological development [8,39]. Huang [40] and other computational biologists have co-opted the Waddington metaphor to represent a "rugged" energy landscape with hills (high energy) and valleys (low energy). Similar to the concept of potential energy in classical mechanics, biological development is described as a downhill process driven by the landscape gradient. According to Huang, "cells placed on top of a mountain top or at a 'watershed' in the epigenetic landscape will roll down into just the few distinct valleys accessible to them, driving the spontaneous separation into discrete fates" [41].

By extrapolation and analogy from Huang in evolutionary developmental biology to the chemoresistant phenotype in oncology, if the crest of the hill is depicted as a drug-resistant state, which is energetically unfavorable, requiring the consumption of ATP for the transcription and translation of proteins, nonessential for proliferation, but essential for resistance, then in the absence of treatment the cancer cell, like a ball rolling downhill, may spontaneously "arrive" at the "low-energy state" [41] associated with drug sensitivity (Figure 16.2).

For this reason, the introduction of treatment holidays, which consist of structured short-term breaks or chemotherapy-free rest periods, may induce reversion to the lowest-energy or baseline drug-sensitive state.

FIGURE 16.2

Waddington's epigenetic landscape as a framework for cancer cell resistance. Huang et al., developmental biologists, have co-opted Waddington's landscape, to explain a developing cell's fate in terms of a ball rolling down a hill which eventually ends in a valley (mature cell type). In this chapter, we apply this analogy to the loss of acquired or *de novo* resistance. Resistant cells are higher upon the epigenetic landscape and, thus, have a greater potential energy (since resistance is energetically expensive) than nonresistant cells. These resistant cells, represented by balls, may sit at the top of a peak on this landscape, due to inertia, but a small deviation will cause them to "roll downhill" and assume a sensitive phenotype.

Resensitization to EGFR tyrosine kinase inhibitors (TKIs) has been anecdotally observed in nonsmall-cell lung cancer (NSCLC) patients [42,43] and genitourinary cancers after a treatment interruption "with reported clinical benefits and good tolerability when sequencing back to first-line therapy" [44].

Similar retreatment responses have been reported for several other anticancer agents including cyclophosphamide, doxorubicin, 5-fluorouracil, vincristine, and methotrexate [45], which suggests that drug resistance is a metastable state, resulting in "backsliding" to drug sensitivity when the exogenous selective pressure is removed. However, these "passive" resensitization events are welcome but one-off exceptions to the general rule in oncology that chemoresistance is insuperable.

For the sake of completeness we mention another "narrow" one-time exception, which is the 2004 pivotal BOND (**B**owel **O**ncology and Cetuximab A**N**tibo**D**y) trial, serving as the basis for the approval of Cetuximab (Erbitux®), the anti-EGFR monoclonal antibody, where the no-resensitization rule was challenged (we are tempted to write shaken but not stirred). In this trial, metastatic colorectal cancer (mCRC) patients previously treated with irinotecan-based regimens were randomized to cetuximab+irinotecan or cetuximab alone. The response rate (RR) and OS for the combination arm of cetuximab+irinotecan were 23% and 8.6 months, respectively, versus 11% and 6.9 months, for cetuximab alone, respectively, which supports resensitization [46].

These cetuximab resensitization results are limited to a specific niche, the *Kirsten rat sarcoma viral oncogene homolog (KRAS)* mutated irinotecan-refractory setting, making it a special nongeneralizable circumstance, since one of the underlying mechanisms of irinotecan resistance is EGFR upregulation [47]. Nevertheless, this exception does not undermine the generality of the rule regarding the stability of resistance: Wadlow et al. [48] published a phase II trial of 20 patients with cetuximab-refractory mCRC treated subsequently with human anti-EGFR monoclonal antibody, panitumumab (Vectibix®), where no responses were observed.

However, where the resistance rule is inverted or turned upside down is in the context of epigenetic priming, which potentially leads to common, generalizable effects across multiple tumor types, as will be discussed later, since it is the central thesis of this chapter. This complementary, hand-in-glove relationship between epigenetic agents and resensitization necessitates a hybrid word, episensitization, to explain it.

16.4.2 METRONOMIC DOSING

The following title excerpt from a manuscript by Hanahan, Bergers, and Bergsland is an apt and pithy description of metronomic chemotherapy in general: "Less is more, regularly" [49]. In contrast to the MTD-associated "cell kill" paradigm, metronomic administration involves low maintenance doses of conventional chemotherapy to circumvent both myelosuppression and drug resistance [50]. The latter is possibly mediated, at least in part, through epigenetic mechanisms. If the drug-resistant state is thermodynamically or energetically unstable, as Huang et al. have suggested, or an "unnatural" one, as Gatenby has claimed, that will spontaneously default to baseline when the treatment selection pressure is removed or attenuated, then ultra-low doses of chemotherapy may forestall the onset of treatment resistance. However, to the extent that is true, it does not prevent it: resistance inevitably develops [51], possibly due to eventual chemotherapy-mediated depletion of sensitive cells, which allows resistant clones to proliferate, leading to tumor recurrence. Nevertheless, Hahnfeldt et al. [52] have described repeated "resensitization" to chemotherapy after initial refractoriness with metronomic dosing, which suggests that in terms of reversal of resistance less really is more.

16.4.3 INTERMITTENT DOSING

In general, the current dominant paradigm to manage metastatic cancer patients involves sequential lines of *continuous* systemic therapy, with a switch to the next line triggered by radiologic/symptomatic progression or intolerable toxicity, followed by and culminating in an admission to hospice after all other options have been exhausted. However, the prevalence of ad lib "hit-and-run" treatment holidays to diminish the burden of cumulative toxicities and restore drug sensitivity suggests that this paradigm should be reevaluated; in other words, that it may not be necessary or even desirable to administer chemotherapy in an uninterrupted fashion.

For example, mCRC is treated with combination regimens of 5-fluorouracil, folinic acid, and oxaliplatin or irinotecan with addition of targeted therapies, such as bevacizumab, cetuximab, and panitumumab. The administration of prolonged chemotherapy frequently results in acquired resistance as well as toxicity, in particular fatigue, hand–foot syndrome, and oxaliplatin-related neuropathy [53], which significantly impacts quality of life and necessitates treatment interruptions, that is, drug holidays. Two clinical trials in mCRC, OPTIMOX-1, and COIN, have investigated whether it is safe to introduce chemotherapy-free periods without impacting OS. They are briefly discussed below from the perspective of resensitization.

The OPTIMOX-1 trial, intended to address the issue of cumulative neurotoxicity with oxaliplatin-based therapy, demonstrated that induction FOLFOX4 treatment for six cycles followed by maintenance 5-fluorouracil/leucovorin (FU/LV) FU/LV only, with no oxaliplatin for 12 cycles and reintroduction FOLFOX4 for six cycles results in similar efficacy with less neurotoxicity compared with continuous FOLFOX7 until disease progression or toxicity [54].

In the COIN trial, patients were randomized to receive continuous oxaliplatin and fluoropyrimidine combination versus 3 months of oxaliplatin and fluoropyrimidine combination followed by a chemotherapy holiday until disease progression at which point the same treatment was restarted. While nonadherence to the protocol compromised the validity of clinical trial results (at some centers the restart rate of oxaliplatin and fluoropyrimidine was $<60\%$, leading to a detriment of around 6 weeks in survival on the intermittent arm), patients at centers that restarted $>60\%$ tended to survive longer [55]; this result is similar to OPTIMOX-1 [56], where patients treated at more compliant centers with regard to reintroduction of leucovorin (folinic acid), fluorouracil, oxalipplatin (FOLFOX) after maintenance also lived longer, which possibly suggests a resensitization effect through re-emersion of sensitive clones.

16.4.4 HYPOMETHYLATING AGENTS

The same "less is more" dictum [57] that characterizes metronomic dosing also underlies the activity of the azanucleotides, AZA, and DAC. At higher doses, the azanucleotides behave as classical pyrimidine analog cytotoxics [58] that inhibit DNA synthesis and induce cell-cycle arrest; in the 1970s and 1980s, high, toxic doses of AZA and DAC (600–1500 mg/m^2 per course) were administered to treat hematologic diseases [59]. This is in contrast to the FDA-approved schedules of 75 mg/m^2 daily for 7 days and 15 mg/m^2 every 8 h for nine doses for AZA and DAC, respectively, in acute myeloid leukemia (AML). It is at these lower doses that the hypomethylating properties of these agents, resulting from covalent trapping of DNMTs [60], which leads to reexpression of silenced tumor suppressor genes, became apparent. Consequently, the azanucleotide-specific phrase "epigenetically targeted dose" has entered the scientific lexicon to differentiate it from an aza-cytotoxic dose [61].

At these epigenetically targeted doses, *systematic* resensitization to previously failed therapies has been observed most notably in platinum-refractory ovarian cancer, which will be discussed later. At present, note that the word systematic is italicized to call attention to the difference between active resensitization with hypomethylating agents, which elicits a predictable pattern of altered gene expression, and spontaneous resensitization (e.g., from treatment interruptions) that occurs inconsistently and unpredictably. This inconsistency and unpredictability is possibly related to the "inertia" of gene-expression states, which may persist in the absence of an active exogenous "push," like a ball at rest on top of a hill that tends to remain at rest unless acted upon by an outside force, according to Newton's first law.

To return to the image of an epigenetic landscape with peaks and valleys, the resistant phenotype is at a maximum rather than a minimum of its potential energy, and, therefore, like the ball perched precariously at the crest of the hill, reversion to drug sensitivity may happen eventually, that is, the ball finally overcomes inertia and rolls downhill, however due to randomness it is evidently impossible to predict when (or if) that event will occur; hence for this reason passive or spontaneous resensitization strategies are unreliable. The different examples of nonepigenetic resensitization are represented below (Figure 16.3). Clinical examples of active resensitization with hypomethylating agents will be presented later in this chapter.

16.4.5 EPIGENETIC CLASSES AND SPECIFIC INHIBITORS

There are two main classes of epigenetic agents: DNA demethylators and HDAC inhibitors (HDACi). Due to space limitations, only a top-level overview of their mechanisms of action, development status, toxicities, and retreatment potential is provided; preclinical compounds, beyond the scope of this chapter, are not included. In particular, since epi-resensitization is a multiple reuse paradigm, necessitating long-term or (hopefully) even lifetime care, close attention is given to the toxicities of these agents, to assess the feasibility of repeated interval retreatment.

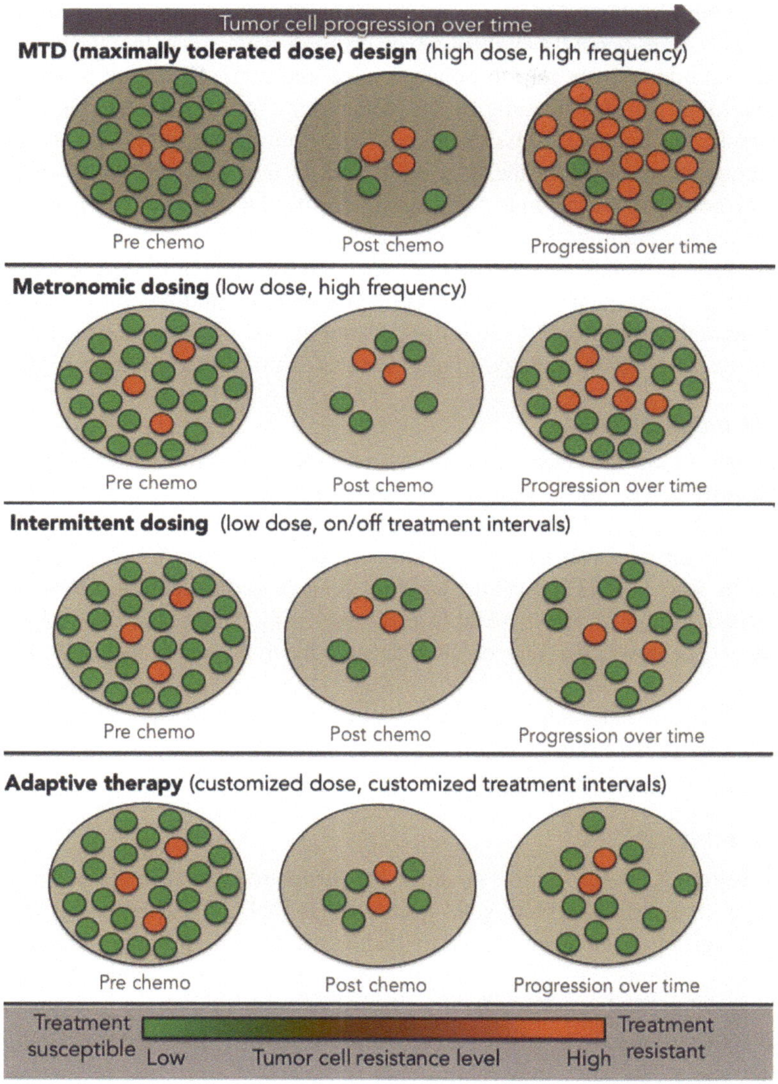

FIGURE 16.3

Nonepigenetic resensitization strategies. From top to bottom. MTD: maximal doses lead to maximal resistance. Metronomic dosing: low doses administered regularly may lead to less selection pressure on the tumor with slower development of resistance. Intermittent dosing: the longer the treatment-free interval, theoretically, the greater the chance for tumors to lose acquired resistance. Adaptive therapy: treatment to contain, not to cure.

As a potential blueprint for a long-term epi-retreatment plan, a consecutive immunization strategy, also known as "prime-boost," which involves priming with a vaccine and boosting it multiple times to maintain immunity may help to conceptualize and mainstream the practice of therapeutic rechallenge. According to this paradigm, an epigenetic "prime" would precede a refractory chemotherapy "boost" in repeat succession to induce a state of perpetual chemosensitivity akin to longlasting antitumor immunity (see the ROCKET trial design). Like Humpty Dumpty, the tumor-specific pattern of gene expression may "fall apart" during the epigenetic "prime" only to eventually reassemble (unlike Humpty Dumpty) during the "boost," at which point, the sequence is reinitiated and repeated, like a nursery rhyme, with the same or different agents, depending on response or tolerability. The goal is to extend OS where patients remain alive and on therapy with, hopefully, less chemotherapy-induced morbidity, assuming the regimen is well tolerated, a prerequisite with an epi-resensitization strategy, given the serial nature of reexposure to chemotherapies.

16.4.5.1 DNMT inhibitors

One of the hallmarks of cancer cells [62] is promoter hypermethylation at CpG islands [63], resulting from overactivity of DNMTs, which contributes to (reversible) transcriptional silencing of tumor suppressor genes. DNMTs catalyze the transfer of a methyl group from the co-factor S-adenosyl-L-methionine to cytosine in CpG dinucleotides. In humans, the primary DNMTs are DNMT1, DNMT3a, and DNMT3b. While possibly an oversimplification, DNA methylation patterns are established by the *de novo* methyltransferases [64], DNMT3A and DNMT3B, and then faithfully propagated to newly replicated DNA strands by the maintenance DNMT, DNMT1 [65] (Table 16.1).

DNMT inhibitors (DNMTi) induce anticancer effects, at least in part, through a hypomethylating effect on genes involved in diverse cellular processes such as apoptosis, cytostasis, differentiation, and tumor angiogenesis (Table 16.2).

16.4.5.2 HDAC inhibitors

Modification of histones by acetylation changes the structure of chromatin [69], which increases the accessibility of transcription factors to nucleosomal DNA. Histone acetyltransferase, which transfers an acetyl group from acetyl-CoA to lysine residues in proteins, and HDAC enzymes, which removes it, maintains the acetylation state of chromatin in a highly dynamic equilibrium. Human HDACs are classified into four families based on their homology to yeast factors. HDACi are divided into four structural classes: (i) hydroxamic acids; (ii) aliphatic acids; (iii) benzamides; and (iv) cyclic tetrapeptides. They may inhibit only some HDACs (HDAC isoform-selective inhibitors) or against all types of HDACs (pan-inhibitors) [69]. While the HDACi are generally characterized by a broad therapeutic window especially compared with AZA and DAC, the most common side effects, nausea/vomiting, fatigue, diarrhea, cardiotoxicity, and transient thrombocytopenia and neutropenia, may predispose to drug intolerability with long-term administration, which is particularly relevant in the context of epi-resensitization.

16.4.5.3 Multiepigenetic inhibitors

To the best of our knowledge, RRx-001, a systemically nontoxic intravenous agent, sourced from the aerospace industry and constructed around a novel dinitroazetidine pharmacophore, in phase II clinical trials, is the only epigenetic agent in the clinic, experimental or approved, that targets

Table 16.1 Overview of DNMTi, Their Mechanisms of Action, Development Status, Clinical Toxicities and Retreatment Potential

Name	Mechanism of Action	Development Status	Clinical Toxicities	Retreatment Potential
AZA (Vidaza™)	Ribonucleoside analog that (i) interrupts mRNA translation and (ii) incorporates into DNA where it inhibits methylation by covalently binding to or "trapping" DNMT1	FDA-approved for treatment of myelodysplastic syndrome (MDS) in 2004	Incorporation of nucleoside analogs into DNA results in a permanent alteration of the genome, raising the possibility of future cancer development. Possibility for life-threatening toxicities such as neutropenia and thrombocytopenia	Low–intermediate
DAC (Dacogen®)	Analog of the nucleoside 2′-deoxycytidine that incorporates into DNA and inhibits methylation by trapping DNMTs, which results in the reduced methylation of cytosines in DNA synthesized after drug treatment	FDA-approved for treatment of MDS in 2006	As above	Low–intermediate
MG98	Oligonucleotide antisense inhibitor of DNMT1	Completed phase I study, demonstrating safety and preliminary evidence of activity	Bone pain, nausea, and fever	High
Procaine and procainamide	Partial competitive inhibitors of DNMT1 that bind to CG-rich DNA and prevent DNMT1 binding [67]	Approved by the FDA for the treatment of cardiac arrhythmias. In preclinical studies	Q–T interval prolongation and hypersensitivity reactions including fever, rash, agranulocytosis and the development of systemic lupus erythematosus [68]	Intermediate
SGI-110	Dinucleotide of DAC and deoxyguanosine that increases the *in vivo* exposure of DAC by protecting it from deamination	Completed phase I and phase II studies in AML	Myelosuppression and grade 1/2 injection site events	Intermediate

Adapted from Gravina et al. [66].

Table 16.2 Overview of Clinical HDACi, Their Chemical Structure, Name, Mechanisms of Action, Development Status, Clinical Toxicities, and Retreatment Potential

Chemical Structure	Name	Mechanism of Action	Development Status	Clinical Toxicities	Retreatment Potential
Hydroxymate	SAHA (vorinostat)	Pan-inhibitor	FDA-approved for treatment of cutaneous T-cell lymphoma (CTCL)	High rate of fatigue, nausea, anorexia, vomiting, and thrombocytopenia	Low–intermediate
Hydroxymate	PXD101 (belinostat)	Pan-inhibitor	Phase I/II in hematological malignancies and solid tumors	Fatigue, diarrhea, atrial fibrillation, and nausea/vomiting	Low–intermediate
Hydroxymate	LBH589 (panobinostat)	Classes I and II	Phase I/II in hematological malignancies and solid tumors	Nausea, diarrhea, vomiting, hypokalemia, loss of appetite, and thrombocytopenia	Low–intermediate
Hydroxymate	4SC-201 (resminostat)	Pan-inhibitor	Phase II orphan status designation in HCC	Reversible gastrointestinal toxicities and fatigue	Low–intermediate
Cyclic peptide	Depsipeptide/FK228 (romidepsin)	Class I	FDA-approved for CTCL phase I/II in hematological malignancies and solid tumors	Fatigue, nausea, anorexia, vomiting, and cardiotoxicity	Low–intermediate
Benzamide	MS-275 (entinostat)	Class I	Phase I/II in hematological malignancies and solid tumors	Reversible neurotoxicity, hypophophatemia, hyponatremia, and hypoalbuminemia	Intermediate
Benzamide	MGCD0103 (mocetinostat)	Class I	Phase I/II in hematological malignancies and solid tumors	Fatigue, nausea, vomiting, cardiotoxicity, and diarrhea	Low–intermediate
Aliphatic fatty acid	Valproic acid	Classes I and IIa	FDA-approved for epilepsy, seizures, mania, bipolar disorders, migraines Phase I/II in hematological malignancies and solid tumors in particular advanced NSCLC	Neurocognitive impairment, somnolence	Intermediate

Adapted from Hrabeta et al. [69].

enzymes of different classes. As an reactive oxygen species/reactive nitrogen species (ROS/RNS)-mediated triple inhibitor of DNMTs, HDACs, and lysine demethylases, RRx-001 has demonstrated preliminary evidence of epi-resensitization to irinotecan in a randomized phase II mCRC trial versus regorafenib fittingly acronymed ROCKET, given the aerospace spin-off, for **R**Rx-001 in COlon Cancer TaKen To Extend Time. The design of this trial, really the first of its kind to systematically explore an alternating "prime-boost" switch-at-progression strategy of epigenetic priming with RRx-001 followed by "boost" irinotecan then RRx-001 then irinotecan, etc., as well as the preliminary results from it, will be discussed below in the section about epi-resensitization trials.

RRx-001 is characterized by a remarkably benign toxicity profile, which makes it uniquely suitable for long-term chronic administration, either singly or combined. The only adverse effect, completely eliminable with longer infusion durations, is reversible pain with intravenous administration. Interestingly, RRx-001 is a radiosensitizer in tumor, while a radioprotector [70] in normal tissue as a result of Nrf2 and p53 upregulation. It may even result in chemoprotection of normal tissue as well, by the same mechanism, although this is speculative, but based on dramatically improved quality of life and performance indices during RRx-001 administration and higher-than-expected patient tolerability of irinotecan-based therapies post-RRx-001. If this indeed turns out to be the case, RRx-001 may improve the side effect profile of formerly intolerable therapies as well as resensitize to them.

In addition to the ongoing metastatic colon cancer ROCKET study, several phase II epi-resensitization-themed clinical trials with RRx-001 are scheduled to start in 2015 in the following tumor types:

- cholangiocarcinoma with sequential resensitization to gemcitabine/cisplatin,
- lung cancer (small-cell lung carcinoma and nonsmall-cell lung carcinoma) with sequential resensitization to platinum doublets,
- hepatocellular carcinoma (HCC) with sequential resensitization to sorafenib,
- multiple myeloma with concomitant resensitization to bortozemib (Velcade),
- allcomers concomitant resensitization to anti-PD-1 inhibitors.

A Google search of epigenetic agents revealed another "triple epigenetic inhibitor" called UVI5008 [71], an analog of the natural product psammaplin A [72], that targets HDACs, sirtuins, and DNMTs and potently "induces cancer cell-selective death ... in several human tumor xenografts and genetic mouse models of human breast cancer in vivo"; however, despite this encouraging preclinical efficacy, UVI5008 has not yet made it to the clinic and is, therefore, as previously mentioned, outside the scope of this chapter.

16.4.5.4 Oncolytic viruses

If epigenetics is operationally defined as reversible changes in the expression patterns of a defined set of genes without modification to the underlying DNA sequence, then viruses are, by definition, innate epigenetic agents. To make the host cell do its bidding that is, serve the viral agenda, which includes synthesizing viral proteins, disrupting the innate antiviral response and producing viral progeny, the virus manipulates the host machinery at the transcriptional level to reprogram and restrict patterns of gene expression, which is the hallmark of epigenetic regulation.

In the same way, the success of so-called oncolytic viruses or OVs (adenovirus, herpes simplex virus, vesicular stomatic virus, reovirus, coxsackie virus, and vaccinia virus), which, in an example of intelligent design [73], have been engineered to infect tumors, depends on their ability to hijack the cellular translational and transcriptional machinery of the cancer cell and downregulate its complex repertoire of protein synthesis to a more restricted set of viral genes and proteins.

Viral vectors have been further altered with therapeutic transgenes to improve both oncolytic potency and replicative selectivity; however, in a repeat of a predictable plot line, a resistant tumor subpopulation will eventually emerge that circumvents the viral tropism and/or the cytotoxic effects of the transgene. One potential solution to this all-too-familiar development is the use of onco*static* viruses that establish slower chronic and/or latent infections, rather than rapid lytic infections, in which the viral genome persists in and disrupts the normal expression of the cancer cell's genes. Although persistent viral infection is highlighted as a drawback or a limitation rather than an advantage [74], because of a failure to induce rapid tumor cell lysis, we hypothesize that oncostasis may lead to episensitization.

In general, viruses are in competition with the host for the cellular translation machinery, prompting them to hinder, in some cases even shut down, host gene expression in favor of their own [75]. Hence, the virus acts an epigenetic "filter," concentrating and confining the "broadband" molecular heterogeneity of the tumor to a narrow spectrum. In cancer cells this highly restricted pattern of gene expression inactivates alternative survival pathways—escape hatches from drug-induced growth inhibition and death, rendering the transcriptionally boxed-in tumor vulnerable to refractory chemotherapies.

ONYX-015 [76], an E1B−55 kDa gene-deleted adenovirus, with a favorable safety profile [77], independent of the route of administration, was evaluated in several combination chemotherapy clinical trials. The first and most significant, a phase II trial in refractory, recurrent head and neck cancer patients with cisplatin and 5-FU, led to an approximately 65% RR, compared to the historic RR of 25−35% in this population. Moreover, no evidence of new tumors or regrowth of treated ones was apparent 6 months after treatment. Because it is not clear how many patients previously received the platinum−5-FU combination, definitive evidence of epi-resensitization is lacking but presumably patients would have been exposed and refractory to one or both of these agents.

Shanghai Sunway Biotechnology, a Chinese company, acquired US development rights to ONYX-015 during phase III and stopped the trial. China's State Food and Drug Administration approved H101, a virus very similar to ONYX-015 (and also owned by Shanghai Sunway Biotechnology), for treatment of head and neck cancer by intratumoral injection in combination with cisplatin and 5-FU chemotherapy. The approval was based on a 79% RR in patients who received intratumoral administration of H101 in combination with chemotherapy versus 40% in controls with cisplatin and 5-FU alone, that is, without virus treatment ($P = 0.001$) [78]. The main side effects reported were fever (45.7%), injection site reaction (28.3%), and influenza-like symptoms (9.8%).

In addition to this remarkably benign safety profile that theoretically allows for repeat administration (assuming protection from neutralizing antibodies) what makes OVs such ideal epi-resensitization agents is their ability to globally restrict gene expression and protein synthesis, which severely limits the escape options of the cancer cell, that is, its ability to access alternative redundant drivers of cellular proliferation and survival.

Finally, these two virotherapy trials, while suggestive of epi-resensitization, were not designed to demonstrate it, unlike the proof-of-concept studies presented in the next section.

16.5 Epi-RESENSITIZATION TRIALS

Now that the theoretical and foundational underpinnings of epi-resensitization have been established we arrive at the much-anticipated Jerry Maguire-esque "moment of truth": show me the proof; evidence of clinical efficacy, which ultimately separates interesting theories and intuition from therapeutically relevant treatment strategies. The null hypothesis assumes that resensitization is an exercise in futility—and toxicity. The alternative hypothesis, ours, posits the contrary. Keeping in mind that the benefit of the doubt goes to the null hypothesis, that is, the status quo, which is correct? To ultimately answer this question, or rather attempt to answer it, since some of the information is piecemeal and preliminary, we turn to the ultimate proving ground of clinical trials, which have formally explored epi-resensitization as a proof-of-concept outcome. Ovarian cancer, HCC, CRC, NSCLC, and cholangiocarcinoma are some of the tumor types where an epi-sensitization strategy has already been explored or is currently under investigation.

16.5.1 OVARIAN CANCER

The combination of a platinum and a taxane, for example, carboplatin and paclitaxel, following cytoreductive surgery remains the standard first-line chemotherapy for advanced ovarian cancer [79]. Nevertheless, despite initial responses to systemic therapy, tumors inevitably develop resistance, at which point they are deemed incurable, accounting for a dismal prognosis, with only 20% surviving at 5 years [80]. For decision-making purposes in recurrent ovarian cancer, patients are somewhat arbitrarily classified [81] (based on >20-year-old definitions) into platinum-refractory, platinum-resistant, and platinum-responsive: the platinum-refractory group is unresponsive and progresses during platinum-based therapy; platinum-resistant disease recurs less than 6 months after completion of treatment; platinum-sensitive disease is characterized by a complete clinical response that lasts at least 6 months from the completion of the initial platinum-based therapy [82]. With the exception of platinum-sensitive disease, which may benefit from retreatment with either cisplatin or carboplatin, resistant/refractory patients do not tend to respond to platinum-based agents [83].

16.5.1.1 Phase II clinical trial of DAC and carboplatin in platinum-refractory/resistant ovarian cancer

Based on preclinical experimental data, which demonstrated enhanced gene expression and drug sensitivity in response to inhibitors of DNMTs (DNMTIs), Matei and Nephew [84] administered daily low-dose DAC for 5 days followed by carboplatin on day 8 in 17 patients with heavily pretreated and platinum-resistant ovarian cancer in a phase II clinical trial. At 8½ months, seven patients were alive, resulting in a 35% objective RR and progression-free survival (PFS) of 10.2 months, with nine patients (53%) free of progression at 6 months; this is compared to the small percentage of short-lived objective responses (<10%) usually induced in this patient population [85], which is indicative of epi-resensitization.

16.5.1.2 Phase I/II study of 5-azacytidine and carboplatin in platinum-refractory/resistant ovarian cancer

In this similar trial, Fu et al. [86] administered AZA 75 mg/m^2 subcutaneously daily for 5 days and carboplatin over 1 h on day 2 every 28 days to 30 patients with platinum-resistant or -refractory ovarian cancer. The ORR was 13.8% (4 of 29 patients) with one patient achieving a complete response and three patients achieving a partial response (PR); the disease control rate (PR plus stable disease (SD)) was 45% (13 of 29 evaluable patients) and the median OS was 14 months, also indicative of epi-resensitization.

16.5.2 ADVANCED HCC

The multikinase inhibitor, sorafenib [87], is the first and, to date, the only systemic agent to statistically improve survival in HCC patients, however, RRs are low and resistance to therapy inevitably develops with no treatment approved for treatment in the second-line setting. Resminostat is an oral pan-HDACi with reversible gastrointestinal toxicities (nausea/vomiting and diarrhea) as the main adverse events.

16.5.2.1 Phase II SHELTER study of resminostat + sorafenib versus resminostat alone in second-line HCC after sorafenib progression in first-line treatment

Data are scant [88] but the OS is reportedly 8.1 months in the resminostat + sorafenib arm versus 4.1 months for resminostat alone. The expected OS in second-line HCC is 5.2 months, resulting in a survival benefit of 3 months in the sorafenib/resminostat epi-resensitization arm.

16.5.3 METASTATIC COLORECTAL CANCER

In metastatic colon cancer, a nuanced "continuum of care" [89] model has supplanted the standard "tick box," straight-line trajectory that characterizes the treatment of other tumor types. While this model involves strategically empiric "mixing and matching" of approved chemotherapy agents, including 5-FU, irinotecan, bevacizumab, capecitabine, panitumumab, cetuximab, and regorafenib (it has been suggested that the actual sequence of regimens may be less important than availability and access to all of them at some point during the course of treatment) [90], patients are not typically rechallenged with formerly failed therapies because of the potential for an unfavorable risk–benefit ratio. Regorafenib is recommended for third-line therapy in patients pretreated with fluoropyrimidines, oxaliplatin, irinotecan, bevacizumab, or anti-EGFR antibodies.

The epigenetics of CRC has been reviewed extensively and a cogent argument for the effectiveness of epigenetic therapy in reversing, for example, 5-FU resistance has been made [91]. Indeed, though vorinostat has been reported to sensitize resistant CRC cells to apoptosis by 5-FU, there are no current clinical results available.

16.5.3.1 Phase II ROCKET study of RRx-001 randomized against current third-line treatment, regorafenib, and followed by irinotecan-based therapies on progression

In this recently initiated randomized phase II clinical trial of RRx-001 versus regorafenib, an oral multikinase inhibitor, in metastatic irinotecan-refractory third/fourth-line colon cancer, reintroduction of irinotecan-based therapy is mandated on progression on both arms, if clinically appropriate. In addition, the RRx-001 patients, if appropriate, are allowed to continue on trial in a perpetual retreatment loop (Figure 16.4). Epi-resensitization is defined by either (i) a drop in carcinoembryonic

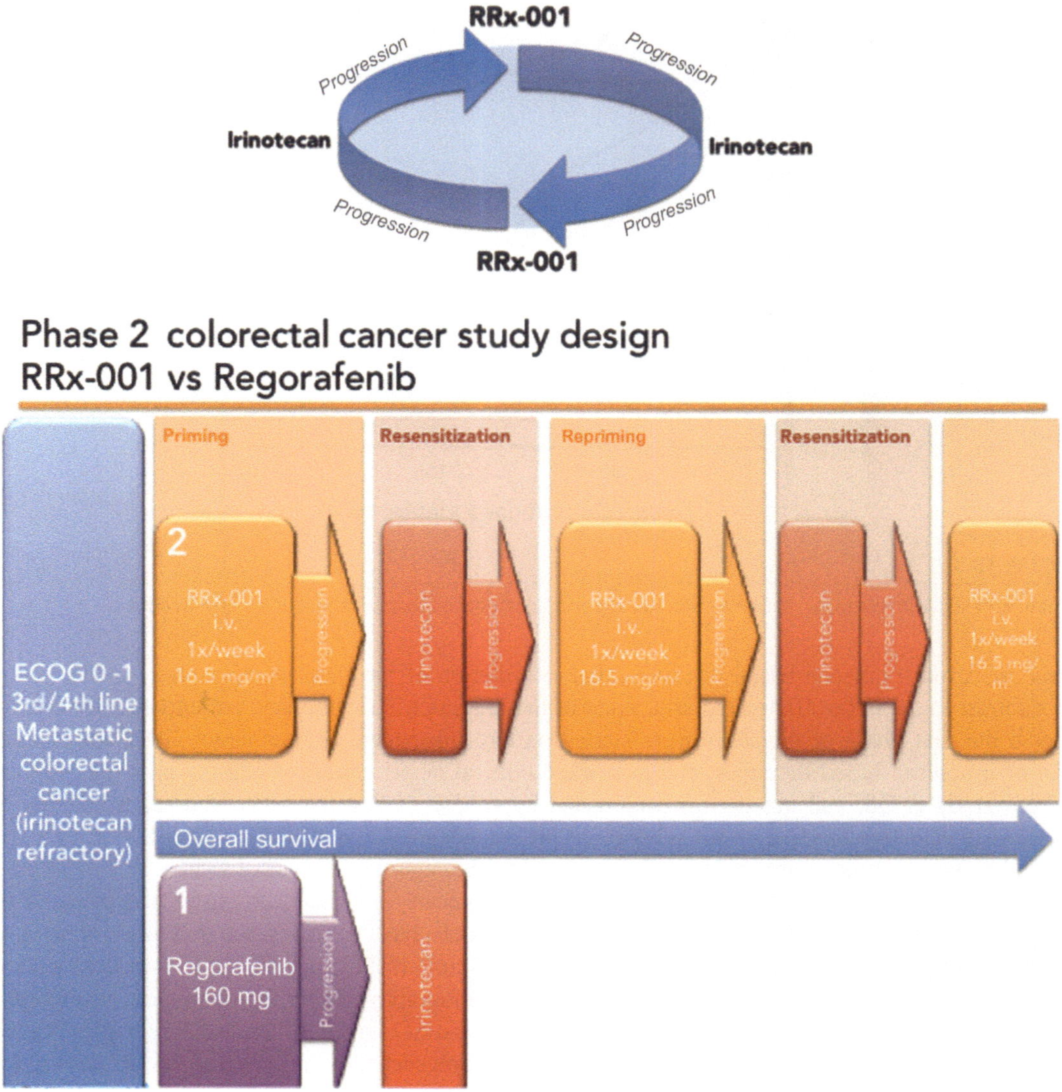

FIGURE 16.4

Clinical trial schema. Left: Clinical trial design of the ROCKET trial demonstrating rechallenge with irinotecan-based therapies after progression on either RRx-001 or regorafenib. Right: The continuing potential resensitization loop of RRx-001-irinotecan-RRx-001-irinotecan, etc., for RRx-001 patients only, not available to regorafenib patients.

antigen (CEA) levels, (ii) clinical and/or radiologic stability, or (iii) prolonged clinical benefit on therapy leading to increase OS. The primary endpoint is OS.

To date, six out of seven (6/7) RRx-001 patients (one is in the hospital for a complicated parapneumonic effusion and, therefore, currently nonevaluable for the purposes of epi-resensitization)

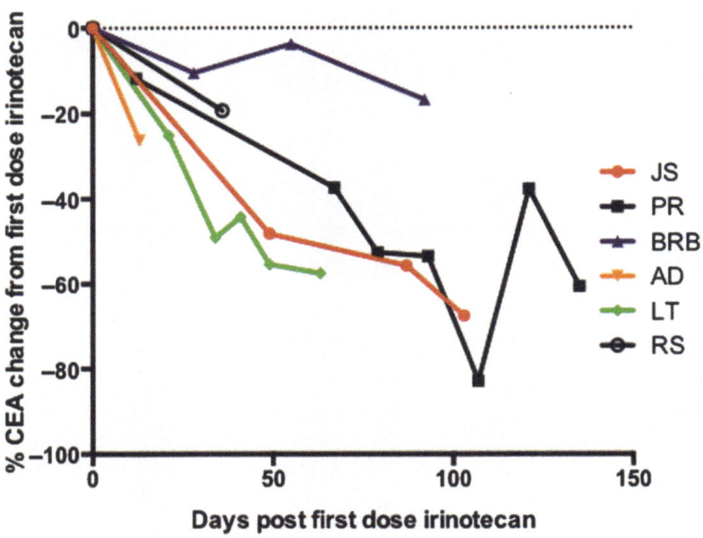

FIGURE 16.5

Decline of CEA levels in six patients after starting irinotecan-based therapies post-RRx-001 treatment.

have restarted irinotecan-based therapies after completion of the first, priming phase of the study. In contrast to the RRx-001-treated arm, none of the regorafenib randomized patients that have completed the first phase of the study were started on any subsequent chemotherapy, let alone irinotecan or irinotecan-based therapies, due to marked clinical deterioration. While the primary endpoint of this study is OS, and not resensitization, the hypothesis is that the successful reintroduction of previously effective therapies with subsequent prolonged stabilization of disease will lead to a statistically significant increase in survival, compared to the regorafenib control arm.

Six of seven (6/7) patients who restarted irinotecan or irinotecan-based therapies (e.g., FOLFIRI+/−bevacizumab or irinotecan plus bevacizumab) have been resensitized as demonstrated by a decline in the tumor-associated marker, CEA (Figure 16.5), improvement in clinical performance status and/or PR/SD on positron emission tomography (PET) or computed tomography (CT). One of the patients had a metabolic PR on PET and SD on CT. These early examples suggest that RRx-001 is effective maintenance therapy that epigenetically primes tumors to re-respond to irinotecan-based therapies. Perhaps even more significantly, these prolonged resensitizations to irinotecan-based therapies are akin to those of treatment-naive mCRC patients, who have never seen chemotherapy before, since the median PFS in second-line metastatic colorectal therapy is 4.5 months [92].

16.5.3.2 Phase I study of SGI-110 combined with irinotecan followed by a randomized phase II study of SGI-110 combined with irinotecan versus regorafenib in previously treated mCRC patients

This is a recently initiated two-part combination phase I clinical trial of the second-generation DNMTi, SGI-110, with irinotecan versus regorafenib in metastatic irinotecan-refractory colon cancer. No data are available.

16.5.4 NONSMALL-CELL LUNG CANCER

The median survival of all patients with advanced NSCLC is 10–12 months. The fraction of patients who are living 1 year after diagnosis has only increased slightly over the past decade. The current first-line chemotherapy standard for patients with advanced NSCLC is one of a number of platinum-based doublets. Second-line treatment for recurrent or progressive disease includes treatment with the chemotherapy agents docetaxel and pemetrexed, or treatment with an oral EGFR antagonist, erlotinib. Currently, patients with advanced NSCLC who have progressed after second-line treatment have limited options. Retrospective analysis using chemotherapy for third-line treatment show RRs of only 2% and median survival of 4 months [93].

16.5.4.1 Phase I/II trial with AZA and entinostat in extensively pretreated recurrent metastatic NSCLC

While technically not epi-resensitization-based, this trial is included since it demonstrates tumor preconditioning or priming prior to subsequent therapies, which is consistent with episensitization. Juergens et al. [94] treated recurrent metastatic NSCLC with AZA and entinostat, inhibitors of DNA methylation and histone deacetylation, respectively. The treatment was well tolerated and achieved a favorable objective RR and survival benefits (>1 year in approximately 20% of the patients and a median OS of 6.4 months) exceeded historical controls (48% expected survival after 6 months). Interestingly, the favorable trend in survival was attributed to an "unusually robust response to subsequent cytotoxic therapies, with which the majority of patients were treated" [12]. The subsequent therapies in the NSCLC trial included pemetrexed, docetaxel, erlotinib, antiprogrammed cell death protein (PD-1) monoclonal antibodies, gemcitabine, irinotecan/bevacizumab, and cisplatin, suggesting that this combination of epigenetic inhibitors reversed the resistant phenotype, making the tumor more widely susceptible to a variety of subsequent chemotherapeutic agents.

16.5.4.2 Pilot three-arm study (TRIPLE THREAT) of RRx-001 administered in small-cell lung cancer, NSCLC, and extrapulmonary neuroendocrine tumors prior to readministration of platinum-based doublet regimens

This trial was scheduled to start recruitment in the first quarter of 2015.

16.5.5 CHOLANGIOCARCINOMA

Cholangiocarcinoma is a heterogeneous group of tumors with an uncommon incidence of 1–2 per 100,000 cases in the United States [95]. For advanced, unresectable patients, who comprise the majority of cases, palliative gemcitabine plus platinum combination chemotherapy is standard practice [96] as first-line treatment, while the benefits and feasibility of second-line chemotherapy are unclear and under investigation.

16.5.5.1 Phase II (EPIC) trial of RRx-001 administered for 6 weeks in second-line therapy prior to administration of gemcitabine/cisplatin

This trial was scheduled to start recruitment in 2015.

16.6 DISCUSSION

The prevailing assumption in oncology that "once resistant always resistant," generally preventing a retrial of previously effective chemotherapy due to the risk of toxicity without benefit, is contradicted by the work of Huang and Gatenby, in particular, discussed earlier, which demonstrates that acquired resistance, energetically expensive to produce and maintain, is context-dependent and reversible. As explained by Gatenby and reiterated above, treatment is a form of environmental induction that exerts diffuse selection pressures on cancer cells, favoring the expression of multiple defense mechanisms, effective against multiple chemotherapies [97]. However, the possession of these defense mechanisms is metabolically costly and compromises available resources for "mission-critical" proliferation; therefore, when the associated costs outweigh the benefits, the tumor will adopt, on balance, a drug-sensitive phenotype. The problem is that due to random fluctuations in the environment and the inertial character of gene expression, resistance may outlast the treatment challenge; therefore, it is unpredictable if, when and how quickly a phenotypic switch will occur. Moreover, the very act of treatment even after a chemotherapy-free interval re-elicits induced resistance.

The apparent solution to this therapeutic "catch 22" is epigenetic modulation, because it *alters the pattern of gene expression that is perfectly or near perfectly matched to environmental conditions.* Genes influence whole "pathways" [11]: thus, it is likely that even a small epigenetically mediated change in transcription, turning on one gene here, silencing another gene there, which alters the epigenome, leads to a "butterfly effect"-like cascade of consequences that affects phenotype.

Here we arrive at a crucial point in our (alternative) hypothesis: because epigenetic inhibition both upregulates and downregulates a panoply of genes, it pushes the tumor out of its "comfort zone," by repressing certain pathways, and/or derepressing others; ritualized countermeasures to reestablish a steady state in response to this epigenetic disruption are energetically expensive and the high cost forces the tumor to "deballast" like a sinking ship jettisoning nonessential cargo; in the process since energy optimization during stress is at a premium the tumor redirects all necessary resources to survival at the expense of *de novo* and acquired resistance, the maintenance of which is also expensive, as discussed earlier.

The basic, essential "purpose" of epigenetic alterations, if it is possible to anthropomorphically ascribe a purpose or a motive to an inanimate epigenetic mechanism (referred to as "pathetic fallacy" in literature), is to reprogram the functions of the cell in response to changes in environmental conditions. It makes teleological sense, therefore, that the knee-jerk adoption of a "survival phenotype" over a resistance phenotype in response to epigenetic inhibition, which is presumably meant to conserve resources and increase the "coping capabilities" of the tumor, leads to resensitization with formerly effective but refractory treatments. While reversal of resistance and resensitization may occur spontaneously during treatment interruption, the localized and sporadic pattern of its occurrence supports the contention that resensitization is more reliably elicited with epigenetic agents; hence, the use of the term epi-resensitization. The desire to piece together this jigsaw puzzle of epigenetics and resensitization extends beyond mere scientific or explanation-seeking curiosity: as a therapeutic paradigm, epi-resensitization has important hands-on clinical and treatment implications and may predict time to progression as well as OS.

An article by Richard Pazdur, the FDA's oncology chief, makes the case that the link between improved RRs, PFS, and survival benefits is tenuous [98], since it has never been definitively established that "increases in survival are directly attributable to decreases in tumor burden produced by the therapy." A confounding issue with OS is the effect of subsequent treatments, which are generally unknown. By contrast, resensitization controls for subsequent therapies, since the treatment plan is essentially predetermined. Moreover, unlike either PFS or RRs, resensitization is a relevant and reliable surrogate for OS because more options equal more time, due to a dearth of choices beyond second or third line in metastatic cancer. While OS may not always be the best or most appropriate endpoint, especially when treatment-related toxicities render quality of life so poor that a few more months time becomes essentially meaningless and counterproductive, resensitization has the potential to increase both survival *and* quality of life, through recycling and reuse of treatments with tolerable side effects.

According to the FDA, OS is the undisputed gold standard "considered the most reliable cancer endpoint, and... it is usually the preferred endpoint." However, patient preferences are typically less black-and-white, more subjectively complex and not as amenable to a simple answer: the most valuable commodity may not be the amount of time *per se*, although that is certainly important, but the quality of time because if the additional few months of survival are accompanied with intolerable treatment-associated toxic effects and a deteriorated quality of life (QOL) then it is a mostly meaningless (and counterproductive) statistic. From the perspective of the medical establishment and the FDA, due to the acutely existential threat of cancer, the severe toxicities of potentially life-saving chemotherapies may and in fact usually do take precedence over the golden rule of "first do no harm" (primum non nocere). In other words, in oncology, the end (i.e., tumor control or eradication) justifies the means, which is a form of "medical Machiavellianism (MM)." For Niccolo Machiavelli, the author of the 1513 political treatise, *Il Principe*, The Prince, "by any means necessary" was a morally permissible expedient because in the end what matters is success, no matter the methods necessary to achieve it. In the MM context of cancer the benefit of OS outweighs quality-of-life considerations.

Nevertheless, in the balance between survival gains versus quality-of-life impact what may matter more or most, from the perspective of the patients, is how well they live in the time remaining and not just for how long, since eradication of metastatic cancer by surgery, radiation, drugs, biotherapy, and so on is nearly impossible. On the other hand, it is clearly to the ultimate benefit of patients, their families, and physicians, if survival is increased, provided quality of life is acceptable.

Episensitization has the potential to reconcile this apparent disconnect between quality and quantity of life, greatly increasing the expectancy of survival from mere months to multiple years, with minimal side effects, as resensitized patients "see-saw" or "boomerang" between epigenetic agents and formerly refractory treatments in consecutive cycles. In the current "been there done that on to the next" oncology treatment algorithm patients advance or "progress" (ironically, progression denotes, not an improvement, but a worse prognosis) through a gauntlet of successive lines of therapy until all standard options have been exhausted, at which point only hospice or clinical trials with, as yet unproven agents, remain. By contrast, episensitization is literally a "revolutionary" approach since, like a wheel, treatment *revolves* (360°) and returns or regresses to the starting point, which blurs the boundaries between lines of therapy.

This "boundary blurring" also extends to the symbolic border separating the inhospitable no man's land of cancer from the geographically distinct territory of chronic, manageable diseases like HIV, multiple sclerosis, chronic obstructive pulmonary disease (COPD), and diabetes, as resensitized metastatic patients attempt to crossover and live out, not a death sentence, but a *life sentence*. The basis for "full citizenship" in this territory is chronicity, which means that the cancer is controlled and contained but not cured over the long term.

At this point, we are tempted to insert the grandiose phrase, "paradigm shift," if only to keep up with the scientific Joneses. Apparently it is de rigueur, based on the increasing frequency with which it appears in the literature, to claim at least one, if not more, paradigm shifts (per paragraph). However, judging by the quality of several scientific manuscripts that overuse the clichéd term to hype their results, the actual importance of a manuscript may be proportional to the frequency with which paradigm shift is invoked, according to an inverse square law, like the effects of gravity! (scientific merit $\propto$ 1/PS [2]). Therefore, instead of a paradigm or a paradigm shift, it may be more appropriate to refer to episensitization as a therapeutic "turning point" or "a new era"—out with the old status quo assumptions about acquired resistance and in with the new, a "changing of the guard" approach that has the potential to decatastrophize cancer and transform it from a terminal to a chronic but noncurable disease.

What a hopeful juxtaposition—cancer and chronic disease! The fact that such a chronic, long-term scenario for a currently lethal condition requiring heroic end-stage medical intercession is even imaginable or possible (hopefully an indication of progress rather than delusion!) has enormous implications for the choice and safety margin of agents that may precede and follow epigenetic therapies. In this chronic context, the safety, cost, convenience, and side-effect profiles of first-line and salvage therapies may (and probably should) take relative precedence over efficacy, provided efficacy is present, whereas currently the emphasis is on the latter. The cumulative and irreversible toxicities associated with long-term exposure to several chemotherapies, potentially resulting in end-organ damage, are not feasible for a prime-boost resensitization continuum because it would render patients too physically and mentally debilitated to continue subsequent treatment.

To return to the example of mCRC, the tyrosine multikinase inhibitor (TKI), regorafenib, also referred to as son of sorafenib [99], was approved despite a marginal survival benefit compared to placebo [100] and *considerable* toxicity including diarrhea, fatigue, rash, and stomatitis/mucosal inflammation simply because it is the "last chance therapy," that is, currently no other options exist. However, in episensitized patients, sorafenib/regorafenib-like agents with marginal benefit–risk ratios are pushed aside in favor of better-tolerated and less toxic regimens as part of a "déjà vu (all over again) strategy" to repeat previously beneficial therapies. As the least toxic and most broad-spectrum pan-epigenetic inhibitors, oncolytic (or potentially oncostatic) viral vectors and RRx-001, may hopefully, 1 day, serve as the "karmic" backbone of the episensitization therapies; the word "karmic" is chosen because of the associated phrase "what goes around comes around," which perfectly encapsulates the epi-resensitization strategy. This begs the question: how do we get from here to there? The answer: one karmic step at a time.

As the spiritual equivalent to the butterfly effect, karma "acts in multiple feedback loops, with the present moment being shaped both by past and by present actions; present actions shape not only the future but also the present." In other words, through episensitization, patients receive and reap the karmic consequences of the circular treatment chain with relatively nontoxic agents. For these present and future cancer patients, after all the impediments, adversities, setbacks, and frustrations during the long, hard road of treatment, it is quite literally about time.

ACKNOWLEDGMENTS

The authors gratefully acknowledge the worthy accomplishments of Major Harry Lybeck M.D., Ph.D, a living 96-year-old Finnish Medical Officer serving under General Mannerheim in the Second World War and our hero who, denying that he is one, continues to touch lives, including our own, with his spirit of selfless dedication and devotion. Now a proud great grandfather to Brooke Shannon Harper, Dr Lybeck always worked in wartime and in peacetime to give his patients a second chance at life, which is what we hope and anticipate will happen with the strategy of episensitization in the fight against cancer.

REFERENCES

[1] Hahnfeldt P, Folkman J, Hlatky L. Minimizing long-term tumor burden: the logic for metronomic chemotherapeutic dosing and its antiangiogenic basis. J Theor Biol 2003;220:545.

[2] Reid T, Dad S, Korn R, Oronsky B, Knox S, Scicinski J. Two case reports of resensitization to previous chemotherapy with the novel Hypoxia-Activated hypomethylating anticancer agent RRx-001 in metastatic colorectal cancer patients. Case Rep Oncol 2014;7(1):79–85.

[3] Birkhead BG, Rankin EM, Gallivan S, Dones L, Rubens RD. A mathematical model of the development of drug resistance to cancer chemotherapy. Eur J Cancer Clin Oncol 1987;23(9):1421–7.

[4] Clynes M. (2013). Multiple Drug Resistance in Cancer 2: Molecular, Cellular and Clinical Aspects, Springer Netherlands.

[5] Stewart BW, Wild CP. World Cancer Report 2014. Lyon: International Agency for Research on Cancer; 2014.

[6] Siegel R, Ma J, Zou Z, Jemal A. Cancer statistics, 2014. CA Cancer J Clin 2014;64(1):9–29.

[7] Haig D. The (dual) origin of epigenetics. Cold Spring Harb Symp Quant Biol 2004;69:67–70.

[8] Waddington C. The epigenotype. Endeavour 1942;1:18–20.

[9] Shipony Z, Mukamel Z, Mendelson Cohen N, Landan G, Chomsky E, Zeliger SR, et al. Dynamic and static maintenance of epigenetic memory in pluripotent and somatic cells. Nature 2014;513(7516):115–19.

[10] Tie F, Banerjee R, Stratton CA, Prasad-Sinha J, Stepanik V, Zlobin A, et al. CBP-mediated acetylation of histone H3 lysine 27 antagonizes Drosophila Polycomb silencing. Development 2009;136(18):3131–41.

[11] Klajic J, Busato F, Edvardsen H, Touleimat N, Fleischer T, Bukholm I, et al. DNA methylation status of key cell-cycle regulators such as CDKNA2/p16 and CCNA1 correlates with treatment response to doxorubicin and 5-Fluorouracil in locally advanced breast tumors. Clin Cancer Res Off J Am Assoc Cancer Res 2014;20(24):6357–66.

[12] Oronsky B, Oronsky N, Scicinski J, Fanger G, Lybeck M, Reid T. Rewriting the epigenetic code for tumor resensitization: a review. Transl Oncol 2014;7(5):626–31.

[13] Merlo LM, Pepper JW, Reid BJ, Maley CC. Cancer as an evolutionary and ecological process. Nat Rev Cancer 2006;6(12):924–35.

[14] Alexander S, Friedl P. Cancer invasion and resistance: interconnected processes of disease progression and therapy failure. Trends Mol Med 2012;18(1):13–26.

[15] Kienle G, Kiene H. From reductionism to holism: systems-oriented approaches in cancer research. Glob Adv Health Med 2012;1(5):68–77.

[16] Kokko H, Lopez-Sepulcre A. The ecogenetic link between demography and evolution: can we bridge the gap between theory and data? Ecol Lett 2007;10(9):773–82.

[17] Stein GS, Zaidi SK, Stein JL, Lian JB, van Wijnen AJ, Montecino M, et al. Genetic and epigenetic regulation in nuclear microenvironments for biological control in cancer. J Cell Biochem 2008;104 (6):2016–26.
[18] Szyf M. DNA methylation signatures for breast cancer classification and prognosis. Genome Med 2012;4(3):26.
[19] Tong M, Han BB, Holpuch AS, Pei P, He L, Mallery SR. Inherent phenotypic plasticity facilitates progression of head and neck cancer: endotheliod characteristics enable angiogenesis and invasion. Exp Cell Res 2013;319(7):1028–42.
[20] Mamane Y, Petroulakis E, LeBacquer O, Sonenberg N. mTOR, translation initiation and cancer. Oncogene 2006;25(48):6416–22.
[21] Lopez-Lazaro M. The warburg effect: why and how do cancer cells activate glycolysis in the presence of oxygen? Anti-Cancer Agents Med Chem 2008;8(3):305–12.
[22] Berg JM, Tymoczko JL, Stryer L. Biochemistry. 5th ed. New York, NY: W H Freeman; 2002 Section 18.6, The Regulation of Cellular Respiration Is Governed Primarily by the Need for ATP; 2002
[23] Tam WL, Weinberg RA. The epigenetics of epithelial-mesenchymal plasticity in cancer. Nat Med 2013;19(11):1438–49.
[24] Leder K, Holland EC, Michor F. The therapeutic implications of plasticity of the cancer stem cell phenotype. PloS One 2010;5(12):e14366.
[25] Yu HA, Riely GJ, Lovly CM. Therapeutic strategies utilized in the setting of acquired resistance to EGFR tyrosine kinase inhibitors. Clin Cancer Res Off J Am Assoc Cancer Res 2014;20(23):5898–907.
[26] Lorenz E.N. The statistical prediction of solutions of dynamic equations. In: Proc internattional symposium on numerical weather prediction, Tokyo; 1962. p. 629–35.
[27] Lorenz EN. Predictability: does the flap of a butterfly's wings in Brazil set off a tornado in Texas? In: 139th Annual Meeting of the American Association for the Advancement of Science (29 Dec 1972), in Essence of Chaos (1995), Appendix 1, 181; 1995.
[28] Lowe EJ. For want of a Nail. Analysis 1980;40(1):50–2.
[29] Smith KT, Workman JL. Histone deacetylase inhibitors: anticancer compounds. Int J Biochem Cell Biol 2009;41(1):21–5.
[30] Oronsky BT, Oronsky NC, Fanger GR, Parker CW, Caroen SZ, Lybeck MM, et al. Follow the ATP: tumor energy production: a perspective. Anti-Cancer Agents Med Chem 2014;14(9):1187–98.
[31] Silva AS, Gatenby RA. A theoretical quantitative model for evolution of cancer chemotherapy resistance. Biol Direct 2010;5:25.
[32] Meacham CE, Morrison SJ. Tumour heterogeneity and cancer cell plasticity. Nature 2013;501 (7467):328–37.
[33] Gatenby RA, Silva AS, Gillies RJ, Frieden BR. Adaptive therapy. Cancer Res 2009;69(11):4894–903.
[34] Gatenby RA, Frieden BR. Inducing catastrophe in malignant growth. J Math Med Biol 2008;25 (3):267–83.
[35] McFarland LV. Normal flora: diversity and functions. Microbial Ecol Health Dis 2000;12:193–207.
[36] Hochster HS. Stop and go: yes or no? J Clin Oncol 2009;27(34):5677–9.
[37] Chibaudel B, Maindrault-Goebel F, Lledo G, Mineur L, Andre T, Bennamoun M, et al. Can chemotherapy be discontinued in unresectable metastatic colorectal cancer? The GERCOR OPTIMOX2 study. J Clin Oncol Off J Am Soc Clin Oncol 2009;27(34):5727–33.
[38] Maughan TS, Adams RA, Smith CG, Meade AM, Seymour MT, Wilson RH, et al. Addition of cetuximab to oxaliplatin-based first-line combination chemotherapy for treatment of advanced colorectal cancer: results of the randomised phase 3 MRC COIN trial. Lancet 2011;377(9783):2103–14.
[39] Goldberg AD, Allis CD, Bernstein E. Epigenetics: a landscape takes shape. Cell 2007;128(4):635–8.

[40] Huang S. Non-genetic heterogeneity of cells in development: more than just noise. Development 2009;136(23):3853–62.

[41] Huang S, Ernberg I, Kauffman S. Cancer attractors: a systems view of tumors from a gene network dynamics and developmental perspective. Semin Cell Dev Biol 2009;20(7):869–76.

[42] Yano S, Nakataki E, Ohtsuka S, Inayama M, Tomimoto H, Edakuni N, et al. Retreatment of lung adenocarcinoma patients with gefitinib who had experienced favorable results from their initial treatment with this selective epidermal growth factor receptor inhibitor: a report of three cases. Oncol Res 2005;15 (2):107–11.

[43] Kurata T, Tamura K, Kaneda H, Nogami T, Uejima H, Asai Go G, et al. Effect of re-treatment with gefitinib ('Iressa', ZD1839) after acquisition of resistance. Ann Oncol Off J Eur Soc Med Oncol/ESMO 2004;15(1):173–4.

[44] Biondani P, Procopio G, Pietrantonio F, de Braud F, Verzoni E. New perspectives in advanced genitourinary malignancies. Tumori 2012;98(2):267–9.

[45] Cara S, Tannock IF. Retreatment of patients with the same chemotherapy: implications for clinical mechanisms of drug resistance. Ann Oncol Off J Eur Soc Med Oncol/ESMO 2001;12(1):23–7.

[46] Cunningham D, Humblet Y, Siena S, Khayat D, Bleiberg H, Santoro A, et al. Cetuximab monotherapy and cetuximab plus irinotecan in irinotecan-refractory metastatic colorectal cancer. New Engl J Med 2004;351(4):337–45.

[47] Petitprez A, Larsen AK. Irinotecan resistance is accompanied by upregulation of EGFR and Src signaling in human cancer models. Curr Pharm Des 2013;19(5):958–64.

[48] Wadlow RC, Hezel AF, Abrams TA, Blaszkowsky LS, Fuchs CS, Kulke MH, et al. Panitumumab in patients with KRAS wild-type colorectal cancer after progression on cetuximab. Oncologist 2012;17 (1):14.

[49] Hanahan D, Bergers G, Bergsland E. Less is more, regularly: metronomic dosing of cytotoxic drugs can target tumor angiogenesis in mice. J Clin Investig 2000;105(8):1045–7.

[50] Lennernas B, Albertsson P, Damber JE, Norrby K. Antiangiogenic effect of metronomic paclitaxel treatment in prostate cancer and non-tumor tissue in the same animals: a quantitative study. Acta Pathol Microbiol Immunol Scand 2004;112(3):201–9.

[51] Chow A, Wong A, Francia G, Man S, Kerbel RS, Emmenegger U. Preclinical analysis of resistance and cross-resistance to low-dose metronomic chemotherapy. Investig New Drugs 2014;32(1):47–59.

[52] Hahnfeldt P, Folkman J, Hlatky L. Minimizing long-term tumor burden: the logic for metronomic chemotherapeutic dosing and its antiangiogenic basis. J Theor Biol 2003;220(4):545–54.

[53] Storey DJ, Sakala M, McLean CM, Phillips HA, Dawson LK, Wall LR, et al. Capecitabine combined with oxaliplatin (CapOx) in clinical practice: how significant is peripheral neuropathy? Ann Oncol Off J Eur Soc Med Oncol/ESMO 2010;21(8):1657–61.

[54] Tournigand C, Cervantes A, Figer A, Lledo G, Flesch M, Buyse M, et al. OPTIMOX1: a randomized study of FOLFOX4 or FOLFOX7 with oxaliplatin in a stop-and-Go fashion in advanced colorectal cancer—a GERCOR study. J Clin Oncol Off J Am Soc Clin Oncol 2006;24(3):394–400.

[55] Adams RA, Meade AM, Seymour MT, Wilson RH, Madi A, Fisher D, et al. Intermittent versus continuous oxaliplatin and fluoropyrimidine combination chemotherapy for first-line treatment of advanced colorectal cancer: results of the randomised phase 3 MRC COIN trial. Lancet Oncol 2011;12 (7):642–53.

[56] de Gramont A, Buyse M, Abrahantes JC, Burzykowski T, Quinaux E, Cervantes A, et al. Reintroduction of oxaliplatin is associated with improved survival in advanced colorectal cancer. J Clin Oncol Off J Am Soc Clin Oncol 2007;25(22):3224–9.

[57] Shen H, Laird PW. In epigenetic therapy, less is more. Cell Stem Cell 2012;10(4):353–4.

[58] Stresemann C, Lyko F. Modes of action of the DNA methyltransferase inhibitors azacytidine and decitabine. Int J Cancer 2008;123(1):8–13.
[59] Jain N, Rossi A, Garcia-Manero G. Epigenetic therapy of leukemia: an update. Int J Biochem Cell Biol 2009;41(1):72–80.
[60] Qin T, Jelinek J, Si J, Shu J, Issa JP. Mechanisms of resistance to 5-aza-2'-deoxycytidine in human cancer cell lines. Blood 2009;113(3):659–67.
[61] Juergens RA, Rudin CM. Aberrant epigenetic regulation: a central contributor to lung carcinogenesis and a new therapeutic target. Am Soc Clin Oncol Educ Book 2013;:. Available from: http://dx.doi.org/10.1200/EdBook_AM.2013.33.e295.
[62] Issa JP. Hypermethylator phenotypes in aging and cancer. DNA alterations in cancer: genetic and epigenetic changes. Natick: Eaton Publishing; 2000. p. 311–22.
[63] Esteller M. CpG island hypermethylation and tumor suppressor genes: a booming present, a brighter future. Oncogene 2002;21(35):5427–40.
[64] Robertson KD. DNA methylation, methyltransferases, and cancer. Oncogene 2001;20(24):3139–55.
[65] Holliday R, Pugh JE. DNA modification mechanisms and gene activity during development. Science 1975;187(4173):226–32.
[66] Gravina GL, Festuccia C, Marampon F, Popov VM, Pestell RG, Zani BM, et al. Biological rationale for the use of DNA methyltransferase inhibitors as new strategy for modulation of tumor response to chemotherapy and radiation. Mol Cancer 2010;9:305.
[67] Lee BH, Yegnasubramanian S, Lin X, Nelson WG. Procainamide is a specific inhibitor of DNA methyltransferase 1. J Biol Chem 2005;280(49):40749–56.
[68] Kim SY, Benowitz NL. Poisoning due to class IA antiarrhythmic drugs. Quinidine, procainamide and disopyramide. Drug Saf 1990;5(6):393–420.
[69] Hrabeta J, Stiborova M, Adam V, Kizek R, Eckschlager T. Histone deacetylase inhibitors in cancer therapy. A review. Biomed Pap Med Fac Univ Palacky, Olomouc, Czech 2014;158(2):161–9.
[70] Ning S, Bednarski M, Oronsky B, Scicinski J, Saul G, Knox SJ. Dinitroazetidines are a novel class of anticancer agents and hypoxia-activated radiation sensitizers developed from highly energetic materials. Cancer Res 2012;72(10):2600–8.
[71] Nebbioso A, Pereira R, Khanwalkar H, Matarese F, Garcia-Rodriquez J, Miceli M, et al. Death receptor pathway activation and increase of ROS production by the triple epigenetic inhibitor UVI5008. Mol Cancer Therap 2011;10(12):2394–404.
[72] Garcia J, Franci G, Pereira R, Benedetti R, Nebbioso A, Rodriguez-Barrios F, et al. Epigenetic profiling of the antitumor natural product psammaplin A and its analogues. Bioorgan Med Chem 2011;19(12):3637–49.
[73] Ottolino-Perry K, Diallo JS, Lichty BD, Bell JC, McCart JA. Intelligent design: combination therapy with oncolytic viruses. Mol Ther J Am Soc Gene Ther 2010;18(2):251–63.
[74] Chia SL, Yusoff K, Shafee N. Viral persistence in colorectal cancer cells infected by Newcastle disease virus. Virol J 2014;11:91.
[75] Schneider R, Shenk T. Impact of virus infection on host cell protein synthesis. Ann Rev Biochem 1987;56:317–32.
[76] Reid T, Galanis E, Abbruzzese J, Sze D, Wein LM, Andrews J, et al. Hepatic arterial infusion of a replication-selective oncolytic adenovirus (dl1520): phase II viral, immunologic, and clinical endpoints. Cancer Res 2002;62(21):6070–9.
[77] Opyrchal M, Aderca I, Galanis E. Phase I clinical trial of locoregional administration of the oncolytic adenovirus ONYX-015 in combination with mitomycin-C, doxorubicin, and cisplatin chemotherapy in patients with advanced sarcomas. Methods Mol Biol 2009;542:705–17.

[78] Xia ZJ, Chang JH, Zhang L, Jiang WQ, Guan ZZ, Liu JW, et al. Phase III randomized clinical trial of intratumoral injection of E1B gene-deleted adenovirus (H101) combined with cisplatin-based chemotherapy in treating squamous cell cancer of head and neck or esophagus. Ai Zheng=Chinese J Cancer 2004;23(12):1666–70.

[79] Raja FA, Chopra N, Ledermann JA. Optimal first-line treatment in ovarian cancer. Ann Oncol Off J Eur Soc Med Oncol/ESMO 2012;23(Suppl. 10):x118–27.

[80] Armstrong DK. Relapsed ovarian cancer: challenges and management strategies for a chronic disease. Oncologist 2002;7(Suppl. 5):20–8.

[81] Ferriss JS, Rice LW. The role of in vitro directed chemotherapy in epithelial ovarian cancer. Rev Obstet Gynecol 2010;3(2):49–54.

[82] Funt SA, Hricak H, Abu-Rustum N, Mazumdar M, Felderman H, Chi DS. Role of CT in the management of recurrent ovarian cancer. AJR Am J Roentgenol 2004;182(2):393–8.

[83] Fung-Kee-Fung M, Oliver T, Elit L, Oza A, Hirte HW, Bryson P. Optimal chemotherapy treatment for women with recurrent ovarian cancer. Curr Oncol 2007;14(5):195–208.

[84] Matei DE, Nephew KP. Epigenetic therapies for chemoresensitization of epithelial ovarian cancer. Gynecol Oncol 2010;116(2):195–201.

[85] Matei D, Fang F, Shen C, Schilder J, Arnold A, Zeng Y, et al. Epigenetic resensitization to platinum in ovarian cancer. Cancer Res 2012;72(9):2197–205.

[86] Fu S, Hu W, Iyer R, Kavanagh JJ, Coleman RL, Levenback CF, et al. Phase 1b-2a study to reverse platinum resistance through use of a hypomethylating agent, azacitidine, in patients with platinum-resistant or platinum-refractory epithelial ovarian cancer. Cancer 2011;117(8):1661–9.

[87] Cainap C, Qin S, Huang WT, Chung IJ, Pan H, Cheng Y, et al. Linifanib versus sorafenib in patients with advanced hepatocellular carcinoma: results of a randomized phase III trial. J Clinl Oncol Off J Am Soc Clin Oncol 2015;33(2):172–9.

[88] Bitzer M, Horger M, Ganten T, Ebert MP, Siveke J, Woerns MA, et al. Clinical update on a phase I/II trial of HDAC inhibitor resminostat in patients with sorafenib-resistant hepatocellular carcinoma (HCC): the shelter study. J Clin Oncol Off J Am Soc Clin Oncol 2011;29:(suppll; abstr e 13603).

[89] Goldberg RM, Rothenberg ML, Van Cutsem E, Benson AB, Blanke CD, Diasio RB, et al. The continuum of care: a paradigm for the management of metastatic colorectal cancer. Oncologist 2007;12 (1):38–50.

[90] Grothey A, Sargent D, Goldberg RM, Schmoll HJ. Survival of patients with advanced colorectal cancer improves with the availability of fluorouracil-leucovorin, irinotecan, and oxaliplatin in the course of treatment. J Clin Oncol Off J Am Soc Clin Oncol 2004;22(7):1209–14.

[91] Bardhan K, Liu K. Epigenetics and colorectal cancer pathogenesis. Cancers 2013;5(2):676–713.

[92] Giessen C, Laubender RP, Ankerst DP, Stintzing S, Modest DP, Schulz C, et al. Surrogate endpoints in second-line treatment for mCRC: a systematic literature-based analysis from 23 randomised trials. Acta Oncol 2014;1–7.

[93] Massarelli E, Andre F, Liu DD, Lee JJ, Wolf M, Fandi A, et al. A retrospective analysis of the outcome of patients who have received two prior chemotherapy regimens including platinum and docetaxel for recurrent non-small-cell lung cancer. Lung Cancer 2003;39(1):55–61.

[94] Juergens RA, Wrangle J, Vendetti FP, Murphy SC, Zhao M, Coleman B, et al. Combination epigenetic therapy has efficacy in patients with refractory advanced non-small cell lung cancer. Cancer Discov 2011;1(7):598–607.

[95] Vogl TJ, Naguib NN, Nour-Eldin NE, Bechstein WO, Zeuzem S, Trojan J, et al. Transarterial chemoembolization in the treatment of patients with unresectable cholangiocarcinoma: results and prognostic factors governing treatment success. Int J Cancer 2012;131(3):733–40.

[96] Valle J, Wasan H, Palmer DH, Cunningham D, Anthoney A, Maraveyas A, et al. Cisplatin plus gemcitabine versus gemcitabine for biliary tract cancer. New Engl J Med 2010;362(14):1273–81.
[97] Kam Y, Das T, Minton S, Gatenby RA. Evolutionary strategy for systemic therapy of metastatic breast cancer: balancing response with suppression of resistance. Women's Health 2014;10(4):423–30.
[98] Pazdur R. Response rates, survival, and chemotherapy trials. J Natl Cancer Inst 2000;92(19):1552–3.
[99] Carr BI, Cavallini A, Lippolis C, D'Alessandro A, Messa C, Refolo MG, et al. Fluoro-Sorafenib (Regorafenib) effects on hepatoma cells: growth inhibition, quiescence, and recovery. J Cell Physiol 2013;228(2):292–7.
[100] Grothey A, Van Cutsem E, Sobrero A, Siena A, Falcone A, Ychou M, et al. Regorafenib monotherapy for previously treated metastatic colorectal cancer (CORRECT): an international, multicentre, randomised, placebo-controlled, phase 3 trial. Lancet 2013;381(9863):303–12.

CHAPTER

THE EMERGING POTENTIAL FOR EPIGENETIC THERAPEUTICS IN NONCANCER DISORDERS 17

Christopher G. Bell[1,2,3,4]

[1]*Department of Twin Research & Genetic Epidemiology, St Thomas' Hospital, King's College London, London, UK* [2]*MRC Lifecourse Epidemiology Unit, University of Southampton, Southampton, UK* [3]*Academic Unit of Human Development and Health, University of Southampton, Southampton, UK* [4]*Epigenomic Medicine, Centre for Biological Sciences, Faculty of Environmental and Natural Sciences, University of Southampton, Southampton, UK*

CHAPTER OUTLINE

17.1 INTRODUCTION

Epigenetics is defined as the chemical modifications of the genome that are able to be inherited though mitosis—but do not alter the underlying genetic code. These factors include molecular additions to the DNA molecule itself, as well as the packaging and wrapping of the strands around proteins, termed histones [1]. Furthermore, various noncoding classes of RNA are viewed to play an integral role in the epigenetic process [2]. The presence of these factors impact and influence gene expression in a time- and tissue-dependent fashion [3]. They are thus vital for the synergistic functioning of the human body.

The chemical modifications of DNA itself include the most well-studied mark, DNA methylation, the addition of a methyl group onto the 5′ carbon of the cytosine base [4]. This classically occurs within a CpG dinucleotide context: a C followed by G in the 5′ to 3′ direction of the DNA strand. Additional rarer chemical modifications, identified more recently, are products of active demethylation and have illuminated this process in more detail. These include the first oxidative produce of methylation, 5-hydroxymethylcytosine (5hmC)—catalyzed by the TET enzymes—and

G. Egger & P. Arimondo (Eds): Drug Discovery in Cancer Epigenetics. DOI: http://dx.doi.org/10.1016/B978-0-12-802208-5.00017-5

the subsequent products of further oxidation of 5hmC to 5-formylcytosine and 5-carboxylcytosine [5]. These modifications may also possess functions in their own right, beyond their role in the demethylation process. This includes generalized features, such as their coordinated role with TET in enhancer modulation [6], as well as potential specific roles such as via methyl-C binding protein 2 (MeCP2) being the major 5hmC-binding protein within the brain [7].

The packaging of DNA around histone protein octamers is also a fundamental epigenetic mechanism, as posttranslational modifications of the tails of these molecules, as well as histone protein variants, can influence gene function [8]. Specific histone tail locations can be modified by methylation, acetylation, phosphorylation, etc. [9]. Currently most data point toward the importance of modifications of lysines within the tail of histone 3, which can be associated with active promoters (H3K4me3), active enhancers (H3K4me1 and H3K27ac), and repressed regions either constitutive (H3K9me3) or facultative (H3K27me3) heterochromatin [10]. Recent evidence of independent transmission of the H3K9me2/3 mark has strengthened the role of histone modifications as pure epigenetic actors [11]. These modifications can be combinatorially processed to "segment" the genome of specific cell-types into demarcated active, poised, transcribed, repressed, etc., regions, such as by ChromHMM [12] or Segway [13] algorithms.

While these epigenetic modifications are vital for tissue-specific functioning, there is also increasing evidence that this level of genomic control may be influenced by the environment and thus act as an environmental interface with the genome [14,15]. Due to the inherent stability, but also the plasticity of the DNA methylation mark—and furthermore with more easier high-throughput analysis possible—the potential to use this modification as a proxy or biomarker to measure accrued lifetime environmental exposures has elicited much interest [16]. This possibility for a more accurate quantitation of epidemiological biological effects, with a potential pathogenic role, has led to an explosion of epigenome-wide association studies (EWAS) [17]. However these studies require considerably different study design and analysis than genetic association studies due to the inherent variability of the epigenome that is cognizant with its function [18].

The first insights into the role epigenetic change may play in disease pathology came more than three decades ago, with the identification of the dramatic DNA hypomethylation that occurs in cancer tissue [19]. Due to this initial strong finding, much of the fundamental knowledge in epigenetics has stemmed from the field of cancer biology [19]. The additional paradigm in epigenetic research is imprinting—or parent-of-origin-specific allele-specific methylation (ASM)—with abnormalities first identified and associated with phenotypes in the early 1990s [20].

However, coarse global measures, or the examination of a discrete locus in a few individuals, were the only methods available until the genomic era. Now with precise genome-wide data, in larger numbers, further specific patterns have been able to be identified, such as increased cancer- [21], tissue- [22], reprogramming- [23], and even species-specific [24] DNA methylation changes occurring within the regions that surround CpG dinucleotide dense CpG islands—termed CpG island shores ($\sim$2 kb up/downstream). Also, the precise delineation of large hypomethylated DNA blocks in cancer has been shown, which can account for half the genome and lead to a dramatic variation in gene expression, and contribute to the global DNA hypomethylation identified decades earlier [25].

Aging-specific changes in the epigenome identified include increased DNA methylation in the promoters of the targets of polycomb-binding proteins [26]. Combinations of these DNA methylation changes can be used as a highly accurate biological clock [27] and furthermore the fascinating observation that a mismatch between biological and an increased epigenetically predicted age is associated with not only decreased physical and cognitive fitness [28], but also increased risk of mortality [29]. The strongest direct environmental influence documented so far on the DNA methylome has been that of tobacco smoking. Distinct CpGs modified across a number of studies have been identified, including both passive and in utero exposure [30–33], such as CpGs within the *AHRR* gene, the aryl-hydrocarbon receptor repressor, itself involved in the tobacco smoke detoxification pathway. Some loci appear to be longstanding, with ex-smokers retaining abnormal DNA methylation levels even up to 10 years after quitting [34]. Animal models have explored epigenetic modifications via toxins and nutritional compounds and small-scale studies have also suggested this in humans [35]. Furthermore, increased or attenuated effects on common genetic risk factors [36,37] may act through these epigenetic mechanisms [14,38], and thus have further driven exploration of the epigenome in common disorders, such as obesity [39] and cardiovascular disease [40].

The strong confounding factor of genetic heterogeneity has robustly pointed toward the advantage of monozygotic twins [41,42] and longitudinal subjects in the study design for these EWAS analyses [43] and much work will need to be done in replicating findings considering the biological complexity of the epigenome.

17.2 EPIGENETIC TARGETS FOR THERAPEUTICS

While epigenetic modifications identified in association studies may simply be passive biomarkers, they can also be intrinsically involved in the pathogenesis of the disease. Moreover, due to the plasticity of the epigenome there is the possibility of remedying these aberrations across the gamut of diseases (Figure 17.1). This potential of epigenetic therapy was highlighted by Egger et al. [44] in the context of cancer and there is considerable biotechnology interest in reversing these pathogenic "epimutations" that impact on disease-associated gene expression [45]. Undoubtedly there is great promise in this approach, but only slow progress has been made to date, due to the intertwined nature of the epigenetic mechanisms involved, that the last decade's more detailed studies have begun to reveal [46]. There is now increasing optimism however, as a more coherent and systematic framework for epigenetic drug discovery is beginning to develop [46]. These novel therapeutics may work via various epigenetic avenues, such as removing or adding DNA modifications, influencing the large number of histone-modifying enzymes acting on histone posttranslational modifications, or through long noncoding RNAs (lncRNAs) that are increasingly seen to play a crucial role as molecular scaffolding coordinating these epigenetic processes within loci. Furthermore they may also be requisitioned to correct for not only an epigenetic abnormality but to compensate for an underlying genetic defect.

One method is through the remodeling of higher-order chromatin structure by genome-wide histone-modifying enzyme modifiers. Histone acetyl transferases, by adding an acetyl to lysine increase accessibility, and histone deacetylases (HDACs) remove acetyl groups from lysine residues leading to a more condensed or repressed formation. HDAC inhibitors (HDACi) are small interfering molecules in histone deacetylation, thus by their action help retain acetylation levels and prevent gene repression.

The proteins involved in epigenomic modification, including the acetylation and methylation networks of the human epigenome, are extremely abundant but also present a large number of

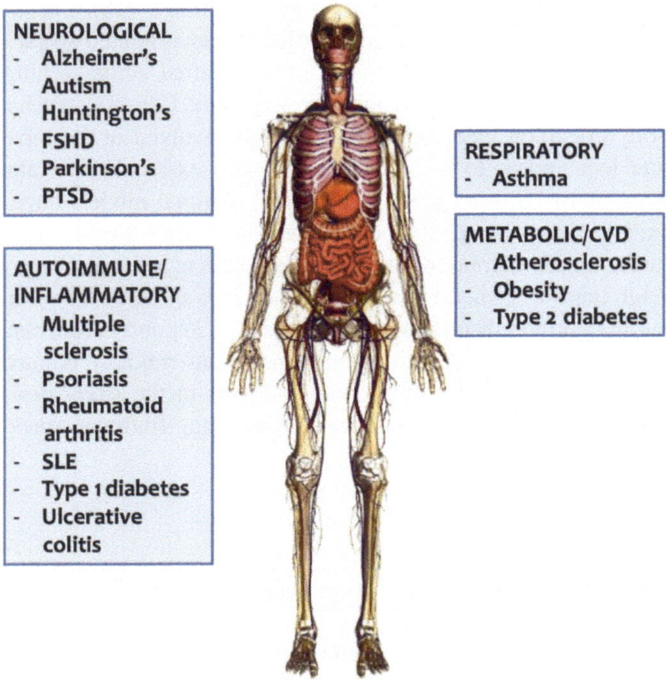

FIGURE 17.1

The potential for epigenetic drugs ranges across all systems of the human body. Examples of diseases where this therapy is now being explored. FSHD, facial-scapular-humoral dystrophy; PTSD, posttraumatic stress disorder; SLE, systemic lupus erythematosus.

druggable targets [47]. These genome-wide mechanisms are nonspecific and therefore can lead to side effects. However there is some evidence that implies that abnormal modifications may respond preferentially, such that demethylating drugs, such as azacytidine, do not reactivate the inactivate X chromosome in females [45]. This is assumed to be due to fact that there are multiple mechanisms of epigenetic regulation and that combination therapy of epigenetic drugs affecting distinct components of the pathway may enable specific reactivation.

Bromodomains are lysine acetylation recognition domains contained within certain proteins involved in regulating gene transcription that include methyltransferases and transcriptional coactivators [48]. Bromodomain-containing proteins are implicated in cancer, and also nonmalignant diseases, such as HIV, inflammation, and neurological disorders, and are seen as a therapeutic target of great potential.

Raynal et al. recently identified that 1% of current US FDA-approved drugs have significant epigenetic activity, by being shown to epigenetically switch off promoters in colon cancer cells [49]. Thus this indicates that a good proportion of these already-approved drugs could also be repurposed for their epigenetic effect.

17.3 NEUROLOGICAL DISEASES

With the extreme complexity and paucity of true functional understanding of the biological processor that is the human brain, the potential for novel insights to be made through epigenetic analysis makes this a highly stimulating field of study. The importance of DNA methylation in higher brain functioning can be clearly displayed by the impact of DNA sequence mutation, and expansion, in both Rett and fragile X syndromes, respectively, due to the downstream DNA methylation changes they produce [50]. While the postnatal cognitive decline in Rett syndrome strongly indicates the functional importance of the MeCP2 in early development, further evidence from a mouse model has shown that it is still required for continued adult brain functioning as well [51]. Particularly of recent interests are the levels of non-CpG DNA methylation and DNA hydroxymethylation (5hmC) that have been shown to be uniquely high within the brain, particularly during early postnatal neuronal development [52].

Epigenetic mechanisms can play a role in molecular memory due to their plasticity. The concept of this epigenetic role, through molecular turnover or addition to DNA, has been a longstanding proposition in long-term memory formation, as discussed by Crick [53]. Evidence has taken longer to accumulate, though examples are increasing, such as rat model experiments that have revealed cortical gene-specific hypermethylation induced by associative learning experiences [54]. Furthermore, abrogation of the methylation machinery by 5-azadeoxycytidine DNA methyltransferase inhibition disturbed this long-term memory formation. This has led to more recent work from Graff et al. [55] exploring the therapeutic possibilities of epigenetic modification in posttraumatic stress disorder. This psychological illness and related anxiety symptoms are hypothesized to be due to the overcompensated hardwired persistence of unpleasant memories of traumatic events. Conventional therapeutic options such as anxiolytics are limited and cognitive-behavioral methods, such as exposure-based therapies, involving memory reconsolidation are more effective. These researchers identified, in mice, that by retaining histone acetylation levels by using a histone deacteylatase 2 targeting inhibitor, the plasticity of remote memories could be increased and thus aid in their loss. This epigenetically primed and increased the expression of neuroplasticity-related genes [55]. Thus these intriguing experimental data indicate the potential to augment current behavioral techniques, by increasing the plasticity of long-term memory recall invoked during memory reconsolidation.

Fascinating evidence of the role that small regulatory noncoding RNAs may play in this epigenetic memory mechanism has come from work in the sea slug *Aplysia*, where neuronal-expressed PIWI-interacting RNAs (piRNAs) were discovered [56]. Previously, piRNAs were thought to be germline-specific. In this model these RNAs, which are so-named because they complex with a class of regulatory protein called PIWI, were shown to enable serotonin-dependent methylation of specific CpGs implicated in synaptic modulation; thus manipulating the long-term state of neurons.

The importance of epigenetics in the brain is starkly clear in imprinting disorders—where neurological symptoms commonly accompany many of these syndromes [57]. Therefore "de-imprinting" and activating the imprinted allele may have significant benefits in some of these individuals, although there will be developmental critical windows that, if missed, cannot be rectified for at a later timepoint. Angelman syndrome is caused by a deficiency in the maternal *UBE3A* transcript—thus those affected still possess an epigenetically silenced paternal *UBE3A* gene. Meng et al. [58]

have recently shown that by reducing the expression of the silencing-inducing antisense long noncoding transcript *Ube3a-ATS*, continued expression of the paternal *Ube3a* in neurons in mice could be achieved. This reduced selected neurological symptoms present in this model and while in this case was induced by an antisense oligonucleotide for the lncRNA, additional DNA or chromatin epigenetic modulation of the system would be assumed to have the same, if not more sustained effects. Likewise similar approaches will be applicable to the maternally imprinted transcripts, critical small nucleolar RNAs, within the same region implicated in Prader–Willi syndrome [59].

Valproic acid was first prescribed as an approved antiepileptic drug in 1967 in France and subsequently over a decade later for bipolar disorder [60], and hence became a mainstay in the therapeutic armamentarium. It was subsequently found to be active as an antineoplastic drug as well. The drug is in fact directly epigenetically active as an HDACi. Valproic acid is also used in the United States in fragile X syndrome to treat the attention deficit hyperactivity disorder symptoms associated with the majority of affected boys with this disorder, although affect on the methylated expansion itself has not been definitively shown [61]. Its treatment remit has further broadened with protective effects in neurodegenerative diseases, such as Alzheimer's, not only reducing behavioral disturbances, but also as a neuroprotective agent by lessening neurofibrillary tangles [62].

Recent EWAS in Alzheimer's disease performed in postmortem brain samples revealed associations with specific changes in 71 CpG's methylation, including in the genes *ABCA7* and *BIN1*, previously genetically implicated in this neurodegenerative disease [63]. Changes were identified presymptomatically in patients with amyloid build-up and thus are potentially early markers pathogenically involved in the disease process. Additionally, regional disease-related changes were found in the ankyrin 1 (*ANK1*) gene in the entorhinal cortex, superior temporal gyrus, and prefrontal cortex, but not in the relatively unaffected cerebellum or in peripheral blood, and suggest its possible role in cortical degeneration [64].

Furthermore, work from the Epigenomic Roadmap consortium investigated chromatin state changes with disease progression in the hippocampus of a mouse model of Alzheimer's-like neurodegeneration [65]. This showed a coordinated repression of promoters and enhancers involved in synaptic plasticity, with a concurrent upregulation of immune-related regulatory regions. The orthologous human regions of these disease-associated active enhancers also possessed immune-specific enhancer signatures, and are also enriched for Alzheimer's Genome-Wide Association Study (GWAS) variants. This thus further supports immune cell dysregulation as the driver in Alzheimer's disease, rather than primary neuronal cell degeneration. An immune role in multiple sclerosis was similarly supported using an analogous integrated genetic and epigenomic approach in a study from Farh et al. [66].

Another severe neurological disease, Huntington's, is a hypermorphic or gain-of-function, adult-onset, and furthermore incurable disorder, that is caused by the mutant huntingtin gene possessing an expanded CAG trinucleotide repeat. This leads to an abnormally long polyglutamine tract at the N-terminus of the corresponding protein. Being a dominant disorder, switching off this mutant allele would have clear therapeutic utility. RNA interference techniques have been shown to be successful in murine models [67], therefore more durable epigenetic modification may also have potential.

Parkinson's disease, with classical symptoms of tremor, stooped gait and facial rigidity is due to the substantial loss of dopaminergic neurons of the substantia nigra within the midbrain. While genetic susceptibilities are known, the influence of the environment on these late-onset diseases is also a factor and thus while no clear epigenetic factors have been so far identified, they have also been proposed as a therapeutic target [68].

The progressive upper limb and facial muscle weakness disorder, facial-scapular-humoral dystrophy (FSHD), is caused by the disrupted regulation of the D4Z4 macrosatellite repeat. A genetically susceptible allele facilitates an epigenetic abnormality, consistent in all pathologies. Furthermore, a chromatin regulator, SMCHD1, causes type 2 (via digenic inheritance [69]) and increases the severity of type 1 FSHD [70]. Considerable differences in disease penetrance and variable expressivity, plus the pathogenic role of SMCHD1, suggest epigenetic drug targets may exist.

Autism is consistently associated with CNVs and *de novo* mutation, however epigenetic insights again have been sorted to explore the pathogenesis, such as H3K4me3 analysis of the prefrontal cortex [71]. To discover further insights into autism, commonality and differences in the autistic symptoms experienced in neurodevelopmental disorders, such as Rett syndrome and fragile X syndrome, have been explored [72]. This has proposed that the knowledge gained through successful animal models of Rett syndrome needs to be mimicked across autism, in order to make significant breakthroughs in understanding. Although this is currently difficult due to the extent of genetic variation associated with the disease. A recent mouse model of Rett syndrome (with autism-like symptoms) identified that MeCP2 represses neuronal-specific long genes by binding to non-CpG methylated CpA sites, which does not occur with the mutant MeCP2 protein [73].

Ziller et al. [74], in a study of brain development, examined the epigenetic priming and lineage commitment of neuronal progenitors. This demonstrated an enrichment of Alzheimer's and bipolar disorder associated GWAS hits within H3K27Ac peaks. Further insights into human neurological and psychiatric disease may also come from the charting of human-specific epigenomic characteristics from comparative epigenomic analysis, with the potential to more fully understand higher cognitive functioning and human-specific vulnerabilities [75–78]. Mutations in epigenetic global modifiers, such as *EHMT1*, a lysine methyltransferase (aka *KMT1D*) implicated in the intellectual disability Kleefstra syndrome [79], further display the importance of these global epigenomics modifiers and chromatin regulators in brain function. Other examples are *MLL1* (*KMT2A*) associated with psychiatric disorders such as schizophrenia [80], and *CHD8* in autism [81]. In exon sequencing analysis these chromatin-modifying genes, as a class, are shown to possess a high mutational load, in human neurodevelopmental and psychiatric disorders [82], as has been previously seen in cancer [83].

17.4 AUTOIMMUNE AND INFLAMMATORY DISEASE

The success of GWAS has been significant in the study of autoimmune diseases with 100s of risk loci now known. Recent work in 21 autoimmune diseases to pinpoint the likely functional variants integrated these genetic results with transcriptional and *cis*-regulatory epigenetic chromatin information from specialized immune cell subsets [66]. This has revealed novel insights into the potential causative disease cell types, with a broad enrichment for these results within immune cell enhancers [66]. Additionally a gain of histone acetylation and evidence of enhancer-associated RNA expression was revealed with immune simulation in these loci. Candidate causal variants within enhancer regions were however indicated to alter only ~10–20% known transcription-factor-binding sites, although they were often in near proximity to these motifs.

Disturbance of the epigenome in autoimmune disease is associated with significant changes in the normal pattern of gene expression, such as in systemic lupus erythematosus with the hypomethylation of regulatory regions of *ITGB2*, *IL4*, and *IL6* in T cells [84]. Thus the use of therapeutics that can modify these epigenetic changes will potentially have strong clinical utility [85].

The use of HDACi's in inflammation is being thoroughly explored [86]. HDACi's proposed immunomodulatory effects are via affecting molecules involved in the processing and presentation of antigens, furthermore influencing the co-stimulatory and adhesion factors [87]. Experimental models are in use across a range of diseases, including psoriasis [88], multiple sclerosis [89], systemic lupus erythematosus [90], ulcerative colitis [91], and rheumatoid arthritis [92]. The evidence to date reveals increased anti-inflammatory production by antigen-producing cells, as well as amplified T-regulatory-cell generation. However results have shown that HDACi effects, while widespread, are strongly tissue-dependent in effect, and also influenced by the context and dose level [87]. As such, pan-HDACi's are found to antagonize the status quo. HDACs themselves rely on *cis*-acting elements that vary from cell to cell, not specific DNA-binding activity, therefore precise targeting and use of specific HDACi's have clear advantages.

In mouse models of colitis, HDACi's have showed modifications across the clinical spectrum, including reductions in weight loss, histological inflammation, colon shortening, diarrhea, and bleeding scores, plus inflammatory markers, such as IL-6 and IFNγ, and also increases in the anti-inflammatory IL-10 [91]. This improvement is thus seen to be due to a limitation of cell differentiation and proliferation, including CD4+ T cells, and a modulation of cytokine production toward reduced inflammation. There is a protective effect limiting epithelial barrier breech of the mucosa by bacteria, which is driven via an increase in human β defensin-2, through HDACi's such as sulforaphane and butyrate [93]. HDACi's have also been shown to influence noncoding RNAs, such as miR-146a expression, involved in innate inflammation [94].

Epigenetic activation of the synovial fibroblasts within the joint, the significant players in the pathogenesis of rheumatoid arthritis, has warranted detailed investigation [92]. These inflammatory cell recruiters invade and destroy the intra-articular cartilage of the joint and this aggressive phenotype is found to even persevere in culture, through multiple cell passages [95]. The DNA methylomes of normal, osteoarthritic and rheumatoid synovial fibroblasts are distinct, with additional differences between early and longstanding rheumatoid disease [96]. Small studies have proposed a role for reduced DNMT1 in rheumatoid arthritis, as well as increased skewing of X-inactivation, linked with the known increased disease risk for women [92]. With respect to the picture of histone modifications, rheumatoid arthritis sufferers were shown to have increased HDAC activity in peripheral blood mononuclear cells (PBMCs) in comparison to healthy individuals. Furthermore an investigation with a nonselective (trichostatin A) and selective (MI192, HDAC-3 selective) HDACi's revealed both regulated cytokine production (TNF-α and IL1-β) from PBMCs, but with the selective inhibitor only affecting the rheumatoid individuals [97]. Thus further supporting the use of these selective inhibitors in disease states.

Current work is indicating the potential of wide-ranging repressive effects even with the broad HDACi's (pan-HDACi), in this case vorinostat (suberanilohydroxamic acid), licensed from Merck for T-cell lymphomas, with HIV/AIDS phase I and II trials underway at University of North Carolina [46]. This study is comparing HIV-infected patients on antiretrovirals, before and after one treatment of the HDACi, and measuring the HIV RNA expression and infection within resting CD4+ cells.

Additional novel targets of posttranslational modifications include citrullination, catalyzed by peptidylarginine deiminase (PAD) enzymes, leading to histone citrullination in activated neutrophils, which has been implicated in a number of inflammatory diseases [98]. Experimental cell-line and mouse model evidence has shown that inhibitors of these PAD enzymes suppress neoplastic growth but also inflammation [99]. This has led to interest in their potential across a number of inflammatory disease targets, including multiple sclerosis [98].

Type 1 diabetes (T1D) was one of the earliest disease phenotypes to undergo EWAS analysis, and was examined by one of the earlier generations of Illumina Infinium arrays, the 27 k array [100]. In this very early exploratory EWAS, this technology was used to compare within T1D those who had and had not developed the significant symptom of diabetic neuropathy, in the surrogate tissue of peripheral blood [101]. As a forerunner of later integrative studies, one of the 19 significant hits was in the promoter of the *UNC13B* gene, which has been genetically implicated in diabetic nephropathy. This was followed by a more specific T1D analysis, however again only with the 27 k array, in monozygotic twins discordant for the disease [102], with the analysis performed in the T1D pathogenesis-related purified CD14+ monocytes. One hundred and thirty-two T1D disease-associated CpGs were identified, but only very small absolute methylation changes were seen, although interestingly some modifications were identified in islet autoantibody-positive individuals years prior to diagnosis. Direct DNA methylation array analysis of the pancreas has also been performed, revealing sex differences [103]. Histone modifications have also been explored in T1D and again the therapeutic potential role for HDACi's is being investigated as well [104].

The complex environmental interplay and the variable history of inflammatory respiratory diseases, such as asthma, lend themselves to thorough epigenetic analysis. The wide range of exposures to potential environmental trigger factors may, through epigenetic changes, show the biological effects of their action [105].

There has been some recent conflicting evidence about the validity of murine models of human inflammation diseases [106,107], most critically because of their low rate in transitioning therapeutic success from the mouse to human [108]. However, with more precise understanding of the biological mechanisms involved, that may be detailed through locus and sequence-specific epigenetics analysis, improved transfer across species may become possible. For instance, certain strains of mouse may be found to be more applicable for certain disease studies, due to good conservation of pathogenic markers, such as has been recently seen in the CK-p25 mouse and Alzheimer's disease [65]. Also, intriguing epigenetic insights may be rendered by simple and noninvasive comparative epigenomic analyses across even more similar species, such as peripheral blood differences identified between primates [24,109], that can highlight human-specific immune differences, as specifically seen in the leukotriene B4 receptor locus, and across the entire leukotriene production pathway, with at present undetermined implications [24,110]. Further fascinating genetic clues from more recent human history are also being uncovered, such as GWAS shared leprosy and inflammatory associations [111].

17.5 METABOLIC AND CARDIOVASCULAR DISEASE

With the current pervasive influence of the "obesogenic environment"—the high intake of high-calorific foods coupled with low physical exercise—that exists increasingly worldwide, obesity and associated diseases, such as type 2 diabetes (T2D), cardiovascular disease, hypertension, and dyslipidemia, are at unprecedented levels [112,113]. Quantitating this environment and its influence on the human physiology by epigenetic means is a tractable avenue to investigate. With regards to the epigenetics of obesity, this has extended from early studies into the imprinting disorder Prader–Willi syndrome, with its morbid obesity phenotype, to more recent explorations of the common trait [39]. Dietary influences on the epigenome with EWAS using precise metabolite data are the start of a fascinating interplay between these -omic data [114], although the strong genetic influences on these factors will need accounting for [115].

The *FTO* locus with its strong, by GWAS standards, influence on a trait, in this case BMI, has garnered much detailed analysis. Facilitative epigenetic changes in the risk haplotype due to variation in the CpG-Single Nucleotide Polymorphism (SNP) density within the genetically implicated region of the *FTO* gene were identified [116]. These colocated with other functional mechanisms, such as chromatin enhancer data, as well as loci implicated in the distal control of the *IRX3* transcription factor gene by Ragvin et al. [117] in a zebrafish model, and then subsequently in humans by Smemo et al. [118].

Changes in DNA methylation with respect to T2D status were also identified within the first intron of the *FTO* gene that was not contributed to by sequence polymorphism within the region [119]. This was a focal region of hypomethylation, although the inverse increased methylation within the entire risk-haplotype was also replicated. More recently this was further explored across both an Israeli Ashkenazi Jewish and Palestinian Arab population, with the latter possessing a higher T2D prevalence. T2D individuals again confirmed the local region of lower methylation. This region also showed consistent hypomethylation with age, but this effect was even more accelerated in the T2D individuals, therefore identifying that this was a premature aging signal in those with the disease state across both populations [120].

With this robust epigenetic finding within *FTO*, further interesting environmental influences on the known genotype–phenotype relationship have been able to be identified, including an increased effect with low physical activity [121] and the inverse attenuating it [37]. Also, a cohort generational effect was shown on the genotype–phenotype relationship, with those born before 1942 having decreased penetrance of the BMI correlation [122].

Analysis of changes in adipose tissue post gastric-bypass has identified variation in or near genes associated with obesity, development, and epigenetic regulation [123]. Further understanding of the role of inflammation in obesity-related disorders and also its role in atheroma plaque formation has also been gathered [124]. Atherosclerosis leads to a substantial risk of cardiac or hemorrhagic disease and there is an emerging role for epigenetics in the inflammation implicated in the etiological process [125]. As such the therapeutic treatment of this inflammation and the potential of a role for HDACi's is clear [110]. Regarding epigenetic abnormalities, a study from 2010 identified a pathogenic change in the methylation of the repetitive element long interspersed nucleotide element-1 in relation to ischemic heart disease in blood [126]. The potential for repetitive elements to play a pathogenic role across diseases, which is unable to be assessed with current array technology, is an underexplored avenue, as latent regulatory effects [127] occurring with age, etc. [128] are potentially unleashed. Analysis of the endothelial epigenome has shown the formation of super enhancers in response to nuclear factor-kappa B causing dramatic changes in gene expression and isoform selection leading to an atherogenic state [129].

Interesting future projects are being proposed such as the EPIOSA study (Epigenetics modifications and Subclinical Atherosclerosis in Obstructive Sleep Apnea) to explore these epigenetic changes occurring over the course of a 5-year period, due to the associated cardiovascular disease risk with this condition [130]. In-depth phenotyping at yearly intervals is planned to assess inflammatory-related changes to the epigenome.

Moreover, in a recently published study it is intriguing to note that the potential for the environment to modulate even severe monogenic disease has been seen, which has strong implications for all research and diagnostic genetics. A study of the maternal-age-associated risk of congenital heart disease found this to be modifiable, in a mouse model, by increased exercise, so that even when the offspring carried the causative mutation this maternal intervention could reduce the risk [131].

17.6 PRECISE MANIPULATION OF THE EPIGENOME

The ability to target base-resolution-specific epigenetic changes, to functionally understand the implications of disease-associated variation and to explore therapeutic opportunities, is an intense area of experimental research [132]. Early work focused on the use of transcription activator-like effector nucleases (TALENs) to influence gene expression, by fusing epigenetic modifiers such as DNMT3A and the TET enzymes (Tet1, Tet2, Tet3) to modify DNA methylation.

In a pioneering study from the groups of George Church and Feng Zhang, optogenic tools were used to develop a method to increase localization and allow precise deployment, using a blue-light-sensitive protein, crytochrome 2 from *Arabidopsis thaliana* [133]. This "optoepigenetics" approach leads to a conformational change and binding of an additional protein (crytochrome-interacting basic-helix-loop-helix 1), enabling the TALEN with a DNA methyltransferase attached to switch on in the presence of the blue light. This then leads to the clinical potential of optic-activation through their use in fiber-implanted cables within internal organs [134].

Clustered regularly interspersed short palindromic repeats (CRISPR) technology is a programmable DNA targeting mechanism for genome editing derived from bacterial DNA that is creating a high degree of excitement in the field of genome editing [135]. These transcribe RNA that recognize foreign DNA via sequence complementarity. This superior RNA guide, compared to the previous protein attachment methods, also binds a nuclease, Cas9, which enables sequence-specific DNA cutting. This method can be modified using a mutant Cas9 to target and anchor epigenetic modifiers. Konermann et al. [136] repurposed a CRISPR/Cas9 system to gather multiple activators, appending the short guide RNA with further sequence to draw in an RNA-binding protein, thereby acting as a molecular scaffold to recruit further transcriptional stimulators (Figure 17.2). Thus this mimics the lncRNA that often act in this fashion within enhancers [138].

A study from 2015 by Hilton et al. [139] has now used targeted acetylation with CRISPR technology to precisely activate specific genes. By fusing the acetyltransferase protein p300, this leads to H3K27ac formation at promoters and enhancers, with the consequence of highly specific gene activation.

17.7 CONCLUSION AND OUTLOOK

Novel findings continue to arrive at a rapid pace across the wide spectrum of epigenetics: from the insights into the mechanism whereby CpG-containing bacterial and viral DNA is recognized by Toll-like receptor 9 as part of the innate immune system [140]; large-scale imputation of epigenomes, enabling the rapid identification of significant outliers [141]; the use of the potent chromatin

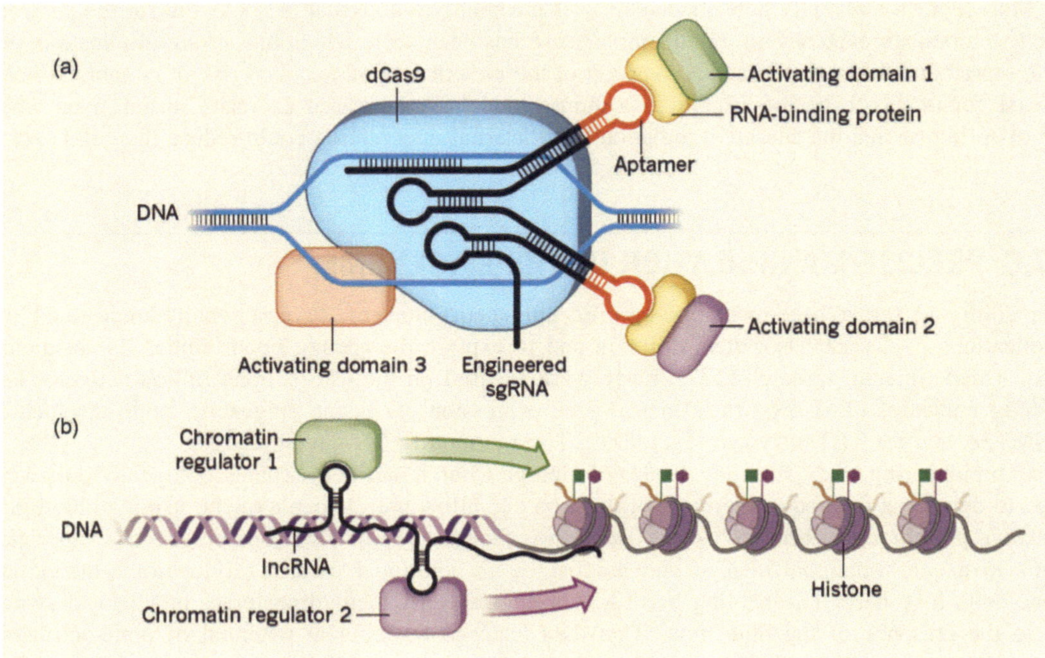

FIGURE 17.2

Epigenome editing with CRISPR technology. (A) Konermann et al. [136] have utilized the short guide RNA in CRISPR to act as a molecular scaffold to enable recruitment of activating proteins for local epigenomic modification. (B) lncRNAs act as a molecule scaffold for epigenomic chromatin modifiers.

Reproduced from Cho and Chang [137] © Nature Publishing Group.

insulator CTCF for less mutagenicity and toxicity in human gene therapy [142]; to the potential for improved drug discovery to develop "therapeutic tools that tune regulation" arising from the Epigenomic Roadmap project [143].

The epigenomic instruments of the future will require precise programming of time and cellular-restricted activity and the selective restoration or repression of individual isoforms implicated in disease pathogenesis [134]. Getting therapeutics to the correct tissue will be a significant factor, with current RNA-based approaches including use of modified antisense oligonucleotides or spherical nucleic acids to target specific RNAs. These may require invasive intrathecal or intracerebroventricular injection justified in the treatment of severe neurodisorders [134], but newer and more nuanced mechanisms will also need to be developed, such as the spatiotemporal control possible with optic tools [144].

Epigenetics also gives us a key to unlock the biology involved with the aging process; through understanding the aging-related epigenetic changes found, and helping toward the goal of healthy aging. Furthermore, the high level of somatic mutation identified with age in peripheral blood in genes known to be epigenomic modifiers, *DNMT3A*, *TET2*, and *ASXL1* [145,146], may give further insights. Understanding the pathway physiology behind the systemic negative health effects, for example, those implicated with *F2RL3* DNA methylation mortality prediction, may also identify novel biology [147].

Due to the many diseases that have an epigenetic component, there is vast potential to develop tools for functional knowledge, cellular engineering, and drug development [132]. As genome sequencing becomes a routine pathological assessment, a method of altering and fine-tuning these genetic findings will become the focus. The prospect that epigenomics will have significant impact on the prognosis, prevention, and treatment of human disease is undoubted, and the coming decades will see epigenomic medicine become a fundamental cornerstone of future healthcare.

ACKNOWLEDGMENTS

CGB is funded via the European Research Council (project number 250157) and TwinsUK. TwinsUK is funded by the Wellcome Trust; European Community's Seventh Framework Programme (FP7/2007-2013); and also receives support from the National Institute for Health Research (NIHR) BioResource, Clinical Research Facility and Biomedical Research Centre based at Guy's and St Thomas' NHS Foundation Trust and King's College London.

REFERENCES

[1] Cedar H, Bergman Y. Linking DNA methylation and histone modification: patterns and paradigms. Nat Rev Genet 2009;10:295–304.

[2] Kowalik KM, Shimada Y, Flury V, Stadler MB, Batki J, Buhler M. The Paf1 complex represses small-RNA-mediated epigenetic gene silencing. Nature 2015;520:248–52.

[3] Roadmap Epigenomics Consoritum, Kundaje A, Meuleman W, Ernst J, Bilenky M, Yen A, et al. Integrative analysis of 111 reference human epigenomes. Nature 2015;518:317–30.

[4] Schubeler D. Function and information content of DNA methylation. Nature 2015;517:321–6.

[5] Branco MR, Ficz G, Reik W. Uncovering the role of 5-hydroxymethylcytosine in the epigenome. Nat Rev Genet 2012;13:7–13.

[6] Hon GC, Song C-X, Du T, Jin F, Selvaraj S, Lee Ah Y, et al. 5mC oxidation by Tet2 modulates enhancer activity and timing of transcriptome reprogramming during differentiation. Mol Cell 2014;56:286–97.

[7] Mellen M, Ayata P, Dewell S, Kriaucionis S, Heintz N. MeCP2 binds to 5hmC enriched within active genes and accessible chromatin in the nervous system. Cell 2012;151:1417–30.

[8] Margueron R, Reinberg D. Chromatin structure and the inheritance of epigenetic information. Nat Rev Genet 2010;11:285–96.

[9] Kouzarides T. Chromatin modifications and their function. Cell 2007;128:693–705.

[10] Zhou VW, Goren A, Bernstein BE. Charting histone modifications and the functional organization of mammalian genomes. Nat Rev Genet 2011;12:7–18.

[11] Audergon PNCB, Catania S, Kagansky A, Tong P, Shukla M, Pidoux AL, et al. Restricted epigenetic inheritance of H3K9 methylation. Science 2015;348:132–5.

[12] Ernst J, Kellis M. Interplay between chromatin state, regulator binding, and regulatory motifs in six human cell types. Genome Res 2013;23:1142–54.

[13] Hoffman MM, Buske OJ, Wang J, Weng Z, Bilmes JA, Noble WS. Unsupervised pattern discovery in human chromatin structure through genomic segmentation. Nat Methods 2012;9:473–6.
[14] Bell CG, Beck S. The epigenomic interface between genome and environment in common complex diseases. Brief Funct Genom 2010;9:477–85.
[15] Feinberg AP. Phenotypic plasticity and the epigenetics of human disease. Nature 2007;447:433–40.
[16] Bock C. Analysing and interpreting DNA methylation data. Nat Rev Genet 2012;13:705–19.
[17] Paul DS, Beck S. Advances in epigenome-wide association studies for common diseases. Trends Mol Med 2014;20:541–3.
[18] Heijmans BT, Mill J. Commentary: the seven plagues of epigenetic epidemiology. Int J Epidemiol 2012;41:74–8.
[19] Feinberg AP, Vogelstein B. Hypomethylation distinguishes genes of some human cancers from their normal counterparts. Nature 1983;301:89–92.
[20] DeChiara TM, Robertson EJ, Efstratiadis A. Parental imprinting of the mouse insulin-like growth factor II gene. Cell 1991;64:849–59.
[21] Irizarry RA, Ladd-Acosta C, Wen B, Wu Z, Montano C, Onyango P, et al. The human colon cancer methylome shows similar hypo- and hypermethylation at conserved tissue-specific CpG island shores. Nat Genet 2009;41:178–86.
[22] Ji H, Ehrlich LI, Seita J, Murakami P, Doi A, Lindau P, et al. Comprehensive methylome map of lineage commitment from haematopoietic progenitors. Nature 2010;467:338–42.
[23] Doi A, Park IH, Wen B, Murakami P, Aryee MJ, Irizarry R, et al. Differential methylation of tissue- and cancer-specific CpG island shores distinguishes human induced pluripotent stem cells, embryonic stem cells and fibroblasts. Nat Genet 2009;41:1350–3.
[24] Wilson GA, Butcher LM, Foster HR, Feber A, Roos C, Walter L, et al. Human-specific epigenetic variation in the immunological Leukotriene B4 receptor (LTB4R/BLT1) implicated in common inflammatory diseases. Genome Med 2014;6:19.
[25] Hansen KD, Timp W, Bravo HC, Sabunciyan S, Langmead B, McDonald OG, et al. Increased methylation variation in epigenetic domains across cancer types. Nat Genet 2011;43:768–75.
[26] Teschendorff AE, Menon U, Gentry-Maharaj A, Ramus SJ, Weisenberger DJ, Shen H, et al. Age-dependent DNA methylation of genes that are suppressed in stem cells is a hallmark of cancer. Genome Res 2010;20:440–6.
[27] Horvath S. DNA methylation age of human tissues and cell types. Genome Biol 2013;14:R115.
[28] Marioni RE, Shah S, McRae AF, Ritchie SJ, Muniz-Terrera G, Harris SE, et al. The epigenetic clock is correlated with physical and cognitive fitness in the Lothian Birth Cohort 1936. Int J Epidemiol 2015.
[29] Marioni RE, Shah S, McRae AF, Chen BH, Colicino E, Harris SE, et al. DNA methylation age of blood predicts all-cause mortality in later life. Genome Biol 2015;16:25.
[30] Joubert BR, Haberg SE, Nilsen RM, Wang X, Vollset SE, Murphy SK, et al. 450K epigenome-wide scan identifies differential DNA methylation in newborns related to maternal smoking during pregnancy. Environ Health Perspect 2012;120:1425–31.
[31] Monick MM, Beach SR, Plume J, Sears R, Gerrard M, Brody GH, et al. Coordinated changes in AHRR methylation in lymphoblasts and pulmonary macrophages from smokers. Am J Med Genet Part B Neuropsychiatr Genet off Publ Int Soc Psychiatric Genet 2012;159B:141–51.
[32] Philibert RA, Beach SR, Brody GH. Demethylation of the aryl hydrocarbon receptor repressor as a biomarker for nascent smokers. Epigenet Off J DNA Methylation Soc 2012;7:1331–8.
[33] Breitling LP, Yang R, Korn B, Burwinkel B, Brenner H. Tobacco-smoking-related differential DNA methylation: 27K discovery and replication. Am J Hum Genet 2011;88:450–7.
[34] Zeilinger S, Kuhnel B, Klopp N, Baurecht H, Kleinschmidt A, Gieger C, et al. Tobacco smoking leads to extensive genome-wide changes in DNA methylation. PLoS One 2013;8:e63812.

[35] Bakulski KM, Fallin MD. Epigenetic epidemiology: promises for public health research. Environ Mol Mutagen 2014;55:171–83.
[36] Qi Q, Chu AY, Kang JH, Jensen MK, Curhan GC, Pasquale LR, et al. Sugar-Sweetened beverages and genetic risk of obesity. New Engl J Med 2012;367:1387–96.
[37] Kilpelainen TO, Qi L, Brage S, Sharp SJ, Sonestedt E, Demerath E, et al. Physical activity attenuates the influence of FTO variants on obesity risk: a meta-analysis of 218,166 adults and 19,268 children. PLoS Med 2011;8:e1001116.
[38] Feinberg AP. Epigenetics at the epicenter of modern medicine. JAMA 2008;299:1345–50.
[39] Bell CG. Epigenomic factors in human obesity. Amsterdam, Boston, MA: Elsevier/AP; 2012.
[40] Baccarelli A, Rienstra M, Benjamin EJ. Cardiovascular epigenetics: basic concepts and results from animal and human studies. Circ Cardiovasc Genet 2010;3:567–73.
[41] Roos L, Spector TD, Bell CG. Using epigenomic studies in monozygotic twins to improve our understanding of cancer. Epigenomics 2014;6:299–309.
[42] van Dongen J, Slagboom PE, Draisma HH, Martin NG, Boomsma DI. The continuing value of twin studies in the omics era. Nat Rev Genet 2012;13:640–53.
[43] Mill J, Heijmans BT. From promises to practical strategies in epigenetic epidemiology. Nat Rev Genet 2013;14:585–94.
[44] Egger G, Liang G, Aparicio A, Jones PA. Epigenetics in human disease and prospects for epigenetic therapy. Nature 2004;429:457–63.
[45] Karberg S. Switching on epigenetic therapy. Cell 2009;139:1029–31.
[46] Hunter P. The second coming of epigenetic drugs: a more strategic and broader research framework could boost the development of new drugs to modify epigenetic factors and gene expression. EMBO Rep 2015.
[47] Arrowsmith CH, Bountra C, Fish PV, Lee K, Schapira M. Epigenetic protein families: a new frontier for drug discovery. Nat Rev Drug Discov 2012;11:384–400.
[48] Conway SJ. Bromodomains: are readers right for epigenetic therapy? ACS Med Chem Lett 2012;3: 691–4.
[49] Raynal NJ, Lee JT, Wang Y, Garriga J, Malouf G, Dumont S, et al. Abstract 380: discovery of new epigenetic drugs among FDA-approved drug libraries. Cancer Res 2014;74:380.
[50] Illingworth RS, Gruenewald-Schneider U, De Sousa D, Webb S, Merusi C, Kerr AR, et al. Interindividual variability contrasts with regional homogeneity in the human brain DNA methylome. Nucl Acids Res 2015;43:732–44.
[51] McGraw CM, Samaco RC, Zoghbi HY. Adult neural function requires MeCP2. Science 2011;333:186.
[52] Kinde B, Gabel HW, Gilbert CS, Griffith EC, Greenberg ME. Reading the unique DNA methylation landscape of the brain: non-CpG methylation, hydroxymethylation, and MeCP2. Proc Natl Acad Sci 2015.
[53] Crick F. Memory and molecular turnover. Nature 1984;312:101.
[54] Miller CA, Gavin CF, White JA, Parrish RR, Honasoge A, Yancey CR, et al. Cortical DNA methylation maintains remote memory. Nat Neurosci 2010;13:664–6.
[55] Graff J, Joseph NF, Horn ME, Samiei A, Meng J, Seo J, et al. Epigenetic priming of memory updating during reconsolidation to attenuate remote fear memories. Cell 2014;156:261–76.
[56] Rajasethupathy P, Antonov I, Sheridan R, Frey S, Sander C, Tuschl T, et al. A role for neuronal piRNAs in the epigenetic control of memory-related synaptic plasticity. Cell 2012;149:693–707.
[57] Ferguson-Smith AC. Genomic imprinting: the emergence of an epigenetic paradigm. Nat Rev Genet 2011;12:565–75.
[58] Meng L, Ward AJ, Chun S, Bennett CF, Beaudet AL, Rigo F. Towards a therapy for Angelman syndrome by targeting a long non-coding RNA. Nature 2015;518:409–12.

[59] Millan MJ. An epigenetic framework for neurodevelopmental disorders: from pathogenesis to potential therapy. Neuropharmacology 2013;68:2–82.
[60] Henry TR. The history of valproate in clinical neuroscience. Psychopharmacol Bull 2003;37(Suppl. 2):5–16.
[61] Torrioli M, Vernacotola S, Setini C, Bevilacqua F, Martinelli D, Snape M, et al. Treatment with valproic acid ameliorates ADHD symptoms in fragile X syndrome boys. Am J Med Genet A 2010;152a: 1420–7.
[62] Monti B, Polazzi E, Contestabile A. Biochemical, molecular and epigenetic mechanisms of valproic acid neuroprotection. Curr Mol Pharmacol 2009;2:95–109.
[63] De Jager PL, Srivastava G, Lunnon K, Burgess J, Schalkwyk LC, Yu L, et al. Alzheimer's disease: early alterations in brain DNA methylation at ANK1, BIN1, RHBDF2 and other loci. Nat Neurosci 2014;17:1156–63.
[64] Lunnon K, Smith R, Hannon E, De Jager PL, Srivastava G, Volta M, et al. Methylomic profiling implicates cortical deregulation of ANK1 in Alzheimer's disease. Nat Neurosci 2014;17:1164–70.
[65] Gjoneska E, Pfenning AR, Mathys H, Quon G, Kundaje A, Tsai LH, et al. Conserved epigenomic signals in mice and humans reveal immune basis of Alzheimer's disease. Nature 2015;518:365–9.
[66] Farh KK-H, Marson A, Zhu J, Kleinewietfeld M, Housley WJ, Beik S, et al. Genetic and epigenetic fine mapping of causal autoimmune disease variants. Nature 2015;518:337–43.
[67] Yu D, Pendergraff H, Liu J, Kordasiewicz HB, Cleveland DW, Swayze EE, et al. Single-stranded RNAs use RNAi to potently and allele-selectively inhibit mutant huntingtin expression. Cell 2012;150:895–908.
[68] Ammal Kaidery N, Tarannum S, Thomas B. Epigenetic landscape of parkinson's disease: emerging role in disease mechanisms and therapeutic modalities. Neurotherapeutics 2013;10:698–708.
[69] Lemmers RJ, Tawil R, Petek LM, Balog J, Block GJ, Santen GW, et al. Digenic inheritance of an SMCHD1 mutation and an FSHD-permissive D4Z4 allele causes facioscapulohumeral muscular dystrophy type 2. Nat Genet 2012;44:1370–4.
[70] Himeda CL, Jones TI, Jones PL. Facioscapulohumeral muscular dystrophy as a model for epigenetic regulation and disease. Antioxid Redox Signal 2014.
[71] Shulha HP, Cheung I, Whittle C, Wang J, Virgil D, Lin CL, et al. Epigenetic signatures of autism: trimethylated H3K4 landscapes in prefrontal neurons. Arch Gen Psychiat 2012;69:314–24.
[72] Percy AK. Rett syndrome: exploring the autism link. Arch Neurol 2011;68:985–9.
[73] Gabel HW, Kinde B, Stroud H, Gilbert CS, Harmin DA, Kastan NR, et al. Disruption of DNA-methylation-dependent long gene repression in Rett syndrome. Nature 2015 (advance online publication).
[74] Ziller MJ, Edri R, Yaffe Y, Donaghey J, Pop R, Mallard W, et al. Dissecting neural differentiation regulatory networks through epigenetic footprinting. Nature 2014 (advance online publication).
[75] Shulha HP, Crisci JL, Reshetov D, Tushir JS, Cheung I, Bharadwaj R, et al. Human-specific histone methylation signatures at transcription start sites in prefrontal neurons. PLoS Biol 2012;10:e1001427.
[76] Bell CG, Wilson GA, Butcher LM, Roos C, Walter L, Beck S. Human-specific CpG "beacons" identify loci associated with human-specific traits and disease. Epigenetics 2012;7:1188–99.
[77] Bell CG, Wilson GA, Beck S. Human-specific CpG 'beacons' identify human-specific prefrontal cortex H3K4me3 chromatin peaks. Epigenomics 2014;6:21–31.
[78] Provencal N, Suderman MJ, Guillemin C, Massart R, Ruggiero A, Wang D, et al. The signature of maternal rearing in the methylome in rhesus macaque prefrontal cortex and T cells. J Neurosci 2012;32:15626–42.
[79] Kleefstra T, Brunner HG, Amiel J, Oudakker AR, Nillesen WM, Magee A, et al. Loss-of-function mutations in euchromatin histone methyl transferase 1 (EHMT1) cause the 9q34 subtelomeric deletion syndrome. Am J Hum Genet 2006;79:370–7.

[80] Cheung I, Shulha HP, Jiang Y, Matevossian A, Wang J, Weng Z, et al. Developmental regulation and individual differences of neuronal H3K4me3 epigenomes in the prefrontal cortex. Proc Natl Acad Sci USA 2010;107:8824–9.

[81] Sugathan A, Biagioli M, Golzio C, Erdin S, Blumenthal I, Manavalan P, et al. CHD8 regulates neurodevelopmental pathways associated with autism spectrum disorder in neural progenitors. Proc Natl Acad Sci USA 2014;111:E4468–77.

[82] Ronan JL, Wu W, Crabtree GR. From neural development to cognition: unexpected roles for chromatin. Nat Rev Genet 2013;14:347–59.

[83] Dawson MA, Kouzarides T, Huntly BJ. Targeting epigenetic readers in cancer. New Engl J Med 2012;367:647–57.

[84] De Santis M, Selmi C. The therapeutic potential of epigenetics in autoimmune diseases. Clin Rev Allergy Immunol 2012;42:92–101.

[85] Szyf M. Epigenetic therapeutics in autoimmune disease. Clin Rev Allergy Immunol 2010;39:62–77.

[86] Dinarello CA, Fossati G, Mascagni P. Histone deacetylase inhibitors for treating a spectrum of diseases not related to cancer. Mol Med 2011;17:333–52.

[87] Woan KV, Sahakian E, Sotomayor EM, Seto E, Villagra A. Modulation of antigen-presenting cells by HDAC inhibitors: implications in autoimmunity and cancer. Immunol Cell Biol 2012;90:55–65.

[88] Zhang P, Su Y, Lu Q. Epigenetics and psoriasis. J Eur Acad Dermatol Venereol 2012;26:399–403.

[89] Handel AE, Ebers GC, Ramagopalan SV. Epigenetics: molecular mechanisms and implications for disease. Trends Mol Med 2010;16:7–16.

[90] Reilly CM, Regna N, Mishra N. HDAC inhibition in lupus models. Mol Med 2011;17:417–25.

[91] Felice C, Lewis A, Armuzzi A, Lindsay JO, Silver A. Review article: selective histone deacetylase isoforms as potential therapeutic targets in inflammatory bowel diseases. Aliment Pharmacol Ther 2015;41:26–38.

[92] Klein K, Gay S. Epigenetic modifications in rheumatoid arthritis, a review. Curr Opin Pharmacol 2013;13:420–5.

[93] Schwab M, Reynders V, Loitsch S, Steinhilber D, Schroder O, Stein J. The dietary histone deacetylase inhibitor sulforaphane induces human beta-defensin-2 in intestinal epithelial cells. Immunology 2008;125:241–51.

[94] Sonkoly E, Ståhle M, Pivarcsi A. MicroRNAs and immunity: novel players in the regulation of normal immune function and inflammation. Semin Cancer Biol 2008;18:131–40.

[95] Whitaker JW, Shoemaker R, Boyle DL, Hillman J, Anderson D, Wang W, et al. An imprinted rheumatoid arthritis methylome signature reflects pathogenic phenotype. Genome Med 2013;5:40.

[96] Ai R, Whitaker JW, Boyle DL, Tak PP, Gerlag DM, Wang W, et al. DNA methylome signature in early rheumatoid arthritis synoviocytes compared with longstanding rheumatoid arthritis synoviocytes. Arthritis Rheum 2015.

[97] Gillespie J, Savic S, Wong C, Hempshall A, Inman M, Emery P, et al. Histone deacetylases are dysregulated in rheumatoid arthritis and a novel histone deacetylase 3-selective inhibitor reduces interleukin-6 production by peripheral blood mononuclear cells from rheumatoid arthritis patients. Arthritis Rheum 2012;64:418–22.

[98] Chirivi RG, Rosmalen JW, Jenniskens GJ, Pruijn GJ, Raats JMH. Citrullination: a target for disease intervention in multiple sclerosis and other inflammatory diseases? J Clin Cell Immunol 2013;4:146.

[99] Mohanan S, Cherrington BD, Horibata S, McElwee JL, Thompson PR, Coonrod SA. Potential role of peptidylarginine deiminase enzymes and protein citrullination in cancer pathogenesis. Biochem Res Int 2012;2012:895343.

[100] Bibikova M, Le J, Barnes B, Saedinia-Melnyk S, Zhou L, Shen R, et al. Genome-wide DNA methylation profiling using Infinium assay. Epigenomics 2009;1:177–200.

[101] Bell CG, Teschendorff AE, Rakyan VK, Maxwell AP, Beck S, Savage DA. Genome-wide DNA methylation analysis for diabetic nephropathy in type 1 diabetes mellitus. BMC Med Genom 2010;3:33.

[102] Rakyan VK, Beyan H, Down TA, Hawa MI, Maslau S, Aden D, et al. Identification of type 1 diabetes-associated DNA methylation variable positions that precede disease diagnosis. PLoS Genet 2011;7: e1002300.

[103] Hall E, Volkov P, Dayeh T, Esguerra JL, Salo S, Eliasson L, et al. Sex differences in the genome-wide DNA methylation pattern and impact on gene expression, microRNA levels and insulin secretion in human pancreatic islets. Genome Biol 2014;15:522.

[104] Dang MN, Buzzetti R, Pozzilli P. Epigenetics in autoimmune diseases with focus on type 1 diabetes. Diabetes Metab Res Rev 2013;29:8–18.

[105] Miller RL, Ho S-m. Environmental epigenetics and asthma. Am J Respir Crit Care Med 2008;177:567–73.

[106] Seok J, Warren HS, Cuenca AG, Mindrinos MN, Baker HV, Xu W, et al. Genomic responses in mouse models poorly mimic human inflammatory diseases. Proc Natl Acad Sci USA 2013;110:3507–12.

[107] Takao K, Miyakawa T. Genomic responses in mouse models greatly mimic human inflammatory diseases. Proc Natl Acad Sci 2015;112:1167–72.

[108] Perrin S. Preclinical research: make mouse studies work. Nature 2014;507:423–5.

[109] Hernando-Herraez I, Prado-Martinez J, Garg P, Fernandez-Callejo M, Heyn H, Hvilsom C, et al. Dynamics of DNA methylation in recent human and great ape evolution. PLoS Genet 2013;9:e1003763.

[110] Charo IF, Taub R. Anti-inflammatory therapeutics for the treatment of atherosclerosis. Nat Rev Drug Discov 2011;10:365–76.

[111] Liu H, Irwanto A, Fu Xa, Yu G, Yu Y, Sun Y, et al. Discovery of six new susceptibility loci and analysis of pleiotropic effects in leprosy. Nat Genet 2015 (advance online publication).

[112] Friedman JM. Obesity: causes and control of excess body fat. Nature 2009;459:340–2.

[113] Gluckman PD, Hanson M, Zimmet P, Forrester T. Losing the war against obesity: the need for a developmental perspective. Sci Transl Med 2011;3:93cm19.

[114] Petersen AK, Zeilinger S, Kastenmuller G, Romisch-Margl W, Brugger M, Peters A, et al. Epigenetics meets metabolomics: an epigenome-wide association study with blood serum metabolic traits. Hum Mol Genet 2014;23:534–45.

[115] Suhre K, Gieger C. Genetic variation in metabolic phenotypes: study designs and applications. Nat Rev Genet 2012;13:759–69.

[116] Bell CG, Finer S, Lindgren CM, Wilson GA, Rakyan VK, Teschendorff AE, et al. Integrated genetic and epigenetic analysis identifies haplotype-specific methylation in the FTO type 2 diabetes and obesity susceptibility locus. PLoS One 2010;5:e14040.

[117] Ragvin A, Moro E, Fredman D, Navratilova P, Drivenes O, Engstrom PG, et al. Long-range gene regulation links genomic type 2 diabetes and obesity risk regions to HHEX, SOX4, and IRX3. Proc Natl Acad Sci USA 2010;107:775–80.

[118] Smemo S, Tena JJ, Kim K-H, Gamazon ER, Sakabe NJ, Gomez-Marin C, et al. Obesity-associated variants within FTO form long-range functional connections with IRX3. Nature 2014;507:371–5.

[119] Toperoff G, Aran D, Kark JD, Rosenberg M, Dubnikov T, Nissan B, et al. Genome-wide survey reveals predisposing diabetes type 2-related DNA methylation variations in human peripheral blood. Hum Mol Genet 2012;21:371–83.

[120] Toperoff G, Kark JD, Aran D, Nassar H, Ahmad WA, Sinnreich R, et al. Premature aging of leukocyte DNA methylation is associated with type 2 diabetes prevalence. Clin Epigenet 2015;7:35.

[121] Andreasen CH, Stender-Petersen KL, Mogensen MS, Torekov SS, Wegner L, Andersen G, et al. Low physical activity accentuates the effect of the FTO rs9939609 polymorphism on body fat accumulation. Diabetes 2008;57:95–101.

[122] Rosenquist JN, Lehrer SF, O'Malley AJ, Zaslavsky AM, Smoller JW, Christakis NA. Cohort of birth modifies the association between FTO genotype and BMI. Proc Natl Acad Sci USA 2015;112:354–9.

[123] Benton MC, Johnstone A, Eccles D, Harmon B, Hayes MT, Lea RA, et al. An analysis of DNA methylation in human adipose tissue reveals differential modification of obesity genes before and after gastric bypass and weight loss. Genome Biol 2015;16:8.

[124] Schulz C, Massberg S. Atherosclerosis—multiple pathways to lesional macrophages. Sci Transl Med 2014;6:239ps2.

[125] Abi Khalil C. The emerging role of epigenetics in cardiovascular disease. Ther Adv Chronic Dis 2014;5:178–87.

[126] Baccarelli A, Wright R, Bollati V, Litonjua A, Zanobetti A, Tarantini L, et al. Ischemic heart disease and stroke in relation to blood DNA methylation. Epidemiology 2010;21:819–28.

[127] Ward MC, Wilson MD, Barbosa-Morais NL, Schmidt D, Stark R, Pan Q, et al. Latent regulatory potential of human-specific repetitive elements. Mol Cell 2012;12:00944–6.

[128] Van Meter M, Kashyap M, Rezazadeh S, Geneva AJ, Morello TD, Seluanov A, et al. SIRT6 represses LINE1 retrotransposons by ribosylating KAP1 but this repression fails with stress and age. Nat Commun 2014;5:5011.

[129] Brown JD, Lin CY, Duan Q, Griffin G, Federation AJ, Paranal RM, et al. NF-κB directs dynamic super enhancer formation in inflammation and atherogenesis. Mol Cell 2014;56:219–31.

[130] Marin JM, Artal J, Martin T, Carrizo SJ, Andres M, Martin-Burriel I, et al. Epigenetics modifications and subclinical atherosclerosis in obstructive sleep apnea: the EPIOSA study. BMC Pulm Med 2014;14:114.

[131] Schulkey CE, Regmi SD, Magnan RA, Danzo MT, Luther H, Hutchinson AK, et al. The maternal-age-associated risk of congenital heart disease is modifiable. Nature 2015;520:230–3.

[132] Keung AJ, Joung JK, Khalil AS, Collins JJ. Chromatin regulation at the frontier of synthetic biology. Nat Rev Genet 2015;16:159–71.

[133] Konermann S, Brigham MD, Trevino AE, Hsu PD, Heidenreich M, Le C, et al. Optical control of mammalian endogenous transcription and epigenetic states. Nature 2013;500:472–6.

[134] Day JJ, Kennedy AJ, Sweatt JD. DNA methylation and its implications and accessibility for neuropsychiatric therapeutics. Ann Rev Pharmacol Toxicol 2015;55:591–611.

[135] Cong L, Ran FA, Cox D, Lin S, Barretto R, Habib N, et al. Multiplex genome engineering using CRISPR/Cas systems. Science 2013;339:819–23.

[136] Konermann S, Brigham MD, Trevino AE, Joung J, Abudayyeh OO, Barcena C, et al. Genome-scale transcriptional activation by an engineered CRISPR-Cas9 complex. Nature 2015;517:583–8.

[137] Cho SW, Chang HY. Genomics: CRISPR engineering turns on genes. Nature 2015;517:560–2.

[138] Wang KC, Chang HY. Molecular mechanisms of long noncoding RNAs. Mol Cell 2011;43:904–14.

[139] Hilton IB, D'Ippolito AM, Vockley CM, Thakore PI, Crawford GE, Reddy TE, et al. Epigenome editing by a CRISPR-Cas9-based acetyltransferase activates genes from promoters and enhancers. Nat Biotech 2015 (advance online publication).

[140] Ohto U, Shibata T, Tanji H, Ishida H, Krayukhina E, Uchiyama S, et al. Structural basis of CpG and inhibitory DNA recognition by Toll-like receptor 9. Nature 2015.

[141] Ernst J, Kellis M. Large-scale imputation of epigenomic datasets for systematic annotation of diverse human tissues. Nat Biotech 2015;33:364–76.

[142] Liu M, Maurano MT, Wang H, Qi H, Song C-Z, Navas PA, et al. Genomic discovery of potent chromatin insulators for human gene therapy. Nat Biotech 2015;33:198–203.

[143] Mullard A. The roadmap epigenomics project opens new drug development avenues. Nat Rev Drug Discov 2015;14:223–5.

[144] Tischer D, Weiner OD. Illuminating cell signalling with optogenetic tools. Nat Rev Mol Cell Biol 2014;15:551–8.
[145] Jaiswal S, Fontanillas P, Flannick J, Manning A, Grauman PV, Mar BG, et al. Age-related clonal hematopoiesis associated with adverse outcomes. New Engl J Med 2014;371:2488–98.
[146] Genovese G, Kähler AK, Handsaker RE, Lindberg J, Rose SA, Bakhoum SF, et al. Clonal hematopoiesis and blood-cancer risk inferred from blood DNA sequence. New Engl J Med 2014;371:2477–87.
[147] Zhang Y, Yang R, Burwinkel B, Breitling LP, Holleczek B, Schottker B, et al. F2RL3 methylation in blood DNA is a strong predictor of mortality. Int J Epidemiol 2014;43:1215–25.

Index

Note: Page numbers followed by "*f*" and "*t*" refer to figures and tables, respectively.

B

F

G

H

I

J

K

L

M

N

O

P

Q

R

Kolibri 360 GmbH
Ferdinand-Jühlke-Straße 7
99095 Erfurt, Deutschland
produktsicherheit@kolibri360.de